AF066469

Air Pollution Modeling and Its Application IV

NATO • Challenges of Modern Society

A series of edited volumes comprising multifaceted studies of contemporary problems facing our society, assembled in cooperation with NATO Committee on the Challenges of Modern Society.

Air Pollution Modeling and Its Application IV

Edited by
C. De Wispelaere
Prime Minister's Office for Science Policy
Brussels, Belgium

Published in cooperation with
NATO Committee on the Challenges of Modern Society

PLENUM PRESS • NEW YORK AND LONDON

Library of Congress Cataloging in Publication Data

International Technical Meeting on Air Pollution Modeling and Its Application (14th: 1983: Copenhagen, Denmark)
Air pollution modeling and its application IV.

(NATO challenges of modern society; v. 7)
"Proceedings of the Fourteenth International Technical Meeting on Air Pollution Modeling and Its Application, held September 27-30, 1983 in Copenhagen, Denmark"—T.p. verso.
Includes bibliographies and index.
1. Atmospheric diffusion—Mathematical models—Congresses. 2. Air—Pollution—Meteorological aspects—Mathematical models—Congresses. I. Wispelaere, C. de. II. North Atlantic Treaty Organization. Committee on the Challenges of Modern Society. III. Title. IV. Series.
QC880.4.D44I57 1983 628.5'3'0724 84-26352

ISBN-13: 978-1-4612-9491-7 e-ISBN-13: 978-1-4613-2455-3
DOI: 10.1007/978-1-4613-2455-3

Proceedings of the Fourteenth International Technical Meeting on Air Pollution Modeling and Its Application, held September 27-30, 1983, in Copenhagen, Denmark

Softcover reprint of the hardcover 1st edition 1985

A Division of Plenum Publishing Corporation
233 Spring Street, New York, N.Y. 10013

PREFACE

In 1969 the North Atlantic Treaty Organization established the Committee on the Challenges of Modern Society. Air Pollution was from the start one of the priority problems under study within the framework of the pilot studies undertaken by this Committee. The organization of a yearly symposium dealing with air pollution modeling and its application is one of the main activities within the pilot study in relation to air pollution.

After being organized for five years by the United States and for five years by the Federal Republic of Germany, Belgium, represented by the Prime Minister's Office for science Policy, became responsible in 1980 for the organization of this symposium.

This volume contains the papers presented at the 14th International Technical Meeting on Air Pollution Modeling and its Application held in Copenhagen, Denmark, from 27th to 30th September 1983. This meeting was jointly organized by the Prime Minister's Office for Science Policy, Belgium, and the National Agency of Environmental Protection, Air Pollution Laboratory, Risø National Laboratory, Denmark. The conference was attended by 103 participants and 43 papers have been presented. The members of the selection committee of the 14th I.T.M. were A. Berger (Chairman, Belgium), W. Klug (Federal Republic of Germany), K. Demerjian (United States of America), L. Santomauro (Italy), H. Van Dop (The Netherlands), H.E. Turner (Canada), C. De Wispelaere (Coordinator, Belgium).

The main topic of this 14th I.T.M. was Parametrization of Transformation and Removal Processes in Air Quality Modeling. On this topic a review paper was presented by Ø. Hov (Norwegian Institute for Air Research). Other topics of the conference were : Coastal Meteorology related to Air Pollution Modeling, Lagrangian Modeling for Synoptic Range Transport, Atmospheric Experiments pertinent to Air Quality Modeling, Practical Applications of Air Quality Modeling.

U. Peching (Department Meterology, San Jose State University, California, U.S.A.) and F.L. Ludwig (SRI International, Menlo Park, U.S.A.) presented a review paper as an introduction to the topic on Costal Meteorology related to Air Pollution Modeling Dr. J. Knox (Lawrence Livermore National Laboratory, California, U.S.A.) gave a review on the topic Lagrangian Modeling for Synoptic Range Transport.

On behalf of the selection committee and as organizer and editor I should like to record my gratitude to all participants who made the meeting so stimulating and the book possible. Among them I particulary mention the chairmen and rapporteurs of the different sessions. Thanks also to the local organizing committee, chaired by Dr. J. Fenger. Thanks also to Mrs. J. Husted and P. Sondergaard, who were the Conference Secretaries. Finally I have the pleasure to record my thanks to Mrs. N. Desees and L. Mille and Miss J. Bertrand for typing and to Mr. D. Poelman for preparing the papers.

C. De Wispelaere

CONTENTS

PARAMETERIZATION OF TRANSFORMATION AND REMOVAL PROCESSES IN AIR QUALITY MODELING

LAGRANGIAN MODELING FOR SYNOPTIC RANGE TRANSPORT

ATMOSPHERIC EXPERIMENTS PERTINENT TO AIR QUALITY MODELING

PRACTICAL APPLICATIONS OF AIR QUALITY MODELING

1: PARAMETERIZATION OF TRANSFORMATION AND REMOVAL PROCESSES IN AIR QUALITY MODELING

Chairmen: A. Berger
Ø. Hov
U. Pechinger

Rapporteurs: J. Van Ham
E. Lyck
P. Zannetti

ASPECTS OF THE PARAMETERIZATION OF TRANSFORMATION AND REMOVAL PROCESSES IN AIR QUALITY MODELLING

Øystein Hov

Norwegian Institute for Air Research
P.O. Box 130
N-2001 Lillestrøm, Norway

1 INTRODUCTION

The air quality is linked to the ambient air concentration of species like ozone (O_3), nitrogen dioxide (NO_2), peroxyacetylnitrate (PAN), aldehydes, carbon monoxide (CO), sulphur dioxide (SO_2), sulphate ($SO_4^=$), nitric acid/nitrate (HNO_3/NO_3^-), soot and many other types of particulate material (containing trace metals, hydrocarbons, chlorinated hydrocarbons etc.). Ozone alone, or in combination with SO_2 or NO_2, is responsible for up to 90% of the crop losses in the U.S. caused by air pollution. An estimate made for the U.S., assuming that all areas just met the current O_3 standard of 120 ppb as hourly average, showed a loss of 2 to 4% of the crop production. A test program, utilizing field chambers where the ozone concentration could be controlled, demonstrated yield reductions in all crops at seasonal 7h/day mean O_3 concentrations of 60 - 70 ppb when compared with a control value of 25 ppb, thought to represent the natural background (Heck et al., 1982).

At the 1982 Stockholm conference on acidification of the environment it was stated in a consensus report that "the recently reported forest damage in an estimated one million hectares of Central Europe seems to be related to (among others) the direct effects of gaseous pollutants and soil impoverishment, and toxicity arising from very large amounts of wet and dry deposition" (Hileman, 1983). For a long time the most noticeable effect from acid rain has been the lowering of pH in thousands of lakes in Scandinavia and eastern North America. A consequence has been a substantial increase in the amount of dissolved aluminium, which has caused a wide spread decimation of the fish population.

Pollutants affecting the air quality are emitted directly or they are derived chiefly from the emissions of nitrogen oxides (NO_x), SO_2 and hydrocarbons (HC). An attempt at characterizing in terms of spatial and temporal scales of dispersion and transformation the most important chemical species which affect the air quality, is shown in Table 1. If the atmospheric fate of a given pollutant is to be computed by a model, it is required that the temporal and spatial domains are at least as indicated in Table 1. If an attempt is made at analysing e.g. ozone generation by constructing a model covering an area extending a few km and over a few hours, it is likely that the computational result is strongly dependent on the boundary conditions.

The chemical species which affect the air quality through their presence in the atmospheric boundary layer (ABL), are in general removed from the ABL on a time scale comparable to the typical time length between ABL break-down situations. At our latitudes, the ABL typically takes 2 - 4 d to reach a break-down situation (Smith and Carson, 1977). Break-down may occur as a result of large scale convective instability where convective elements may carry significant amounts of heat, moisture and pollution upwards into layers otherwise not affected by the underlying surface. Break-down may occur at fronts through upsliding motion, in mountainous regions where the vertical mixing may be considerable, and by continuous synoptic-scale vertical motion in combination with the diurnal cycle in the depth of the boundary layer which causes the contents of the boundary layer to be pumped upwards in the middle of the day, where it is left behind to be acted upon by the steady synoptic upward motions. Water soluble species may be rained out by precipitation associated with ABL break-down. The ABL vertical pumping effect distributes species like N_2O, CFC's, CH_4, O_3, CO and others into the free troposphere, to be acted upon on a global spatial scale.

The design of a particular air quality model has to be chosen on the basis of the problem to be analysed. The calculation of street canyon concentrations of a primary pollutant like CO requires different chemical and physical/meteorological considerations compared to the modelling of the generation, transport and loss of ozone or sulphate in the ABL. The time between ABL break-downs defines important temporal and spatial scales for ABL air quality modelling, however, and in this paper emphasis will be put on chemical transformation and removal of HC, SO_2 and NO_x over several days.

It is to be understood that parametrization of atmospheric transformation and removal processes of chemical species involves assigning numerical values to the significance of the known pathways for transformation and removal. In a broader sense, parametrization also includes the mathematical framework in which the numerical values for physical and chemical processes are to be embedded, and how to solve the mathematical equations which arise.

Table 1. Characterization in terms of spatial and temporal scales of dispersion and transformation of chemical species affecting the air quality.

Species	Scales for dispersion and transformation: Space	Time	Comments
O_3	$\leq$ 1000 km	$\leq$ 10 d	
PAN	$\leq$ 1000 km	$\leq$ 5 d	
SO_2	$\leq$ some 100 km	$\leq$ 2 d	
$SO_4^{=}$	$\leq$ 1000 km	$\leq$ 5 d	
NO_2	$\leq$ 1 km	$\leq$ 1 h	Scales appropriate for traffic. NO_2 may also be a dominating oxidized nitrogen species in long-range transported air (Grennfelt, 1979).
HNO_3 (NO_3^-)	$\leq$ 1000 km	$\leq$ 10 d	
CO	$\leq$ 1 km	$\leq$ 1 h	Scales appropriate for traffic. There is also a significant anthropogenic influence on the tropospheric global CO budget.
HCHO (aldehydes)	$\leq$ 10 km	$\leq$ 10 h	
C_6H_6 (benzene)	$\leq$ 1000 km	$\leq$ 10 d	
N_2O, CH_4, CFC, CO_2, O_3	global	years	Anthropogenic influence on the global tropospheric budgets may affect the future global climate.

2 MATHEMATICAL FORMUALTION OF AN AIR QUALITY MODEL

Mass conservation of each chemical species is a common requirement in air quality modelling. This can be written as

$$\frac{\partial c_i}{\partial t} + \nabla \cdot \vec{v}c_i = \nabla \cdot D_i \nabla c_i + R_i \tag{1}$$

where c_i denotes the concentration of species i, $\vec{v}$ is the wind field (stochastic variable), D_i molecular diffusion coefficient for species i, while R_i includes the effects of atmospheric transformation, removal and emission. In most air quality models the ensemble averaged atmospheric diffusion equation is applied

$$\frac{\partial \langle c_i \rangle}{\partial t} + \nabla \cdot \bar{\vec{v}}\langle c_i \rangle = \nabla \cdot (K \cdot \nabla \langle c_i \rangle) + R_i(\langle c_1 \rangle, \ldots, \langle c_p \rangle), \quad i = 1, \ldots, p \tag{2}$$

where $\langle c_i \rangle$ is the ensemble averaged concentration (the brackets will be omitted in the following), $\bar{\vec{v}}$ is the deterministic wind velocity and K the tensor introduced to approximate the turbulent transport term by the mean concentration gradient, p is the number of components (see e.g. Seinfeld, 1975, McRae et al., 1982). In this paper the R_i in eq. (2) will be examined, since this term describes the effects of chemistry, removal and emission on the time development of the concentration of species no. i.

In Figure 1 is shown a diagram of the factors which make up R_i. Eq. (2) is often taken as the general form of the continuity equation for a chemical species. In practical applications, less general forms of the equation are used depending on the problem to be analysed, by reducing the dimensionality to two or one, by transformation into a trajectory model, etc. Recent reviews are published by McRae et al., 1982, and Johnson, 1983. In practical applications, the mathematical formulation is a compromise based on many factors: The quality and detail of the meteorological data, emission fields, data describing chemical transformations, availability of computational and economical resources, and availability of air quality data against which the model can be compared.

The numerical solution of eq. (2) or some simplified form of eq. (2) is not straight forward. The component of the equation system which describes the chemical transformation is in general non linear, and exhibits stiff properties. This means that there is a wide range of chemical decay times involved. Special techniques have to be applied to keep the computer cost down and obtain

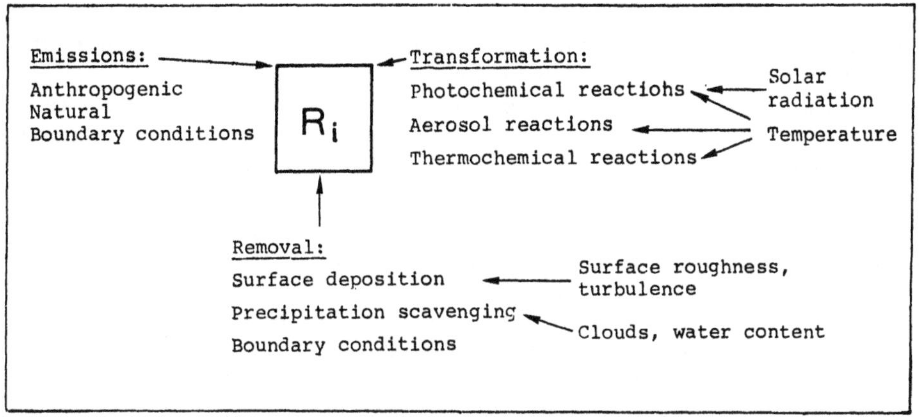

Figure 1. Factors in the term R_i in eq. (2).

solutions with acceptable numerical accuracy (Gear, 1971, Whitten, 1976). A so called quasi-steady state approximation technique has been developed by Hesstvedt et al., 1978, which cuts down the computer cost for the integration of kinetic equations, but which depends on other methods (e.g. Gear, 1971) for cross checks to assess the numerical accuracy (see also Hov et al., 1978a and Derwent and Hov, 1979).

Recently several workers have applied operator splitting techniques to solve particular versions of eq. (2). It is then possible to apply special numerical procedures to solve the various parts of the equation, where the basic elements are those due to transport and those due to chemistry (see McRae et al., 1982, Hov, 1983a).

3 PARAMETRIZATION OF CHEMICAL TRANSFORMATIONS IN THE GAS PHASE

3.1 Composition of atmospheric hydrocarbons

The emissions of organic material into the atmosphere have an exceedingly complex composition. Even though the mass of natural emissions dominates on a global basis, it is generally believed that natural organic emissions play a minor role in the formation of pollutants which affect the air quality (Dimitriades, 1981). Attention will therefore be paid here to the chemical transformation of anthropogenic hydrocarbon emissions.

Vehicle traffic is a dominating source of a long range of hydrocarbons. In a survey of the gas-phase hydrocarbons $\geq C_5$ generated by motor vehicles in highway operation, approximately 400 vehicle-generated compounds were detected in the Allegheny Mountain Tunnel in Pennsylvania, U.S. (Hampton et al., 1982. This paper also contains a comprehensive list of references dealing with the chemical composition of exhaust gas). The hydrocarbon composition as measured in road tunnels or close to busy roads, is representative also of the composition of the emissions. The large variation in reactivity among the anthropogenic hydrocarbons quickly causes a change in the concentration distribution of a given air sample with time or distance from the source. In Table 2 is shown the average ratios of hydrocarbons to acetylene as measured in the Lincoln Tunnel in New York. In Table 3 is shown the concentrations of nine hydrocarbons detected in clean air in the remote troposphere far away from anthropogenic sources. The numbers can be viewed on the background of the reactivity distribution of anthropogenic hydrocarbons illustrated in Figure 2.

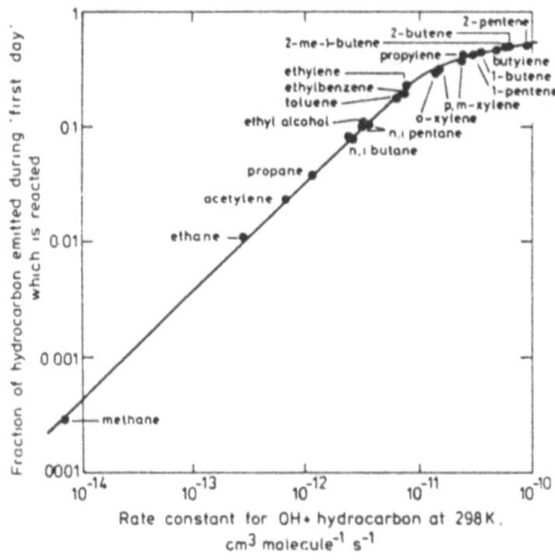

Figure 2. The fraction of each hydrocarbon emitted into an Eulerian urban box model for London which was reacted in a day as a function of the OH reaction rate coefficient. The continuity equation (2) was simplified to

$$\frac{dc}{dt} = R - (c-c_b)\left(\frac{1}{h}\ \frac{dh}{dt} + \frac{v}{L}\right)$$

where c_b is the background concentration, h the mixing height, v mean wind (2.4 m/s) and L horizontal dimension of the model box (50 km) (Derwent and Hov, 1979).

Table 2. Average ratios of hydrocarbons to acetylene in the Lincoln Tunnel in New York, calculated from component concentrations in ppbC (compiled by Killus and Whitten, 1983).

Component	Ratio of component to C_2H_2 and standard deviation
Ethylene	1.33 ± 0.14
Isobutane	0.34 ± 0.05
n-Butane	0.97 ± 0.12
Propylene	0.61 ± 0.07
Isopentane	1.25 ± 0.14
Isobutylene	0.34 ± 0.04
Sum of C_4 olefins	0.60 ± 0.07
n-Pentane	0.62 ± 0.07
Sum of C_5 olefins	0.53 ± 0.08
Cyclopentane	0.76 ± 0.08
3-Methylpentane	0.34 ± 0.04
n-Hexane	0.36 ± 0.05
2,4-Dimethylpentane	0.34 ± 0.04
2,2,4-Trimethylpentane	0.27 ± 0.23
Toluene	1.27 ± 0.23
Ethyl benzene	0.22 ± 0.03
p-Xylene	0.25 ± 0.03
m-Xylene	0.70 ± 0.15
o-Xylene	0.28 ± 0.04
Sum of C_8 aromatics	1.44 ± 0.25
3 & 4-Ethyl toluene	0.38 ± 0.05
sec-Butyl benzene	0.40 ± 0.06
Sum of paraffins	6.81 ± 0.92
Sum of olefins	3.24 ± 0.32
Sum of aromatics	3.87 ± 0.58
Total nonmethane hydrocarbons	13.9 ± 1.5
Carbon monoxide	63.4 ± 6.1

Table 3. Hydrocarbons in clean air (compiled by Penkett, 1982).

Compound	Concentration pptv		Source
	N.H.*)	S.H.*)	
CH_4	1575×10^3	1510×10^3	n (a)*)
C_2H_6	1030	240	n a
C_2H_4	80	100	n
C_2H_2	160	20	a
C_3H_8	85		n a
C_3H_6	53		n
C_4H_{10}	10		a
C_6H_6	64		a
C_7H_8	17		a

*) S.H. and N.H.: Southern and northern hemispheres, n denotes natural origin, a anthropogenic, ( ) minor contribution.

Northern hemispheric ratios of nonmethane hydrocarbons to acetylene on a ppbC basis:

paraffins/acetylene	7.4	
olefins/acetylene	1.0	(.0 if anthropogenic influence only is considered)
aromatics/acetylene	1.6	

Acetylene is a slowly reacting substance of anthropogenic origin. The characteristic chemical lifetime is approximately 1 month with a mean hydroxyl concentration of 1×10^6 molecules/cm^3. Ethane is even less reactive, and from Table 3 it can be seen that it is the dominant paraffin species in clean air. The more reactive paraffins have been depleted during transport (cpr. Figure 2 and Table 2). The shift in the hydrocarbon composition from the polluted site (Table 2) to the remote troposphere (Table 3) is further demonstrated by the strong decline in the olefins/acetylene and aromatics/acetylene ratios. This is due to the in general high reactivity of olefins and aromatics. These species are removed with characteristic chemical lifetimes of the order of 1 - 100 h, with the exception of benzene which has a chemical lifetime of approximately 10 d against the attack of hydroxyl with a mean concentration of 1×10^6 molecules/cm^3.

The change in the hydrocarbon composition with time from the source region is further demonstrated in Figure 3. Paraffins and slowly reacting species like acetylene and acetone were calculated to make up the bulk of the hydrocarbons a few days downwind of the U.K. sources, and among the paraffins ethane dominated. The change with time in the composition of the hydrocarbons in an ageing, polluted air mass was further discussed by Derwent and Hov, 1982. See Table 4 for specification of model HC-mixture.

The hydrocarbons do not affect the air quality, with the possible exception of formaldehyde, benzene (and PAH, chlorinated hydrocarbons etc.) close to the sources. It is their role in the formation of secondary pollutants which dictates the treatment of hydrocarbons in air quality modelling.

The discussion above has shown that over a time period of several days there is a significant shift in the hydrocarbon composition of a polluted air mass. Species which do not contribute significantly to the oxidant formation on an urban time scale (a few hours), can be important precursors on a regional spatial scale where a polluted air mass undergo chemical transformations over several days under the action of sunlight.

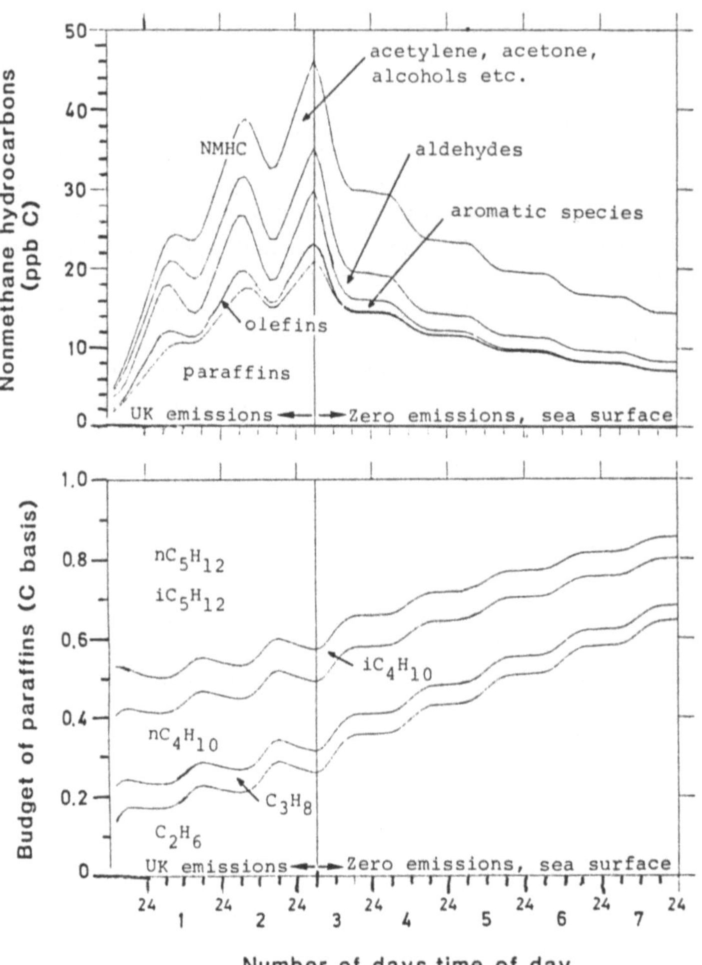

Figure 3. Development with time of the nonmethane hydrocarbons (NMHC) and their composition (upper diagram), and of the relative composition of the paraffins (on a C-atom basis), lower diagram. Lagrangian type box model:

$$\frac{Dc}{dt} = \frac{E}{h} + P_{ch} - \left(L_{ch} + \frac{v_d}{h}\right) c$$

where h is the mixing height (1300 m). Emissions of hydrocarbon were distributed among 35 different species. Average U.K. emissions of NO_x, HC and SO_2 were emitted for 2½ days, whereafter transport over a sea surface was modelled (Hov, 1983c).

Table 4. Species which were emitted in the model development by Derwent and Hov (1979) for the U.K.

NO	iC_5H_{12}	CH_3CHO	CH_3OH
SO_2	C_2H_4	C_2H_5CHO	C_2H_5OH
CO	C_3H_6	C_3H_7CHO	1-butene
CH_4	C_2H_2	iC_3H_7CHO	2-butene
C_2H_6	toluene	C_4H_9CHO	2-pentene
C_3H_8	o-xylene	CH_3COCH_3	1-pentene
nC_4H_{10}	m-xylene	$CH_3COC_2H_5$	2-methyl-1-butene
iC_4H_{10}	p-xylene	methylpropylketone	3-methyl-1-butene
nC_5H_{12}	ethylbenzene	methyl-i-propylke-tone	2-methyl-2-butene
	HCHO		butylene
			benzaldehyde

3.2 Hydrocarbons in atmospheric photochemistry

NO_2, NO and O_3 take part in a cyclic set of reactions:

$$NO_2 + h\nu \xrightarrow{k_1} NO + O(^3P) \qquad \text{(R1)}$$

$$O(^3P) + O_2 + M \rightarrow O_3 + M \qquad \text{(R2)}$$

$$NO + O_3 \rightarrow NO_2 + O_2 \qquad \text{(R3)}$$

M denotes an air molecule. The calculation or assessment of the dissociation rate coefficients (e.g. k_1 for NO_2) is an important part of any photochemical air quality model, since the solar energy drives the chemistry. In general,

$$k_i(z,t) = \int_{\lambda_1}^{\lambda_2} \phi_i(\lambda,z)\ \sigma_i(\lambda,z)\ F(t,\lambda,z)d\lambda \qquad (3)$$

where λ_1 and λ_2 denote the wavelength interval of absorption for species i, $\phi_i(\lambda,z)$ denotes the wavelength and height dependent quantum yield, $\sigma_i(\lambda,z)$ the wavelength and generally also height dependent absorption crossection, and $F(t,\lambda,z)$ denotes the time of day, wavelength and height dependent actinic irradiance (solar flux). Most dissociation rate coefficients are calculated theoretically. Several parametrization schemes are in use for this purpose. (Leighton, 1961, Luther and Gelinas, 1976, Peterson, 1976, Isaksen et al., 1976, Schere and Demerjian, 1977). Absorption by O_3 and O_2, Rayleigh scattering, reflection by the surface or clouds, and absorption and scattering by aerosols affect the solar flux in the atmosphere.

Reaction (R3) followed by (R1) and (R2) does not affect the ozone concentration. NO may be converted to NO_2 without any loss of O_3, however:

$$HO_2 + NO \rightarrow NO_2 + OH \qquad (R4)$$

$$RO_2 + NO \rightarrow NO_2 + RO \qquad (R5)$$

which followed by (R1) and (R2), lead to the formation of O_3. RO_2 denotes an peroxyalkyl radical (e.g. CH_3O_2). Hydroperoxy and peroxyalkyl radicals arise from the oxidation of hydrocarbons, which occurs through reactions of the form

$$HC_i + OH \xrightarrow{k_{i,6}} \text{products} \qquad (R6)$$

This reaction is usually the rate determining step in a chain of reactions where the hydrocarbon (HC_i) eventually is decomposed into stable end products (CO, CO_2) or intermediate species are formed, like aldehydes. Several reactions of the types (R4) or (R5) may occur during the decomposition, depending on the number of carbon and hydrogen atoms in the original molecule.

Hydrocarbons containing double bonds (olefins) also react with ozone. This is an important point to bear in mind in air quality modelling covering several days, since O_3-hydrocarbon reactions dominate at night when hydroxyl vanishes.

The hydroxyl radical is assumed to drive the major part of the hydrocarbon degradation. It is important to keep track of its production and loss mechanisms. The major production in polluted air comes from the degradation of hydrocarbons or intermediate species like formaldehyde where hydroperoxy radicals are formed and converted to hydroxyl through (R4). Photolysis of hydrogenperoxide also leads to hydroxyl formation. In clean air the most important source of hydroxyl is through the reactions

$$O_3 + h\nu \rightarrow O(^1D) + O_2 \qquad \lambda < 3100Å \qquad (R7)$$

followed by

$$O(^1D) + H_2O \rightarrow 2OH \qquad (R8)$$

Important loss of hydroxyl and hydroperoxy radicals takes place through

$$NO_2 + OH \rightarrow HNO_3 \qquad (R9)$$

$$SO_2 + OH \rightarrow HSO_3 \qquad (R10)$$

$$HO_2 + CH_3O_2 \rightarrow CH_3O_2H + O_2 \qquad (R11)$$

$$HO_2 + HO_2 \rightarrow H_2O_2 \qquad (R12)$$

$$HO_2 + NO_2 \rightarrow HO_2NO_2 \qquad (R13)$$

The degree of loss of radicals through reaction (R10) is determined by the fate of the HSO_3 radical. One investigator believed that hydroperoxy radicals are formed through the subsequent decomposition of HSO_3, in which case the sum $OH + HO_2$ is not affected and reaction (R10) is not a real sink of radicals containing H atoms (Cox, 1974). Photolysis of CH_3O_2H, H_2O_2 and HO_2NO_2 and thermal decomposition of HO_2NO_2 also lead to formation of OH or HO_2 in which case these species only serve as a temporary storage for radicals. The species are water soluble, however, and the reactions (R11) - (R13) followed by removal of the products in water droplets, serve as radical sink reaction pathways.

3.3 Modelling of the photochemistry of an urban atmosphere

Most model studies of photochemical transformation of air pollutants have focused on the urban spatial scale and on a time scale of a few hours, i.e. "first day" episodes where carry-over effects from previous days have been disregarded or only crudely parametrized. Most of these models have been applied in California (Los Angeles - San Francisco).

There are several different strategies currently in use to parametrize the chemical transformation of hydrocarbon precursors. The most straight forward procedure is the explicit representation of the chemical degradation of each individual hydrocarbon thought to play a role in the chemical turnover. In practical work it is not possible to model all the hydrocarbons known to be emitted, and the explicit representation of each individual hydrocarbon is replaced by a surrogate mechanism, where organic species in a particular class, e.g. alkenes, are represented by one or more members of that class. In Table 5 the representation of the hydrocarbons in some surrogate mechanisms is listed.

Table 5. Representation of hydrocarbons in some surrogate mechanisms.

Problem studied	Hydrocarbons included	Investigator
Urban photochemistry, Northern New Jersey	Propene, formaldehyde, acetaldehyde, proprion-aldehyde	Graedel et al., (1976).
Urban photochemistry, Oslo	C_2H_4, C_3H_6, n-C_4H_{10}, hexane, m-xylene, C_2H_2, CH_4, HCHO, CH_3CHO	Hov et al., (1978b).
Urban photochemistry, urban plume, regional photochemistry	Listed in Table 4 (35 different species)	Derwent and Hov, (1979, 1980a).
Long-range transport	C_2H_6, n-C_4H_{10}, C_2H_4, m-xylene, C_3H_6	Eliassen et al. (1982).
Boundary layer photochemistry	C_2H_6, n-C_4H_{10}, C_2H_4, m-xylene, C_3H_6	Hov (1983b).
Reactive plume	Ethane, ethene, propane, propene, n-butane, i-pentane, benzene, toluene, m-xylene	Hov and Isaksen (1981).
Reactive plume, U.K.	35 different species, listed in Table 4	Hov and Isaksen (1981).

It should be strongly emphasized that the usefulness of any chemical mechanism applied in a model, rests heavily upon the following factors: a) The data for hydrocarbon emissions must have the same detail as the model representation. b) The certainty by which the chemical degradation pathways, reaction rate coefficients and end products of each individual hydrocarbon is known. c) The availability of air quality data where the hydrocarbons are measured with the temporal and spatial resolution and speciation required by the model. d) Availability of computer resources. Among these four points, the first three represent serious obstacles for further progress in the modelling of the transformation of organic species in the atmosphere. If a ranking of the areas where the deficiencies of knowledge are most serious, the lack of appropriate emission data (a) and air quality data (c) come first. An extensive database now exists where detailed mechanisms for the photooxidation of hydrocarbons have been modelled and the results kept together with smog chamber simulation data for model validation. Examples can be found in the work by Demerjian et al. (1974), Carter et al. (1979, 1981), Killus and Whitten (1982, 1983), Atkinson et al. (1980, 1982), Akimoto et al. (1977) and Glasson and Wendschuh (1977). Smog-chamber studies usually cover the first few hours of development of a chemical mixture.

Using a surrogate mechanism is one way of representing a complex hydrocarbon mixture in a model. Other approaches involve lumping the hydrocarbons by classes according to a common basis such as structure or reactivity and using a generalized reaction mechanism for these classes. In Table 6 some lumped mechanisms are reviewed. In a lumped photochemical mechanism the atmospheric hydrocarbon chemistry is represented by reactions of the form (McRae et al., 1982)

$$HC_j + X_m \xrightarrow{\overline{k^m_j}} \text{products} \qquad \text{(R14)}$$

where X_m, typically atomic oxygen, hydroxyl radical or ozone and the j'th hydrocarbon class react. Each class is composed of many different species, and the lumped reaction rate coefficient ($\overline{k^m_j}$) is composition dependent. It can be written as

$$\overline{k^m_j} = \sum_{i=1}^{N_j} k^m_i n_i \Big/ \sum_{i=1}^{N_j} n_i \qquad (4)$$

Table 6. Lumped mechanisms for hydrocarbon representation in some photochemical smog models.

Problem studied	Partition of organic compounds	Investigator
Urban atmosphere	HC	Reynolds et al., (1973).
Urban atmosphere (Los Angeles)	Low reactivity hydrocarbon (HC) and a high reactivity hydrocarbon (HC_2).	Reynolds et al., (1973).
Urban atmosphere (Houston)	Olefins, paraffins, aldehydes, aromatics (4 classes).	Demerjian and Schere (1979).
Urban (regional) San Francisco Bay Area (LIRAQ)	Alkenes (HC1), alkanes (HC2), aldehydes (HC4)	Mac Cracken et al. (1978), based on Hecht et al. (1974).
Urban atmosphere (Los Angeles)	Alkanes, olefins, aromatics, oxygenated compounds (aldehydes)	McRae et al. (1982), based on Falls and Seinfeld (1978).
Urban atmosphere (ELSTAR)	Alkanes, ethene, other alkenes, aromatic hydrocarbons, formaldehyde, acetaldehyde	Lloyd et al. (1979).
Urban atmosphere	Propane, $\geq C_4$ alkanes, ethene, 1-alkenes, internal alkenes, benzene, monoalkylbenzenes, dialkylbenzenes, formaldehyde, acetaldehyde, higher aldehydes and ketones	Atkinson et al. (1982).

where N_j is the number of species in the lumped hydrocarbon class j, while k_i^m is the reaction rate coefficient of the individual hydrocarbon i in class j with X_m, and n_i is the concentration of hydrocarbon i in class j. It is obvious that the composition of hydrocarbons in each lumped class j will change with time, thus $\overline{k_j^m}$ will change. The more reactive hydrocarbons will be consumed first, which means that $\overline{k_j^m}$ in general will decrease with time. This may not be a significant problem in the modelling of urban photochemistry on a time scale of a few hours with continuous emissions. On a regional scale with computations over several days, the evaluation of $\overline{k_j^m}$ will become a serious problem, however. The lumped mechanism does not allow keeping track of the development of the composition of each HC class. In Figure 3 it was shown how the composition of the paraffins, typically taken together in lumped mechanisms, was calculated to change drastically with time in a long-range transport situation. This will obviously affect $\overline{k_j^m}$. Falls and Seinfeld (1978) stated that lumped mechanisms by necessity involve parameters which must be chosen within theoretically defined bounds to provide the best fit of data and predictions. The art of developing a mechanism is to deal with the inherent uncertainties in an optimal manner.

Lumped mechanisms have not been applied in the modelling of photochemical transformations over several days and covering a spatial scale of the order of 1000 km, to this author's knowledge. It seems difficult to resolve the evaluation of $\overline{k_j^m}$ in such studies, and it may be recommended to use surrogate mechanisms under such circumstances. It also seems as if the carbon-bond mechanism may then be better suited than lumped mechanisms.

The carbon-bond mechanism (CBM) is a generalized kinetic mechanism in which most carbon atoms with similar bonding are treated similarly, regardless of the molecules in which they occur. It is based on the concept of grouping carbon atoms with similar chemical bonding, rather than grouping hydrocarbon molecules of similar chemical type (Whitten et al., 1980, Killus and Whitten, 1983). In its current form (CBM-III) described by Killus and Whitten (1983), the CBM treats the reactions of six types of carbon atoms: (1) single-bonded carbon atoms, whose principal constituent is paraffinic carbon molecules (PAR), (2) relatively reactive double-bonded carbon (OLE), (3) slow double bonds, which are almost exclusively ethene (ETH), (4) reactive aromatic rings (ARO), (5) carbonyl compounds such as aldehydes and ketones (CARB), and (6) highly photolytic α-dicarbonyl compounds such as methyl glyoxal and biacetyl (DCRB). Other types of carbon atoms can also be treated within this frame work. An extensive tabulation of recommended CBM fractions for a variety of organics is also provided. There are several advantages with the CBM approach over the lumped mechanism approach. No data on average molecular weights are necessary, and the range of reaction rate constants to be averaged can be reduced (Whitten et al., 1980). In a lumped mechanism, the accurate simulation of a mixture of propene, butene and pentene using a single "olefin" species would require careful selection of an average molecular weight. During the simulation, the average molecular weight would change due to differences in reactivity of the different species of the olefin group. It would not be possible to calculate this change appropriately, and some parametrization would have to be involved.

Whitten et al. (1980) illustrated the second advantage of the CBM mechanism (the range of reaction rate constants to be averaged is reduced) as follows: the paraffins n-butane and isooctane react with OH with rate constants of about 3800 and 11500 ppm^{-1} min^{-1}, respectively. If they are grouped by numbers of single-bonded carbon atoms, these rate constants become 950 and 1440 ppm^{-1} min^{-1}, which considerably narrows the range of rate constants to be averaged. A third advantage of CBM over the lumped mechanisms, is that it is designed to conserve the total carbon with time. This is an important property when the budget of hydrocarbons is to be assessed over time periods typical of regional oxidant studies: a few days. The CBM mechanism has been used in several regional-scale air quality models (Liu and Reynolds, 1983, Killus et al., 1983, Scherer and Stern, 1982). Scherer and Stern applied the CBM 3-D air quality dispersion model developed by Systems Applications Inc., to study the photochemistry occurring within a grid of size approximately 160×160 km^2 in the Rhine-Ruhr area, only to find that the spatial and temporal scales of the problem frequently were such that the boundary conditions strongly influenced the computations.

The chemical mechanisms reviewed above, are all based on kinetic, mechanistic and product data taken from the literature. Reaction rate constants and mechanisms are usually consistent with recent evaluations of NBS (Hampson and Garvin, 1978), NASA (1979, 1982), CODATA (Baulch et al., 1980) and Atkinson et al. (1979).

3.4 Composition of nonmethane hydrocarbon emissions in surrogate mechanisms

The most reasonable approach in model work may be to represent the hydrocarbon precursor mixture by just a few of the most abundant species. The way in which the selection of such a representative hydrocarbon mixture is made, however, has a significant influence on the overall model results. This will be discussed in this section for cases on a temporal scale of some days.

Six simplified compositions were selected, two of them taken from EPA publications (mixtures No. 4 and 6, see legend of Table 7). Four others were constructed, usually with five hydrocarbons. For some mixtures, the dividing lines between the amounts attributed to each species were drawn on the basis of the reactivity distribution of the original U.K. inventory. Reactivity of a hydrocarbon is only one aspect of the secondary pollutant generation, however. Other important factors describe the stoichiometry, that is how much of a given pollutant is generated per hydrocarbon degraded, and inventory, that is how much is emitted of a given hydrocarbon species.

Comparison of the performance of the various mixtures was done for a situation where the development of each mixture was followed for 7½ days. The total hydrocarbon emissions were identical for all mixtures on a mass basis, equal to the average U.K.-1975 release.

In Table 7 is given the peak or near-peak concentration of ozone, PAN and hydroxyl for every day for the U.K. mixture (No. 1), and for all the other 6 mixtures relative to the U.K. case. In Table 7 is also given the geometric mean of the ratios for the whole period.

The performance relative to the U.K. mixture is seen to be quite dependent on time, with the largest discrepancies on days 2 and 3. For ozone and hydroxyl, the results fell quite well in line for all mixtures on days 5 - 8. The prediction of PAN showed more spread. These results were expected when it is kept in mind that the ozone and hydroxyl production is comparable for most hydrocarbons. PAN is more sensitive to the structure of the hydrocarbon precursor.

Ozone production requires the production of peroxy radicals in the hydrocarbon degradation, a condition which is generally satisfied. PAN can only be formed from a particular hydrocarbon if it produces an acetylperoxy radical in its degradation. Methane and acetylene do not have the required structure, for example. The 35 nonmethane hydrocarbons in the model by Derwent and Hov, 1979 have a fairly even distribution with respect to ozone production, while a much more uneven distribution is observed for PAN (Derwent and Hov, 1980b). This means that a correct composition of the hydrocarbon emissions in the model is very important for a satisfactory calculation of the absolute concentration of PAN, while it is less critical for ozone.

Mixtures 4 and 5 gave the poorest result compared to the U.K. case, while 2, 3 and 7 were quite similar. All mixtures except no. 6 overestimated secondary pollutant yield compared to the U.K. case. Mixture no. 6 comprised as many as eight different hydrocarbons.

Table 7. Model calculations for 8 days of ozone, PAN and hydroxyl. All calculations were done with identical emissions on a mass basis, equal to the U.K. release for 1975 (Derwent and Hov, 1979). The calculations were done with 7 different compositions of the hydrocarbon emissions, and started at tropospheric background concentrations at 12 a.m. on day No. 1.

Mixture No.	*Composition*
1	*UK emissions for 1975, the hydrocarbons distributed among 35 species.*
2	*20% of total hydrocarbon on each of C_3H_8, nC_4H_{10}, C_2H_4, C_3H_6 and m-xylene.*
3	*C_3H_8, nC_4H_{10}, C_2H_4, C_3H_6 and m-xylene distributed according to reactivity to represent the total NMHC spectrum as estimated for mixture 1 (Hov and Derwent, 1981).*
4	*25% as C_3H_6, 75% as nC_4H_{10}, 2% as CH_3CHO and 1% as HCHO on a C basis (Dodge, 1977).*
5	*UK mixture distributed among C_3H_6, nC_4H_{10}, HCHO and CH_3CHO according to reactivity.*
6	*"Urban mix" as recommended by Glasson and Wendschuh, 1977 (9.2% as C_2H_6, 18.3% as C_3H_8, 23% as nC_4H_{10}, 10.3% as iC_5H_{12}, 11.6% as C_2H_4, 8.2% as C_3H_6, 13.6% as toluene, 5.3% as m-xylene).*
7	*30% as C_2H_6, 10% as nC_4H_{10}, 20% as C_2H_4, 10% as C_3H_6, 30% as m-xylene.*

Ozone at 6 pm

Hydrocarbon mixture No.	1	2	3	4	5	6	7
	ppb	Concentration ratio relative to mixture 1					
Day No.							
1	8.2	1.13	1.05	1.16	1.33	0.91	1.07
2	35.7	1.56	1.33	1.37	1.80	.85	1.46
3	74.5	1.28	1.21	1.25	1.39	.89	1.23
4	108.6	1.12	1.10	1.14	1.19	.99	1.09
5	129.8	1.08	1.08	1.12	1.13	1.03	1.05
6	144.3	1.05	1.07	1.10	1.10	1.04	1.03
7	154.1	1.04	1.06	1.10	1.09	1.05	1.02
8	161.1	1.04	1.06	1.09	1.08	1.05	1.02
Geometric mean of ratios		1.15	1.12	1.16	1.25	.97	1.11

(continued)

Table 7. Aspects of the parameterization of Transformation (Continued)

PAN at 6 pm

Hydrocarbon mixture No.	1	2	3	4	5	6	7
	ppb	Concentration ratio relative to mixture 1					
Day No.							
1	.02	1.69	1.37	1.74	2.85	.73	1.46
2	.18	2.62	2.23	2.49	3.86	.75	2.41
3	.68	1.52	1.69	2.15	2.24	.84	1.36
4	1.20	1.22	1.44	1.90	1.82	1.12	1.06
5	1.52	1.17	1.41	1.87	1.74	1.17	1.01
6	1.76	1.14	1.39	1.85	1.69	1.18	.99
7	1.93	1.13	1.38	1.83	1.66	1.18	.98
8	2.05	1.11	1.37	1.81	1.63	1.17	.98
Geometric mean of ratios		1.39	1.51	1.94	2.08	1.00	1.22

Hydroxyl at noon

Hydrocarbon mixture No.	1	2	3	4	5	6	7
	*	Concentration ratio relative to mixture 1					
Day No.							
1							
2	5.91	1.51	1.26	1.37	1.95	.81	1.36
3	7.15	1.47	1.38	1.52	1.43	.81	1.37
4	9.49	.95	.96	.94	.93	.97	.96
5	8.28	1.00	.99	.99	1.00	.96	1.00
6	7.67	1.01	1.01	1.01	1.02	.99	1.01
7	7.24	1.02	1.02	.99	1.04	1.00	1.02
8	6.91	1.03	1.02	1.04	1.05	1.00	1.02
Geometric mean of ratios		1.11	1.07	1.09	1.14	.94	1.08

* Concentration in 10^6 molecules cm^{-3}

It can be seen that mixture 7 predicted approximately 40% more ozone and hydroxyl, and a factor two higher PAN, than the U.K. mixture on day no. 2, while the results fell in line on days 5 - 8. It cannot be claimed that these numbers represent the uncertainty of model predictions with respect to choice of composition of the hydrocarbon emissions. A substantial error is linked to this choice, however.

As the number of days increased, the daily increase in ozone approached zero. This means that the production of ozone balanced the loss due to ground deposition and chemistry. The time scale for reaching a near steady state level could be measured by the reactivity of the slowest-reacting nonmethane hydrocarbons. It turned out that after 7 - 10 days, all hydrocarbons which were attacked by hydroxyl only, reached a fairly steady level on a diurnal average basis. Olefins also react with ozone, and were suppressed due to the increasing ozone levels. The assumption of identical mass of HC emissions was in the end sufficient to reach almost the same ozone level in all cases regardless of the composition of the hydrocarbon emissons.

The calculated concentration of PAN was strongly dependent on the choice of composition of the hydrocarbon emissions. A difference of more than a factor of 4 was found in the concentrations of PAN for equal emissions. This difference was much less pronounced for ozone or hydroxyl.

3.6 Hydrocarbon mechanisms and control strategy predictions

The U.S. EPA kinetic model and ozone isopleth package (OZIPP, Whitten and Hogo, 1978) has been developed to relate photochemical oxidants (expressed as ozone) downwind of an urban area to organic compounds and oxides of nitrogen. The physical model underlying the empirical kinetic modelling approach (EKMA) is based on the concept of a column of air transported along an assumed trajectory. Fresh emissions may be added as the column moves. The kinetic model is based on a chemical scheme for the decomposition of a mixture of n-butane, propene, and NO_x. EKMA is best used to determine the sensitivity of maximum hourly O_3 concentrations observed within or downwind of a city to changes in ambient levels of nonmethane hydrocarbons, NO_x precursors, and concentrations of O_3 and precursors transported from upwind areas.

The urban plume model for London described by Derwent and Hov (1979) also described the chemical transformations in a column of air transported along an assumed trajectory. The kinetic model was much more complex, however, cpr. Table 4. Approximately 160 different species were computed, involving nearly 300 chemical and photochemical reactions (Derwent and Hov, 1979).

Ozone isopleths diagrams have been calculated for London applying both models (Hov and Derwent, 1981). In Figure 4 is shown the diagram obtained with the EKMA model and the other model (AERE Harwell model). The formulation of the AERE Harwell urban plume model is shown in Figure 4 as well. The ozone isopleths diagram for London using the AERE Harwell model was constructed by varying the urban emissions, and then taking the 0900 h concentrations of NMHC and NO_x as representative of the maximum urban burden. The maximum downwind concentrations of ozone were taken to construct the isopleths diagram.

It can be seen that the choice of hydrocarbon mechanism had a significant impact on the model result. The AERE Harwell model predicted that HC control always reduces O_3, while NO_x control in some cases may even increase O_3. A combined HC and NO_x control strategy may leave O_3 near the same level, or increase O_3 if the NO_x control dominates over the HC control. If HC is controlled efficiently, O_3 may drop. These results do not agree very well with those of OZIPP, where NO_x control or a combined HC and NO_x control in most cases reduces O_3 (Hov and Derwent, 1981).

There are significant difference in the conclusions drawn from the two model calculations described above. It is therefore important to be aware that the assumptions and parametrizations made may constrain the computations to a large extent. The results may become model specific rather than physically realistic.

3.7 Parametrization of night-time chemistry and variable sun

A model for photochemistry of an urban atmosphere (Oslo) was described by Hov et al. (1978b). The formulation of the continuity equation was similar to that of the London urban model, compare the caption of Figure 2.

In Figure 5 is shown the effect of assuming a "dead" night-time chemistry, i.e. the initial values at sunrise were equal to the values at sunset. The ozone maximum was lowered by about 10%, compared to the case when night-time chemistry was included, while the development before noon deviated strongly and there was a false

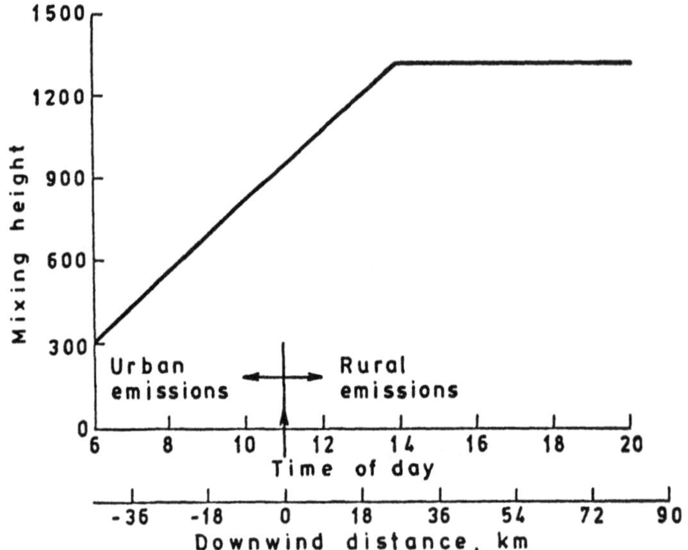

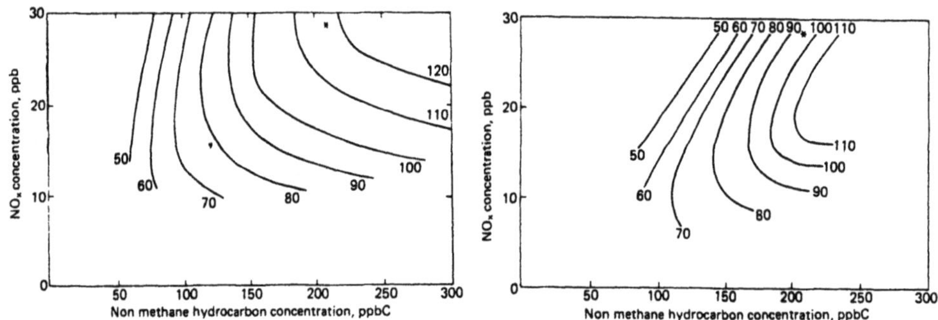

Figure 4. (Top) Formulation of the urban plume model for London (Derwent and Hov, 1979).
(Bottom left) Ozone isopleth diagram for London constructed using OZIPP. Isopleth ozone concentrations for 1800 h in ppbv.
(Bottom right) Ozone isopleth diagram for London calculated with AERE Harwell fixed point urban model. NMHC and NO_x concentrations are taken for 0900 h and the ozone contours in ppbv represent the daily maixmum values.

*refers to present quality.

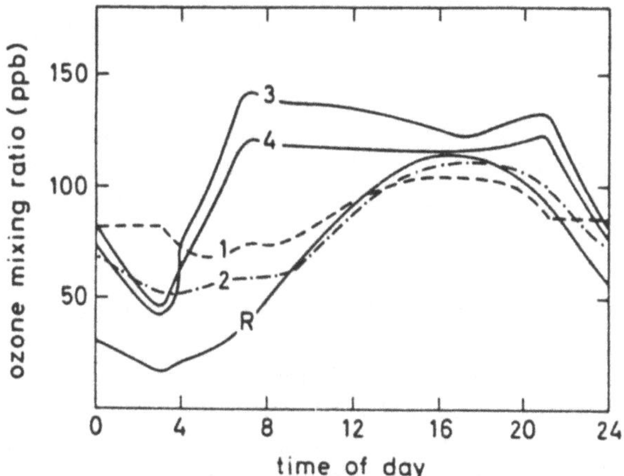

Figure 5. Diurnal variation of ozone in different urban model sensitivity studies. R denotes the standard ozone curve. For curve 1 a "dead" night-time chemistry was assumed, i.e. the concentrations at sunrise were assumed equal to the concentrations at sunset. Curve 2: Low emissions during the night and higher during the day. Curve 3: Noon values of the dissociation rate coefficients during the sunlit hours. Curve 4: Same as curve 3, except 3 p.m. dissociation rate coefficients were used (Hov et al., 1978b).

minimum introduced about two hours after sunrise. The diurnal mean in ozone concentration was considerably overestimated if the effect of night-time chemistry was neglected. Other compounds with daily maxima showed deviations similar to those of ozone, while the effect on compounds with nocturnal maxima (e.g. NO_3 and hydrocarbon peroxy radicals) could not be accounted for as long as the day/night shifting was not modelled.

To investigate whether the nocturnal minimum in ozone is caused by gas chemical or physical processes (deposition, dilution), a simulation run was made where all parameters were unchanged except the pollutant emissions. They were assumed to have a diurnal variation with a low rate at night (about 20% of the daily average) and a higher rate during the day (Figure 5). The nocturnal minimum of ozone was much less marked. This was caused by the difference in NO concentrations in the two cases, and it indicated that gas chemical processes are responsible for the nocturnal ozone minimum often observed in urban areas.

Use of constant dissociation rates or a fixed solar zenith angle cannot be recommended in the modelling of a day/night system. The results exhibited obvious weaknesses (Figure 5). With constant dissociation rates, a considerable shift in time and size of ozone maximum occurred, with false maximum values at sunrise and sunset. The use of a constant 3 p.m. value for dissociation rates was in principle as bad an approximation as application of the noon values.

4 PARAMETRIZATION OF WET PHASE CHEMISTRY AND REMOVAL PROCESSES

The processes for removal of chemical species from the atmosphere are divided into two main groups: Wet deposition comprises the incorporation of chemical compounds in cloud droplets (rainout) and removal by falling precipitation (washout), while dry deposition denotes the direct collection of gaseous and particulate species on land or water surfaces (Garland, 1978). No attempt will be made to review the extensive literature on how to model wet phase chemistry and wet and dry removal from the atmosphere. Only some pertinent references will be given.

Rate data and chemical schemes for the aqueous phase oxidation of SO_2 in the presence of ammonia to ammonium sulphate were published by Scott and Hobbs (1967) and McKay (1971). Mechanisms for the oxidation of SO_2 by ozone in droplets were published by Penkett (1972), Penkett and Garland (1974), Erickson et al. (1977), Larson et al. (1978), Hegg and Hobbs (1978) and Penkett et al. (1979). Later the role of hydrogen peroxide in wet phase oxidation has been assessed by Penkett et al. (1979). Recently the free radical chemistry of cloud droplets, aqueous atmospheric aerosols and raindrops has been assessed by Chameides and Davis (1982) and Graedel and Weschler (1981). Chameides and Davis suggested that OH and HO_2 are formed in the gas phase during daytime within a warm cloud. Through heterogeneous scavenging, the radicals formed in the gas phase can represent a major source of free radicals to cloud water. H_2O_2 may then be formed in the cloud droplets, leading to a rapid oxidation of sulphur species. Graedel and Weschler arrived at similar conclusions for aqueous aerosols and raindrops.

A useful review of processes affecting the formation of atmospheric acidity, has been published by Cox and Penkett (1982).

Hales (1972, 1978, 1981, 1982a, 1982b) has published papers on the theory and modelling of wet removal of atmospheric gases and aerosols. General reference may also be given to the special issue of Atmospheric Environment on precipitation chemistry (Hales, 1982c). A parametrization scheme for sulphate washout has also been published by Hegg (1983).

Reviews on dry deposition have been published by Garland (1978, 1979) and Sehmel (1980). The parametrization of dry deposition in air quality models is carefully reviewed by McRae et al. (1982).

5. SENSITIVITY ANALYSIS

There is a long range of chemical and physical processes which are parametrized in air quality models: Chemical degradation pathways of hydrocarbons, interaction with inorganic nitrogen-hydrogen-oxygen-sulphur chemistry, choice of reaction rate coefficients, absorption cross sections and quantum yields of species which are photodissociated, calculation of solar fluxes, choice of composition of hydrocarbon precursor emissions, intensity of precursor emissions of hydrocarbons, nitrogen oxides and sulphur dioxide, input fields and boundary conditions of the model components, efficiency of ground removal processes and gas-particle interactions, meteorological parameters like temperature, relative humidity, air pressure, and transport processes (mean flow and diffusion). It is inevitable that there is a significant uncertainty inherent in most of the data applied to describe these processes and parameters. An assessment of the model response to perturbations of boundary conditions, initial conditions or any parameter whithin the model, may be done indirectly by solving the model several times at distinct values of the input parameters. It may also be done directly. Dunker (1980, 1981) described a direct method where the sensitivity coefficients, which are the partial derivatives of the solution with respect to the parameters in question, can be calculated from the set of partial differential equations which define the model.

Dunker (1980) reported calculations done in a direct way of the response of the model developed by Reynolds et al. (1973) to changes in input parameters and functions. In a model involving non-linear chemical reactions, the output concentrations will not respond linearly for changes of arbitrary magnitude in the input functions. However, the response may be linear for input changes of restricted magnitude. Dunker (1980) suggested that changes in the initial conditions, emission rates or boundary conditions yield a response which is linear locally, given that the changes were restricted, though still sizable.

The sensitivity coefficients provide direct information on the effect on the individual state variables of small variations in each parameter about its normal value. An analysis that accounts for simultaneous parameter variations of arbitrary magnitude can be termed a global sensitivity analysis (Tilden and Seinfeld, 1982). In a global sensitivity analysis it is possible to consider the average sensitivity of simultaneous parameter variations over the actual expected ranges of uncertainty. Tilden and Seinfeld (1982) presented a method called Fourier amplitude sensitivity test method (FAST) to perform a global sensitivity analysis. They found that the predictions of a mathematical model for photochemical air pollution were most sensitive to uncertainties (±50%) in mixing height, photolysis rate constants, initial conditions and emission intensities.

Kramer et al. (1982a) described a computer code that computes first-order sensitivity coefficients for constant temperature and pressure chemical kinetics problems. The sensitivity equations were solved using the Analytically Integrated Magnus method. This is a general method for the calculation of first and second order sensitivities, and initial condition sensitivitites, for systems of first order ordinary differential equations. Second order sensitivities provide information on non-linearity and parametric interactions ($\partial^2 y_i(t)/\partial\alpha_j\partial\alpha_k$ where y_i is the i'th dependent variable, α_j the j'th system parameter, and t is the variable of integration), Kramer et al. (1982b).

The question of sensitivity analysis is also reviewed by Edelson (1981).

REFERENCES

Akimoto, H., Hoshino, M., Inoue, G., Okuda, M. and Washida, N. (1977) Photooxidation of the toluene-NO_2-O_2-N_2 system in a small smog chamber. EPA-600/3-77-001b, pp. 737-744.

Atkinson, R., Darnall, K.R., Lloyd, A.C., Winer, A.M. and Pitts, J.N., Jr. (1979) Kinetics and mechanisms of the reaction of the hydroxyl radical with organic compounds in the gas phase. Adv. Photochem. 11, 375-488.

Atkinson, R., Carter, W.P.L., Darnall, K.R., Winer, A.M. and Pitts, J.N., Jr. (1980) A smog chamber and modeling study of the gas phase NO_x-air photooxidation of toluene and the cresols. Int. J. Chem. Kin. 12, 779-836.

Atkinson, R., Lloyd, A.C. and Winges, L. (1982) An updated chemical mechanism for hydrocarbon/NO_x/SO_2 photooxidations suitable for inclusion in atmospheric simulation models. Atmospheric Environment 16, 1341-1355.

Baulch, D.L., Cox, R.A., Hampson, R.F., Kerr, J.A., Troe, J. and Watson, R.T. (1980) Evaluated kinetic and photochemical data for atmospheric chemistry. J. Phys. Chem. Ref. Data 9, 295-471.

Carter, W.P.L., Lloyd, A.C., Sprung, J.L. and Pitts, J.N., Jr. (1979) Computer modeling of smog chamber data: Progress in validation of a detailed mechanism for the photooxidation of propene and n-butane in photochemical smog. Int. J. Chem. Kin. 11, 45-101.

Carter, W.P.L., Atkinson, R., Winer, A.M. and Pitts, J.N., Jr. (1981) Evidence for chamber-dependent radical sources: Impact on kinetic computer models for air pollution. Int. J. Chem. Kin. 13, 735-740.

Chameides, W.L. and Davis, D.D. (1982) The free radical chemistry of cloud droplets and its impact upon the composition of rain. J. Geophys. Res. 87, 4863-4877.

Cox, R.A. (1974) The photolysis of nitrous acid in the presence of carbon monoxide and sulphur dioxide. J. Photochem. 3, 291-304.

Cox, R.A. and Penkett, S.A. (1982) Formation of atmospheric acidity. Proc. CEC workshop on "acid deposition", Berlin 9 September 1982. D. Reidel, Dordrecht, pp. 58-83.

Demerjian, K.L., Kerr, J.A. and Calvert, J.G. (1974) The mechanism of photochemical smog formation. Adv. Environmental Sci. Technol. 4, 11-262.

Demerjian, K.L. and Schere, K.L. (1979) Applications of a photochemical box model for O_3 air quality in Houston, Texas. Proc., Specialty conference on ozone/oxidants: interactions with the total environment. Houston, Texas 14-17/10-1979.

Derwent, R.G. and Hov, Ø. (1979) Computer modelling studies of photochemical air pollution formation in North West Europe. AERE-R9434, HMSO, London.

Derwent, R.G. and Hov, Ø. (1980a) Computer modelling studies of the impact of vehicle exhaust emission controls on photochemical air pollution formation in the United Kingdom. Environ. Sci. Technol. 14, 1360-1366.

Derwent, R.G. and Hov, Ø. (1980b) A simplified numerical method for estimating the potential for photochemical air pollution formation in the United Kingdom. AERE R-9682, Her Majesty's Stationery Office, London.

Derwent, R.G. and Hov, Ø. (1982) The potential for secondary pollutant formation in the atmospheric boundary layer in a high pressure situation over England. Atmospheric Environment 16, 655-665.

Dimitriades, B. (1981) The role of natural organics in photochemical air pollution. Issues and research needs. JAPCA 31, 229-235.

Dodge, M. C. (1977) Combined use of modeling techniques and smog chamber data to derive ozone precursor relationships. EPA-600/3-77-001b, pp. 881-889.

Dunker, A. (1980) The response of an atmospheric reaction-transport model to changes in input functions. Atmospheric Environment 14, 671-679.

Dunker, A. (1981) Efficient calculation of sensitivity coefficients for complex atmospheric models. Atmospheric Environment 15, 1155-1161.

Edelson, D. (1981) Computer simulation in chemical kinetics. Science 214, 981-986.

Eliassen, A., Hov, Ø., Isaksen, I.S.A., Saltbones, J. and Stordal, F. (1982) A lagrangian long-range transport model with atmospheric boundary layer chemistry. J. Appl. Met. 21, 1645-1661.

Erickson, R.E., Yates, L.M., Clark, R.L. and McEwen, D. (1977) The reaction of sulfur dioxide with ozone in water and its possible atmospheric significance. Atmosperic Environment 11, 813-817.

Falls, A.H. and Seinfeld, J.H. (1978) Continued development of a kinetic mechanism for photochemical smog. Environ. Sci. Technol. 12, 1398-1406.

Garland, J.A. (1978) Dry and removal of sulphur from the atmosphere. Atmospheric Environment 12, 349-362.

Garland, J.A. (1979) Dry deposition of gaseous pollutants. Proc. WMO symposium Sofia 1-5 Oct. 1979, WMO No. 538, Geneva, pp. 95-103.

Gear, C.W. (1971) The automatic integration of ordinary differential equations. Comm. A.C.M. 14, 176-179.

Glasson, W.A. and Wendschuh, P.H. (1977) Multiday irradiation of NO_x-organic mixtures. EPA-600/3-77-001b, pp. 677-685.

Graedel, T.E., Farrow, L.A. and Weber, T.A. (1976) Kinetic studies of the photochemistry of the urban troposhere. Atmospheric Environment 10, 1095-1116.

Graedel, T.E. and Weschler, C.J. (1981) Chemistry within aqueous atmospheric aerosols and raindrops. Rev. Geophys. Space Phys. 19, 505-539.

Grennfelt, P. (1979) Oxidized nitrogen compounds in long-range transported polluted air masses. Proc., WMO-symposium, Sofia 1-5 Oct., 1979, WMO No. 538, Geneva, pp. 199-206.

Hales, J.M. (1972) Fundamentals of the theory of gas scavenging by rain. Atmospheric Environment 6, 635-659.

Hales, J.M. (1978) Wet removal of sulfur compounds from the atmosphere. Atmospheric Environment 12, 389-399.

Hales, J.M. (1981) Pluvius: a generalized one-dimensional model of reactive pollutant behaviour, including dry deposition, precipitation formation, and wet removal. Battelle Pacific Northwest Laboratory, Richland, Washington 99352.

Hales, J.M. (1982a) The role of NO_x as a precursor of acidic deposition. Air pollution by nitrogen oxides, Eds. T. Schneider and L. Grant, Elsevier Scientific Publishing Co., Amsterdam, pp. 61-77.

Hales, J.M. (1982b) Mechanistic analysis of precipitation scavenging using a one-dimensional, time-variant model. Atmospheric Environment 16, 1775-1783.

Hales, J.M. (1982c) Precipitation chemistry, special issue, ed. Hales, J.M. Atmospheric Environment 16, 1603-1794.

Hampson, R.F. and Garvin, D. (1978) Reaction rate and photochemical data for atmospheric chemistry - 1977, National Bureau of Standards Special Publication 513.

Hampton, C.V., Pierson, W.R., Harvey, T.M., Updegrove, W.S. and Marano, S. (1982) Hydrocarbon gases emitted from vehicles on the road. 1. A qualitative gas chromatography/mass spectrometry survey. Environ. Sci. Technol. 16, 287-298.

Hecht, T.A., Seinfeld, J. and Dodge, M.C. (1974) Further development of a generalized kinetic mechanism for photochemical smog. Environ. Sci. Technol. 8, 327-339.

Heck, W.W., Taylor, O.C., Adams, R., Bingham, G., Miller, J., Preston, E. and Weinstein, L. (1982) Assessment of crop loss from ozone. JAPCA 32, 353-361.

Hegg, D.A. and Hobbs, P.V. (1978) Oxidation of sulfur dioxide in aqueous systems with particular reference to the atmosphere. Atmospheric Environment 12, 241-253.

Hegg, D.A. (1983) The sources of sulfate in precipitation. I. Parameterization scheme and physical sensitivities. J. Geophys. Res. 88, 1369-1374.

Hesstvedt, E., Hov, Ø. and Isaksen, I.S.A. (1978) Quasi-steady state approximation in air pollution modelling: Comparison of two numerical schemes for oxidant prediction. Int. J. Chem. Kinet. 10, 971-994.

Hileman, B. (1983) Outlook: 1982 Stockholm conference on acidification of the environment. Environ. Sci. Technol. 17, 15A-18A.

Hov, Ø., Isaksen, I.S.A. and Hesstvedt, E. (1978a) A numerical method to predict secondary air pollutants with an application on oxidant generation in an urban atmosphere. Proc., WMO-Symposium, Norrköping, 19-23 June, 1978, WMO No. 510, Geneva, pp. 219-226.

Hov, Ø., Isaksen, I.S.A. and Hesstvedt, E. (1978b) Diurnal variation of ozone and other pollutants in an urban area. Atmospheric Environment 12, 2469-2479.

Hov, Ø. and Derwent, R.G. (1981) Sensitivity studies of the effects of model formulation on the evaluation of control strategies for photochemical air pollution formation in the United Kingdom. JAPCA 12, 1260-1267.

Hov, Ø. and Isaksen, I.S.A. (1981) Generation of secondary pollutants in a power plant plume: a model study. Atmospheric Environment 15, 2367-2376.

Hov, Ø. (1983a) Numerical solution of a simplified form of the diffusion equation for chemically reactive atmospheric species. Atmospheric Environment 17, 551-562.

Hov, Ø. (1983b) One-dimensional vertical model for ozone and other gases in the atmospheric boundary layer. Atmospheric Environment 17, 535-549.

Hov, Ø. (1983c) Modelling of the long-range transport of peroxyacetylnitrate to Scandinavia. Submitted for publication.

Isaksen, I.S.A., Midtbø, K.H., Sunde, J. and Crutzen, P.J. (1976) A simplified method to include molecular scattering and reflection in calculation of photon fluxes and photodissociation rates. Geophysica Norvegica 31, 11-26.

Johnson, W.B. (1983) Interregional exchanges of air pollution: Model types and applications. JAPCA 33, 563-574.

Killus, J.P. and Whitten, G.Z. (1982) A mechanism describing the photochemical oxidation of toluene in smog. Atmospheric Environment 16, 1973-1988.

Killus, J.P., Morris, R.E. and Liu, M.K. (1983) Application of a regional oxidant model to the northeast United States. Paper presented at EPA-OECD international conference on long-range transport models for photochemical oxidants and their precursors, EPA, Research Triangle Park, N.C. 12-14 April, 1983.

Killus, J.P. and Whitten, G.Z. (1983) A new carbon-bond mechanism for air quality simulation modeling. SYSAPP-83/048. Systems Applications, Inc., San Rafael, Calif.

Kramer, M.A., Kee, R.J. and Rabitz, H. (1982a) CHEMSEN: A computer code for sensitivity analysis of elementary chemical reaction models. Princeton University, Dep. of chemistry, Princeton, N.J. 08544.

Kramer, M.A., Calo, J.M., Rabitz, H. and Kee, R.J. (1982b) AIM: The analytically integrated Magnus method for linear and second order sensitivity coefficients. Princeton University, Dep. of chemistry, Princeton, N.J. 08544.

Larson, T.V., Horike, N.R. and Harrison, H. (1978) Oxidation of sulfur dioxide by oxygen and ozone in aqueous solution: A kinetic study with significance to atmospheric rate processes. Atmospheric Environment 12, 1579-1612.

Leighton, P.A. (1961) Photochemistry of air pollution. Academic Press, New York.

Liu, M.K. and Reynolds, S.D. (1983) Development of a regional-scale air quality model. Paper presented at EPA-OECD international conference on long-range transport models for photochemical oxidants and their precursors, EPA, Research Triangle Park, N.C. 12-14 April, 1983.

Lloyd, A.C., Lurmann, F.W., Godden, D.A., Hutchins, J.F., Eschenroeder, A.Q. and Nordsieck, R.A. (1979) Development of the ELSTAR photochemical air quality simulation model and its evaluation relative to the LARPP data base. ERT document No. P-5287-500, ERT, Inc., Westlake Village, CA 91361.

Luther, F.M. and Gelinas, R.J. (1976) Effects of molecular multiple scattering and surface albedo on atmospheric photodissociation rates. J. Geophys. Res. 81, 1125-1132.

MacCracken, M.C., Wuebbles, D.J., Walton, J.J., Duewer, W.H. and Grant, K.E. (1978) The Livermore regional air quality model: I. Concept and development. J. Appl. Met. 17, 254-272.

McKay, H.A.C. (1971) The atmospheric oxidation of sulphur dioxide in water droplets in presence of ammonia. Atmospheric Environment 5, 7-14.

McRae, G.J., Goodin, W.R. and Seinfeld, J.H. (1982) Development of a second-generation mathematical model for urban air pollution - I. Model formulation. Atmospheric Environment 16, 679-696.

NASA (1979) Chemical kinetics and photochemical data for use in stratospheric modeling. Evaluation Number 2, JPL 79/27. Jet Propulsion Lab., Pasadena.

NASA (1982) Chemical konetics and photochemical data for use in stratospheric modeling. Evaluation Number 5, JPL 82/57, Jet Propulsion Lab., Pasadena.

Penkett, S.A. (1972) Oxidation of SO_2 and other atmospheric gases by ozone in aqueous solution. Nature 240, 105-106.

Penkett, S.A. and Garland, J.A. (1974) Oxidation of sulphur dioxide in artificial fogs by ozone. Tellus 26, 284-290.

Penkett, S.A., Jones, B.M.R., Brice, K.A. and Eggleton, A.E.J. (1979) The importance of atmospheric ozone and hydrogen peroxide in oxidising sulphur dioxide in cloud and rainwater. Atmospheric Environment 13, 123-137.

Penkett, S.A. (1982) Non-methane organics in the remote troposphere. Atmospheric chemistry, ed. E.D. Goldberg, Dahlem Konferenzen 1982. Springer Verlag, Berlin, pp. 329-355.

Peterson, J.T. (1976) Calculated actinic fluxes (290-700 mm) for air pollution photochemistry application. U.S. Environmental Protection Agency Report EPA-600/4-76-025.

Reynolds, S.D., Roth, P.M. and Seinfeld, J.H. (1973) Mathematical modeling of photochemical air pollution - I. Formulation of the model. Atmospheric Environment 7, 1033-1061.

Schere, K.L. and Demerjian, K.L. (1977) Calculation of selected photolytic rate constants over a diurnal range. U.S. Environmental Protection Agency Report EPA-600/4-77-015.

Scherer, B. and Stern, R. (1982) Analysis of a photochemical smog episode and preparation of the meteorological input data for a three dimensional air quality dispersion model. Proc. Second European Symposium on Physico-chemical behaviour of atmospheric pollutants, Varese, Italy 29 Sept. - 1 Oct. 1981. D. Reidel, Dordrecht, pp. 561-571.

Scott, W.D. and Hobbs, P.V. (1967) The formation of sulfate in water droplets. J. Atm. Sci. 24, 54-57.

Sehmel, G.A. (1980) Particle and gas dry deposition: a review. Atmospheric Environment 14, 983-1011.

Seinfeld, J.H. (1975) Air Pollution: Physical and chemical Fundamentals, McGraw-Hill, Inc., New York, 523 pp.

Smith, F.B. and Carson, D.J. (1977) Some thoughts on the specification of the boundary layer relevant to numerical modelling. Boundary Layer Met. 12, 307-330.

Tilden, J.W. and Seinfeld, J.H. (1982) Sensitivity analysis of a mathematical model for photochemical air pollution. Atmospheric Environment 16, 1357-1364.

Whitten, G.Z. and Meyer, J.P. (1976) CHEMK: A computer modelling scheme for chemical systems. Systems Applications Inc., San Rafael, California.

Whitten, G.Z. and Hogo, H. (1978) User's manual for kinetics model and ozone isopleth plotting package. EPA-600/8-78-014a.

Whitten, G.Z., Hogo, H. and Killus, J.P. (1980) The carbon-bond mechanism: a condensed mechanism for photochemical smog. Envir. Sci. Technol. 14, 690-700.

AN EXAMINATION OF LINEAR PARAMETERIZATIONS IN A STATISTICAL LONG-RANGE TRANSPORT MODEL

Akula Venkatram and Jonathan Pleim

Environmental Research & Technology, Inc.
696 Virginia Road,
Concord, Massachusetts 01742

ABSTRACT

The assumptions underlying linear models for long-range transport and deposition are examined. We see that the linear parameterization of the wet removal of the SO_2 cannot be justified. This suggests that linear models might be inadequate for the examination of emission control scenarios. However, we show that by assuming "efficient" removal of SO_2 we can construct a model which avoids explicit incorporation of the wet removal process. The validity of this model is demonstrated by comparing model predictions with observations of annually averaged wet deposition at 62 stations in Eastern North America (ENA).

INTRODUCTION

The most common type of long-term model (Fisher, 1978; Venkatram et al, 1982) for the transport and deposition of sulfur represents transformation and removal processes with parameters that do not depend on the concentration of sulfur. Although these models yield long-term deposition predictions that compare well with corresponding observations, they have been attacked on the ground that these "linear" parameterizations cannot represent inherently non-linear processes such as chemistry and wet removal. It is often suggested that these models work only because they have been "tuned" against data; in effect, they represent linearizations of a non-linear system about a state corresponding to the wet deposition observations used to tune the model. This of course implies that such models cannot be used to examine the effect of emission patterns very different from that used in the tuning exercise. In this paper, we examine the validity of these criticisms of linear models. We then construct a model that is free of the major shortcomings of other linear models. The usefulness of this model is demonstrated by comparing model results against observations of wet deposition made during 1980 at 62 stations covering the Eastern United States and Canada.

Linear Models

The recent (linear) long-term models for transport of sulfur (Fisher, 1978; Smith, 1981) use the concept of wet and dry regions. A wet region is one in which there is precipitation occurring while a dry region contains no rain. Pollutants embedded in these moving synoptic-scale regions can be accordingly classified as wet and dry pollutants. For a detailed discussion of these concepts, the reader is referred to Smith (1981). It can be shown that the Lagrangian evolution of a unit mass of wet or dry pollutant emitted from a source is described by the following set of differential equations,

$$\frac{dG_d}{dt} = -\lambda_d G_d - \frac{1}{\tau_d} G_d - k_d G_d + \frac{1}{\tau_w} G_w \tag{1}$$

$$\frac{dG_w}{dt} = -\lambda_w G_w - \frac{1}{\tau_w} G_w - k_w G_w + \frac{1}{\tau_d} G_d \tag{2}$$

$$\frac{dS_d}{dt} = k_d G_d - \bar{\lambda}_d S_d - \frac{1}{\tau_d} S_d + \frac{1}{\tau_w} S_w \tag{3}$$

$$\frac{dS_w}{dt} = k_w G_w - \bar{\lambda}_w S_w - \frac{1}{\tau_w} S_w + \frac{1}{\tau_d} S_d \tag{4}$$

In the above equation, G refers to unit mass of SO_2 while S refers to sulfate. The subscripts 'w' and 'd' denote wet and dry respectively. The total scavenging rate is denoted by λ, and k is the gas phase oxidation of SO_2 and SO_4. The removal rate λ_w of SO_2 during rain includes the heterogeneous in-cloud oxidation of SO_2 to SO_4. The rates of conversion of wet to dry periods and vice versa are determined by the average wet and dry durations (in the Lagrangian sense) τ_w and τ_d.

Equations 1 and 4 can be solved readily if we assume that the parameters λ and k are independent of G and S. Is this assumption of linearity justified? It is well accepted (See Acid Deposition, 1983) that the concentration of the OH radical controls the gas phase oxidation of SO_2 to SO_4. A recent study (Stockwell and Calvert, 1983) shows that because the $OH\text{-}SO_2$ reaction is not a chain terminating step in a NO rich atmosphere, the OH concentration is insensitive to the SO_2 concentration. Therefore, there is justification in taking k to be independent of G in using the model for Eastern North America (ENA). In our study we will assume that k is a constant of the order of a percent an hour. This value is probably high for an annually averaged value. However, as we will show in a later section, model results are relatively insensitive to k.

Everything we presently know about incloud oxidation of SO_2 (See Acid Deposition, 1983) indicates that we cannot assume that λ_w is independent of G. Therefore it would seem that a linear parameterization for the wet removal of SO_2 would be useless if we wanted to examine emission control scenarios. However, we can make progress if we use the observation that SO_2 is removed very efficiently in rain events. This is equivalent to the statement $\lambda_w \tau_w >> 1$, a condition which holds if we accept the values $\lambda_w \approx 10^{-4} s^{-1}$ and $\tau_w = 10$ hr given in the literature (Fisher, 1978; Smith 1981).

This implies that the "destruction" terms in Eqs. (2) and (4) are dominated by wet removal. Because the time scales of production of G_w and S_w, τ_d is much larger than the removal time scales $1/\lambda_w$ and $1/\bar{\lambda}_w$ $(\lambda_w \tau_d >> 1)$ we can assume that G_w and S_w are governed by quasi-steady equations. We then find,

$$G_w \approx \frac{1}{\lambda_w \tau_d} G_d \tag{5}$$

and $$S_w \approx \frac{1}{\bar{\lambda}_w \tau_d} S_d \tag{6}$$

We see immediately that the "wet" terms in Eqs. 1 and 3 can be neglected and we find that G_d and S_d are governed by the simple equations

$$\frac{dG_d}{dt} = -\left(\lambda_d + \frac{1}{\tau_d} + k_d\right)G_d \tag{7}$$

and $$\frac{dS_d}{dt} = -\left(\bar{\lambda}_d + \frac{1}{\tau_d}\right)S_d + k_d\, G_d \tag{8}$$

Current research (See Hicks, Critical Assessment Document, USEPA) indicates that the dry removal rates λ_d and $\bar{\lambda}_d$ are insensitive to the concentrations of SO_2 and SO_4. This means that we are left with the set of equations 7 and 8 in which the parameters are essentially independent of G and S. Assuming that these parameters do not vary over the modeling region of interest we can integrate the equations to yield,

$$G_d(t) = G_d(0) \exp(-\ell_g t) \tag{9}$$

and $$S_d(t) = S_d(0)e^{-\ell_s t} + \frac{k_d G_d(0)}{(\ell_g - \ell_s)} \left(e^{-\ell_s t} - e^{-\ell_g t} \right) \tag{10}$$

where the loss terms ℓ_s and ℓ_g are defined by

$$\ell_g = \lambda_d + \frac{1}{\tau_d} + k_d \tag{11a}$$

$$\ell_s = \bar{\lambda}_d + \frac{1}{\tau_d} \tag{11b}$$

Assuming that trajectories are essentially straight for distances of the order of 1,000 km, the expression for the wet deposition of sulfur associated with a point source can be written as

$$D_w = \frac{f_\theta}{2\pi r u} \left(G_d + S_d \right) \frac{1}{\tau_d} \tag{12}$$

In Eq. 12, r is the distance from source to receptor, f_θ is the frequency with which the large scale wind blows in the sector containing the source and receptor, and u is the average sector wind speed. The next section describes the use of Eq. 12 to estimate wet deposition.

Application of the Model - Input Data

The simple model was used to estimate annually averaged deposition of sulfur. "Background" depositions of sulfur in rain are typically 2 kg/hec/yr (Fisher 1978), a value which is significant compared to the observations in the Western United States and Canada. To avoid accounting for the unknown background deposition, we confined our analysis to the Eastern United States and Canada where wet deposition is dominated by sources within the modeling region. The observations correspond to annually averaged deposition of sulfur measured in 1980 at the 62 stations listed in Table 1.

TABLE 1

COMPARISON OF OBSERVATIONS OF ANNUALLY AVERAGED WET DEPOSITION OF SULFUR WITH MODEL PREDICTIONS FOR 1980

NO.	STATION NAME		PREDICTED	OBSERVED	OBS/PRED
1	LONG POINT		14.1	11.6	0.8
2	CHALK RIVER		7.2	8.2	1.1
3	KEJIMKUJEK		4.5	6.0	1.3
4	GANDER		1.1	7.1	6.6
5	TRURO		3.3	8.9	2.7
6	ST. JOHN		8.3	11.7	1.4
7	CHARLO		3.4	9.0	2.6
8	SEPT-ILES		2.1	5.8	2.8
9	GOOSE		0.8	3.4	4.4
10	FORT CHIMO		0.6	1.1	1.9
11	NITCHEQUON		1.4	2.6	1.9
12	QUEBEC		5.7	14.9	2.6
13	CHIBOUGAMAU		3.3	5.1	1.6
14	ST. HUBERT		11.2	11.5	1.0
15	MANIWAKI		6.4	10.1	1.6
16	KINGSTON		9.2	13.4	1.5
17	PETERBOROUGH		9.4	12.5	1.3
18	HARROW (WINDSOR)		16.4	13.8	0.8
19	DORSET		8.8	9.1	1.0
20	MOOSONEE		2.6	2.7	1.0
21	WHITEFACE		8.6	12.1	1.4
22	ITHACA		13.3	9.7	0.7
23	PENN STATE		17.8	8.6	0.5
24	VIRGINIA		11.9	5.6	0.5
25	ILLINOIS		10.3	7.5	0.7
26	BROOKHAVEN		12.8	8.5	0.7
27	LEWES		11.2	6.4	0.6
28	OXFORD		12.7	7.0	0.6
29	CARIBOU	ME	3.9	5.8	1.5
30	GREENVILLE	ME	5.1	5.9	1.2
31	HUBBARD BROOK	NH	7.8	6.8	0.9
32	HUNTINGTON WILDLIFE	NY	9.1	9.3	1.0
33	AURORA RESEARCH FARM	NY	12.6	13.3	1.0
34	KNOBIT	NY	12.3	10.5	0.8
35	STILLWELL LAKE-WEST PT	NY	13.7	9.5	0.7
36	LEADING RIDGE	PA	17.8	9.4	0.5
37	KANE EXPER. FOREST	PA	16.8	11.9	0.7
38	PARSONS	WV	18.4	14.4	0.8
39	HORTON'S STATION	VA	12.2	9.5	0.8
40	LEWISTON	NC	7.3	8.6	1.2
41	RESEARCH TRIANGLE PARK	NC	7.6	9.9	1.3
42	FINLEY (RALEIGH)	NC	7.4	9.4	1.3
43	CLINTON CROPS RES. STA	NC	6.3	7.5	1.2
44	PIEDMONT RES. STATION	NC	8.6	9.9	1.1
45	COWEETA	NC	8.6	8.5	1.0
46	CLEMSON	SC	7.7	7.8	1.0
47	GEORGIA STA.	GA	7.1	8.1	1.1
48	BRADFORD FOREST	FL	4.2	6.0	1.4
49	AUSTIN-CARY FOREST	FL	3.8	8.5	2.3
50	WALKER BRANCH WATERSHD	TN	13.0	13.4	1.0
51	CALDWELL	OH	19.9	16.1	0.8
52	WOOSTER	OH	16.5	12.3	0.7
53	DELAWARE	OH	12.1	13.3	1.1
54	MERIDIAN	MS	5.6	6.3	1.2
55	SIU	IL	11.1	7.2	0.6
56	SALEM	IL	10.7	9.1	0.9
57	BONDVILLE	IL	10.3	9.9	1.0
58	ARGONNE	IL	12.5	12.8	1.0
59	KELLOGG BIOLOGICAL STA	MI	9.9	10.0	1.0
60	WELLSTON	MI	6.8	7.5	1.1
61	UMI BIOLOGICAL STA	MI	5.6	6.3	1.1
62	TROUT LAKE	WS	4.7	5.1	1.1

The emission inventory, compiled by ERT for the Canadian Electrical Association, is a composite of information from the EPS (Canada), NEDS and individual state inventories, and is nominally representative of 1980. This inventory agrees well with that prepared by Working Group 3B under the U.S. - Canada Memorandum of Intent (MOI 1982). For use in the model, the emissions were aggregated into 519 point sources whose distribution is shown in Figure 1.

The wind speed u and f_θ correspond to a 600 m wind rose derived from rawinsonde observations made during the period 1972-1976 (Niemann 1982). Observations from 67 stations were combined into 14 wind roses corresponding to different regions in the U.S. and Canada. A typical wind rose, shown in Figure 2, indicates the dominance of westerly, south-westerly flows in Eastern North America (ENA).

The average length of dry periods τ_d were taken from Henmi and Reiter (1978) who made the calculations from rainfall data at 61 stations in 1974. Strictly, τ_d should be a Lagrangian statistic. However, Hamrud and Rodhe (1981) have shown that there is little difference between Eulerian and Lagrangian precipitation statistics. We realize that there is some difficulty in defining, and therefore in measuring τ_d. The values of τ_d used in our calculations ranged from 60 to 75 hours in the Eastern United States. For convenience, wet deposition was calculated with the τ_d value at the relevant receptor.

Comparison of Model Estimates with Observations

Figure 3 shows a comparison of observations of sulfur deposition with model estimates made using the parameters

$1/\lambda_d$ = 60 hrs; Background deposition = 2 kg/hec/yr
k_d = 0%/hr
τ_d = 60 - 75 hrs (Henmi and Reiter 1978)
f_θ = Wind roses at 600 m, 1972 - 1976

Note that sulfate production is neglected.

It is seen that the model does tend to overpredict deposition. However, considering the inevitable uncertainty in model parameters and the emission inventory, the agreement between model predictions and observations is more than adequate. This comparison provides considerable support for the basic hypothesis that the removal of SO_2 (through in-cloud oxidation) in rain is rapid enough to allow us to neglect the details of the wet scavenging process. Notice that because model estimates do not use rainfall information we have avoided taking advantage of the observed high correlation between rainfall and wet deposition.

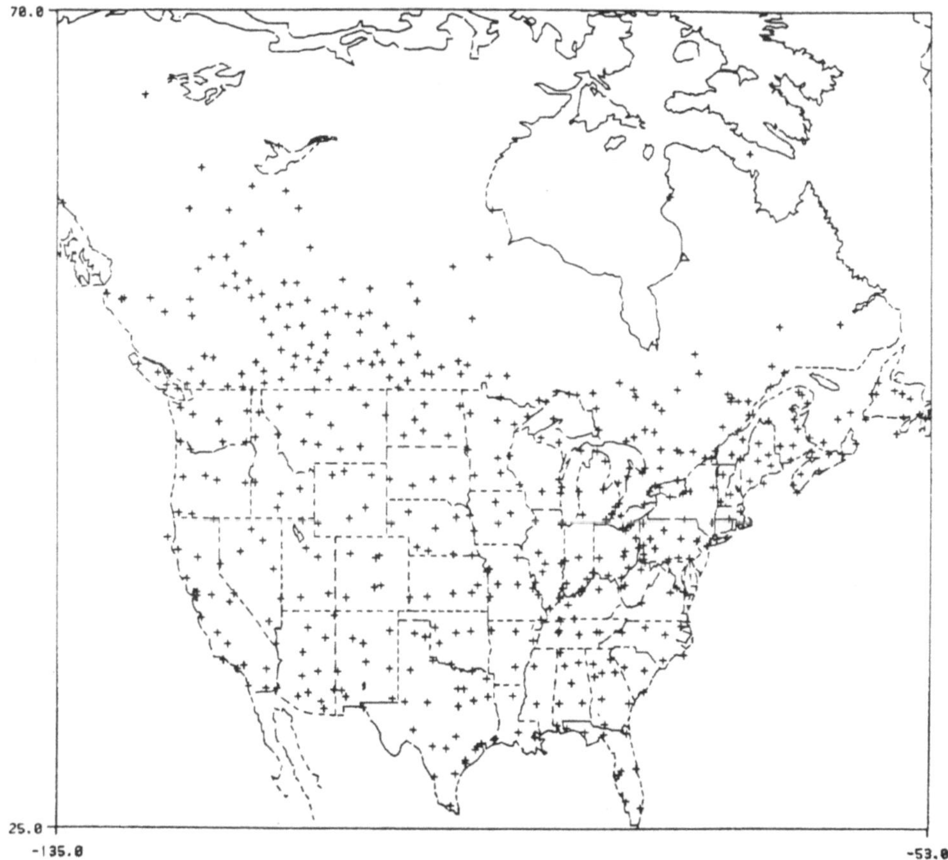

Figure 1 Distribution of Emissions Used in the Study

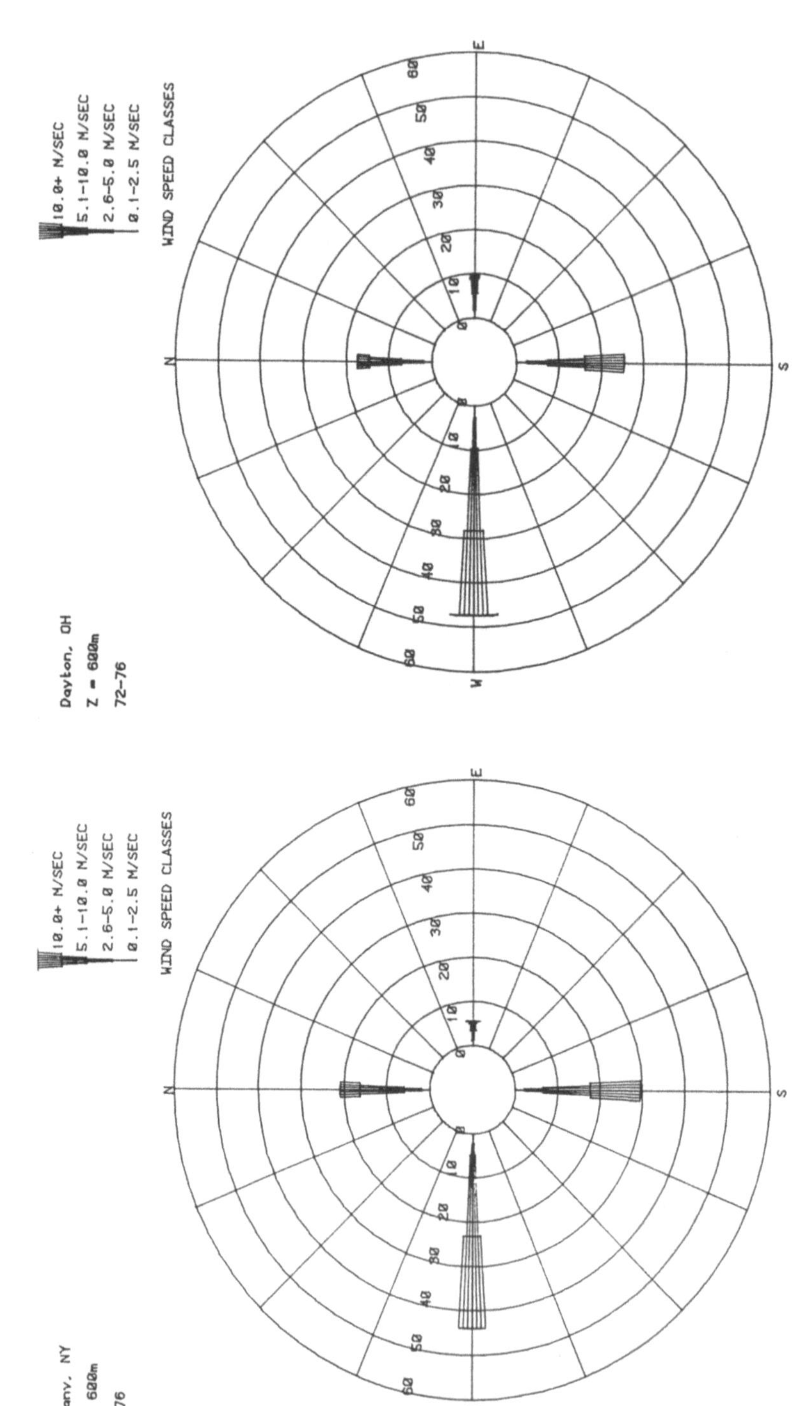

Figure 2 Typical 600m Wind Roses

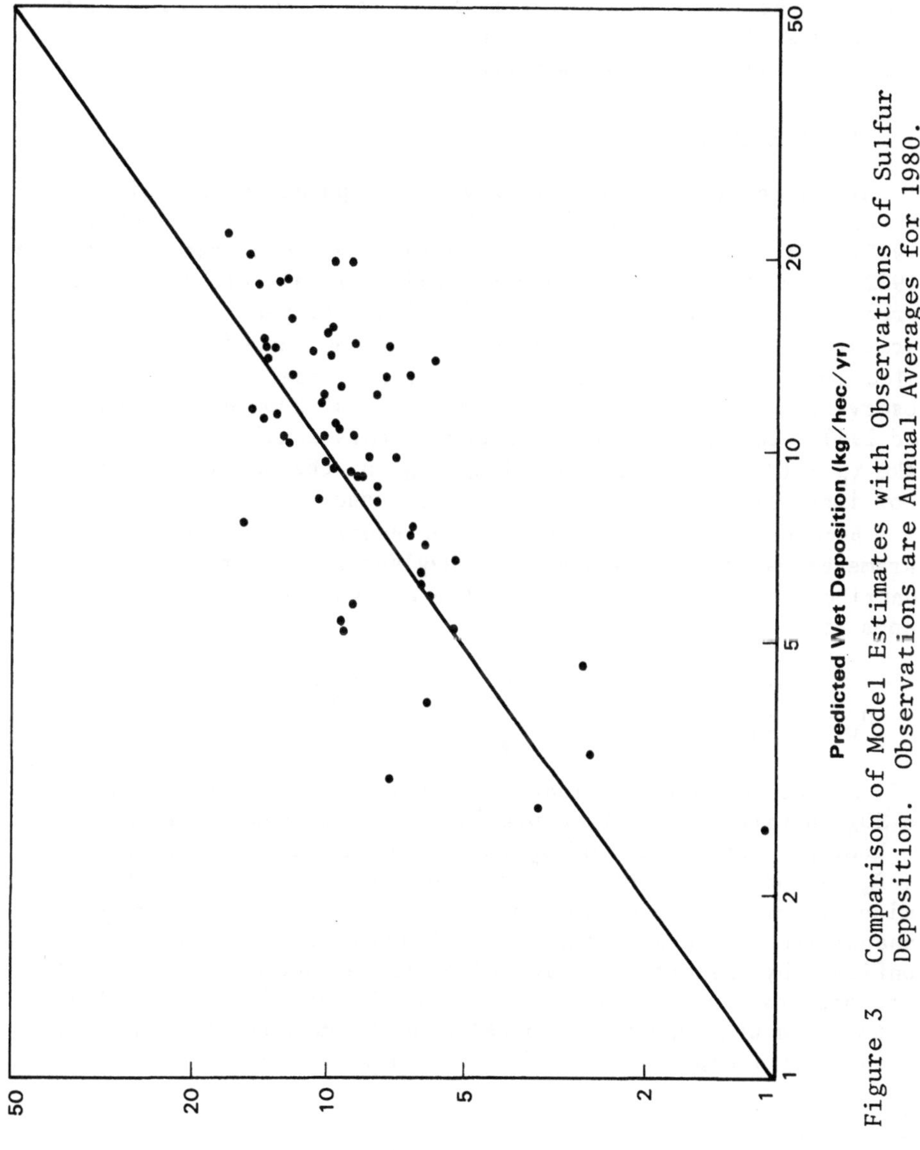

Figure 3 Comparison of Model Estimates with Observations of Sulfur Deposition. Observations are Annual Averages for 1980.

Figure 4 presents model results based on the assumption that SO_2 is not removed by rain; all the sulfur in rain is associated with sulfate formed by gas phase oxidation of SO_2 at a rate of 1%/hr. We see that this assumption underpredicts observations by large margins even though we have conservatively taken the emitted sulfate to be 10% of the total emission for each source. Clearly, secondary sulfate contributes very little to sulfur in rain falling in Eastern North America.

Sensitivity Studies

Model results depend strongly on the parameter τ_d which is difficult to determine in the first instance. Henmi and Reiter (1978) indicate that $\tau_d \simeq 70$ hrs in 1974, while Niemann's analysis (1982) shows that $\tau_d \simeq 120$ hrs for 1978. In all probability this large variation of τ_d is the result of actual year to year variability as well as the difference in the method used to calculate τ_d. The comparison of model estimates with observations is bound to reflect our lack of knowledge of τ_d for 1980. In Figure 5, we demonstrate this sensitivity; if we take $\tau_d = 100$ hrs for the modeling region, the model results show improvement over that of the base case.

Because the model is simple, the sensitivity of the model to parameter values is transparent. Neglecting the small contribution of primary sulfate, the expression for the wet deposition of sulfur is

$$D_w = \frac{Qf_\theta}{2\pi ru\tau_d} e^{-\ell_g t} \left(1 + \frac{k_d}{(\ell_g - \ell_s)} \left(e^{-(\ell_s - \ell_g)t} - 1\right)\right) \tag{13}$$

Because k_d could include the nonlinear conversion of SO_2 to SO_4 in fair weather (non-precipitating) clouds, it is useful to know the sensitivity of D_w to k_d. We notice that for small t, D_w is independent of k_d. This suggests, that in a high emission density region such as ENA, the wet deposition rate is insensitive to the dry conversion of SO_2 to SO_4. This is confirmed by the virtually identical comparison of model results with observations for k_d=0 and 2%/hr. The implication of this is that the wet deposition in ENA depends on parameters which are essentially independent of the concentration levels.

Discussion

We have examined the assumptions underlying linear models for the long-range transport and deposition of sulfur. The major source of non-linearity in such models is the wet removal rate of SO_2 which includes the incloud conversion of SO_2 to SO_4. We show how we can get around the need to account explicitly for wet removal by assuming that the time scale of wet scavenging $1/\lambda_w$

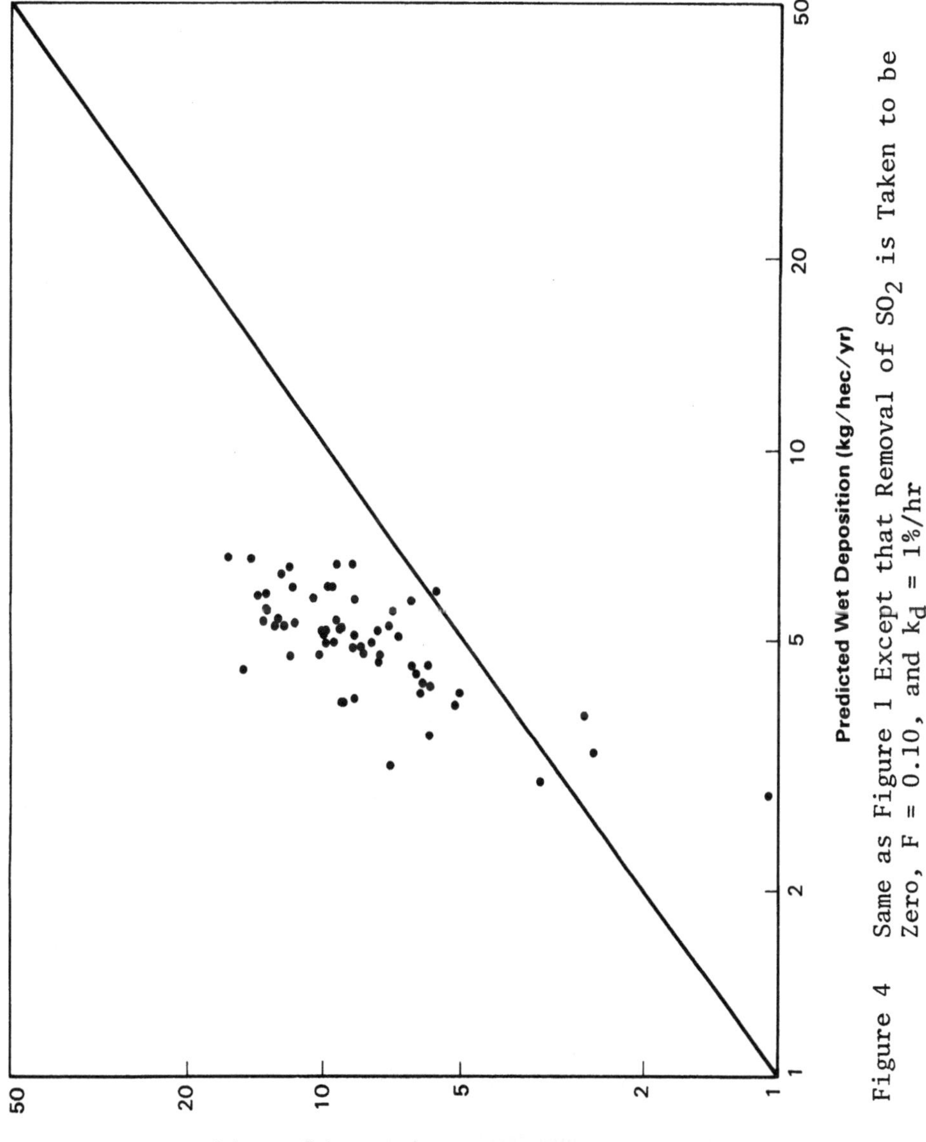

Figure 4 Same as Figure 1 Except that Removal of SO_2 is Taken to be Zero, F = 0.10, and k_d = 1%/hr

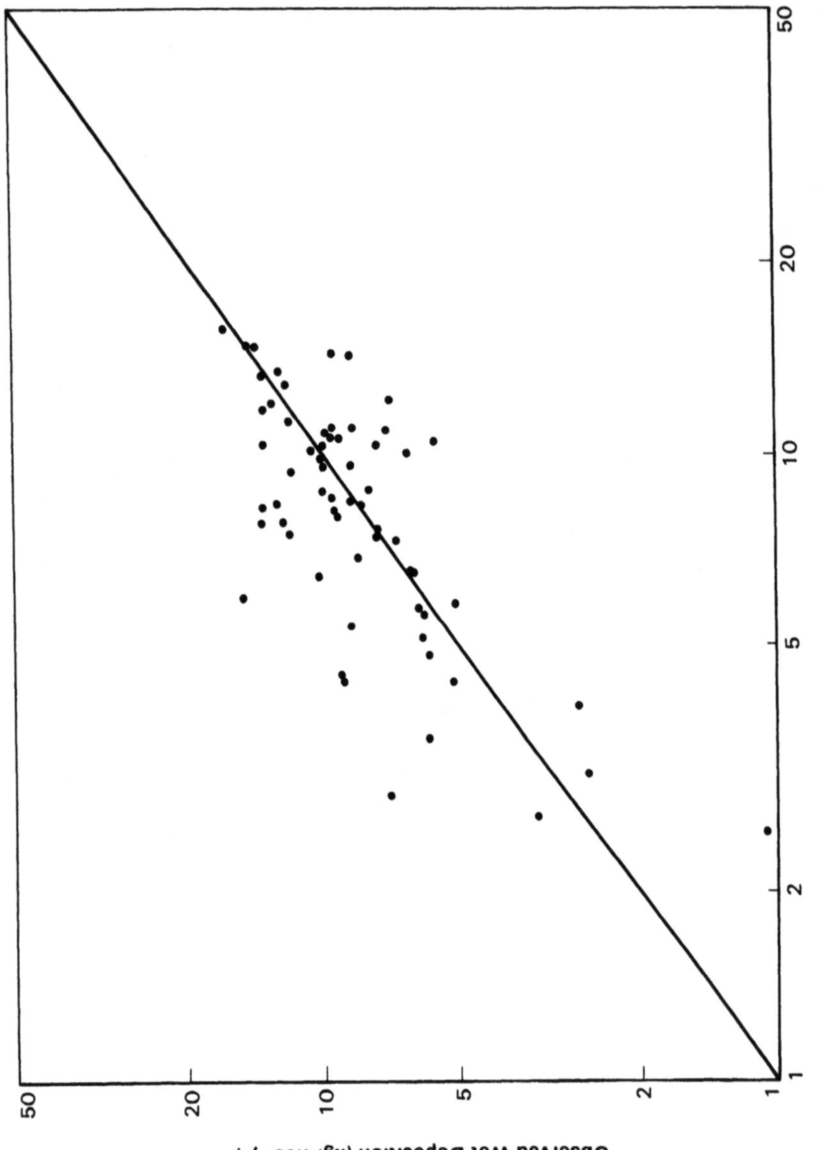

Figure 5 Same as Figure 3 Except that τ_d = 100 Hours for the Whole Modeling Region

is much smaller than the time scale τ_d that determines the rate at which SO_2 is supplied from dry periods to wet periods. Also $\lambda_w \tau_w >> 1$, so that very little SO_2 leaves a wet period. This implies that τ_d controls the rate of wet removal and the details of the wet scavenging process (including in-cloud oxidation) can be neglected. Notice that this assumption is fundamentally different from that used in other models (Scriven and Fisher, 1975) based on a similar mass balance. In these models wet deposition is accounted through a scavenging coefficient which is a product of a removal rate (based on rainfall rate) and the fraction of time it rains between source and receptor. Although τ_d is likely to be related to annual rainfall, our formulation for wet scavenging is not an explicit function of rainfall or the length of the wet period τ_w.

Model estimates from this simple model compare very well with observations of annual sulfur deposition at 62 stations distributed over the United States and Canada. We believe that this provides strong support for the structure of this admittedly empirical model. Note that this paper does not discuss dry deposition of SO_2 which contributes significantly to the problem of acid deposition.

REFERENCES

Acid Deposition: Atmospheric Processes in Eastern North America, 1983; National Academy Press, 2101 Constitution Avenue, NW, Washington, D.C., 20418.

Fisher, B.E.A. 1978. The calculation of long-term sulphur deposition in Europe. Atmospheric Environment, 12:489-501.

Hales, J.M. and T.M. Dana 1979. Precipitation scavenging of urban pollutants by convective storm systems. J. Appl. Met., 18:244-316.

Hamrud, M. and H. Rodhe 1981. A numerical comparison between Lagrangian and Eulerian rainfall statistics. Tellus, 33:235-241.

Henmi, T. and E.R. Reiter 1978. Regional residence time of sulfur dioxide over the Eastern United States. Atmospheric Environment, 12:1489-1496.

MOI, 1982: United States - Canada Memorandum of Intent, 1982. Emissions, Costs and Engineering Assessment - Final Report of Working Group 3-B.

Niemann, B.L. 1982. The 1980 Data Set for further evaluation of regional air quality/acid deposition simulation models. EPA report, write to G. Foley, USEPA, ORD, Washington, D.C. 20460.

Rodhe, H. and J. Grandell, 1972: On the removal time of aerosol particles from the atmosphere by precipitation scavenging. Tellus, 24, 5, 442-454.

Scriven, R.A. and B.E.A. Fisher 1975. The long range transport of airborne material and its removal by deposition and washout - 1. General considerations. Atmospheric Environment, 9:49-58.
Smith, F.B., 1981: The significance of wet and dry synoptic regions on long-range transport of pollution and its deposition. Atmospheric Environment, 15,863-873.
Stockwell, W.R. and J.G. Calvert, 1983: The mechanism of the HO-SO_2 reaction. Atmospheric Environment, In press.
Venkatram, A., B.E. Ley and S.Y. Wong 1982. A statistical model to estimate long-term concentrations of pollutants associated with long-range transport. Atmospheric Environment, 16:249-258.

DISCUSSION

A. ELIASSEN I completely agree with Dr. Venkatram that the long-term wet deposition of sulfur can be treated as a linear problem, since the residence time for SO_2 or SO_4 with respect to wet deposition is basically determined by the Lagrangian dry time. Dry deposition, however, may still be non-linear, since it depends on the ratio of SO_2 to SO_4 and therefore on the transformation rate of SO_2 which may be concentration dependent. Have you investigated this part of the problem ?

A. VENKATRAM Both the dry and wet deposition of sulfur depend on the transformation rate of SO_2. As you point out, the dry deposition is more sensitive to this rate because of the large difference between the dry deposition velocities of SO_2 and SO_4. If k is associated primarily with homogeneous gas phase oxidation, nonlinearity should not be of concern because of the insensitivity of the OH concentration to the concentrations of the co-pollutants. However, if the transformation rate, as it could well be, is related to in-cloud (fair weather) conversion of SO_2 to SO_4, non-linearity has to be important. Although I have not yet done so, I do intend to look at this problem.

A. BERGER Have you used only one value for τ_d for all U.S. ? To which extent is that parameter a tuning or a physical factor ?

A. VENKATRAM T_d could well be an empirical parameter. However, the conclusions of this paper are not changed as long as the parameter is not a function of the concentration level.

A MULTI-LAYERED, LONG-RANGE TRANSPORT, LAGRANGIAN TRAJECTORY MODEL: COMPARISON WITH FULLY MIXED SINGLE LAYER MODELS

M. Trevor Scholtz and Boris Weisman

MEP Company, Markham, Ontario, Canada

INTRODUCTION

The Lagrangian modeling approach for Long-Range Transport and Deposition is a practical technique for integrating a number of the important processes controlling the budget of SO_2 emitted into the atmosphere and it's distribution into the several reservoirs in the environment.

In order to test how this distribution depends on the degree of detail with which the major processes are treated, several versions of the TRANS model (Weisman 1980, 1983) have been generated, and used to model all North American sources on a seasonal basis.

The various models were designed to examine the influence of a diurnally varying vs constant mixing height, fully mixed vs vertically distributed mass distribution, differentiation of emission height vs instantaneous mixing, dry removal from surface layer vs fully mixed column, and simulated enhanced in-cloud oxidation vs constant oxidation rate.

This paper presents some of the results of a detailed comparison between model predictions both for the standard parameters, as well as more sensitive measures generated from the model output.

MODEL DESCRIPTION

The basic TRANS (Transport of Anthropogenic Nitrogen and

Sulphur) model for long-range transport and deposition simulation has been described by Weisman (1980, 1983). Briefly, the Lagrangian model comprises four layers with heights of 0.1, 0.25, 0.5 and 1.0 times the maximum mixing height. Material in each layer is similarly advected with a vertically uniform wind field. The SO_2 and SO_4 budgets in each of the four layers are computed along trajectories taking into account first order chemical transformation as well as wet and dry deposition. For dry deposition only the mixed layer undergoes depletion, while for wet deposition all layers participate in the wet removal process.

The emissions in each source area are distributed in the vertical among the four layers according to the distributions given in the Interim Report of Work Group 2, Memorandum of Intent on Transboundary Air Pollution (1981), so that elevated emissions during night-time periods do not contribute to the surface concentration and dry deposition near the source.

In order to determine the sensitivity of derived concentration and deposition fields to the model structure, three versions of the TRANS model were constructed and evaluated using actual 1978 meteorology and emissions for Eastern North America. Version 1 (LM) is the basic TRANS model. Table 1 summarizes the model versions.

Version 2 (LMF)

This is the fullest version of the layered model. There is considerable evidence that the rate of conversion of SO_2 to sulphate is greatly enhanced in-cloud over the homogeneous dry rate. This enhanced rate is generally overlooked by Lagrangian, linear chemistry models. However, since the wet enhanced conversion is generally accompanied by precipitation, which has a higher efficiency for wet removal of particulate sulphate as compared to SO_2, it is anticipated that simulation of the wet enhancement process would significantly alter the derived wet sulphur deposition rates. In order to simulate the enhanced wet conversion and deposition processes, the fraction of each plume segment estimated to be subjected to wet processes (f) is treated separately from the dry fraction of the plume segment. For the dry fraction the transformation and dry deposition is as for the basic model. For the wet fraction, a distinction between summer and winter conditions is made. During winter when convective activity is weak, an enhanced rate of 5%/hr. is applied to the upper model layer (500-1000 m) in which the cloud is assumed to exist. During the summer when convective activity is strong an enhanced rate of 10%/hr. is applied to all layers for fully mixed conditions while for limited mixing the enhanced rate is used only in the upper

Table 1. Summary of Layered and Fully Mixed Models

	Model Version			
	LMF	LM	FM1	FM2
Time Resolution	3 hr.	3 hr.	3 hr.	3 hr.
Number of Layers	4	4	1	1
Seasonal Variation of Mixing Depth (h_m)	Yes	Yes	Yes	Yes
Winter Mean $\bar{h}_m$	463 m	463 m	500 m	500 m
Summer Mean $\bar{h}_m$	694 m	694 m	750 m	750 m
Diurnal Variation of h_m (0.2x, 0.2x, 0.5x, 1x, 2x, 2x, 1x, 0.5x$\bar{h}_m$)	Yes	Yes	No	No
Winter Maximum h_m	1000	1000	-	-
Winter Minimum h_m	100	100	-	-
Summer Maximum h_m	1500	1500	-	-
Summer Minimum h_m	150	150	-	-
Vertically mixed or distributed emission heights	Distributed		Mixed	
Laterally mixed or partitioned wet and dry transformation and deposition processes.	Partitioned	Mixed		

layer (750-1500 m). Only the wet fraction of the plume participates in wet removal processes.

The fraction of each plume segment which is subject to wet processes was estimated from the fraction of stations reporting precipitation under the plume width ($2\sigma_y$).

Version 3 (FM1)

In order to assess the differences between the layered model and a fully mixed model, the FM1 version is the basic model except that a single mixed layer is used with a seasonal variation of the mixing depth.

MODEL COMPARISON STRATEGY

The objective in comparing the models was to determine firstly if the various model structures lead to significantly different overall results and secondly to see if the underlying source/receptor relationships which one would infer from these various models would differ.

All three models were exercised for a one year period (1978) using the same meteorology. The Environment Canada emissions inventory prepared for the MOI (Memorandum of Intent) modeling intercomparison study was used for the simulation. Two runs were carried out with the FM1 model: Run FM2 uses double the conversion and dry deposition rates used for other model runs. Table 2 gives the model parameters used for each simulation.

In order to assess the behaviour of the models in both high and low emissions density regions, four receptor points were selected for study. These receptors are shown in Figure 1. Also shown in this figure, surrounding each receptor, are the source regions chosen for evaluating the influences of medium-range transport as opposed to the long-range influence at each receptor.

RESULTS

Ambient SO_2 and SO_4 Concentrations

Figure 2 compares the results of the four similations for annual average SO_2 and SO_4 surface concentrations at each of the four receptors. The lowest SO_2 and highest SO_4 concentrations are given by the FM2 model due to the doubled conversion rate. The two

Table 2. Model Parameters Used for Simulations

	Model Version			
	LMF	LM	FM1	FM2
Fraction of Plume Subjected to Wet Processes	f	1	1	1
Dry Transformation Rate (%/h)				
Summer	1.0(1-f)	1.0		2.0
Winter	0.5(1-f)	0.5		1.0
Wet Transformation Rate (%/h)				
Summer (all layers)	10.0 f			
Winter (500 to 1000 m only)	5.0 f			
SO_2 Dry Deposition Velocity (cm/s)				
Diurnal Variation	Yes	Yes	No	No
Summer Mean	0.56	0.56	0.6	1.2
Maximum	1.2	1.2	-	-
Minimum	0.12	0.12	-	-
Winter Mean	0.19	0.19	0.2	0.4
Maximum	0.4	0.4	-	-
Minimum	0.04	0.04	-	-
SO_4 Dry Deposition Velocity (cm/s)				
Diurnal Variation	Yes	Yes	No	No
Summer Mean	0.1	0.1	0.1	0.2
Winter Mean	0.1	0.1	0.1	0.2
SO_4 Washout Coefficient (hr^{-1})				
Summer	0.3 f	0.3	0.3	0.3
Winter	0.3 f	0.3	0.3	0.3
SO_2 Washout Coefficient (hr^{-1})		Barrie (1981)		

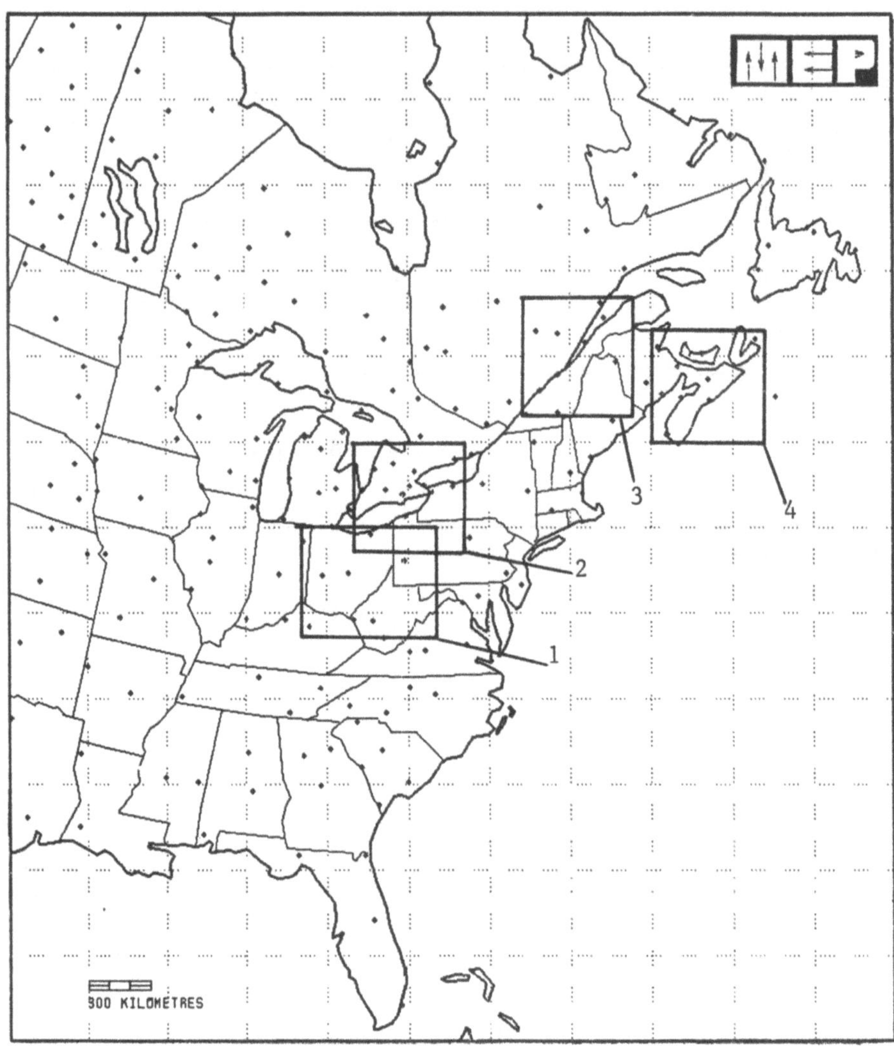

Figure 1. Total Geographical Coverage of Simulations. Meso-Regions Surrounding Receptors: (1) Duncan Falls, Ohio; (2) Simcoe, Ontario; (3) Quebec City, Quebec; (4) St. Margarets Bay, Nova Scotia. Precipitation Observing Stations (♦).

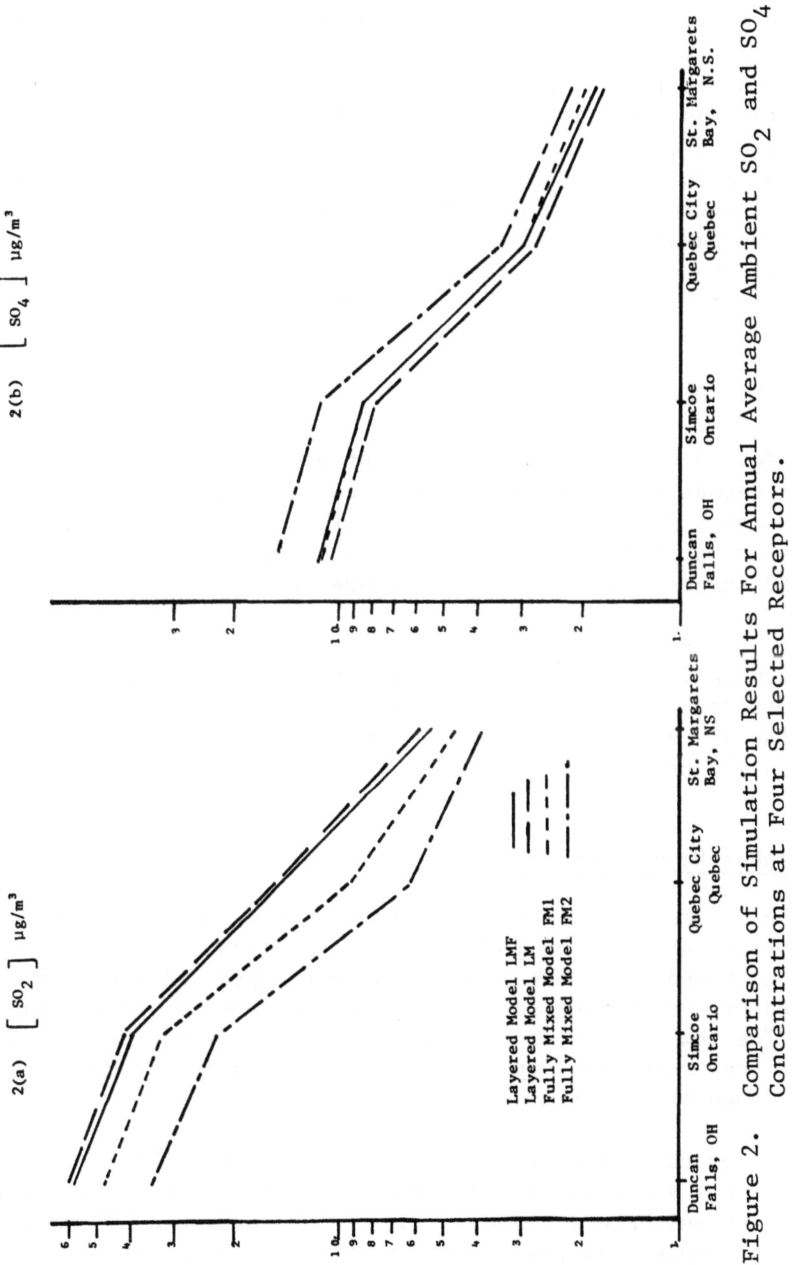

Figure 2. Comparison of Simulation Results For Annual Average Ambient SO_2 and SO_4 Concentrations at Four Selected Receptors.

layered models give approximately 30% higher SO_2 concentrations than the FM1 model. This is mainly due to the limited mixing of near surface emissions during night periods: The FM model mixes these through the full layer at all times. The reverse effect has been noted for simulations near major elevated point sources with the FM model yielding higher SO_2 concentrations than the layered model. The same distinction does not appear in the SO_4 comparison where both the layered models and FM1 give quite similar concentrations. This is attributed to the relatively slow conversion to sulphate and in the course of one diurnal cycle with the layered models, the SO_2 is uniformly mixed through all layers. The small difference between the two layered models is attributed to the 'by-passing' of SO_4 in the dry fraction of the plume in the LMF model so avoiding the very efficient washout process and leading to a 10% increase in ambient sulphate levels.

Dry Sulphur Deposition

Figure 3 compares the results for primary and secondary sulphur deposition. The FM2 model gives the highest deposition rate in both cases due to the doubled deposition velocities, which in the case of primary dry deposition, more than compensates for the lower SO_2 concentration shown in Figure 2.

The dry deposition rates for the layered models and FM1 are quite similar. The significantly higher SO_2 concentrations for the layered models do not reflect strongly in primary dry depositions since the dry deposition velocity is much reduced during the more stable night periods.

Wet Sulphur Deposition (Figure 4)

The primary wet deposition given by the layered and FM1 models is very similar with the LMF model giving slightly lower values than the LM model (except at the Simcoe receptor) due to the enhanced wet conversion rate which depletes the SO_2 in raining areas. The low results of the FM2 model are a direct result of the doubled conversion and dry deposition rates which deplete SO_2 more rapidly than the other models.

The secondary wet deposition results shown in Figure 4 show that all four simulations give significantly different results. The highest wet sulphate deposition rate is shown by the LMF model even though the ambient sulphate concentrations shown in Figure 2 are considerably lower than FM2 and are comparable to the FM1 and LM models. This is due to the simulated enhanced wet oxidation rate

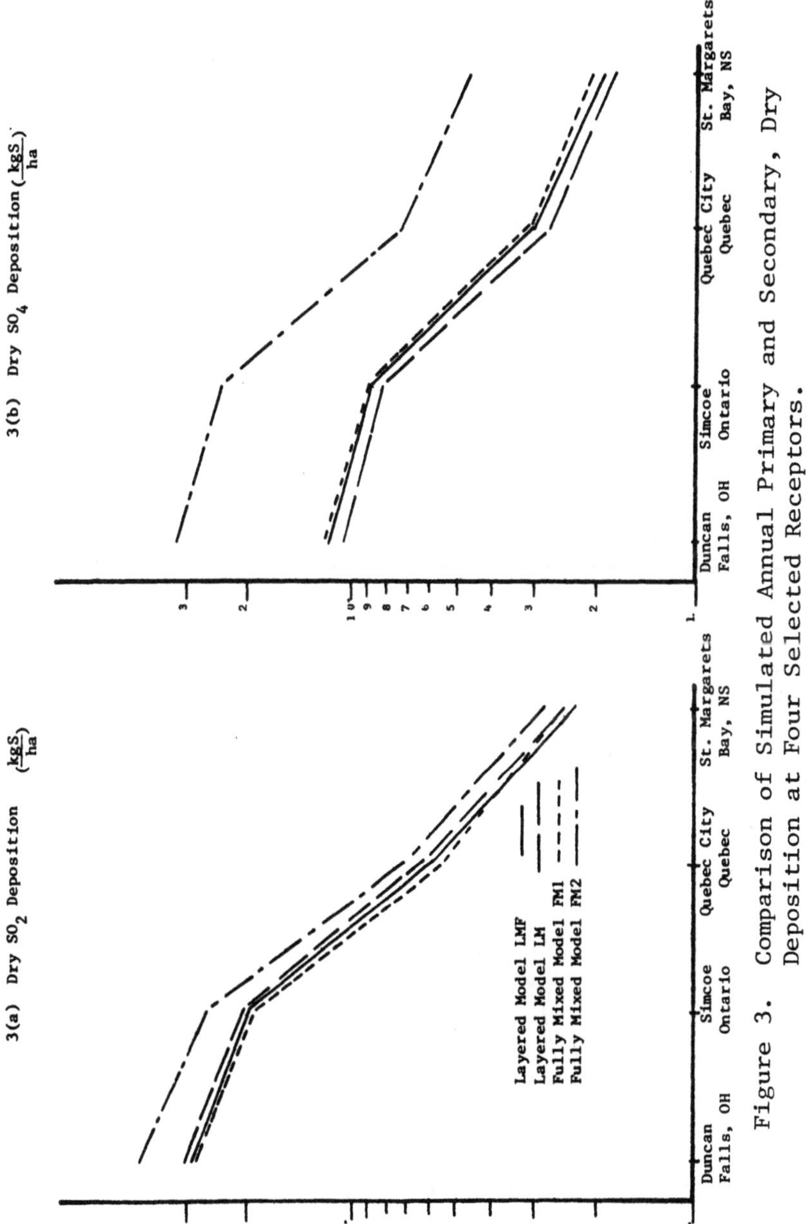

Figure 3. Comparison of Simulated Annual Primary and Secondary, Dry Deposition at Four Selected Receptors.

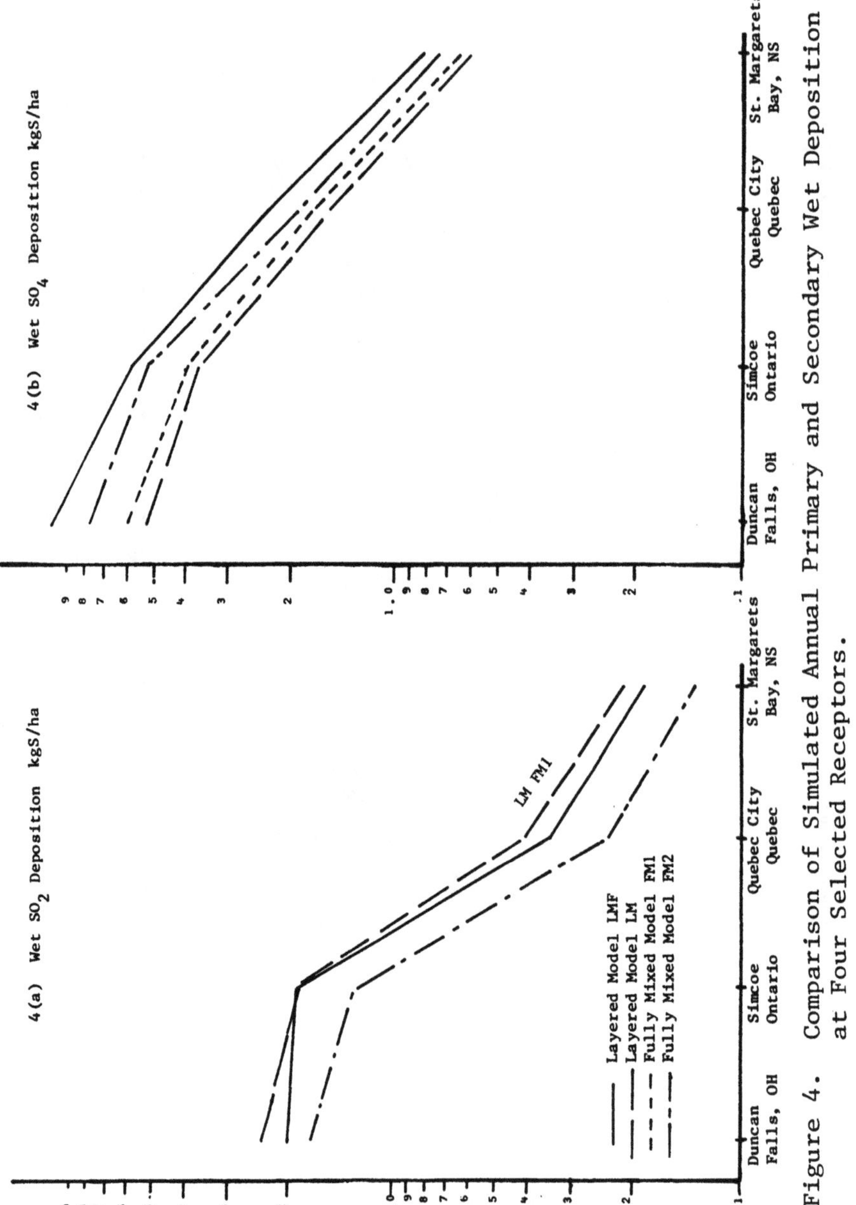

Figure 4. Comparison of Simulated Annual Primary and Secondary Wet Deposition at Four Selected Receptors.

used in the LMF model. Since sulphate is very efficiently washed out of the plume, the controlling rate step tends to be the rate at which sulphate is produced. This is highlighted by comparing the two layered models which give very similar ambient sulphate levels while the secondary wet deposition rates differ by a factor of almost two. The FM2 model result shows that doubling the overall oxidation rate in an attempt to artificially account for enhanced wet oxidation significantly increases the wet deposition. This is to a minor extent attributable to increased sulphate accumulation during dry periods which does not occur with the LMF model due to the wet removal which accompanies the enhanced rate (See Figure 2). The lowest rates are given by the LM and FM1 models due to depletion of available sulphate by washout and slow replenishment by oxidation of primary sulphur.

Total Sulphur Deposition

In these simulations, dry sulphur deposition is predominantly primary sulphur while wet sulphur deposition is dominated by secondary sulphur and the corresponding plots (not shown) give similar relationships between the models to Figures 3(a) and 4(b). Total sulphur deposition is shown in Figure 5. The FM2 model shows the highest rates due mainly to the doubled deposition velocities. The differences among the remaining three simulations are of the order of 10%.

Impact of the Emissions in the Medium-range (approximately <300 km) and the Long-range (approximately >300 km)

Since one of the primary goals of long-range transport modeling is to derive regional source/receptor relationships for control strategy evaluation, the models have been compared on the basis of the impact of sources within approximately 300 km of the receptor (medium-range transport) and of sources at long-range (i.e. all sources outside the medium-range).

Tables 3 and 4 show the results for ambient and deposited sulphur for the LMF layered model with simulated enhanced wet oxidation and the fully mixed FM1 model versions. Also shown are the percentage differences between the two model results.

Medium-range Impact (Table 3). From Table 3 it is seen that there are significant differences between the two models for ambient SO_2 concentrations and wet sulphate depositions, the layered model giving a greater general impact in the medium-range than the fully mixed model. The largest differences between the

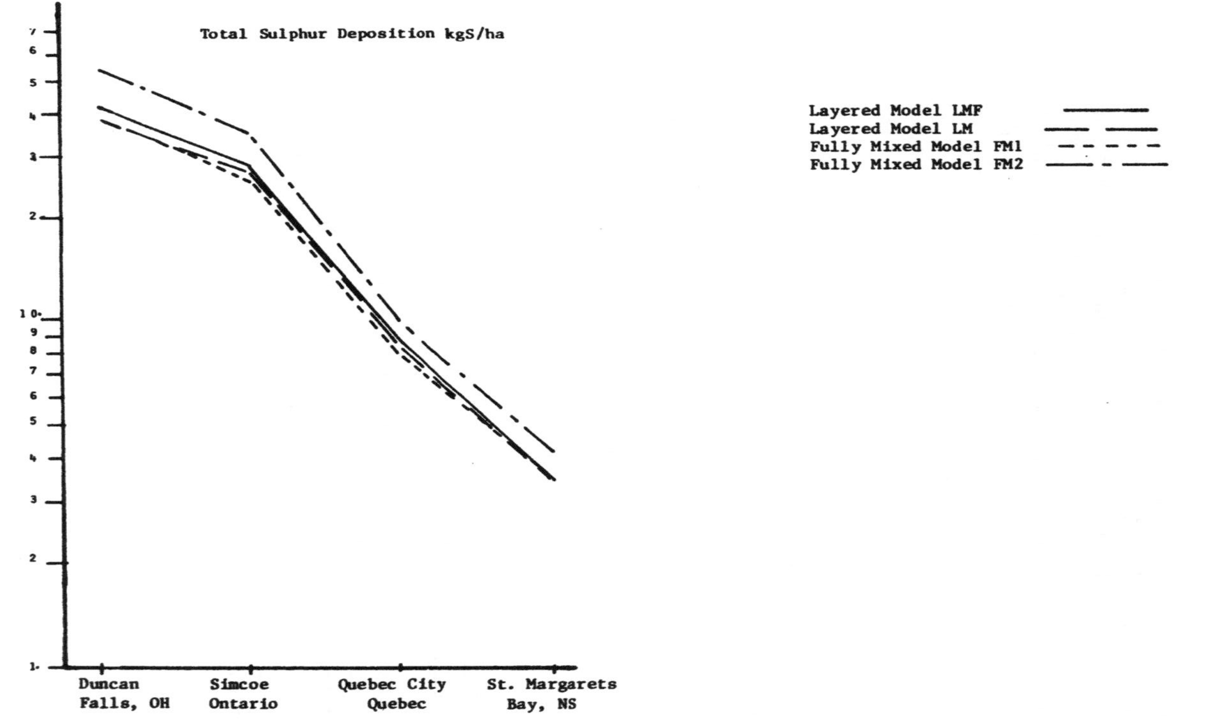

Figure 5. Comparison of Simulated Annual Total Sulphur Deposition at Four Selected Receptors.

Table 3. Medium-Range (<300 km) Impact of Sulphur Emissions

	Emissions MT SO_2/y	Concentration $\mu g/m^3$		Dry Deposition KgS/ha		Wet Deposition KgS/ha		Total Sulphur Deposition KgS/ha
		SO_2	SO_4	SO_2	SO_4	SO_2	SO_4	
Duncan Falls, Ohio	5.21							
LMF		48.3	7.2	24.2	0.75	1.57	8.02	34.6
FM1		37.6	7.0	23.9	0.74	1.81	5.02	31.5
Δ%		+28	+3	+1	+1	-13	+60	+10
Simcoe, Ontario	2.21							
LMF		22.9	2.8	10.8	0.29	0.72	1.59	13.4
FM1		15.7	2.8	9.7	0.29	0.61	1.24	11.8
Δ%		+46	0	+11	0	+18	+28	+14
Quebec City, P.Q.	0.31							
LMF		10.2	0.49	3.5	0.050	0.070	0.27	3.9
FM1		4.3	0.46	2.6	0.050	0.076	0.16	2.9
Δ%		+137	+7	+35	0	-8	+70	+34
St. Margarets Bay, N.S.	0.31							
LMF		3.0	0.31	1.2	0.032	0.053	0.17	1.5
FM1		1.9	0.32	1.1	0.033	0.059	0.12	1.3
Δ%		+58	-3	+9	-3	-10	+42	+15

models are for SO_2 concentration, particularly in the weaker source regions (+137% for the LMF at Quebec City) and wet sulphate deposition (+20 to +60%): The dry SO_2 and total sulphur depositions are accordingly also higher (up to +35%) for the layered model. The lower SO_2 concentration and dry depositions of the fully mixed model are due to the mixing of low elevation emissions into the full daytime mean mixing depth even under limited mixing conditions. The enhanced wet sulphate deposition of the layered model is a direct result of in situ production of sulphate at an enhanced rate in the wet fraction of the plumes. In the fully mixed model with a lower oxidation rate, even during rain periods, the wet sulphate washout tends to be limited by the production rate of sulphate.

The sulphate concentrations and dry depositions for the two models are comparable.

Long-range Impact (Table 4). Table 4 compares the absolute contributions due to long-range transport given by the layered (LMF) and fully mixed (FM1) models. The receptor values in this table represent the contribution due to all sources excluding the meso region surrounding the receptor. Other than for wet sulphate deposition, this table shows that the ambient and deposited sulphur values attributed to long-range transport are lower for the layered model. While the LMF model gives considerably higher total SO_2 concentrations at all receptors (Figure 2(a), Table 4 shows that this is mainly a medium-range effect rather than due to long-range transport. Long-range transport of ambient sulphate is very similar for the two models.

The portion of dry sulphur deposition attributed to long-range transport follows the trend of the SO_2 and SO_4 concentrations when allowance is made for the 7.5% lower diurnially averaged deposition velocities with the layered model (See Table 2). For the LMF layered model, wet sulphur deposition of primary sulphur is lower as would be expected from the trend of the SO_2 concentration while the wet deposition of sulphate is greatly enhanced in the layered model relative to the fully mixed model. This increase is again attributed to the simulated enhanced in-cloud oxidation of the LMF model. Since wet periods represent a relatively minor fraction of the total annual period, the greatly enhanced wet sulphate removal rate does not markedly affect the plume budgets of SO_2 and SO_4 at long-range. It therefore appears that both the long- and medium-range contributions to wet sulphate deposition are sensitive to the manner in which wet processes are handled by the model and neglect of these processes can lead to an underestimate of the long-range effects on wet sulphur deposition. The results for the fully mixed FM2 model with double the oxidation rate, (not in the

Table 4. Long-Range (>300 km) Impact of Sulphur Emissions

	SO_2 $\mu g/m^3$	SO_4 $\mu g/m^3$	Dry SO_2 Dep kgS/ha	Dry SO_4 Dep kgS/ha	Wet SO_2 Dep kgS/ha	Wet SO_4 Dep kgS/ha	TSD kgS/ha
Duncan Falls, Ohio							
LMF	8.9	4.2	4.2	0.43	0.44	1.9	6.9
FM1	9.9	4.3	5.1	0.45	0.58	0.90	7.0
Δ%	-10	-2	-18	-4	-24	+110	-1
Simcoe, Ontario							
LMF	17.5	5.8	9.0	0.60	1.2	4.2	15.0
FM1	16.6	5.0	9.7	0.61	1.3	2.8	14.4
Δ%	+5	0	-7	-2	-10	+53	+4
Quebec City, P.Q.							
LMF	4.5	2.5	2.3	0.25	0.28	2.1	4.9
FM1	5.1	2.5	2.9	0.26	0.33	1.6	5.0
Δ%	-12	-1	-21	-4	-16	+32	-2
St. Margarets Bay, N.S.							
LMF	2.5	0.16	1.1	0.16	0.14	0.66	2.0
FM1	2.8	0.16	1.4	0.18	0.16	0.53	2.2
Δ%	-11	0	-21	-11	-15	+25	-9

table) show a high ambient SO_4 transport at long-range but the corresponding contribution to wet sulphate deposition is none-the-less lower than the layered model with simulated in-cloud processes (Figure 4(b)).

Relative Amounts of Primary and Secondary Sulphur (Table 5)

Table 5 compares the relative importance of primary and secondary sulphur in ambient and deposited sulphur levels as derived from the FM1 and LMF models.

The ratios of primary/secondary sulphur in ambient air and dry deposited sulphur are generally quite similar with both models except for ambient concentrations in the medium-range where the LMF model gives higher ratios at all receptors particularly in the weaker source regions of Nova Scotia and Quebec. Wet sulphur depositions are generally lower in primary sulphur with the layered model, especially in the Ohio region for long-range transport, and in the weaker source regions for both transport ranges. In the Ohio region total deposited sulphur is much richer in sulphate with the layered model for both long- and medium-range contributions to the receptor; this is due to the relatively strong emissions region which surrounds the medium-range region so ensuring a ready supply of primary sulphur for the enhanced wet oxidation in both ranges.

At the other receptors, the greater fraction of secondary sulphur deposition with the layered model is noted mainly for long-range transport. The higher secondary sulphur depositions are a direct result of the simulation of enhanced wet oxidation in the layered model.

SUMMARY AND CONCLUSIONS

A comparison of the annual average ambient and deposited sulphur results for a one year simulation period (1978) has been made using fully mixed and four-layered versions of the Lagrangian TRANS model. The objective of these comparisons was to determine if the more detailed structure of the layered model would lead one to significantly different conclusions regarding the medium- and long-range impact of sulphur emissions. One version of the layered model includes the simulation of enhanced in-cloud oxidation of primary sulphur species in the fraction of the plumes subjected to wet processes.

The results of this study indicate that in the medium-range, (<300 km) fully mixed models give lower estimates for the annual

Table 5. Ratios of Primary/Secondary Sulphur At Medium-Range (MR) and Long-Range (LR)

	$[SO_2]/[SO_4]$		Dry SO_2 Dep / Dry SO_4 Dep		Wet SO_2 Dep / Wet SO_4 Dep		Tot SO_2 Dep / Tot SO_4 Dep	
	MR	LR	MR	LR	MR	LR	MR	LR
Duncan Falls, Ohio								
LMF	6.7	2.1	32.	9.8	0.20	0.23	2.9	2.0
FM1	5.4	2.3	32.	11.0	0.23	0.64	4.5	4.2
Simcoe, Ontario								
LMF	8.2	3.0	37.	15.	0.45	0.29	6.1	2.1
FM1	5.6	3.3	33.	16.	0.50	0.46	6.7	3.2
Quebec City, P.Q.								
LMF	21.	1.8	70.	9.2	0.26	0.13	11.	1.1
FM1	9.4	2.0	52.	11.	0.48	0.21	13.	1.7
St. Margarets Bay, N.S.								
LMF	9.7	16.	38.	6.9	0.31	0.21	6.2	1.5
FM1	5.9	18.	33.	7.8	0.50	0.30	7.6	2.2

average ambient surface SO_2 concentrations and primary sulphur depositions by mixing near-surface emissions into the full model layer even under limited mixing conditions; the reverse would be true in areas where strong elevated sources are predominant. Secondary sulphur concentrations and dry depositions in the medium-range and long-range (>300 km) are comparable with both models. At long-range the fully mixed model gives higher SO_2 concentrations and higher primary sulphur deposition (both wet and dry). The four-layered model with simulated enhanced wet SO_2 oxidation gives significantly higher wet sulphate deposition in both the medium-range and long-range. This is attributed to the more rapid in situ generation of sulphate in the wet fraction of the plume in the layered model, and does not depend to a great extent on the ambient levels of primary and secondary sulphate which are generally lower at long-range with the layered model. This study shows that since wet sulphate deposition is limited by the production rate of sulphate, the manner in which wet processes are handled in Lagrangian long-range transport models can significantly affect the results for wet sulphate deposition at both medium- and long-range transport distances.

REFERENCES

Barrie, L.A., 1981, 'The Prediction of Rain Acidity and SO_2 Scavenging in Eastern North America', Atmos. Env., 15:31.

Weisman, B., 1983, 'Application of Trajectory Model To Regional Characterization', Air Pollution Modeling and its Application II, Volume 3, Editor C. De Wispelaere, Plenum Press 1983, pp 245-264.

Weisman, B., 1980, 'Long-Range Transport Model for Sulphur', 73rd Annual Meeting of the Air Pollution Control Association, Montreal.

Work Group 2, Memorandum of Intent on Transboundary Air Pollution, Phase 2 Interim Report, July, 1981.

DISCUSSION

P.J.H. BUILTJES

Are there any field data available which could be used for a comparison between calculations and field data ?

T. SCHOLTZ

Yes. We compared the LMF model results with the 1978 CANSAP data for wet sulfur deposition, and SURE data for ambient sulphate concentrations. This was a screened data set used by Work Group 2 of the MOI for model evaluation.

The geometric mean ratio of observed to predicted was 1.19 for ambient sulphate and 1.98 for wet sulfur deposition with correlation coefficients of 0.86 and 0.92 respectively. A biogenic background of 2 KgS/ha was added to the modeled wet deposition results. We also compared our wet deposition results with the 1979/80 NADP data and in this case the corresponding geometric mean ratio was 0.94.

ANALYSIS OF ACID DEPOSITION DUE TO SULFUR DIOXIDE EMISSIONS FROM OHIO

Stanley Mermall and Ashok Kumar

Civil Engineering
University of Toledo
Toledo, Ohio

INTRODUCTION

Atmospheric transport of emissions over hundreds of kilometers is termed long range transport. Examples of this phenomenon are: the continental advection of smoke generated by volcanos and forest fires, high tropospheric levels of ozone in regions where the precursors of ozone production are not found in high quantities, radioactive fallout in regions distant from the nuclear release, and the transport and deposition of sulfur and nitrous oxides from industrial sources on areas hundreds of kilometers away.

The detailed investigation of long range transport of emissions began when the public was interested in the effects of radioactive contamination due to nuclear explosions and releases (Hicks and Shannon, 1979). As the open air releases of radioactive materials were reduced so did the concern over radioactive fallout. More recently the transport, transformation, and deposition of sulfur and nitrous oxides has received most of the attention. Commonly known as acid deposition this phenomenon has become one of the most controversial and emotional issues of our time. Acid deposition touches our society politically, economically, and environmentally. Acid deposition has been blamed for causing the destruction of fish life in sensitive lakes in Canada, northeastern United States and Europe. It has become a source of tension between the relatively non-industrial regions in the northeastern United States and southeastern Canada and the industrial midwest (Robert, 1982, Jacobson, 1981).

There are two general ways to determine the contribution of a source region on acid deposition. The first scheme, called the receptor mode, seeks to explain observed concentrations by following the contaminant backwards in time until a source of the observed pollution is determined. This method has several drawbacks: (i) the placement of the source becomes more uncertain the further back in time the analysis goes. As a result the source region of the observed pollutant can not always be determined with certainty, and (ii) this analysis could be biased since it is trying to find a cause for an observed result. A second scheme, source mode, follows the contamainant leaving a source.

Over the last decade Ohio has received considerable attention concerning its contribution or impact upon the acid deposition phenomenon in North America. This is due to the following:

(i) Ohio has the largest sulfur dioxide emission rate of any state in the United States of America. Ohio contributes 10.7% of the total U.S. sulfur dioxide emissions with utilities making up to 79.3% of this amount (U.S.-Canada, 1982).

(ii) Ohio is located in a region where its' pollutants can be carried, by the prevailing wind flow, to the northeastern United States and southeastern Canada. These regions are thought to be sensitive to acid deposition and have been the subject of political discussions between the United States and Canada.

In this paper the relative impact of utility emissions from individual source regions of Ohio is investigated using an improved event type model (in the source mode) during the period of July 3 through July 8, 1978. The period was chosen because during this interval the eastern half of the United States was dominated by a stagnated high pressure system and thus the potential of an acid deposition long range transport episode existed.

MODEL STRUCTURE

A long range transport model should include dispersion of pollutants as well as physical processes such as dry deposition, wet deposition, and chemical decay. According to Johnson (1983), long range transport models can be classified as Lagrangian trajectory models, Eulerian grid type, or statistical models. Since this study is concerned with a pollution episode, either a Lagrangian trajectory model or a Eulerian grid type model is suitable. The proposed event model utilizes the features from both model types.

The equations used in our model to calculate concentrations of sulfur dioxide and sulfate were derived from the following mass

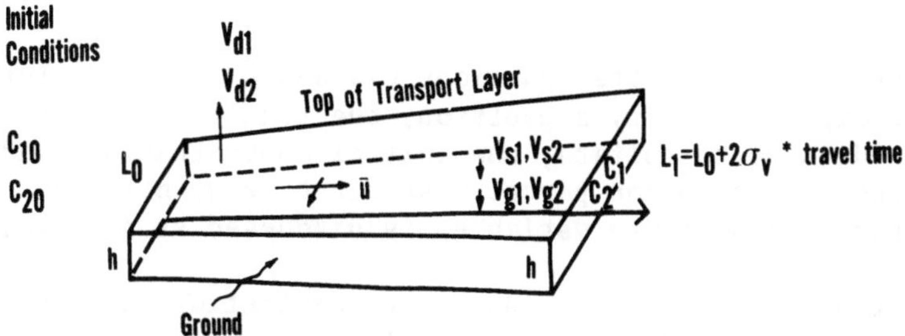

Fig. 1. Mass balance for concentration equations

$$\bar{u} \cdot h \frac{d}{dx}(LC_1) = - [V_{g1} + fV_{S1} + Vd_1] \cdot LC_1 - wC_1 \cdot Lh \qquad (1)^*$$

$$\bar{u} \cdot h \frac{d}{dx}(LC_2) = - [V_{g2} + fV_{S2} + Vd_2] \cdot LC_2 + 1 \cdot 5\, wC_1 \cdot Lh \qquad (2)$$

By assuming that the mean wind speed, initial concentration and plume width, wet and dry deposition velocities, transformation rate and transport height terms are known. Equations (1) and (2) were solved to give equations (3) and (4) which are used to calculate the concentrations of SO_2 and SO_4 respectively.

$$C_1 = \frac{C_{10} \cdot L_0}{L_0 + 2\sigma_v \frac{x}{\bar{u}}} \cdot \exp\left[- \frac{R_1}{h} \cdot \frac{x}{\bar{u}}\right] \qquad (3)$$

$$C_2 = \frac{C_{20} \cdot L_0}{L_0 + 2\sigma_v \frac{x}{\bar{u}}} \cdot \exp\left[- \frac{R_2}{h} \cdot \frac{x}{\bar{u}}\right] + 1 \cdot 5\, w \frac{h}{R_1 - R_2} \cdot$$

$$\cdot \left\{ C_1 \left[\exp\left[\frac{R_1 - R_2}{h} \cdot \frac{x}{\bar{u}}\right] - 1\right]\right\} \qquad (4)$$

* Nomenclature is given at the end of the paper.

Equations (3) and (4) are similar to the equations of Henmi and Reiter (1979).

In order to solve the equations (3) and (4), the depletion due to precipitation, dry deposition, the transformation rate of sulfur dioxide to sulfate, the initial concentration, and the path of the air pollution parcel must first be quantified. The parameterization of these variables is discussed as follows.

The removal of pollutant due to precipitation was expressed in this model in terms of the precipitation scavenging velocity which is defined as:

$$V_S = k/\chi \cdot P \tag{5}$$

and

$$f = \frac{\tau_p}{\tau_d + \tau_p} \tag{6}$$

The duration of wet and dry periods and the precipitation rate was calculated from data provided by the stations located in 31 regions of homogenous land use.

The ratio k/χ is commonly defined as the washout ratio. In this model the washout ratio was assumed to be 5×10^4 for sulfur dioxide and 1×10^5 for sulfate. The scavenging velocity for all receptors within the plume was averaged. This average was then used as input for equations (3) and (4).

The removal of pollutants due to dry deposition was expressed in this model in terms of the dry deposition velocity. The dry deposition velocity is defined as:

$$V_g = -F/\chi \tag{7}$$

In the computer program either a constant or a variable dry deposition velocity may be used. The variable dry deposition velocity option allows the dry deposition velocity to vary as a function of land use, pollutant, season, and atmospheric stability using results obtained from research performed by Sheih (1979). (See Table 1) If the constant dry deposition velocity option was used, the dry deposition velocity was assumed to be:

Table 1.

Values of dry deposition velocity of SO_2 and SO_4 as a function of landuse and stability in cm/sec.

	1	2	3	4	5	6	7	8	9
Stability	Crop land Pasture	Crop and Woodland grazing land	Irrigated crops	Forest wood land grazed	Forest wood land ungrazed	Subhumid grass land & semi arid grass land	Open wood land grazed	Dessert Shrubland	Swamp
A SO_2	.65	.65	.45	.85	.85	.55	.65	.45	.95
SO_4	.75	.85	.65	.95	.95	.75	.75	.85	.75
B SO_2	.75	.75	.65	.85	.85	.65	.75	.45	.85
SO_4	.85	.85	.75	.95	.95	.75	.85	.85	.85
C SO_2	.75	.75	.65	.85	.85	.75	.75	.45	.05
SO_4	.85	.85	.75	.95	.95	.85	.85	.85	.85
D SO_2	.35	.35	.25	.35	.35	.35	.35	1.0	.95
SO_4	.85	.85	.75	.95	.95	.85	.85	.85	.85
E SO_2	.05	.05	.05	.05	.05	.05	.05	.05	.55
SO_4	.65	.75	.55	.85	.85	.65	.65	.75	.65
F SO_2	.55	.65	.45	.85	.85	.45	.55	.05	.55
SO_4	.45	.45	.35	.55	.55	.35	.45	.05	.45

Dry Deposition Velocity of SO_2
Daytime 2 cm/sec
Nighttime 1 cm/sec

Dry Deposition Velocity of SO_4
Daytime .2 cm/sec
Nighttime .1 cm/sec

The model accounts for the transformation of SO_2 to Sulfate through the following relationship

$$\frac{d[H_2SO_4]}{dt} = \frac{d[SO_4]}{dt} = w[SO_2] \tag{8}$$

Verbally, equation (8) states that the formation rate of H_2SO_4 is equal to the rate of SO_4 formation which is equal to a constant (w) times the concentration of SO_2 present (U.S. Canada, 1982). There is considerable evidence that the transformation rate is not constant and that it varies as a function of sunlight intensity, relative humidity, temperature, and the mixture and concentration of substances making up the air mass. Because we do not know enough about the effects of these variables to justify a variable transformation rate most models assume a constant rate. In this model a constant transformation rate was assumed to be:

$w = .1$ %/hr Daytime

$w = .01$ %/hr Nighttime

Emission input into the model is in the form of concentration. The annual emission rates were obtained from the emissions inventory maintained by the United States Environmental Protection Agency and is the one used by the United States-Canadian study on acid deposition. The sulfur dioxide emissions from Ohio's utilities were divided into six regions (see Figure 2). The emissions from each region were aggregated and placed as nearly as possible in the center of the individual sources. The initial concentration which is required as a boundary condition for equation (3) and (4) was calculated following Kumar, 1978:

$$C_o = q/(\overline{u}* \ z* \ y) \tag{9}$$

Equations (3) and (4) require a value for horizontal dispersion parameter be determined. This was approximated using

$$\sigma_y = 0.5 \ t \tag{10}$$

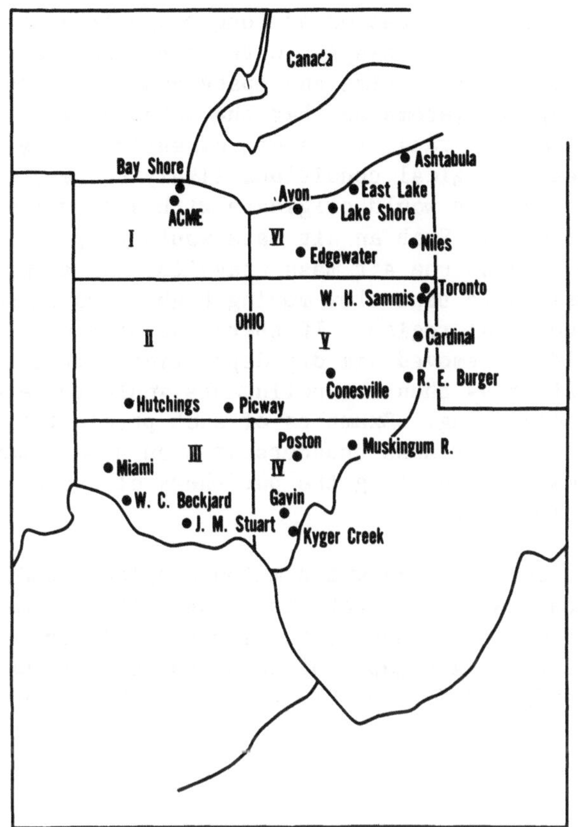

Fig. 2. Individual utility sources in Six Source Regions of Ohio.

This equation was found to give a reasonable estimate for the horizontal diffusion over several days duration. (Heffter 1975) The use of this type of dispersion term allows the model to run when a detailed knowledge of the transport layer is not known. It also uses a dispersion factor which is accepted by many researchers who are involved in the study of long range transport.

The path of a parcel of air in time is called a trajectory. In this study trajectories were obtained using Heffter's (1980) model.

METEOROLOGY

To understand the processes of long range transport it is useful to investigate the changes in the concentrations of sulfur pollution as a function of time and space and relate those concentrations to synoptic systems so that the meteorological conditions favorable for the long range transport of sulfur oxides may be identified. Meteorological conditions likely to result in a long range transport episode would originate with a slow moving air mass over a source region. Such an air mass would trap pollutants and then transport them as the air mass traveled. An example of this type of air mass would be a slow moving high pressure zone traveling over an industrial region. If there was no precipitation the pollutants would be removed via dry deposition. If precipitation occurs in the air mass then the pollutants would be removed by precipitation scavenging. Examples of such precipitation producing situations would be air mass thunderstorms on a warm day, low pressure systems passing along the periphery of the air mass, or orographic precipitation.

One way to characterize which meteorological conditions are favorable to long range transport is to describe air masses in a region in terms of homogeneous temperature, humidity, cloud cover, and source regions. An example of this type of work was illustrated in the study performed for the Electric Power Research Institute

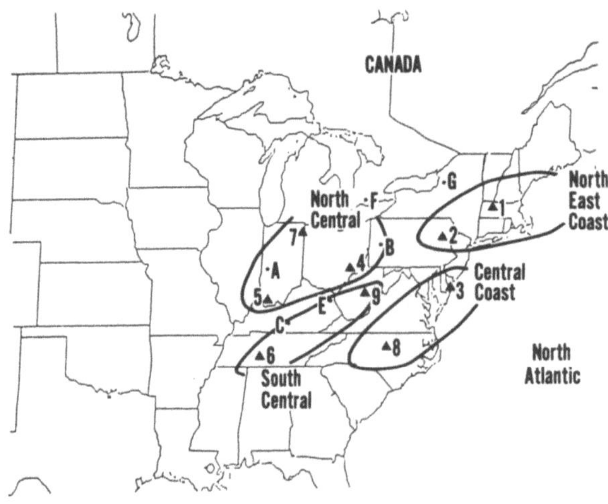

Fig. 3. Four subregions and locations of SURE stations.

(EPRI) (SURE, 1982). In this study the eastern half of the United States was divided into the four subregions (North Central, Northeast coast, Central coast and South Central) (See Figure 3) and the meteorological conditions were described in terms of five air mass categories; continental polar warm (cPw), continental polar colder (cPk), continental polar (cP), transitional (Tr) and maritime (mT).

Using a technique similar to the EPRI research, we compared the spatial and temporal sulfur dioxide and sulfate concentrations observed in the SURE network to the synoptic conditions over the period of July 2 to July 8, 1978. Figure 4 indicates that during the period studied, the highest observed sulfate levels were associated with mT and the lowest with cPk or Tr air masses. The maximum sulfate concentrations in the north central and north coastal regions, which occured when these regions were influenced by mT air masses, can be explained by the effect of stagnated air masses over the midwest and the associated north eastward wind pattern which carries pollution into the two subregions. Sulfate levels associated with mT air masses were less prevelant in the central coastal and south central region where the air flow did not pass over high emission source regions. Secondary sulfate concentration maxima were associated with cPw air masses.

The relationship between sulfur dioxide concentrations and meteorology for each region is shown in Figure 5. This figure indicates that the observed sulfur dioxide concentration pattern is complex. This complexity can be explained by the influence of local sulfur dioxide emissions. This pattern does not exist in the sulfate data because sulfate concentrations are influenced less by local sources than is sulfur dioxide. Moreover, the sulfate data was averaged over a 24 hour period while the sulfur dioxide data represented an average of the hourly observations.

In the northeast coast, south central and north central subregions, the highest observed concentrations of sulfur dioxide were associated with cPw and mT air masses. Secondary maxima were associated with Tr air masses in the north central region. The minimum observed sulfur dioxide concentrations in the south central, northeast coast, and north central subregions were associated with cPk air masses.

The EPRI study showed that sulfate levels were highest in maritime tropical regions and lowest in continental polar, colder and transitional air masses. Secondary maximum concentrations of sulfate were found to exist in Illinois, Indiana, Ohio, Pennsylvania and the Carolinas when these regions were under the influence of continental polar warm air masses. Similar relationships were also observed by Chung in Southeastern Canada (Chung, 1978).

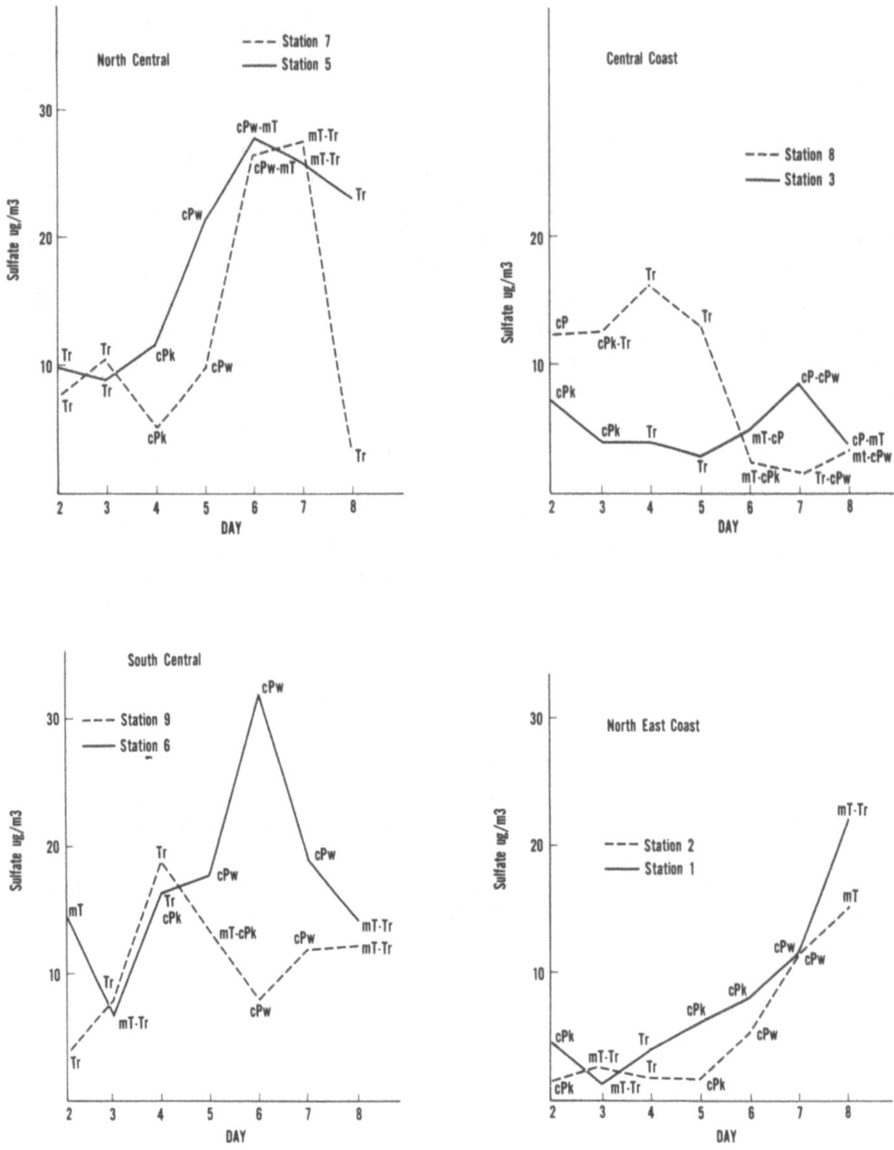

Fig. 4. Indicates that during the period studied, the highest observed sulfate levels were associated with mT and the lowest with cPk or Tr air air masses.

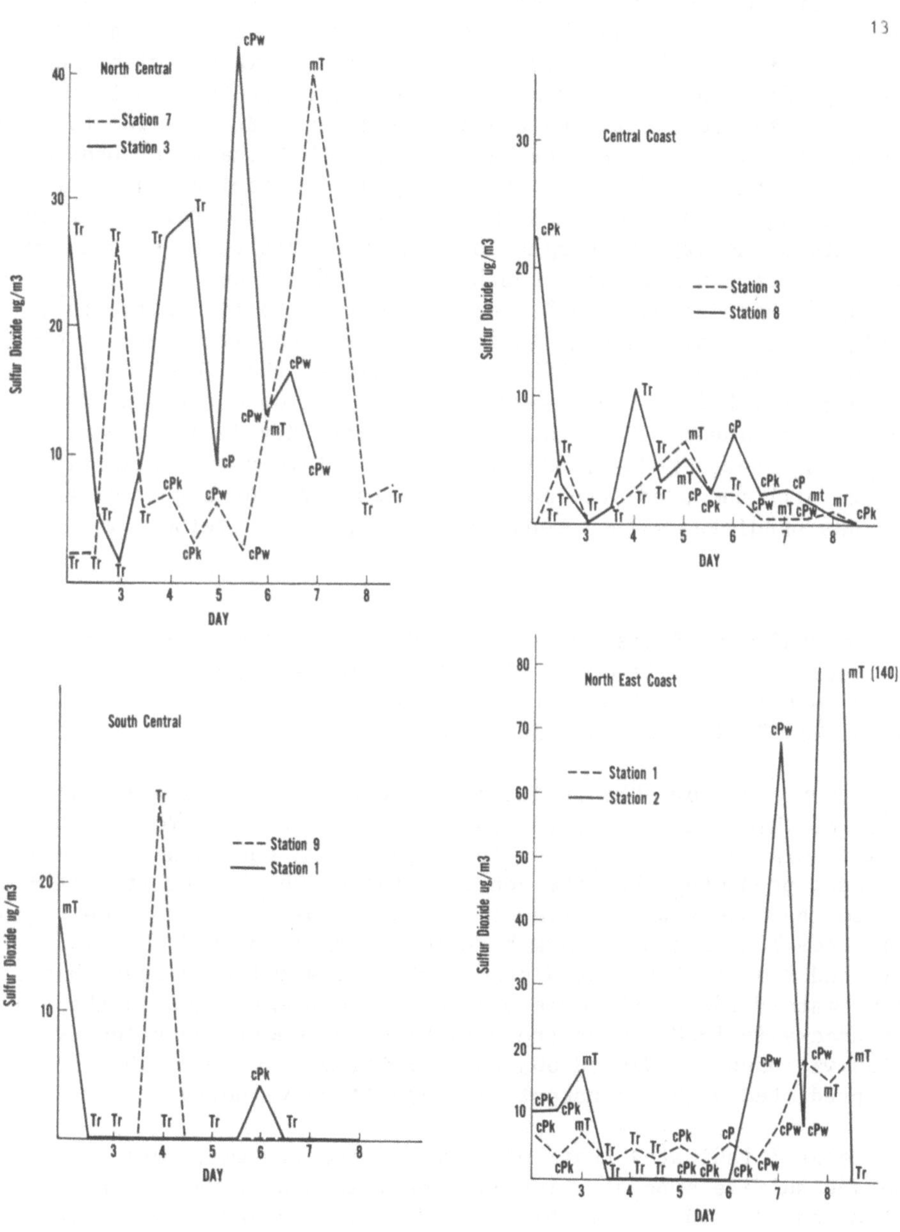

Fig. 5. The relationship between sulfur dioxide concentrations and meteorology for each region shown.

RESULTS AND DISCUSSION

The results of the model runs performed from July 3 to July 8 for the six Ohio emission subregions are summarized in Figure 6 and Tables 2 and 3. In these tables the observed concentration for SO_2 and SO_4 at EPRI's SURE network is given for comparison with the models predictions. Note that the observed concentration is due to all the sources while the predicted concentration is due to only Ohio utility sources. Figure 6a indicates that the model predicted Ohio, northern West Virginia and Pennsylvania received the largest amount of sulfur dioxide from Ohio's emission. According to these predictions the region thought to be most sensitive to acid depo- osition (northern New York, southeastern Ontario and the New England states) received less than 25% of its sulfur dioxide from Ohio's utility emissions. This is in general agreement with results obtained by Endlich et al. (1982) who predicted that Ohio's emissions accounted for 15% of the observed sulfur dioxide concentration in New York in January 1977. When the variable dry deposition velocity option was used, predicted SO_2 contaminent levels were slightly higher, with the predicted maximum levels covering a greater area than was predicted with a constant dry deposition velocity.

According to Table 2 Region V had the greatest impact of the SO_2 contaminent levels observed in the acid sensitive region. Predicted contributions from this source region ranged from 20% in central New York to 1% in southern Ontario.

Figure 6c shows the predicted concentrations of sulfate as a result of Ohio's utility emissions. Southwestern New York, western Pennsylvania and Ohio received the highest SO_4 contamination levels. The model predicted that the contribution of Ohio's utility emission to the observed contaminent levels in the acid sensitive region range from 20% in central New York to 5% in Quebec and Ontario. Small and Samson (1982) predicted that Ohio's emission contributed approximately 18% of the observed sulfate concentration in the Adirondacks in 1978. When the variable dry deposition velocity option was used, predicted SO_4 concentration were comparable to that predicted with a constant dry deposition velocity.

Table 3 indicates that Region V had the greatest impact on the observed sulfate concentrations in the acid sensitive region. Predicted contributions from this source region ranged from 10% in central New York to .5% in Southern Quebec.

Table 2.

Predicted Amount of Sulfur Dioxide Contamination Accounted for by SURE Network

SURE Station	Observed Concentration	Total Predicted Concentration due to Ohio		Percent of Observed Concentration at the Station					
	$\mu g/m^3$	$\mu g/m^3$	%	REGION I	REGION II	REGION III	REGION IV	REGION V	REGION VI
1	6.92	2.16	31.2	.7	.3	1.1	1.9	20.8	6.4
2	25.6	15.3	59.6	.1	.2	1.3	1.8	56.3	0
3	3.04	.56	18.4	0	.3	1.3	15.8	1.00	0
4	15.5	22.93	148.1	1.4	3.0	32.3	81.3	4.1	.32
5	44.16	0	0	0	0	0	0	0	0
6	2.79	.06	2.2	0	0	0	0	0	2.2
7	8.78	.59	6.8	3.0	0	3.8	0	0	0.6
8	2.99	.15	5	0	0	1.0	1.7	2.3	0
9	.68	1.20	176.5	0	1.5	10.3	163.	2.9	0

Table 3.

Predicted Amount of Sulfate Contamination Accounted for by SURE Network

SURE Station	Observed Concentration	Total Predicted Concentration due to Ohio		Percent of Observed Concentration at the Station					
	$\mu g/m^3$	$\mu g/m^3$	%	REGION I	REGION II	REGION III	REGION IV	REGION V	REGION VI
1	8.28	3.15	38	.8	.6	.2	4.11	15.9	4.59
2	10.6	3.75	35	.5	.8	7.5	8.2	9.2	0
3	6.29	1.04	17	0	.2	23	9.9	.5	0
4	17.3	4.16	24	.3	.4	10	4.5	2.0	.3
5	18.13	.03	0	0	0	0	0	0	0
6	16.19	.08	.5	0	0	0	0	.1	0
7	11.98	.21	1.7	.3	0	.8	0	0	0
8	8.69	.58	6.7	0	0	.8	1.0	2.0	0
9	12.23	.44	3.6	.1	.1	.4	2.0	.2	.1

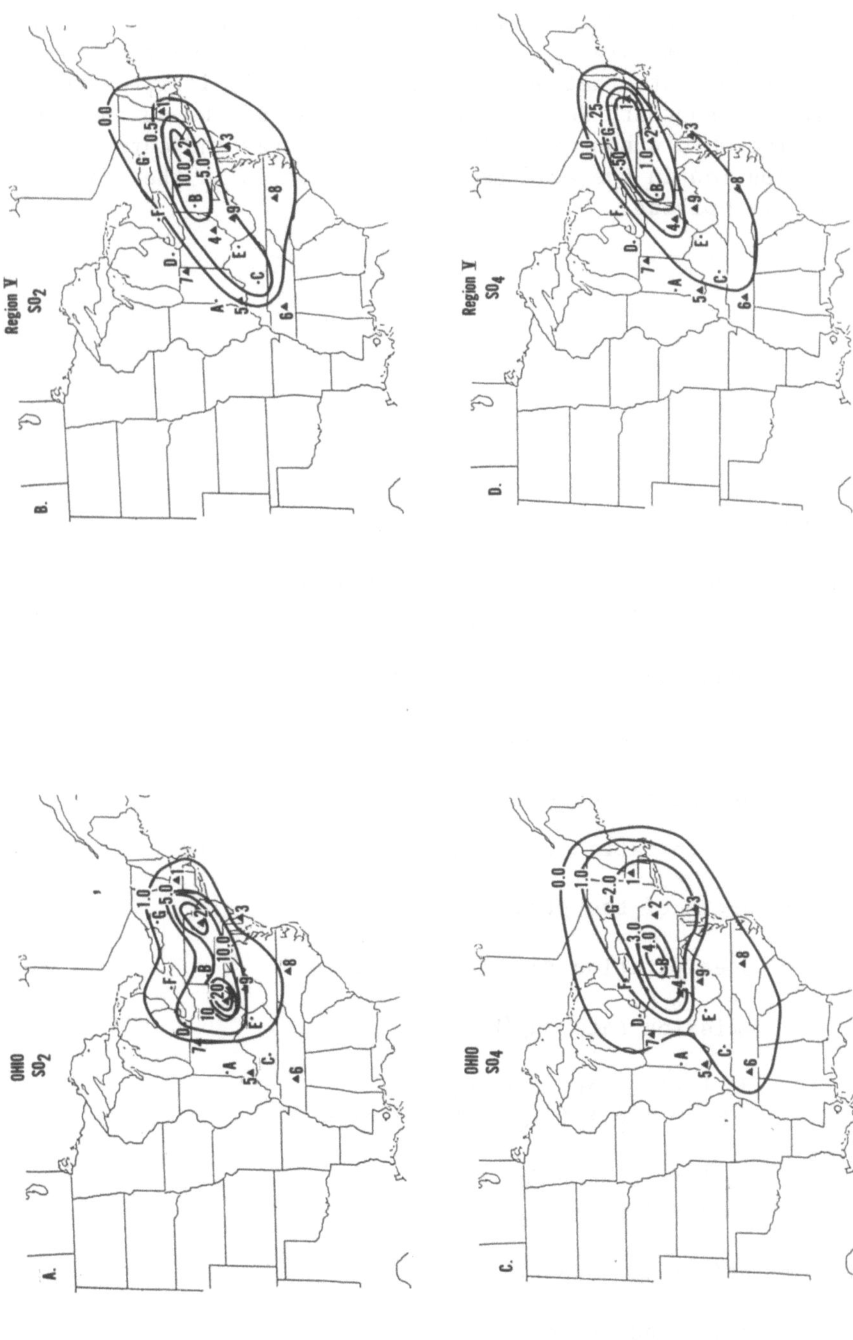

Fig. 6. The results of the model runs performed from July 3 to July 8 for the six Ohio emission subregions are summarized.

CONCLUDING REMARKS

An event type long range transport model has been applied to study acid deposition resulting from Ohio's utility sources for an episode occuring from July 3, 1978 to July 8, 1978. The results of the model runs indicate that Ohio, northern West Virginia, western Pennsylvania and southern New York received most of the sulfur contamination from Ohio's sources. Indirect estimates indicate that Ohio's contribution in the acid sensitive region is in the range of 20% for this episode.

NOMENCLATURE

English Symbols

$C1,2$ - concentration of sulfur dioxide and sulfate in the transport layer

f - frequency of precipitation events

F - flux of the pollutant

h - height of the transport layer

k - concentration of the pollutant in the precipitation

L - width of the plume

P - precipitation rate (meters/hour)

q - emission rate (u_g/hr)

R_1 - depletion term for SO_2
$R_1 = V_{g1} + f\ V_{s1} + V_{d1} + w \cdot h$

R_2 - depletion term for SO_4
$R_2 = V_{g2} + f\ V_{s2} + V_{d2}$

t - travel time (sec)

$\bar{u}$ - mean wind spead

Vd - transport of the pollutant through the top of the transport layer

V_g - dry deposition velocity

V_s - precipitation scavenging velocity

w - transformation rate of sulfur dioxide to sulfate

x - downwind distance from the source

y - width of plume

z - thickness of the transport layer

Greek Symbols

σ_v - horizontal dispersion parameter

τ_d - duration of dry periods

τ_p - duration of wet periods

χ - concentration of the pollutant in the air at the receptor

REFERENCES

Chung, Y. S., 1978, The Distribution of Atmospheric Sulfates in Canada and its Relationship to Long Range Transport of Air Pollution, Atmospheric Enviornment, Vol 12, pp. 1471-1480.

Endlich, R. M., Bhumralkar, C. M., and Brodzinsky, R., 1982, ENAMAP-1 Long Term Air Pollution Model. Refinement of Transformation and Deposition Mechanisms. Presented at 75th Annual Meeting of the Air Pollution Control Association, New Orleans, Louisana, June 1982.

Heffter, J., 1980, Air Resources Laboratories, Atmospheric Transport and Dispersion (ARL-ATAD), Air Resources Laboratories, Silver Springs, Maryland, NOAA. Tech. Memo. ERL-ARL-81.

Heffter, J., 1975, A Regional-Continental Scale Transport, Diffusion-Deposition Model: Part I Trajectory Model, Part II Diffusion-Deposition Model, Air Resources Laboratories. Silver Springs, Maryland, June 1975.

Henmi, T., and Reiter R., 1979, Long Range Transport and Transformation of Sulfur Dioxide and Sulfate, Environmental Sciences Research Laboratories, Office of Research and Development, U.S.E.P.A., Research Triangle Park, N.C. 27711.

Hicks, B. B., and Shannon, J. D., 1979, A Method for Modeling the Deposition of Sulfur by Precipitation Over Regional Scales, Journal of Applied Meteorology, Vol. 18, pp. 1415-1420.

Jacobson, J. S., 1981, Acid Rain and Environmental policy, Air Pollution Control Association Journal, Vol. 31, pp. 1071-1073.

Johnson, W., 1983, Interregional Exchanges of Air Pollution: Model Types and Applications, Air Pollution Control Association Journal, Vol. 33, pp. 563-574.

Kumar, A., 1978, Pollutant Dispersion in the Planetary Boundary Layer, Syncrude Canada LTD., Professional Paper 1978-1.

Roberts. 1982, Solving The Acid Rain Equation Keynote Address, 75th A.P.C.A. Annual Meeting and Exhibition, Air Pollution Control Association Journal, Vol. 32, pp. 925-928.

Small, M., and Samson, P., 1982, Mathematical Simulation of Lagrangian Precipitation and Associated Sulfur Wet Deposition. Proceeding Atmospheric Deposition Specialty Conference, Air Pollution Control Association, November 1982.

SURE, 1982, EPRI Sulfate Regional Experiment Results and Implications, EPRI EA-2165-SY-LD Project 862-2, Summary Presentation, December 1982.

U.S. Canada, 1982, United States-Canada Memorandum of Intent on Transboundary Air Pollution: Final Report, Atmospheric Sciences and Analysis Work Group 2.

DISCUSSION

T. LAVERY

You overestimated SO_2 levels at Duncan Falls; you apparently underestimated SO_4 levels there. What component of the model do you attribute that to ?

S. MERMALL

The formulation of the model is based on several assumptions. The results are heavily influenced by transformation rate of SO_2 to sulfate. The values in the paper were based on the work of Henmi and Reiter (1979). Current literature suggests a higher value. This will resolve the problem you mentioned.

G. DEN HARTOG You used two scenarios in your paper, a constant dry deposition velocity to 1 to 2 cm/sec and a variable deposition velocity considerably smaller. Since you state the results were similar in the two cases is the model not sensitive to dry deposition velocity.

S. MERMALL The results are similar in nature. The model is sensitive to the dry deposition velocity and some variations in SO_2 concentrations were observed. However, enough numerical experiments were not carried out to make a mathematically precise statement.

DEVELOPMENT OF AN ACID RAIN IMPACT ASSESSMENT MODEL

R. Yamartino, J. Pleim and W. Lung

Environmental Research and Technology, Inc.
696 Virginia Road
Concord, MA 01742

INTRODUCTION

Numerous studies have implied the significance of long-range transport of air pollutants in the acidification of lakes in the Eastern United States and Canada (National Research Council, 1983). The present study was designed to address the question: Do current levels of SO_x and NO_x emissions from electric utility power plants in New York State contribute significantly to acid deposition and acidity in lakes and streams in acid-sensitive regions of New York, New England and the Canadian Maritime Provinces? To put this question into perspective, the Empire State Electric Energy Research Corporation (ESEERCO) contracted with ERT to develop a model to quantify the acid rain impacts due to all sources of SO_x and NO_x in Eastern North America including the New York utility power plants' emissions and to assess changes in acid deposition and acidity of surface waters in the above areas resulting from changes in selected emissions.

In order to focus this impact assessment, four acid-sensitive lakes in the Adirondacks, New Hampshire, Central Massachusetts and southwest New Brunswick were selected for detailed evaluation. These areas have been generally identified as sensitive to acid inputs and are located predominantly downwind of the New York utilities under prevailing southwesterly to westerly wind patterns. In addition, one lake with a large buffering capacity was selected for comparison purposes.

Attainment of the above stated objective required a multidisciplinary effort to assemble, evaluate, and exercise the following key elements of the program:

- Establishment of a comprehensive set of lake selection criteria based on all, presently known, variables affecting the lake acidification process.

- Assembly of 1980 base year emissions from New York utilities and from major external emissions regions.
- Development of a coupled atmospheric transport, deposition, and lake water chemistry model consisting of three basic components: a trajectory generator; a Lagrangian atmospheric dispersion, statistical transformation and acidic deposition module; and a two-layer, lake acidification model to compute alkalinity and CO_2 acidity.

This paper will focus on the development and application of the atmospheric transport and acidic deposition models. The lake water chemistry model, as well as the lake selection criteria and emissions inventory, are described in Yamartino et. al. (1983).

ATMOSPHERIC TRANSPORT, DISPERSION, CHEMICAL TRANSFORMATION, AND DEPOSITION MODELING

In many types of atmospheric dispersion models, the simulation of the advection, diffusion, chemical transformation, and deposition processes is performed simultaneously. This results from the fact that such simultaneous application of operators is either trivial, as in the case of Gaussian plume modeling, or imperative, as in the case of Eulerian grid models incorporating non-linear chemistry. In the present modeling effort we are fortunate that the chemistry is dealt with in a linear way through the use of first-order rate constants. This linearity enables one to split the computationally expensive transport operator, which must compute the trajectories followed by parcels of air through interpolation of observed meteorological data, from the relatively inexpensive computation of dispersion, linearized chemistry, and deposition.

In the following subsections, the theoretical bases of these two operator groupings, referred to programatically as ARL-ATAD and DISCDEP, are considered.

The ARL-ATAD Trajectory Model

The Air Resources Laboratories Atmospheric Transport and Dispersion Model (ARL-ATAD; Heffter, 1980) is a regional scale Lagrangian trajectory model. ARL-ATAD calculates four air parcel trajectories per day from any number of source locations for durations of up to five days. Each trajectory is composed of linear segments which are the products of a three-hour time step and the advecting winds, which are vertically and horizontally weighted averages of nearby rawinsonde wind observations.

The depth of the layer through which the rawinsonde wind observations are averaged, the Transport Layer Depth (TLD), is calculated in two ways depending on whether the trajectory starts in the daytime or the nighttime.

i) The daytime technique involves scanning the temperature profiles for a critical inversion. The TLD is then defined as the layer between the ground and the inversion. This method assumes that the plume instantly mixes throughout the transport layer.

ii) For nighttime trajectory starts, the TLD is defined by a Gaussian dispersion function. Vertical growth proceeds proportional to the square root of the travel time for the duration of the night. After the first night, the TLD is calculated by the daytime method.

The advective velocity is calculated from the vertically averaged layer winds from all reporting stations within a prescribed radius. The station winds are spatially averaged using the weighting factor $1/R^2$, where R is the distance between the station and the trajectory segment midpoint. Also, a trigonometric alignment factor is applied such that stations off to either side are weighted only half as much as stations in alignment with the trajectory segment.

Trajectories are terminated when sufficient rawinsonde data is not available within a prescribed radius or when the user selected maximum trajectory duration ($\leq$5 days) is reached.

The meteorological input data for this model consists of upper air wind and temperature data collected from the North American rawinsonde station network by the United States Air Force (USAF-ETAC, 1972) and available through the National Climatic Center (NCC, 1975). The model extracts wind and temperature observations from the surface up to the 500 mb level for the dates and region specified by the user. These data are generally available only at 0Z and 12Z, although a few stations also report at 6Z and 18Z.

In addition to the standard ARL-ATAD output, the program now generates a more concentrated output file that contains trajectory segment endpoint coordinates as well as transport layer depths. This file serves both as input to the Dispersion-Chemical Transformation Deposition model (DISCDEP) and as a data file for graphical software. An example of a trajectory plot made from these files is presented in Figure 1.

The Dispersion-Chemical Transformation-Deposition Model (DISCDEP)

The Dispersion-Chemical Transformation-Deposition Model (DISCDEP) combines the Statistical Acid Deposition Model (STADMOD; Venkatram et. al., 1982) with a puff dispersion model of the type used by Heffter (1980). The condensed output file from ARL-ATAD, containing trajectory segment endpoints and transport layer depths, is input to DISCDEP. The output from DISCDEP is the average daily deposition fluxes at each receptor from each source group on a unit emission basis.

The dispersion model within DISCDEP simulates plume dispersion by considering Gaussian pollutant puffs which follow ARL-ATAD generated trajectories. Puffs are released once per day

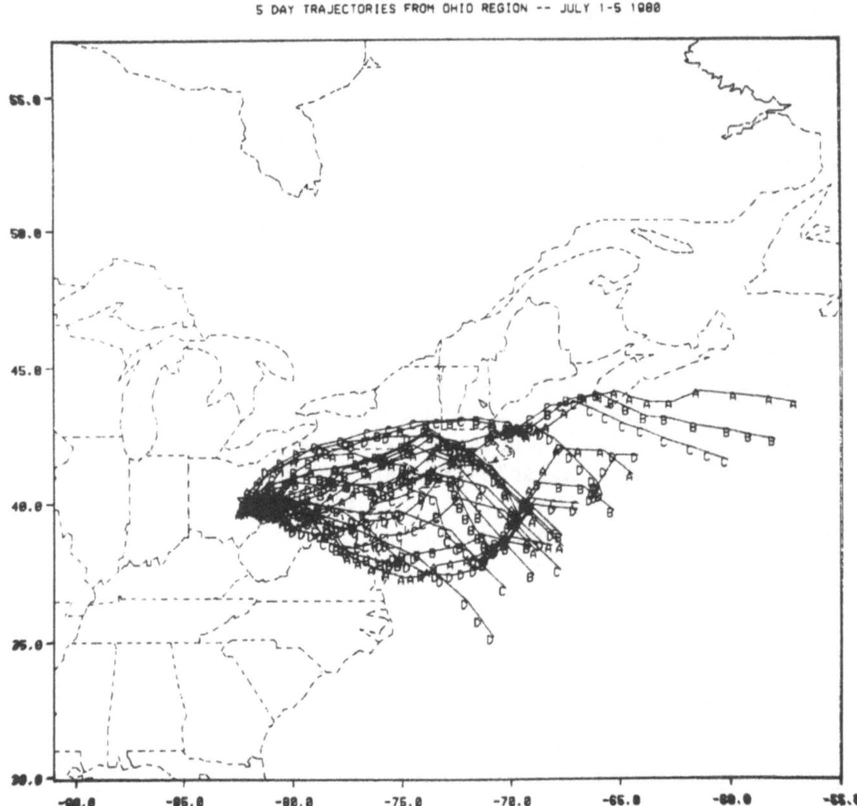

Figure 1 Examples of trajectories generated by ARL-ATAD for the five day period July 1-5, 1980. Four trajectories per day, starting at 0Z, 6Z, 12Z, and 18Z and represented by the letters A, B, C, and D respectively at six-hour intervals along the trajectory, are shown originating from Ohio.

at random hours assuming the entire day's emissions are contained in the one puff. All calculations are based on unit emissions (1 gm/day), in order to allow for greater flexibility when evaluating various emissions scenarios. Puffs released at hours in-between trajectory start times follow interpolated trajectories. Concentrations are calculated at model determined time-steps as

$$C = Q/(2\pi\sigma_h^2 \Delta Z) \exp(-r^2/2\sigma_h^2) \quad (1)$$

where,

Q = emission mass per puff,
σ_h = horizontal standard deviation of puff,
ΔZ = mixed layer depth,
r = distance from the puff center, and

where σ_h is given as

$$\sigma_h(m) = .5\ t(sec) + \sigma_{ho} \quad (2)$$

and σ_{ho} is an initial plume size based on the size of the aggregated source region. The 0.5 m/sec growth rate is based on experiments summarized by Heffter (1965).

Puff concentrations are calculated at discrete time intervals along each 3-hour trajectory segment. The length of the time increments are determined from puff size and advection speed such that there will be sufficient overlap between successive puffs to accurately simulate a continuous plume. The number of time steps per 3-hour trajectory segment, $N = R/(.5\ \sigma_h)$ where R is the length of the trajectory segment, was arrived at through numerical convergence experiments.

The mixed layer depth is generally taken as the transport layer depth (TLD) as calculated by ARL-ATAD for each trajectory segment. However, if TLD decreases with time, the mixing depth is maintained at its previous value, thus preventing concentrations from increasing. Concurrently with the concentration calculation, the Statistical Acid Deposition Model is called to simulate chemical transformation and plume depletion by means of wet and dry deposition.

The chemical transformation of NO_x and SO_x emissions to nitrates and sulfates and deposition of these chemical species is simulated on a long-term average statistical basis by STADMOD. The basic premise of STADMOD is that long-term deposition patterns are insensitive to short-term meteorological fluctuations. Therefore, the physical processes of chemical transformation, scavenging, and deposition can be statistically parameterized.

The rates at which NO_x and SO_2 emissions are transformed to nitrate (NO_3^-) and sulfate (SO_4^{2-}) depend on meteorological conditions and other chemical constituents present in the plume and atmosphere. The natural removal of primary and secondary species through dry deposition at the surface varies with the contaminant species, meteorology, and vegetation parameters. Natural removal is also achieved during precipitation through nucleation and incloud scavenging. STADMOD was developed to address these removal processes in order to calculate ambient pollutant concentration and deposition fields on a long-term average basis. Since the modeling of sulfur and nitrogen are substantially identical in this methodology, differing only in input parameters such as transformation rates and wet and dry removal rates, the model description is presented in terms of sulfur compounds only.

On a long-term average basis the time of transport from source to receptor is composed of both wet and dry periods. A wet period is when the polluted parcel is in an area of precipitation and a dry period is when the polluted parcel is not in an area of precipitation. A basic assumption of this model is that the physical and chemical processes involved in the transformation of SO_2 to SO_4^{2-} and the deposition of both SO_2 and SO_4^{2-} must be separately parameterized during wet and dry periods. Therefore, the model computes the concentrations of four atmospheric species, wet and dry SO_2 and wet and dry SO_4^{2-}. Figure 2 shows schematically the evolution of the modeled species, where G represents SO_2, S represents SO_4^{2-}, and the subscripts d and w indicate dry and wet respectively. The rate of conversion of dry SO_2 or SO_4^{2-} to wet SO_2 or SO_4^{2-} is given by the inverse of the average duration of dry periods in a Lagrangian sense ($1/\tau_d$), while the rate of conversion of wet SO_2 or SO_4^{2-} to dry SO_2 or SO_4^{2-} is given by the inverse of the average duration of wet periods in a Lagrangian sense ($1/\tau_w$). The dry and wet oxidation rates of SO_2 are k_d and k_w whereas λ and λ represent SO_2 and SO_4^{2-} scavenging coefficients respectively. The differential equations which describe this system (Venkatram et. al., 1982) are

$$\frac{dG_d}{dt} = -\lambda_d G_d - k_d G_d - \frac{1}{\tau_d} G_d + \frac{1}{\tau_w} G_w \tag{3a}$$

$$\frac{dG_w}{dt} = -\lambda_w G_w - k_w G_w - \frac{1}{\tau_w} G_w + \frac{1}{\tau_d} G_d \tag{3b}$$

$$\frac{dS_d}{dt} = -\bar{\lambda}_d S_d + k_d G_d - \frac{1}{\tau_d} S_d + \frac{1}{\tau_w} S_w \tag{3c}$$

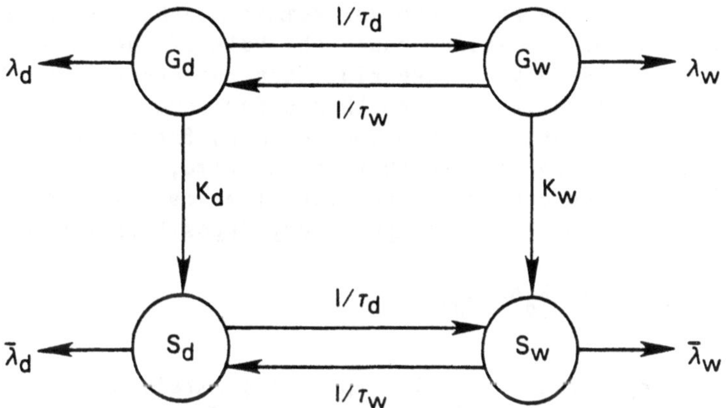

Source: Venkatram et. al. 1982.

Figure 2 Schematic evolution of modeled species. G and S denote gaseous and solid species whereas the subscripts d and w denote dry and wet respectively. The various transition (i.e., interspecies) and removal pathways and associated rate sonstants are described in the text.

$$\frac{dS_w}{dt} = -\bar{\lambda}_w S_w + k_w G_w - \frac{1}{\tau_w} S_w + \frac{1}{\tau_d} S_d \tag{3d}$$

subject to the initial conditions

$$G_d(0) = f_d Q_{SO_2}; \quad G_w(0) = f_w Q_{SO_2}$$

where f_d and f_w are the Eulerian dry and wet fractions.

There was some concern that the termination of trajectories at five days would act selectively to reduce the impact of sources very distant from the receptor lakes. For this reason the mass fractions of each airborne chemical species were computed as a function of travel time for a typical summer month. As can be seen in Figure 3 for the sulfur case, the bulk of the material is deposited within five days, suggesting that the error induced by trajectory termination is not more than a few percent.

Following computation of the species mass fractions G_d, G_w, S_d, and S_w for a particular puff travel time, concentrations at the candidate receptor site are computed using Equation (1). The model then computes dry deposition fluxes as

$$F_d = V_d \cdot C_d = \lambda_d \Delta Z \; C_d, \tag{4a}$$

where the dry deposition velocity, V_d is separately determined for sulfur and nitrogen species in gaseous and particulate forms to the watershed and lake surfaces. Wet deposition fluxes are modeled by first noting that the wet removal rate can be expressed as $\lambda_w = \alpha(R_s/R_o)$ as suggested by Maul (1978), where α is the scavenging rate for a reference rainfall rate R_o and R_s is the mean rainfall rate at the trajectory origin (i.e., at the source). This λ_w is then assumed to characterize wet removal along the entire trajectory; however, the wet deposition flux is then computed as

$$F_w = C \; \lambda_w \Delta Z \; (R_r/R_s) \tag{4b}$$

where R_r is the mean rainfall rate at or near the receptor of interest. While this factor (R_r/R_s) creates some conceptual problems, in that, like the local V_d values, it does not properly feedback into the mass budget equations, it is intended to correct local wet fluxes with the local rainfall values that may differ significantly from the value governing plume depletion over most of the trajectory. In annual average deposition modeling using STADMOD (Lague et. al., 1983), such a local rainfall correction factor resulted in substantially better agreement with annual wet deposition measurements for sulfur.

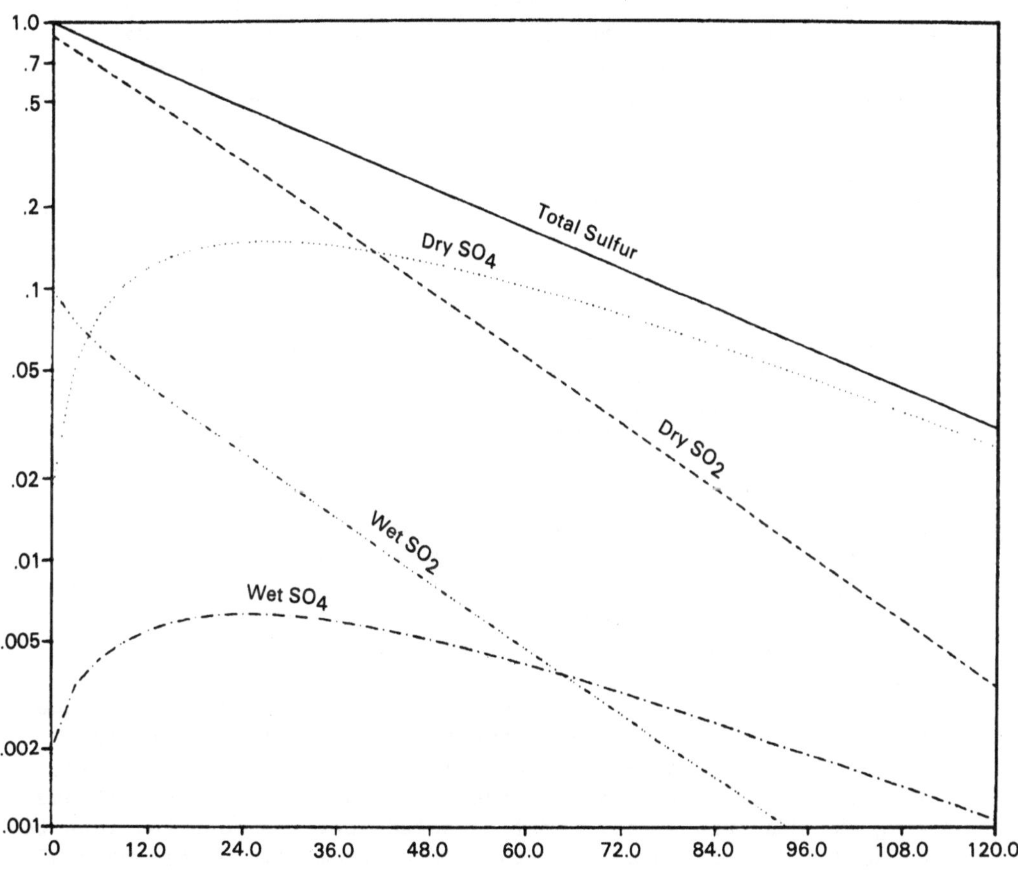

Figure 3 Sulfur Plume mass remaining in air versus travel time (hr).

MODEL APPLICATION AND PARAMETER SELECTION

As the ARL-ATAD computation of the air parcel trajectories is the most computer intensive portion of this modeling effort, it was decided to aggregate the several hundred source aggregates of the basic Canadian Electric Association (CEA) inventory (Lague et. al., 1983) into a few major aggregates. The aggregate regions were chosen to conform to geographical and political boundaries as closely as possible given the limitation of the simple latitude and longitude boundaries shown in Figure 4. The source aggregates were represented in the model as sources located at the source strength (SO_2) weighted mean latitude and longitude with an initial lateral dispersion defined by the second moment of the SO_2 source distribution. This initial puff spread is defined as:

$$\sigma_{ho} = (\Sigma Q_i R_i^2 / \Sigma Q_i)^{1/2} \tag{5}$$

where:

- Σ indicates the summation over each source, i, within the region
- Q_i is the SO_2 source strength of source i
- R_i is the distance between source i and the center of emissions mass for the region.

Single unit sources, not representing an aggregate of many sources, such as individual N.Y. power plants and the Sudbury smelter, are given an initial lateral dispersion coefficient of zero. Sudbury is included as an individual source because its 1980 SO_2 emissions are similar in magnitude to many entire source regions.

Sources west of 91° were not considered in this study, since their inclusion would require computation of significantly more trajectories and a greater area of meteorological data coverage, without significantly changing results. Ancillary investigations using the CEA statistical model revealed that the contributions to the acidic deposition at the study lakes from all sources west of 91° longitude are less than 5% of the total; thus, justifying their exclusion from this study.

The ARL-ATAD trajectory generator was then used to produce trajectories from each of the 22 source aggregate locations for each day in 1980. The resulting data file of trajectories, each lasting a maximum of five days, was segregated into twelve monthly files and served as a principal input to the DISCDEP model. The Statistical Acid Deposition module, contained in DISCDEP, requires the specification of several deposition and transformation related parameters and, as the Lake Acidification Model requires monthly average atmospheric depositions as input, these parameters are

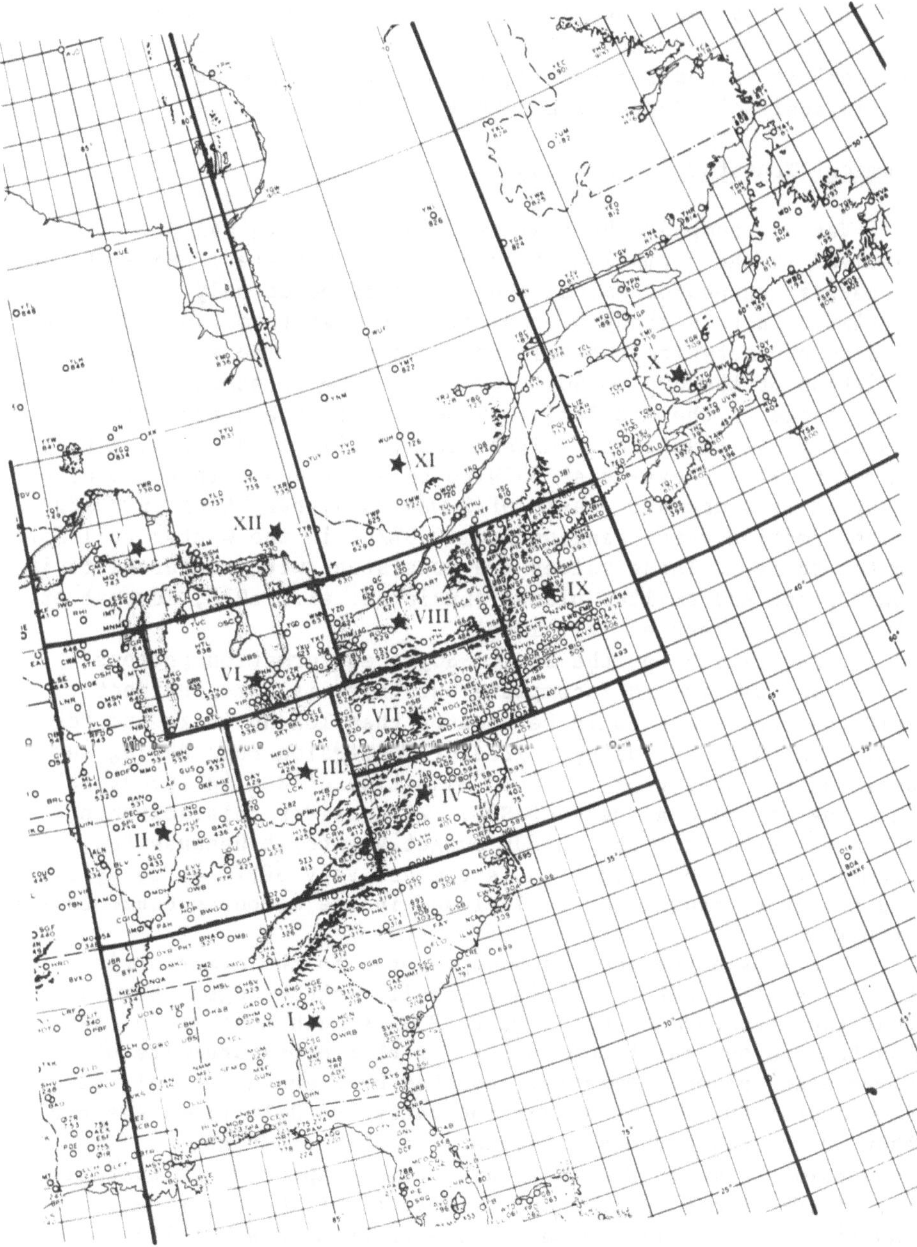

Figure 4 Source aggregations used in this modeling study. In addition to the twelve major source regions shown here, ten additional sources were designated to represent individual power plants or clusters of power plants within New York state.

allowed to vary on a monthly basis. For most parameters, however, values are available, at best, on a seasonal-average basis, so that monthly values must be inferred from simple linear interpolations between seasonal values. In the following subsections values for the requisite input parameters for the Statistical Acid Deposition module are considered.

Chemical Transformation Rates

Various complex chemical reactions and physical processes are involved in the formation of sulfates and nitrates. Many of these processes are photochemical and therefore occur only in the daytime. In this study diurnal average transformation rates are used to parameterize the relevant processes. The seasonal variation of the SO_2 to SO_4 conversion rate is expected to be quite large in the mid-latitudes because of the dependence on photochemical reactions. Altshuller (1979) found, for example, summer to winter ratios at 35°N to be about 4-5. In the present study, a less dramatic ratio of 3 was used, resulting in sulfur transformation rates ranging from 0.5%/hr in the winter to 1.5%/hr in the summer. These values are consistent with measurements in power plant and smelter plumes (Lusis et. al., 1978; Forrest et. al., 1979; Forrest et. al. 1980).

Understanding of the SO_2 to SO_4 transformation process during wet periods is incomplete but there is some indirect evidence that rates could be greater than 10%/hr (Hegg and Hobbs, 1981). Rather than using a greater conversion rate during wet periods, the effect of in-cloud oxidation of SO_2 is incorporated into the wet removal rate of SO_2. We consider this approach more realistic from the sulfur budget viewpoint because SO_4 formed by in-cloud oxidation is removed very efficiently by rain and therefore should not be included in the airborne concentration of SO_4.

Far less attention has been devoted to the atmospheric formation of nitrates. It is generally acknowledged, however, that the transformation of NO_x to NO_3 is much more rapid than the transformation of SO_2 to SO_4. Airborne measurements of power plant plume chemistry, reported by Richards et. al. (1981), showed that NO_x to NO_3 conversion rates are 3-10 times greater than SO_2 to SO_4 conversion rates. In this study the conversion of NO to NO_2 is assumed to be rapid enough that all NO_x emissions can be replaced by equivalent NO_2 emissions. In addition, all oxidized products are treated as NO_3. The transformation rates of NO_2 to NO_3 chosen for this study are 7.5%/hr in the summer and 2.5%/hr in the winter (i.e., 5 times the SO_2 conversion rate). These are within the range of values found by Forrest et. al. (1981) using filter pack measurements within a coal-fired, power plant plume.

Dry Deposition to Land

The dry deposition rate of a contaminant onto the ground or vegetative surface, depends on its ground level ambient concentration and the ability of the surface to retain that particular species. A deposition velocity is often used as an empirical measure of this capability: that is,

$$V_d = F_d/C$$

where both the deposition flux F_d and the concentration C are measured at the same height. Major factors controlling deposition velocities are atmospheric turbulence and surface resistance, and both of these depend on land use characteristics. Therefore, two sets of dry deposition velocities were specified for this study, one for the deposition to the receptor lakes and one for the deposition to the watersheds.

The values chosen to describe deposition to the watersheds of the receptor lakes are the same values used for dry plume depletion along the trajectory paths, since both the watersheds in particular and eastern North America in general, are primarily covered with forests of a mixed deciduous-coniferous nature. Atmospheric turbulence in forests is, on average, relatively high due to mechanical turbulence caused by the large roughness length of the trees. The surface resistance is largely a function of the stomatal activity of the leaves. In deciduous forests both factors will vary with season as the presence of leaves not only affects surface resistance but also increases aerodynamic roughness. The seasonality of dry deposition velocities is also influenced by the seasonal variation in radiative heating, and thus, of atmospheric stability.

Values for the dry deposition velocity of SO_2, reported experimentally for a variety of surfaces by several authors, typically range from 0.4 cm/sec to 0.8 cm/sec. Sheih et. al. (1979) calculated deposition velocities of SO_2 to forests of from 0.6 cm/sec to 1.0 cm/sec for unstable conditions and from 0.1 cm/sec to 0.4 cm/sec for stable and neutral conditions. McMahon and Denison (1979) reported a daytime deposition velocity to "countryside" of 0.8 cm/sec. At night, dry deposition velocities are generally less than 0.2 cm/sec due to stable atmospheric conditions and increased stomatal resistance. In light of these reports, 0.6 cm/sec was chosen to represent a diurnal average deposition velocity of SO_2 during the summer months, which represents the upper limit of the annual range.

In winter the watersheds will usually be snow covered. Measurements of deposition velocities to snow range from 0.1 cm/sec, during low wind speed, stable conditions (Garland, 1976), to 0.3-0.4 cm/sec as back-calculated from average ground-level concentrations and depositions (Chamberlain, 1980). Thus, a diurnal average SO_2 dry deposition velocity of 0.3 cm/sec was used for January and February.

Comparitively little data is available on the dry deposition velocity of NO_2. It is reasoned, however, that its seasonal variation should be similar to that of SO_2. Also, since NO_2 is less soluble than SO_2, its dry deposition velocity should be less by some factor. Beilke (1970) suggested that the dry deposition velocity of NO_2 should be one quarter that of SO_2. Judeikis and Wren (1978), experimenting with soil and cement surfaces, came up with a factor of about 0.5. Sehmel (1980) reported a factor of about 2/3, based on experiments involving alfalfa.

The controlling factor for the daytime deposition of gaseous contaminants to a heavily vegetated surface is most probably the canopy resistance. Canopy resistances to NO_2 of three times those for SO_2 have been reported by Hicks (1983) for daytime measurements. Since nighttime deposition velocities of any gas are much smaller than daytime values, the diurnal average dry deposition velocity of NO_2 is assumed to be about 1/3 that of SO_2.

Dry deposition velocities of particulate aerosols are particularly difficult to quantify. Neither experimental nor theoretical approaches have yielded sufficiently certain results. Some of the many factors involved in the process of dry particle deposition are surface characteristics, atmospheric stability, solubility, wind speed, and particle size distribution.

Particulate aerosols are removed from the air by mechanical impaction on surface vegetation and other surface features, and the processes by which they are brought into contact with the ground are also a function of particle size. The bulk of sulfate aerosol mass is thought to be in the 0.1 to 1.0 µm range. Theoretical estimates concerning particles in this size range suggest deposition velocities of 0.1 cm/sec or less (Slinn, 1977; Droppo, 1979).

Experimental measurements of particulate deposition velocities have yielded widely varying results. The experimental studies that have been made over vegetated surfaces, however, generally suggest deposition velocities between 0.1 and 1.0 cm/sec. For example, Hicks, et. al. (1982), using eddy correlation techniques, found deposition velocities to grass as high as 0.7 cm/sec in daytime with a long term average of about 0.2 cm/sec. In forested areas, deposition velocities are expected to be somewhat higher due to a greater roughness length. Hicks and Wesely (1980), using eddy flux measurements made over a loblolly pine plantation, calculated an average dry deposition velocity of sulfate of about 0.7 cm/sec. Sheih, et al. (1979) suggested that deposition velocities for sulfates over forests should be similar in magnitude to those for SO_2.

Given the current uncertainty with respect to the dry deposition velocities of small particles, a value of 0.3 cm/sec was chosen for the diurnal-average, summer sulfate deposition velocity. This value, which is half the respective value for SO_2, is intended as a compromise between those who claim that the deposition velocity of sulfates is 10 times smaller than that of SO_2 and those who suggest that the two are comparable.

The seasonal variation in the dry deposition velocity of sulfates is expected to be greater than for that of SO_2 because of the greater dependence on the turbulence inducing characteristics of deciduous trees. Also, during winter months the ground is usually snow covered and particulate deposition velocities become quite small. Resistances to sulfate deposition for smooth surfaces such as snow have been reported by Hicks (1983) to be of order 15 s/cm, which translates to a deposition velocity of only 0.06 cm/s. During January and February the lakes are assumed to be snow or ice covered all the time so that 0.06 cm/sec was used as the dry deposition velocity of sulfate to the lakes. The average winter deposition velocity to the watersheds should be somewhat greater because of the presence of coniferous trees and periods during which the ground is bare. Consequently, 0.08 cm/sec was chosen for the winter average dry deposition velocity of sulfate.

Dry deposition velocities of secondary nitrates are more difficult to establish. Nitrates in power plant plumes initially take the form of gaseous nitric acid (HNO_3). Due to the high vapor pressure of nitric acid, condensation is not a significant mechanism for the formation of particulate nitrate (Middleton and Kiang, 1979). However, dissolution of gaseous nitrate onto wetted aerosol surfaces and subsequent reactions with dissolved cations, such as ammonium (NH_4^+), are more important. Since the solubility of nitric acid increases with decreasing temperature, the ratio of particulate to gaseous nitrate is expected to be, on average, highest in the winter. Particulate nitrate is generally found in about the same size range as particulate sulfate and is therefore likely to be deposited at the same rate. Nitric acid, like SO_2 is very soluble, and therefore should be deposited at a similar rate.

Present data are insufficient to quantify the relative amounts of gaseous and particulate nitrates in typical long-range transport plumes. However, it has been reported (Stelson et.al. 1979) that the amount of gaseous HNO_3 is of the same order as the amount of particulate nitrate. Therefore, it was assumed that the proportion of particulate nitrate ranges from 25% in the summer to 75% in the winter. Assuming that gaseous nitric acid deposits at the same rate as SO_2 and that particulate nitrates deposit at the same rate as sulfate, the average winter value for the deposition velocity of dry nitrate was calculated to be 0.15 cm/sec and the average summer value to be 0.50 cm/sec.

Dry Deposition to Lakes

Since direct deposition to lakes is also of concern in this study, a separate attempt was made to estimate the dry deposition velocities over water to better represent this component.

In January and February both lakes and forests are usually snow covered in the receptor areas of interest in this study. Therefore, the dry deposition velocities to the lakes are assumed to be slightly less than to the watersheds due to lower mechanical turbulence.

During months when the lakes are unfrozen, deposition is largely influenced by air-water temperature differences. Therefore, the dry deposition velocities for the summer months are assumed to be smaller over the lakes than the watersheds because of the stratifying influence of the cooler water surface. Similarly, the lake deposition velocities in the fall are assumed to be greater since the water surface is generally warmer than the air, resulting in greater instability of the surface layer.

For smooth surfaces, such as water, the deposition velocity of sulfate is generally reported to be about 0.1 cm/sec (Garland 1978). Using this as a median value for the unfrozen months, the deposition velocity was varied upward in the fall to a maximum monthly average of 0.13 cm/sec during October and downward to a value of 0.07 cm/sec in the summer. The same reasoning is applied to the deposition velocity of SO_2 and, using a median value of 0.5 cm/sec (Garland 1976; and Owens and Powell 1974), yields a maximum value of 0.65 cm/sec in the fall and a minimum of 0.35 cm/sec in the summer.

The dry deposition velocities of NO_2 to lakes are again assumed to be 1/3 of the corresponding value for SO_2. The dry deposition velocities of nitrate are calculated the same way as the watershed values, using the appropriate SO_2 and sulfate values to lakes.

Wet Removal Rates

The removal of gaseous contaminants by precipitation is a complex process that is governed by such factors as initial precipitation pH, background ambient contaminant concentration, precipitation type, and precipitation rate. The dominant process by which SO_2 is removed during wet periods is probably the incloud oxidation of SO_2 to SO_4 and subsequent rainout. This process is included in the wet removal rate of SO_2 rather than in the SO_2 to SO_4 transformation rate, as mentioned previously. The wet removal rates used in this study are given by:

$$\lambda_w = \alpha(R_s/R_o)$$

as suggested by Maul (1978), where R_s is the monthly average rainfall rate characteristic of the particular source region and R_0 is the reference rainfall rate of one mm/hr and α = $3.0 \times 10^{-5}\ s^{-1}$ for SO_2. This value for α assumes rain with a pH in the range of 4-5. For SO_4, $\alpha = 1.0 \times 10^{-4}\ s^{-1}$ as suggested by Garland (1978).

The chemistry and scavenging of NO_x is less well understood. Wet deposition of NO_2 and NO_3 are treated in a manner analagous to wet sulfur deposition. The values of α, $7.5 \times 10^{-6}\ s^{-1}$ and $5.0 \times 10^{-5}\ s^{-1}$ for NO_2 and NO_3 respectively, take the relative solubility of each species into account.

In general, snow is less efficient at removing airborne pollutants. There is evidence, however, that nitrate is more efficiently removed by snow than sulfate (Bowersox and DePena 1980). The removal rates by frozen precipitation used in this study are $0.1 \times 10^{-5}\ s^{-1}$ for SO_2 (Summers, 1977), $3.0 \times 10^{-5}\ s^{-1}$ for SO_4 and NO_3 and zero for NO_2 (Scire and Lurmann, 1983). Monthly averaged values were computed from estimated ratios of frozen to liquid precipitation that ranged from 0.0 in summer to 0.6-0.8 in winter.

Lagrangian Wet and Dry Period Durations

The average durations of wet and dry periods along the path of an air parcel trajectory (τ_w and τ_d) are Lagrangian parameters which cannot be determined from routine Eulerian meteorological observations. Estimates of these parameters are therefore rather tenuous. However, both Venkatram et al. (1982) and Smith (1981) found that model results were relatively insensitive to τ_w and τ_d. The values used in this study are τ_w = 7 hrs and τ_d = 46 hrs. These estimates were made by Slinn et. al. (1979) based on actual Lagrangian trajectories.

PARAMETER SENSITIVITY ANALYSIS

The discussion accompanying parameter selection indicates that most of the parameters used in the statistical model for pollutant transformation and removal actually represent a consensus of experimentally observed or inferred results. In the case of most of these parameters a variation, or shift, of a factor of two in either direction would still leave one in a viable range for that parameter. Choosing a factor-of-two (i.e., either halving or doubling) variation from the normal operational value for the initial evaluation of each parameter's sensitivity, the percentage changes in wet, dry, and total, sulfur plus nitrogen deposition at each of the four lake regions during July 1980 were computed. July was chosen because of the large magnitude of deposition predicted and observed in that month combined with the continued, relatively high sensitivity of the lake's chemistry and aquatic life in early summer.

The results indicated that, from the point of total acidic deposition, the wet removal rates of the primary and secondary species are the most sensitive parameters, followed by the dry deposition velocity of primary species and the chemical transformation rate. The dry deposition velocity of secondary species was found to be the least sensitive variable, probably because secondary species are predominantly removed by wet processes.

A more thorough sensitivity analysis of the kinetic parameters used in acid deposition models of this type was performed by Golumb et. al. (1983). They found that, in a model assuming continuous rain, the wet deposition of sulfate was far more sensitive to the oxidation rate than to the wet removal rate. This was because in their model wet removal proceeded at a much faster rate than oxidation and wet removal of the primary species was not considered; therefore, oxidation was the rate determining step. In the current model, however, wet deposition is a function of long-term average precipitation statistics. This greatly reduces the wet removal rate compared to a continuous rain scenario, and thus makes it more of a rate determining process.

COMPARISON OF DEPOSITION PREDICTIONS WITH OBSERVATIONS

Model predictions of wet deposition fluxes have been compared to wet deposition fluxes measured at Woods Lake by Altwicker and Johannes, (1983), for the first 8 months of 1980. The observations were presented in terms of concentrations of sulfate and nitrate in precipitation. These values were converted to fluxes by multiplying by the monthly rainfall amounts. Observed and predicted wet fluxes are presented in Table 1 along with the observed to predicted ratio, a quantity less susceptible to bias induced by the known, high correlation between rainfall and sulfur deposition.

While drawing strong conclusions about the model's behavior seems premature given the amount of data available for comparison, examination of these data does, however, suggest some general characteristics of the curent model. For example, the general trend of underprediction in the winter and overprediction in the summer suggests that the seasonal variation of the wet deposition parameters may be overestimated. The underprediction in January and February, in particular, may be an artifact of underestimating the scavenging rate during frozen precipitation events. The wet removal rates for these months are simply averages of liquid and frozen precipitation removal rates weighted by an estimate of the fraction of frozen precipitation. Frozen precipitation generally originates from liquid clouds, so that incloud conversion of primary to secondary should still occur during these events. In this model this process was included in the wet removal rate of the primary species rather than as an increased transformation rate during wet periods. This assumption or the values used for the fractions of liquid and frozen precipitation and the frozen precipitation removal rate may not adequately account for this process.

Another possible explanation for the differences between model predictions and observations is the assumed linear dependence of the wet removal rate on the rainfall rate. There is evidence (Scott, 1982) that the efficiency of wet removal diminishes with increasing rainfall intensity. This may be because intense rainstorms will scavenge the majority of available contaminants in the first few minutes, leaving the subsequent rain relatively clean. It should be noted, however, that the statistical transformation and deposition portion of this model was designed to simulate these processes on a long-term average basis, i.e., annual or seasonal. On such a basis rainfall rates exhibit far less variation, so that the suspected non-linear relation between wet removal rate and rainfall rate is less important. Therefore, when applying this model on a monthly-average basis, it may be desirable to modify it to take this possible nonlinearity into account.

Figure 5 shows log of the observed/predicted deposition rates plotted versus the log of the monthly rainfall. The significantly negative slopes, -0.42 for sulfur and -0.84 for nitrogen, suggest that the exponent for the rainfall dependence in the removal rate, λ_w, or in the wet flux, F_w, should be substantially smaller than unity and possibly species dependent (e.g., 0.15 for nitrogen and 0.6 for sulfur). A modification of the model formulation in this direction awaits a more comprehensive evaluation using wet deposition data from a number of sites.

DEPOSITION MODELING RESULTS

Using the twelve, one-month trajectory files produced by ARL-ATAD for 1980 and the emissions aggregates discussed previously, the DISCDEP model was run for five candidate lakes and watersheds of interest (i.e., Woods and Panther Lakes in the Adirondacks, Bickford Reservoir in Massachusetts, Caldwell Pond in New Hampshire and Mosquito Lake in New Brunswick, Canada). Total deposition fluxes of sulfur and nitrogen species were computed along with combined total acidic deposition, expressed as equivalents per day, to the lakes and their watersheds. Deposited fluxes were found to be greatest at Woods and Panther Lakes. The major contribution to the higher total deposition flux at Woods and Panther lakes was the wet deposition component, and these lakes received significantly more rainfall than Bickford Reservoir or Caldwell Pond in 1980. The tendency of rainfall to be greater

Table 1 Observed and predicted wet deposition fluxes ($\mu eq/m^2$), at Woods Lake, NY during 1980

SULFUR

MONTH	PRED	OBS	OB/PR	RAIN(MM)
JAN	1528.	2240.	1.47	62.2
FEB	1394.	1422.	1.02	28.4
MAR	6155.	2688.	.44	103.4
APR	7399.	4702.	.64	90.4
MAY	2459.	2652.	1.08	30.5
JUN	15613.	7896.	.51	119.6
JUL	25463.	16287.	.64	198.6
AUG	7974.	7548.	.95	71.9

RATIO OF AVERAGED OBSERVED TO PREDICTED = 0.67

NITROGEN

MONTH	PRED	OBS	OB/PR	RAIN(MM)
JAN	724.	1929.	2.66	62.2
FEB	518.	2418.	4.67	28.4
MAR	2052.	2688.	1.31	103.4
APR	2555.	3255.	1.27	90.4
MAY	826.	1311.	1.59	30.5
JUN	5374.	2991.	.56	119.6
JUL	9330.	6157.	.66	198.6
AUG	2964.	3019.	1.02	71.9

RATIO OF AVERAGED OBSERVED TO PREDICTED = 0.98

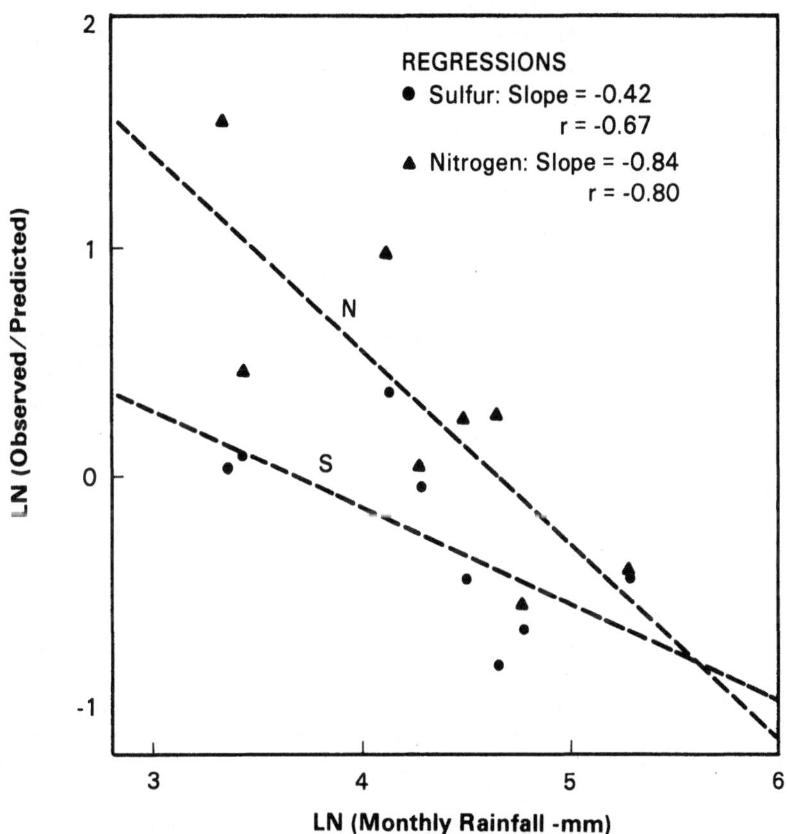

Figure 5 Natural logarithm of observed/predicted wet deposition fluxes at Woods Lake versus log of local rainfall (mm/month).

in this region is, however, not peculiar to 1980, since Woods and Panther Lakes are located on the windward side of the Adirondack Mountains and to the lee of Lake Ontario: an ideal location for the occurrence of orographically induced precipitation.

The annual acidic deposition contributed by all New York utility (NYPP) emissions was found to be 5% of the total annual deposition at Woods and Panther Lakes and less at the more distant lakes. Thus, emissions changes of ±20% by these sources suggest deposition changes of only ±1% on the closest, acid-sensitive region of interest, the Adirondacks. These assessments were also found to be in good agreement with annual predictions made (Lague et. al., 1983) using the same statistical transformation and deposition module coupled to annual climatological probability functions for rainfall and wind direction.

It should be noted that this model is designed for regional scale assessments and is therefore inappropriate for simulations involving near field receptors which are within several hours of transport, or about 100 km, of a source. Since the model treats all emissions as ground level sources, dry deposition predictions from elevated point sources will be incorrect for such near field receptors. Woods and Panther lakes, however, are sufficiently remote from all modeled sources that long-term average predictions of the relative contributions of the New York utilities as a group to both wet and dry deposition should be realistic.

CONCLUSIONS AND STUDY FINDINGS

In this project a Lagrangian atmospheric transport scheme has been linked to a long-term average statistical model for acidic deposition. The deposition predictions are then input to a lake model which, while not as complicated as some could envision, is presently data limited. The predictions of this set of linked models is in general agreement with simpler modeling and it is doubtful that the next step in modeling sophistication, given current understanding of the mechanisms, will yield qualitatively different results. In addition, this "next step" might involve coupling a large Eulerian grid model to a water model as comprehensive as TetraTech's (1983), a task that would involve one or two orders of magnitude greater effort and that would be even more hampered by current data limitations.

The results of this model development and application effort can be summarized as follows:

- While wet and dry removal of primary (i.e., emitted) species tends to peak deposition close to the source, conversion to secondary species creates a plume component which peaks, in terms of airborne mass, after about one day of plume travel. This secondary plume is longer lived than the primary plume and hence is responsible for distant sources (i.e., several days of transport away) contributing to deposition fluxes.

- After five days of transport, 97% of a plume's total mass is deposited out. This result is, of course, sensitive to the values of conversion and deposition parameters.
- New York utility emissions account for about 5% of annual acidic deposition to lakes in the Adirondacks. This result is obtained both from annual average modeling and from the trajectory based model presented in this paper.

While not described in this paper, findings based upon the application of the Lake Acidification Model include:

- The Lake Acidification Model (LAM) was able to reproduce the observed trends of transient alkalinity and pH reductions in the epilimnion of two Adirondack lakes (Woods Lake and Panther Lake). These transients are associated with springtime snowpack release of accumulated acidic depositions to the lakes.
- The Lake Acidification Model was also used to analyze the seasonal dynamics of alkalinity consumption and generation during the summer stratification period of the Bickford Reservoir. Through model calibration, LAM was able to quantify the seasonal kinetics in the epilimnion and hypolimnion by matching the model calculations with the observed alkalinity and pH levels.
- The calibrated LAM models for Woods Lake, Panther Lake and Bickford Reservoir were used to predict the changes in alkalinity and pH in these lakes that would result from ±20% changes in the emissions of New York utilities. The 1% changes in predicted deposition for the Adirondack lakes do not generate any appreciable changes in the alkalinity and pH levels in Woods Lake or Panther Lake over an annual cycle. The concurrent impacts on the other lakes are even less significant.

ACKNOWLEDGMENTS

The authors are indebted to Dr. Peter Coffey of the Empire State Electric Energy Research Corporation for his valuable technical input. This project was funded by ESEERCO.

REFERENCES

Altshuller, A.P., 1979. Model predictions of the rates of homogeneous oxidation of sulfur dioxide to sulfate in the troposhere. Atm. Env. 13:1653-1661.

Altwicker, E.R. and A.H. Johannes, 1983. Wet and dry deposition into Adirondak watersheds. In: The Integrated Lake-Watershed Acidification Study: Proceedings of the ILWAS Annual Review Conference. EPRI EA-2827.

Beilke, S., 1970. Laboratory investigations of washout of trace gases, Proc. Symp. on Precip. Scavenging, 1970. USAEC Symp. Services No. 22, pp. 261-269.

Bowersox, W.C., R.G. Depena, 1980. Analysis of precipitation chemistry at a central Pennsylvania site. J. Geophys. Research 85:5614-5620.

Chamberlain, A.C., 1980. Dry deposition of sulfur dioxide, pp. 185-197 of Atmospheric Sulfur Deposition (D.S. Shriner, C.R. Richmond, and S.E. Lindberg, eds.), Ann Arbor Press, 568 pp.

Droppo, J.G., 1979. Experimental techniques for dry deposition measurements. In: Atmospheric Sulfur Deposition (D.S. Shriner, C.R. Richmond, and S.E. Lindberg, eds.), Ann Arbor Science Inc., Ann Arbor, pp. 209-222.

Forrest, J., S.E. Schwartz, and L. Newman, 1979. Conversion of sulfur dioxide to sulfate during the Da Vinci flights. Atm. Env. 13:157-167.

Forrest, J., R. Garber, and L. Newman, 1981. Conversion rates in power plant plumes based on filter pack data--Part I: The Coal-Fired Cumberland Plume, Atm. Env. 15: 2273-2282.

Garland, J.A., 1976. Dry deposition to a snow surface: Discussion. Atm. Env. 10:1033.

Garland, J.A. 1978. Dry and wet removal of sulfur from the atmosphere, Atm. Env., 12, 349-362.

Golomb, D.S. Batterman, J. Gruhl, and W. Labys 1983. Sensitivity analysis of the kinetics of acid rain models. Atm. Env. 17:645-653.

Heffter, J.L., 1965. The variation of horizontal diffusion parameters with time for travel periods of one hour or longer. J. Appl. Meteorol., 4(1): 153-156.

Heffter, J.L. 1980. Air Resources Laboratories atmospheric transport and dispersion model (ARL-ATAD). NOAA. Tech. Memo ERL-ARL-81, Air Resources Laboratory, Silver Springs, MD 20910.

Hegg., D.A. and P.V. Hobbs, 1981. Cloud water chemistry and the production of sulfates in clouds. Atm. Env. 15:1597-1604.

Hicks, B.B., 1983. Dry deposition. In: Critical Assessment Document on Acid Deposition. Air Resources Atmospheric Turbulence and Diffusion Laboratory, National Oceanic and Atmospheric Administration, Oak Ridge, TN 37830.

Hicks, B.B., and M.L. Wesely, 1980: Turbulent transfer process to a surface and interaction with vegetation, pp. 199-207 of Atmospheric Sulfur Deposition (D.S. Shriner, C.R. Richmond and S.E. Lindberg, eds.) Ann Arbor Press, 568 pp.

Hicks, B.B., M.L. Welely, R.L. Coulter, R.L. Hart, J.L. Durham, R.E. Speer, and D.H. Stedman, 1982: An experimental study of sulfur deposition to grassland, Proceedings, Fourth International Conferrence on Precipitation Scavenging, Dry Deposition, and Resuspension, Santa Monica, California, 29 November - 3 December, in press.

Judeikis, H.S. and A.G. Wren, 1978. Laboratory measurements of NO and NO_2 depositions onto soil and cement surfaces. Atm. Env. 12:2315-2319.

Lague, J.S., A. Venkatram, J. Young, K. Ganesan, R. Dowd, P.A. Hansen, B. Doyle, J. Pleim, 1983. Acid deposition effects of projected North American emission trends. Report for the Canadian Electrical Association, by Environmental Research and Technology, Inc., 2625 Townsgate Road, Westlake Village, CA 91361.

Lusis, M.A., K.G. Anlauf, L.A. Barrie, and H.A. Wiebe, 1978. Plume chemistry studies at a northern Alberta power plant. Atm. Env. 12:2429-2437.

Maul, P.R., 1978. Preliminary estimates of washout coefficient for sulfur dioxide using data from an East Midlands ground level monitoring network. Atm. Env. 12, 2515-2517.

McMahon, T.A. and P.J. Denison, 1979. Empirical atmospheric deposition parameters--a survey. Atm. Env. 13:571-585.

Middleton, P. and C.S. Kiang, 1979. Relative importance of nitrate and sulfate aerosol production mechanisms in urban atmospheres. In: Nitrogeneous Air Pollutants (D. Grosjean, ed.), Ann Arbor Science, Ann Arbor, pp. 269-288.

National Climatic Center, 1975: USAF. Global Weather Center surface data. NOAA. EDS. NCC., Asheville, NC 28801.

National Research Council, 1983. Acid Deposition-Atmospheric Processes in Eastern North America. National Academy Press, Washington, D.C. 375 pp.

Owers, M.J., and A.W. Powell, 1974: Deposition velocity of sulphur dioxide on land and water surfaces using a ^{35}S method, Atm. Env. 8:63-67.

Richards, L.W., J.A. Anderson, D.L. Blumenthal, A.A. Brandt, J.A. McDonald, N. Waters, E.S. Macias, and P.S. Bhardwaja, 1981. The chemistry, aerosol physics and optical properties of a western coal-fired power plant plume. Atm. Env. 15:2111-2134.

Scire, J.S. and F.W. Lurmann, 1983. Development of MESOPUFF-II dispersion model. Presented at the AMS sixth Symposium on Turbulence and Diffusion. March 22-25, 1983, Boston, Mass.

Scott, B.C., 1982. Theoretical estimates of the scavenging coefficient for soluble aerosol particles as a function of precipitation type, rate, and altitude. Atm. Env. 16:1753-1762.

Sehmel, G.A., 1980. Particle and gas dry deposition: A review. Atm. Env. 14:983-1011.

Sheih, C.M., M.L. Wesely, and B.B. Hicks, 1979: Estimated dry deposition velocities of sulfur over the Eastern United States and surrounding regions, Atm. Env., 13, 1361-1368.

Slinn, W.G. 1977. Some approximations for the wet and dry removal of particles and gases from the atmosphere. Water Air and Soil Pollution, 7, 513-543.

Slinn, W.G.N., P.C. Katen, M.A. Wolf, W.D. Loveland, L.F. Radke, E.L. Miller, L.J. Ghannam, B.W. Reynolds, and D. Vickers, 1979. Wet and dry and resuspension of AFCT/TFCT Fuel processing radionuclides. U.S. Department of Energy Final Report, SR-0980-10, Air Resources Center, Oregon State University, Corvallis, Oregon.

Smith, F.B., 1981. The significance of wet and dry synoptic regions on long-range transport of pollution and its deposition. Atm. Env., 15, 863-873.

Stelson, A.W., S.K. Friedlander, and J.H. Seinfeld, 1979. A note on the equilibrium relationship between ammonia and nitric acid and particle ammonium nitrate. Atm. Env. 13:369-372.

Summers, P.W., 1977. Note on SO_2 Scavenging (R.G. Semonin and R.W. Beadle, eds.), Tech. Information Centre, ERDA, pp. 88-94.

Tetra Tech, Inc., 1983. The Integrated Lake-Watershed Acidification Study: Proceedings of the ILWAS Annual Review Conference. EPRI EA-2827, Project 1109-5.

USAF-ETAC., 1972: DATSAV-SYNFILE. Upper air meteorological data, Washington Navy Yard Annex, Washington, D.C.

Venkatram, A., B.E. Ley and S.Y. Wong, 1982. A statistical model to estimate long-term concentrations of pollutants associated with long-range transport. Atm. Env., 16, 249-257.

Yamartino, R.J., J.E. Pleim, W. Lung, 1983. Development of an Acid Rain Impact Assessment Model. Final report to Empire State Electric Energy Research Corporation, 1271 Avenue of the Americas, New York, NY 10020.

DISCUSSION

A. BERGER What is the range of values that you have used for b, the exponent of R in the wet flux formula?

R. YAMARTINO We have assumed that the wet deposition flux is linear in the rainfall measured at the receptor (i.e., b = 1); however, the results presented in Figure 5 indicate that b-values less than unity would provide better estimates of the monthly wet fluxes. We are in the process of further refining this model.

J. SHANNON Did you estimate the effect of non-utility New York sources on deposition in the lakes?

R. YAMARTINO Yes, they were included in source aggregate VIII as indicated in Figure 4 of the paper.

G. DEN HARTOG If New York power utilities are responsible for less than 5% of the acid deposition to the Adirondack lakes, what is the major source of this deposition?

R. YAMARTINO Rather than a single major source, the modeling suggests that there are many small contributors. Numerical values for these contributions have not been released pending further model modification.

ESTIMATION OF NORTH AMERICAN ANTHROPOGENIC SULFUR DEPOSITION AS A FUNCTION OF SOURCE/RECEPTOR SEPARATION

Jack D. Shannon

Environmental Research Division
Argonne National Laboratory
Argonne, IL 60439

INTRODUCTION

The relative contributions to acidic or acidifying deposition in North America from local sources and from long-range sources are contentious matters. Indeed, there is not even agreement on the terms of reference. To some, the term local implies that the source and the receptor are in the same political or administrative entity, such as a state or Environmental Protection Agency Region. To others, local and mesoscale transport are out to 500 km, with transport greater than 500 km termed continental or long-range. Since states and provinces encompass a wide range of shapes and areas, and since this research effort is aimed toward investigation of rather fundamental aspects of transport and deposition rather than who is doing what to whom, we here combine all U.S. and Canadian emissions of anthropogenic sulfur, and calculate separately the deposition contributions from sources located within 200 km, from 200 to 500 km, from 500 to 1000 km, and more than 1000 km from each receptor in turn. For convenience, we will call the source/receptor separation categories local, regional, long-range, and continental, respectively.

The separation classification involves a moving radius or annulus of influence for each point in a grid of receptors with all sources outside the circle or annulus ignored. Thus, for each receptor location the emission subsets are unique.

THE MODEL

The model of transport and deposition used for these simulations is the Advanced Statistical Trajectory Regional Air

Pollution (ASTRAP) model.[1] The ASTRAP model simulates long-term (monthly to seasonal) concentrations of oxides of sulfur and of nitrogen and their respective products. The model combines calculation of trajectory distribution statistics, representing a series of bivariant normal puffs fitted to endpoint ensembles as a function of time since release for each of a grid of points across the U.S.A. and Canada, with a one-dimensional vertical integration for each of six potential initial effective emission layers extending to 800 m depth. The total depth through which pollutants eventually can be diffused is as much as 1800 m, depending upon the atmospheric stability profile. Wet removal is a function of the half power of the 12-hour precipitation amount; horizontal distributions of simulated trajectory tracers deposited by wet processes are summarized statistically in a manner similar to that described above. In the vertical integration, dry deposition velocities, vertical eddy diffusivity profiles, and linear chemical transformation rates are given diurnal and seasonal variations. All calculations described above are for unit emissions.

The ASTRAP model is periodically adjusted to utilize input data with different temporal and spatial resolution and to parameterize better the results of recent field data. The model is not adjusted merely to reproduce limited observations more closely, as there are too many degrees of freedom for confidence in model tuning. Most parameterizations have diurnal and seasonal variations; the average parameterization rates for summer and winter and meteorological input are summarized in Table 1.

Table 1 : ASTRAP model profile.

Input data		
Emissions	seasonal averages usually gridded on 1/3 NMC scale (100-125 km) six layers to 800 m primary sulfate function of fuel type	
Winds	Hemispheric analyses of 1000 mb and 850 mb wind fields every six hours on NMC scale	
Precipitation	Continental analyses of 24-hr totals every 12 hours on 1/3 NMC scale	
Parameterization averages	Summer	Winter
SO_2 deposition velocity	0.45 cm s^{-1}	0.25 cm s^{-1}
SO_4 deposition velocity	0.22 cm s^{-1}	0.12 cm s^{-1}
SO_2 transformation rate	2 % hr^{-1}	0.5 % hr^{-1}
Transport wind depth	1500 m	1000 m
Afternoon mixing depth	1800 m	1000 m
Bulk wet removal	$\min \begin{cases} P^{1/2} \\ 2/3 \end{cases}$	$\min \begin{cases} P^{1/2} & \text{(south)} \\ 0.5\, P^{1/2} & \text{(north)} \\ 2/3 \end{cases}$
	P precipitation in cm/12 hr	

In order to calculate concentrations or deposition, the emissions of each cell of an emission inventory gridded horizontally and vertically are used to scale the horizontal distribution statistics for the appropriate trajectory starting point (with the necessary horizontal bias between trajectory starting point and virtual emission source taken into consideration) multiplied by the one-dimensional surface concentration or dry deposition increment for the appropriate initial effective stack height. The wet deposition horizontal distribution times the emission rate is multiplied by the vertically integrated pollutant mass rather than by the surface concentration. The result of all of this multiplication is an overlapping set of weighted puffs that can be evaluated at each receptor point. The trajectory and vertical distribution statistics need be calculated only once per season, regardless of how many emission scenarios are examined.

RESULTS

The emission inventory used in these simulations, based upon the inventories developed at Brookhaven National Laboratory and in the research efforts related to the Memorandum of Intent on Transboundary Air Pollution between the United States and Canada, is summarized in Table 2.

Table 2 : Total Anthropogenic Sulfur Emissions in the United States and Canada Used in ASTRAP Simulations.

Winter	Spring	Summer	Autumn	Total
4265	3520	3534	3636	14955

(units kT S)

The integrated deposition totals calculated for the meteorological conditions experienced during 1980 and 1981 are given in Tables 3 and 4.

Table 3 : Cumulative annual sulfur deposition (kT) for 1980 as a function of source/receptor separation.

Deposition Region		Source Category LO	RE	LR	CO	Total
E. U.S.A.	dry	1084	981	489	115	2669
	wet	522	860	631	206	2219
E. Canada	dry	148	160	166	130	604
	wet	91	198	377	400	1066
W. U.S.A.	dry	344	267	148	51	810
	wet	55	83	95	75	308
W. Canada	dry	73	77	52	26	228
	wet	21	39	36	45	141

(LO = 0 to 200 km, RE = 200 to 500 km, LR = 500 to 1000 km, CO = more than 1000 km)

Table 4 : Cumulative annual sulfur deposition (kT) for 1981 as a function of source/receptor separation.

Deposition Region		Source Caterogy LO	RE	LR	CO	Total
E. U.S.A.	dry	1083	951	468	114	2616
	wet	530	784	584	214	2112
E. Canada	dry	156	168	185	165	674
	wet	97	188	379	448	1112
W. U.S.A.	dry	333	258	146	56	793
	wet	60	81	96	71	308
W. Canada	dry	75	82	55	30	242
	wet	21	40	38	42	141

The difference in transport and deposition meteorology for the two years is surprisingly small. Simulation of other meteorological periods, such as January and July of 1978, has indicated marked changes (factors of five to ten in extreme cases) in deposition for the same month of different years, but comparison of the annual accumulations for 1980 and 1981 do not reflect this sensitivity.

Here, east is defined as east of the Mississipi River and the Ontario/Manitoba border. As the scale of the source/receptor separation category increased, the annuli contain larger areas and generally more total emissions. Yet, the cumulative dry deposition from local sources is the equal of that from regional sources, and in the United States, almost twice that from long-range and continental-range sources. Dry deposition in Canada receives significant relative contributions from sources in all distance separation categories. This difference from results for the U.S.A. is due to the large portion of the area of Ontario and Quebec that is remote from sources. The dry deposition amounts simulated for those remote areas are low in absolute terms, however.

As one might expect, analysis of cumulative wet deposition contributions reveals a shift, relative to the dry deposition pattern, from local dominance toward the larger scales. While dry deposition is a continuous process as soon as a plume is mixed down to the surface (although the rate can approach zero when atmospheric and surface resistances are high), wet deposition is episodic. Plumes may thus travel for several days before encountering a precipitation system. The wet deposition per unit area from a particular source decreases as one moves farther downwind, but more sources are included in the regional and long-range categories because the areas of the annuli increase. The largest contributions to integrated wet deposition are regional sources for the eastern U.S.A. and western Canada, long-range sources for the western U.S.A., and continental-range transport for eastern Canada.

Some seasonal differences in the source/receptor distance relationships are examined in Tables 5 and 6. Since transport winds are generally stronger in winter and since removal mechanisms are more effective for both dry and wet processes during summer, the relative local source contribution is generally greater during summer. The local source contribution to wet deposition in Canada during the summer may be unusually low due to drought conditions near major source areas.

Table 5 : Cumulative sulfur deposition (kT) for winter 1980 as a function of source/receptor separation.

Deposition Region		Source Category LO	RE	LR	CO	Total
E. U.S.A.	dry	250	244	142	42	678
	wet	55	123	106	40	324
E. Canada	dry	32	40	47	43	162
	wet	6	21	56	75	158
W. U.S.A.	dry	79	68	45	20	212
	wet	11	16	15	11	53
W. Canada	dry	18	22	18	9	67
	wet	2	5	6	7	20

Table 6 : Cumulative sulfur deposition (kT) for summer 1980 as a function of source/receptor separation.

Deposition Region		Source Category LO	RE	LR	CO	Total
E. U.S.A.	dry	312	262	105	18	697
	wet	188	292	206	62	748
E. Canada	dry	37	38	37	24	136
	wet	32	54	85	112	283
W. U.S.A.	dry	102	74	34	7	217
	wet	13	18	23	19	73
W. Canada	dry	19	17	9	3	48
	wet	10	16	14	11	51

UNCERTAINTY

At this point the range of uncertainty about the model estimates should be addressed. Monitoring networks for dry deposition are not yet operating, although research efforts to develop the monitoring methods are underway. Monitoring networks for wet deposition do exist and are expanding, but measurements combine the contributions from all sources and thus do not directly address source apportionment.

When the ASTRAP model was run with 1980 data and results were compared with annual observed wet deposition of sulfate in the major U. S. and Canadian networks, results were as summarized in Table 7.

Table 7 : ASTRAP verification statistics for annual wet sulfate deposition 1980.

REGION	N	$\overline{OB}$	$\overline{MODEL}$	$\overline{RESIDUAL}$	ρ	VARIANCE EXPLAINED
U.S.A. and Canada	95	18.1	20.6	- 2.5	0.85	72 %
East	66	23.5	29.0	- 5.5	0.75	56 %
West	29	5.9	1.7	4.2	0.35	12 %

Here the observations have been reduced by 2.5 micromoles sulfate per liter of precipitation, as a crude estimate of the unmodeled natural contribution. It was found that changing the estimate of natural source contribution within reasonable limits affected the bias but had almost no effect on variance explanation. The model performs better when all observations are considered because the basic pattern of a deposition peak in the upper Ohio Valley/lower Ontario region and low values in the west contributes strongly to both observed and modeled variance. This "first harmonic" is mostly absent when the eastern or western subsets are modeled. The poor results for the western subset are believed due to the much larger relative contribution from natural sources than in the east. The observation adjustment used here is apparently too crude to account properly for the natural source contribution.

Some sources of modeling uncertainty can be examined as to their potential effect upon the source/receptor relationships :

1. Dry deposition velocities : if deposition velocities are underestimated, relative local contribution is underestimated.
2. Wet removal rate : a wet removal rate that is too low will underestimate the relative local contribution, but less so than an equivalent bias in the dry deposition velocities, because of the episodic nature of wet deposition.
3. Chemical transformation rate : since wet removal in ASTRAP is for bulk sulfur while the dry deposition velocities for SO_2 in ASTRAP are about twice those for sulfate, an underestimate of the transformation rate will produce an overestimate of the relative contribution of local sources to dry deposition.

4. Mixing depth : if the mixing depths implied by the eddy diffusivity profiles are too low, surface concentrations are too high and thus relative local source contribution is overestimated for dry deposition.
5. Transport depth : since wind speeds generally increase with height, an underestimate of the transport depth will produce an overestimate of the local source contribution to both dry and wet deposition.
6. Emission rates : experience with different inventories intended to represent the same area and the same time leads to a rough estimate of about 25 % uncertainty. Since the ASTRAP model is linear in the relationship between emissions and deposition, the uncertainty would be reflected in the category to which a source is assigned for a particular receptor.
7. Meteorological variability : the uncertainty is thought to be much larger than that revealed by the simulations for 1980 and 1981 meteorology, which may be unusual in their similarity.
8. Meteorological analyses : resolution problems are most likely to affect the calculation of the local source contribution. The lack of precipitation analyses over the ocean should lead to a slight overprediction of long-range and continental-scale contributions, since recurving trajectories are insufficiently depleted.

An additional source of modeling uncertainty results from the gridding of sources on a 1/3 National Meteorological Center (NMC) scale, about 100-125 km for the latitudes of interest, and the assignment of a virtual source to the center of each cell. The assignment to a separation category is based upon the horizontal separation between emission centroid and receptor point. A cell whose centroid is 175 km away might contain some sources 200 to 230 km away, but all of the emissions will be assigned to the local category. This resolution problem should not create a significant bias in results, but it does contribute to increased variance.

SUMMARY

Local sources have been shown to have a much greater relative contribution to dry deposition than to wet deposition. This is due to the fact that dry deposition, for a given surface and atmospheric stability, is directly proportional to surface concentration, which for a primary pollutant steadily decreases away from sources due to depletion and vertical dilution, while wet deposition is episodic and removes pollutants from a greater depth of the atmosphere.

The influence of distant sources is seen to be relatively greater in eastern Canada than in the eastern U.S. This is due both to the less dense emission field in Canada and to the fact that much of eastern Canada is northeast of the major U.S. and Canadian sources and thus downwind in the sense of the prevailing flow, particularly during summer.

Results indicate that sulfur deposition results from a wide distribution of sources, particularly in the case of wet deposition. It can be inferred that a policy to reduce total sulfur deposition by a specified proportion in sensitive areas would not need to be as broadly based as a program to reduce wet deposition by that proportion since the relative local source contribution to dry deposition appears to be greater than that for wet deposition.

Acknowledgements

Although the research described in this article has been funded wholly or in part by the United States Environmental Protection Agency (EPA) through DW930060-01-0 to DOE, it has not been subject to EPA review and therefore does not necessarily reflect the views of EPA and no official endorsement should be inferred.

REFERENCES

Shannon, J. D., 1981 : A model of regional long-term average sulfur atmospheric pollution, surface removal, and net horizontal flux. Atmos. Environ. 15, 689-701.

DISCUSSION

A. ELIASSEN

You pointed out that the elements of your transfer matrix varied considerably from year to year. If one uses models to study the effect of various emission reduction scenarios, one might therefore be completely misled if one restricts the study to a few short episodes. The current "supermodels" under development are so complex than they probably cannot be run on several years of data. I would therefore assume that their value in emission reduction studies is very limited. Could you comment on this ?

J.D. SCHANNON

It is critically important that multiple years of meteorology be used in models applied in policy assessment.

A joint project between Dr. Samson at the University of Michigan and myself is attempting to examine the meteorological variability of long-range transport and deposition over a 15-year period. The primary use of the elaborate Eulerian models may be to aid in analysis of field experiments via diagnostic applications and to upgrade the parameterizations of the simpler models. The interchange of scientific understanding between models is an area in which insufficient effort has been spent thus far.

A. VENKATRAM If rainfall were uniform over the grid, do you expect the sulfur wet deposition pattern to be similar to the SO_2 emission pattern ?

J.D. SHANNON Obviously the wet deposition pattern would be shifted downwind from the emission pattern, with the shift a function of the transport speeds and the hypothetical uniform precipitation pattern.

CHEMISTRY OF SULFATE AND NITRATE FORMATION

Christian Seigneur, Pradeep Saxena, and Philip M. Roth

Systems Applications, Inc.
101 Lucas Valley Road
San Rafael, CA 94903

INTRODUCTION

The atmospheric deposition of acidic species, i.e., sulfates and nitrates, has been known to occur for many decades. In recent years, concern about the ecological damage that acidic deposition may cause to streams and lake ecology, soils, forests, and materials has considerably increased (e.g., NRC, 1983; EPA, 1983; OTA, 1982). At the first meeting of the Convention on Long-Range Transboundary Air Pollution of the U.N. Economic Commission for Europe, held in Geneva in June 1983, Nordic countries proposed a 30 percent reduction in sulfur emissions to be implemented by 1993. In North America, the U.S./Canada Memorandum of Intent on Transboundary Air Pollution was set to define a common policy for the United States and Canada. Canada calls for a 50 percent reduction in SO_2 emission in both countries. In the United States, several reports acknowledge the contribution of sulfur dioxide (SO_2) and nitrogen oxide (NO_x) anthropogenic emissions to acidic deposition in northern America and the need for emission control (NRC, 1983; OSTP, 1983; OTA, 1982; EPA, 1983).

There is little controversy about the existence of a cause-effect relationship between SO_2 and NO_x emissions and acid deposition. However, the spatial and temporal distribution of the acid deposition is a complex function of atmospheric chemical and transport processes, and other atmospheric variables such as cloud distribution and sunlight. Moreover, the oxidation of SO_2 and NO_x

to sulfates and nitrates requires the presence of oxidants that are primarily produced through an intricate chemistry of NO_x and reactive hydrocarbons (RHC). Therefore, the effects of changes in atmospheric concentrations of SO_2, NO_x, and RHC on acidic species formation are interrelated and depend on several factors related to the atmospheric chemistry of these compounds.

The complexity of the problem makes it difficult to define effective control strategies that can mitigate acid deposition in the most sensitive areas, while minimizing the cost of emission control. In order to approach this issue in a rigorous and efficient way, it is imperative to integrate the available information in a quantitative structured model.

In this paper, we address one part of the acid deposition problem; namely, the formation of sulfate and nitrate species in the atmosphere. We first describe a mathematical model that was recently developed at Systems Applications, Inc. and that treats the formation of acidic species in both the gas phase and liquid phase in a detailed manner. Next, we use this model to investigate the relative importance of the various gaseous and aqueous oxidation pathways that lead to sulfate and nitrate formation under various ambient conditions.

DESCRIPTION OF THE MODEL

The acid rain chemistry model comprises gas-phase kinetic mechanisms, interfacial equilibria/transport between the gas and liquid phases, liquid-phase equilibria, and liquid-phase kinetic mechanisms. The liquid- and gas-phase reaction kinetics are coupled, i.e., rate equations are now solved simultaneously rather than sequentially (Seigneur, Saxena, and Roth, 1982).

The gas-phase chemistry module describes the reactions among chemical species occurring in the atmosphere. The input to this submodel consists of initial concentrations of the chemical species and the time-varying photolysis rate constants for the photochemical reactions. The output of this module is the time-dependent concentrations of the chemical species in the atmosphere.

The liquid-phase chemistry module describes the chemistry of a liquid droplet or liquid-coated aerosol. The initial droplet radius, the initial concentrations of liquid-phase chemical species, and the time-varying moisture content of the air parcel or cloud constitute the initial input to this module. This submodel yields the liquid-phase concentrations of chemical species.

The acid rain chemistry model is a box model that conserves the mass over the two phases as chemical reactions and interphase transport occur. The model is capable of treating time-varying emission rates of gaseous species. Species such as iron, manganese, sodium, chloride, and magnesium, commonly identified in the chemistry of aerosols and acid precipitation, are included in this model.

Currently, the liquid-phase-chemistry model considers only one droplet size at any given time. However, the droplet radius is changed to account for time-dependence of the moisture or cloud-water content. The model can be extended to include a size distribution of droplets and aerosols along with their dynamic behavior. Such coupling of aerosol dynamics and chemistry has been previously implemented for aerosols (e.g., Bassett, Gelbard, and Seinfeld, 1981) and can be applied to fog or cloud droplets in a similar fashion. However, for the purpose of this study, the assumption of a monosize droplet distribution satisfies our needs without affecting the conclusions of the study. A complete description of the acid rain chemistry model is presented in the following sections.

Gas-Phase Kinetic Mechanisms

The Carbon-Bond Mechanism (CBM) III is used to simulate gas-phase chemistry (Killus and Whitten, 1983). This mechanism comprises the reactions among NO_x, reactive hydrocarbons, and O_3. The Carbon-Bond Mechanism is a photochemical kinetic mechanism that has been developed expressly to provide a reasonable compromise between chemical realism and computational efficiency. The CBM is designed to meet stringent validation standards in the simulation of laboratory smog-chamber studies; it has also been successfully applied to many atmospheric studies.

It appears, according to recent reviews of SO_2 gas-phase chemistry, that the reaction of SO_2 with OH radicals is the major pathway of the gas-phase oxidation of SO_2 (Atkinson and Lloyd, 1981; Burton et al., 1982). The following reactions were added to treat the oxidation of SO_2 by OH radicals :

$$SO_2 + OH \xrightarrow{O_2} HSO_5, \qquad k_1 = 1500\ ppm^{-1}\ min^{-1}$$

$$HSO_5 \longrightarrow HO_2 + SO_3, \qquad k_2 = 10{,}000\ min^{-1}$$

$$SO_3 + H_2O \longrightarrow H_2SO_4, \qquad k_3 = 10{,}000\ ppm^{-1}\ min^{-1}$$

Nitric acid is formed rapidly from the OH reaction with NO_2. During daylight hours this reaction appears to be the major pathway:

$$OH + NO_2 \rightarrow HONO_2, \qquad k_4 = 1.60 \times 10^4 \text{ ppm}^{-1} \text{ min}^{-1}$$

However, the concentration of OH is zero at night, and nitric acid generation must be handled by other pathways (Richards, 1983). One pathway believed to be important involves N_2O_5, which is held in an equilibrium concentration by

$$NO_3 + NO_2 \rightleftarrows N_2O_5 \quad .$$

The forward rate constant is 1280 ppm^{-1} min^{-1} and the reverse constant is 2.7 min^{-1} according to an evaluation by NASA (1982). The reaction of N_2O_5 with vapor phase H_2O is believed to generate nitric acid :

$$N_2O_5 + H_2O \xrightarrow{\text{aerosol surface}} 2\ HONO_2 \quad .$$

This reaction has been studied by Harker and Strauss (1981). Their experimental results suggest a pseudo-first-order kinetics with respect to N_2O_5 that is as follows :

$$k_5 = 6.9 \times 10^{-2} \times \frac{\text{aerosol surface}}{\text{volume of air}} \ (\text{min}^{-1}), \text{ for dry aerosols}$$

$$k_6 = 2.07 \times \frac{\text{aerosol surface}}{\text{volume of air}} \ (\text{min}^{-1}), \text{ for liquid-coated aerosols or droplets}$$

where the aerosol surface/volume of air is in m^{-1}.

For the chemistry of H_2O_2, the following gas-phase reactions may become important removal mechanisms in addition to the reaction with SO_2 in the liquid phase :

$$H_2O_2 \xrightarrow{h\nu} 2OH, \qquad k_7(h\nu)$$

$$H_2O_2 + OH \longrightarrow HO_2 + OH, \qquad k_8 = 2500 \ \text{ppm}^{-1} \text{ min}^{-1}$$

These reactions were included in this model. The rate constant for H_2O_2 dissociation is dependent on solar radiation. The overall gas-phase chemistry is illustrated in Figure 1.

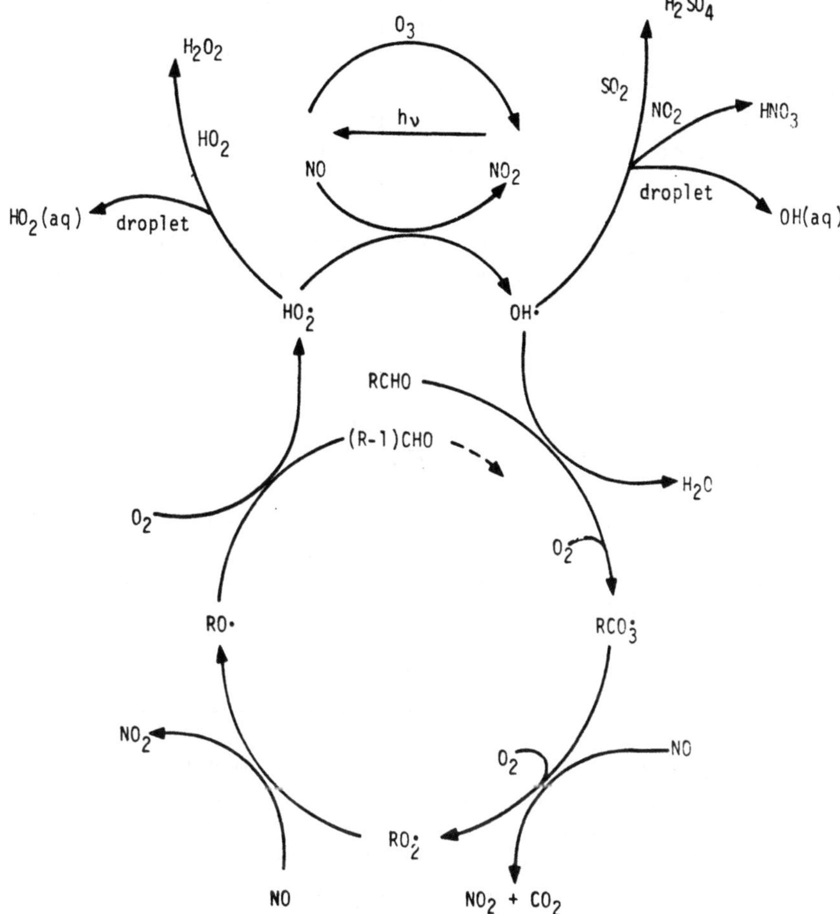

Fig. 1. Simplified description of gas-phase chemistry. (After Burton et al., 1983)

Liquid-Phase Chemistry

The liquid-phase chemistry module is largely derived from a multiphase aerosol model developed recently (Saxena, Seigneur, and Peterson, 1983). The gas-phase/liquid-phase equilibria, liquid-phase chemical kinetics, and chemical equilibria within the liquid phase are described in this module. The atmospheric chemistry of the aqueous phase has been reviewed in great detail by several authors (Graedel and Weschler, 1981; Chameides and Davis, 1982; Calvert, 1983); all chemical reactions presently considered important in atmospheric aqueous chemistry of sulfate and nitrate formation were incorporated. Table 1 presents a detailed description of the liquid-phase chemical reactions.

The oxidation of SO_2 in solution occurs primarily by reaction with H_2O_2, O_3, O_2, NO_x, and OH. The oxidation by O_2 is slow unless catalyzed by trace metal ions such as Mn^{2+} and Fe^{3+}. Synergism between the catalytic properties of Mn^{2+} and Fe^{3+} may exist (Barrie and Georgii, 1976; Martin, 1983) and is included. The interaction of aldehyde chemistry with sulfur chemistry is also considered.

The oxidation of NO_2 in solution occurs primarily by autooxidation. Formation of aqueous nitrate may also occur by decomposition of peroxyacetylnitrate (PAN) but was not considered here.

Interfacial Mass Transport

The transport of chemical species from the gas-phase to the liquid-phase involves two sequential steps :

(1) Diffusion from the bulk phase to the droplet surface

(2) Diffusion at the droplet surface from the gas-phase to the liquid-phase.

For gases with a high vapor pressure, the liquid-phase concentrations are determined by the liquid/gas-phase equilibrium and not by the mass transfer rate. Under these conditions, the gas-phase concentration at the droplet surface may be assumed equal to the bulk gas-phase concentration (Peterson and Seinfeld, 1980). This is the case for SO_2, O_3, H_2O_2, NO_2, HNO_3, HCHO, NH_3, HNO_2, and CO_2. For these species, the liquid-phase concentration is calculated from the gas-phase concentration by means of Henry's law.

For gases with a low vapor pressure, the gas-phase concentration at the droplet surface is very low, and the rate-limiting process is the diffusion from the bulk gas-phase to the droplet surface. This is the case for H_2SO_4. The saturation vapor pressure of H_2SO_4, for example, has been reported to be less than 10^{-7} atm at 25°C (Roedel, 1979); Morrison and Chu, 1979). Gas-phase diffusion is also the rate-limiting process in the case of radical (e.g., OH and HO_2) scavenging by aerosols or droplets. The kinetics of this diffusion-limited process is governed by the Fuchs-Sutugin equation (Fuchs and Sutugin, 1971) :

$$\frac{d\, m_i}{dt} = \frac{4\pi\, r\, D_i}{RT\ (1 + L\ Kn)}\, P_i\ ,$$

where

m_i = moles of species i,
t = time,

Table 1. Chemical Reaction of the Liquid-Phase Model

Reaction	Kinetic Parameter	Value at 25°C	Reference
$H_2O(g) \rightleftarrows H_2O(aq)$	$\frac{1}{P_{sat_w}} = \frac{a_w}{P_{H_2O}}$	$31.99\ atm^{-1}$	Perry and Chilton (1973)
$H_2O(aq) \rightleftarrows H^+ + OH^-$	$K_w = \frac{[H^+][OH^-]\gamma_{H^+}\gamma_{OH^-}}{a_w}$	$1.008 \times 10^{-14}\ mol^2\ \ell^{-2}$	Robinson and Stokes (1965)
$H_2SO_4(aq) \rightleftarrows H^+ + HSO_4^-$	$K_{1SA} = \frac{[H^+][HSO_4^-]\gamma_{H^+}\gamma_{HSO_4^-}}{[H_2SO_4(aq)]}$	$1\ mol\ \ell^{-1}$	Cotton and Wilkinson (1980)
$HSO_4^- \rightleftarrows H^+ + SO_4^{2-}$	$K_{2SA} = \frac{[H^+][SO_4^{2-}]\gamma_{H^+}\gamma_{SO_4^{2-}}}{[HSO_4^-]\gamma_{HSO_4^-}}$	$0.01\ mol\ \ell^{-1}$	Eigen et al. (1964)
$SO_2(g) \rightleftarrows SO_2(aq)$	$K_{HS} = \frac{[SO_2\ (aq)]}{P_{SO_2}}$	$1.24\ mol\ \ell^{-1}\ atm^{-1}$	Johnstone and Leppla (1934)
$SO_2(aq) \rightleftarrows H^+ + HSO_3^-$	$K_{1S} = \frac{[H^+][HSO_3^-]\gamma_{H^+}\gamma_{HSO_3^-}}{[SO_2(aq)]}$	$0.0127\ mol\ \ell^{-1}$	Yui (1940)
$HSO_3^- \rightleftarrows H^+ + SO_3^{2-}$	$K_{2S} = \frac{[H^+][SO_3^{2-}]\gamma_{H^+}\gamma_{SO_3^{2-}}}{[HSO_3^-]\gamma_{HSO_3^-}}$	$6.24 \times 10^{-8}\ mol\ \ell^{-1}$	Yui (1940)
$S(IV)(aq) \xrightarrow[Fe^{3+}]{O_2,\ Mn^{2+}} SO_4^{2-}$	(See Note 1)	(See Note 1)	Martin (1983)

(continued)

Table 1 (continued)

Reaction	Kinetic Parameter	Value at 25°C	Reference
$Fe^{3+} + 3\ OH^- \rightleftarrows Fe(OH)_3$	$K_{sm} = \frac{[Fe^{3+}]}{[Fe^{3+}] + [Fe(OH)_3]}$	(See Note 2)	Hoffmann and Jacob (1983)
$HSO_3^- + H_2O_2(aq) \rightarrow SO_4^{2-}$	(See Note 3)	(See Note 3)	Martin and Damschen (1981) Hoffmann and Edwards (1975)
$H_2O_2(g) \rightleftarrows H_2O_2(aq)$	$K_{HHP} = \frac{[H_2O_2(aq)]}{P_{H_2O_2}}$	7.1×10^4 mol ℓ^{-1} atm^{-1}	Martin and Damschen (1981)
$SO_2(aq) + O_3(aq) \rightarrow SO_4^{2-}$	k_{A2}	5.9×10^2 mol^{-1} ℓ s^{-1}	Overton, Aneja, and Durham (1979)
$HSO_3^- + O_3(aq) \rightarrow SO_4^{2-}$	k_{A3}	3.1×10^5 mol^{-1} ℓ s^{-1}	Erickson et al. (1977)
$SO_3^{2-} + O_3\ (aq) \rightarrow SO_4^{2-}$	k_{A4}	2.2×10^9 mol^{-1} ℓ s^{-1}	Erickson et al. (1977)
$O_3(g) \rightleftarrows O_3(aq)$	$K_{HO} = \frac{[O_3\ (aq)]}{P_{O_3}}$	1.14×10^{-2} mol ℓ^{-1} atm^{-1}	Kosak-Channing and Helz (1983)
$HNO_3(g) \rightleftarrows HNO_3(aq)$	$K_{HN} = \frac{[HNO_3(aq)]}{P_{HNO_3}}$	1.983×10^5 mol ℓ^{-1} atm^{-1}	Tang (1980)
$HNO_3(aq) \rightleftarrows H^+ + NO_3^-$	$K_{1N} = \frac{[H^+]\ [NO_3^-]\ \gamma_{H^+}\ \gamma_{NO_3^-}}{[HNO_3(aq)]}$	15.4 mol ℓ^{-1}	Tang (1980)

Reaction	Kinetic Parameter	Value at 25°C	Reference
$NO_2(g) \rightleftarrows NO_2(aq)$	$K_{HND} = \frac{[NO_2(aq)]}{P_{NO_2}}$	7.0×10^{-3} mol ℓ^{-1} atm^{-1}	Lee and Schwartz (1981)
$2NO_2(aq) \rightarrow 2H^+ + NO_2^- + NO_3^-$	K_{A6}	1.0×10^{8} mol^{-1} ℓ s^{-1}	Lee and Schwartz (1981)
$NH_3(g) \rightleftarrows NH_3(aq)$	$K_{HA} = \frac{[NH_3(aq)]}{P_{NH_3}}$	57 mol ℓ^{-1} atm^{-1}	Morgan and Maahs (1931)
$NH_3(aq) \rightleftarrows NH_4^+ + OH^-$	$K_{1A} = \frac{[NH_4^+][OH^-]\gamma_{NH_4^+}\gamma_{OH^-}}{[NH_3(aq)]}$	1.774×10^{-5} mol ℓ^{-1}	Robinson and Stokes (1965)
$CO_2(g) \rightleftarrows CO_2(aq)$	$K_{HCD} = \frac{[CO_2(aq)]}{P_{CO_2}}$	3.4×10^{-2} mol ℓ^{-1} atm^{-1}	Morgan and Maahs (1931)
$CO_2(aq) \rightleftarrows H^+ + HCO_3^-$	$K_{1c} = \frac{[H^+][HCO_3^-]\gamma_{H^+}\gamma_{HCO_3^-}}{[CO_2(aq)]}$	4.45×10^{-7} mol ℓ^{-1}	Robinson and Stokes (1965)
$HCO_3^- \rightleftarrows H^+ + CO_3^{2-}$	$K_{2c} = \frac{[H^+][CO_3^{2-}]\gamma_{H^+}\gamma_{CO_3^{2-}}}{[HCO_3^-]\gamma_{HCO_3^-}}$	4.68×10^{-11} mol ℓ^{-1}	Robinson and Stokes (1965)
$4O_3(aq) \xrightarrow{OH^-} H_2O_2(aq)$	(See Note 4)	(See Note 4)	Gurol and Singer (1982)

(continued)

Table 1 (continued)

Reaction	Kinetic Parameter	Value at 25°C	Reference
$HCHO(g) \rightleftarrows HCHO(aq)$	$K_{HF} = \frac{[HCHO(aq)]}{P_{HCHO}}$	7.04×10^3 mol ℓ^{-1} atm^{-1}	Dasgupta et al. (1980)
$HCHO(aq) + HSO_3^- \rightleftarrows HOCH_2SO_3^-$	$K_F = \frac{[HOCH_2SO_3^-]\ \gamma_{HMSA^-}}{[HCHO(aq)]\ [HSO_3^-]\ \gamma_{HSO_3^-}}$	7.41×10^4 mol^{-1} ℓ	Ledbury and Blair (1925)
$2HO_2(aq) \rightarrow H_2O_2(aq) + O_2(aq)$	k_{R1}	7.5×10^5 mol^{-1} ℓ s^{-1}	Behar et al. (1970)
$H_2O_2(aq) \xrightarrow{h\nu} 2OH(aq)$	k_{R2}	(See Note 5)	Livingston and Zeldes (1966)
$HSO_3^- + OH(aq) \rightarrow 0.5\ SO_4^{2-} + 0.5\ SO_4^-$	k_{R3}	9.5×10^9 mol^{-1} ℓ s^{1}	Farhataziz and Ross (1977)
$SO_4^- + H_2O_2(aq) \rightarrow HSO_4^- + HO_2(aq)$	k_{R4}	1.2×10^7 mol^{-1} ℓ s^{-1}	Ross and Neta (1979)
$OH(aq) + H_2O_2(aq) \rightarrow HO_2(aq) + H_2O(aq)$	k_{R5}	4.5×10^7 mol^{-1} ℓ s^{-1}	Farhataziz and Ross (1977)
$S(IV)(aq) + NO_2^- \rightarrow SO_4^{2-} + 0.5\ N_2O(aq)$	See text	--	Martin, Damschen, and Judeikis (1981)
$HNO_2(g) \rightleftarrows HNO_2(aq)$	$K_{HN2} = \frac{[HNO_2(aq)]}{P_{HNO_2}}$	47.6 mol ℓ^{-1} atm^{-1}	Martin, Damschen and Judeikis (1981)

Reaction	Kinetic Parameter	Value at 25°C	Reference
$HNO_2(aq) \rightleftarrows H^+ + NO_2^-$	$K_{2N} = \frac{[H^+][NO_2^-]\gamma_{H^+}\gamma_{NO_2^-}}{[HNO_2(aq)]}$	5.9×10^{-4} mol ℓ^{-1}	Martin, Damschen and Judeikis (1981)
$O_3(aq) + H_2O_2(aq) \rightarrow OH(aq) + HO_2(aq) + O_2(aq)$	K_{R6}	2.0 mol^{-1} ℓ s^{-1}	Damschen (1982)

(1) $$\frac{d[H_2SO_4(aq)]}{dt} = \frac{4.7\,[Mn^{2+}]^2}{[H^+]} + \frac{0.82[Fe^{3+}][S(IV)]}{[H^+]}\left(1 + \frac{1.7 \times 10^3\,[Mn^{2+}]^{1.5}}{6.31 \times 10^{-6} + [Fe^{3+}]}\right), \quad \text{mol } \ell^{-1}\ s^{-1}$$

(2) $K_{sm} = 1$, $\quad pH < 3.6$

$K_{sm} = 374.3021 - 319.1337 \times pH + 102.088 \times (pH)^2 - 14.517 \times (pH)^3 + 0.7741 \times (pH)^4$, $\quad 3.6 < pH < 5$

$K_{sm} = 0$, $\quad pH > 5.$

(3) $$\frac{d[H_2SO_4(aq)]}{dt} = \frac{5.2 \times 10^6\,[HSO_3^-][H_2O_2(aq)][H^+]}{0.1 + [H^+]}, \quad \text{mol } \ell^{-1} s^{-1}$$

(4) $$\frac{d[O_3]}{dt} = -\,8.5 \times 10^4\,[OH^-]^{0.55}\,[O_3]^2, \quad \text{mol } \ell^{-1}\ s^{-1}, \quad pH > 4.$$

$$\frac{d[O_3]}{dt} = -\,0.27[O_3]^2, \quad \text{mol } \ell^{-1}\ s^{-1}, \quad pH < 4.$$

(5) The rate constant is dependent on solar radiation.

r = aerosol radius,

D_i = gas-phase diffusion coefficient of species i,

R = universal gas constant,

T = temperature,

P_i = bulk gas-phase concentration of species i,

$Kn = \frac{r}{\lambda}$ = Knudsen number,

λ = mean free path of species i,

and L, the noncontinuum correction factor, is a function of Kn and is defined by (Fuchs and Sutugin, 1971) :

$$L = \frac{4/3 + 0.71\ Kn^{-1}}{1 + Kn^{-1}} \quad .$$

For free radicals (OH, HO_2), an empirical accomodation coefficient is included in the rate equation to account for the non-unity efficiency of the scavenging process. We used a value of 10^{-2} for this efficiency (Davis, 1983).

CASE STUDIES

The formation of sulfate and nitrate species in dry atmospheres via gas-phase chemical reactions has been extensively investigated and has been reviewed in two recent articles (Burton et al., 1983; Richards, 1983). The gas-phase chemistry of sulfate and nitrate species may be summarized as follows :

> Sulfuric acid is formed primarily by the reaction of SO_2 with OH radicals and occurs therefore during daytime only. Other routes such as the reactions of SO_2 with O atoms and Criegee intermediates are believed to be minor.
>
> Nitric acid is formed during daytime by reaction of NO_2 with OH. The OH radicals react about 10 times faster with NO_2 than with SO_2. At night, if NO concentrations are low, NO_2 will react with O_3 to form NO_3; this radical reacts with NO_2 to form N_2O_5 that can then be hydrolyzed to HNO_3 on the surface of liquid-coated aerosols. The NO_3 radical may also dissolve in liquid-coated aerosol and produce nitrate.

In this section we focus on the chemistry of sulfate and nitrate formation in wet atmospheres, i.e., in fogs or clouds. A few modeling studies have analyzed the relative importance of various oxidation pathways of SO_2 and NO_2 for specific conditions, e.g., urban aerosols (Middleton et al., 1980) or urban fog (Jacob and Hoffmann, 1983). There has been no attempt, however, to evaluate a general mechanism of the gas-phase and liquid-phase chemistry of acidic species formation under a variety of atmospheric conditions. We address this problem by applying the mathematical model described in Section 2 to three case studies that are characterized by very different conditions. These three case studies have been selected because experimental data were available to evaluate the model thus providing a sound basis for our theoretical investigation of the mechanisms of sulfate and nitrate formation. The three case studies are :

Summer nonraining cloud in the Adirondacks, northeastern United States,

Fall raining cloud in the Ohio River Valley, midwestern United States, and

Nighttime urban fog in the Los Angeles Basin.

The initial conditions used in the model simulations were deduced to the extent possible from ambient measurements and are presented in Table 2.

Nonraining Cloud-Adirondacks

The Adirondacks are characterized by high levels of acidic deposition and lakes that have a small buffering capacity. Atmospheric measurements have been conducted over the past years in the Adirondacks, at Whiteface Mountain, to investigate the chemistry of clouds and fogs (Mohnen, 1983). A cloud event starting at 5:00 a.m. in the summer season was simulated for three hours. Initial concentrations of SO_2 and NO_x are low; sulfates and nitrates are present in a mole ratio of 1.5. The cloud water content is constant throughout the simulation.

Simulation results for sulfate, nitrate, H_2O_2, and cloud pH are shown in Figure 2. It is clear that SO_2 oxidation is dominated by the liquid-phase H_2O_2 reaction, whereas NO_2 oxidation occurs primarily by gas-phase reaction with OH radicals. The chemistry of this cloud event may be summarized as follows. At 5:00 a.m. initial sulfate aerosols and gaseous nitric acid are dissolved into the cloud almost instantaneously and the cloud pH is about 4.1.

Table 2. Initial Conditions for Model Simulations

Variable	Non-Raining Cloud[a] (Adirondacks)	Raining Cloud[b] (Ohio River Valley)	Urban Fog[c] (Los Angeles Basin)
SO_2 (ppb)	1	10	20
NO_x (ppb)	1	6	150
O_3 (ppb)	45	100	20
NH_3 (ppb)	1	5	5
H_2O_2 (ppb)	1	1	1
RHC (ppb)	15	30	1700
CO_2 (ppm)	320	320	320
Sulfate ($\mu g \cdot m^{-3}$)	3	10	15
Nitrate ($\mu g \cdot m^{-3}$)	1.3	2.6	20
Fe^{3+} ($\mu g \cdot m^{-3}$)	--	--	1
Mn^{2+} ($\mu g \cdot m^{-3}$)	--	--	0.05
SO_2 Emissions ($ppb \cdot min^{-1}$)	--	--	0.05
NO_x Emissions ($ppb \cdot min^{-1}$)	--	--	0.11
NH_3 Emissions ($ppb \cdot min^{-1}$)	--	--	0.005
Cloud water content ($g \cdot m^{-3}$)	0.5	1	0.1
Droplet Radius (μm)	7.5	7.5	7.5
Cloud Transmissivity	0.9	0.5	--
Time	5 a.m.	3 p.m.	5 a.m.
Month	June	November	December

[a] Mohnen (1983)
[b] Lazrus et al. (1983)
[c] Jacob and Hoffman (1983)

At this pH the high solubility of H_2O_2 and its fast reaction with HSO_3^- lead to a rate of SO_2 oxidation in the aqueous phase of more than 100 percent per hour. The H_2O_2 concentration profile shows that gas-phase H_2O_2 decreases very fast, first by dissolution in the cloud droplets, then by aqueous reaction. Around 7:00 a.m. photochemical reactions start to produce HO_2 radicals that lead to H_2O_2 formation at a rate that exceeds the rate of H_2O_2 consumption in the droplets. At pH 4.1 reaction of SO_2 with O_3 is too slow (less than 0.5 percent of SO_2 per hour) to contribute notably to sulfate formation. Gas-phase oxidation of SO_2 is slow at 5:00 a.m. (about 0.2 percent per hour) and reaches 1.5 percent per hour at 8:00 a.m.; this pathway would contribute more to total sulfate formation later in the day when photochemical reactions maintain high OH radical concentrations.

Nitric acid is formed in the gas phase primarily by reaction with OH radicals, since this pathway accounts for 94 percent of total nitrate formation. The remaining 6 percent originates from the reaction of NO_2 with NO_3 radicals.

There is good agreement between the model calculations and measurements obtained at Whiteface Mountain (Mohnen, 1983). Measured concentrations of sulfate are on the order of 4 $\mu g/m^3$ and those of nitrate 2 $\mu g/m^3$, with an average mole ratio of 1.4. The predicted sulfate to nitrate mole ratio is 2.4. This

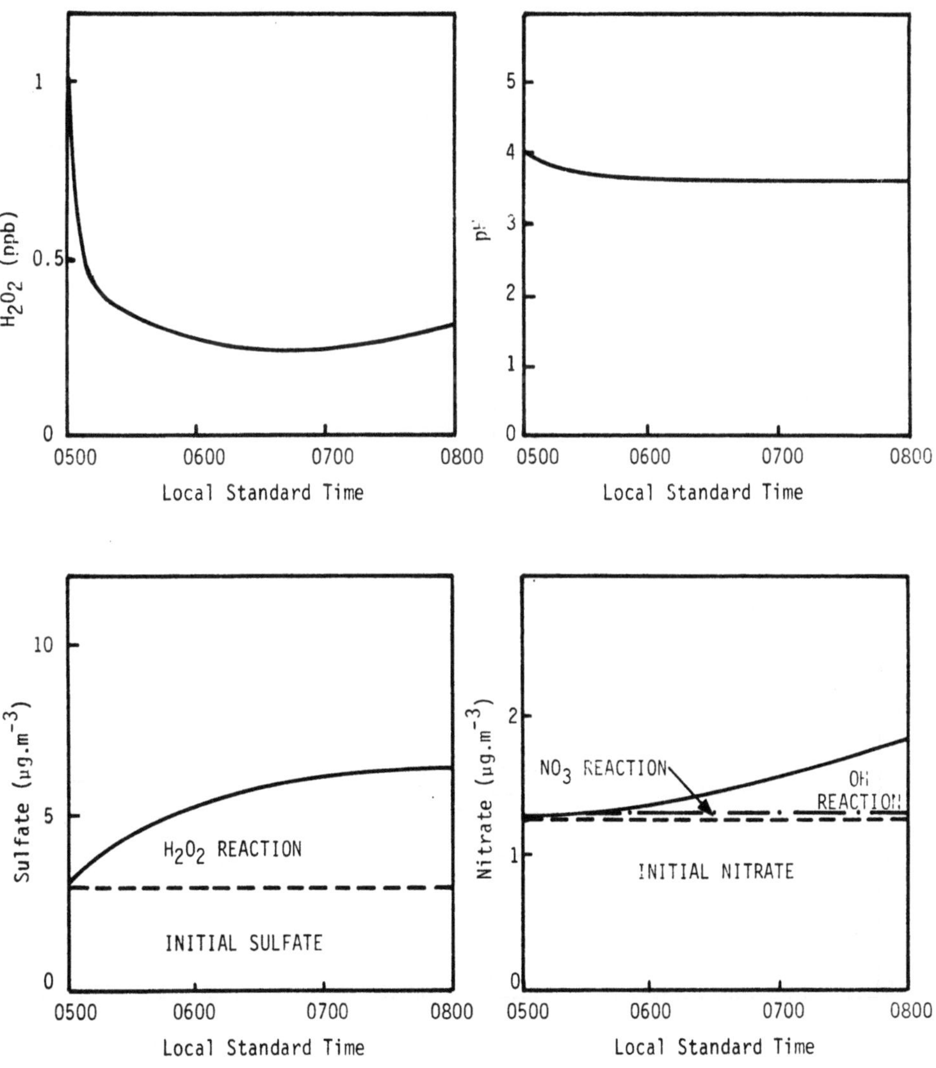

Fig. 2. Nonraining cloud simulation-Adirondacks.

higher value is due to the fact that the model simulates early morning chemistry where nitric acid formation is still slow. The predicted pH value decreases from 4.1 to 3.6; the average measured pH value is 3.8.

Raining Cloud-Ohio River Valley

Measurements were conducted in dry air and clouds during a warm frontal precipitation event in the Ohio River Valley during November

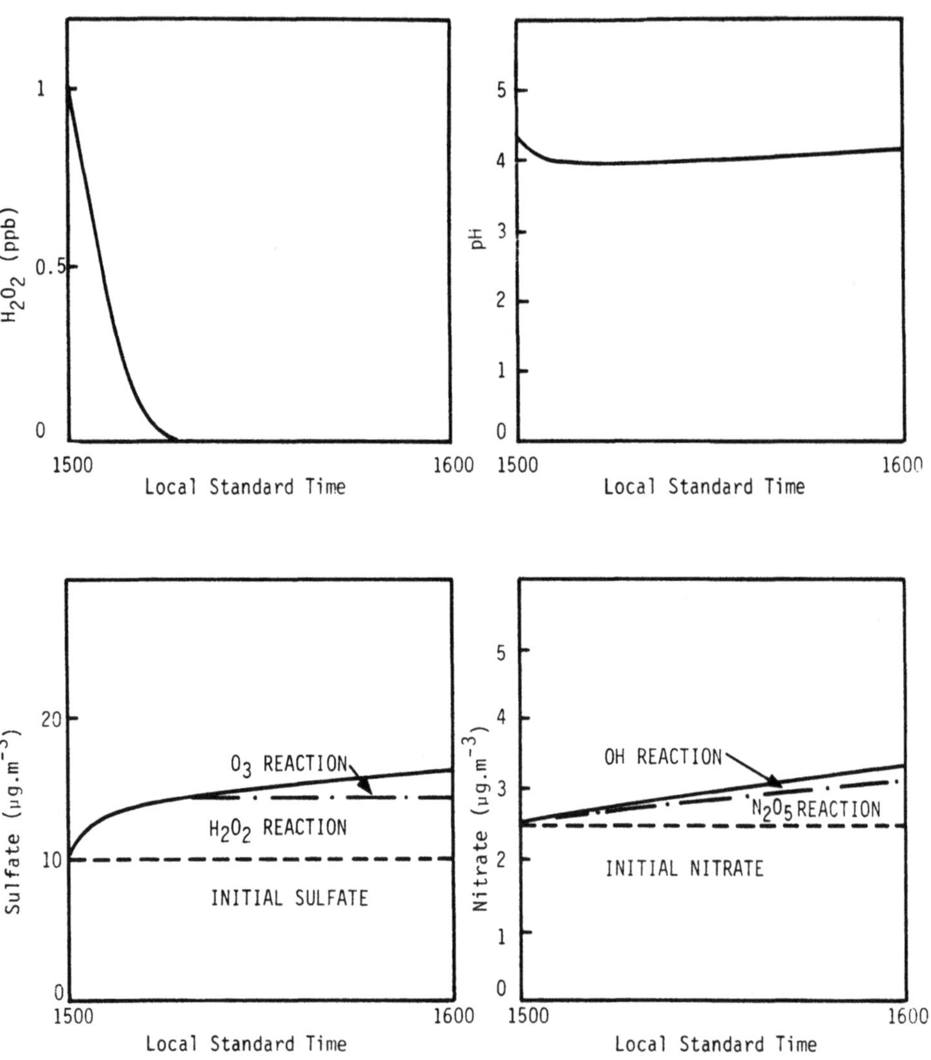

Fig. 3. Raining cloud simulation-Ohio River Valley.

1979 (Lazrus et al., 1983). The model does not presently treat cloud dynamics, and the simulation of chemistry in a raining cloud was approximated as follows. The initial cloud water content was 1 $g.m^{-3}$, which was assumed to increase according to the average water vapor condensation rate of 1.5 $g.m^{-3}.h^{-1}$ (Lazrus et al., 1983).

The results of the one-hour simulation are shown in Figure 3. Notable differences appear between this simulation and the previous case study. Although sulfate formation is again dominated by the H_2O_2 aqueous reaction, H_2O_2 is depleted very fast and the corresponding SO_2 oxidation rate decreases from 140 percent per hour at the beginning to 1.5 percent per hour after one hour. The rate of SO_2 oxidation by O_3 ranges from 6 to 10 percent per hour. Concentrations of OH radicals are low because of the low solar irradiations (November evening with a cloud ransmissivity of 0.5). The rate of SO_2 oxidation by OH radicals is less than 1 percent per hour.

Formation of nitric acid by reaction of NO_2 with OH is not dominant because of the low photochemical activity within the cloud, and the dark reaction of NO_2 with NO_3 accounts for 80 percent of nitrate formation. The pH drops from 4.3 to about 4.0 at the beginning of the simulation as sulfate is rapidly formed by the reaction of SO_2(aq) with H_2O_2. As the cloud water content increases, the pH tends to slightly increase due to dilution in spite of acidic species formation.

Qualitatively, the simulation results agree well with the ambient measurements obtained at the cloud base. The predicted liquid-phase concentrations of sulfate are in the range of 6.0×10^{-5} to 1.2×10^{-4} M, and those of nitrate in the range of 2.0×10^{-5} to 4.0×10^{-5} M. These concentrations compare well with the measured values of $3.3. \times 10^{-5}$ to 4.2×10^{-5} M for sulfate and 2.0×10^{-5} to 4.1×10^{-5} M for nitrate. The rates of acidic species formation estimated from the cloud measurements are 1.2 ppb/h for sulfate and 0.9 ppb/h for nitrate. Model calculations give a sulfate formation rate of 1.7 ppb/h and a nitrate formation rate of 0.3 ppb/h. The predicted pH varies from 4.3 to 4.0, which compares well with the measured values of 3.9 to 4.1.

Urban Fog--Los Angeles Basin

Measurements of fog chemical composition have been conducted in the Los Angeles Basin and have shown the presence of high concentrations of sulfate, nitrate, and trace metals in fog droplets (Munger et al., 1983). The simulation conditions used by Jacob and Hoffman (1983) in their simulation of acid fog were selected. A varying fog water content was chosen to represent the formation, stagnation, and evaporation of a nighttime fog during a three-hour period.

Simulation results are presented in Figure 4. Sulfate formation by reaction of HSO_3^- with H_2O_2 occurs in the first 20 minutes. At night, O_3 is not produced and is consumed in a few minutes by reaction with NO. Therefore, no oxidation of SO_2 by O_3 occurs. Trace metals--iron and manganese--catalyze the oxidation of SO_2 by O_2 and account for most of the sulfate formation.

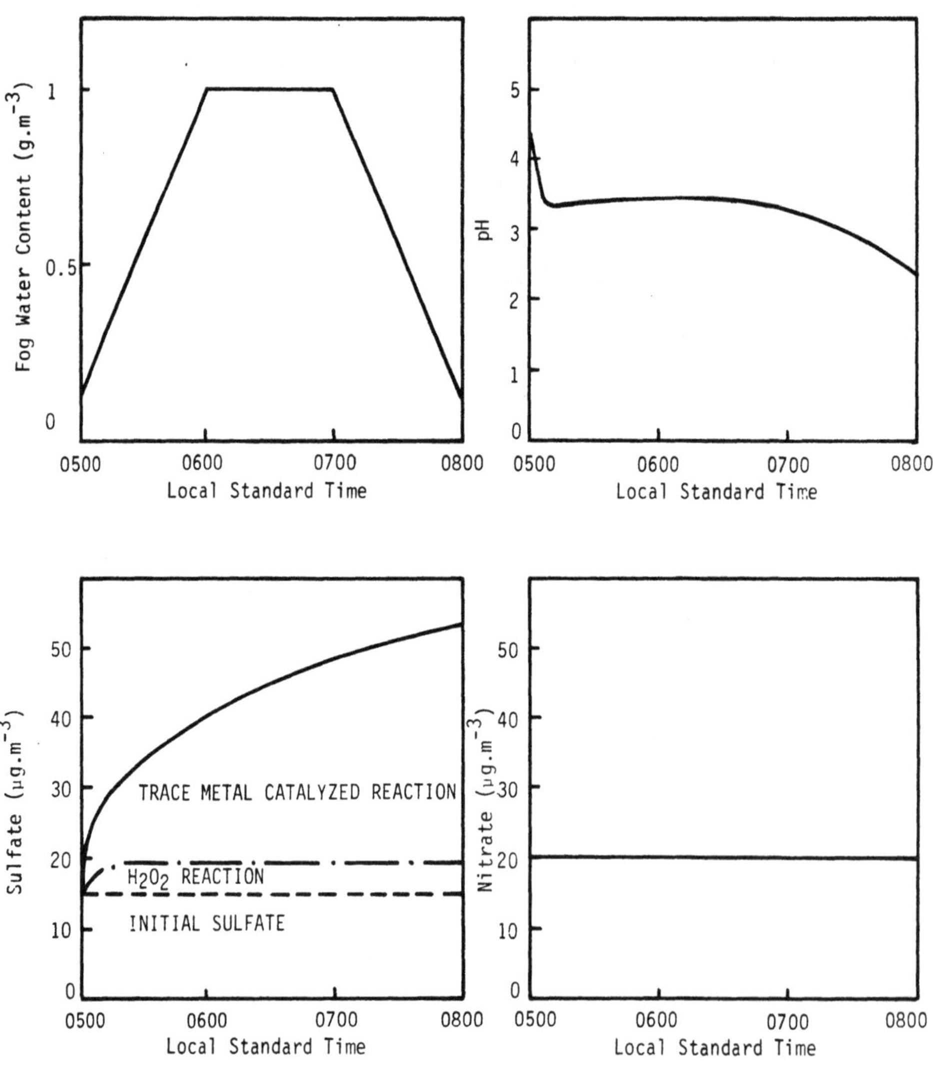

Fig. 4. Urban fog simulation-Los Angeles Basin.

The formation of bisulfite ion complexes by formaldehyde in solution to form hydroxymethanesulfonic acid (HMSA) has little effect on fog chemistry. The concentration of HSO_3^- ions is governed by the pH of the solution and the gas-phase SO_2 concentrations unless the HMSA concentration becomes comparable to the SO_2 concentrations. In this simulation, HMSA concentrations are about 50 times lower than SO_2 gas-phase concentrations and have therefore negligible effect on HSO_3^- concentrations and sulfate formation. Negligible effect of HMSA on sulfate formation was also predicted in the two other case studies.

No nitrate formation occurs, since the OH radical concentrations are negligible at night and no O_3 is available to react with NO_2 to form NO_3 radicals. Aqueous auto-oxidation of NO_2 is too slow to produce any notable amount of nitrate. Initially the pH drops as sulfate is formed very rapidly. As the fog water content increases, the pH tends to increase slightly and finally drops sharply to a low value of 2.3 as the fog dissipates.

Munger and co-workers (1983) reported sulfate concentrations of 2.5×10^{-4} to 2.5×10^{-3} M. Model predictions of sulfate levels range from 4.0×10^{-4} to 5.6×10^{-3} M. The predicted pH varies from an initial value of 4.3 to a final value of 2.3. This compares well with measured pH values that range form 2.2 to 5.8.

CONCLUSION

A model that describes in detail the gas- and liquid-phase chemistry of sulfate and nitrate formation has been used to simulate the formation of acidic species in clouds and fog. Good qualitative agreement was obtained between model predictions and ambient measurements. From the results of these model simulations, we may summarize the atmospheric chemistry of sulfate and nitrate species as follows.

In a dry atmosphere SO_2 oxidation occurs by gas-phase reaction with OH radicals only during daytime. Nitric acid is formed during daytime by reaction of NO_2 with OH radicals. If NO concentrations are low, O_3 will be present at night and nitric acid will form by reaction of NO_2 with NO_3 and hydrolysis of the reaction product N_2O_5.

If clouds of fog are present, liquid-phase oxidation of SO_2 becomes the primary pathway for sulfate formation. When gaseous and aerosol species enter a cloud, the cloud pH drops as sulfate aerosols and nitric acid are dissolved. At acidic pH values, reaction with H_2O_2 is the major SO_2 oxidation mechanism with a rate that can be on the order of 100 percent per hour. Generally, H_2O_2 is consumed very fast, and O_3 may contribute notably to sulfate formation if the cloud pH is not too low, i.e., pH ⩾ 4 with SO_2 oxidation rates between 1 and 10 percent per hour. Gas-phase oxidation of SO_2 may occur at a rate of about 1 percent per hour and will generally not contribute much to sulfate formation in clouds. Liquid-phase reactions, such as oxidation by OH radicals and HNO_2, are too slow to contribute notably to sulfate formation in fog or clouds. In fog or clouds, where trace metals are present in high levels, catalytic oxidation of SO_2 by dissolved O_2 may become the dominant mechanism. The formation of bisulfite ion complexes by formaldehyde does not appear to affect sulfate formation notably.

Formation of nitric acid occurs primarily by oxidation of NO_2 by OH when photochemical activity is high. In cases when photochemistry takes place at a slow rate, such as in optically thick clouds or winter clouds, reaction of NO_2 with NO_3 becomes the main mechanism for nitric acid formation. Decomposition of peroxyacetylnitrate in cloud droplets is an alternate pathway for nitrate formation (Holdren et al., 1982; Richards et al., 1983) and its contribution to nitrate levels should be investigated.

The chemistry of sulfate and nitrate species formation involves several pathways that vary in importance according to the atmospheric conditions, season, time of day, and background air quality. A rigorous approach to the study of acidic species formation therefore requires a detailed description of the major kinetic processes.

ACKNOWLEDGMENTS

This work was supported by Southern California Edison Company, Consolidation Coal Company, Peabody Coal Company, and Systems Applications, Inc. We thank Dr. Mohnen (New York State University), Mr. Mirabella (Southern California Edison), Mr. Kerch (Consolidation Coal Company), Mr. Wootten (Peabody Coal Company), and Drs. Burton and Whitten (Systems Applications, Inc.) for helpful discussions. We thank also Ms. Judy Rodich and Mr. Howard Beckman for their excellent technical editing.

REFERENCES

Behar, D., Czapski, G., Rabani, J., Dorfman, L. M., and Schwartz, H. A., 1970, The acid dissociation constant and decay kinetics of the perhydroxyl radical, J. Phys. Chem., 74:3209-3213.

Burton, C. S., Liu, M. K., Roth, P. M., Seigneur, C., and Whitten, G. Z., 1983, Chemical transformation in plumes, Air Pollution Modeling and Its Application--II, C. de Wispelaere, ed., pp. 3-58, Plenum Press, New York, New York.

Cotton, F. A., and Wilkinson, G., 1980, "Advanced Inorganic Chemistry," 4th ed., John Wiley, New York, New York.

Atkinson, R., and Lloyd A. C. 1981, Evaluation of kinetic and mechanistic data for modeling of photochemical smog, J. Phys Chem. Ref. Data, 10.

Barrie, L. A., and Georgii, H. W., 1976, An experimental investigation of the absorption of sulfur dioxide by water drops containing heavy metal ions, Atmos. Environ., 10:743-749.

Bassett, M. E., Gelbard, F., and Seinfeld, J. H., 1981, Mathematical model for multicomponent aerosol formation and growth in plumes, Atmos. Environ. 15:2395-2406.

Calvert, J. G., ed. 1983, "Acid Precipitation," Ann Arbor Science Publishers, Ann Arbor, Michigan, in press.

Chameides, W. L., and Davis, D.D., 1982, The free radical chemistry of cloud droplets and its impact upon the composition of rain, J. Geophys. Res., 83:4863-4877.

Damschen, 1982, private communication.

Dasgupta, P.K. Decesare, K., and Ulbrey, J. C., 1980, Determination of atmospheric sulfur dioxide without tetrachlorrmercurate (11) and the mechanism of the Schiff reaction, Anal. Chem., 52:1912-1922.

Davis, D.D., 1983, private communication.

EPA, 1983, "The Acidic Deposition Phenomenon and Its Effects--Critical Assesment Review Papers. "EPA-600/800/8-83-016, U.S. Environmental Protection Agency, Washigton, D.C.

Eigen, M., Kruse, W., Maass, G., and DeMager, L., 1964, Rate constants of protolytic reactions in aqueous solution, in "Progress in Reaction Kinetics," G. Porter, ed., Macmillan, New York, New York.

Erickson, R. E., Yates, L. M., Clark, R. L., and McEwan, D., 1977, The reaction of sulfur dioxide with ozone in water and its possible atmospheric significance, Atmos. Environ., 11:813-817.

Farhataziz, and Ross, A. B., 1977, "Selected Specific Rates of Reactions of Transients From Water in Aqueous Sollution, III: Hydroxyl Radical and Perhydroxyl Radical and Their Radical Ions," NSRDS-NBS 59, U.S. Department of Commerce, Washington, D.C.

Fuchs, N. A., and Sutugin, A. G., 1971, High-dispersed aerosols, in Topics in Current Aerosol Research," G. M. Hidy and J. R. Brock, eds., Pergamon Press, New York, New York.

Graedel, T. E., and Weschler, C. J., 1981, Chemistry within aqueous atmosperic aerosols and raindrops, Rev. Geophys. Space Phys., 19:505-539.

Gurol, M. D., and Singer, P. C.,1982, Kinetics of ozone decomposition: a dynamic approach, Environ. Sci. Technol., 16:337-383.

Harker, A. B., and Strauss, D.R., 1981, "Kinetics of the Heterogeneous Hydrolysis of Dinitrogen Pentoxide over the Temperature Range 214-263 K," Rockwell International Science Center (Federal Aviation Administration, Publication FAA-EE-81-3).

Hoffmann, M. R., and Jacob, D. J., 1983, Kinetics and mechanisms of the catalytic oxidation of dissolved sulfur dioxide in aqueous solution : an application to nighttime fog water chemistry, in "Acid Precipitation," J. G. Calvert, ed., Ann Arbor Science Publishers, Ann Arbor, Michigan (in press).

Holdren, M. W., Ward, G. F., Keigley, G. W., and Spicer, C.W., 1982, "Preliminary Investigation of the Effects of Peroxyacetylnitrate Precipitation Chemistry," Battelle Pacific Northwest Laboratories, Seattle, Washington.

Jacob, D. J., and Hoffmann, M. R., 1983, A dynamic model for the production of H^+, NO_3^-, and SO_4^{2-} in urban fog, J. Geophys. Research, in press.

Johnstone, H. F., and Leppla, P. W., 1934, J. Am. Chem. Soc., 56:2233.

Killus, J.P., and Whitten, G. Z., 1983, "A New Carbon-Bond Mechanism for Air Quality Simulation Modeling," 81245, Systems Applications, Inc., San Rafael, California.

Kosak-Channing, L. F., and Helz, G. R., 1983, Solubility of ozone in aqueous solutions of 0-0.6 M ionic strength at 5-30°C, Environ. Sci. Technol., 17:145-149.

Lazrus, A. L.,Haagenson, P. L., Kok, G. L., Huebert, B. J., Kreitzberg, C. W., Likens, G. E., Mohnen, V. A., Wilson, W. E., and Winchester, J. W., 1983, Acidity in air and water in a case of warm frontal precipitation, Atmos. Environ., 17:581-591.

Ledbury, W., and Blair, E. W., 1925, The partial formaldehyde vapor pressures of aqueous solution of formaldehyde, part II, J. Chem. Soc., 127:2832-2839.

Lee, Y.N., and Schwartz, S. W., 1981, Reaction kinetics of nitrogen dioxide with liquid water at low partial pressure, J. Phys. Chem., 85:840-848.

Livingston, R. and Zeldes, H., 1966, "Paramagnetic Resonance Study of Liquids During Photolysis--Hydrogen Peroxide and Alcohols," J. Chem. Phys., Vol. 44, pp. 1245-1259.

Martin, L. R., 1983, Kinetic studies of sulfite oxidation in aqueous solution, in "Acid Precipitation," J. G. Calvert, ed., Ann Arbor Science Publishers, Ann Arbor, Michigan (in press).

Martin L. R., and Damschen, D. E., 1981, Aqueous oxidation of sulfur dioxide by hydrogen peroxide at low pH, Atmos. Environ., 15:1615-1621.

Martin, L. R., Damschen, D. E., and Judeikis, M. S., 1981, The reaction of nitrogen oxide with SO_2 in aqueous aerosols, Atmos. Environ., 15:191-195.

Middleton, P., Kiang, C.S., and Mohnen, V.A., 1980, theeretical estimates of the relative importance of various urban sulfate aerosol production mechanisms, Atmos. Environ., 14:463-472.

Mohnen, V. A., 1983, private communication.

Morgan, O. M., and Maahs, O., 1931, Can.J. Res., 5:162.

Morrison, R.A., and Chu, K. S., 1979, Measurement and calculation of the total pressure in equilibrium with highly concentrated sulfuric acid, Symposium on Environmental and Climatic Impact of Coal Utilization, 17-19 April, Williamsburg, Virginia.

Munger, J. W., Jacob, D. J., Waldman, J. M., and Hoffman, M.R., 1983, Fogwater chemistry in an urban atmosphere, J. Geophys. Research, in press.

NAPAP, 1982, "Annual Report 1982 to the President and Congress," National Acid Precipitation Assessment Program, Washington, D.C.

NASA, (1982), "Chemical Kinetics and Photochemical Data for Use in Stratospheric Modeling: Evaluation Number 5," 82-57, Jet Propulsion Laboratory, National Aeronautics and Space Administration, Pasadena, California.

NRC, 1983, "Acid Deposition--Atmospheric Processes in Eastern North America," National Research Council, National Academy Press, Washington, D.C.

OSTP, 1983, "Press Advisory and Interim Report from OSTP's Acid Rain Peer Review Panel," Office of Science and Technology Policy, Executive Office of the President, Washington, D.C.

OTA, 1982, "The Regional Implication of Transported Air Pollutant: An Assessment of Acidic Deposition and Ozone," Office of Technology Assessment, Congress of the United States, Washington, D.C.

Overton, J. H., Aneja, V.P., and Durham, J. L., 1979, Production of sulfate in rain and raindrops in polluted atmospheres, Atmos. Environ., 13:355-367.

Perry, J. H., and Chilton, C. H., 1973, "Chemical Engineer's Handbook," Mc Graw-Hill, New York, New York.

Peterson, T. W., and Seinfeld, H. J. 1980, Heteorogeneous condensation and chemical reaction in droplets--application to the heteorogeneous atmospheric oxidation of SO_2, Adv. Environ. Sci. Technol., 10:125-180.

Richards, L. W., 1983, Comments on the oxidation of NO_2 to nitrate--day and night, Atmos. Environ., 17:397-402.

Richards, L. W., Anderson, J. A., Blumenthal, D L., Duckhorn, S. L., and McDonald, J. A., 1983, "Characterization of Reactants, Reaction Mechanisms, and Reaction Products Leading to Existence of Acid Rain and Acid Aerosol Conditions in Southern California," California Air Resources Board, Sacramento, California.

Robinson, R. A., and Stokes, R. H., 1965, "Electrolyte Solutions", Butterworthe, London, England.

Roedel, W., 1979, Measurements of sulfuric acid saturation vapor pressure : Implications for aerosol formation by heteromolecular nucleation, J. Aerosol Sci., 10:375.

Ross, A. B., and Neta, P., 1979, "Rate Constants for Reactions of Inorganic Radicals in Aqueous Solution, NSRDS-NBS 65, U.S. Department of Commerce, Washington, D.C.

Saxena, P., Seigneur, C., and Peterson, T. W., 1983, Modeling of multiphase atmospheric aerosols, Atmos. Environ., 17:1315-1329.

Seigneur, C., Saxena, P;, and Roth, P. M., 1982, Preliminary results of acid rain chemistry modeling," Proc. Atmospheric Deposition Specialty Conference, pp. 330-341, Air Pollution Control Association, Pittsburgh, Pennsylvania.

Tang, I. N., 1980, On the equilibrium partial pressures of nitric acid and ammonia in the atmosphere, Atmos. Environ., 14:819-828.

Yui, T., 1980, Tokyo Inst. Phys. Chem. Res. Bull., 19:1229.

DISCUSSION

A. VENKATRAM — Could you comment on the relative importance of Fe, Mn catalysis and free radical chemistry in the aqueous phase ?

Chr. SEIGNEUR — Fe and Mn catalysis is very important for SO_2 oxidation when these trace metal concentrations are high as it is the case for example in the Los Angeles Basis. According to our model simulations, free radical chemistry did not affect sulfate and nitrate formation notably for the cases considered.

S. SHANNON — Have you extended your model to simulate deposition via fog or cloud impaction ?

Chr. SEIGNEUR — Not yet. That is planned future work.

T. LAVERY — Based on your simulations, what is the role of NO_x in controlling the pH of the Los Angeles fog ?

Chr. SEIGNEUR — According to our simulations no nitrate is formed in the fog. Thus, all the nitrate present in this fog was formed previously (e.g. daytime gas-phase chemistry). The presence of nitrate leads to lower pH values. Lower pH values may in turn lead to faster SO_2 oxidation in the liquid phase and to a faster decrease of the pH.

A. VENKATRAM — How did you fix the initial concentrations in the aqueous phase ?

Chr. SEIGNEUR The initial concentrations in the aqueous phase are from measured aqueous phase concentrations (e.g. sulfate) or are calculated by the model from the gas phase concentrations according to Henry's law equilibrium.

THE USE OF FIELD DATA IN AVERAGE WET DEPOSITION MODELING

Nadezda Sinik, Edita Loncar and Sonja Vidic

Republican hydrometeorological institute of Croatia
Centre for meteorological research
41000 Zagreb, Gric 3, Yugoslavia

INTRODUCTION

In view of today's energy production it appears important to incorporate into long term pollution modeling the rate of pollutants removal from the air by rain scavenging. The degree of an air concentration decrease by washout processes can be estimated on the basis of various mathematics and physics in dependence on pollutant solubility and on size distributions of aerosol-rain systems. Greenfield (1957) modeled washout by means of inertia forces in the subcloud layer in connection with prior coagulation and diffusion processes inside a cloud. His and several earlier theoretical approaches have been very instructively discussed by Makhon'ko (1965) who pointed out that washout processes are about ten times less efficient than the rain out. Still, the washout itself and consequently the wet deposition makes a comparable removal mechanism for soluble pollutants on both a short and a long term basis (Scriven and Fisher, 1975). The latter is dependent on the frequency of rainy periods.

No matter what the physics of a process is, all wet deposition modelers try to determine the rate of precipitation scavenging λ. It is generally a function of many various factors. Chamberlain (see Scriven and Fisher, 1975) takes $\lambda = 10^{-4}\ I^{1/2}$ (I is the precipitation intensity), for Rodhe and Grandel (1972) λ ranges between 3×10^{-5} to 5×10^{-4} etc. Besides, many authors emphasize the importance of statistical distributions of periods with precipitation for λ determination (eg. Davies 1976, Rodhe 1980, Smith 1981, etc.).

Barrie (1981) suggested the use of a dimensionless washout ratio W instead of λ. He found that W is rather strongly dependent on the rain pH and on the mean air temperature T in the subcloud layer. This relationship is governed by the chemistry of a given pollutant. Also, the rain scavenging coefficient λ for the case of SO_2 has been correlated to W, to a rain intensity, I, and to a depth of the box in an one layer model, H.

If the data exist on the pollutant amount which is wet deposited during cloud passages or during a given time period in connection with trajectory models, one can do various elaborated studies on rain scavenging efficiency. This is usually the case for greater regions, say Europe, (see Eliassen 1978, Mac Mahon, Denison and Fleming 1976, Smith and Hunt 1978 etc.). Unfortunately, this is very often not the case with smaller regions. Here, one usually can make use of the most common field data on precipitation (amounts and number of days) to get some idea of that region's natural ability with respect to rain scavenging efficiency without being able to check the results with appropriate measurements. This was the problem in the middle coastal region of Yugoslavia which we had to solve - and our paper describes our endeavors and results.

The study was aimed at determining λ on the basis of precipitation data at all weather stations inside a small region (diameter of about 130 km). The resulting monthly average spatial distribution of the washout coefficient has been presented. It was tested indirectly by means of pH measurements at several rain stations inside the region using Barrie's empirical relations.

MODELING

Washout coefficient λ has been generally defined as

$$\lambda = -\frac{1}{\chi}\frac{d\chi}{dt}\ \sec^{-1} \tag{1}$$

where χ is pollutants air concentration and $-\frac{d\chi}{dt}$ the rate of air concentration decrease by washout. If we neglect droplets evaporation, turbulence, electrical and other possible effects we can use a general expression

$$\lambda = C\ I^{\alpha}\ \sec^{-1} \tag{2}$$

which postulates λ to be a function of rainfall intensity, I. Values of C and α have to be determined. Relations of a general form, as (2), can be applied to any of the gases. Still, since the wet deposition depends in a great extent on a solubility of a given gas (Engelmann et al. 1966, Marsh 1978) one should determine C and α corresponding to the material which is supposed to be influenced by rain scavenging. We considered the gases with a good solubility for which there exists and experimentally found relationship of λ and I, given in (Slade 1968). On the basis of Slade's results a set of C and α values has been calculated. It came out that both C and α vary with rain intensity, C between 3.1×10^{-4} and 4.0×10^{-4} and α between 0.8 and 1.0. Next, for the same set of λ and I values and for given values of α a set of C values has been computed. As a result, coefficient C appeared to be function of rain intensity only for $\alpha = 1.0$. For all the other values of α ($0.8 \leq \alpha < 1.0$) C is quasiconstant having a mean value of 3.3×10^{-4} hour mm^{-1} sec^{-1} with corresponding mean value of $\alpha = 0.9$. Finally, the wet deposition coefficient λ is given by the equation

$$\lambda = 3.3 \times 10^{-4} I^{0.9} sec^{-1} \qquad (3)$$

Performing the calculations of λ by the eq. (3) and also by Chamberlain's equation one obtains comparable results for rainfall intensities less than 1.0 mm $hour^{-1}$. Our values of C and α agree well with Ritchie, Brown and Wailand's (1978) results for cases of a heavy rain. They gave two values of C : for a stable and for an unstable atmosphere. Our value is between those two and corresponds to a neutral atmosphere. Since we wanted to apply equation (3) to an average state of the atmosphere we supposed its neutral value to be quite a good approximation.

The most common long term precipitation data are precipitation amounts and number of days with precipitation ($\geq$ 1.0 mm). Therefore, the simplest approach for determining an average value of λ is

a/ - Compute the rainfall intensity

$$I = \frac{RR}{24\ N} \text{ mm hour}^{-1} \qquad (4)$$

at every raingauge station in the considered region. Here, RR and N are taken to be the monthly amount and number of days with precipitation, respectively.

b/ - Compute λ by means of equation (3).

For the same value of I the rain scavenging will be more efficient when the rain is more frequent. Therefore, λ should be multiplied by the relative frequency of rainy days, N/D, at every raingauge station (D = duration of the considered time period, say month, in days). One obtains

$$\bar{\lambda} = \frac{N}{D} \lambda \qquad (5)$$

which represents a rough estimate of long term washout coefficient.

A relative frequency or probability of rainy days is used here to substitute much more relevant parameter of "wet" periods duration. Smith (1981) determines the rain probability in a wet region as a whole. Here we take into account the spatial variability of this quantity and subsequently of $\bar{\lambda}$ inside a given region.

$\bar{\lambda}$ can be applied to any of long term diffusion and transport models. One has to compute the nearground pollutant concentration at any locality in the surroundings of an emission source and then to apply the local $\bar{\lambda}$ value.

Resulting from the eq. (1) the precipitation scavenging has an exponential form, so one has to multiply the average nearground concentration $\bar{\chi}$ by a "wet correction" factor

$$r = e^{-\bar{\lambda}\Delta t} \tag{6}$$

at a given point inside the region under consideration. Here Δt should represent the average duration of rainfall, in seconds. However, in air pollution modeling Δt is restricted by the time interval over which the pollutant concentration is calculated. The corrected nearground air concentration of a soluble pollutant is then

$$\bar{\chi}_r = r.\ \bar{\chi} \tag{7}$$

Almost the same result can be obtained by the equation

$$\bar{\chi}_r = \bar{\chi}\ (1 - \bar{\lambda}\Delta t) \tag{8}$$

which contains the linear term in a series expansion of (6), or by the discretization of (1). Similar approach is given in Pack (1978).

APPLICATION

The described procedure was applied to an area in the coastal part of Yugoslavia. Monthly average washout coefficients $\bar{\lambda}$ in August (period 1969-1978) have been computed at every raingauge station in that area. They vary between 0.3×10^{-4} and $0.8 \times 10^{-4} s^{-1}$. The isolines on the figure 1. give a spatial distribution of $\bar{\lambda}$ and enable to obtain its value at any part of the region.

The greatest wet correction factor has been computed in the middle and northeast part of the region and it is $r = 0.866$ for half-an-hour concentrations. Monthly means of half-an-hour SO_2 nearground air concentrations in this region range up to 0.165 mg m^{-3}. Rain scavenging may lessen this amount to 0.143 mg m^{-3}.

There was no possibility to test our model by direct measurements of wet deposition. Fortunately, Barrie (1981) gave the following empirical relation for SO_2 :

$$\lambda = \frac{W_{SO_2} I}{H}$$

where the washout ratio W can easily be determined using Barrie's graph as well as corresponding equations as a function of pH and T (the meaning of all the parameters is given in the "Introduction").

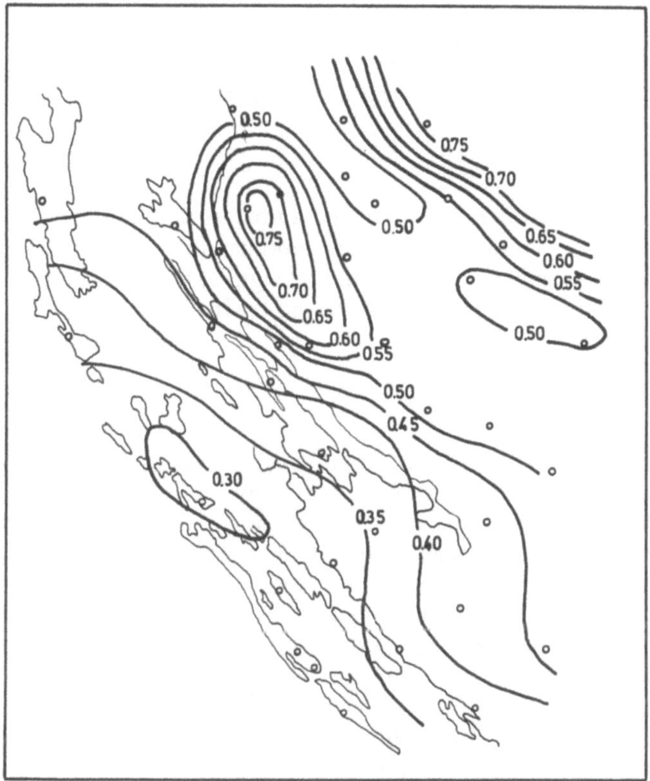

Fig. 1. Isolines of $\bar{\lambda}$ for the month of August (period 1969-78)-Yugoslavia, middle Adriatic coast.

Mean monthly rain intensities in our region varied between 0.2 mm $hour^{-1}$ and 0.65 mm $hour^{-1}$. Three stations with pH measurements inside the region indicated a spatial quasi-constancy with a characteristic value of pH = 6.0 for the whole region. Further, the parameter H in Barrie's equation was given a value of the monthly mean cloud base height in this region. A special analysis of relevant cloud date gave the values of H in the range 800 m and 1200 m. The air layer mean temperature in August is T = 293°K. Using the mentioned graph in Barrie (1981) we have obtained $W = 5.9 \times 10^5$. With all this values we have obtained

a/ for the smallest intensity I and the greatest mean cloud base height H

$$\lambda = 0.41 \times 10^{-4} sec^{-1}$$

b/ for the greatest I and the smallest H

$$\lambda = 0.89 \times 10^{-4} sec^{-1}$$

The computed values of λ agree quite well with $\bar{\lambda}$ on our figure 1. This fact encourages to apply our method of $\bar{\lambda}$ calculations with more certainty at least for the case of SO_2.

CONCLUSION

A spatial distribution of an average washout coefficient can be determined by means of the most common precipitation data. The described simple method gives just rough estimates, still it can be usefull in determining a degree of rain scavenging efficiency in a given region.

REFERENCES

1. L. A. Barrie, The prediction of rain acidity and SO_2 scavenging in Eastern North America, Atm. Env. 15:31 (1981).
2. T. D. Davies, Precipitation scavenging of sulphur dioxide in an industrial area, Atm. Env. 10:879 (1976)
3. A. Eliassen, The OECD study of long range transport of air pollutants: long range transport modelling, Atm. Env. 12:479 (1978).
4. R. J. Engelmann, Scavenging Prediction Using Ratios of Concentrations in Air and Precipitation, J. Appl. Meteor. 10:493 (1971).
5. S. M. Greenfield, Rain scavenging of radioactive particulate matter from the atmosphere, J. Met. 14:115 (1957).
6. T. A. Mc Mahon, P. J. Denison and R. Fleming, A long-distance air pollution transportation model incorporating washout and dry deposition components, Atm. Env. 10:751 (1976).
7. K. P. Makhon'ko, Simplified theoretical notion of contaminant removal by precipitation from the atmosphere, Tellus XIX (3): 467 (1967).
8. A. R. W. Marsh, Sulphur and nitrogen contributions to the acidity of rain, Atm. Env. 12:401 (1978).
9. D. H. Pack, Meteorology of long-range transport, Atm. Env. 12:425 (1978).
10. L. T. Ritchie, W.D. Brown, J. R. Wailand, Effects of Rainstorms and Runoff on Consequences of Atmospheric Releases from Nuclear Reactor Accidents, Nuclear Safety 19(2):220 (1978).
11. H. Rodhe and J. Grandell, On the removal time of aerosol particles from the atmosphere by precipitation scavenging, Tellus 24:442 (1972).
12. H. Rodhe, Estimate of wet deposition of pollutants arround a point source, Atm. Env. 14:1197 (1980).

13. R. A. Scriven and B. E. A. Fisher, The long range transport of airborne material and its removal by deposition and washout - I. general consideration - II. the effect of turbulent diffusion, Atm. Env. 9:49 (1975).
14. D. H. Slade, "Meteorology and Atomic Energy", U.S. Atomic Energy Commission, TID-24190 (1968).
15. F. B. Smith and R. D. Hunt, Meteorological aspects of the transport of pollution over long distances, Atm. Env. 12:461 (1978).
16. F. B. Smith, The significance of wet and dry synoptic regions on long-range transport of pollution and its deposition, Atm. Env. 15:863 (1981).

DISCUSSION

A. BERGER What is the seasonal variation of α ?

N. SINIK α is a function of rain intensity and varies seasonally as much as rain intensity does. According to data, given in Slade's "Meteorology and Atomic energy" α varies between 0.8 and 1.0.

MODELING OF CHEMICAL TRANSFORMATIONS OF SO_x AND NO_x IN THE POLLUTED ATMOSPHERE - AN OVERVIEW OF APPROACHES AND CURRENT STATUS

N. V. Gillani

Mechanical Engineering Department
Washington University
St. Louis, Missouri

ABSTRACT

Two principal approaches are identified in the modeling of chemical transformations of SO_x and NO_x in the polluted atmosphere. The fundamental approach involves simulation of the detailed chemical kinetics of the SO_x-NO_x-HC system ; in the empirical approach, relatively simple parameterizations of the bulk rates of conversion of precursors to secondary products are sought in terms of environmental factors which, based on laboratory and field measurements, appear to control these rates. In this paper, the principal features of both approaches are described in the form of an overview, with examples of the main results. An assessment is made of the current state of development of both methods, particularly within the context of their applicability in regional transport-transformation models. A brief overview is also presented of the controversial question of linearity of the transformation module with respect to the precursor emissions.

INTRODUCTION

Most of the harmful effects of atmospheric sulfur and nitrogen compounds, viz. acidic depositions, visibility degradation, material damage, climatic change, and human health effects, are more directly associated with the secondary products (sulfates and nitrates) of chemical transformations of the precursor emissions (SO_2, NO, NO_2) into the atmosphere. In the past two or three decades, a great deal of research was focused on these transformation processes, specifically to understand the detailed mechanisms and to obtain quantitative estimates of bulk rates of the conversion of SO_2 to sulfates, and of NO to NO_2 to nitrates, based on laboratory simulations of the polluted atmosphere as well as on field measurements in the real polluted atmosphere. Attempts were also being made to formulate mathematical and numerical models for simulation of the transformation processes. Most recently, a great deal of interest has been aimed at the formulation of the transformation module in meso- and synoptic scale coupled transport-transformation-deposition models. In this paper, an overview is presented of the current state of development of this module.

Secondary products of chemical transformations of SO_x and NO_x emissions are generally more acidic than their precursors. In the context of acidification of lakes, vegetation and soil, however, the chemical form in which the deposition arrives at the surface is of relatively little significance (because precursor depositions are rapidly converted to the secondary forms following deposition) compared to the fact that the rate of the deposition process itself depends strongly on its chemical form. Thus, for example, sulfate particles are believed to have a considerable longer average atmospheric residence than SO_2, and hence a larger range of impact. Nitric acid, on the other hand, is likely to be removed from the atmosphere more efficiently and rapidly than its precursors. Consequently, it is necessary for transport-deposition models to distinguish between primary and secondary pollutants, and to facilitate atmospheric chemical transformations through appropriate modules.

The chemical transformation module is an integral part of the overall transport-transformation-removal model. The framework within which the larger model is formulated and solved may be Lagrangian (trajectory), or Eulerian (grid), or some hybrid scheme. Langrangian or trajectory models simulate the changing concentration field within a given polluted air parcel (e.g., a puff or plume release) as a result of the combined effects of dilution, chemistry, and depositions. Typically, the concentration field as well as meteorological variables are assumed to be homogeneous within the air parcel. Recent attempts have also been made to obtain simulations with finer spatial resolutions within the air parcel. Lagrangian models may be tailored for simulations of pollutant kinetics at the plume scale. Regional Lagrangian simulations are commonly based on simple linear superpositions of individually-calculated concentrations of multiple plumes. Individual plumes may be referred to point sources or area sources. For the modeling of nonlinear processes in multiple interacting plumes over regional scales, Eulerian grid models are more appropriate. They are based on the solution of coupled transport-transformation-removal mass balance equations of individual species over specified two- or three-dimensional spatial grids. Typical grid sizes in regional models vary from 50 to 100 km to a side. Within each grid cell, pollutant concentrations, as well as meteorological variables, are assumed to be uniformly distributed. In a pure grid model, emissions within a grid cell are considered in an aggregate sense, and are instantaneously homogenized over the entire cell volume. The error of this approximation is particularly severe in two-dimensional grid models which lack vertical resolution. The effects of sub-grid scale processes are sometimes included in terms of bulk parameterizations. Alternately, a hybrid scheme may be used in which individual plumes may be modeled in a Lagrangian sense and detail until they acquire the spatial dimensions of the Eulerian grid size, and subsequent simulation is within the Eulerian framework. The output from a grid model is an evolving series of snapshots of the deposition field over the entire modeled region. This is clearly very desirable in regional modeling. Grid models, however, require far more extensive input information, computations and computational resources than trajectory models, and are generally quite expensive to implement. The chemical transformation module does not depend, per se, on the framework of the larger model formulation. However, its validity does depend on the spatial-temporal resolution of the simulation, and on the facility for accommodating nonlinear processes and plume interactions with its chemically different environment. The remainder of this section is focused on the transformation module.

An objective of this section is to review and assess briefly our present ability to predict the rates of chemical transformations of primary emissions of SO_x and NO_x to secondary acidic products (sulfates and nitrates) during atmospheric transport. Such predictions are based on transformation models, which are mathematical formulations relating secondary pollutant formation rates to concentrations of the precursor gases (e.g., SO_2, NO), and to any other chemical and meteorological factors considered to contribute to the transformation processes. The principal approaches in formulating such models are discussed for S and N compounds, for power plant and urban plumes, and for each of the major conversion mechanisms believed to be important. Specific formulations of practical interest are reviewed briefly along with their applications, and major outstanding problem areas are identified. An overall assessment is presented of our present standing in terms of the desired goals of transformation modeling. Emphasis is placed on formulations believed to be suitable for inclusion as transformation modules in current long-range transport-transformation models aimed at simulating regional-scale acidic depositions.

The atmospheric transformation processes are very complex, involving multiple parallel pathways (mechanisms) of physical diffusion and homogeneous and heterogeneous chemical reactions of a wide variety of reactants and catalysts. The reactants may be of primary or background origin or intermediate or secondary products of concurrent reactions. A variety of meteorological factors-- UV radiation, temperature, relative humidity, clouds, fogs, atmospheric turbulence, and others -- also have important influence on atmospheric transformation processes. Many of these factors are interdependent ; e.g., UV radiation, temperature, clouds, and turbulent mixing are closely related to insolation. Furthermore, a given factor may simultaneously have opposite effects on different chemical reactions ; e.g., the effect of plume dispersion should be to "quench" reactions between coemitted species (Schwartz and Newman 1978), but also to promote reactions of primary emissions with background species (Wilson 1978, Gillani and Wilson 1980). Given the complex array of reactants and their reactions influenced in a complicated manner by interdependent environmental factors, one must recognize that no single and simple mathematical expression can describe adequately the transformation processes of a given pollutant. Realistic transformation models should be capable of distinguishing among the different conversion mechanisms and, for each mechanism, should reasonably reflect the dependence of the conversion rate on current plume, background, and environmental conditions.

Historically, the science of transformation modeling is young. As recently as 1977, the state of the art was such that in a widely acclaimed regional monitoring and modeling program, the conversion rate of SO_2 to SO_4^{2-} was represented by a single constant number over a regional scale, regardless of time of day, season, or prevailing meteorological conditions (OECD 1977). Even today, such practice is not uncommon in regional models, perhaps with some justification. Since 1977, however, significant progress has been made in developing transformation modules appropriate for regional models, particularly for the gas-phase mechanism of S conversions. Applicable models for the liquid-phase mechanism are still rare and primitive. Current transformation models for N compounds are generally complex, requiring extensive computational resources even for mesoscale applications. Their validations are limited.

APPROACHES TO TRANSFORMATION MODELING

Basically two approaches to transformation modeling exist - a fundamental approach and an empirical approach.

The fundamental approach

The fundamental approach consists of the so-called "explicit mechanisms method" and its simplified counterparts. In theory, the explicit mechanisms method involves consideration of all significant reactants and their elementary reactions involved in each mechanism of sulfate or nitrate formation. Concentration changes by all chemical reactions are calculated simultaneously for all species at short-term intervals (typically a few seconds). Reactants include not only the precursors (e.g., SO_2, and NO), their principal oxidizing agents (e.g. OH, HO_2, and RO_2 in the gas-phase mechanism, and O_2, O_3 and H_2O_2 in the liquid-phase mechanism), and the secondary products of concern (e.g., H_2SO_4 and HNO_3)

but also catalysts and significant intermediate species involved in the mechanisms. Of particular significance are the multitude of reactive HC species and their derivatives involved in gas-phase chain reactions that contribute to photochemical smog formation, as well as to sulfate and nitrate formation. In a spatially homogeneous system (well-mixed plume) consisting of n species, a total of 2n first-order, nonlinear, ordinary differential equations must be solved simultaneously at each time step to evaluate the changing species concentrations in the plume and in the background with which the plume interacts. Plume-background interactions must be facilitated in the model. If spatial inhomogeneities are important and need to be resolved, the system of equations becomes substantially larger. Also, because a wide range of reaction-time scales are generally involved, computations for the equations' solutions at each time step are quite involved, time-consuming, and expensive.

Implementation of the explicit mechanisms method has many associated problems. The list of possible reactants is long, and sometimes there is even disagreement about what the products are in given individual reactions. Values of many elementary reaction rate constants have either not been measured or are not quite reliable. Model input requirements also include specification of initial concentrations of all species in the plume and in the background. While primary emissions from major point sources are reasonably well characterized, area source emissions are not. This is particularly true for the hydrocarbons. Also, the spatial-temporal resolution of the current area source emissions inventories is generally inadequate to verify model performance based on the available mesoscale field data of power plant and urban plume transport and transformations. Atmospheric measurements are either rare or nonexistent for many short-lived species, some of crucial importance (e.g., OH, HO_2, RO_2, and H_2O_2). Detailed HC and aldehyde measurements in the atmosphere are not common. Input specifications and model validations are thus only partial and very approximate.

Perhaps the best example of an attempt to simulate smog chemistry by explicit mechanisms is the work of Demerjian et al. (1974), which incorporated more than 200 species, the great majority of them arising from the explicit use of specific reactive HC and corresponding organic intermediates and sinks. Despite this model's comprehensiveness, the authors warn that its representation of the real atmosphere, which undoubtedly contains hundreds of organic compounds, may be an oversimplification. Such complex chemical modeling is currently impractical for application in regional models. Simplifications and further approximations are necessary. The key is to achieve a reasonable condensation of the vast number of HC and aldehydes, and their reactions, while adequate representation is maintained. Such condensation is attempted either by "lumping" groups of species by some common criterion and then treating each group as a single species in the model, or by substituting a single surrogate species either for all HC (e.g., propylene by Greadel et al. 1976, "nonmethane HC" by Miller et al. 1978) or for a particular lumped group of HC (e.g., xylene for aromatics, by Hov et al. 1977). Two principal methods of "lumping" have been developed : the HSD method (Hecht et al. 1974), and the carbon bond mechanism (CBM) method (Whitten et al. 1980). In the HSD method, organic species of like reactivities are grouped into four main classes : paraffins, aromatics, olefins, and aldehydes. Many models use a modification of this in which the following six lumped classes are used after Falls and Seinfeld (1978) and Falls et al. (1979) : ethylene, higher molecular weight olefins, paraffins, aromatics, formaldehyde, and higher molecular weight aldehydes. In the CBM method, similarly bonded C atoms are lumped into four or more classes. In principle, the CBM is closer to the explicit mechanism and is also easier to use in conjunction with measured data than is the HSD mechanism. Such formulations have been further condensed in specific simulations by reducing the number of species modeled through the use of surrogate reactions and rate coefficients which effectively include the role of the omitted species (Levine and Schwartz 1982).

Validation of simulations performed by detailed chemical models has, to date, been generally based on matching calculated concentrations of certain key aspects of photochemical smog formation (e.g., HC loss, and OH or O_3 formation) with those measured in controlled smog chamber studies in the laboratory. The roles of such meteorological variables as sunlight, temperature, and relative humidity are simulated directly in the experiments and included in the calculations through the dependence of elementary reaction rates on them. The role of other meteorological variables such as turbulence and inhomogeneous mixing generally is not simulated in laboratory experiments. This is probably a serious limitation.

In the real polluted atmosphere, the deficiency of certain key reactive ingredients in a primary emission may well be overcome through entrainment of such ingredients from the back-ground air. The formation of ozone and sulfates in HC-poor power plant emissions in the eastern United States during summer afternoons is thus almost as rapid as in HC-rich urban emissions (Fig. 1, and Gillani and Wilson 1980). Appropriate background characterization and treatment of plume-background interaction can be of critical importance in realistic modeling of transformation processes.

An important positive feature of detailed chemical models is that nonlinear chemical couplings between species, including the coupling between sulfur and nitrogen chemistry, is explicitly retained. In this sense, the same model can, in principle, perform simulations of SO_x and NO_x transformations, as well as of urban or power plant plume chemical evolution. With appropriate spatial-temporal resolution, the effect of plume-plume and plume-background interactions can also be performed.

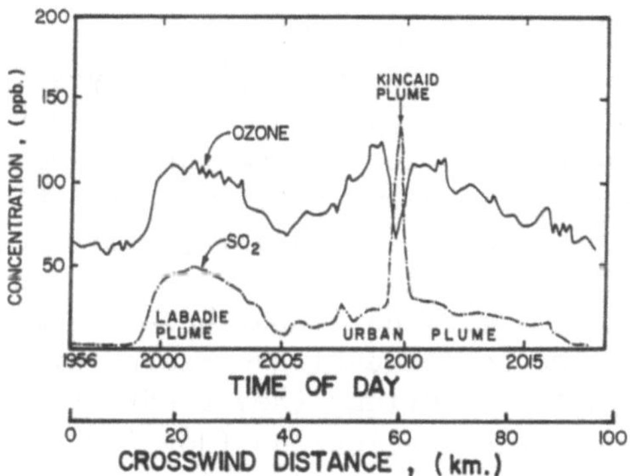

Fig. 1. Crosswind profiles of primary (SO_2) and secondary (O_3) pollutants in the Labadie power plant plume and in the urban plume of St. Louis (including the emissions from a major petroleum refinery complex). The data were collected during an aircraft traverse through the plumes at about 450 m AGL. The plumes were released around mid-day ; the traverse was made around 2000. Observe the comparable levels of excess ozone in both plumes. The Kincaid power plant plume represents a relatively fresh emission into the aged urban plume. Figure is from Gillani et al. (1978)

One of the major undesirable features of the detailed chemical approach is the necessity of performing extensive computations. However, considerable differences exist in amounts of computation necessary depending on choice of numerical method and degree of chemical approximations involved. The number of species in the chemical schemes commonly used varies between 10 and 100. The amount of computations increases nonlinearly and rapidly with increasing number of species. For any given chemical scheme of smog simulation, the main numerical problem arises from the fact that the various chemical reactions occur at speeds which vary by several orders of magnitude. This wide range of time scales involved in this problem makes the corresponding set of differential equations quite "stiff". Standard techniques for integrating sets of differential equations (e.g., the Runge-Kutta Method) cannot provide stable solutions of such stiff systems at realistic cost. Special techniques such as those developed by Gear (1971) provide much more efficient numerical integrations by performing automatic time and error control, and are capable of providing accurate numerical solutions, albeit at considerable cost and requiring the use of large high-speed computers. The Gear technique has been used widely in simulations of photochemical smog. Other attempts to reduce computations have resorted to the use of quasi-steady-state assumptions for certain very reactive species. Such assumptions are not always justified and have been shown to lead to large inaccuracies not only under polluted conditions but also in relatively clean background conditions (Farrow and Edelson 1974, Dimitriades et al. 1976, Jeffries and Saeger 1976, Hesstvedt et al. 1978). Judiciously invoked steady-state approximations (QSSA) based on continuous monitoring of pollutant time scales during on-going simulations, can permit locally analytical solutions (Hesstvedt et al. 1978) and even locally linearized analytical solutions (Hov 1983a). Such numerical techniques can provide solutions comparable in accuracy to the Gear solutions at a fraction of the cost, and can be implemented on smaller computers.

Examples of specific detailed chemical model calculations for atmospheric applications are considered in Section 4.1. A recent review paper by (Hov 1983c) is also recommended for those interested in further details pertaining to the fundamental approach to transformation modeling.

The empirical approach

Given the substantial uncertainties and gaps in the input information needed for detailed chemical models, and given the discrepancies in reported transformation rates of SO_x and NO_x, the use of detailed kinetic models continues to be questioned, and simpler empirical rate expressions are often favored. A great deal of experimental research on chemical transformations has been directed at obtaining estimates of the conversion rates of SO_2 to sulfates, and of NO_2 to nitrates in the laboratory and in the field. In recent years, some success has been achieved in relating field estimates of the conversion rates to specific conversion mechanisms and to specific, measured influencing factors. A large number of source-related and environmental factors have been implicated as influencing transformations. They include the time and height of source release, the nature and amounts of the acid precursors, other coemitted species, the reactivity of the air mass in which emissions are trans ported, as well as such meteorological factors as sunlight, temperature, absolute humidity, clouds and fogs, and atmospheric stability.

In the empirical approach, an attempt is made to identify the rate-controlling factors for each mechanism and to formulate and validate an overall rate expression for measured sulfate or nitrate formation by each mechanism directly in terms of these factors, which are also measured. In other words, the effect of the multiple elementary reactions is parameterized in terms of pertinent, measurable chemical and meteorological factors. Such parameterizations of the conversion rate are simple rate expressions, which can be inserted directly into regional models as the transformation module. They entail very few computations and require inputs that are, for the most part, relatively readily available even on a regional scale. In spite of their simplicity, they often yield quite reliable estimates of actual atmospheric formations of such final products as sulfates. This is particularly true when their formulation is based directly on field data and their application is based on measured input data. Their principal disadvantage is that they lack generality, being applicable mainly under environmental conditions reasonably close to those for which they have been successfully validated. In specific applications for which relevant parameterizations are available, their simplicity and reliability make them immensely valuable.

The existing empirical parameterizations of sulfur chemistry are largely based on mesoscale plume data. At least three important practical implications of this limitation may be identified. First of all, given the dominance of source-specific characteristics in mesoscale plume transport, empirical parameterizations which are mesoscale in origin must be considered to be specific to the source type (e.g., power plant plumes versus urban-industrial plumes) for which they were developed. Secondly, since the characteristic time scales of the atmospheric residence of secondary pollutants such as sulfates and ozone are significantly longer than mesoscale, it must be presumed that the parameterizations for plumes would be sensitive to boundary conditions. In fact, empirical observations have shown that sulfate and ozone formation rates in power plant as well as urban plumes are strongly sensitive to the chemical condition of the background air, and to the rate of plume dilution by entrainment of this background air (Gillani and Wilson 1980 ; Miller and Alkezweeny 1980). Plume-background interactions can sometimes even obscure the initial chemical distinctions between a power plant plume and a petroleum refinery plume (see Fig. 1). Finally, one must question the validity of empirical parameterizations of meso-scale origin in synoptic scale applications. On the positive side, however, it has been demonstrated empirically that pollutant plumes evolve to the chemical maturity characteristic of regional air masses within only a few hours of transport during sunny convective conditions typical of summer days in the eastern United States (Gillani and Wilson 1980). At least under such conditions, chemical parameterizations derived from data of chemically aged plumes may have validity even during long range transport.

The reactions governing SO_2 oxidation have the general form :

$$SO_2 + O_x + (M) \rightarrow \text{products} \rightarrow SO_4^{2-} \qquad (1)$$

where O_x represents the principal oxidizing agents ; i.e., OH and possibly HO_2 and RO_2 for gas-phase oxidation (Calvert et al. 1978), and H_2O_2, O_3, and O_2 for liquid-phase oxidation (Penkett et al. 1979) ; (M) represents the catalysts, if and when any are involved. With the possible exception of catalyzed reactions (Freiberg 1974), the rate of sulfate formation, r_s, may be expressed as :

$$r_s = \frac{\partial}{\partial t}(SO_4^{2-}) = k_s \cdot (SO_2) \qquad (2)$$

where the fractional conversion rate, k_s, depends on O_x, the oxidizing species. Such a relationship is valid as long as SO_2 is not in stoichiometric excess. The validity and linearity of this equation are further discussed in section 3. Parameterization of k_s which is the goal of empirical transformation models, is thus a representation of the weighted contributions of factors which effectively determine the O_x concentrations. It may be broken down by mechanisms into :

$$k_s = k_{s_G} + k_{s_L} + k_{s_{HET}} \qquad (3)$$

where components on the right hand side represent, respectively, the fractional conversion rates by gas-phase, liquid-phase, and heterogeneous aerosol surface reaction mechanisms. No parameterizations have been attempted for the heterogeneous mechanism, partly because reliable and particular atmospheric data are lacking and partly because the mechanism generally is not considered important on the regional scale. Specific parameterizations of S conversions are most developed for k_{s_G}, and efforts to parameterize k_{s_L} have just begun. These are discussed in the next section.

Similarly, the formation of the two principle secondary nitrates (HNO_3 and PAN) are largely governed by the reactions

$$NO_2 + OH \rightarrow HNO_3 \quad (4a)$$

and

$$NO_2 + RCOO_2 \rightarrow PAN \,. \quad (4b)$$

Hence, their formation rates may be expressed as :

$$r_{HNO_3} = k_{HNO_3} \cdot (NO_2) \quad (5a)$$

$$r_{PAN} = k_{PAN} \cdot (NO_2) \,, \quad (5b)$$

where the fractional conversion rates, k_N (N = HNO_3, PAN), depend on the concentrations of OH and $RCOO_2$, respectively. The parameterizations of k_N would represent the weighted contributions of the factors which effectively determine these free radical concentrations. Empirical parameterizations of k_N based on field data have not been formulated or tested. Sensitivity of k_N to the HC - NO_x mix has been studied in smog chamber experiments. Some of the most recent specific results and their implications will be discussed in a later section.

THE QUESTION OF LINEARITY

A much debated matter, and one of considerable practical importance in terms of regional modeling and control strategy, is the question of linearity (or the lack of it) in the source-receptor relationship between emissions of SO_x and NO_x and their depositions. An important subset of this larger question pertains to the linearity of relationships between r_s and SO_2, and r_N and NO_x. In this section, the discussion is limited to the question of linearity of the chemical transformation processes. If the transformation chemistry is nonlinear, certain common modeling practices based on the assumption of linearity must be viewed with caution. For example, regional models typically have a spatial resolution over grids of 50 to 100 km to a side. The assumption of uniform species concentrations within a grid cell which includes concentrated emissions sources may give erroneous transformation estimates unless some appropriate parameterization of sub-grid scale processes is included. Distinctions in the chemical mix of different source emissions are also presumably important in the case of nonlinear chemistry. Linear superpositions of species concentrations, calculated for individual plumes assumed to be isolated, will also give erroneous estimates of nonlinear secondary formations in regions with multiple plume interactions. The validity of the linearity assumption is also crucial to the success of attempts to control secondary pollutants by a strategy of linear rollback of precursor emissions.

The lack of consensus on the question of linearity, particularly with respect to sulfur chemistry, is probably due to different interpretations of the definition of the term linear relationship. By definition, the relationship between r_s and SO_2 is linear if it can be stated in the form of Equation (2), and if k_s is independent of SO_2. Clearly, k_s is variable through its dependence on species such as the OH free radical which are responsible ultimately for the oxidation of SO_2. Therefore, the critical question is whether these oxidizing agents are themselves dependent on SO_2. There is no doubt that in a fresh plume with high concentration of SO_2, OH level is significantly controlled by SO_2 itself, and the oxidation of SO_2 is a nonlinear process. Such conditions, however, are short-lived. Subsequently, if there are no further fresh injections of SO_2 into this plume, the formation of OH will be governed by the NO_x-HC chemistry in the plume and by entrainment from the background of OH itself and of other reactive species contributing to OH formation. The direct dependence of plume NO_x-HC chemistry on local SO_2 concentration is very weak in this stage of plume transport. Consequently, one commonly finds in the published literature explicit or implicit statements about linear sulfur chemistry under such conditions. If the mathematical definition of linearity is to be interpreted strictly, such statements are correct within the context of the transport of a particular plume release. In the broader context of modeling of longer-term averages or continuous emissions, possibly varying with time, and with inevitable plume-plume and plume-background interactions, an indirect form of nonlinearity does exist because of the correlation between SO_2 emissions and the co-emissions of NO_x and HC. A broader definition of linearity which requires k_s to be independent not only of SO_2 but also of co-emitted species is implicit in the works of Cahir et al. 1982 and Hidy 1982.

The significance of the role of the co-emitted species is illustrated in the following practical example. Suppose we wish to answer the following question : "Will a 50 percent reduction of SO_2 emission from source A (or region A) result in a corresponding 50 percent decrease in downwind sulfate formation ?". There is no unique answer to this question. First, the manner in which the emission reduction is achieved is important. If source A is a coal-fired power plant, and the reduction in SO_2 emission is achieved by a 50 % reduction in the amount of fuel burned, there may be an accompanying reduction in NO_x emissions which, in turn, will cause k_s to be different. The answer to the question, therefore, is "no". The cause of this apparent of effective nonlinearity is the indirect dependence of k_s on SO_2 through the correlation between co-emitted SO_2 and NO_x. The 50 percent reduction in SO_2 emission could also have been achieved by the use of fuel of 50 percent lower sulfur content or by scrubbing SO_2 from the combustion products prior to stack emission. To the extent that these latter procedures may not have changed NO_x emissions, k_s will remain unchanged except during initial transport, and the downwind sulfate formation would be expected to decrease by about 50 percent, all other conditions being the same. The answer to the question is therefore "yes".

A second factor which will profoundly influence downwind sulfate formation is the composition of the air which the plume encounters during mesoscale and long range transport. There is field evidence to suggest that the role of co-emitted species may be substantially enhanced, or overwhelmed, by the role of the background air which the plume entrains by mixing processes. Like the co-emitted species, a polluted background can also serve as the source of the oxidizing agents. Figure 1 shows an example of the side-by-side transport of two St. Louis plumes of very different emission composition, yet comparable secondary formations. The Labadie power plant emission is characterized by a very low HC/NO_x ratio. The urban plume of St. Louis, including the emissions from a large petroleum refinery complex, by contrast is much richer in reactive HC emissions. The secondary formation of ozone in large plumes on summer days is closely related to the formation of sulfates (White et al. 1976, Gillani and Wilson 1980). The formation of ozone and sulfates in power plant plumes at rates comparable to those in urban plumes is due to the entrainment of polluted background air. During long-range transport, the role of the background air may well predominate as a source of reactive species which oxidize SO_2. In laboratory measurements with no role of a variable background, a first order dependence of sulfate formation on SO_2 concentrations has been observed for gas-phase reactions (Miller 1978) as well as liquid-phase reactions (Penkett et al. 1979). Mesoscale field measurements are also generally consistent with pseudo-first-order dependence between r_s and SO_2, except during early transport.

A common practice in detailed models of sulfur chemistry in the atmosphere is to represent the SO_2 + OH reaction as a terminal reaction, effectively leading to the formation of H_2SO_4 and depletion of the OH radical concentration. This dependence of OH on SO_2 therefore contributes to nonlinearity of the sulfur transformation process. It has been pointed out recently (Stockwell and Calvert 1983) that the SO_2-OH reaction may initiate a chain of reactions which may lead not only to formation of H_2SO_4, but also to regeneration of OH in the presence of NO_x and hydrocarbons. Such recycling of OH would have the effect of weakening the nonlinearity of sulfur chemistry. An important conclusion of the recent NAS report on acid deposition (NAS 1983) was indeed that nonlinearity of sulfur chemistry is probably quite weak, and that even the broader coupling between sulfur emissions and depositions may be substantially linear on a long-term average basis over the spatial scale of eastern North America.

Based on theoretical considerations, the relationship between r_N and NO_x is expected to be nonlinear, since k_N depends on OH, for example, which depends directly on the NO_x chemistry. Results of recent smog chamber experiments suggest, however, that the nonlinearity of r_N may also be shortlived relative to the time scale of long range transport (Spicer 1983). Pseudo-first-order parameterizations of r_N may be justifiable, but k_N may also need to reflect the make-up of the air which an emission encounters during transport.

SOME SPECIFIC MODELS AND THEIR APPLICATIONS

Detailed Chemical Simulations

Detailed chemical modules based on the explicit mechanisms approach have been used within Eulerian as well as Lagrangian formulations, and in model applications at the plume scale as well as the regional scale. Such transformation modules differ principally in terms of their representations of the hydrocarbons, and in the methods used for the numerical solution of the set of nonlinear differential equations describing the species concentration changes by chemical reactions. The following discussion outlines some specific representative models, and is not intended as an extensive review of chemical models.

The LIRAQ model (McCracken et al. 1978, Duewer et al. 1978) is an example of a two-dimensional grid model (single well-mixed vertical layer). The transformation module attempts to simulate photochemical smog formation based on the HSD scheme (Hecht et al. 1974), and the numerical solution is based on the Gear technique. The SAI Airshed Model (Reynolds et al. 1979) is a three-dimensional grid model which permits initial isolation of elevated point sources from surface sources. It uses the carbon bond mechanism of photochemical smog simulation (Whitten and Hogo 1977), and numerical solution is by a finite difference technique (SHASTA) developed by Boris and Book (1973). An ambitious three-dimensional regional grid model currently under development at EPA (Lamb 1981) presently uses the chemical scheme of Demerjian and Schere (1979) which uses four hydrocarbon classes of different reactivities. In some regional models (e.g., McRae et al. 1979), point source plumes are simulated in a Lagrangian sense within the framework of an Eulerian grid network until they attain the dimensions of the grid cell. Thereafter, the simulation is continued in the Eulerian frame.

On a global basis, the troposphere is presumed to be clean and the organic species most relevant to smog formation are carbon monoxide (CO) and methane (CH_4). Recently, a two-dimensional global model was employed by Fishman and Drutzen (1978) to predict the global distribution of OH, HO_2, and CH_3O_2 radical concentrations. Predicted OH concentrations were reasonably comparable with recent, measured atmospheric concentrations (Sheppard et al. 1978).

Altshuller (1979) used this model for OH to investigate the variability of the sulfate formation rate by the homogeneous gasphase mechanism with respect to latitude and altitude. His results showed that in the clean environment, OH is the principal oxidizing agent, and that, at higher latitudes, e.g., in the northeastern United States, Canada, and northern Europe, large seasonal differences in sulfate formation by this mechanism are to be expected. Very little sulfate formation is likely in winter by gas-phase mechanisms.

The regional model of Carmichael and Peters (1979) is based on the chemistry of a clean background in which the only organic species are CO and CO_2. They invoke the pseudo-steady-state assumption for the oxidizing species OH, HO_2, H_2O_2, and O_3, and use their iterative solution for these species in first order expressions for the oxidation of SO_2 to estimate the sulfate formation rate.

Most plume simulations are based on trajectory-type models. Calculations made for polluted industrial regions and urban areas have simulated certain observed phenomena related particularly to O_3 behavior (Greadel et al. 1978) but at the same time have yielded conflicting results concerning important control strategies. Results by Graedel et al. (1978) suggest OH levels to be directly proportional to NO_2 levels, implying that reduction of NO_x emissions would help control nitrate and sulfate production. Miller (1978) showed rather that NO_x emissions tend to delay SO_2 oxidation and that the ratio ($NMHC/NO_x$) of initial concentrations of nonmethane HC's and NO_x's dominates the SO_2 oxidation rate. Miller's conclusions were verified experimentally. Actually, as suggested by Miller (1978), precursor effects may significantly differ in the first several hours of daytime plume transport from their effects during subsequent regional transport.

Detailed chemical calculations also have been applied to simulate sulfate and nitrate formation in urban plumes (Isaksen et al. 1978, Miller and Alkezweeny 1980, Bazzell and Peters 1981) and in power plant plumes (Miller et al. 1978, Bottenheim and Strausz 1979, Levine 1980, Hov and Isaksen 1981, Steward and Liu 1981). In these calculations, proper simulations of the changing background air and of plume-background interactions were necessary for at least qualitative agreement with field observations. Levine (1980) neglected plume-background interactions and, as a result, his conclusion that power plant plume dilution inhibits sulfate formation is contrary to field observations in moderately polluted regions (Gillani and Wilson 1980). Hov and Isaksen (1981), on the other hand treated crosswind spatial inhomogeneities in sulfate formation resulting from plume-background interaction and succeeded in simulating, at least qualitatively, many features of the crosswind plume data of Gillani and Wilson. Stewart and Liu (1981) similarly provided cross-wind resolution and plume-background interactions with their reactive plume model which was based on the carbon bond mechanism for the simulation of chemical kinetics. Recently, Hov (1983b) performed a plume simulation in which vertical stratification of the concentration field was considered. In general, plume simulations have indicated that O_3 and aerosol formation are greater when the background is polluted, that OH is the dominant oxidizing species for SO_2 and NO_2, and that OH and peroxy radical (HO_2, RO_2) concentrations, which play an important role in O_3 formation, peak at midafternoon in polluted regions.

In all of the above simulations, only the homogeneous gas-phase chemistry was included. Rodhe et al. (1979) added reactions of SO_2 and NO_2 with H_2O_2 in the presence of "clouds" to a highly lumped gas-phase chemistry model. H_2O_2 generation was calculated based on the gas-phase reactions. The authors recognized qualitatively that the effective rate constants for cloud reactions must include not only the effect of the liquid-phase transformations occurring in cloud droplets and in precipitating clouds, but also exchange rates of the reacting species between the droplets and the surrounding air, and the frequency and occurrence of clouds and precipitation. They then proceeded to choose rate constant values such that overall gas- and liquid-phase oxidation rates of SO_2 became comparable and the liquid phase oxidation of NO_2 became relatively insignificant compared to its gas-phase counterpart. This procedure for the liquid-phase mechanism represents a highly parameterized approach, with parameter values assumed rather subjectively. Their calculations were applied regionally to the European industrial environment under summertime conditions. The relative contributions of gas-phase and liquid-phase mechanisms to sulfate and nitrate formation, of course, reflected their assumptions.

Parameterized Models

For many years, no consensus could be reached concerning the relative importance of the many chemical and meteorological factors implicated as influencing gas-to-particle S conversion. Most transport-transformation models used constant pseudo-first-order rates for the oxidation of SO_2. Documentation of sunlight as a dominant environmental factor governing sulfate formation in power plant plumes (Gillani et al. 1978) has since been verified and widely accepted and used. In particular, in a recent review of field data on sulfate formation in power plant plumes during all seasons in the United States, Canada, and Australia, Wilson (1981) observed that the outstanding common pattern in this broad data base was the diurnality of the sulfate formation directly related to solar radiation. Such a role of sunlight is also consistent with the observed distinct summer peak in regional SO_4^{2-} distribution in the eastern United States (Husar and Patterson 1980), even though corresponding SO_2 emissions are distributed fairly uniformly over all seasons (DOE 1979).

A sunlight-dependent model of the form $k_s \propto R_T$, the total incoming solar radiation flux at ground level, was used by Gillani (1978) in a diagnostic mesoscale plume model and by Husar et al. (1978) in a multiday plume S budget study. A similar parameterization has been used by Shannon (1981) and by others. Gillani found that such a model based only on sunlight could not simulate the observed day-to-day variation in sulfate formation. Evidently, factors other than sunlight must be included. Also, the manner in which sunlight influences the conversion process must be more carefully considered. As Wilson (1981) noted, observed correlations of the conversion rate with sunlight, or with air temperature (Eatough et al. 1981), do not imply the direct role of these factors in the underlying mechanism. These two factors are highly correlated, as are both to turbulent mixing, convective cloud formation, and a number of other factors, which alone can exert rate-controlling influences on specific conversion mechanisms. Accordingly, formulation of meaningful parameterizations must be based on mechanistic considerations.

Gillani et al. (1981) recently advanced a parameterization of the gas-to-particle S conversion by the gas-phase mechanism based on plume data collected during the summer in the Midwest (Missouri and Tennessee). The motivation for their gas-phase parameterization was derived from their earlier identification of a current pattern of O_3 and aerosol generation in power plant plumes, which evidently involved participation of reactive species entrained from the background (Gillani and Wilson 1980). Gillani et al. argued that accelerated photochemical generation of the radical species OH, HO_2 and RO_2 that oxidize gas-phase SO_2 would be facilitated by reactions involving NO_x emissions and entrained reactive HC and free radical species. Consequently, the quality of the background air and the extent of plume dilution by its entrainment were judged to be important contributing factors, in addition to sunlight which powers the photochemical reactions. Given the lack of detailed data of the oxidizing species, the authors resorted to using O_3 as a surrogate for, or an indicator of, air mass reactivity. Vertical plume spread, Δz_p, was chosen as a measure of the extent of plume dilution. The resulting gas-phase parameterization is :

$$k_{s_G} \cong (.03 \pm .01) R_T \cdot (\Delta z)_p \cdot (O_3)_o \, , \qquad (6)$$

where k_{s_G} is in percent hr^{-1}, R_T is in kW m^{-2}, $(\Delta z)_p$ is in meters, and background ozone, $(O_3)_o$, is in ppm. The coefficient 0.03 ± 0.01 was chosen on the basis of the best fit between the calculated (Equation 6) and measured values of k_{s_G}. The measured values were for dry (relative humidity < 75 percent), cloudless conditions when gas-phase reactions may safely be assumed to predominate. The parameterization was validated successfully by data collected in the plumes of three large central power generating stations in Missouri and Tennesse during two different summers. The emperical coefficient (0.03) thus pertains to such large power plant plumes in which the initial NO_x/SO_2 ratio is about 1:3.

The above parameterization is believed to provide good estimates of the gas-phase sulfate formation rate under the moderately polluted conditions characteristic of the eastern United States in summer and appears to be valid even under more polluted conditions during stagnation episodes. Its validity in winter, even in this region, remains to be tested. Its performance in clean regions such as the Southwest, and in extremely polluted areas such as Los Angeles, CA, on a smoggy day is also unproven. Furthermore, the parameterization has no validity for urban plumes and possibly also plumes from small power plants owing to substantially different composition of the emissions. In spite of these restrictions, the parameterization is of practical significance. Its input requirements are minimal and can be satisfied presently over a regional scale in the eastern United States. Its explicit inclusion of plume-background interactions and air mass conditions probably gives it some validity even during long-range transport when the role of the background is expected to be dominant. Application of the parameterization based on 1976 St. Louis, MO, data of the input variables yields the diurnal and seasonal pattern of k_{s_G} as shown in Figure 2. The magnitudes and temporal variations shown are plausible and consistent with available field data, as well as with expectations based on detailed chemical calculations (Calvert et al. 1978, Altshuller 1979). The results predict that in the Midwest, gas-phase mechanisms may be expected to convert about 10 to 20 percent of the SO_2 in a power plant plume to SO_4^{2-} during an average summer day, while corresponding conversion in winter may be about an order of magnitude smaller. By comparison, measured values of SO_2 to SO_4^{2-} conversion by all mechanisms range between 15 and 35 percent for summer conditions in the same region (Gillani and Wilson 1983). It may be inferred, therefore, that liquid-phase mechanisms may convert about 5 to 15 percent of the SO_2 to SO_4^{2-} per day during summer in the Midwest.

Gillani et al. (1983) have recently also made a first attempt to formulate a parameterization of liquid-phase SO_4^{2-} formation resulting from plume-cloud interactions. The formulation explicitly recognizes that the overall conversion rate, k_{s_L}, depends not only on the chemical reaction rate within cloud droplets, k_{s_L}, but also on the physical extent of plume-cloud interactions. Because clouds are discrete entities in space and time, and plume-cloud interactions are somewhat random events, the authors choose to describe plume-cloud interactions in probabilistic terms. The overall formulation has the general form

$$k_{s_L} = P.\ K_{S_L} \tag{7}$$

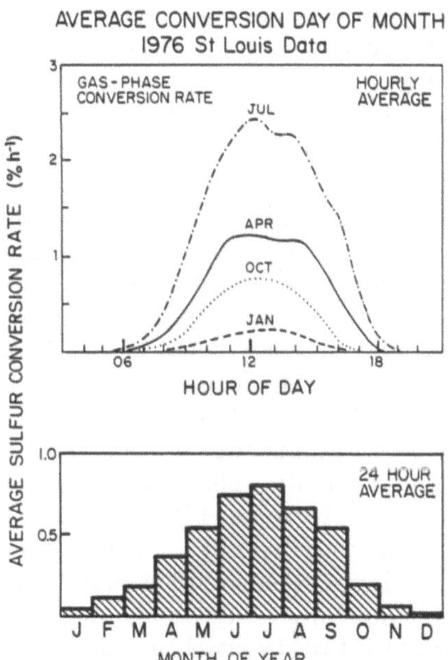

Fig. 2. Diurnal and seasonal patterns of the gas-to-particle conversion of sulfur by gas-phase mechanisms, according to calculations based on the parameterization of Gillani et al. (1981).

where P represents a measure of the probability and extent of plume-cloud interactions. All three quantities in the equation are time dependent. The dependence of P on local plume and cloud dimensions has been derived explicitly (details given in original reference), and its values are determined during an actual power plant plume model run based on current, calculated plume dimensions and local cloud data from surface weather observations of the National Weather Service network of stations, as well as on local lidar and aircraft measurements. P represents a measure of the fraction of a given plume volume which is in contact with the liquid phase.

The authors did not attempt to parameterize K_{S_L}. It depends on such variables as liquid water concentration, droplet pH, and concentrations of dissolved S, oxidizing agents (H_2O_2, and O_2), and catalysts (Fe and Mn). No data were available for such cloud chemical composition. The authors did, however, obtain an average daytime estimate for K_{S_L} under typical summertime fair-weather convective cloud conditions in the Kentucky-Tennessee area. The inferred value of K_{S_L} (summer daytime average conversion rate within clouds) was 12 ± 6 percent hr^{-1}. This value compares with values of 0 to 104 percent hr^{-1} estimated by Hegg et al. (1980), based on ambient SO_2 and SO_4^{2-} measurements in wave cloud situations and with predicted values ranging from 10 to 20 percent hr^{-1} in large storm cloud systems in the summer based on an indirect mass balance technique (Scott 1981). Also, the value of P averaged over 24 hr is expected to be significantly less than 0.1 during summer as well as winter. In other words, the average bulk plume conversion rate by liquid-phase mechanisms is likely to be less than the local droplet-phase conversion rate by more than an order of magnitude. All of these estimates involve several assumptions and approximations and must be used with caution. Values of K_{S_L} at night and in winter are believed to be substantially smaller as a result of lower concentrations of the photochemically generated oxidizing species, O_3 and H_2O_2.

Based on the above parameterizations and St. Louis, MO data, it is estimated that the 24-hr average, overall sulfate formation rates in July are likely to be 0.8 ± 0.3 percent hr^{-1} by gas-phase reactions and at least 0.4 ± 0.2 percent hr^{-1} by liquid-phase reactions. Winter rates by gas-phase reactions are estimated to be an order of magnitude smaller than in summer and by liquid-phase reactions are estimated to be comparable during the two seasons.

A variety of empirical data suggests that liquid-phase conversions in wetted aerosols may be significant at relative humidity between 75 and 100 percent (Dittenhoefer and de Pena 1980, McMurry et al. 1981). Winchester (1983) has formulated the following empirical parameterization of k_s which highlights the role of absolute humidity and temperature :

$$k_s \propto (P_{H_2O})^{3.08} \; (P_{H_2O,sat})^{1.213} \qquad (8)$$

where P_{H_2O} denotes the partial pressure of water vapor, and $P_{H_2O,sat}$ denotes the saturation vapor pressure of water vapor (a measure of temperature).

No comparable parameterizations of NO_x transformations have been formulated. Summertime plume measurements suggest that NO_3 formation is primarily in the form of HNO_3 vapor (Forrest et al. 1979, 1981 ; Hegg and Hobbs 1979 ; Richard et al. 1981) and that oxidation of NO_2 to HNO_3 may proceed about three times faster than does oxidation of SO_2 to H_2SO_4 (Forrest et al. 1981, Richards et al. 1981). Gas-phase mechanisms of HNO_3 formation are believed to predominate in the summer.

Whitby recently used a simple model assuming the total accumulation mode aerosol formation rate to be directly proportional to UV radiation intensity, to simulate observations of aerosol formation in the St. Louis, MO, urban plume of 18 July 1975. He estimated that about 1000 tons of secondary fine aerosol may be produced in the St. Louis plume in one summer irradiation day (Whitby 1980). For the same plume transport, Isaksen et al. (1978) used a detailed chemical model to simulate the measured data of O_3 and $SO_4{}^{2-}$ formation presented by White et al. (1976) and estimated peak H_2SO_4 and HNO_3 formation rates of 5 and 20 percent hr^{-1}, respectively, to occur in the early afternoon. Alkezweeny and Powell (1977) also measured the St. Louis plume and estimated afternoon $SO_4{}^{2-}$ formation rates to be 10 to 14 percent hr^{-1}. Miller and Alkezweeny (1980) measured $SO_4{}^{2-}$ formation rates in the Milwaukee urban plume, particularly related to the quality of the background air mass, to range from 1 to 11 percent hr^{-1}.

Spicer (1977) estimated the NO_2-to-products transformation rate in the Los Angeles urban plume as 10 ± 5 % hr^{-1}. In more recent measurements downwind of Los Angeles (Spicer et al. 1979), the observed lower limit of NO_x conversion rates ranged from 1 to 16 % hr^{-1}, with typical rates in the 5 to 10 percent hr^{-1} range. Spicer (1980) estimated NO_x transformation/removal rate for the Phoenix urban plume to be less than 5 percent hr^{-1}, while data for Boston showed rates in the 14 to 24 percent hr^{-1} range. Transformation products of NO_x transformations include not only inorganic nitrate (e.g., HNO_3), but also organic species (e.g. PAN). Spicer attributes the low conversion rate in Phoenix at least partly to thermal decomposition of PAN and its analogs at the high ambient temperatures of the desert area.

Recently, Middleton et al. (1980) performed a model study of relative amounts of sulfate production in wetted aerosols in a polluted environment by two different mechanisms : condensation of SO_2 gas-phase oxidation products, and catalytic and noncatalytic SO_2 oxidation in the liquid phase. The microphysical vapor transfer to the aerosols and the chemical conversion within the aerosols were treated as coupled kinetic processes. Concentrations of the oxidizing species (e.g., OH, and H_2O_2) and of the catalysts (e.g. Fe, Mn and soot) were assumed known, and representative values for day and night and summer and winter were used. The study concluded that in the daytime, photochemical reactions and liquid-phase oxidation by H_2O_2 are likely to predominate, with particle acidity playing a minor role. At night, sulfate production rates are low, being principally by catalytic liquid-phase mechanisms involving O_3 and O_2. The daytime H_2O_2 reaction rate was enhanced by the lower winter temperatures.

SUMMARY AND CONCLUSIONS

Transformation models can, at best, be only as good as our understanding of the transformation processes. Significant gaps in this understanding remain, particularly with respect to physical and chemical kinetics of the liquid-phase processes. The validity and extrapolation of laboratory results to real atmospheric conditions are often questionable. Field measurements, in simultaneous physical and chemical measurements pertaining to plume-cloud interactions are almost nonexistent.

Detailed chemical models are not yet practical for application in regional models to predict acidic product formation and deposition. Many individual pieces of information - microphysical pathways and chemical reactions - must be put together correctly and we are still struggling to assemble an adequate information base about the individual pieces. To complicate matters, important couplings exist between the different major mechanisms of sulfate and nitrate formation (e.g., H_2O_2 formed by gas-phase photochemistry is of paramount importance in liquid phase chemistry), and significant interdependences exist among the major influencing environmental factors.

Considerable progress has been made in transformation modeling in recent years. Significant gaps remain, however, in our ability to predict transformation rates of SO_x and NO_x under atmospheric conditions. The following observations summarize the current status of the principal aspects of transformation modeling :

- It is now possible to simulate the principal features of the smog chamber chemistry of the SO_x-NO_x-HC system rather accurately by detailed modeling of the chemical kinetics based on lumped representations of the hydrocarbons, even though details of the chemical mechanisms are not fully understood.

- Detailed chemical models of plume transformations under atmospheric conditions have successfully simulated many qualitative features of field observations, including some details of crosswind profiles influenced by plume-background interactions. These simulations are mainly restricted to gas-phase chemistry.

- The principal current limitations in detailed chemical modeling are probably related to inadequate characterization of the emission field and of the ambient polluted regional background. Improved and more detailed inventories of the emissions of SO_x, NO_x, and HC from major sources including the urban area sources, and reliable measurements of reactive species (e.g., OH, RO_2, H_2O_2) in the ambient atmosphere are needed before reliable conclusions concerning the regional-scale transformation processes can be made. The relative importance of co-emissions vs background entrainment as sources of oxidizing agents (OH, RO_2, H_2O_2, etc.) is not fully understood at the present time.

- Current detailed chemical models generally do not include liquid-phase chemistry. Quantitative descriptions of the liquid-phase environment (e.g., cloud dynamics, plume-cloud interaction, etc.) are not adequately incorporated into transformation models. Cloud and fog chemistry measurements are sparse and much needed. Coupled modeling of gas- and liquid-phase chemistry is necessary, particularly under summer conditions. First steps in this direction have been taken.

- For the near future, it appears that transformation modules based on empirical parameterizations will continue to predominate in operational regional models. All models, to varying degrees use parameterizations based on laboratory and field data. Currently, regional models mostly employ pseudo-first-order or constant first order bulk conversion rates. Preliminary parameterizations have been developed only for sulfate formation in power plant plumes, and will undoubtedly continue to be improved. No practical parameterizations exist yet for nitrate formation or for urban plumes.

- The basis for refining these estimates to reflect at least the gross diurnal and seasonal variations, and even the role of a changing background, exists. Increasingly, new models are incorporating such empirical expressions, which are constantly being improved. The state-of-the-art of such parameterizations will be further advanced as more data are obtained and analyzed. Detailed chemical models can serve to improve our understanding and basis for the formulation of empirical parameterizations which reflect the underlying physical-chemical processes rather than merely expressing statistical correlations. Adherence to mechanistic considerations is recommended in formulating the parameterizations. More, and more reliable, measurements of such important variables as the atmospheric concentrations of OH, H_2O_2, NH_3, HC's, SO_4^{2-} and NO_3^- and of cloud dimensions and cloud chemical composition are needed direly.

H_2SO_4 and HNO_3 formations apparently peak during daytime and in summer. Gas-phase mechanisms are believed to contribute a larger share, on the average, to these secondary formations under warm, sunny conditions. Typically, on a summer day (24 hr) in the eastern United States, about 25 $\pm$ 10 percent of the airborne SO_2 in power plant plumes is likely to be converted to sulfates. Nighttime conversion is a small part (about 5 percent or less). S transformations may be somewhat higher than these in the southeastern United States. HNO_3 formation rate in power plant plumes is about three times as fast as the sulfate formation rate by gas-phase mechanisms. Aerosol NO_3^- formation rate is apparently very small, at least in the summer. Both sulfate and nitrate formation are faster in urban plumes.

At this time, the major sources of uncertainty in determining atmospheric transport ranges of pollutants are probably associated with transport and deposition processes rather than with transformation processes.

AKNOWLEDGMENT

The preparation of this paper was funded, in part, by the United States Environmental Protection Agency through Co-operative Agreement CR-809713. The paper has not been subjected to the Agency's peer or policy review, and therefore does not necessarily reflect the views of the Agency, and no official endorsement should be inferred.

REFERENCES

Alkezweeny A.J. and Powell D.D. (1977). Estimation of transformation rate of SO_2 to SO_4 from atmospheric concentration data. Atmos. Envir. 11, 1979-182.

Altshuller A.P. (1979). Model predictions of the rates of homogeneous oxidation of sulfur dioxide to sulfate in the troposphere. Atmos. Envir. 13, 1653-1661.

Bazzell C.C and Peters L.K. (1981). The transport of photo-chemical pollutants to the background troposphere. Atmospheric Environment 15, 957-968.

Boris J.P. and Book D.L. (1973). Flux corrected transport. I. SHASTA, an algorithm that works. J. Comp. Phys. 11, 38-69.

Bottenheim J.W. and Strausz O.P. (1979). The effect of a polluting source on the air quality downwind of prestine northern areas. Atmospheric Environment 13, 1085-1089.

Cahir J.J., Pitts J.N., Ross J., Twomey S.A., and Wiesenfeld J.R. (1982). The source-receptor relationship in acid precipitation : Implications for generation of electric power from coal. Physical Dynamics, Inc. Report No. PD-LJ-82-268R on Workshop of January 18-22.

Calvert J.C., Su F., Bottenheim J.W. and Strausz O.P. (1978). Mechanism of the homogeneous oxidation of sulfur dioxide in the troposphere. Atmos. Envir. 12, 197-226.

Carmichael G.R. and Peters L.K. (1979). Numerical simulation of the regional transport of SO_2 and sulfate in the eastern United States. Preprint, 4th. AMS Symp. Turbulence, Diffusion, and Air Pollution. Reno, NV. January 15-18.

Clark W.C., Landis D.A. and Harker A.B. (1976). Measurements of the photochemical production of aerosols in ambient air near a freeway for a range of SO_2 concentrations. Atmospheric Environment 10, 637-644.

Demerjian K.L., Kerr J.A. and Calvert J.G. (1974). The mechanism of photochemical smog formation. In Advances in Environmental Science & Technology, Vol. 4, 1-262, Wiley, N.Y.

Demerjian K.L. and Schere K.L. (1979). Applications of a photochemical box model for ozone air quality in Houston, Texas. Proc. APCA Conference on Ozone/Oxidants : Interactions with the Total Environment II, Houston, TX. October 14-17.

Dimitriades B., Dodge, M.C., Bufalini, J.J., Demerjian K.L., and Altshuller A.P. (1976). Letter to the Editor. Environ. Sci. Technol. 10, 934-936.

Dittenhoefer A.C. and de Pena R.G. (1980). Sulfate aerosol production and growth in coal-fired power plant plumes. J. Geophys. Res. 85, No. C-8, 4499-4506.

Duewer W.H., McCracken M.C. and Walton J.J. (1978). The Livermore Regional Air Quality Model : II. Verification and sample application in the San Francisco Bay area. J. Appl. Met. 17, 273-311.

U.S. Department of Energy, Monthly Energy Review, DOE/E1A-0035/10(79), October 1979.

Eatough D.J., Richter B.E., Eatough N.L. and Hansen L.D. (1981). Sulfur chemistry in smelter and power plant plumes in the western U.S. Atmos. Envir. 15, 2241-2254.

Falls A.H. and Seinfeld J.H. (1978). Continued development of a kinetic mechanism for photochemical smog. Environ. Sci. Technol. 12, 1398-1406.

Falls A.H., Mc Rae G.J. and Seinfeld J.H. (1979). Sensitivity and uncertainty of reaction mechanisms for photochemical air pollution. Int. J. Chem. Kinetics 11, 1137-1162.

Farrow L.A. and Edelson D. (1974). The steady-state approximation : Fact or fiction ? Int. J. Chem. Kinetics 6, 787-800.

Fishman J. and Crutzen P.J. (1978). The distribution of hydroxyl radical in the troposphere. Atmos. Sci. Paper No. 284, Colorado State University.

Forrest J., Garber R. and Newman L. (1979). Formation of sulfate, ammonium and nitrate in an oil-fired power plant plume. Atmos. Environ. 13, 1287-1297.

Forrest J., Garber R. and Newman L. (1981). Conversion rates in power plant plumes based on filter pack data - Part I : The coal-fired Cumberland plume. Atmos. Environ. 15, 2273-2282.

Freiberg J. (1974). Effects of relative humidity and temperature on iron-catalyzed oxidation of SO_2 in atmospheric aerosols. Environ. Sci. Technol. 8, 731-734.

Gear C.W. (1971). Chapter 11 in Numerical Initial Value Problems in Ordinary Differential Equations. Prentice-Hall, Englewood Cliffs, NJ.

Gillani N.V. (1978). Project MISTT : Mesoscale plume modeling of the dispersion, transformation, and ground removal of SO_2. Atmos. Environ. 12, 569-588.

Gillani N.V., Husar R.B., Patterson D.E. and Wilson W.E. (1978). Project MISTT : Kinetics of particulate sulfur formation in a power plant plume out to 300 km. Atmos. Environ. 12, 589-598.

Gillani N.V. and Wilson W.E. (1980). Formation and transport of ozone and aerosols in power plant plumes. Annals N.Y. Acad. Sci. 338, 276-296.

Gillani N.V., Kohli S. and Wilson W.E. (1981). Gas particle conversion of sulfur in power plant plumes : I. Parameterization of the conversion rate for dry, moderately polluted ambient conditions. Atmos. Environ. 15, 2293-2313.

Gillani N.V. and Wilson W.E. (1983). Gas-to-particle conversion of sulfur in power plant plumes : II. Observations of liquid phase conversions. Atmos. Environ. 17, 1739-1752.

Gillani N.V., Colby J.A. and Wilson W.E. (1983). Gas-to-particle conversion of sulfur in power plant plumes : III. Parameterization of plume-cloud interactions. Atmos. Environ. 17, 1753-1764.

Graedel T.E., Farrow L.A. and Weber T.A. (1976). Kinetic studies of the photochemistry of the urban troposphere. Atmos. Environ. 10, 1095-1116.

Graedel T.E., Farrow L.A. and Weber T.A. (1978). Urban kinetic calculations with altered source conditions. Atmos. Environ. 12, 1403-1412.

Hecht T.A., Seinfeld J.H. and Dodge M.C. (1974). Further development of generalized kinetic mechanism for photochemical smog. Environ. Sci. Technol. 8, 327-339.

Hegg D.A. and Hobbs P.A. (1979). Some observations of particulate nitrate concentration in coal-fired power plant plumes. Atmos. Environ. 13, 1715-1716.

Hesstvedt E., Hov O. and Isaksen I. (1978). Quasi-steady-state approximations in air pollution modeling : Comparisons of two numerical schemes for oxidant prediction. Int. J. Chem. Kinetics 10, 971-994.

Hidy G.M. (1982). Potential fallacy in assuming linear proportionality between SO_2 emissions and acid deposition. Paper presented at Second National Symposium on Acid Rain, Pittsburgh, PA, October 6-7.

Hov O. (1983a). Numerical solution of a simplified form of the diffusion equation for chemically reactive atmospheric species. Atmos. Environ. 17, 551-562.

Hov O. (1983b). One-dimensional vertical model for ozone and other gases in the atmospheric boundary layer. Atmos. Environ. 17, 535-550.

Hov O. (1983c). Aspects of the parameterization of transformation and removal processes in air quality modeling. Paper presented at the 14th NATO/CCMS Technical Meeting on Air Pollution Modeling and Its Applications, Copenhagen, Sept. 24-27.

Hov. O. and Isaksen I.S.A. (1981). Generation of secondary pollutants in a power plant plume : A model study. Atmos. Environ. 15, 2367-2376.

Hov. O., Isaksen I.S.A. and Hesstvedt E. (1977). Diurnal variations of ozone and other pollutants in an urban area. Report No. 24, Institute for Geophysics, Univ. of Oslo.

Husar R.B., Patterson D.E., Husar J.D., Gillani N.V. and Wilson W.E. (1978). Sulfur budget of a power plant plume. Atmos. Environ. 12, 549-568.

Husar R.B. and Patterson D.E. (1980). Regional scale air pollution : sources and effects. Annals N.Y. Acad. Sci. 338, 399-417.

Isaksen I.S.A., Hesstvedt E. and Hov O. (1978). A chemical model for urban plumes : test for ozone and particulate sulfur formation in St. Louis urban plume. Atmos. Environ. 12, 599-604.

Jeffries H.E. and Saeger M. (1976). Letter to the Editor. Environ. Sci. and Technol. 10, 936-937.

Lamb R.G. (1981). A regional scale (1000 km) model of photochemical air pollution : I. Theoretical formulation. U.S. EPA Technical Report. In press.

Levine S.Z. (1981). A model for stack plume reactions with atmospheric dilution (SPREAD). Atmos. Environ. 15, 2573-2581.

Levine S.Z. and Schwartz S.E. (1982). Construction and testing of a Surrogate CHemical MEchanism (SCHEME) for tropospheric photochemical reactions. Chap. 11. In Trace Atmos. Const., S.E. Schwartz, ed. Vol. 12 in Advances in Environ. Sci. and Technol. John Wiley & Sons, Inc., New York.

McCracken M.C., Wuebbles D.J., Walton J.J., Duewer W.H., and Grant K.E. (1978). The Livermore Regional Air Quality (LIRAQ) model : I. Concept and development. J. Appl. Met. 17, 254-272.

McMurry P.H., Rader D.J., and Stith K. (1981). Growth laws for secondary aerosols in power plant plumes : Implications for chemical conversion mechanism. Atmos. Environ. 15, 2315-2327.

McRae G.J., Gooding W.R. and Seinfeld J.H. (1979). Development of a second-generation airshed model for photochemical air pollution. Proc. 4th AMS Symp. on Turbulence, Diffusion and Air Pollution, Reno, Nevada, January 15-19.

Middleton P., Kiang C.S. and Mohnen V.A. (1980). Theoretical estimates of the relative importance of various urban sulfate aerosol production mechanisms. Atmos. Environ. 14, 463-472.

Miller D.F. (1978). Precursor effects of SO_2 oxidation. Atmos. Environ. 12, 273-280.

Miller D.F., and Alkezweeny A.J., Hales J.M. and Lee R.N. (1978). Ozone formation related to power plant emissions. Sci. 202, 1186-1188.

Miller D.F. and Alkezweeny A.J. (1980). Aerosol formation in urban plumes over Lake Michigan. Annals. N.Y. Acad. Sci. 338, 219-232.

National Academy of Sciences (1983). Report on Acid Deposition - Atmospheric Processes in Eastern North America, National Academy Press, Washington D.C.

OECD (1977). The OECD programme on long range transport of air pollutants : measurements and findings. Final Report. Organization for Economic Co-operation and Development, Paris 1977.

Penkett S.A., Jones B.M.R., Brice K.A. and Eggleton A.E.J. (1979) : The importance of atmospheric ozone and hydrogen peroxide in oxidizing sulfur dioxide in cloud rainwater. Atmos. Environ. 13, 123-137.

Reynolds S.D., Tesche T.W. and Reid L.E. (1979). An introduction to the SAI airshed model and its usage. SAI Technical Report EF 79-31. Systems Applications, Inc., San Rafael, CA.

Richards L.W., Anderson J.A., Blumenthal D.L., Brandt A.A., Mc Donald J.A., Watus N., Macias E.S. and Bhardwaja P.S. (1981). The chemistry, aerosol physics, and optical properties of a western coal-fired power plant. Atmos. Environ. 15, 2111-2134.

Rodhe H., Crutzen P. and Vanderpol A. (1979). Formation of sulfuric and nitric acid in atmosphere during long range transport. Proc. WMO Symposium on Long Range Transport of Pollutants, Sofia, Bulgaria, October 1-5, 1979.

Schwartz S.E. and Newman L. (1978). Processes limiting oxidation of sulfur dioxide in stack plumes. Environ. Sci. Technol. 12, 67-73.

Scott B.C. (1981). Predictions of in-cloud conversion rates of SO_2 to SO_4 based upon a simple chemical and dynamical model. Atmos. Environ. (in press).

Sheppard J.C., Campbell M.J. and Au B. (1978). Boundary layer hydroxyl measurements by a ^{14}C tracer technique. Presented before the Div. Environmental Chemistry, American Chemical Society, Miami Fl., September 11-14.

Spicer C.W. (1977). The fate of NO_x in atmosphere. In Adv. Environ. Sci. Technol. 1, 163-261. (Pitts J.N. and Metcalf R.L., eds. Wiley).

Spicer C.W., Joseph D.W., Sticksel P.R., Sverdrup G.M. and Ward G.F. (1979). Reactions and transport of nitrogen oxides and ozone in the atmosphere. Battelle-Columbus Report to EPA, 1979.

Spicer C.W. (1980). The rate of NO_x reaction in transported urban air. In Atmospheric Pollution 1980, Proc. 14th Colloq. Paris, May 5-8. M.M. Benarie (ed.), Elsevier Scientific Publishing Company, Amsterdam.

Spicer C.W. (1983). Smog chamber studies of NO_x transformation rate and nitrate-precursor relationship. Environ. Sci. and Tech. 17, 112-120.

Stewart D.A. and Liu M.K. (1981). Development and application of a reactive plume model. Atmos. Environ. 15 (in press).

Stockwell W.R. and Calvert J.G. (1983). The mechanism of the $HO\text{-}SO_2$ reaction. Atmos. Environ. (in press).

Whitby K.T. (1980). Aerosol formation in urban plumes. Annals N.Y. Acad. Sci. 338, 258-275.

White W.H., Anderson J.A., Blumenthal D.L., Husar R.B., Gillani N.V., Husar J.D. and Wilson W.E. (1976). Formation and transport of secondary air pollutants : ozone and aerosols in the St. Louis urban plume. Sci. 194, 187-189.

Whitten G.Z., Hogo H. and Killus J.P. (1980). The Carbon Bond Mechanism : A condensed kinetic mechanism for photochemical smog. Environ. Sci. Technol. 14, 690.

Whitten G.Z. and Hogo H. (1977). Mathematical modeling of simulated photochemical smog. U.S. EPA Technical Report, EPA-6003-77-011.

Wilson W.E. (1978). Sulfates in the atmosphere : a progress report on Project MISTT. Atmos. Environ. 12, 537-547.

Wilson W.E. (1981). Sulfate formation in point-source plumes : A review of recent field studies. Atmos. Environ. 15, 2573-2581.

Winchester J.W. (1983). Sulfur, acid aerosols, and acid rain in the eastern United States. Ch. 6 in Adv. Environ. Sci. Technol., 12, John Wiley and Sons, New York.

DISCUSSION

A. VENKATRAM

In suggesting your empirical formulation for SO_2 to $SO_4^=$ conversion, are you implying that it is easier to predict O_3 than $SO_4^=$? It does not make sense to specify the O_3 concentration pattern in a modeling exercise.

N. GILLANI

My parameterization for sulfate formation in power plant plumes by the gas-phase mechanism requires knowledge of ozone in the back-ground air mass, not in the plume. Since gas-phase chemistry is important during the convective period in the daytime, ground-level measurements of ozone during that time are representative of ozone throughout the mixing layer. Ground level ozone concentrations are routinely monitored at many (several dozen) ground stations in the eastern U.S. Such ozone data from rural sites are quite appropriate indicators of regional ozone background, and hence suitable, necessary and adequate inputs for my parameterization. It may be added that this minimal chemical input compares with the input requirement in detailed chemical models of not only ozone in the background but a large number of other species, including hydrocarbons, OH, and other rarely measured reactive species. The detailed models further require initial values in the plume of all of these species.

T. LAVERY

What measure of mixing is used in the parameterization ?

N. GILLANI

The vertical spread of the plume is used as a surrogate for overall plume dilution.

T. LAVERY So, in a prognostic sense, you need to predict the depth of the mixed layer (or vertical plume dimension) and the O_3 concentration to be able to predict $SO_4^=$ concentration ?

N. GILLANI Yes. I repeat here that ozone concentration is needed only for the background air-mass. The chemical reactivity of the background air, and entrainment of the background air into the plume are critical factors in determining plume chemistry. My parameterization recognizes this critical influence of background reactivity and of atmospheric mixing, and facilitates their inclusion by using the most elementary surrogates of these factors. No regional model worth credence can avoid mixing height as an input or calculated variable. I am further contending that regional models which include plume chemistry but ignore the condition of the prevailing background must be viewed with caution.

PLUME MODEL FOR NITROGEN OXIDES

Christer Persson

Swedish Meteorological and Hydrological Institute

Norrköping, Sweden

ABSTRACT

A local scale dispersion plume model including atmospheric chemistry for nitrogen oxides is developed.

The purpose of the study is to obtain a practical air quality model which can be used in environmental planning for e.g. coal-fired power plants. Problems concerning both high concentrations of NO_2 in the air and deposition of different nitrogen compounds to the ground are considered.

In the numerical model the instantaneous plume dilution is described - assuming total mixing within the plume - as a function of emission parameters and meteorological conditions. The interplay of emissions, instantaneous plume dilution, entrainment of polluted ambient air, meteorological conditions and atmospheric chemistry processes determines the production or loss of the different compounds. On a local scale only six chemical reactions have to be included. On a somewhat larger scale the number of reactions needed increase substantially. For the numerical solution we have used the method by Gear.

Instantaneous plume concentrations and the NO_2 fraction of the total NO_x in the plume at different conditions are obtained directly from the above mentioned calculations. In order to obtain 1-h mean values of different nitrogen oxides and long-term deposition values conventional gaussian respectively K-formulation procedures are used. Meteorological input data are obtained from climatological observations and also generated by a boundary layer model. From the results it is obvious that the meteorological conditions and the ambient ozone concentration are of great importance for the relative amount of NO_2 in the plume.

INTRODUCTION

The environmental impact on a local geographical scale of NO_x-emission from a power plant depends both on the concentration and the chemical nature of the species produced by the emitted material, when it reaches the receptor site. The NO_x-emission consists to 90-95 % of NO but it will more or less rapidly be oxidized in the atmosphere. In the first step it is oxidized mainly to NO_2, then further to e.g. HNO_2 and HNO_3. This has to be considered as the negative effects on health by NO_2 is much more serious than by NO. Also NO_2 is more directly related to the acidification of ground since it is deposited - through dry deposition to vegetation mainly - much more efficiently than NO. First when HNO_2 and HNO_3 are produced the wet deposition becomes important. In studying these problems theoretical models including both atmospheric transport and dispersion and necessary chemical reactions are of interest.

Plume models for nitrogen oxides have been presented by e.g. White (1977) and Peters and Richards (1977). They include only three reactions and assume photostationary state. Carmichael and Peters (1981) apply chemical calculations on the NO-oxidation in an observed plume. Cocks and Fletcher (1982) present a model including a comprehensive chemical schema compled with a very simple method for calculating the plume dispersion.

The present study is aiming at development of a practical air quality model for different nitrogen oxides, which can be used in environmental planning.

PLUME MODEL FOR NITROGEN OXIDES

The instantaneous plume dilution and the simultaneous chemical reactions in the plume are described in a lagrangian coordinate system following the wind i.e. following a trajectory. We start with a segment of the plume, which is moved by the wind away from the source and at the same time the concentrations of different compounds are changed through dilution, entrainment of ambient polluted air and chemical reactions within the plume.

The following basic assumptions are made :

- The distribution of the pollutants is homogeneous ("top hat" concentration distribution) within each cross section of the instantaneous plume.

- Wind speed is constant with height within each cross section of the instantaneous plume.

- The effect of very short time concentration fluctuations on the chemical transformations can be neglected.

In order to describe the changes in concentrations in the instantaneous plume, we start with a simple expression for mass balance for each of the considered compounds. A balance in mass flux for a compound "i" through two cross sections of the plume, see figure 1, separated with a distance Δx can be expressed as :

$$\frac{\Delta C_i}{\Delta t} = P_i + C_{i0} \frac{2R\Delta R + \Delta R^2}{(R + \Delta R)^2} \times \frac{u}{\Delta x} - S_i - C_i \frac{2R\Delta R + \Delta R^2}{(R + \Delta R)^2} \times \frac{u}{\Delta x} \quad (1)$$

where

C_i (x) = concentration of compound "i" in the instantaneous plume at distance x downwind from the source

C_{i0} = background i.e. concentration of compound "i" in the ambient air

P_i = production of compound "i" per unit volume through chemical reactions

S_i = sink of compound "i" per unit volume through chemical reactions

R = plume cross section radii

t = time

u = wind speed at plume level

Thus the transport time between the two cross sections can be calculated from the expression $\Delta x = u \times \Delta t$.

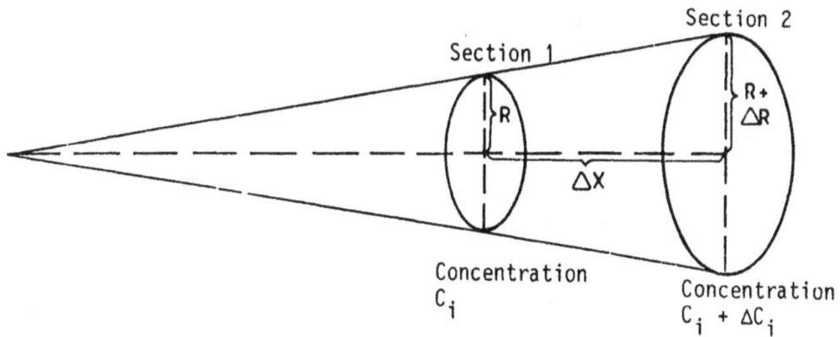

Fig. 1 - A balance in mass flux for a compound "i" through two cross sections of the plume.

If the instantaneous plume cross section area (A) is not circular, R in the expression above can be calculated through $R = \sqrt{A/\pi}$. Thus the concept "radii" still can be used.

Plume dispersion

With the assumption of "top hat" concentration distribution the problem with describing the instantaneous plume dispersion can be reduced to a description of how the plume cross section area and thus R varies. R is a function of the distance from the source and depends on the inherent turbulence within the plume (especially at an initial stage) and the atmospheric turbulence (at larger distances). Different methods have been used to determine R according to table 1. During the initial phase the rate of growth has been calculated from the simple relation $R = \beta \times \Delta h$, where Δh is the plume rise. The equations have been presented by Briggs (1975) and Högström (1978). The constant β has been set equal to 0.65 for unstable conditions, 0.60 for neutral and 0.55 for stable conditions, cf. discussions by Briggs (1975) and Hanna (1982).

Information about the instantaneous plume dispersion parameters were obtained from Högström (1964). On larger distances than 7 km, very little information about instantaneous plume dispersion was found. Therefore σ-values based on ordinary gaussian dispersion studies, i.e. an averaging time of 10 min or more, had to be used. However, at these rather large distances the length of sampling time is probably of somewhat minor importance.

x_t goes to infinity when u goes towards 0. The same is true for x_{ins} and x_s when the stratification becomes close to neutral. This is of course not realistic and has to be handled by means of introducing certain restrictions. At special combinations of emission patterns and meteorological conditions occasionally illogical differences can arise between the sizes of the calculated plume cross-sections at different stabilities, since no general calculation method exist. In the model these problems are - for the time being - handled through comparison between neutral and stable or unstable. If the stratification is stable the smallest R-value is chosen, while the largest value is chosen when the stratification is unstable.

Chemical reactions

The reaction scheme first utilized in the model is presented in appendix. This scheme consists of about 30 reactions with a few "summary" reactions, i.e. incomplete reactions since the intermediate stages are not specified. Calculations however showed that on a local scale (0 - ∿ 20 km) only six reactions were of importance for the NO_x-chemistry. These are given in table 2 together with their rate constants. The photo dissociation rate has been

Table 1. Theoretical expressions used in order to describe the instantaneous plume dilution.

	Unstable	Neutral	Stable
Initial phase (only inherent turbulence)	$0<x\leq x_{ins}$ $x_{ins}=4.44\,\frac{F^{2/5}\times u^{3/5}}{q^{3/5}}$ $q=\frac{W_*^{\,3}}{H}$ $R=\beta\Delta h=\beta\left[\frac{8.33\,F_m\,x}{u^2}+\frac{4.17\,F\,x^2}{u^3}\right]^{1/3}$ (x_{ins} max 4.5 km)	$0<x\leq x_t$ $x_t=\frac{3000\,F}{u^3}$ $R=\beta\Delta h=\left[\frac{8.33\,F_m\,x}{u^2}+\frac{4.17\,F\,x^2}{u^3}\right]^{1/3}$ (x_t max 4.5 km)	$0<x\leq x_s$ $x_s=\frac{4.5\,u}{\omega}$ $\omega=(\frac{g}{\Theta_a}\frac{d\Theta a}{dz})^{1/2}$ $R=\beta\Delta h=\beta\,2.4\,(\frac{F}{uw^2})^{1/3}\times(1+(1+(\frac{wF_m}{F})^2)^{1/2})^{1/3}\times(\frac{x}{x_s})^{1/3}$ $F_m=\frac{\rho_0}{\rho_a}\times\frac{w_0V_0}{\pi}$ (x_s max 4.5 km)
	$x>x_{ins}$	$x>x_t$	$x>x_s$
Continuing phase (inherent turbulence + atmospheric turbulence)	$R^2=R^2(x_{ins})+2\sigma_{zr}\sigma_{yr}$ or $R=\sqrt{\frac{H\,W}{\pi}}$ where $W=\sqrt{2\pi}\sigma_{yr}$	$R^2=R^2(x_t)+2\sigma_{zr}\sigma_{yr}$ or $R=\sqrt{\frac{H\,W}{\pi}}$ where $W=\sqrt{2\pi}\sigma_{yr}$	$R^2=R^2(x_s)+2\sigma_{zr}\sigma_{yr}$

Nomenclature to table 1 :

- x = horizontal coordinate along the plume
- y = horizontal coordinate across the plume
- z = vertical coordinate
- F = bouancy flux of the source = $\frac{V_o\,g}{\pi}\times\frac{T_o-T_a}{T_o}$
- V_o = source volume flux
- T_o = absolute temperature of gases emitted from stack
- T_a = absolute temperature of ambient air
- g = gravitational acceleration
- q = $\frac{w_*^{\,3}}{H}$
- w_* = scaling velocity for a convective mixing layer
- w_* = $[g/T_a\times\overline{w'\Theta'}\times H]^{1/3}$
- $\overline{w'\Theta'}\rho C_p$ = sensible heat flux at the ground
  - ρ = air density
  - C_p = spec. heat of the air
- w = vertical speed of the air
- Θ = potential temperature

prime indicates fluctuations

- H = mixing height
- u = air speed along the x-axis
- Δh = plume rise
- ω = $(\frac{g}{\Theta}\times\frac{d\Theta_a}{dz})^{1/2}$ Brunt-Väisälä frequency
- Θ_a = potential temperature of ambient air
- β = entrainment constant
- σ_{zr} = gaussian vertical dispersion parameter for instantaneous plumes
- σ_{yr} = gaussian horizontal dispersion parameter for instantaneous plumes
- F_m = momentum flux of source
- ρ_o = density of effluent
- ρ_a = density of ambient air
- W_o = efflux velocity of effluent

determined for 60^0N at different times of the day and year according to a method developed by Isaksen et al (1977).

TABLE 2.

Chemical reactions considered in the model (see also appendix). Reaction rates for three component reactions are given in cm^6 $molecules^{-2}$ s^{-1}, for two component reactions in cm^3 $molecules^{-1}$ s^{-1} and dissociation rates in s^{-1}.
T = temperature of the plume (^{0}K).

		Reaction rates
R1	$NO_2 + h\nu \rightarrow NO + O(^3P)$	$j_1 = 0 - 6.2\ 10^{-3}$
R2	$O(^3P) + O_2 + M \rightarrow O_3 + M$	$k_2 = 1.1\ 10^{-34}\ e^{510/T}$
R3	$O_3 + NO \rightarrow NO_2 + O_2$	$k_3 = 2.1\ 10^{-12}\ e^{-1450/T}$
R4	$2NO + O_2 \rightarrow 2NO_2$	$k_4 = 1.5\ 10^{-40}.e^{1780/T}$
R5	$NO + NO_2 + H_2O \rightarrow 2HNO_2$	$k_5 = 6.0\ 10^{-38}$
R6	$2HNO_2 \rightarrow NO + NO_2 + H_2O$	$k_6 = 1.9\ 10^{-11}\ e^{-5000/T}$

The reactions R1-R4 are on this scale by far the most important, which also has been shown in a study by Cocks and Fletcher (1982), where a very large number of reactions were included.

Explicit expression for $(\frac{dc_i}{dt})_{chem} = P_i - S_i$, with notations according to equation 1, were obtained from the reaction scheme. Equation 1 was then integrated numerically using a version of Gear's method given by Hindmarsh (1974). The method is using variable time steps.

Calculations of one hour mean concentration values

Normally in practical applications, one hour mean values of the concentration at fixed geographical locations around a power plant are of prime interest. Those values can be obtained through a combination of :

- information concerning the instantaneous plume, given by equation 1

- information of the concentration in the ambient air

- calculation routines from a conventional gaussian model for one hour sampling time.

The probability, k, for the point (x, y, z) to be within the instantaneous plume is

$$k = \frac{\pi R^2}{2\pi\sigma_y\sigma_z} \cdot \left[e^{\frac{-(z-h_{eff})^2}{2\sigma_z^2}} + e^{\frac{-(z+h_{eff})^2}{2\sigma_z^2}} \right] \cdot e^{\frac{-y^2}{2\sigma_y^2}}$$

where h_{eff} is the effective chimney height and $\sigma_y(x)$ and $\sigma_z(x)$ are dispersion parameters for one hour sampling time. The one hour mean concentration $\bar{c}_i$ is obtained from

$$\bar{c}_i(x, y, z) = k \cdot c_i \text{ (instantaneous plume)} + (1-k)c_{i0} \text{ (ambient)}.$$

Deposition

It is possible to estimate deposition values for different nitrogen oxides rather easily by combining the solution of equation 1 with a conventional K-model. K here stands for the eddy diffusivity.

From the calculated concentrations, $c_i(x)$, we can obtain $Qi(x) = u\iint c_i(x)dydz$, which indicates the "effective source strength" for the compound "i" at the distance x. Subsequently the K-model gives the deposition. When dealing with cases, where the deposition is important, equation 1 should be complemented with a term that subtracts the deposited part given by the K-model.

RESULTS FROM APPLICATIONS TO POWER PLANTS

In the calculations we have assumed the following ambient concentrations, which are typical for rural parts of southern and middle Sweden.

TABLE 3.

	Warm summer	Other periods
O_3	90 ppb	30 ppb
NO	0.5 ppb	0.5 ppb
NO_2	2.0 ppb	2.0 ppb
HNO_2	0.1 ppb	0.1 ppb

In figures 2, 3 and 4 some results are summerized from calculations for a 270 MW coal-fired power plant. The emissions from this power plant was assumed to be

TABLE 4.

	Winter	Spring	Summer
NO	26 g/s	19 g/s	7 g/s
NO_2	1.9 g/s	1.5 g/s	0.6 g/s

This corresponds to a NO_x-concentration of about 170 ppm in the stack (10). Some examples in the figures also refer to a concentration of about 500 ppm (30). Meteorological data typical for southern Sweden is assumed.

During daytime the traditional steady state from reactions R1, R2 and R3 is successively achieved. From figure 2 we see that in this case it is reached after about 4 km during a normal summer day. At night time on the other hand, we get a total oxidation of NO. This occurs after about 10 km transport in the summer, see figure 2, and at a very much greater distance in winter. The slow oxidation in winter is due to the stable stratification and also to some degree to a lower O_3 concentration in the ambient air.

The calculated 1-hour mean concentrations in figure 3 show that the largest NO_2-concentrations are not obtained close to the source during unstable conditions at daytime, which is typical for inert emissions, but rather at night and at a distance of 10-15 km from the source.

Figure 4 illustrates a possible urban fumigation situation during a warm summer night. Outside the city the stratification is very stable while over the urban area, due to the heat island effect, the stratification is slightly unstable.

In figures 5 and 6 some example of results are given from calculations, where the effect of different fuels and sizes of power plants were studied. The presented results were obtained with the assumption of normal meteorological conditions in springtime for southern Sweden and with the emission data in table 5.

The results in figures 5 and 6 refer to daytime conditions only, when the photochemical equilibrium eventually is reached. We can see that this happens at a distance of less than or about 1 km for the very small plants of 0.1 or 1.0 MW. For the 100 MW power plant the calculations indicate that photochemical

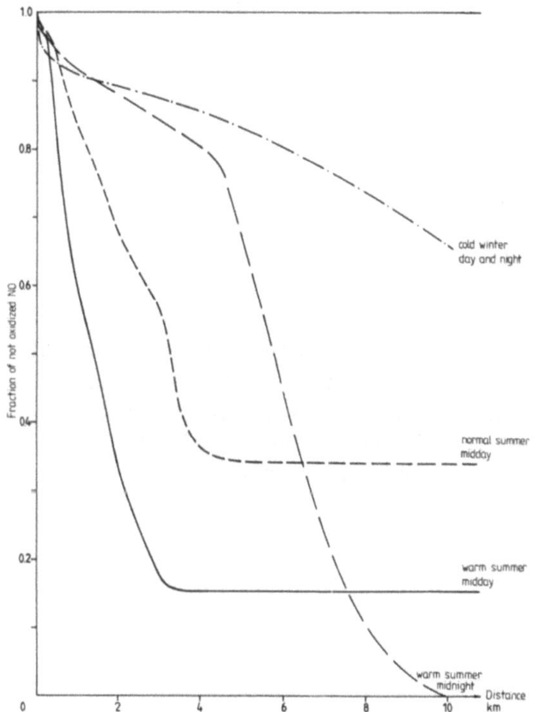

Figure 2. Calculated fraction of not oxidized NO in the emitted plume given as a function of distance from the source. The curves refer to a 270 MW coal-fired power plant for some different meteorological situations.

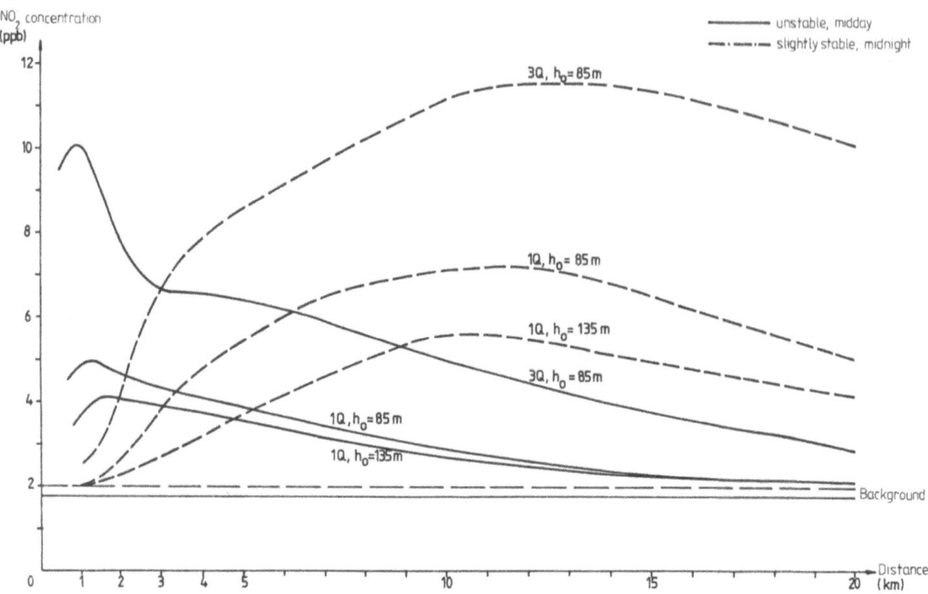

Figure 3. Calculated 1-hour mean concentration at ground level just below the plume given as a function of distance from the source. The results refer to a 270 MW power plant for normal meteorological conditions at springtime in southern Sweden.

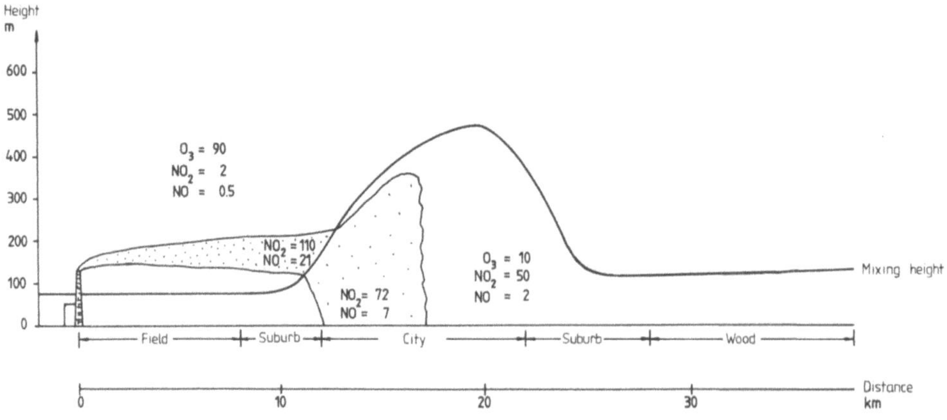

Figure 4. An attempt to simulate urban fumigation of a power plant plume during a summer morning. Information of the mixing height was obtained from a boundary layer model.

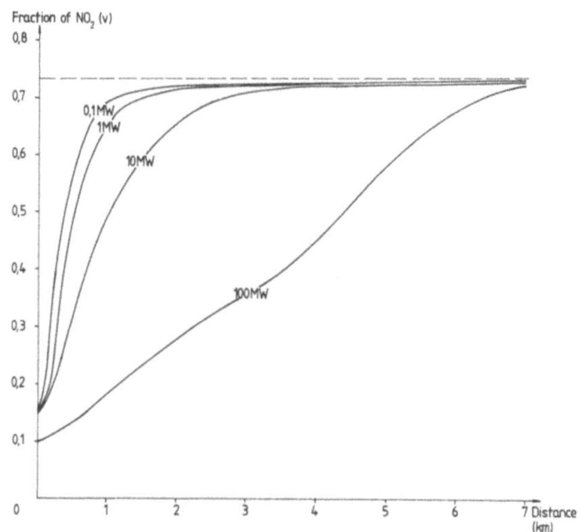

Figure 5. Calculated NO_2-fraction of the total emitted NO_x in the plume as a function of distance from the source. The curves refer to coal-fired plants of 0.1, 1.0, 10 and 100 MW respectively. The calculations were performed with the assumptions of normal meteorological conditions for southern Sweden at midday in springtime.

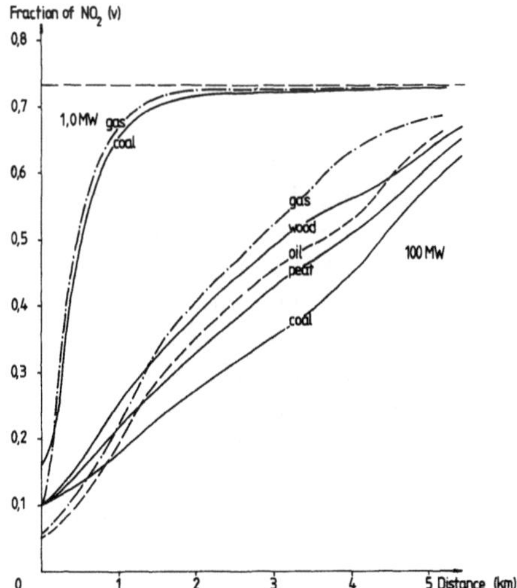

Figure 6. Calculated NO_2-fraction of the total emitted NO_x in the plume as a function of distance from the source. The curves refer to power plants of 100 MW and in two cases of 1.0 MW. The calculations were performed for the following five different fuels : coal, peat, wood, oil and gas.

TABLE 5.

	Emission			
	NO	NO_2	O_2	$H_2O(v)$
100 MW				
Coal	14.5 g/s	2.48 g/s	4 %	7 %
Peat	10.4 "	1.76 "	4 "	22 "
Wood	7.75 "	1.32 "	4 "	25 "
Oil	7.75 "	0.626 "	1 "	10 "
Gas	5.22 "	0.420 "	1 "	17 "
10 MW				
Coal	0.985 "	0.263 "	5 "	6 "
1.0 MW				
Coal	0.111 "	0.030 "	8 "	5 "
0.1 MW				
Coal	0.124 "	0.0034 "	10 "	4 "

equilibrium is reached at a distance of 7 km or more. This is quite far beyond the distance where the maximum concentration at ground level is reached for inert compounds.

At daytime with slightly unstable conditions the reaction $NO+O_3$ is by far the most important for the formation of NO_2 in the plume. Eventually the photochemical equilibrium is reached, which however takes some time due to the successive entrainment of ozon. At nighttime with stable stratification the reaction $NO+NO+O_2$ normally is very important for the NO_2 formation, although the shortage of O_2 in the emission can reduce the importance of this reaction during the first seconds after emission. At larger distances, when some dilution and entrainment of O_3 has taken place, the $NO+O_3$ reaction dominates also in stable conditions.

From the results it is obvious that emission conditions, meteorological factors and ambient ozone concentration are of great importance for the relative amount of NO_2 in the plume.

VERIFICATIONS

The model has, for the time being, not been tested against measured concentrations. A study is planned where concentration data from a mobile lidar system shall be used. Attempts are also made to obtain continuous NO, NO_2 and O_3-registrations from areas close to major power plants.

REFERENCES

Briggs G., 1975, Plume rise predictions, "AMS workshop on meteorological and environment assessment", Boston Ma.

Carmichael G. and Peters L., 1981, Application of the mixing-reaction in series model to NO_x-O_3 plume chemistry, Atm. Env. 15, 1069-1074.

Cocks A. and Fletcher I., 1982, Possible effects of dispersion on the gas phase chemistry of power plant effluents, Atm. Env. 16, 667-678.

Hanna S., 1982, Review of atmospheric diffusion models for regulatory applications, WMO Tech. Note No 177.

Hindmarsh A.C., 1974, Gear : Ordinary differential equation system solver, Lawrence Livermore Laboratory Report UCID-30001.

Högström U., 1964, An experimental study on atmospheric diffusion, Tellus, 16, 205-251.

Högström U., 1978, "Sulfur in the environment : Part I", John Wiley & Sons, Inc.

Isaksen I., Midtbø K., Sunde J. and Crutzen P., 1977, A simplified method to include molecular scattering and reflection in calculations of photon fluxes and photodissociation rates. Geoph. Norv. 31, 11-26.

Peters L. and Richards W., 1977, Extension of atmospheric dispersion models to incorporate fast reversible reactions. Atm. Env. 11, 101-108.

Rodhe H., Crutzen P. and Vanderpol A., 1981, Formation of sulfuric and nitric acid in the atmosphere during long range transport. Tellus 33, 132-141.

White W., 1977, NO_x-O_3 photochemistry in power plant plumes : Comparison of theory with observation. Env. Science & Tech. 11, 995-1000.

APPENDIX LARGER REACTION SCHEME

Below is given the reaction scheme first utilized in the model. The reactions R10-R15 are "summary" reactions, i.e. incomplete reactions since the intermediate stages are not specified (see Rodhe et al, 1981). Calculations showed, however, that on a local scale only reactions R1-R6 had to be included.

R 1 $NO_2 + h\nu \rightarrow NO + O(^3P)$
R 2 $O(^3P) + O_2 + M \rightarrow O_3 + M$
R 3 $O_3 + NO \rightarrow NO_2 + O_2$
R 4 $2NO + O_2 \rightarrow 2NO_2$
R 5 $NO + NO_2 + H_2O \rightarrow 2HNO_2$
R 6 $2HNO_2 \rightarrow NO + NO_2 + H_2O$

R 7 $O_3 + h\nu \rightarrow O_2 + O(^1D)$
R 8 $O(^1D) + M \rightarrow O(^3P) + M$
R 9 $O(^1D) + H_2O \rightarrow 2OH$
R10 $CO + OH \rightarrow - - \rightarrow HO_2 + CO_2$
R11 $CH_4 + OH \rightarrow - - - \rightarrow CH_2O + HO_2$
R12 $C_2H_4 + OH \rightarrow - - - \rightarrow HO_2 + 2CH_2O$
R13 $CH_2O + OH \rightarrow - - - \rightarrow CO + HO_2$
R14 $CH_2O + h\nu \rightarrow - - - \rightarrow CO + 2HO_2$
R15 $CH_2O + h\nu \rightarrow - - - \rightarrow CO + H_2$
R16 $2HO_2 \rightarrow H_2O_2 + O_2$
R17 $HO_2 + OH \rightarrow H_2O + O_2$
R18 $H_2O_2 + OH \rightarrow HO_2 + H_2O$
R19 $H_2O_2 + h\nu \rightarrow 2OH$
R20 $NO + HO_2 \rightarrow OH + NO_2$
R21 $NO_2 + OH \rightarrow HNO_3$
R22 $O(^3P) + NO_2 \rightarrow O_2 + NO$
R23 $O_3 + h\nu \rightarrow O(^3P) + O_2$
R24 $O_3 + NO_2 \rightarrow O_2 + NO_3$
R25 $O_3 + OH \rightarrow O_2 + HO_2$
R26 $O_3 + HO_2 \rightarrow 2O_2 + OH$
R27 $NO + NO_3 \rightarrow 2NO_2$
R28 $NO + OH \rightarrow HNO_2$
R30 $OH + HNO_2 \rightarrow NO_2 + H_2O$
R31 $HNO_2 + h\nu \rightarrow NO + OH$

DISCUSSION

P. BUILTJES

You have two basic assumptions : homogeneous mixing and the neglect of short time concentration fluctuations. Which assumption is to your opinion the most critical one ?

G. OMSTEDT

These two assumptions are dependent on each other and dependent on the time-scales involved. The few tentative comparisons on the NO_2/NO_x-fraction between calculations and measurements that have been made, do not indicate that these assumptions are too critical.

DRY DEPOSITION OF FINE PARTICLES TO CITY SURFACES

N.O. Jensen

Meteorology Section
Risø National Laboratory
DK-4000 Roskilde, Denmark

INTRODUCTION

A literature study has revealed that very little, if any, data exist for the dry deposition of fine particles to city surfaces. Dry deposition to a typical large city, however, with its relatively smooth surfaces of concrete tile and asphalt, would very likely be less than to a vegetated fibrous surface. In an investigation of the deposition velocity of caesium-137 to building surfaces, Roed (1983) finds small values indeed. In making reference to model studies there certainly seems to be an effect. However, it is not at all clear how to extrapolate to full scale, i.e., what scaling factors to use. Presumably they would be combinations of physical parameters, such as: aerosol size, d; surface roughness, z_0; and turbulence, u_* (friction velocity), but lack of data prevents adequate guidance to modellers in this field.

Some work on deposition to cities has been done in the past. A study by Andersen et *al.* (1978) deals with comparisons of vegetation uptake and total funnel collection (including wet deposition). Fig. 1 shows isopleths of funnel collection amounts from this study and the distribution of areal use in the city of Copenhagen.

The reason for such studies to be difficult to interpret is not only due to the mix of dry and wet deposition but also due to the general unrepresentativeness of collection agent (i.e., non typical spots in the general surroundings). It depends of course on what the usage of deposition estimates are: in some respects a large deposition is a conservative estimate, in other the contrary. But in connection with aerosols and respiratory effects, it would be a

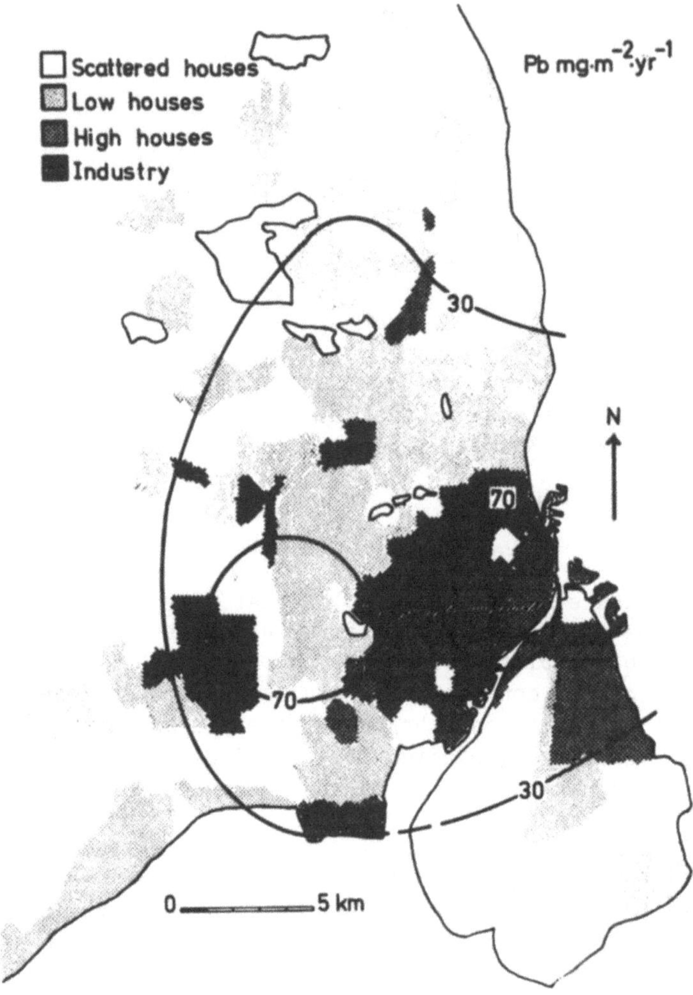

Fig. 1. Total (dry and wet) funnel collection results for Pb in the Copenhagen area, from Andersen et al. (1978).

wrong strategy to overestimate deposition. Hence areally representative estimates are called for.

The paper describes, mostly in qualitative terms, the deposition of particles and how the mechanics depend on particle size. It goes on to discuss the effect of rough surfaces. It is concluded that knowledge on the subject, at relevant large Reynolds numbers, is indeed lacking. Various methods for measurements of deposition

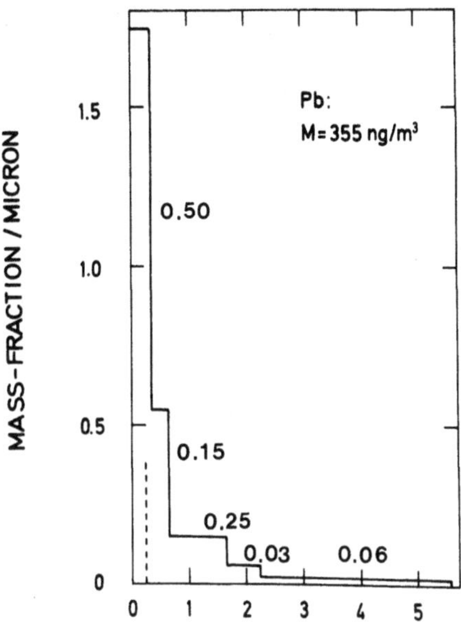

Fig. 2. Size distribution of Pb in a 2-hour cascade impactor sample from Copenhagen (from Flyger et al., 1976). The relative mass fractions are noted on the figure. The dotted line marks the mass median diameter.

is mentioned and in its last paragraph the paper gives some general ideas on how a suitable full scale experiment should be laid out in order to produce some data on the problem of dry deposition to city surfaces.

PARTICLES IN THE URBAN ATMOSPHERE

Figs. 2 and 3 show the size distribution of Lead and Calcium particles from measurements in Copenhagen area (Flyger et al., 1976). The suspended load of these two components are seen to be about the same in this case (~ 0.3 $\mu g/m^3$) but the shapes of the distribution are very different.

Thus for Pb, about 50% of the mass fall in the smallest category (cascade impactor range corresponding to diameters, $d \lesssim 0.3$ μm). The origin of these particles is thought to be automobile exhaust from engines burning leaded petrol. The three stages of the impactor collecting particles up to about 1.7 μm account for ~ 90% of the mass. For a "fresh" aerosol, however, where agglomeration

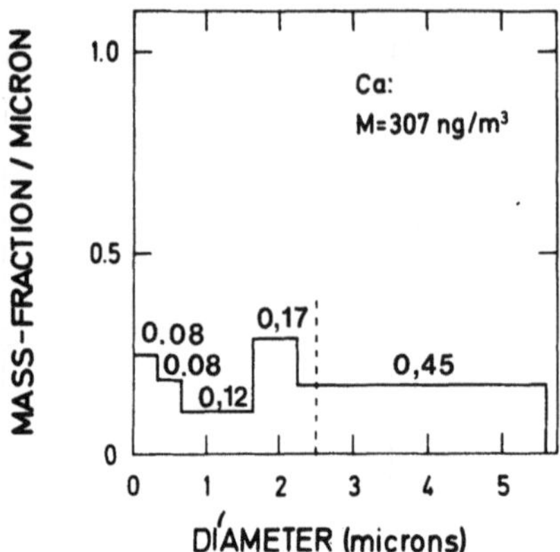

Fig. 3. As fig. 2 but for Ca.

has not yet had much chance to occur, the arosol is even smaller. Thus Whitley et al. (1975) find that 30 m from a freeway the aerosol is predominantly below 0.15 μm and exhibits a strong combustion mode at about 0.02 μm. With winds from the sampling station towards the freeway the aerosol was rather larger (~ 0.2 μm) in agreement with the above and with findings from other cities, e.g. St. Louis (Stampfer and Andersen, 1975; Alkezweeny, 1978).

Fig 3, however, which deals with Calcium shows a quite different and more even distribution with about 50% of the mass associated with particles larger than 2 μm. This is explained by the expectation that Calcium derives from wind blow dust.

Thus the total aerosol distribution in a city is not likely to be monotonic or describable with a simple mathematical expression as Junge's equation (Jaenicke and Davies, 19769), but may rather appear as double peaked function.

Although, as will be mentioned below, the 1 μm size particles may be the least depositing, they may be the least abundant in the city atmosphere anyway, since the main production occurs on either side of this size.

DEPOSITION OF PARTICLES IN GENERAL

An often used analogy is to consider the deposition velocity v_d as the inverse of a resistance (the larger the resistance, the smaller the deposition) and then sum the various individual physical processes limiting the deposition as a series of resistances. If the total resistance, or inverse transport capacity, consists of:

r_a = the resistance in connection with the turbulent boundary layer in general

r_B = the resistance in connection with transport through the laminar (or molecular diffusive) boundary layer near surfaces

r_s = the resistance in connection with the very process of surface uptake,

then

$$1/v_d = r_a + r_B + r_s \; . \qquad (1)$$

The inertia of sub-micron particles is so small that they behave like a gas. There is little tendency for these particles to rebound at a surface. Once they touch the surface their small momentum is overcome by intermolecular forces. Thus any surface is an efficient sink, and the surface resistance r_s is practically zero (Garland, 1980).

The laminar sublayer resistance, r_B, determined by Brownian diffusion is dependent on particle size. Thus the diffusivity,

$$D_p \propto \frac{kT}{\mu d} \, , \qquad (2)$$

where k is the Boltzman constant, T is the absolute temperature, μ is the viscosity of air and d is the particle diameter. This diffusivity is much less than diffusivities of gases. Consequently r_B is larger for particles than for gases. Thus while r_s in addition to the aerodynamic resistance is the limiting factor for deposition of gases, this role is in the case of particles played by r_B. The smaller the particles the larger the diffusivity (see Eq. (2)) and therefore the deposition increases with decreasing particle size.

For larger particles ($d \gtrsim 1$ µm) the inertia becomes significant: these particles are mostly deposited by interception and impaction. The larger the particles the more pronounced are these effects.

Particles of a few microns may be captured largely by interception. Thus the fine structure of vegetation consists of hairs and microscopic roughness elements, and particles following the flow and passing within one particle radius of these elements will be captured. Of course the efficiency of this process depends on the details of the surface. For still larger particles the mechanism may chiefly be impaction: When stream lines curve around macroscopic details of the surface these particles continue straight forward and may coast through the laminar sublayer (i.e., shortcut r_B). Also pure gravitational settling is an increasing function of particle size (increases with d^2) but is only significant for fairly large particles (for d ~ 3 μm the settling velocity is ~ $3 \cdot 10^{-2}$ cm/s) in situations where v_d is fairly small for other reasons.

In the range ~ 0.1 to ~ 1 μm neither diffusion nor interception/impaction are particularly efficient. Thus depositon is at a minimum for these particles.

For all sizes, the deposition is also dependent on the turbulence in the flow at distance from the surface elements. In Eq. (1) this is represented by r_a. Thus the stronger the turbulence (represented by the friction velocity u_*) the more deposition will occur - except for very large particles which may show a reversed effect due to bounce-off after impact, this being more likely the larger the impaction velocity is.

DEPOSITION TO ROUGH SURFACES IN GENERAL

The aerodynamic part, r_a, in Eq. (1) can be expressed as $u(z)/u_*^2$, where u is the wind speed at height z. It sets the maximum possible value for v_d (when $r_B = r_s = 0$). For $r_B + r_s$, Chamberlain (1966) has introduced the definition $(Bu_*)^{-1}$ where B may be recognized as the dimensionless sublayer Stanton number of Owen and Thomson (1963). Thus for $r_B + r_s \gg r_a$, i.e. for a low affinity between the surface and the depositing material, we have

$$v_d \simeq Bu_* \,, \tag{3}$$

and the discussion of deposition can then concentrate on the size of B for various surfaces.

For a range of different surfaces characterized by the aerodynamic surface roughness z_o relative to the laminar sublayer thickness ν/u_* (ν is the kinematic viscosity of air) which would exist over a smooth surface under the same flux conditions, diffusion results for various gases, mainly water vapour, have been used to obtain values of B (Garratt and Hicks, 1973). The results are shown in Fig. 4. Above $Re \equiv z_o/(\nu/u_*) \simeq 10^2$, the data seem to split according to surface texture: deposition to fibrous surfaces tends towards saturation (independent of Re), whereas B for deposition to surfaces consisting of smooth, regular roughness elements continues to

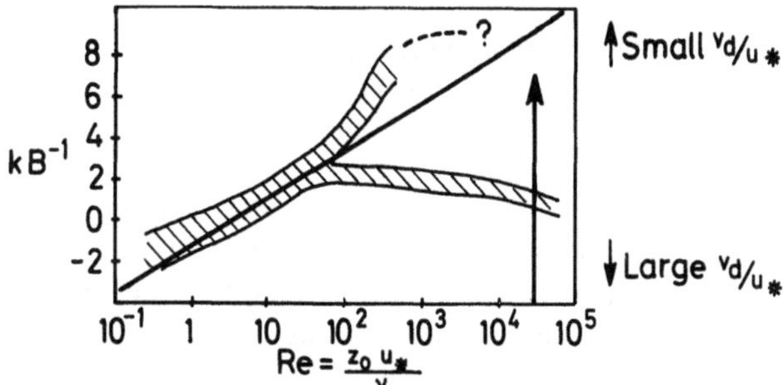

Fig. 4. The overall behaviour of kB^{-1} (defined in the text) in diffusion of water vapour to various surfaces. Lower branch: Fibrous elements. Upper branch: Smooth rougness elements (from Garland and Hicks, 1973). The large vertical arrow denotes a typical Reynolds number for a large city.

decrease with increasing Re. This implies a decreasing deposition with increasing roughness for surfaces of the latter type, in which category we may put cities. Results at the relevant Re-number do not exist, however, so application will require extrapolation.

DEPOSITION OF PARTICLES ONTO ROUGH SURFACES

In scaling of model results (e.g. Sehmel, 1973) as well as in theoretical developments (e.g. Owen, 1969, Wood, 1981) the depostion velocity is often expressed in terms of particle diameter, relative to viscous sublayer thickness:

$$\frac{v_d}{u_*} \propto f_1 \left(\frac{d}{\nu/u_*}\right) \quad , \tag{4}$$

which seems to appear regardless of whether a stopping length hypothesis is involved or not. For geophysical flows, this is however not likely to be a workable proposition, since in the developments leading to Eq. (4) it is always assumed that z_0 is of the order of ν/u_* or less ($Re \lesssim 1$).

It is quite likely that a proper scaling in very rough flows, where $z_0 \gg \nu/u_*$ (large Reynolds number) would involve a correlation of the sort described in the previous paragraph, where for gases

$$\frac{v_d}{u_*} \propto f_2 \left(\frac{z_0}{\nu/u_*}\right) \quad . \tag{5}$$

But it is also likely that an equivalent expression for particle deposition would have to show an additional dependence on d. For small particles, a proposition would be the following.

Consider a layer of thickness δ over which a concentration difference $\Delta\chi = \chi - \chi_0$ exists. By the flux-gradient hypothesis (Fick's law) we have that flux equals $\kappa(\Delta\chi/\delta)$. Assuming, that $\chi_0 \ll \chi$ (relevant for particles where $r_s \sim 0$) we get by direct definition

$$v_d = \kappa/\delta \ , \tag{6}$$

where κ is the appropriate diffusivity. Further, having a viscous (sub-) sublayer developiong along each roughness element of scale z_0, its thickness will be $\delta \sim \sqrt{\nu t} \sim \sqrt{\nu z_0/u_*}$ where the advective time of development is estimated as z_0/u_* (u_* being a relevant velocity in the roughness flow layer) whereby Eq. (6) may be written as

$$\frac{v_d}{u_*} = \frac{\kappa}{\nu} \left(\frac{z_0}{\nu/u_*}\right)^{-\frac{1}{2}} , \tag{7}$$

in accordance with Eq. (5). The structure of Eq. (7) is similar to the formula given by Owen and Thomson (1963) for gases and later confirmed by Chamberlain (1968) whereby κ/ν is given to the power 1.25 and the Reynolds number $z_0 u_*/\nu$ to the power $-$ 0.45. Using Eq. (7) with κ substituted for the particle diffusivity according to Eq. (2) results in

$$v_d \propto \sqrt{u_*/z_0}/d \tag{8}$$

where the proportionality factor contains all the physical constants. This formula combines Eqs. (4) and (5) in the case of sub-micron particles. For large particles in the impaction range we have presently no theoretical proposition.

Knowledge regarding the deposition of particles is indeed very meager. Of course, in agreement with the above development, very fine particles can always be postulated to behave as gases, but the argument is not really appropriate for particles of size > 1µm. Thus, deposition of the latter particles will probably not scale as simply as the deposition of gases but rather be dependent on the surface geometry itself (i.e. whether surfaces with the same aerodynamic z_0 consist of densely packed or more spread elements): thus there will be less diffusion of small particles to a surface of less area which presumably means that the roughness elements are widely spread and of simple geometry, but more impaction of large particles to the protrusions of a surface with such widely spread elements). Very little guidance regarding parameterization of this problem can presently be given.

RESUSPENSION

After deposition particles adhere with great tenacity to solid surfaces (Corn, 1961) to the extent that even vigorous blowing on a surface dislodge only few particles. Even washing seems inefficient in this respect. The only way resuspension seems to arise is when dislodgement of larger particles, predominantly soil grains, occur as a result of saltation (Bagnold, 1941). Such wind blown dust can have adhered to it earlier deposited material. Material deposited in a city will be less prone to this process.

METHODS OF MEASUREMENTS OF DRY DEPOSITION

In addition to the methods mentioned in the introduction, i.e. analyzing the amount of material present on artificial or natural samples with the related question of areal representativeness (besides collecting wet deposits, moss is hardly representative of the major part of a city surface in regard to dry deposition) a number of micrometeorological alternatives excists which directly or indirectly measure the downward flux of material in question in the air above the surface.

One such method is the gradient method in which the flux, F, of material is determined from

$$F = K \frac{\partial \chi}{\partial z} , \tag{9}$$

where K is an eddy diffusivity in the turbulent flow which may be assumed proportional to $u_* z$, and $\partial\chi/\partial z$ is the gradient of particle concentration. Thus measurements of profiles of wind and concentration is needed.

Another method, the socalled eddy correlation method, measures the flux directly. Defining χ' as the instantaneous deviation from the average concentration and w' as the instantaneous vertical velocity (the mean may be taken as zero)

$$F = \overline{\chi' w'} , \tag{10}$$

where the overbar means an average over a suitable length of time (e.g. 10 min). Compared to the previous method, the latter requires very fast responding concentration measuring equipment (measurements of w' is no problem). A few suitable instruments exist for this, based on various techniques in connection with light scattering off a continuously flowing aerosol sample.

A special version of the eddy correlation method is the so-called eddy accumulation technique (Desjardins, 1977) where sampling during occasions with upward respectively downward velocity fluctuations is done separately. The outstanding feature of this method, which has not been emphasized much in the contemporary literature, is that the sampling can be made onto filters, which later in the laboratory can be analyzed for any chemical element.

Use of both of the above methods in a city would require spatial averaging (over several house blocks) as well, which conveniently might only be obtained by flying the instruments from an aircraft.

Indirectly deposition can be obtained by using some sort of a budget method which again for the present problem would require aircraft concentration sampling. An exception is in case of a well mixed boundary layer with a capping inversion, where a few tower measurements from moderate height would suffice. In this case great precision of the concentration measuring equipment is required (Williams, 1982).

The main problem with traditional micrometeorological methods, however, is that local sources and horizontal gradients in the concentration field in general contaminate the measurements.

SUGGESTION FOR AN EXPERIMENTAL LAY OUT

The problem of the overall horizontal gradient over a city could be overcome by flying cross-wind, but the presence of local sources makes the fluxes spurious no matter what measurement technique is used including upwind/downwind budget methods.

The only alternative seems to be the use of an artificial or tracer aerosol, which can be destinguished from the background. It should be of a well defined particle size; it should be possible to produce in various size ranges in order to investigate deposition over the relevant spectrum and above all it should be non toxic as it probably must be applied in a sizeable dose to enable significant pick-up. Only a few laboratories in the world are probably able to taylor such a tracer. Their interest and cooperation is hoped for.

Eddy correlation measurements of this tracer would probably be difficult as fast instruments concentration measurements are not specific to chemical composition. However, one could use a version of the eddy-accumulation technique where samples were collected on filters for later conventional laboratory analysis.

However, if a second non depositing tracer (e.g. SF_6) is released and sampled simultaneously with the particle tracer, such that measurements of relative concentration is enabled, budget estimates can be made more readily. Thus it is no longer necessary to care for inhomogeneous areal influence or representativeness of measurement points as the technique directly gives the integrated effect of the surface in question.

A particularity simple lay out for an experiment then emerges where everything is sampled at ground level as average concentrations over suitable lengths of time. For a full answer comparison experiments would have to be conducted over agricultural land, however. The advantage of using an aircraft that remains is that much more sampling flexibility can be obtained relative to a given upwind release; further that it might be possible to conduct the comparison measurements at the same time during that release; and lastly that some advantage may be obtained by letting the aircraft release the agents in a suitable configuration upwind (e.g. crosswind or more realistically in an alongwind direction simulating a continuous elevated source) which will lead to a complete indenpendence of wind direction vagaries.

SUMMARY

Regarding dry deposition of fine particles to city surfaces, or to aerodynamically rough surfaces consisting of relatively smooth elements at sufficiently high Reynolds number, it has been concluded that substantial knowledge is absent. To remedy this we have suggested conduction of a relatively simple experiment, micrometeorologically speaking. The main difficulty in principle is the advent of a suitable particle tracer. Further a theoretical formula for deposition of small particles to the surfaces in question has been suggested.

ACKNOWLEDGMENT

This study was supported by Nordisk Ministerråd under the project MIL4, "Various sources' relative contribution of pollutants in the environment".

REFERENCES

Alkezweeny, A.J., 1978, Measurements of Aerosol particles and trace gases in METROMEX, J. Appl. Meteor, 17:609.
Andersen, A., Hovmand, M.H., and Johnsen, I., 1978, Atmospheric heavy metal deposition in the Copenhagen area, Environ. Pollut., 17:133.
Bagnold, R.A., 1941, The physics of blown sand and desert dunes, Methuen London.
Chamberlain, A.C., 1966, Transport of gases to and from grass and grass-like surfaces, Proc. Roy. Soc., A 290:236.
Chamberlain, A.C., 1968, Transport of gases to and from surfaces with bluff and wave-like roughness elements, Quart. J. Roy. Met. Soc., 94:318.
Corn, M., 1961, The adhesion of solid particles to solid surfaces - I. A review. J. Air Pollut. Control Ass., 11:523.
Desjardins, R.L., 1977, Description and evaluation of a sensible heat flux detector, Boundary-Layer Meteorol., 11:147.
Flyger, H., Palmgren Jensen, F., and Kemp, K., 1976, Air Pollution in Copenhagen, Part I. Element analysis and size distribution of aerosols. Risø Report No. 338. 70 pp.
Garland, J.A., 1980, Surface deposition from radioactive plumes. In: Seminar on Radioactive Releases and Their Dispersion in the Atmosphere Following a Hypothetical Reactor Accident, Risø, April 22-25, 1980. Edited by F. van Hoeck and P. Recht, Commission of the European Communities, Luxembourg, Vol. 1:293.
Garratt, J.R., and Hicks, B.B., 1973, Momentum heat and water vapour transfer to and from natural and artificial surfaces, Quart. J. Roy. Met. Soc., 99:680.
Jaenicke, R., and Davies, C.N., 1976, The mathematical expression of the size distribution of atmospheric aerosol, J. Aerosol Sci., 7:255.
Owen, P.R., and Thomson, W.R., 1963, Heat transfer across rough surfaces, J. Fluid Mech., 15:321.
Owen, P.R., 1969, Pneumatic transport, J. Fluid Mech., 39:407.
Roed, J., 1983, Deposition velocity of caesium-137 on vertical building surfaces, Atmosph. Env., 17:663.
Sehmel, G.A., 1973, Particle eddy diffusitivies and deposition velocities for isothermal flow and smooth surfaces, Aerosol Science, 4:125.
Stampfer, J.F., and Anderson, J.A., 1975, Locating the St. Louis Urban Plume at 80 and 120 km and some of its characteristics, Atmos.Env., 9:301.
Whitby, K.T., Clark, W.E., Marple, V.A., Sverdrup, G.M., Sem. G.J., Willeke, K., Liu, B.Y.H., and Pui, D.Y.H., 1975, Characterisation of California aerosol - I. Size distributions of freeway aerosol, Atmos. Env., 9:463.

Williams, R.M., 1982, Uncertainties in the use of box models for estimating dry deposition velocity, Atmos. Env., 16:2707.
Wood, N.B., 1981, A simple method for the calculation of turbulent deposition to smooth and rough surfaces, J. Aerosol Sci., 12:275.

2. COASTAL METEOROLOGY RELATED TO AIR POLLUTION MODELING

Chairmen: J. Knox
F. L. Ludwig

Rapporteurs: P. Builtjes
C. Ludwig

A REVIEW OF COASTAL ZONE METEOROLOGICAL PROCESSES IMPORTANT TO THE MODELING OF AIR POLLUTION

F. L. Ludwig

Atmospheric Science Center
SRI International
Menlo Park, California 94025

INTRODUCTION

Coastal areas are frequently favored as sites for cities, industrial activity and power plants. Therefore, it is important to understand the meteorological conditions that prevail in coastal zones and how these conditions affect the transport, transformation and diffusion of materials emitted into the atmosphere. A brief review of the atmospheric processes that govern pollutant concentration will provide a basis for discussing shoreline conditions and their effects on concentration. The major factors determining atmospheric concentrations are the

- Wind field,
- Turbulent diffusion,
- Chemical transformations, and
- Removal processes.

All of these are affected in one way or another by the presence of a shoreline.

Of the items listed above, the wind field is probably the most important. Its most obvious influence is on the transport of materials once they leave their source. It is the wind that determines where emissions go, and with what other emissions they interact. Wind field changes determine whether the emissions continue to move from the source or recirculate. The wind determines the initial dilution; the initial concentration tends to be inversely proportional to wind speed because, other things being equal, higher wind speeds provide more air for dilution. For buoyant gases, this same diluting effect quickly reduces buoyancy so that plumes may not rise as high as in still air.

The second most important effect on concentrations is turbulent diffusion. The origins of turbulent motion are quite complex, but the thermal stratification of the atmosphere is one of the more important determinants. Stable stratification inhibits vertical motion and turbulence generation, while unstable thermal stratification will promote such turbulent motion. Wind shear is another major source of turbulent energy. The shoreline affects both the thermal stratification and the wind shear, so it is not surprising that the resultant atmospheric dilution is different from that over uniform land or water surfaces.

Not all materials are inert and remain unchanged after release into the atmosphere. Some materials undergo chemical reactions which may be influenced by the presence of water in either the vapor (humidity) phase or the liquid (droplet) phase; so the concentrations of such materials are likely to be affected by the presence of a shoreline.

The final factor listed earlier, i.e. removal, manifests itself in at least three ways:

- Dry removal
- Below-cloud removal of gases and aerosols by precipitation
- In-cloud removal of gases and aerosols by precipitation

Although there are undoubted shoreline effects on these processes, e.g., coastal influences on the occurrences of clouds and fog, they will not be a major focus of this review because these effects are somewhat secondary.

The behavior of these atmospheric factors that influence air pollution is very complex in the vicinity of a shoreline, and represents a severe challenge to any boundary layer model that hopes to describe their behavior. Pollution models will not be adequate if they treat only uniform, steady meteorological conditions. They must treat fully three-dimensional, time-varying meteorological conditions and air motions. This review focuses on observational studies of boundary layer behavior in coastal zones and their interpretation. The ways in which boundary layer models have been used to address the complexities of the coastal zone will be reviewed by Pechinger (1983) in a companion paper.

The two most important factors governing the behavior of the atmosphere in coastal areas are the temperature differences between land and water, and the larger scale flow patterns, i.e. offshore flow versus onshore flow. These two factors have guided the organization of this review.

OBSERVED TURBULENCE PATTERNS

Onshore Flow with Land Warmer than Water

Onshore flow, whether caused by synoptic scale weather systems or induced locally by differential surface heating, is the most important situation from the pollution standpoint because such flow is most likely to produce air pollution impacts on people. Furthermore, onshore effects are generally more dramatic than those during offshore flow; e.g., as the air flows across the boundary from water toward land, an internal boundary layer forms. The importance to pollutant dispersion of this shoreline-induced internal boundary layer has been recognized for at least 20 years, e.g., Bierly and Hewson (1962) emphasized the importance of the heating discontinuity in strengthening the internal boundary layer that forms at the roughness discontinuity. Lyons and Cole (1973) have referred to this combined effect as the thermal internal boundary layer (TIBL) and the term TIBL is widely used to refer to the thermally-augmented internal boundary layers that occur over land during shoreward flow over warmer land.

Figure 1, patterned after diagrams that have been used by many authors (e.g., Lyons, 1975; Raynor et al., 1980), is a schematic depiction in the vertical plane of the important meteorological zones during flow from cooler water to warmer land. The different regions can have radically different characteristics. The uppermost region (I) is the free atmosphere above the highest levels to which mixing reaches at inland locations. The lowest region over water (III) is a stable layer which has been cooled by the underlying cold water. The lowest region over land (IV) is the TIBL whose depth increases gradually as the air moves inland. Region II is a transition layer between the free atmosphere and the underlying layers. Any boundary layer model or pollution model must be able to account for these very different turbulent regimes and the sharp boundaries between them, if it is to describe pollutant behavior accurately.

Figure 2 patterned after a figure given by Lyons (1975) shows how a plume imbedded in the relatively stable layer over the water interacts with the TIBL. The greater turbulent intensity within the TIBL disperses the plume rapidly, mixing relatively high concentrations suddenly downward, causing the ground to be

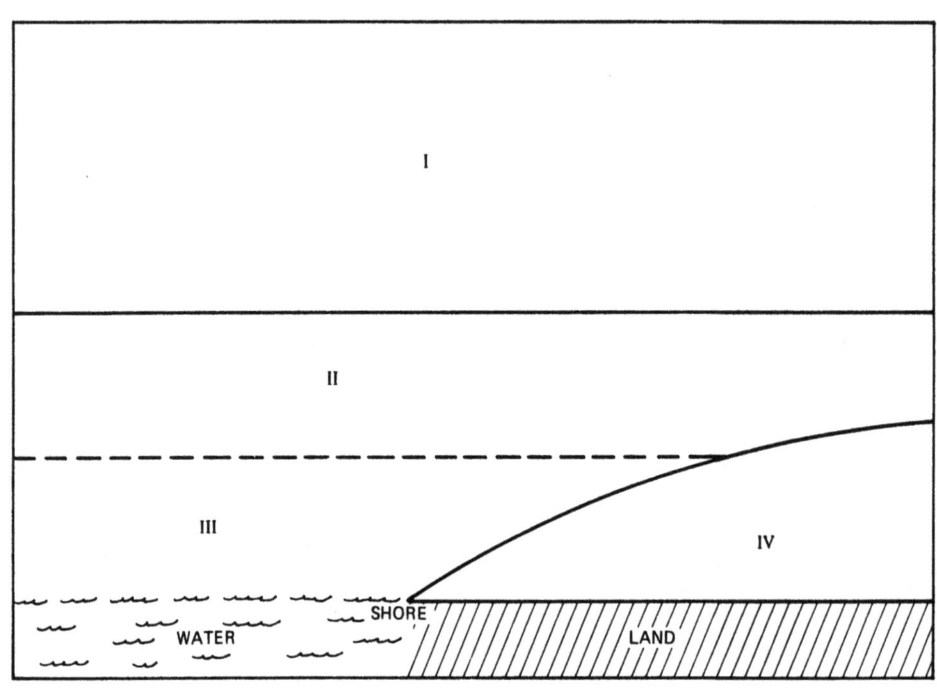

Fig. 1. Major atmospheric regimes during onshore flow

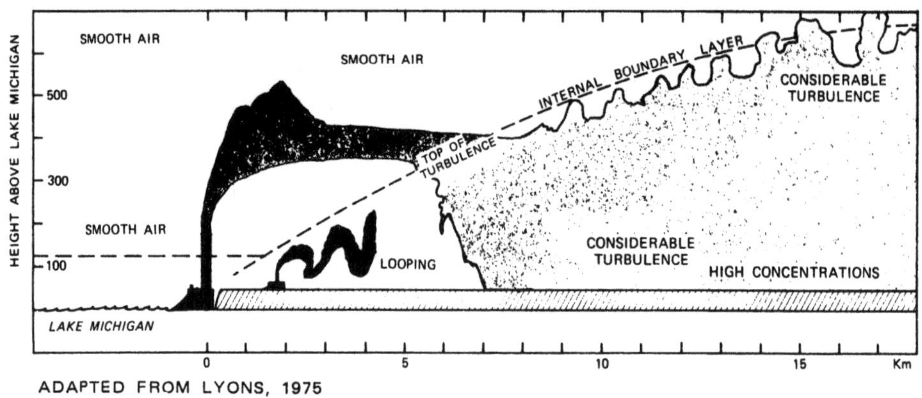

Fig. 2. Schematic diagram of fumigation of a plume entering the tibl

"fumigated." The shoreline fumigation is nearly steady-state and will affect the same general area for long periods of time. This is unlike the short-lived inland fumigation caused by diurnal heating that persists only during the period when the top of the deepening mixing layer is just above the plume and disappearing later when the mixed layer deepens.

The depth of the layer as a function of inland distance was studied by Raynor et al. (1980) who found that a simple model proposed by Venkatram (1977) fit the observations quite well. Venkatram's approach is similar to that used by Summers (1965) to describe boundary layer development over an urban heat island. Figure 3 (derived from a Venkatram figure) shows the typical potential temperature profile in the TIBL. The following discussion has been abbreviated considerably from the original Venkatram (1977) paper, in which he follows Tennekes (1973) and assumes vertical motion to be small compared to the entrainment velocity, $(-\overline{w'\beta'})_h/\Delta$, so that the rate of inversion rise, dh/dt, is proportional to the entrainment velocity:

$$\frac{dh}{dt} = -\frac{(w'\theta')_h}{\Delta} \qquad (1)$$

The numerator on the right side is the inversion heat flux at the top of the mixed layer. Venkatram (1977) assumes that the temperature jump at the inversion, Δ, is proportional to mixing layer depth (h) and the lapse rate above the TIBL (γ), i.e.,

$$\Delta = F\gamma h \qquad (2)$$

where the proportionality constant F is less than one. He further assumes inversion heat flux, $(\overline{w'\theta'})_h$ to be proportional to the surface heat flux and $(\overline{w'\theta'})_s$, as parameterized by Deardorff (1972):

$$(\overline{w'\theta'})_h = \left(\frac{-F}{1-2F}\right)(\overline{w'\theta'})_s = \frac{-FC_\theta U_*}{(1-2F)}(\theta_1 - \theta_m)$$

$$= \frac{-FC_\theta C_u U_m}{(1-2F)}(\theta_1 - \theta_m) \qquad (3)$$

where C_θ is the heat transfer coefficient; C_u is the friction coefficient (i.e., the ratio of the friction velocity U_* to the wind speed, U_m in the mixed layer). The surface and mixed layer temperatures are θ_1 and θ_m, respectively. Substituting in Equation 1 gives:

$$hdh = \frac{C_\theta C_u U_m(\theta_1 - \theta_m)}{\gamma(1-2F)}\, dt \qquad (4)$$

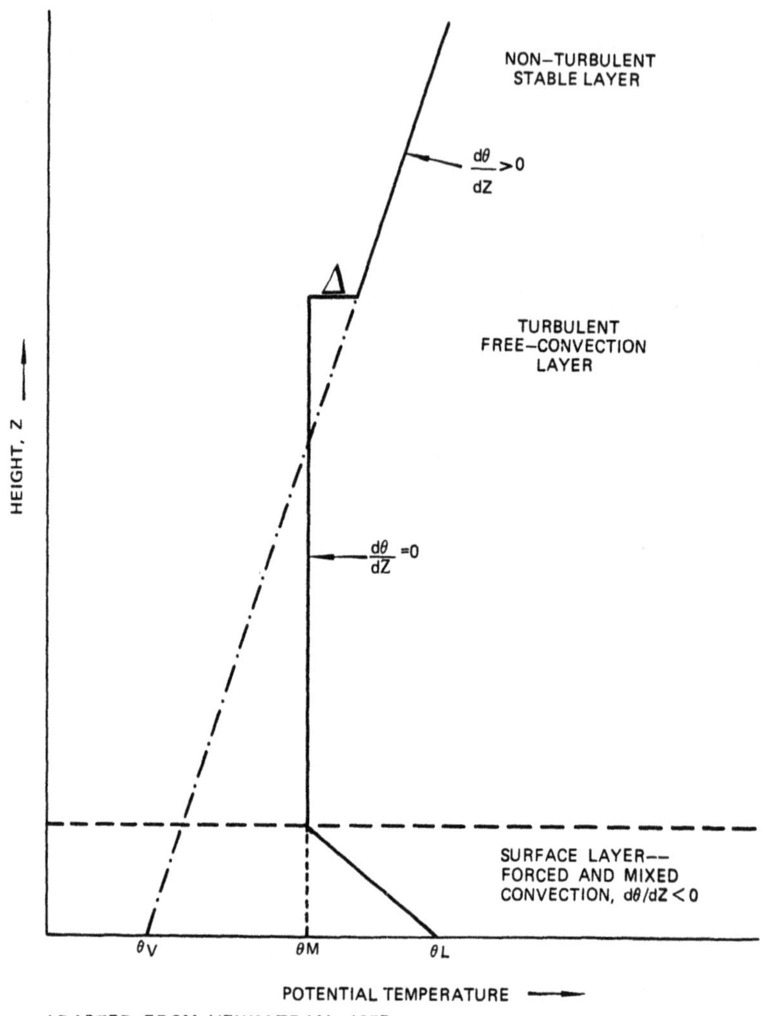

Fig. 3. Schematic diagram of the temperature profile in an idealized TIBL

If the right hand terms are constant and we substitute $U_m dt = dx$ (where x is distance traveled inland), integration gives:

$$h = \left[\frac{2C_\theta C_u(\theta_1 - \theta_m)x}{\gamma(1-2F)} \right]^{1/2} \tag{5}$$

Assuming that $C_u \approx C_\theta \approx U_x/U_m$ and $\theta_m \approx \theta_w$ gives:

$$h = \frac{U_*}{U_m}\left[\frac{2(\theta_1 - \theta_w)x}{\gamma(1-2F)}\right]^{1/2} \tag{6}$$

The water and mixed layer potential temperatures, θ_w and θ_m, have been equated under the assumption that γ is representative of the vertical potential gradient upwind over water and that the proportionality constant F, has been redefined accordingly.

Thus, Equation 6 shows the height of the TIBL to have a square root dependence on: distance inland, land-water temperature difference, and the reciprocal of overwater potential temperature gradient. The height varies directly with friction velocity and the reciprocal of the wind speed. Such a relationship has been found to fit the observed data reasonably well (Raynor et al., 1980; DiVecchio et al., 1976). Kerman et al. (1982) pointed out that the equilibrium internal boundary layer depth, h_e, may not always apply because inland surface temperatures at the point of presumed equilibrium (Peters, 1975; Venkatram, 1977) may be continually changing in response to solar radiative heating.

Lyons (1975) based the qualitative descriptions of turbulence shown on Figure 2 on pilot reports obtained during many field studies. Raynor et al. (1979) reported several-fold increases in turbulence levels (expressed as a ratio of the standard deviation of wind speed fluctuations, σ_u, to the average wind speed, $\bar{U}$) in the internal boundary layer versus those measured outside it; the averages (of 18 cases) were 0.12 at the coast and 0.33 inland. The coastal and inland standard deviations of the horizontal wind direction fluctuations, σ_θ, were about 7° and 48°, respectively, while vertical directional fluctuations, σ_ϕ, were about 1.6° and 15°.

Zanetti et al. (1982) released tracer gases off the California coast during onshore flow conditions and determined the size of the plume about 800 m off the coast and about 5 to 8 km inland. Using measured plume dimensions near the shore, they inferred a stability class (and hence turbulent diffusion characteristics; c.f. Gifford, 1976) and compared it with one inferred from plume dimensions farther inland. Typically, both horizontal and vertical plume dimensions at the inland site suggested that conditions were one Pasquill-Gifford class less stable than at the offshore location, although the inferred stability classes were not necessarily the same for the two dimensions. If the offshore measurements represent the integrated effects of marine turbulence over a distance of several kilometers, and inland observations portray the added effect of the TIBL for several more kilometers, then the observed differences in stability classes are a measure of turbulent intensity differences in the two regimes. Expressed in these terms, typical differences are somewhat greater than "one Pasquill-Gifford stability class."

Ludwig and Dabberdt (1976) examined the effects of a change in urban surface on fluctuations in horizontal wind direction, and found effects similar to the shoreline case. Since many urban areas are built at the coast, this raises the question of what the combined urban and coastline effects on turbulence patterns might be. Bornstein et al. (1983) reported that the range of horizontal wind directions at stations in the New York City area increased during onshore flow from 40° at the shoreline to more than 130° over urbanized coastal areas; in some of the less urbanized parts of New Jersey and Staten Island the range reached 60 to 70°. Their results indicate that urbanization does enhance the effects of the TIBL.

The eddy dissipation rate ($\varepsilon^{1/3}$) was measured by Keen and Lyons (1978) as a turbulence indicator during September lakebreezes on the western shore of Lake Michigan. They reported values greater than 3 $cm^{2/3}s^{-1}$ within the thermal internal boundary layer but less than 10 percent of that value outside it. Lyons (1975) reported similar values for a June day in this same area.

Offshore Flow from Warmer Land to Colder Water

When the larger scale circulations cause air flow from warmer land masses out over colder water, turbulent conditions will be quite different from those described above, especially in those cases where the temperature difference between the two surfaces is large and a very stable layer forms at the surface in response to cooling from below. Raynor et al. (1975) derived an empirical formula for estimating the depth of this conduction inversion. In some respects, the equation is similar to the one given above for the depth of the internal boundary over warm land. The expression is:

$$H = (U_*/\overline{U}) \left[\frac{x(T_1 - T_w)}{-dT/dz} \right]^{1/2} \tag{7}$$

Figure 4 (patterned on a figure given by Raynor et al., 1975) shows a developing inversion layer after the air moves from the warmer, rougher land out over the cool, smooth water. The formation of another internal boundary layer is shown at the right of the figure as the air returns over warm, rough land; the top of the TIBL rises more steeply over land than does the stable layer top over the cold, smooth water. This slow deepening produces a stable layer that is much shallower than the turbulent layer over the land. Lyons (1975) reports that Taylor (1915) studied tropical air moving over cold waters off the shore of Newfoundland. Using kite

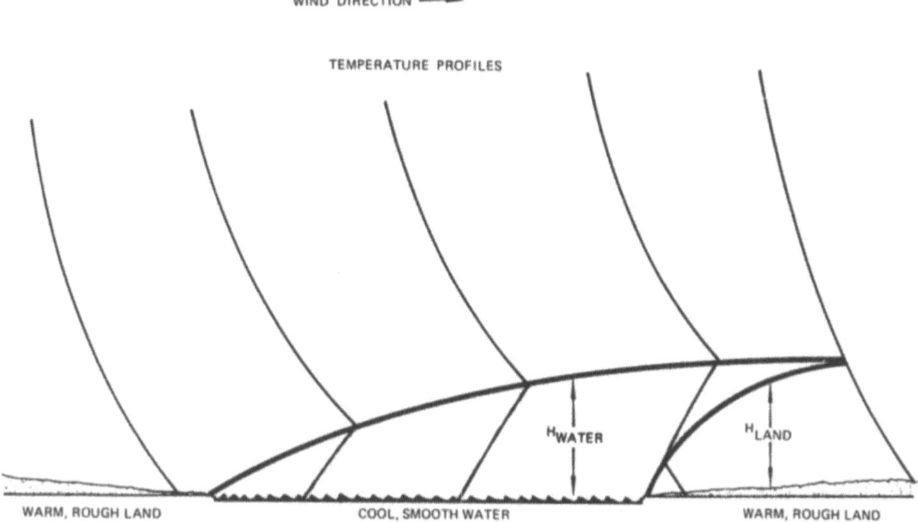

Fig. 4. Schematic diagram of lapse rates and internal boundary layers during flow from land (warm) to water (cool) to land (warm).

soundings he found that the top of the conduction inversion only reached a height of 375 m after several days of travel over cold water. By contrast, a TIBL will reach that depth within a few kilometers from shore. Lyons (1975) also reported that Bellaire (1956) observed a 16°C inversion in the lowest 12 m, only 8 km from shore over Lake Michigan when the land-water temperature difference was about 25°C. The depth of the inversion never exceeded 100 m even at distances as great as 115 km from shore.

Lyon's (1975) own observations were somewhat less dramatic than those described above because they only involved temperature differences of about 10°C. Otherwise they were similar in that he found that the top of the inversion rose only to 150 m during the traverse of the entire width of Lake Michigan. He reported a large difference in the turbulence, as expressed by the eddy dissipation rate; values within the TIBL commonly being about 2 $cm^{2/3}$ s^{-1}, but only about 0.3 $cm^{2/3}$ s^{-1} in the air over the colder water. Lyons (1975) reported that Roll (1965), found eddy diffusion coefficients of around 10^3 cm^2 s^{-1} in over-water inversions, and about 2 orders of magnitude greater over heated land.

More recently, Dabberdt et al. (1982) presented results obtained during low level (30 m) aircraft traverses off the Louisiana Gulf Coast; their data indicate a very abrupt transition in the turbulent regime at the coastline. Figure 5 shows measured

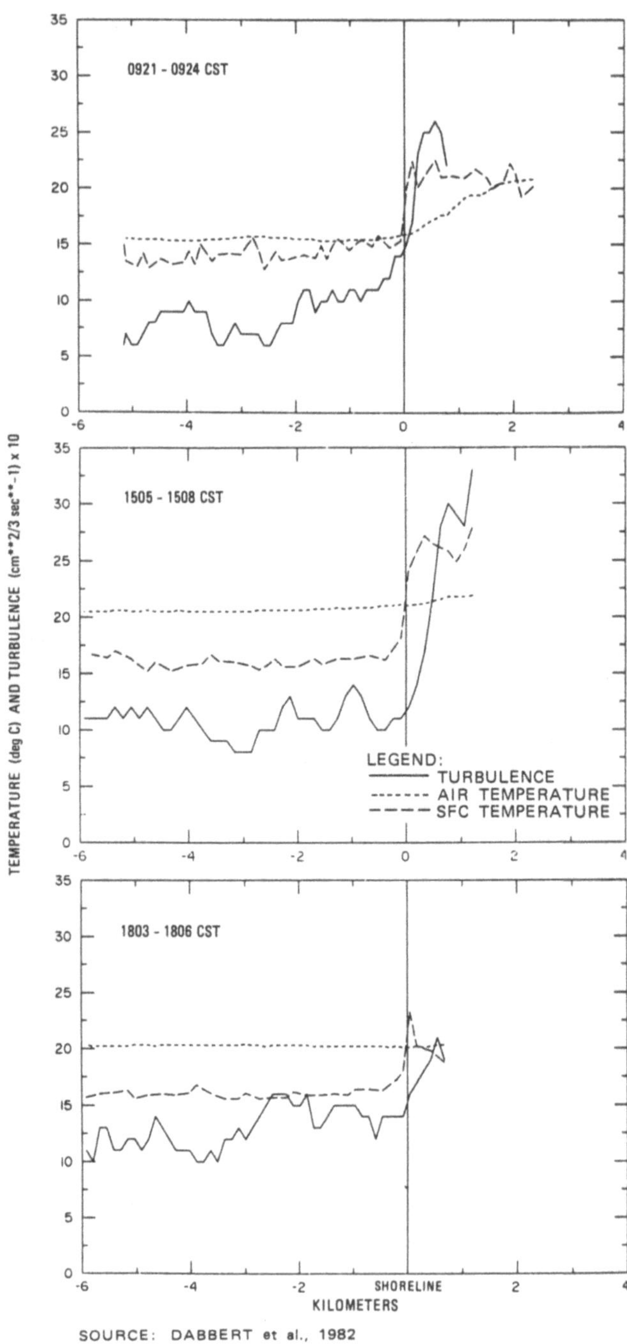

Fig. 5. Offshore-inland meteorological traverses at 30m 23 february 1982

turbulence (dissipation rate, $cm^{2/3}\ s^{-1}$), air temperature and radiative surface temperature (°C) measured on 23 February 1982. During the morning, turbulence rose by about a factor of 3 at the shore in response to a surface temperature change of about 6°C. By mid-afternoon the surface temperature change was about 10°C and, correspondingly, the dissipation rate increased by about a factor of 3. Finally, in the early evening, the surface temperature change at the coast was slightly less than 5°C and the turbulence increase amounted to only about 33 percent.

Onshore Flow Over Cool Land and Offshore Flow to Warmer Water

Conditions during onshore flow from warmer water to colder land are similar to those described in the preceding section, with the roles of the two surfaces reversed. The major differences are accounted for by surface roughness effects. The empirical formula suggested by Raynor et al. (1975), i.e., Equation 7, suggests that the cooled layer depth is proportional to the friction velocity, U_*, which in turn is a measure of the surface roughness. Therefore, cooling should extend to greater heights in air traveling over colder land than it does over water. Air flow from warm water to cold land occurs most frequently at night, and has been of considerably less interest than the more typical daytime sea breeze and nighttime land breeze conditions so that there are few reports of such observations in the literature.

Offshore flow from colder land to warmer water also tends to be a nighttime phenomenon with minor air pollution effects, and consequently there has been little motivation to study it. The fact that such flow involves transfer of sensible heat and also considerable input of latent heat in the form of water vapor leads to some extreme weather conditions e.g., fog, freezing temperatures, rough water and aircraft icing. Such conditions discourage outdoor measurements. Using one of the few data sets available, Lyons (1975) noted that wind statistics obtained by Argonne National Laboratory at a tower 1.6 kilometers offshore from Muskegon, Michigan, were characteristic of much less lateral diffusion, even with warmer water, than was found for similarly unstable conditions over land. Researchers at Brookhaven National Laboratory (Michael et al., 1973; Raynor et al., 1978; Raynor et al., 1975) measured the width of tracer plumes released about 1.5 to 5 km offshore and found that even under unstable conditions, presumably associated with water that is warmer than the overlying air, the width of the plume was similar to that observed under very stable conditions over land. Vertical plume dimensions were more consistent with values for unstable land conditions. Thus, colder air traveling over warmer water seems to have smaller lateral turbulence than would be observed under similar conditions over land, but vertical components are more nearly like those over land. Maas and Harrison

(1977) reached similar conclusions from data collected off the California coast.

Although it has been generally presumed that a sea surface smoother than the land contributes to reduced over-water turbulence, SethuRaman (1979) has shown that this is not necessarily an immutable condition. His measurements indicate that turbulent fluctuations increase approximately linearly with wind speed up to about 10 or 12 m s^{-1}. Above that critical wind speed, the longitudinal and lateral velocity fluctuations increase even more rapidly. According to SethuRaman the increased turbulence is not caused by increased surface roughness at high speeds because there appeared to be no appreciable change in vertical wind component fluctuations. Based on numerous field and laboratory observations, SethuRaman hypothesizes that longitudinal roll vortices form, and introduce the observed turbulence during neutral conditions.

COASTAL CIRCULATIONS

Warmer Land and Cooler Water

Figure 6, which was based on a diagram given by Keen and Lyons (1978) is a schematic representation in cross-section of a well-developed lake- or sea-breeze circulation during mid-afternoon when the synoptic pressure gradients are small. There will frequently be a component of flow parallel to the shoreline so that the circulation shown in Figure 6 will in fact be helical, spiraling into or out of the plane of the diagram. There can be considerable difference in the depth of the inflowing air from one case to another; the diagram indicates a nominal range from about 100 to 1000 m. The inflowing air forms a "front" with pronounced temperature gradients separating the cool ocean air from the warmer air over the land. This front initially advances rather slowly--1 or 2 m s^{-1}. Speed increases as the front advances--to speeds of 3 to 5 m s^{-1}. The front eventually stalls 10 to 20 kilometers inland.

Considerable convergence (up to about 300 x 10^{-5} s^{-1}) and upward vertical motions (around 1 m s^{-1}) accompany the advancing frontal zone. The convergence and vertical motions are in a zone 1 to 2 km wide, parallel to the coast. Rising air may encounter an inversion aloft that will determine the height of the return flow. The return flow is frequently distributed through a layer that is deeper than the inflow layer by a few hundred to 1 or 2 thousand meters, so wind speeds in the return flow are only a few m s^{-1}; the inflow speed reaches 5 to 7 m s^{-1}. The air circulation is completed a few tens of kilometers offshore in a region of weak

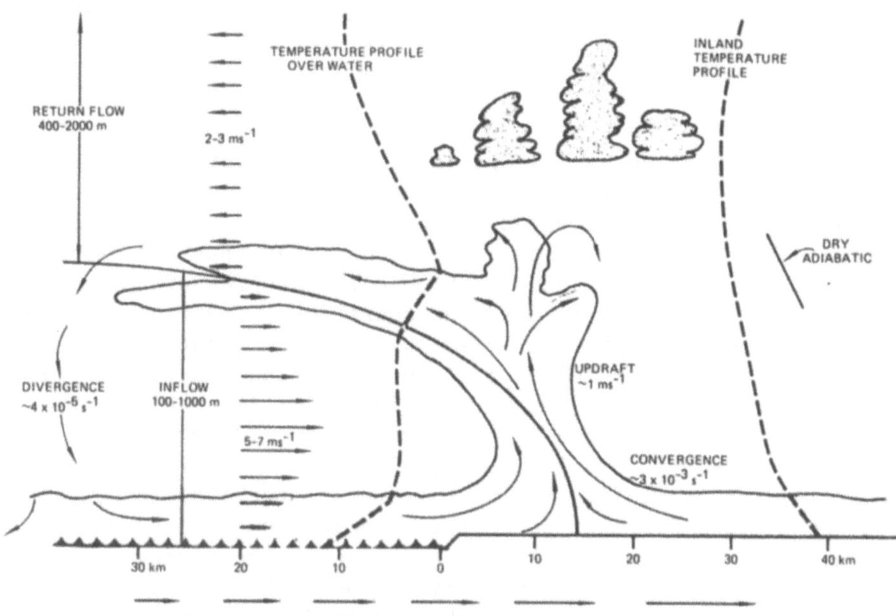

Fig. 6. Schematic diagram of a well developed lake (sea) breeze during mid afternoon

low-level divergence and downward motion. This divergence is about 2 orders of magnitude weaker than the convergence over land.

The subsiding air over the water, in combination with the cooler water itself, causes a temperature inversion to form near the surface. This inversion is eroded by the warmer land as the air moves onshore, as discussed earlier in connection with the TIBL. The TIBL has not been shown in Figure 6 in order to provide a somewhat clearer presentation of other features of the coastal circulation.

Figure 6 shows schematically how atmospheric pollution might be distributed along an urbanized coast. The relatively shallow smoke layer shown over the water surface would not normally be produced there (exceptions might be found where there was off-shore industrial or oil production activities). Instead, such polluted layers are the result of land-based emissions from the preceding evening that were transported offshore by a land breeze (to be discussed later). As the sea breeze develops, this pollution from the preceding day moves onshore and is augmented by new emissions as it moves further inland. At the convergence zone they may encounter smoke traveling in the opposite direction from sources farther inland, and be lifted into the return flow aloft. This circulation leaves a region of relatively clean air between the haze and pollution in the return flow, and that near the water surface.

Figure 6 provides an idealized picture of a typical, well-developed lake breeze for a simple, uncomplicated shoreline. Olsson et al. (1968) collected data during a June afternoon at the eastern shore of Lake Michigan under conditions that corresponded quite well to this idealized, light wind, weak pressure gradient case in a region of nearly straight shoreline. Their data showed the onshore component in the lower layers below about 300 m, penetrating about 7 km inland by 1400 local standard time (LST) and to nearly 15 km by 1700 LST. During this time period, the return flow was found in a layer between about 300 m and 3000 m. Thus, their observations fit the idealized picture of Keen and Lyons (1978) quite well. During the course of the afternoon, the flow developed a distinct northerly component so that by mid-afternoon the developing lake breeze was from the northwest, rather than perpendicular to the generally north-south coastline. By 1900 LST the air had penetrated more than 15 km inland, but flow in the offshore direction had begun to develop over the lake in the lowest layers near the shore, and the return flow had begun to weaken. The close agreement between the idealized sea breeze circulation and actual observations discussed above is not unique; similar events have been documented on the Oregon coast by Johnson and O'Brien (1973) where penetration as far as 60 km inland was found.

Keen et al. (1979) have calculated 3-dimensional trajectories for a lake breeze event on the western shore of Lake Michigan. They found trajectories originating within the coastal zone penetrated about 25 km over land. The Keen, Lyons, and Schuh study used calculated trajectories based on detailed 3-dimensional observations. Johnson et al. (1983) have performed an analogous study where smoke and tracer gases were released in the coastal zone of western Lake Michigan. They used an airborne lidar (laser radar) to plot vertical cross sections of smoke plume behavior. They were able to observe both the inflowing plume and the outflowing portion aloft. The fact that it was in fact the same plume in the return flow was confirmed by the presence of a tracer material that had been released with the smoke.

Uthe et al. (1980) have also obtained lidar cross sections that show a return flow above the marine layer. Figure 7 shows two of their cross sections obtained over the Los Angeles Basin during the morning and afternoon of the same day. The polluted aerosol layer is evident in the low layers in the morning. In

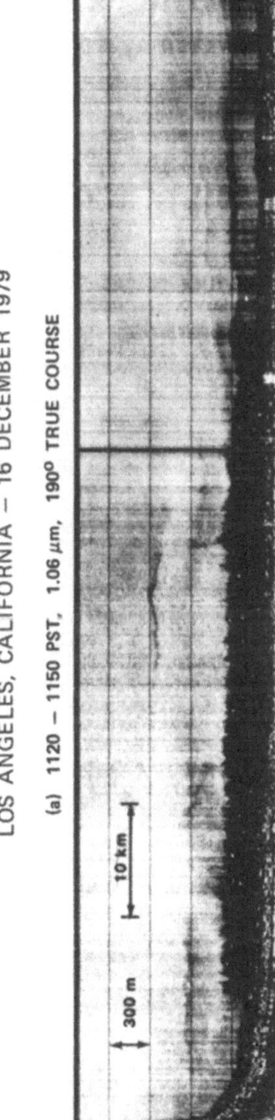

Fig. 7. Los Angeles boundary layer structure derived from Alpha-1 magnetic tape records, 16 december 1979

the afternoon, the cross section shows that the aerosol layer has been entrained in the return flow and is streaming seaward again.

To this point the discussion has focused on idealized cases and observations that correspond to them; but of course coastlines are not always straight and synoptic scale winds are not always light. If the synoptic scale winds are opposite in direction to the sea breeze, then the sea breeze must overcome those prevailing winds. Schroeder (1981) analyzed data from Hawaii where the sea breeze penetration must overcome the prevailing trade winds and found that mid-afternoon penetrations of about 12 km or less are typical; the sea breeze from one side of the island conflicts with trade winds entering from the other side.

Peninsulas,such as Florida may have sea breezes developing at opposite coasts and converging somewhere over the land (e.g., Pielke, 1974). Other complex shoreline shapes and coastal topographies can also lead to interactions between sea breeze circulations from different parts of the coast. Edinger and Helvey (1961) examined such a situation in Southern California, where sea breezes from two parts of the coastline were diverted by an intervening range of mountains, produced a zone of very strong convergence, and eventually met. It appears that such situations may be more likely to occur on coasts where the prevailing wind augments the sea breeze.

Other complications occur at urbanized shorelines with large roughness differences from one region to another. The roughness effects may be compounded by the shape of the shoreline, as in the New York metropolitan area where Bornstein and his coworkers found that the flow patterns show pronounced effects from complexities of shape and roughness. For example, Figure 8 (Bornstein, 1983) shows the patterns of convergence and divergence, and the late afternoon position of the sea breeze front on 9 March 1966. Sea breezes converge from both sides of Long Island and there is strong divergence over Long Island Sound. The angular coastline is reflected in the shape of the sea breeze front which appears to penetrate farther over the low lying, smoother regions between Long Island and Northern New Jersey than over the rougher urbanized areas of Lower Manhattan and Brooklyn. The retarding effect of the urbanized areas is shown clearly during the last few hours of the sequence of successive sea breeze frontal positions displayed by Figure 9 (Bornstein and Thompson, 1981). The retardation by the rougher urban surface also causes the leading slope of the sea breeze front to be steeper in the urban areas than it is in suburban areas; Anderson and Bornstein (1980) estimated frontal slopes using tetroon locations, applying a technique developed by Angell and Pack (1965), and found slopes of 1:125 over Northern New Jersey and 1:25 over Manhattan. Angell and Pack (1965) found a slope about mid-way between these two over Atlantic City, New Jersey.

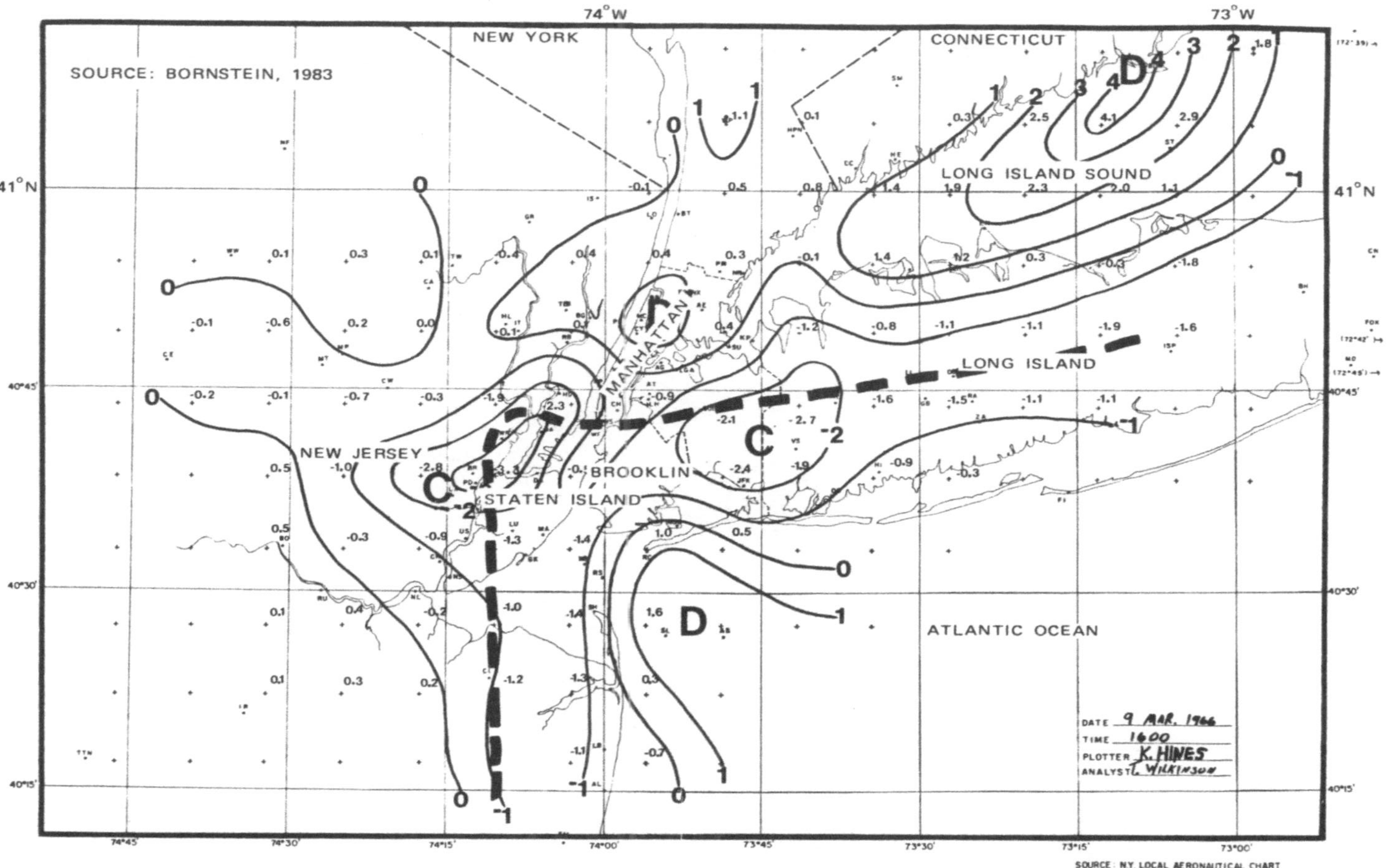

Fig. 8. Convergence (C), Divergence (D) Patterns and the location of the sea breeze front at 1600 on March 1966

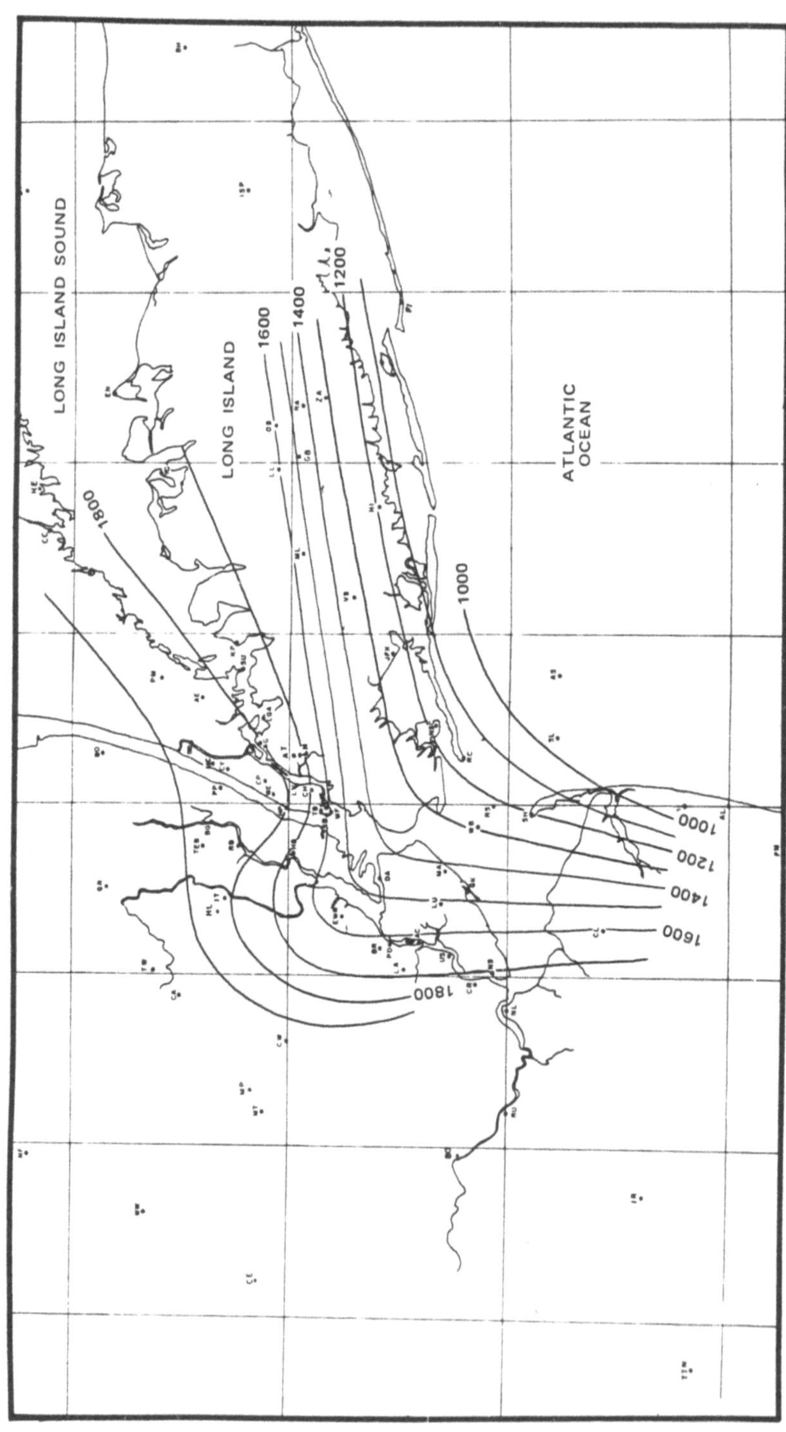

SOURCE: BORNSTEIN AND THOMPSON, 1981

Fig. 9. Positions of the sea breeze front over the New York area at different hours on 9 March 1984

Several attempts have been made to develop methods that would predict sea breeze development and inland penetration. Biggs and Graves (1962) and Lyons (1972) used variations of the same approach. Lyons (1972) defined a dimensionless index for the occurrence of a lake breeze, I:

$$I = V_g^2/(C_p \Delta T) = 12.71 \times 10^6/\Delta N^2(T_1 - T_w)$$

where V_g is geostrophic wind speed (ms^{-1}), ΔN is the 4-mb isobar spacing (km), and T_1 and T_w are the maximum inland air temperature for the day and the water surface temperature (^oC), their difference being ΔT. The specific heat of air is C_p. Lyons (1972) found that the critical value of the index for Chicago was 10; a lake breeze was unlikely to occur for higher values (on 90 to 95 percent of such occasions, if some additional criteria relating to the presence of nonconvective clouds over land were introduced).

The formula developed by Biggs and Graves (1962) used a measured wind from a site "uninfluenced by the lake breeze effect and therefore, representative of the ... wind velocity of the local area" instead of the geostrophic wind. Applying their formula to the western shore of Lake Erie, they found that a critical index value of 3 separated lake breeze events from non-breeze events. This formulation is consistent with Lyons' if the geostrophic wind used by Lyons is greater by about a factor of 1.8 than the airport wind speed.

Remembering that C_p, the specific heat for air has a value* of about 1 $j\ g^{-1}\ {}^oK^{-1}$, we get the following expression:

$$(\Delta T)_{crit} = 0.1\ V_g^2 = 0.3\ V^2$$

for the critical land/sea temperature difference needed to offset the effects of large-scale flow and produce a lake breeze; V is the wind speed ($m\ s^{-1}$) at a "representative" inland site. Thus, a land-sea temperature difference of about 10^oC is required to overcome a geostrophic counterflow of about 10 $m\ s^{-1}$, or a surface wind of about 5 or 6 $m\ s^{-1}$.

Even when the temperature difference between land and water is not great enough to induce a seabreeze, there will still be convergence tendencies and upward motions in shore areas with maximum temperature gradients. A land-sea temperature difference will also accentuate an onshore synoptic flow, producing low-level

---

*Biggs and Graves (1962) and Lyons (1972) used a mixed set of units when they chose $j\ g^{-1}\ {}^oK^{-1}$ for C_p and $m\ s^{-1}$ for wind speed.

convergence regions over land that are not well defined because they lack diametrically opposite wind directions of the idealized case, being marked only by a deceleration of the wind as it moves inland.

Cooler Land/Warmer Water

Figure 10 (derived from the work of Keen and Lyons, 1978) shows the typical pattern associated with a well-developed nighttime land breeze. In the lowest few hundred meters, the flow accelerates as it moves over the water, but at a slower rate than the inflowing daytime sea breeze (see Figure 6). Somewhere over the warmer water (which retains its warmth because of its great heat capacity), the air encounters slower moving air (or a light counterflow) and a convergence zone develops, producing upward motions of a few tenths of a meter per second. The cycle is completed by the landward return flow aloft and subsidence a few tens of kilometers inland. Figure 10 shows how polluted air is transported by the land breeze in a pattern that is more-or-less the mirror image of the daytime case shown in Figure 6.

Land breezes are not as well studied as sea breezes for several reasons. First, they are weaker and harder to measure. Second, they generally happen at night which complicates field measurement programs. Finally, transport away from pollution sources toward

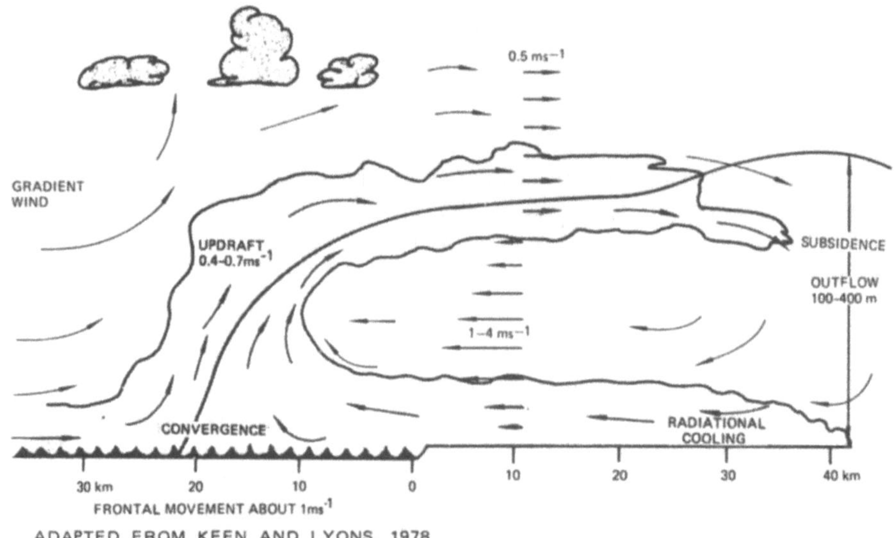

Fig. 10. Schematic diagram of a well developed land breeze at night

unpopulated bodies of water does not seem to interest funding agencies as much as other kinds of circulations do. Olsson et al. (1968) were most interested in sea breezes, but they found some evidence of land breeze circulations during the early and late hours of their study. Meyer (1971) used ultrasensitive 10.7-cm wavelength radar at Wallops Island, Virginia, to show that land breezes at the beach were characterized by a shallow layer of cool air with a depth less than 100 m. For the two cases observed by Meyer (1971), the speed of movement of the land breeze front out to sea was about 1 m s^{-1}; as it moved seaward, the shallow cold layer was modified by convection and increased its thickness to about 600 m in the vicinity of the convergence zone. The maximum seaward penetration of the observed land breeze fronts was about 30 km.

The Importance of the Along-Shore Wind Component

Land and sea breeze circulations are not strictly closed systems; they are very complex with their recirculations aloft and their diurnal reversals of flow. Although the circulations are not closed, it is certainly not safe to assume that pollutants, once they have traveled away from land, will never have an effect on a populated area. The component of the wind parallel to the shore determines where pollutants travel after they have left the land, and where they will once again return to the land. Lyons and Cole (1976) presented evidence that on some occasions, ozone monitored in Milwaukee had its origins in precursors that were emitted in the Chicago area. If one imagines a component of motion into (or out of) Figures 6 or 10, then the resulting streamlines will be spirals, with air moving along the shore--going out over the water and back over the land at different places along the shoreline. In the case of the Chicago-Milwaukee transport, Lyons and Cole (1976) hypothesized that there had been a gradient flow from land to the cooler lake during the morning, with precursors overriding the cool stable layer at the surface. The precursors continue to ride over this relatively clean, cold air, moving northward until a lake breeze develops, by which time the precursors and the ozone formed from them are far northward, off the Milwaukee coast. The developing lake breeze moves the ozone/precursor mixture landward where it becomes entrained in the TIBL. Fumigating to the ground and causing high ozone concentrations at the surface. Westberg et al. (1976) confirmed the findings of Lyons and Cole (1976) with aircraft and ground-based measurements of a wide variety of pollutants. They found that the fumigation associated with the TIBL played an important role in determining pollutant concentration, both for pollutants emitted near the shoreline and those transported from over the water.

Cass and Shair (1980) studied the same phenomena for Southern California. They used calculated surface wind trajectories for the months of September and October to estimate that there is nearly a 90 percent probability that material released into the early morning (0200 LST) land breeze at one location (Redondo Beach, California) would recross the Los Angeles County coast within about 19 hours. A sulfur hexafluoride tracer released from a power plant stack indicated that similar recirculations accounted for nearly all the tracer.

OTHER COASTAL PROCESSES INFLUENCING AIR POLLUTANT CONCENTRATIONS

Convection and Cloud Formation

The zones of convergence and divergence associated with sea and land breeze circulations (see Figures 6 and 10) have their effect on cloud formation, with the convergence zone favoring convective type clouds and the divergence areas tending to suppress them. If clouds do form, they can have a feedback effect on the differential heating and cooling that drives the circulations. In the daytime, the formation of clouds over the land may have a negative effect, inhibiting heating and reducing the temperature differences. At night when the land is radiatively cooling, divergence may inhibit cloud formation and enhance the land's cooling while clouds in the convergence zone over the water slow the cooling and enhance the temperature differences.

Fog and Stratus

Pilié et al. (1979) and Noonkester (1979), among others, have shown that the heating and cooling that takes place at the top of a fog or stratus layer can be important to the thickening and dissipation of those clouds. Pilié et al. (1979) cite four different mechanisms that are important to fog formation in coastal waters;

(1) Fog caused by instability and mixing over patches of warm water;

(2) Radiation fog advected to sea by the nocturnal land breeze,

(3) Fog that propagates downward by stratus thickening,

(4) Fog associated with low-level convergence.

According to Pelié (1979), air traveling over cold water develops an extreme stability in the surface-based inversion that prevents mixing to higher levels of moisture evaporating from the surface. This produces a thin, saturated layer of cool air at the surface. If this layer encounters a patch of warm water, the air's stability is removed so that it can mix with the nearly saturated cool air aloft to cause condensation and fog formation. After the fog forms, radiation cools its top and promotes further instability and turbulence at the upper levels causing the fog layer to get thicker as long as the air continues to pass over warm water.

The second mechanism in the above list represents a similar process. In this case, a shallow fog forms in a saturated stable layer over the nocturnally-cooling land surface. Then, that shallow fog layer is advected over the warmer water by the land breeze, and the mixing and deepening process described earlier is initiated.

The third category of fog formation, i.e., the development of fog through the lowering of stratus, proceeds as follows: Stratus (which frequently forms at the base of a subsidence inversion in coastal areas) is cooled in its uppermost layers by radiational heat loss at night; this cooling introduces instability beneath the inversion and leads to turbulent transport of cool air and cloud droplets downward. The cooling may also enhance the growth of droplets, which then settle from the cloud and evaporate in the clear, cool air below, increasing its humidity. These processes cause the level at which saturation occurs to progress steadily downward.

Finally, convergence may cause air that is nearly saturated to be lifted and to produce stratus clouds at the base of the inversion. Once formed, the stratus layers cool radiatively and thicken by the process described above. This stratus can be advected inland with the return flow, subsiding and evaporating some distance inland. The low level land breeze will then be more humid than it would be otherwise. If the land surface cools sufficiently before the stratus moves inland, then a low level radiation fog may form.

The presence of pollution in coastal zones may serve to enhance the cloud formation processes somewhat. Hung and Liaw (1981) suggest that the sulfate, nitrate and ammonium salts found in industrial aerosol emissions provide favorable sites for condensation, causing advection fogs to form at humidities less than 100 percent. If this were so, it would have its effect on radiative cooling and stability in the lowest layers of the atmosphere, which in turn would affect the nocturnal land-breeze circulations. It should be emphasized that these hypothesized effects have not been verified.

SUMMARY OF IMPORTANT ATMOSPHERIC PROCESSES IN THE COASTAL ZONE AND THEIR RELEVANCE TO AIR POLLUTION MODELING

Table 1 summarizes some of the more important effects that are related to each of the four major processes. The table emphasizes the idealized sea-land breeze patterns when the coast is straight with minimal topographic complexity. The effects of such complexities are not well understood, and provide the motivation for the development of the 3-dimensional boundary layer models discussed in the companion paper (Pechinger, 1983). With these models it will be possible to account for the interactions among all the processes.

Even without the benefit of the boundary layer models, it is apparent from the observations that the problems of pollutant transport and diffusion in coastal zones are fully 3-dimensional and time-varying. This makes both the boundary layer modelers' and the air pollution modelers' tasks much more difficult. They must be able to account for pollutant trajectories that reverse themselves, and emissions that interact with effluents emitted hours earlier from that same source. The models must be able to treat pollutants that travel with minimal dilution over long distances through stable air, and then are brought shoreward and fumigated down at some remote location. It may eventually be necessary to treat convection and condensation processes.

Once adequate boundary layer models are available to describe the circulation in complex coastal regions, it will no longer be necessary to rely on inadequate observations of atmospheric behavior in such regions in order to determine the behavior of pollutants. Air pollution models are almost always diagnostic in nature, using as inputs the temporal and spatial variations of the wind, turbulence, temperature, humidity, and so forth. This information must either be supplied by observations or generated through modeling. Typical pollution models introduce their own simplifications when describing transport, diffusion and transformations of pollutants. In the simplest models, steady-state, spatially-uniform conditions are assumed and transformations ignored. The most complicated models attempt to address all the conditions found in coastal regions. However, they are usually Eulerian in their formulation, and will often have rather poor spatial resolution which can cause difficulties in describing the initial phases of transport and diffusion from major point sources.

The obvious solution to this problem is to treat point source problems in a Lagrangian fashion, approximating them with a sequence of individual elements whose motions are determined by the large-scale air motions. There are two basically different approaches to Lagrangian modeling. One uses dimensionless mass points to represent plume motions, frequently introducing an

Table 1 Summary of Important Atmospheric Processes in Costal Zones

Process	Water/Land Temperature Relationship	Direction of Large-Scale Flow Relative to the Coast		
		Toward Land	Toward Water	Calm or Parallel
Transport and air motions	Water colder than land	1. Increased on-shore flow in the lower BL 2. Decreased (or reversed) onshore flow aloft 3. Convergence zone and upward flow over land in the lower BL 4. Divergence zone and downward flow over water in the lower BL	1. Decreased offshore flow in lower BL 2. Enhanced offshore flow aloft 3. Convergence zone and upward flow over land in the lower BL 4. Divergence zone and downward flow over water in the lower BL	1. Sea breeze flow likely to develop and produce the conditions associated with flow toward land
Transport and air motions	Water warmer than land	1. Decreased onshore flow in the lower BL 2. Convergence and upward motion over coastal zone in the lower BL	1. Increased offshore flow in the lower BL 2. Divergence and downward motion over coastal zone in the lower BL	1. Land Breeze flow likely to develop and produce conditions associated with flow toward water
Turbulence and diffusion	Water colder than land	1. TIBL forms at shore and inland in response to: o increased heating from below o increased surface roughness 2. Continuous fumigation of plumes imbedded in stable layer aloft, where top of TIBL intersects plume	1. Very shallow, very stable layer forms at water surface, effectively isolating surface from air above the stable layer 2. Gradual reduction of turbulent intensity in air above stable layer unless generated from the large-scale flow	1. Sea breeze may develop and produce conditions associated with flow toward land

(continued)

Table 1 (continued)

Process	Water/Land Temperature Relationship	Direction of Large-Scale Flow Relative to the Coast		
		Toward Land	Toward Water	Calm or Parallel
Turbulence and diffusion	Land colder than water	1. Stable surface layer forms over land--not as stable, but deeper than layer which forms over cold water	1. Mixing from below and formation of a deepening turbulent layer (similar to TIBL) over water. Somewhat less pronounced than TIBL because water is generally smoother than land 2. Fumigation similar to onshore flow over warmer land	1. Land breeze may develop and produce the conditions associated with offshore flow over colder water
Large scale convection	Water colder than land	1. Organized convection may develop inland due to surface heating will generally occur where TIBL is high or has merged with planetary BL	1. Cooling from below may cause pre-existing convective activity to weaken, or stop as air passes over colder water	1. A sea breeze may develop and lead to conditions described for landward flow
Large scale convection	Land colder than water	1. Cooling and reduced moisture inputs may cause pre-existing convective activity to dissipate as air moves over land. Some offsetting effects from convergence and upward motion in coastal zone	1. Organized convection may develop over warmer water once any surface stable layers have eroded. Convection over warmer water may be more vigorous than corresponding flow over warmer land because of significant input of moisture (latent heat).	1. A land breeze may develop and lead to conditions described for flow toward the water

Process	Water/Land Temperature Relationship	Direction of Large-Scale Flow Relative to the Coast		
		Toward Land	Toward Water	Calm or Parallel
Fog and cloud formation	Water colder than land	1. Advection fog or stratus moving from water to land and dissipating with time and distance over land	1. Formation of a shallow fog layer at surface of cold water	1. Light sea fog may form and drift onshore where it will dissipate if a seabreeze develops
Fog and cloud formation	Land colder than water	1. Formation of a shallow advection fog or stratus over land	1. Preexisting advection fog or stratus may be maintained.	1. A radiation fog may form over land and drift to sea if a land breeze develops

artificial "diffusion velocity" to account for turbulent dispersion (e.g., Lange, 1973; Sklarew et al., 1972). This is the "particle-in-cell" (PIC) approach. It requires very large computer resources because of the many points required to describe the pollutant field accurately, especially when relatively sharp gradients are found, as could well be the case in the highly variable coastal zone flows.

The second approach to Lagrangian modeling uses fewer elements, each with a finite volume. These elemental volumes can be "puffs" (e.g., Roberts et al., 1970; Start and Wendell, 1974; Ludwig et al., 1977; Zannetti et al., 1982) or more or less cylindrical segments aligned with the plume axis (e.g., Benkley and Bass, 1979). The major difficulty with Lagrangian models that use volume elements is that as the elements grow larger, the conditions at their center will not adequately represent conditions throughout the volume element. This can be particularly troublesome when there is a strong wind shear, or turbulent discontinuities (such as at the top of the TIBL). Thus, the most important coastal processes are exactly the ones that tend to make this kind of model unsuitable.

Sheih (1978) presented an approach where the puffs are described in terms of several points. This approach allows the model to treat the effects of wind shear, but at the expense of more calculations. Variations of this approach have been suggested (Ludwig, 1983), but are still in the developmental stages. It is not clear whether any of these techniques can be made to describe pollutant behavior accurately in the complicated coastal flows.

The Eulerian approaches are demonstrably workable (e.g., Ames et al., 1978) for treating time- and space-varying meteorological fields, as well as chemically interacting species. This approach tends to be expensive, but not prohibitively so, if one does not insist on fine spatial resolution. However, fine resolution will be required to resolve the behavior of emissions from point sources, and even more importantly, in order to treat pollutant behavior in the vicinity of the meteorological discontinuities that characterize the coastal atmosphere.

If one is only interested in simulating behavior of emissions of inert materials (or those that only undergo first-order chemical reactions) from major point sources, then the modeling approach will almost certainly be Lagrangian, although modification will be necessary to make the model capable of taking into account wind shear, and abruptly-varying turbulence fields. If the problem is with inert materials from a mixture of point and area sources, then an argument can still be made for the Lagrangian approach (along the lines used by Roberts et al., 1970). However, a combination of Lagrangian (for the point sources) and Eulerian (for the area sources) seems more reasonable. The plumes from point source emissions would then be transferred to the Eulerian system once their size becomes comparable with the cell size of the Eulerian modeling system. At least one existing model uses a variation of this approach (Liu et al., 1982).

The hybrid Lagrangian-Eulerian approach appears feasible and will meet many of the needs for modeling air pollution in coastal zones. A segmented plume might be used for treating systems with chemical reactions; the species in the plume reacting with each other and with the species concentrations in the Eulerian cell within which the plume segment is found. Eventually, when the plume concentrations become sufficiently dilute they should be transferred to the Eulerian system. Other hybrid approaches also seem possible, where materials from a number of individual, closely-spaced puffs are assigned temporarily, for purposes of simulating the chemical reactions to an Eulerian volume, but this approach will require some rational scheme for redistributing the reactants and the products from the chemical reactions back to their original Lagrangian elements so that the transport and diffusion calculations can be continued.

It is obvious from the observations and the foregoing discussion of their implications for air pollution and boundary layer modelers, that we do not need to worry about solving all the problems too soon and running out of work.

REFERENCES

Ames, J., Myers, T.C., Reid, L.E., Whitney, D.C., Golding, S.H., Hayes, S.R., Reynolds, S.D., 1978, "User's Manual for the SAI Airshed Model," SAI, Inc., San Rafael, California (draft).

Anderson, S.F. and Bornstein, R.D., 1980, "Sea Breeze Frontal Slopes and Vertical Velocities over New York City," Preprints Second Conf. on Coastal Meteorology, Los Angeles, January 30-February 1.

Angell, J.K. and Pack, D.H., 1965, "A Study of the Sea Breeze at Atlantic City Using Tetroons as Lagrangian Tracers," Mon. Wea. Rev., 93:475.

Bellaire, F.R., 1956, "The Modification of Warm Air Moving Over Cold Water," Proc. 8th Conf. on Great Lakes Research.

Benkley, C.W. and Bass, A., 1979, "Development of Mesoscale Air Quality Models, Vol. 2: User's Guide to MESOPLUME Model," Contract Report National Oceanic and Atmos. Admin. Contract 03-6-02-35254, Environmental Research and Technology, Inc., Concord, Massachusetts.

Bierly, E.W. and Hewson, E.W., 1962, "Some Restrictive Meteorological Conditions to be Considered in the Design of Stacks," J. Appl. Meteorol., 1:383.

Biggs, W.G. and Graves, M.E., 1962, "A Lake Breeze Index," J. Appl. Meteorol., 1:474

Bornstein, R.D., 1982, "Transport and Dispersion in an Urban Coastal Environment During Sea Breeze Frontal Conditions," Proc. 1st Internat. Conf. on Meteorol. and Air/Sea Interact. of the Coastal Zone, The Hague, Netherlands.

Bornstein, R.D., 1983, Personal communication.

Bornstein, R.D. and Thompson, W.T., 1981, "Effects of Frictionally Retarded Sea Breeze and Synoptic Frontal Passages in Sulfur Dioxide Concentrations in New York City," J. Appl. Meteorol., 20:843.

Cass, G.R. and Shair, F.H., 1980, "Transport of Sulfur Oxides Within The Los Angeles Sea Breeze/Land Breeze Circulation System," Proc. Second Joint Conf. on Applications of Air Pollution Meteorology, New Orleans, American Meteorological Society.

Dabberdt, W.F., Brodzinsky, R., Cantrell, B.K., Ruff, R.E., Dietz, R., and SethuRaman, S., 1982, "Atmospheric Dispersion over Water and in the Shoreline Transition Zone," Final Report, Vol. I for American Petroleum Institute, SRI International (Project 3450).

Deardorff, J.W., 1972, "Parameterization of the Planetary Boundary Layer," Mon. Wea. Rev., 100:93.

DiVecchio, R.A., Smith, D.B., and Martin, G., 1976, "Performance of a Recent Formulation for Rate of Growth of Boundary Layers Near Shorelines," Proc. Conf. on Coastal Met., September, Virginia Beach, Virginia, American Meteorological Society, Boston, Massachusetts.

Edinger, J.G. and Helvey, R.A., 1961, "The San Fernando Convergence Zone," Bull. Amer. Meteorol. Soc., 42:626.

Gifford, F., 1976, "Turbulent Diffusion Typing Schemes: A Review," Nuclear Safety, 17:68.

Hewson, E.W. and Olsson, L.E., 1967, "Lake Effects on Air Pollution Dispersion," J. Air Pol. Cont. Assoc., 17:757.

Hung, R.J. and Liaw, G.S., 1981, "Advection Fog Formation in a Polluted Atmosphere," J. Air Poll. Cont. Assoc., 31:55.

Johnson, A. and O'Brien, J.J., 1973, "A Study of an Oregon Sea Breeze Event," J. Appl. Meteorol., 12:1267.

Johnson, W.B., Dickson, C.R., Coulter, R.L., and Kornasiewicz, R.A., 1983, "Preliminary Results from SEADEX-I," 14th Inter. Tech. Meeting on Air Poll. Modeling and Its Applic., Copenhagen, Denmark, 27-30 September.

Keen, C.S. and Lyons, W.A., 1978, "Lake/Land Breeze Circulations in the Western Shore of Lake Michigan," J. Appl. Meteorol., 17:1843.

Keen, C.S., Lyons, W.A., and Schuh, J.A., 1979, "Air Pollution Transport Studies in a Coastal Zone Using Kinematic Diagnostic Analysis," J. Appl. Meteorol., 18:606.

Kerman, B.R., Mickle, R.E., Portelli, R.V., Trivett, N.B., and Misra, P.K., 1982, "The Nanticoke Shoreline Dispersion Experiment, June 1978--II, Internal Boundary Layer Structure," Atmos. Env., 16: 423.

Lange, R., 1978, "ADPIC--A Three-Dimensional Particle-in-Cell Model for the Dispersal of Pollutants and Its Comparison to Regional Tracer Studies," J. Appl. Meteorol., 12:320.

Liu, M-k., Stewart, D.A., and Henderson, D., 1982, "A Mathematical Model for the Analysis of Acid Deposition," J. Appl. Meteorol., 21:859.

Ludwig, F.L., 1983, "Air Quality Models Suitable for Use With A Coastal Boundary Layer Model," Review Paper for Stanford University, Civil Engineering Dept., Stanford, California.

Ludwig, F.L. and Dabberdt, W.F., 1976, "A Comparison of Two Atmospheric Stability Classification Schemes in An Urban Application," J. Appl. Meteorol., 15: 1172.

Ludwig, F.L., Gasiorek, L.A., and Ruff, R.E., 1977, "Simplification of a Gaussian Puff Model for Real-Time Minicomputer Use," Atmos. Envir., 11:431.

Lyons, W.A., 1972, "The Climatology and Prediction of the Chicago Lake Breeze," J. Appl. Meteorol., 11:1259.

Lyons, W.A., 1975, "Turbulent Diffusion and Pollutant Transport in Shoreline Environments," Chapter 5 of Lectures on Air Pollution and Environmental Impact Analyses, American Meteorological Society, Boston, Massachusetts.

Lyons, W.A. and Cole, H.S., 1973, "Fumigation and Plume Trapping on the Shores of Lake Michigan During Stable Onshore Flow," J. Appl. Meteorol., 12:494.

Lyons, W.A. and Cole, H.S., 1976, "Photochemical Oxidant Transport: Mesoscale Lake Breeze and Synoptic Aspects," J. Appl. Meteorol., 15:733.

Maas, S.J. and Harrison, P.R., 1977, "Dispersion Over Water: A Case Study of a Nonbuoyant Plume in the Santa Barbara Channel, California," Proc. Conf. on Applications of Air Poll. Meteorol., Salt Lake City, American Meteorological Society, Boston, Massachusetts.

Meyer, J.H., 1971, "Radar Observations of Land Breeze Fronts," J. Appl. Meteorol., 10:1224.

Michael, P., Raynor, G.S., and Brown, R.M., 1973, "Atmospheric Diffusion from an Off-Shore Site," Proc. Phys. Behav. of Radioactive Contam. in the Atmosphere, Vienna, Austria.

Noonkester, V.R., 1979, "Coastal Marine Fog in Southern California," Mon. Wea. Rev., 107:830.

Olsson, L.E., Cole, A.L., and Hewson, E.W., 1968, "An Observational Study of the Lake Breeze on the Eastern Shore of Lake Michigan, 25 June 1965," Proc. 11th Conf. on Great Lakes Research.

Pechinger, U., 1983, "Review of Selected Three-Dimensional Sea-Breeze Models," 14th Inter. Tech. Meeting on Air Pol. Modeling and Its Application, Copenhagen, Denmark, 27-30 September.

Peters, L.K., 1975, "On the Criteria for the Occurrence of Fumigation Inland from a Large Lake," Atmos. Env., 9:806.
Pielke, R.A., 1974, "A Three Dimensional Numerical Model of the Sea Breezes over South Florida," Mon. Wea. Rev., 102:115.
Pilié, R.J., Mack, E.J., Rogers, C.W., Katz, U., and Kocmond, W.C., 1979, "The Formation of Marine Fog and the Development of Fog-Stratus Systems Along the California Coast," J. Appl. Meteorol., 18:1275.
Raynor, G.S., Brown, R.M., and SethuRaman, S., 1978, "A Comparison of Diffusion from a Small Island and an Undisturbed Ocean Site," J. Appl. Meteorol., 17:129.
Raynor, G.S., Michael, P., Brown, R.M., and SethuRaman, S., 1975, "Studies of Atmospheric Diffusion from a Near Shore Site," J. Appl. Meteorol., 14:1080.
Raynor, G.S., Michael, P., and SethuRaman, S., 1980, "Meteorological Measurement Methods and Diffusion Models for Use at Coastal Nuclear Reactor Sites," Nuclear Safety, 21:749.
Roberts, J.J., Croke, E.S., and Kennedy, A.S., 1970, "An Urban Atmospheric Dispersion Model," Proc. Symp. on Multiple-Source Urban Diffusion Models, Air Poll. Cont. Office Pub. No. AP86.
Roll, H.U., 1965, Physics of the Marine Atmosphere, Academic Press, New York.
Schroeder, T.A., 1981, "Characteristics of Local Winds in Northwest Hawaii," J. Appl. Meteorol., 20:874.
SethuRaman, S., 1979, "Structure of Turbulence over Water During High Winds," J. Appl. Meteorol., 18:324.
Sheih, C.M., 1978, "A Puff Pollutant Dispersion Model with Wind Shear and Dynamic Plume Rise," Atmos. Environ., 12:1933.
Sklarew, R.C., Fabrick, A.J., and Prager, 1972, "Mathematical Modeling of Photochemical Smog Using the PICK Method," J. Air Poll. Cont. Assoc., 22:265.
Start, G.E. and Wendell, L.L., 1974, "Regional Dispersion Calculations Considering Spatial and Temporal Meteorological Varitions," National Oceanic and Atmos. Admin. Technical Memo ERL ARL-44.
Summers, P.W., 1965, "An Urban Heat Island Model and Its Application to Montreal," Paper Presented at 1st Canadian Conf. on Micrometeorol., Toronto, Canada.
Taylor, G.I., 1915, "Eddy Motion in the Atmosphere," Trans. Roy. Soc., London, 215:1.
Tennekes, H., 1973, "A Model for the Dynamics of the Inversion Above a Convective Boundary Layer," J. Atmos. Sci., 30:558.
Uthe, E.E., Nielsen, N.B., and Jimison, W.L., 1980, "Airborne Lidar Plume and Haze Analyzer (ALPHA-1)," Bull. Amer. Meteorol. Soc., 61:1035.
Venkatram, A., 1977, "Internal Boundary Layer Development and Fumigation," Atmos. Environ., 11:479.

Westberg, H., Sexton, K., and Roberts, E., 1981, "Transport of Pollutants Along the Western Shore of Lake Michigan," J. Air Poll. Contr. Assoc., 31:385.

Zannetti, P., Wilbur, D.M., and Schacher, G.E., 1982, "Characterizing Coastal Atmospheric Diffusion from Three-Dimensional Monitoring of SF_6 Tracer Releases," Proc. 1st Int. Conf. on Meteorol. and Air/Sea Interact. of the Coastal Zone, The Hague, Netherlands.

DISCUSSION

J. KNOX

Remote sensing has great potential for defining structure and imputs to local to regional scale model, provided the sometimes "massive" amount of data can be principally analyzed. Could you comment on the prognose in this direction ?

F.L. LUDWIG

The first step has been taken in that data are available in digital form, as well as in pictorial form. To date, their use has generally been limited to the definition of mixing lager depth, plume location and plume dimension. To some extent, progress has been hampered by old theory in that a lot of effort has been spent estimating σ values to go with Gaussian plume models. In our group we have recently started to use the data to study the effects of wind shear on plume behavior. We are also trying to adapt some pattern recognition techniques that have proved useful in satellite cloud studies to the problem of analyzing the fine structure of diffusing materials. Lidar is not different from other radically new measurement techniques in that it will take a while before researchers come up with the analytical tools necessary to make best use of the measurements.

T. MIKKELSEN

Consider the situation shown in Fig. 4. Now place an elevated source (as in Fig. 2) at the left hand land/water intersection so that it emits pollution towards the stable boundary layer growing up over the water. Are you able to comment on the decay-time for the transition of the plume diffusion from unstable to stable regime and to put this decay-time in relation to the obviously shorter fumigation timescale for the opposite transition envisioned in Fig. 2.

F.L. LUDWIG There are not many directly relevant data available to answer this question, but Lyons and Cole (1976) have provided at least one illuminating example. In their case, photochemical ozone was mixed through a relatively deep layer that moved out over a cooler lake. Transfer of ozone to the surface must have proceeded very slowly because very steep gradients developed and concentrations aloft remained quite high. The transfer was slow enough that the ozone could be transported 150 to 200 km with only slight diminution of concentration. By contrast, when a plume encounters a TIBL, the mixing and dilution occurs over distances of a few kilometers and time spans of tens of minutes.

REVIEW OF SELECTED THREE-DIMENSIONAL NUMERICAL SEA BREEZE MODELS

Ulrike Pechinger

Department of Meteorology
San José State University
San José, California 95192

and

Zentralanstalt fuer Meteorologie
und Geodynamik
Vienna, Austria

ABSTRACT

A review of three-dimensional sea breeze models is presented, with emphasis on model formulation and application, including simulation of transport and diffusion processes in coastal environments. Problem areas are discussed, and recommendations made for future model development.

INTRODUCTION

Three-dimensional mesoscale planetary boundary layer (PBL) models have been developed for a variety of applications, including simulation of flow over mountains, synoptic cold fronts, sea breeze fronts, and urban circulations. This review concentrates on selected three-dimensional sea breeze models in the open literature. The first model is presented in great detail in order to describe typical features of such models. Only singnificant changes in formulation or solution techniques are discussed for subsequent models, while a summary of selected features from each model is given in Tables 1 to 4.

MODELS

The first three-dimensional numerical sea breeze circulation simulation was carried out by McPherson (1970) for a rectangular shaped bay in a straight coastline like that of Galveston Bay.

Model equations were based on those of the two-dimensional model of Estoque (1961, 1962), in which orography and moisture were neglected. Atmospheric motions were assumed hydrostatic, a valid assumption as the horizontal sea breeze circulation scale is larger than its vertical scale (Dutton and Fichtl, 1969).

The vertical derivative form of the continuity equation was used. While this allows for an upper boundary condition on vertical velocity, mass is no longer conserved.

The model consisted of a 3.8 km deep numerical transition layer above a 50 m deep surface boundary layer (SBL), in which turbulent heat and momentum fluxes were assumed constant with height. Analytical wind and temperature profiles were obtained for three different convective regimes defined by values of the SBL Richardson number Ri. Such formulations have been shown by Clarke (1970) and Taylor and Delage (1971) to reduce inaccuracies associated with finite differencing to the surface.

However, McPherson (and others) used a larger critical Richardson number Ri_c (at which turbulence dies out) than found from SBL observations, i.e., 0.2 versus 33. This is necessary as both the finite difference Ri and Ri_c are directly proportional to vertical grid spacing (Shir and Bornstein, 1977), and the 200 m model spacing was greater than tower heights used in most observational SBL studies. Transition layer eddy viscosity coefficients were assumed equal for heat and momentum, independent of local conditions, and specified to decrease exponentially from their values at the top of the SBL.

At the onshore and offshore boundaries, horizontal wind velocity and the normal gradients of vertical wind velocity, pressure, and temperature were assumed to vanish. Zero gradient conditions normal to the lateral boundaries perpendicular to the coastline were imposed on all parameters, while at the upper boundary steady state conditions were assumed.

A simple sinusoidal surface temperature wave with a maximum amplitude of 10C and a constant surface temperature over water were specified. A uniform roughness length of 1 cm was used and the atmosphere was assumed calm and stably stratified at the initial time of 0800 LST.

A forward in time and centered in space integration scheme with a 5 min time step was utilized in the 56 by 272 km model domain with equal grid spacings of 4 km in the horizontal and 200 m in the vertical. Stable results were only obtained after a 25 point space filter was applied to the horizontal wind and temperature fields at each time step. This filter significantly reduced the amplitudes of waves with two and four grid interval wave lengths.

Horizontal wind and temperature fields at the 250 m level and vertical velocity fields at the 850 m level (only results given) showed clearly thermal effects on the circulation due to the distorted coastline. Two symmetric centers of strongest upward motion initially developed, and a single center of subsidence appeared over the bay. After six hours, the temperature and wind fields became asymmetric due to Coriolis turning, with maximum upward velocities reaching 60 cm s^{-1}. Maximum horizontal onshore velocities of 7 m s^{-1} occurred near the tenth hour.

Differential atmospheric cooling due to smaller wind speeds and reduced turbulence levels on the landward side of the convergence zone prolonged the onshore flow at low levels for several hours beyond the time when the surface temperature gradient had reversed. By 2200 LST, vertical motions had mostly vanished and the horizontal flow was mainly parallel to the coast due to the Coriolis turning.

Information on the vertical extent of the sea breeze circulation, depth of the inflow and return flow layers, intensity of the return flow, and atmospheric thermal stratification were not presented. McPherson had only limited observational data to compare with results, but his suggestion that simulated zones of maximum convergence might coincide with observed zones of enhanced cumulus convection was later confirmed by Pielke and Mahrer (1978) and Cotton et al. (1976).

The model of Pielke (1974) included a discontinuity in surface roughness, constant large-scale synoptic forcing, and moisture (to estimate cloud base heights). The assumption of incompressible flow which filters out sound waves, is valid as the vertical scale height of the sea breeze circulation is less than the scale height of density (Dutton and Fichtl, 1969).

While a constant surface roughness length was prescribed over land, it was calculated over water according to Clarke (1970) as a function of surface stress, as was SBL depth following Blackadar and Tennekes (1968). The depth of the PBL was assumed 25 times that of the SBL.

Wind and temperature profiles in the SBL were dependent on Monin-Obukhov length following Yamamoto and Shimanuki (1966), with those for heat and moisture assumed identical. Two profile functions were distinguished, and thermal stratification was determined by the difference between potential temperature at the first numerical grid point and that at the surface.

Vertical eddy diffusivities above the SBL were evaluated following O'Brien (1970). This formulation produces a maximum value at a height of about one-third the PBL depth. A horizontal diffusion coefficient (from Leith, 1969), given as a function of horizontal grid spacing and flow deformation, was included as a numerical smoother.

The initial velocity profile was determined from a steady state base flow consisting of a geostrophic wind at the top of the PBL and Ekman flow within the transition layer. Potential temperature and specific humidity vertical gradients were specified for each simulation.

To account for the vertical spreading of mass due to the addition of heat into a column, Pielke used a material surface (where local pressure changes were assumed zero) as the model top. This was in contrast to the rigid top used by McPherson. Material surface heights were obtained by integrating the continuity equation from the top grid point to the previous height of this surface. This method produced only negligibly small local pressure changes at the material surface. It was subsequently found by Mahrer and Pielke (1975) that the method reduced the model height (from 20 to 10 km) necessary to simulate two-dimensional flow over a 1900 m high ridge.

The semi-implicit finite-difference scheme on a staggered variable grid was forward in time and centered in space, except for the advection terms which were upstream differenced because this highly dispersive scheme selectively dampens short wavelenghts. The maximum land-sea surface temperature difference was given as 10C.

Two numerical experiments studied the influence of typical southeasterly and southwesterly synoptic winds on development and movement of sea breeze convergence zones over southern Florida. In both cases, a convergence zone formed first along the windward coast due to its larger horizontal thermal gradient. The zone resulted from a change in speed rather than direction. Maximum upward vertical velocities associated with the zone as it moved inland were about 60 cm s^{-1}.

A second region of upward motion subsequently formed along the leeward coast, but remained essentially stationary. Coastal curvature accentuated the convergence pattern, leading to local vertical velocity maxima. Lake Ocheechobee had a substantial influence on convergence patterns in its vicinity, as its lake breeze reinforced the convergence caused by the (larger scale) sea breeze.

Predicted convergence patterns compared well with those of McPherson and with observed locations of convective clouds (from ATS photographs) and shower locations (from radar maps). This agreement indicated that mesoscale (sea breeze) moisture convergence is important in determining convective shower location over southern Florida.

Convergence generated by surface heating alone was substantially larger than that caused solely by the roughness change between land and water. However, surface friction affected surface convergence by increasing the intensity of vertical heat and momentum fluxes.

Further development of the Pielke model was carried out by Mahrer and Pielke (1976) using terrain following height coordinates. They simulated flow over Barbados, a hilly island of about 20 by 30 km and a 240 m maximum elevation. Vertical profiles in the SBL were changed to the newer formulation of Businger (1973) and virtual potential temperature replaced potential temperature. PBL depth was predicted as a function of surface heat and momentum fluxes, vertical motion, and free atmospheric thermodynamic stability following Deardorff (1974).

To avoid numerical problems, the mountain grew during the first 30 minutes of the simulation to its maximum height following a procedure developed by Deaven (1974). Surface temperatures, based on observations, specified a (maximum) land-sea difference of 8C around 1400 LST and a difference of -3C around 0400 LST. An easterly geostrophic wind of 10 m s^{-1} was assumed.

Over the eastern slope, a stationary cell of upward motion was present at all elevations during both daytime and nighttime hours. A wide cell of downward motion occured above the center of the island and above the western slope during the same period. However, off the west coast, a more complex vertical velocity pattern resulted from advection of heated island air.

Agreement was found between simulated vertical velocities and those from tetroon trajectories in the layer between 250 and 750 m, but no observations in upper layers were available.

An extension of this work by Cotton et al. (1976) linked output from a Florida sea breeze simulation to a cumulus cloud model. Results showed precipitating clouds produced by the sea breeze convergence to be significantly deeper and longer lived than those produced by local heating.

In a second study of sea breeze development over southern Florida, Pielke and Mahrer (1978) added a heat balance equation to predict surface temperatures over land. The equation included solar and longwave radiation, as well as latent, sensible, and soil heat fluxes. A soil sublayer was added with 10 levels and a constant grid spacing of 5 cm.

Changes of air temperature due to shortwave and longwave radiative flux divergence were calculated following Atwater and Brown (1974). Atmospheric heating by shortwave radiation was due to water vapor, while both carbon dioxide and water vapor contributed to longwave cooling.

Advection was modeled by an upstream cubic splines formulation, while vertical diffusion was evaluated by a modified Crank-Nicholson scheme as suggested by Paegle et al. (1976) which allowed larger time steps with almost identical solutions. Previous comparisons of the old and new advection schemes by Mahrer and Pielke (1978) had shown little difference in results over flat terrain (like that of Florida).

The new convergence zone patterns agreed well with those from the older model, although convergence magnitudes increased. Results were compared to radar maps, satellite pictures, and surface wind and temperature observations. While most observed showers occured in predicted convergence zones, much of the convergence region was not covered by precipitating clouds. Sea breeze induced moisture fluxes were well correlated with observed cloud base locations during most of the day.

Observed surface winds and temperatures were compared with calculated values at the 3 m level. Differences between observed and predicted surface winds off the eastern coast were attributed to increased mechanical turbulence due to urbanization, a feature not included in the model. Additional disagreement occured in the western half of the peninsula due to local showers which substantially perturbed the observed wind field.

Particle streak lines were calculated for this simulated sea breeze circulation by NcNider and Pielke (1979). Particles were released along the eastern and western coasts at 10 min intervals at heights of 50 and 300 m.

Table 1. General Features

Author(s)	Location	Results	Evaluation	Special Features
McPherson (1970)	Galveston Bay	V, w, T as f(x,y)	None	first three-dimensional simulation
Pielke (1974)	South Florida	V(x,y) w(x,y)	satellite, radar	roughness difference; prevailing flow
Mahrer & Pielke (1976)	Barbados	V(x,y), u(z), $\theta_e(z)$; w, θ, q as f(x,z)	tetroons	orography; relative height coordinates
Pielke & Mahrer (1978)	South Florida	V(x,y), w(x,y), Q(t), V(x,t), θ(x,t)	V(x,y) V(t) satellite, radar	radiative flux divergence
McNider & Pielke (1979)	South Florida	streaklines	None	trajectories
Lyons & Schuh (1979)	Lake Michigan	V, w, H as f(t)	V, w, H as f(t)	good evaluation
Segal et al. (1981)	Tidewater section of Chesapeake Bay	V, H, χ as f(x,y)	None	Eulerian diffusion model

(continued)

Table 1. General Features (Continued)

Author(s)	Location	Results	Evaluation	Special Features
Segal et al. (1982)	Chesapeake Bay	V, w, H as f(x,y) V, w, T as f(x,z) V(x,t), V(z), T(z) streaklines	V(t), V(z), T(z)	space varying prevailing flow
Hjelmfelt & Braham (1983)	Lake Michigan	V, w, p, r, T, T_d, θ_e as f(x,y) u(x,z), w(x,z)	p, V, T, T_d, θ_e as f(x,y) clouds & precipitation	condensation precipitation
Tapp & White (1976)	South Florida	V(x,y), w(x,y)	None	non-hydrostatic
Warner et al. (1978)	Chesapeake Bay	V(x,y), H(x,y,z) θ(x,y), fronts	None	convective adjustment
Estoque & Cross (1981)	Lake Ontario	V(x,y), w(x,y)	None	orography σ-coordinates
Yamada (1980)	Lake Michigan	V(x,y) u, w, K as f(x,z) hodographs	None	second-moment turbulence-closure
Liu et al. (1979)	Tampa Bay	V(x,y), w(x,y)	None	lant $T_o(x,y)$

Author(s)	Location	Results	Evaluation	Special Features
Takano (1983)	Tokyo	V, w, θ as f(x,y) w(x,z), θ(x,z)	None	urban effects
Kikuchi et al. (1981)	Kanto, Japan	V(x,y) V, θ, w as f(x,z)	None	orography; relative height coordinates
Bornstein & Klotz (1983)	New-York City	V, w and θ as f(x,z)	streamlines & isotachs	vorticity mode

List of Symbols in Tables 1-4

- h SBL-height
- H PBL-height
- K vertical eddy diffusivity
- K_H horizontal eddy diffusivity
- p pressure
- q moisture
- Q moisture flux
- r precipitation rate
- s material surface
- t time
- T temperature
- T_o surface temperature
- T_d dewpoint temperature
- u east-west wind component
- v north-south wind component
- $\mathbf{V}$ wind velocity vector
- V horizontal wind speed
- V_N normal wind component
- V_T tangential wind component
- w vertical wind component
- w* vertical (z*) wind component
- x east-west cartesian coordinate
- y north-south cartesian coordinate
- z vertical cartesian coordinate
- z* vertical terrain following coordinate
- z_o roughness length
- z_G ground elevation
- Δ horizontal grid spacing
- Δx minimum horizontal grid spacing
- Δz minimum vertical grid spacing
- π Exner function
- σ relative pressure coordinate
- θ potential temperature
- θ_e equivalent potential temperature
- θ_v virtuel potential temperature
- χ pollutant concentration
- ? unknown

Streak lines from release points on the west coast, where the sea breeze was counter to the synoptic flow, indicated helical trajectories of the type observed in Lake Michigan breezes by Lyons and Olsson (1973). Horizontal convergence occurred for particles released from the east coast.

A comparison of observed and computed lake breeze wind fields on the western shore of Lake Michigan with no topography was carried out by Lyons et al. (1979) using the same model. Observed data consisted of pibal wind profiles, aircraft cross sections of turbulence and pollution, and tetroon trajectories.

Analyses of hourly vertical and horizontal wind profiles showed that the computed lake breeze began approximately one hour after it was observed. Both breezes penetrated about 30 km inland with approximately the same propagation rate. Computed inflow depth was shallower than observed during the the first two hours, about the same during mid-day, and higher in late afternoon hours.

A three-dimensional numerical diffusion model was linked with the Pielke sea breeze model to investigate CO concentration patterns over the Tidewater region (Chesapeake Bay) of Virginia by Segal et al. (1980). Daily average CO-emission data included only surface level sources.

Predicted afternoon concentration patterns, affected by the southeasterly sea breeze, showed CO transport toward the north-west. Intensive vertical mixing over the land caused a gradual reduction of values from morning peak concentrations. Verification of simulated patterns was not possible due to a lack of observations.

Segal et al. (1982) also applied the Pielke model to the greater Chesapeake Bay area, incorporating a spatial variation of an initial westerly synoptic wind field. The sea breeze front moved about 25 km inland and predicted vertical velocities in the sea breeze convergence zones were unexplainably small, with maximum values reaching only 8 cm s^{-1}

A sharp gradient of PBL height developed along the coastal zones around 0900 LST, when the initially uniform 250 m height increased inland to 800 m in response to surface heating. PBL heights inland continued to increase during the day to a maximum of 2100 m.

Due to the advection term in the prognostic PBL height equation, PBL heights over the water increased during the day as the return flow aloft strengthened. Even though the lower part of the marine PBL was stable, it was claimed that higher turbulence intensities aloft in the return flow region would result in increased diffusion of materials in this layer.

Streak lines of emitted particles behaved as expected in a sea breeze circulation. However, diffusive effects can not be reproduced by this method. Effective (advective plus diffusive) velocities found in particle in cell models can incorporate diffusion, but such models require a significantly greater number of particles.

Hjelmfelt and Braham (1983) simulated modification of cold air moving over warm Lake Michigan using the Pielke model and compared results with observations of a précipitat lake effects snow storm. They included a precipitation scheme based on work by Asai (1965) and McCumber (1980) in which excess water vapor freezes and precipitates. Latent heating effects were also included.

Surface temperatures were specified from surface observations, and initial temperature, specific humidity, and wind profiles were based on soundings taken at Green Bay with northwesterly wind speeds increasing from 5 m s^{-1} near the surface to 15 m s^{-1} at 3 km.

The predicted near steady state mesoscale cyclonic circulation at the 250 m (below cloud) level after a 15 hr simulation agreed well with aircraft measurements at that level. Simulated and observed convergence zones were both found over the southeastern shore of the lake, but maximum calculated upward velocities of 20 cm s^{-1} were much smaller than the observed 100 m s^{-1} velocities. As an observed large scale horizontal wind shear was not incorporated into the model, an observed southwesterly flow around the southern end of the lake was not correctly simulated.

Calculated precipitation areas over the eastern half of the lake were in general agreement with aircraft and radar observations. Simulations without latent heat release produced similar (but less intense) circulation patterns. Details of the lake surface temperature distribution had only small effects on the simulated flow patterns.

Table 2. Physical Assumptions

Author(s)	Continuity Equation	SBL	K	K_H	h (m)	H (km)	Top (km)
McPherson (1970)	$\frac{\partial}{\partial z}(\nabla \cdot \mathbf{V}) = 0$	free, forced & near zero	exponential decrease	25-point smoother	50	3.85	3.85
Pielke (1974)	$\nabla \cdot \mathbf{V} = 0$	Yamamoto & Shimanuki (1966)	O'Brien (1970)	f(Δ, V)	Blackadar and Tennekes (1968)		4.82
Mahrer & Pielke (1976)	"	Businger (1973)	"	"	"	"	6
Pielke & Mahrer (1978)	"	"	"	filter of Long (1977)	$\frac{H}{25}$	Deardorff (1974)	"
McNider & Pielke (1979)	"	"	"	"	"	"	"
Lyons & Schuh (1979)	"	"	"	"	"	"	3.5
Segal et al. (1981)	"	"	"	"	"	"	7
Segal et al. (1982)	"	"	"	"	"	"	"

Author(s)	Continuity Equation	SBL	K	K_H	h (m)	H (km)	Top (km)
Hjelmfelt & Braham (1983)	$\nabla \cdot \mathbb{V} = 0$	Businger (1973)	O'Brien (1970)	filter of Long (1977)	$\frac{H}{25}$	Deardorff (1974)	14
Tapp & White (1976)	Not Used	Clarke (1970)	"	$f(\Delta, V)$	75	?	4.82
Warner et al. (1978)	"	Businger (1973)	Busch (1965)	"	10	$f(\frac{\partial\theta}{\partial Z})$	500 mb
Estoque & Gross (1981)	"	Busch (1973)	O'Brien (1970)	"	100	12	12
Yamada (1980)	$\nabla \cdot \mathbb{V} = 0$	Businger (1966) Dyer & Hicks (1970)	2nd moment closure	"	14	None	2.62
Liu et al. (1979)	"	Businger (1971)	O'Brien (1970)	None	Monin-Obukhov length	$f(\frac{\partial\theta}{\partial Z})$	$\sim$ 2

(continued)

Table 2. Physical Assumptions (Continued)

Author(s)	Continuity Equation	SBL	K	K_H	h (m)	H (km)	Top (km)
Takano (1983)	$\nabla \cdot \mathbf{V} = 0$	Yamamoto (1975)	Mellor & Yamada (1974)	$f(\Delta, V)$ or $f(\Delta, \theta)$	50	None	3
Kikuchi et al. (1981)	"	free, forced & near zero	exponential decrease	$f(\Delta, V)$	50	2	2.8
Bornstein & Klotz (1983)	"	Monin-Obukhov (1954) Priestley (1959)	O'Brien (1970)	Constant	50	1	2

Tapp and White (1976) used a dry, non-hydrostatic model to simulate a 30 hour sea breeze circulation over southern Florida, results of which were compared to those of Pielke (1974). Within a 75 m deep SBL, fluxes of heat and momentum were determined as a function of Monin-Obukhov length following Clarke (1970). Other model features, such as eddy diffusivity profiles above the SBL, surface temperature variation, roughness length, horizontal diffusion coefficients, region of integration, and initial conditions were similar to those of Pielke (1974).

Velocity components normal to all lateral boundaries and potential temperature and tangential wind components at inflow boundaries were equal to constant synoptic scale values associated with a 6 m s^{-1} southeasterly flow. At outflow boundaries, potential temperature tendencies were calculated by upstream advection, while tangential wind tendencies were equal to values one grid point inside the boundary. Pressures at the lateral boundaries were determined from gradients normal to the boundary, and horizontal diffusion was increased by a factor of three at the three grid points nearest the lateral boundaries.

Surface fluxes were horizontally smoothed to avoid sudden changes across the coast which tended to produce double grid length waves. A semi-implicit scheme was used on a staggered grid in which terms involving horizontally and vertically propagating sound waves were treated implicitly, with all other terms explicitly differenced.

The convergence zone along the west coast developed further inland than in the Pielke model with a weaker upward motion pattern showing some double grid length waviness. Along the east coast, the flow accelerated in general agreement with results from the Pielke model.

The smaller upward velocities were attributed to the non-hydrostatic system, which allows the vertical pressure gradient to absorb vertical momentum. However, in contrast to the Pielke results and observations, considerable subsidence occured onshore of all coastlines and the effect of Lake Ocheechobee was significantly reduced due to the smoothed surface fluxes.

Warner et al. (1978) also simulated development of a sea breeze circulation in the Chesapeake Bay area using the PBL model of Anthes and Warner (1978). They did not include synoptic forcing, moisture, or radiative flux divergence.

The model used sigma (relative pressure) coordinates with the equations of motion in flux form. The sophisticated PBL parameterization of Busch et al. (1976) was used. Transition layer eddy diffusivities were dependent on local vertical gradients of wind and potential temperature, as well as a prognostic mixing length, itself a function of PBL height and surface heat and momentum fluxes. PBL was defined as the first level above the ground at which the vertical gradient of potential temperature exceeded $1.3^{o}C\ km^{-1}$.

Anthes and Warner (1978) had introduced a dry convective adjustment scheme to prevent numerically produced instabilities from growing uncontrollably in a staticly unstable PBL. The adjustment scheme accounts for vertical convective heat transport due to organized local overturning of convectively unstable layers. This overturning falls in between the vertical scale of sea breeze circulations and that of turbulent fluxes treated by gradient transport theory.

Lateral boundary pressures and temperatures were extrapolated and velocities set to zero. Surface land temperatures were predicted from the energy budget equation of Blackadar (1976). This formulation is less complex than that in the Pielke model as it excludes evaporation and uses a simple linear soil heat flux and constant atmospheric absorptivity. Land and water surface roughnesses were specified, while calm winds and a stable early morning temperature sounding served as initial conditions.

An explicit, centered in space finite difference scheme was used with a time integration scheme developed by Shuman and Hovermale (1968) and generalized by Brown and Campana (1978). This scheme gave the same accuracy as the leapfrog scheme for time steps nearly twice the leapfrog scheme, and it was 1.6 times faster (Anthes and Warner, 1978).

Predicted winds showed the general features of a sea breeze circulation, with inland penetration from both the Atlantic and the Bay and return flows aloft. Zero velocity lateral boundary conditions caused erroneously large westerly wind components, near the eastern model boundary.

PBL depth near the shoreline increased during the morning due to local heating. In the afternoon, advection of cold air behind the fronts significantly affected near shoreline static stability by stabilizing the lower layers and reducing PBL depth. Temperatures along the east coast of the peninsula at 80 m were 7C cooler than along the west coast and about 11C

Table 3. Boundary Conditions, Initial Conditions, and Solution Techniques

Author(s)	Surface Temperature	Lateral Boundary Conditions	Initial Conditions	Numerical Scheme
McPherson (1970)	Prescribed	Onshore and offshore : $u, v, \frac{\partial}{\partial x}(w, \pi, \theta) = 0$ Perpendicular to coastline : zero gradient	Calm; stable (idealized)	Forward time, centered space
Pielke (1974)	"	$v, w, \theta, q, \frac{\partial}{\partial x,y}(\pi, S) = 0$	Ekman; stable	Semi-implicit forward in time upstream
Mahrer & Pielke (1976)	"	$\frac{\partial s}{\partial t}, \frac{\partial}{\partial x,y}(\theta_v, q, \pi, H),$ $w^*, \frac{\partial}{\partial x,y}(u, v) = 0$ inflow : u, v constant	Ekman; near neutral	Semi-implicit forward in time upstream DuFort and Frankel
Pielke & Mahrer (1978)	Heat balance equation	$\frac{\partial}{\partial x,y}(\theta, \pi, Z_G, S, q), w^* = 0$ outflow : $\frac{\partial}{\partial x,y}(u, v) = 0$ inflow : u, v constant	Ekman; stable	"

(continued)

Table 3. Boundary Conditions, Initial Conditions, and Solution Techniques (Continued)

Author(s)	Surface Temperature	Lateral Boundary Conditions	Initial Conditions	Numerical Scheme
McNider & Pielke (1979)	Heat balance equation	$\frac{\partial}{\partial x,y}(\theta, \pi, Z_G, S, q)$, $w = 0$ Outflow : $\frac{\partial}{\partial x,y}(u, v) = 0$ Inflow : u, v constant	Ekman; stable	Semi-implicit forward in time upstream DuFort and Frankel
Lyons & Schuh (1979)	"	"	Observed v; neutral, with stable above	"
Segal et al. (1981)	"	Same as Pielke & Mahrer (1978)	Ekman; observed T & q	Same as Mahrer & Pielke (1976)
Segal et al. (1982)	"	"	Observed V; stable	"
Hjelmfelt & Brahm (1983)	Prescribed	$\frac{\partial}{\partial x,y}(\theta, \pi, Z_G, S, q)$, w^*, $\frac{\partial}{\partial x,y}(u, v) = 0$	Observed V; convectively unstable	"

Author(s)	Surface Temperature	Lateral Boundary Conditions	Initial Conditions	Numerical Scheme
Tapp & White (1976)	Prescribed	V_N from NWP Inflow : θ & V_T from NWP Outflow : $\frac{\partial}{\partial N}(\theta, V_T) = 0$ Neumann for p	Ekman; neutral with stable above	Semi-implicit
Warner et al. (1978)	Heat balance equation	Extrapolated	Calm; stable (idealized)	Brown and Campana (1978) centered in space
Estoque & Gross (1981)	"	Cyclic	Eckman; stable	Matsuno (1966) time integration Grimmer and Shaw (1967)
Yamada (1980)	"	Inflow : one-dimensional solution Outflow : extrapolated	Logarithmic V-profile; neutral with stable above	Peaceman and Rachford (1955)

(continued)

Table 3. Boundary Conditions, Initial Conditions, and Solution Techniques (Continued)

Author(s)	Surface Temperature	Lateral Boundary Conditions	Initial Conditions	Numerical Scheme
Liu et al. (1979)	Specified	$\frac{\partial}{\partial x,y}$ (u, v, θ) = 0	Ekman; idealized θ	Semi-implicit predictive mode : upwind diagnostic mode : centered
Takano (1983)	Heat balance equation	Inflow : (V, θ) = constant Outflow : extrapolated	Ekman; stable	Matsuno (1966) time integration; Arakawa (1972)
Kikuchi et al. (1981)	Specified	Radiation conditions on velocity	Idealized T; Geostrophic	Box method of Bryan (1966) Matsuno (1966) time integration
Bornstein & Klotz (1983)	Specified	Radiation conditions on velocity and θ	Ekman balance; adiabatic	Donor cell forward in time centered in space (for diffusion)

cooler than inland. Cold air advection from the sea breeze reduced PBL depth at 1630 LST from about 2.5 km at inland sites, to 1.5 km at the west coast, and to 0.25 km at the east coast. Warming due to subsidence over the water decreased static stability in the layer between 0.5 and 2.0 km, while static stability at lower levels was not changed significantly.

The model of Estoque and Gross (1981), which simulated development of land-sea breeze circulations over Lake Ontario, included orographic effects of terrain features with heights up to 300 m. The model was similar to that of Anthes et al. (1971) and was originally developed by Gross (1978). It did include synoptic scale forcing, moisture, evaporative surface cooling (following Deardorff, 1978), transition layer eddy diffusivities of O'Brien (1970), the SBL formulation of Estoque (1973), a stability dependent von Kármán constant as suggested by Busch (1973), and cyclic lateral boundary conditions.

The finite-difference scheme for horizontal derivatives was a combination of the Grimmer and Shaw (1967) and Matsuno (1966) forward-backward scheme. Initial temperature and humidity distribution were equal to average vertical observed profiles on 3 October 1972 and surface temperatures were initailly 10C over land and 15C over water.

With no prevailing flow over flat terrain the daytime lake breeze circulation contained irregular flow patterns in areas away from the lake, while during the night an organized land breeze circulation was not simulated. Inclusion of orography generated daytime upslope winds in qualitative agreement with observations, However, observations were not available to verify the computed 30 cm s^{-1} maximum vertical velocities. Inclusion of both a southerly synoptic flow of 4 m s^{-1} and orography produced easterlies over the lake and the southern coastal regions in the afternoon in agreement with observations.

Yamada (1980) used second moment turbulence-closure equations to parameterize turbulence fluxes in a 12 hour sea breeze simulation over Lake Michigan. Prognostic equations for turbulent kinetic energy and length scale were solved and the remaining turbulence second moments were computed from a set of algebraic equations following Mellor and Yamada (1977).

Turbulence levels in coastal environments vary greatly in space and time. Thus, this more sophisticated turbulence parameterization scheme might be expected to produce more realistic turbulence intensity patterns than gradient tranport theory (Cederwall, 1982). Such results would provide more accurate turbulence values when linked with air pollution models.

With upper west-southwesterly geostrophic flow of 2 m s^{-1} and a maximum surface temperature difference of 12C, a daytime lake breeze front developed along the western shore and moved 10 to 20 km inland. The maximum computed turbulent vertical exchange coefficient upwind of the lake was 250 m^2 s^{-1}, while the corresponding value downwind of the lake was only 175 m^2 s^{-1} due to advective cooling of the lower atmosphere in that region by the lake. Turbulence over the lake was considerably reduced in qualitative agreement with plume observations. More observations of turbulence in a coastal environment are needed for an evaluation of the model results.

A 24 hr simulation, including development and decay of a land and sea breeze circulation, was carried out by Liu et al. (1979) for Tampa Bay, using a model with features similar to those already described above in the various models. Temporal and spatial surface temperature variations were specified from empirical data, and initial temperature and wind profiles were based on observations.

The model produced the essential features of a land-sea breeze circulation with inland flow during the day and offshore flow during the night. Convergence zones arising from coastal curvature with maximum upward velocities around 4 cm s^{-1} formed during the day, while divergence zones occured during the night with maximum downward velocities of 2 cm s^{-1}. Comparison with observations was not attempted due to lack of observational data.

Takano (1983) simulated land-sea breeze circulations in the area of Tokyo neglecting orography but including the "level 2" turbulence closure model of Mellor and Yamada (1974). Turbulent heat and momentum fluxes were still parameterized by eddy diffusion coefficients similar to those developed by Busch et al. (1976) and used by Warner et al. (1978). The main difference involved replacement of the previous prognostic equation for mixing length by a diagnostic equation. The new formulation has mixing length as a function of the eddy dissipation rate, which is obtained from a steady state form of the turbulent kinetic energy equation. Urban effects were included by specification of the proper physical parameters in the surface heat balance equation and by inclusion of a constant (with time) anthropogenic heat input to the lowest model layer.

Results showed formation of a daytime urban heat island of about 2C at 50 m and strong upward motion (40 cm s^{-1}) over the city, in agreement with observations. Sea breeze penetration over the city was inhibited for several hours due to the urban

Table 4. Model Grid

Author(s)	Domain (km)	Grid Points	Δ x (km)	Δ z (m)	Simulation (hr)
McPherson (1970)	276 x 56	70 x 50 x 20	4	200	18
Pielke (1974)	572 x 605	33 x 36 x 6	11	50	13
Mahrer & Pielke (1976)	165 x 150	27 x 18 x 10	5	50	29
Pielke & Mahrer (1978)	572 x 605	33 x 36 x 11	11	47.5	24
McNider & Pielke (1979)	"	"	"	"	"
Lyons & Schuh (1979)	290 x 390	29 x 39 x 10	10	40	22
Segal et al. (1981)	300 x 360	30 x 36 x 12	"	10	∿ 12

(continued)

Table 4. Model Grid (Continued)

Author(s)	Domain (km)	Grid Points	Δ x (km)	Δ z (m)	Simulation (hr)
Segal et al. (1982)	300 x 360	30 x 36 x 12	10	10	12
Hjelmfelt & Braham (1983)	440 x 520	36 x 46 x 20	8	52	15
Tapp & White (1976)	363 x 396	33 x 36 x 7	11	595	30
Warner et al. (1978)	290 x 290	30 x 30 x 12	10	∿ 160	12
Estoque & Gross (1981)	190 x 280	26 x 26 x 5	"	∿ 600	24
Yamada (1980)	250 x 330	25 x 25 x 8	"	103	12

Author(s)	Domain (km)	Grid Points	Δ x (km)	Δ z (m)	Simulation (hr)
Liu et al. (1979)	90 x 90	30 x 30 x ?	3	25	24
Takano (1983)	248 x 248	32 x 32 x 21	8	100	48
Kikuchi et al. (1981)	225 x 225	30 x 30 x 12	7.5	50	21
Bornstein & Klotz (1983)	40 x 40	16 x 16 x 16	1	25	20

thermal low which produced a heat island flow counter to the sea breeze on the inland side of the city. When the inland pressure force finally overcame the urban effects, the front moved past the city. The observational study of Bornstein and Fontana (1979) attributed observed stalling of sea breeze fronts over New York City during non-heat island periods to increased frictional drag of the urban surface.

A numerical study of the effects of mountains on land and sea breeze circulations in the Kanto district of Japan was carried out by Kikuchi et al. (1981). The model utilized a material surface and relative height coordinates.

Results without orography showed sea breeze circulations restricted to within 40 km of the shore. When orography was included, mountain and valley winds appeared earlier than did sea and land breeze circulations.

A three-dimensional vorticity-mode sea breeze model has been developed by Bornstein and Klotz (1983). The model is an extension of the two-dimensional URBMET urban boundary layer model of Bornstein (1975) and is called COASTAL-URBMET. It contain progrostic equations for the x- and y- vorticity components and diagnostic equations for two stream functions. The model is Boussinesq and uses the "hydrostatic vorticity," equal to the vertical gradient of the horizontal wind.

In a first test, the model was applied to a right angle coastline approximately the shape of that in the New York City area. A westerly geostrophic flow of 3 mps was specified and the simulation began at 1800 LST. After 20 hours of simulated time, when the land-sea surface temperature difference was 8C, a well developed sea breeze convergence zone formed on the north-south shoreline of New Jersey. The computed near surface streamline pattern showed remarkable agreement with one constructed using data from the NYU/NYC surface aneomometer network (Bornstein et al., 1977).

CONCLUSION

All models described in this review have shown themselves capable of simulating general features of the spatial and temporal variations of mesoscale wind and temperature patterns associated with the development of sea breeze circulations over flat terrain with (real) irregular coastlines. In addition, some models described herein have gone further by:

(1) Simulating the full life cycle of sea-land breeze circulations (Liu et al., 1979; Takano, 1983),
(2) Including orography (Mahrer and Pielke, 1976; Estoque and Gross, 1981; Kikuchi et al., 1981),
(3) Including space varying geostrophic wind (Segal et al., 1982),
(4) Including urban effects (Takano, 1983),
(5) Using higher order turbulence closure (Yamada, 1980), or
(6) Including condensation and precipitation processes (Hjelmfelt and Braham, 1983).

While future simulations should seek to incorporate all or some of these features when appropriate, certain problem areas still exist in some or all models. One is determination of PBL height for use in formulations of transition layer eddy viscosity coefficients. Most models specify this level arbitrarily as either an interior vertical grid level or the model top.

Only Warner et al. (1978) and Segal et al. (1982) use sophisticated methods to estimate this parameter. Both methods produce similar daytime results at inland sites not affected by sea breeze flow. However, they yield dramaticly different PBL growth rates in areas affected by sea breeze circulation.

In the Segal et al. model, large mostly uniform predicted PBL heights over land are advected offshore with the return flow aloft. This causes large (but decreasing) PBl heights offshore within approximately 70 km of the coastline. In the Warner et al. model, cold air advection from water to land stabilizes the lower layers over land and reduces onshore PBL heights, while ocean areas retain their initially specified zero values.

The Warner et al. reduced onshore PBL depths are consistent with reduced vertical eddy diffusivities found downwind of a lake by Yamada (1980) with his second moment turbulence closure model. Thus, the diagnostic vertical temperature gradient method works better than the prognostic local rate of change equation in the near coastal onshore zone.

However, the zero PBL depths predicted by this diagnostic temperature gradient method within the stably stratified marine layer are not reasonable. On the other hand, the large advective produced PBL heights in the Segal et al. application of the prognostic Deardorff equation also do not seem reasonable because of theoretically low turbulence levels within the stable marine layer below. Since the prognostic equation was originally developed for unstable and neutral stratifications, it seems that it should not be applied during stable conditions.

Thus, it appears a different method should be applied within a stable layer when the mixing depth is small. For example, the PBL height (top of the layer of surface influence) could be determined as the top of the surface based radiation inversion. This level would be above the zero value of Warner et al., but below the advective produced value of Segal et al.

The plethora of lateral boundary conditions utilized in models reviewed in this paper, as well as problems with some of the soulutions at the lateral boundaries, indicate that this aspect of numerical PBL modeling is more "an art than a science". However, given the large degree of internal forcing associated with sea breeze flows, this problem is not as serious as with other types of mesoscale flows (Warner et al., 1978).

Another problem associated with all PBL models is selection of numerical solution techniques which minimize numerical errors, such as numerical diffusion and double grid length waves. This problem is discussed in a review of mesoscale modeling by Pielke (1981).

Selection of a proper numerical scheme is one method of reducing computer storage requirements and decreasing model execution time. Because of these limitations, many models reviewed in this paper used too few vertical grid levels, and thus too large vertical grid spacings, to properly simulate vertical details of sea breeze circulations near the surface.

Too few sea breeze modelers present sufficient results to fully describe simulated phenomena. Model results should be presented over a large enough area so that the entire circulation, including zones of subsidence over water, are shown. Various horizontal and vertical cross sections of wind, temperature, and moisture should be presented. Information on inflow and return flow depths, thermal internal boundary layer heights, PBL heights, convergence zone locations, and turbulence levels at different times should be given. All illustrations should be presented legibly so as to allow for qualitative evaluation of model results.

Most current models predict spatial and temporal variations of land surface temperature form energy balance equations involving long and short wave radiation and latent, sensible and soil heat fluxes. This method implicitly allows for surface temperature changes due to air mass advection.

Future surface energy balance equations might also incorporate effects of cloud shadows, cooling of the ground by precipitation (Bhumralkar, 1973), vegetation (Deardorff, 1978), more varied land usage patterns (Myrup and Morgan, 1972) soil moisture effects on subsurface thermal parameters (Santhanam and Bornstein, 1981), and anthropogenic moisture fluxes (Bornstein and Robock, 1976). Other future developments should utilize time and space varying boundary conditions obtained from outputs of larger scale models (Perkey and Kreitzberg, 1976) and include simulation of fog and stratus formation (Lewellen et al., 1983).

Previous linkages between sea breeze and air quality simulation models have either involved surface area CO sources or streak lines from elevated point sources. Future models should simulate the fate of multiple urban point and area sources during sea breeze circulations. Problems associated with air quality modeling in coastal zones are discussed by Ludwig (1983).

The Workshop on Coastal Transport Processes (SethuRaman, 1982) recommended coordinated field studies to gather data on the life cycle of land-sea breeze circulations. Such mean flow and turbulence data are necessary to properly evaluate existing and future sea breeze models.

Acknowledgement

This review was carried out as part of the Coastal Meteorology Research Program at San Jose State University under the Electric Power Research Institute Grant No. RP 1630-13. The author would like to thank Professor Robert D. Bornstein of the Department of Meteorology at San Jose State University for his many suggestions and careful review of this manuscript. The author would also like to thank Ms. Donna Hurth for her careful typing.

REFERENCES

Anthes, R. A., S. Rosenthal and J. Trout, 1971: Preliminary results from an asymmetrical model of the tropical cyclone. Mon. Wea. Rev., 99, 744-758.

Anthes, R. A., and T. Warner, 1978: Development of hydrodynamic models suitable for air pollution and other meso-meteorological studies. Mon. Wea. Rev., 106, 1045-1078.

Arakawa, A., 1972: Design of the UCLA general circulation model. Department of Meteor., Univ. of Calif. Los Angeles, Techn. Rep. No. 6, 116 pp.

Asai, T., 1965: A numerical study of the airmass transformation over the Japan Sea in winter. J. Meteor. Soc. Japan, 43, 1-15.

Atwater, M. A., and P. Brown, Jr., 1974: Numerical calculation of the latitudinal variation of solar radiation for an atmosphere of varying opacity. J. Appl. Meteor., 13, 289-297.

Bhumralkar, C. M., 1973: An observational and theoretical study of atmospheric flow over a heated island: part II. Mon. Wea. Rev., 101, 731-745.

Blackadar, A. K., 1976: Modeling the nocturnal boundary layer. Preprint, Third Symp. Atmospheric Turbulence, DIffusion, and Air Quality, Raleigh, Amer. Meteor. Soc., 46-49.

Blackadar, A. K., and H. Tennekes, 1968: Asymptotic similarity in neutral barotropic planetary layers. J. Atmos. Sci., 25, 1015-1020.

Bornstein, R. D., and A. Robock, 1976: Effects of variable and unequal time steps for advective and diffusive processes of the urban boundary layer. Mon. Wea. Rev., 104, 260-267.

Bornstein, R. and S. Klotz, 1983: Development of the three-dimensional COASTAL-URBMET model. Quarterly Technical Report No. 7. to EPRI from SJSU, Dept. of Meteorology.

Brown, J., and K. Campana, 1978: An economical time-differencing system for numerical weather prediction. Mon. Wea. Rev., 106, 1125-1136.

Busch, N. E., 1973: On the mechanics of atmospheric turbulence. Workshop on Micrometeorology, D. A. Haugen, Ed., Am. Meteor. Soc., Boston, pp. 1-65.

Busch, N. E., S. Chang and R. Anthes, 1976: A multi-level model of the planetary boundary layer suitable for use with mesoscale dynamic models. J. Appl. Meteor., 15, 909-919.

Businger, J. A., 1973: Turbulenct transfer in the atmospheric surface layer. Workshop in Micrometeorology, S. A. Haugen, Ed., Am. Meteor. Soc., Boston, Chap. 2.

Cederwall, R. T., 1982: Review of algebraic stress models for simulating atmospheric turbulence in a three-dimensional sea-breeze model. Suppl. Quarterly Rep. EPRI-RP1630-13 from SJSU, Dept. of Meteorology.

Clarke, R. H., 1970: Recommended methods for the treatment of the boundary layer in numerical models. Aust. Meteor. Mag., 18, 51-73.

Cotton, W. R., R. Pielke and P. Gonnou, 1976: Numerical experiment on the influence of the mesoscale circulation on the cumulus scale. J. Atmos. Sci., 33, 252-261.

Deardorff, J. W., 1974: Three-dimensional numerical study of the height and mean structure of a heated planetary boundary layer. Bound.-Layer Meteor., 7, 81-106.

Deardorff, J. W., 1978: Efficient prediction of ground surface temperature and moisture with inclusion of a layer of vegetation. J. Geophys. Res., 83, 1889-1904.

Deaven, D. G., 1974: A solution for boundary problems in isentropic coordinate models. Ph.D. dissertation, The Pennsylvania State University, 136 pp.

Dutton, J. A., and G. Fichtl, 1969: Approximate equation of motion for gases and liquids. J. Atmos. Sci., 26, 241-254.

Estoque, M. A., 1961: A theoretical investigation of the sea breeze. Quart. J. R. Meteor. Soc., 87, 136-146.

Estque, M. A., 1962: The sea breeze as a function of the prevailing synoptic situation. J. Atmos. Sci., 19, 244-250.

Estoque, M. A., 1973: Numerical modeling of the planetary boundary layer. Workshop in Micrometeorology, D. A. Haugen, Ed., Am. Meteor. Soc., Boston, pp. 217-268.

Estoque, M. A., 1981: Further studies of a lake breeze. Part I: Observational study. Mon. Wea. Rev., 109, 611-618.

Estoque, M. A., and J. Gross, 1981: Further studies of a lake breeze. Part II: Theoretical study. Mon. Wea. Rev., 109, 619-634.

Fontana, P. H., and R. Bornstein, 1979: Observations of frictional retardation of sea breeze fronts. San Jose State Univ. report to NSF, ATM 77-21467, 57 pp.

Grimmer, M., and D. Shaw, 1967: Energy-preserving integrations of the primitive equations on the sphere. Quart. J. R. Meteor. Soc., 93, 337-349.

Gross, J. M., 1978: Lake-effect storms on Lake Ontario. Ph.D. Thesis, University of Miami, Coral Gables, 210 pp.

Hjelmfelt, M. R., and R. Braham, Jr., 1983: Numerical simulation of the airflow over Lake Michigan for a major lake-effect snow event. Mon. Wea. Rev., 111, 205-219.

Kikuchi, Y. et al., 1981: Numerical study of the effects of mountains on the land and sea breeze circulation in the Kanto district. J. Meteor. Soc. Japan, 59, 723-737.

Leith, C. E., 1969: Two-dimensional eddy viscosity coefficients. Proc. WMO/IUGG Symp. Numerical Wea. Prediction, Meteor. Soc. of Japan, Tokyo, Nov. 26-Dec. 4, 1968.

Lewellen, W. S., R. Sykes and D. Oliver, 1983: Further developments of the A.R.A.P. model for the atmospheric marine environment. ARAP #488. Avail. from NTIS, 169 pp.

Long, P., 1977: Personal communication to Pielke and Mahrer.

Liu, M. K., T. Myers and J. McElroy, 1979: Numerical modeling of land and sea breeze circulation along a complex coastline. Mathematics and Computers in Simulation, 21, 359-367.

Ludwig, F. L., 1983: A review of coastal zone meteorological processes important to the modeling of air pollution. Preprints, NATO CCMS 14th ITM on air pollution modeling and its application. Copenhagen, Denmark, Sept. 27-30, 1983.

Lyons, W., and L. W. Olsson, 1973: Detailed mesometeorological studies of air pollution dispersion in the Chicago lake breeze. Mon. Wea. Rev., 101, 387-403.

Lyons, W., J. Schuh and M. McCumber, 1979: Comparison of observed mesoscale lake breeze wind fields to computations using the University of Virginia mesoscale model. Preprints, Amer. Meteor. Soc. 4th Symposium on Turbulence, Diffusion, and Air Pollution, Reno, Nev., Jan 15-18, 1979.

Mahrer, Y., and R. A. Pielke, 1975: A numerical study of the air flow over mountains using the two-dimensional version of the University of Virginia mesoscale model. J. Atmos. Sci., 32, 2144-2155.

Mahrer, Y., and R. A. Pielke, 1976: Numerical simulation of the airflow over Barbados. Mon. Wea. Rev., 104, 1392-1402.

Mahrer, Y., and R. A. Pielke, 1978: A test of an upstream spline interpolation technique for the advective terms in a numerical mesoscale model. Mon. Wea. Rev., 106, 818-830.

Matsuno, T., 1966: Numerical integrations of the primitive equations by a simulated backward difference method. J. Meteor. Soc. Japan, 44, 76-84.

McCumber, M. C., 1980: A numerical simulation of the influence of heat and moisture fluxes upon mesoscale circulation. Ph.D. Thesis, University of Virginia. Also issued as Rep. UVA-ENV-SCI-MESO-1980-2, Dept. Enviorn. Sci., University of Virginia, 255 pp.

McNider, R. T., and R. Pielke, 1979: Application of the University of Virginia Model to air pollutant transport. Preprints, Amer. Meteor. Soc. Fourth Symposium on Turbulence, Diffusion, and Air Pollution, Reno, Nev., Jan 15-18, 1979.

McPherson, R. D., 1968: A three-dimensional numerical study of the Texas coast sea breeze, Report No. 15, NSF Grant GA-367X, Univ. of Texas at Austin, College of Engineering, Atmospheric Science Group, 252 pp.

Mellor, G. L., and T. Yamada, 1974: A hierarchy of Turbulence closure models for planetary boundary layer. J. Atmos. Sci., 31, 1791-1806.

Mellor, G. L., and T. Yamada, 1977: A turbulence model applied to geophysical fluid problems, Proc. Symp. on Turbulence Shear Flows, Pennsylvania State University, State College, PA, April 18-20, 1977.

Myrup, L. O., and D. Morgan, 1972: Numerical model of the urban atmosphere. U. C. Davis Report No. 4, 237 pp.

O'Brien, J. J., 1970: A note on the vertical structure of the eddy exchange coefficient in the planetary boundary layer. J. Atmos. Sci., 27, 1213-1215.

Paegle, J., W. G. Zdunkowski and R. M. Welch, 1976: Implicit differencing of predictive equation of the boundary layer. Mon. Wea. Rev., 104, 1321-1324.

Perkey, D. J., and C. Kreitzberg, 1976: A time-development lateral boundary scheme for limited-area primitive equation models. Mon. Wea. Rev., 104, 744-755.

Peaceman, D. W., and H. H. Rachford, Jr., 1955: The numerical solution of parabolic and elliptic differential equations, SIAM J. Appl. Math., 3, 28-41.

Pielke, R. A., 1974: A three-dimensional numerical model of the sea breeze over the Gulf of Florida. Mon. Wea. Rev., 102, 115-139.

Pielke, R. A., 1981: Mesoscale numerical modeling. Adv. in Geophys., 23, 185-344.

Pielke, R. A., and Y. Mahrer, 1975: Representation of the heated planetary boundary layer in mesoscale models with coarse vertical resolution. J. Atmos. Sci., 32, 2288-2308.

Pielke, R. A., and Y. Mahrer, 1978: Verification analysis of the University of Virginia three-dimensional mesoscale model prediction over south Florida for 1 July 1973. Mon. Wea. Rev., 106, 1568-1589.

Santhanam K. and R. Bornstein, 1981: One-dimensional simulation of temperature and moisture in atmospheric and soil boundary layers. Preprint Vol., AMS Symposium on Turbulence, Diffusion, and Air Pollution, Atlanta, GA, 94-98.

Seagal, M., R. McNider, R. A. Pielke and D. McDougal, 1982: A numerical model simulaltion of the regional air pollution meteorology of the Greater Chesapeake Bay area--Summer day case study. Atmos. Environ., 16, 1381-1398.

Segal, M., R. A. Pielke and Y. Mahrer, 1980: Quantitative assessment of air quality in the Greater Chesapeake Bay area using a three-dimensional atlmospheric model. Proc. Symp. on Intermediate Range Atmos. Transport Processes and Technol. Assess., Gatlinburg, Tenn. Oct. 1-3, 1980.

SethuRaman, S., 1982: Proceedings of the workshop on coastal atmospheric transport processes. BNL-report 51666, Brookhaven, 43 pp.

Shir, C. C., and R. Bornstein, 1977: Eddy exchange coefficients in numerical models of the planetary boundary layer. Bound. Layer Meteor., 11, 171-185.

Takano, K., 1983: Three-dimensional numerical modeling of the land and sea breezes and the urban heat island in the Kanto Plain. Submitted to Bound. Layer Meteor.

Tapp, M. C., and P. W. White, 1976: A non-hydrostatic mesoscale model. Quart. J. R. Meteorol. Soc., 102, 277-296.

Taylor, P. A., and Y. Delaze, 1971: A note on finite-difference schemes for the surface and planetary boundary layers. Bound. Layer Meteor., 2, 108-121.

Warner, T. T., R. Anthes and A. McNab, 1978: Numerical simulations with a three-dimensional mesoscale model. Mon. Wea. Rev., 106, 1079-1099.

Yamada, T., 1980: Numerical simulation of mesoscale atmospheric circulations over the Lake Michigan. ASME Annual Meeting, Nov. 16-21, 1980, Chicago, Ill.

Yamamoto, G., 1975: Generalization of the KEYPS formula in diabolic conditions and related discussion on the critical Richardson number. J. Meteor. Soc. Japan, 53, 189-195.

Yamamoto, G., and A. Shimanuki, 1966: Turbulent transfer in adiabatic conditions. J. Meteorol. Soc. Japan. 44, 301-307.

DISCUSSION

A. BERGER — Are these 3-D models practically capable of simulating a full diurnal cycle ? Or are they limited only to "snapshot" because of computational cost ?

U. PECHINGER — Some of the models have successfully simulated a full diurnal cycle (Liu et al. 1979; Takano, 1983).

J. KNOX — Could you please clarify if the Tapp and White model (1976) is the only non-hydrostatic model among those reviewed ?

U. PECHINGER — Yes.

J. KNOX — The hydrostatic assumption is considered to be invalid if $\left|\frac{\partial u}{\partial z}\right| \cong \left|\frac{\partial w}{\partial x}\right|$. Did Tapp and White examine solutions sensitivity to zone size in view of this above consideration.

U. PECHINGER — No, they only used a II km horizontal grid spacing in the reviewed paper.

We have seen good results for the cases of sea breezes ; do you think reliability can also be archieved in land breezes due to the difficulty to model turbulence in a stable layer ?

U. PECHINGER

Liu et al. (1979) have simulated the development of a land breeze circulation. Flow patterns of this simulation seem reasonable. However, observations to evaluate model results were not available. Takano (1983) also has simulated land breeze circulation in the Tokyo area with a second order turbulence closure model and found general agreement of model results with observations. Turbulence in a O'Brien type K-model is reduced at night, but still non-zero. Therefore, with a reversed temperature gradient even equal in magnitude to the daytime gradient any simulated land breeze will be weaker than a sea breeze. In a higher order turbulence closure model turbulence levels probably will be more reasonably simulated, i.e. more reduced. Therefore the most appropriate model formulation to be utilized in answering this question is one in which there is no prevailing flow. In addition, results from both, a higher order turbulence closure scheme and a K-formulation should be compared to observations.

SIMULATIONS OF A TRACER EXPERIMENT IN THE ØRESUND REGION

Leif Enger

Meteorological Department
Uppsala University
S-751 20 Uppsala, Sweden

Sven-Erik Gryning

Physics Department
Risø National Laboratory
DK-4000 Roskilde, Denmark

Erik Lyck

Air Pollution Laboratory
National Agency of Environmental Protection
DK-4000 Roskilde, Denmark

Ulf Widemo

Studsvik Energiteknik AB
S-611 82 Nyköping, Sweden

ABSTRACT

Main results are presented from a mesoscale tracer experiment that was carried out over a land-water-land area (the Øresund region). In the experiment a tracer was released from a meteorological mast on the coast of Skåne in southern Sweden. After having travelled about 21 km over the Øresund water surface, the plume encountered the coast of Denmark. Tracer samples were taken at two crosswind series, one along the Danish coast and one about 3.5 km inland. During the experiment the atmosphere in the lowest hundred meters was neutral or weakly stable stratified.

A second-order closure dispersion model is presented. The results from the one successful experiment were compared with those from simulations with this model. Very good agreement with the experimental results was obtained.

Simulations were also carried out with the classical Gaussian plume formula, using various schemes for σ_y and σ_z. These models underpredict the measured maximum concentrations by a factor of approximately 2.

INTRODUCTION

From an air pollution point of view, the Øresund region (Øresund is the strait between Denmark and Sweden) is of specific interest because of its rather high population density and the presence of a number of power plants along the coast on both sides of the strait. The atmospheric dispersion is influenced by the water surface in two ways: the water surface will be aerodynamically smoother than the land surface, and their temperatures can substantially differ from each other. We investigated the atmospheric dispersion in this area by carrying out tracer experiments. A technical description of the experimental set-up as well as the measured data from a successful experiment are given in Gryning et al. (1983). In this paper we discuss some model simulations of the experiment.

SITE DESCRIPTION

The Øresund region (with meteorological masts) is shown in Fig. 1a. The part of the region where the tracer experiments were carried out, including the possible tracer sampling unit positions, is shown in Fig. 1b.

The tracer was released from a meteorological mast on a minor (2×2 km^2) peninsula at the Swedish coast of Øresund. The mast is surrounded by a small pine forest. About 400 m west of the mast and directly at the water's edge the Barsebäck nuclear power plant is situated. East of the mast stretches Skåne about 80 km with gently rolling farmland and forests. The width of Øresund on an east-west line through the mast is 21 km. Directly west of Barsebäck (Fig. 1b) the Danish coast region is an almost flat suburban area of Copenhagen. To the south the coast region becomes heavily built-up; to the north park and forest areas show up.

In the Øresund strait lies the island of Saltholm. It is flat (highest point ~ 1 m above mean sea level) and covered with grass and a few trees.

MODELS

The results from the tracer investigation were simulated by a

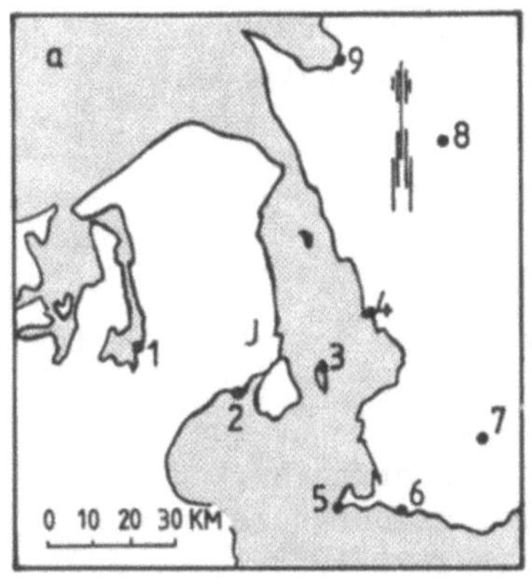

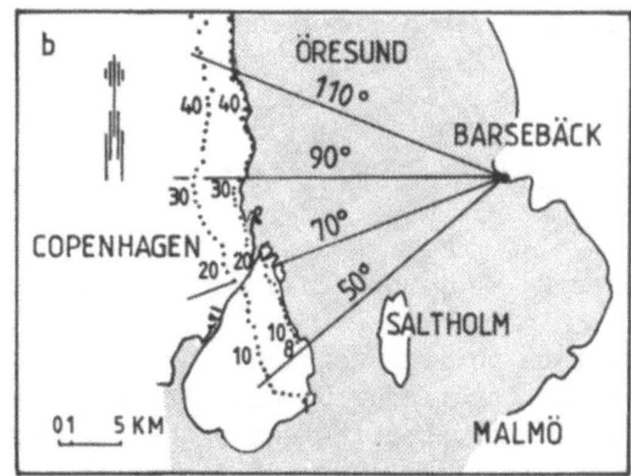

Fig. 1. a) The Øresund region. The positions of the meteorological masts are indicated by •: 1) Risø; 2) Avedøre; 3) Saltholm; 4) Barsebäck; 5) Falsterbo; 6) Maglarp; 7) Sturup; 8) Ljungbyhed and 9) Ängelholm. The radiosonde-station at Jægersborg is indicated by J.

b) The experimental site. Sampling-unit positions available for these experiments are indicated by •, also shown is the numbering of the positions in the individual series.

second-order closure model, and by various versions of the classical Gaussian plume models.

Second-order closure model

The dispersion of a gas from a point source depends on the mean wind and turbulence field. We obtained the wind, turbulence, and temperature field by applying the two-dimensional second-order closure model by Enger (1983a). This model will not be described here. Having determined the meteorological conditions in the area, we simulated the downwind variation of the crosswind-integrated concentration by a two-dimensional second-order dispersion model; the crosswind concentration distribution was assumed Gaussian, and a method to determine σ_y is included. The second-order dispersion model will be described here briefly (a detailed description can be found in Enger (1983b)).

We start with the diffusion equation

$$\frac{\partial C}{\partial t} + U\frac{\partial C}{\partial x} + V\frac{\partial C}{\partial y} + W\frac{\partial C}{\partial z} = -\frac{\partial}{\partial x}(\overline{uc}) - \frac{\partial}{\partial y}(\overline{vc}) - \frac{\partial}{\partial z}(\overline{wc})$$

where C is the concentration, U, V, and W are the mean wind in the x-, y-, and z-direction, respectively, and $\overline{uc}$, $\overline{vc}$, and $\overline{wc}$ are the turbulent fluxes of matter in the three directions. Integrating the equation over y and assuming that the streamwise diffusion, $\partial(\overline{uc})/\partial x$, is much less than the advection, $U\ \partial C/\partial x$, gives

$$\frac{\partial C_y}{\partial t} + U\frac{\partial C_y}{\partial x} + W\frac{\partial C_y}{\partial z} = -\frac{\partial}{\partial z}\int_{-\infty}^{\infty}\overline{wc}\,dy$$

where $C_y = \int_{-\infty}^{\infty} C dy$. In order to solve this equation we introduce the equation for the vertical turbulent flux of matter, $\overline{wc}$, and that for the covariance between temperature and concentration, $\overline{c\theta}$, where θ is the fluctuating and Θ the mean potential temperature. After some simplifications and parameterizations (Enger, 1983b), we end up with the equations:

$$\frac{\partial\overline{wc}_y}{\partial t} + U\frac{\partial\overline{wc}_y}{\partial x} + W\frac{\partial\overline{wc}_y}{\partial z} = -\overline{uw}\frac{\partial C_y}{\partial x} - \overline{w^2}\frac{\partial C_y}{\partial z} + \frac{g}{\Theta}(1-\alpha_2)\,\overline{c\theta}_y + A\frac{\partial}{\partial z}q\cdot\ell\cdot\frac{\partial\overline{wc}_y}{\partial z} - \alpha_1\frac{q}{\ell}\overline{wc}_y$$

and

$$\frac{\partial\overline{c\theta}_y}{\partial t} + U\frac{\partial\overline{c\theta}_y}{\partial x} + W\frac{\partial\overline{c\theta}_y}{\partial z} = -\overline{u\theta}\frac{\partial C_y}{\partial x} - \overline{w\theta}\frac{\partial C_y}{\partial z} - \frac{\partial\Theta}{\partial z}\overline{wc}_y - \alpha_3\frac{q}{\ell}\overline{c\theta}_y + A\frac{\partial}{\partial z}q\cdot\ell\frac{\partial\overline{c\theta}_y}{\partial z}$$

where $\overline{wc}_y = \int_{-\infty}^{\infty}\overline{wc}\,dy$, $\overline{c\theta}_y = \int_{-\infty}^{\infty}\overline{c\theta}\,dy$, $\frac{1}{2}q^2$ is the turbulence energy, ℓ the turbulence length scale taken from the wind model (Enger, 1983a), and α_1, α_2, α_3 and A are constants. According to Enger (1983b), the model coefficients are (α_1, α_2, α_3, A) = (0.36, 0, 0.17, 1.0).

The equations are solved numerically on a grid that is equally spaced in both the x- and z-directions. In these simulations the distance between grid points in the x-direction are 1000 m, and the first point is 100 m downwind from the source. In the z-direction the distance between grid points is 5 m, the lowest point is at the surface and the top of the grid is taken as 500 m. As the plume width is small near the source we use a Gaussian distribution in the two first downstream grid points. The upstream boundary and initial concentration field is taken to be Gaussian:

$$C_y(x,z) = \frac{Q}{\sqrt{2\pi}\cdot\sigma_z(x)\cdot U(h)}\left\{\exp\left(-\frac{(h-z)^2}{2\sigma_z^2(x)}\right) + \exp\left(-\frac{(h+z)^2}{2\sigma_z^2(x)}\right)\right\}$$

where $\sigma_z(x)$ is the vertical standard deviation at the distance x from the source, h the source height, U(h) the wind speed at source height, z the height above the surface and Q the source strength. The boundary conditions at the top of the grid are taken as

$$\frac{\partial C_y}{\partial z} = \frac{\partial \overline{wc}_y}{\partial z} = \frac{\partial \overline{c\theta}_y}{\partial z} = 0$$

The lower boundary conditions are:

$$\frac{\partial C_y}{\partial z} = - \frac{\overline{uw}}{\overline{w^2}} \frac{\partial C_y}{\partial x} \quad , \overline{wc}_y = \overline{c\theta}_y = 0$$

The lateral concentration profile is taken to be Gaussian, an approach which is well supported by dispersion experiments (i.e. Gryning, 1981). The standard deviation of the plume is obtained from the lateral velocity spectrum (Eulerian) for the specific meteorological conditions. The turbulent part of the spectrum is taken from Kaimal et al. (1972), Caughey (1977), and Caughey (1982); the synoptic part is taken from Hess and Clarke (1973) and Smedman-Högström and Högström (1975). After having determined the lateral velocity spectrum, σ_y is determined by use of Taylor's formula.

Gaussian models

Simulations of the tracer experiment have been carried out with 3 versions of the classical Gaussian plume model: the model by Högström (1964), that by Martin and Tikvart (1968), and the version used by Risø (Thykier-Nielsen, 1979).

In Högström's model, σ_y and σ_z have been determined from smoke puff experiments at a coastal site in Sweden (Studsvik), which represents a surface roughness discontinuity from water to slightly hilly terrain, partly covered with wood and partly cultivated. From the measurements σ_y and σ_z have been determined as functions of downwind distance, stability and height at release. The formulas contain terms that take into account the effect of the surface roughness. The sampling time refers to about 1 hour.

The formulas for σ_y and σ_z suggested by Martin and Tikvart (1968) are based on fittings to the Pasquill-Gifford curves. They refer to about 10 min sampling time and apply to flat, open-country conditions.

The model that is being used at Risø (Thykier-Nielsen, 1979) utilizes the 10-min averages of σ_y and σ_z given by Turner (1969). These are roughly the same as the Pasquill-Gifford curves. The conversion between the σ_y-value from the model and a σ_y-value of sampling time t was carried out according to the formula

$$(\sigma_y)_t = (\sigma_y)_{model} \, (t/1800)^{1/3}$$

with t in sec. Under neutral conditions the inversion height is assumed to be 500 m.

EXPERIMENT

A successful tracer experiment was carried out on June 23, 1982 (Gryning et al. 1982).

Experimental

The tracer, sulphurhexafluoride (SF_6), was released without buoyancy at a height of 95 m from the meteorological mast at the nuclear power station at Barsebäck. The tracer was released at a constant flow rate of 6.34 g/s. The time between the start of tracer release and that of tracer sampling was approximately 3.5 x/U(h) where x is the distance to the most distant series. On the experimental day the tracer sampling units were distributed in the series according to the wind direction at Barsebäck. Positions for tracer-sampling, separated by 1.5° when seen from Barsebäck, were marked out in advance; 23 tracer sampling units were used in each series. Tracer sampling was controlled by radio. Three consecutive air samples were taken, each collected over 30 min so that the total averaging time of the three samples was 1½ hour. The samples were analysed for their content of SF_6 by means of a gas chromatograph equipped with an electron capture detector. The tracer concentrations are believed to be known with an uncertainty of 20%. The uncertainty leads to a systematic error that influences all concentrations identically. The reproducibility of the tracer standards within the time necessary to analyse the measurements from one experiment was about 2%; this uncertainty affects the measured concentrations randomly. The calibration results show that SF_6 concentrations could be detected down to about 12 ng/m^3 with a signal-to-noise ratio of 4. A description of the experimental technique can be found in Gryning (1981).

Meteorological measurements

A number of meteorological masts exist in the Øresund region (Fig. 1a). In Sweden, functioning masts at a height of 100 m or more are situated at Barsebäck and Maglarp, and in Denmark at Risø. In addition, smaller meteorological masts are situated at Sturup, Ljungbyhed, Ängelholm, and Falsterbo in Sweden, and at Avedøre in Denmark. No meteorological masts existed on any of the islands in Øresund; therefore, a small (10 m) meteorology mast was put into operation at Saltholm. The mast was situated on the coast of the eastern part of Saltholm in order to have measurements that reflect easterly wind conditions over Øresund. The meteorological mast at Barsebäck is instrumented for routine measurements of wind speed

Table 1. Characteristic times, tracer release rate and Øresund water temperature for the experiment on June 23, 1982.

	TIME (MET)
Start of tracer release	8:30
Tracer sampling units	
run 1	10:48 - 11:18
run 2	11:18 - 11:48
run 3	11:48 - 12:18
Time of launches from Jægersborg:	
radiosondes	10:54 and 12:05
pilot balloons	06:00 and 18:00
Stop of tracer release	12:17

Tracer release rate was 6.34 g/s. The water temperature in Øresund between Copenhagen and Malmö was measured to 14.7 and 15.1 ^{o}C.

(12, 25, 98 m), direction (25, 98 m) and temperature (2.5, 12, 25, 50, 98 m). In addition to these measurements the three-dimensional turbulent wind velocity fluctuations were measured during the experiment at the height of tracer release. The instruments that were used for the turbulence measurements are described in Gryning and Thomson (1979). Radiosondes were routinely launched (12:00 and 00:00 MET) (Mid-European Time; MET = GMT + 1 hour) at Jægersborg, Denmark (Fig. 1a) and pilot balloons were launched every 6 hours. In addition to these routine measurements a radiosonde was launched from Jægersborg especially for this experiment.

Experiment on June 23, 1982

Characteristic times for the experiment are given in Table 1. The synoptic conditions were dominated by an occluded front that stretched from a low over southern England, the North Sea, and the southern part of Denmark. Over Skåne the front was split into a cold and a warm front. Directly north of the front in Øresund the wind speed was rather high (~ 11 m/s at 100 m height) and easterly, that is parallel to the front. On the day of the experiment it was overcast (but no rain) in the morning. At 09:30 the front associated rain began, and it rained continuously during the rest of the experiment. Thus,the experiment was carried out in Pasquill stability class D with rain.

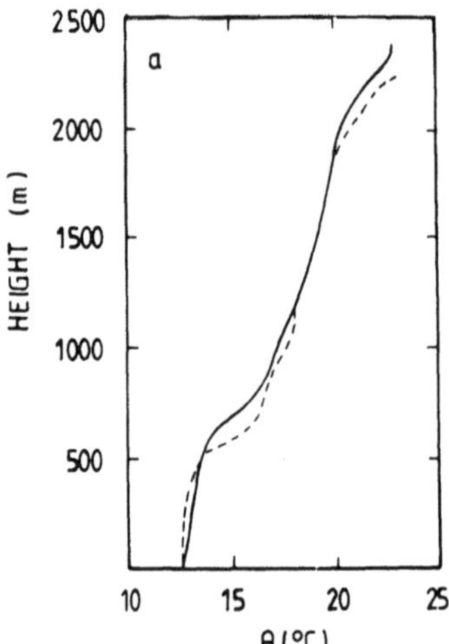

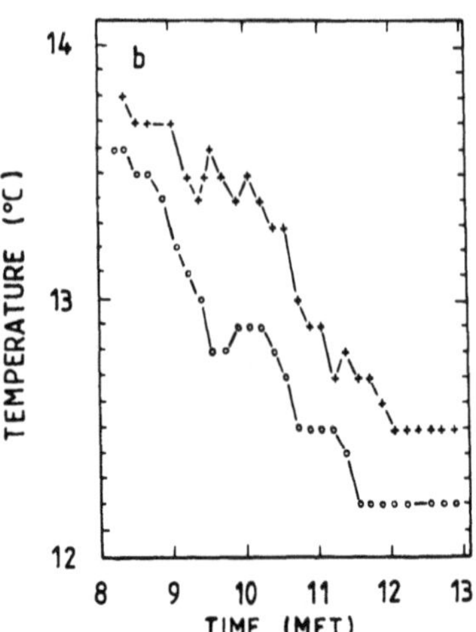

Fig. 2. a) Profiles of potential temperatures at Jægersborg June 23, 1982. The full-line represents the temperature profile at 10:54 and the dashed-line represents the temperature profile at 12:05.
b) Time variation of the observed temperature at the 2 m level on the island of Saltholm (crosses) and at Avedøre (rings).

The measured time variation of the temperature at Saltholm and at Avedøre is illustrated in Fig. 2b. Figure 2a illustrates the measured temperature profile obtained from the 2 radiosonde launches from Jægersborg. The wind measurements obtained from pilot balloons are shown in Fig. 3.

The 1½-h averaged tracer concentration profiles that were measured in Copenhagen at the coastline series (21 km from Barsebäck) and at the inland series (24.5 km from Barsebäck) are illustrated in Fig. 4. We calculated the standard deviation of the lateral tracer concentration profile, σ_y, by use of the formula

$$\sigma_y^2 = \frac{\Sigma C y^2}{\Sigma C} - \left(\frac{\Sigma C y}{\Sigma C}\right)^2$$

where y is crosswind distance. The increase of y at each summation step, Δy, was kept constant and small ($\Delta y \sim 1$ m), and the concentrations between the measuring points were estimated by linear inpolation. The crosswind-integrated concentration, C_y, was estimated in a similar way as $C_y = \Sigma C \Delta y$. At 21.5 km distance, $\sigma_y = 660$ m and $C_y = 1.676 \cdot 10^{-3}$ g/m^2; at 24 km distance $\sigma_y = 743$ m and $C_y = 1.585 \cdot 10^{-3}$ g/m^2.

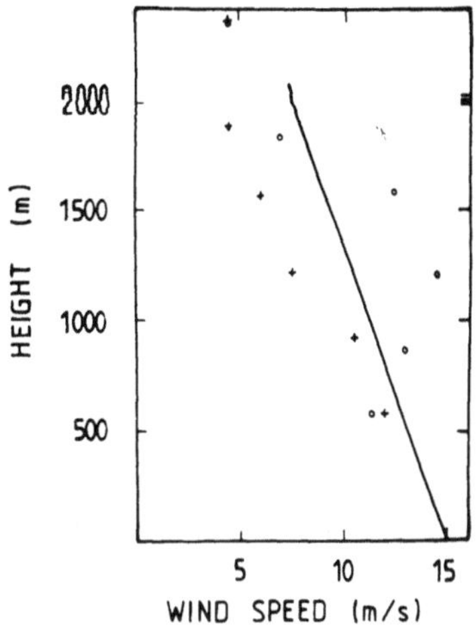

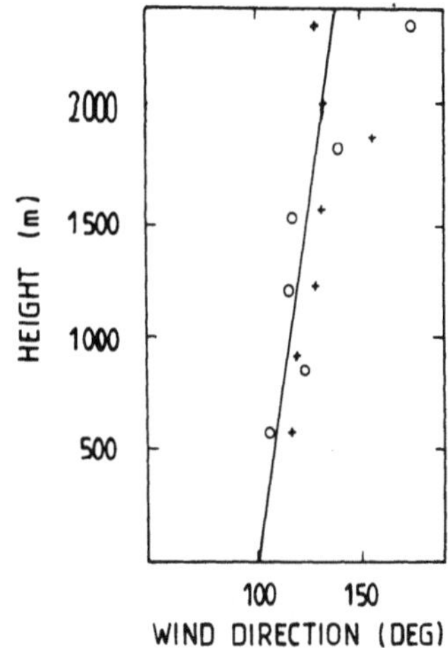

Fig. 3. Measured wind direction and -speed at Jægersborg; crosses represent measurements carried out at 06:00, and rings measurements at 18:00, both on June 23, 1982. The full line represents the relationship that was applied in the simulation of the experiment.

SIMULATIONS OF THE TRACER EXPERIMENT

First we compare the measured data from the tracer experiment with simulations carried out with the higher-order closure models.

The meteorological conditions during the experiment were simulated with Enger's (1983a) boundary layer model. The model requires knowledge of the geostrophic wind. In our simulations we have assumed that the geostrophic wind above 500 m can be approximated by the measured wind. Unfortunately, there were no wind measurements during the dispersion experiment. Therefore, we approximate the geostrophic wind by a straight line between the measured wind profile above 500 m, taken from the pilot-balloon launches at Jægersborg at 06:00 and at 18:00 MET, and then extrapolate this line down to the surface (Fig. 3b). The geostrophic wind direction has been approximated in the same manner. We simulated the experiment using both the temperature profile from the radiosounding at 10:54 and that at 12:05 (Fig. 2). The temperature at 2 m height was 12.5 ^{o}C during the experiment according to the measurements at Saltholm and Avedøre (Fig. 2a). A surface roughness length of 0.2 m was used for grid points in Sweden. Over the water between Sweden and Denmark a length of 0.00024 m was used and in Denmark (Copenhagen) the roughness length was taken as 0.8 m. The distances between grid points are 1000 m. We used 3 grid points in Sweden, 21 grid points over the water, and 6 in Denmark.

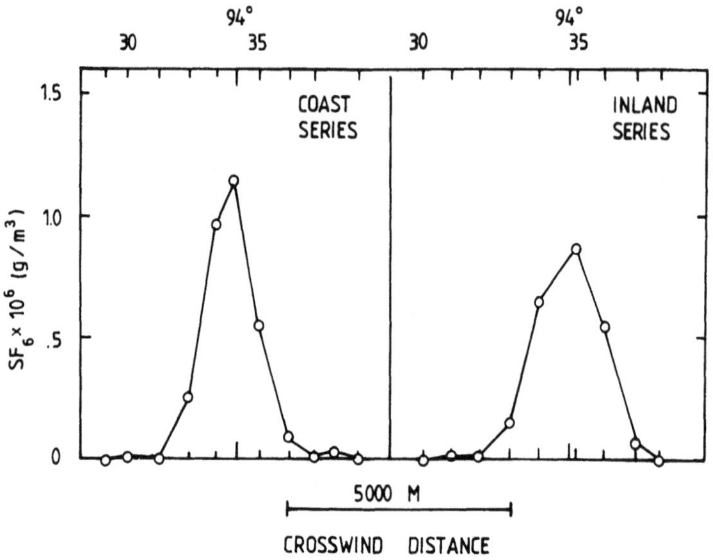

Fig. 4. Experiment on June 23, 1982. 1.5-h mean values of the measured tracer concentrations. The positions have been projected on a line perpendicular to the mean wind direction ($\sim 94^{\circ}$). The positions of the sampling units are indicated on the abscissa; two positions in each crosswind series are numbered according to the numeration in Fig. 1b.

Figure 5 shows the simulated wind profiles at three distances from the source, corresponding to the wind conditions at Barsebäck, Saltholm, and Avedøre, together with the measured values of the wind speed at these sites. In the lowest 50 m of the boundary layer the model seems to underestimate the wind speed by about 1.5 m/s. This difference is probably caused by our use of too low a geostrophic wind speed.

After having determined the meteorological conditions in the areas, we simulated the spread of the tracer by the second-order closure dispersion model. The wind speed that is used in the upstream conditions was taken as the measured wind speed at the source height. The simulations of the ground-level concentrations at the sampling positions are shown in Fig. 6. We note from the figure that the model simulates the concentration distribution very well. We also note that the concentration is dependent very much on the stability. The temperature profile from the sounding at

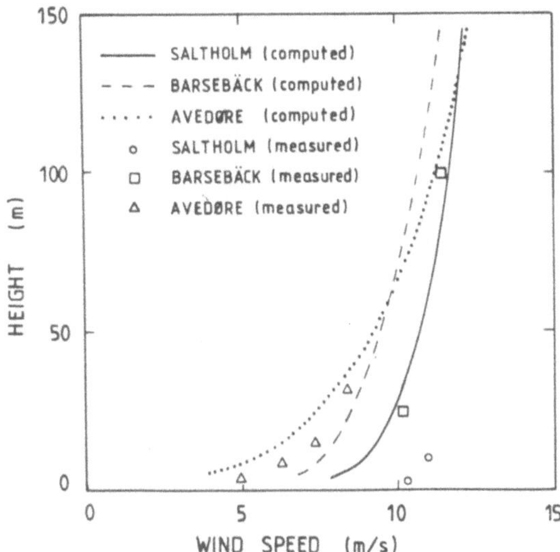

Fig. 5. Measured and computed wind speed profiles at Saltholm, Avedφre, and Barsebäck.

10:54 represents weakly stable conditions in the surface layer, and the temperature profile at 12:05 represents near neutral conditions. The calculations also show that at these distances σ_y essentially depends on the synoptic part of the spectrum.

The experimental data were also compared with the results from calculations carried out with the classical Gaussian models using schemes for σ_y and σ_z; the results are shown in Fig. 7. All of the applied σ-schemes are seen to produce substantially larger values of σ_y than were actually measured in this experiment, and the predicted maximum concentration is substantially lower than the measured ones, typically of the order of a factor of 2.

CONCLUSIONS

The atmospheric dispersion process over a land-water-land area was investigated by carrying out tracer experiments. The results from a successful experiment are reported here. The tracer sulphurhexafluoride was released at a height of 95 m from the meteorological tower at the Barsebäck nuclear power station, situated on the west coast of Sweden at the Øresund. Tracer sampling units were placed in Copenhagen in two semi-circles, one very close to the waterfront and the other about 3.5 km inland. The water stretch

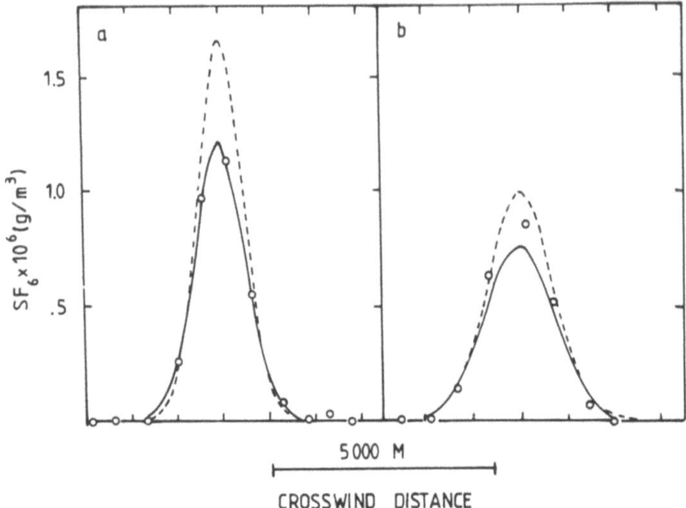

Fig. 6. a) Measured (rings) and calculated tracer concentrations for the coastline series, 21 km from the release point. The simulations were carried out with the second-order closure model. The full and dashed lines represent the concentrations that were calculated using the temperature profiles from the radiosoundings at 10:54 and 12:05, respectively.
b) As in a), except that this is for the inland-series, 24.5 km from the release point.

between the release point and the first semi-circle is about 21 km. Measurements of temperature, wind speed and -direction were performed at several places at meteorological towers by use of radiosondes and pilot balloons.

A numerical higher-order closure dispersion model is presented. The experiment was simulated with this model. The model requires the solution of the laterally integrated diffusion equation, the partial differential equation for the vertical turbulent flux of matter, and for the covariance between temperature and concentration. The mean wind profiles and turbulence quantities are taken from a second-order closure atmospheric boundary layer model developed by Enger (1983a). Very good agreement between the results from the simulation and the experiment was obtained.

Simulations were also carried out with the classical Gaussian

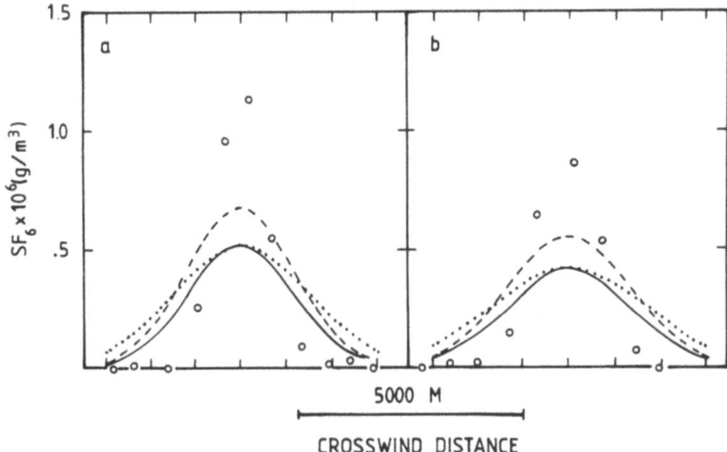

Fig. 7. a) Measured (rings) and calculated tracer concentrations for the coastline series. The simulations were carried out with the usual Gaussian model, using various schemes for σ_y and σ_z:
1) Högström (1964) (full line)
2) Martin and Tikvart (1968) (dashed line)
3) The model used at Risø (Thykier-Nielsen, 1979) (dotted line).

b) As in a), except that this is for the inland series.

plume model, using various commonly used schemes for σ_y and σ_z. These schemes are seen to overpredict the lateral spread of the plume; they also underpredict the maximum concentration by roughly a factor of 2. However, it should be recalled that this comparison rests on the results from only one experiment, and that the uncertainty of these models is often stated to be about a factor of 2.

ACKNOWLEDGEMENTS

A number of people have taken part in the experimental campaign. Arent Hansen from the Meteorology Section, Risø National Laboratory, and Hans Ahleson, Knud A. Hansen, and Morten Hildan from the Air Pollution Laboratory are acknowledged for their skilled technical assistance.

Thanks are due to Arne Svensson and Conny Hellström from the Barsebäck Nuclear Power Station for practical support, and for allowing us to use the facilities at the meteorology mast at Barsebäck. We appreciate the assistance from the Danish Meteorological

Institute; E.W. Nielsen supplied radiosonde data, and the Weather Service prepared forecasts of the wind field in the Øresund region. We further thank the Danish Department of Post and Telecommunication for providing its antenna at the TV-tower at Gladsaxe during these experiments.

The Air Pollution Laboratory of the Danish National Agency of Environmental Protection has developed and maintained the tracer instrument system for dispersion studies. A special electron capture gas chromatograph for tracer analyses was designed by Erling Lund Thomsen. The authors greatly appreciate his work.

REFERENCES

CAUGHEY, S.J., 1977, Boundary layer turbulence spectra in stable conditions. Boundary-Layer Meteorol., 11, 3-14.

CAUGHEY, S.J., 1982, Observed characteristics of the atmospheric boundary layer, in "Atmospheric turbulence and air pollution modelling", F.T.M. Nieuwstadt and H. van Dop, Editors. D. Reidel Publishing Company, Dodrecht, Holland, 107-158.

ENGER, L., 1983a, Numerical boundary layer modeling with application to diffusion. Part 1: A two dimensional higher order closure model. Department of Meteorology, University of Uppsala, Sweden, Report No 70. 54 pp.

ENGER, L., 1983b, Numerical boundary layer modeling with application to diffusion. Part 2: A higher order closure dispersion model. Department of Meteorology, University of Uppsala, Sweden, Report No 71, 45 pp.

GRYNING, S.E. and THOMSON, D.W., 1979, A tall-tower instrument system for mean and fluctuating velocity, fluctuating temperature and sensible heat flux measurements. J. Appl. Meteorol., 18, 1674-1678.

GRYNING, S.E. 1981, Elevated source SF_6-tracer dispersion experiments in the Copenhagen area. Risø-R-446, 187 pp.

GRYNING, S.E., LYCK, E., and WIDEMO, U., 1983, A tracer investigation of the atmospheric dispersion in the Øresund region. Data and Technical report. Studsvik report, STUDSVIK/NW-83/452.

HESS, G,D, and CLARKE, R.H., 1973, Time spectra and cross-spectra of kinetic energy in the planetary boundary layer. Quart. J. Roy. Meteor. Soc., 99, 130-153.

HÖGSTRÖM, U., 1964, An experimental study of atmospheric diffusion. Tellus, 16, 205-251.

MARTIN, D.O. and TIKVART, A., 1968, A general atmospheric diffusion model for estimating the effects on air quality of one or more sources. Paper presented at the 61st annual meeting of the Air Pollution Control Association.

SMEDMAN-HÖGSTRÖM, A-S. and HÖGSTRÖM, U., 1975, Spectral gap in surface layer measurements. J. Atmos. Sci., 32, 340-350.

THYKIER-NIELSEN, S., 1979, Risø model til beregning of konsekvenser at frigørelse af radioaktivt materiale til atmosfären. Internal Report Risø-M-2149, 1979.

TURNER, D.B., 1969, Workbook of atmospheric dispersion estimates. Public Health Service, Div. of Air Pollution, Cincinnati, Ohio. HPS-Pub-999-AP-26. PB-191482. 88 pp.

DISCUSSION

S. VOGT

It is not surprising to me that you are underpredicting the calculated concentrations with the Gaussian model. The reason is that you are applying σ_y and σ_z-values derived at other sites and perhaps other synoptic situations. Do you plan in the future to derive directly σ_y and σ_z from you experiments ?

L. ENGER

Next year new tracer experiments will be carried out in this area. During these experiments a lot of measurements will be done for example turbulence measurements at different heights.

A. GHOBADIAN

In modelling the correlation between pressure fluctuations and concentration gradient fluctuations, how did you fix the constant ?

L. ENGER

To get the constant α_1 in the differential equation for $\overline{wc}$ I have assumed that the underlying surface is flat and homogeneous, that the concentration field is horizontally homogeneous, that stationary conditions apply and that we have neutral conditions. During these conditions the equation is reduced to

$$- \overline{wc} = \frac{\ell}{\alpha_1 q} \overline{w^2} \frac{\partial c}{\partial z} = K_z \frac{\partial c}{\partial z}$$

According to Nieuwstadt and van Ulden

$$K_z = 1.35 K_M = K_H$$

in the surface layer under neutral conditions, which means that

$$\frac{\ell}{\alpha_1 q} \overline{w^2} = 1.35 u_* kz$$

In neutral conditions $q^2 = B_1^{2/3} u_*$ $\overline{w^2} \cong 1.7 u_*^2$ $\ell = kz$

which gives us $\alpha_1 = 0.36$ if $B_1 = 42.5$

α_3 in the equation for $\overline{C\theta}$ has been determined numerically by comparing the results from Willis and Deardorff's water tank experiment for a convective boundary layer with the source in the middle of the convective boundary layer.

If we choosed $\alpha_3 = 0.17$ the development of the plume was similar to Willis and Deardorff's experiment.

B. VANDERBORGHT In your second-order closure model the lateral concentration profile is taken to be Gaussian, and σ_y is obtained from the lateral velocity spectrum. If you should use this σ_y value in the bi-Gaussian model, how should then be the comparison between the measured concentrations and the bi-Gaussian simulation ?

L. ENGER The agreement will be better. But when the plume cross the coastline the surface roughness length will change, causing a large change in the vertical mixing. This cannot be simulated with simple Gaussian models. In other situations with larger temperature differences between land and water surface or in situations with a strong unstable stratification a simple Gaussian assumption for the vertical spread can not be used. The dispersion model used here can simulate the descending centerline of the plume from an elevated source and the ascending plume from a ground level source as predicted by Willis and Deardorffs experiments in a convective boundery layer. These results can not be simulated with a conventional Gaussian model.

THE SHORELINE ENVIRONMENT ATMOSPHERIC DISPERSION EXPERIMENT (SEADEX)

Warren B. Johnson and Edward E. Uthe
Atmospheric Science Center
SRI International
Menlo Park, California 94025

C. Ray Dickson and Gene Start
Air Resources Laboratory - Idaho
National Oceanic and Atmospheric Administration
Idaho Falls, Idaho 83401

Richard L. Coulter
Radiological and Environmental Research Division
Argonne National Laboratory
Argonne, Illinois 60439

Robert A. Kornasiewicz
Office of Nuclear Regulatory Research
U.S. Nuclear Regulatory Commission
Washington, D.C. 20555

NATURE AND OBJECTIVES OF SEADEX

SEADEX (*) is a major field experimental program sponsored by the U.S. Regulatory Commission (NRC) for the purpose of acquiring a comprehensive, high-quality data base for evaluating models of dispersion within coastal zones. A secondary objective of the program is to provide data to help determine which meteorological measurements are most appropriate for emergency preparedness at nuclear power plant sites.

The first field study in this program, SEADEX-I, was conducted in the vicinity of Kewaunee, Wisconsin, on the western shore of Lake Michigan, during the period 28 May - 8 June 1982. SRI International

(*) Shoreline Environment Atmospheric Dispersion Experiment

was responsible for the design and direction of the field study. Other participants in SEADEX-I included the National Oceanic and Atmospheric Administration (NOAA), Argonne National Laboratory (ANL), and Mesomet, Inc.

Analysis of the SEADEX-I data is still in progress at this time, and only very limited results are currently available. The purpose of this brief initial report is to acquaint the scientific community with the design of the SEADEX-I study and with the data base resulting from it. To partially illustrate the type and quality of the data collected, a selected example of the data will also be presented.

DESCRIPTION OF THE FIELD SITE

Figure 1 shows the location of the field site. The study area was centered around Kewaunee Nuclear Power Plant, which is located approximately 27 mi (43 km) southeast of Green Bay, Wisconsin. This particular site was selected primarily for logistical convenience, in that a 72-m tower was available there to serve as a platform for meteorological measurements and for release of the tracer materials.

The land within the primary field study area, defined as within 10 mi (16 km) of the tracer release point, is mostly flat to gently rolling. The highest elevation in the area is approximately 300 m, which is 100 m above the mean water level of Lake Michigan. Since the land in this area is primarily used for dairy farming, it consists mostly of pasture or cropland, with occasional small stands of trees. An extensive road network along section lines is available to facilitate field operations. The shoreline is generally characterized by a low bluff between 6 and 10 m in height.

FIELD STUDY EXPERIMENTAL DESIGN

Field Test Elements and Design Parameters

The two major objectives of the measurement program were to characterize the following as completely as possible during test periods :

- Atmospheric dispersion, and
- Meteorological conditions influencing the dispersion.

For the purpose of dispersion characterization, both a gas tracer (SF_6) and a particulate tracer (oil fog) were used. SF_6 concentrations near the surface were obtained primarily by an array of 130 multiple-bag air samplers. The spacing of these samplers increased with increasing distance from the tracer release point. In addition, a continuous SF_6 analyzer furnished by Brookhaven National Laboratory was used on a Cessna 337 meteorological aircraft to obtain real-time SF_6 concentration measurements at various altitudes above the surface.

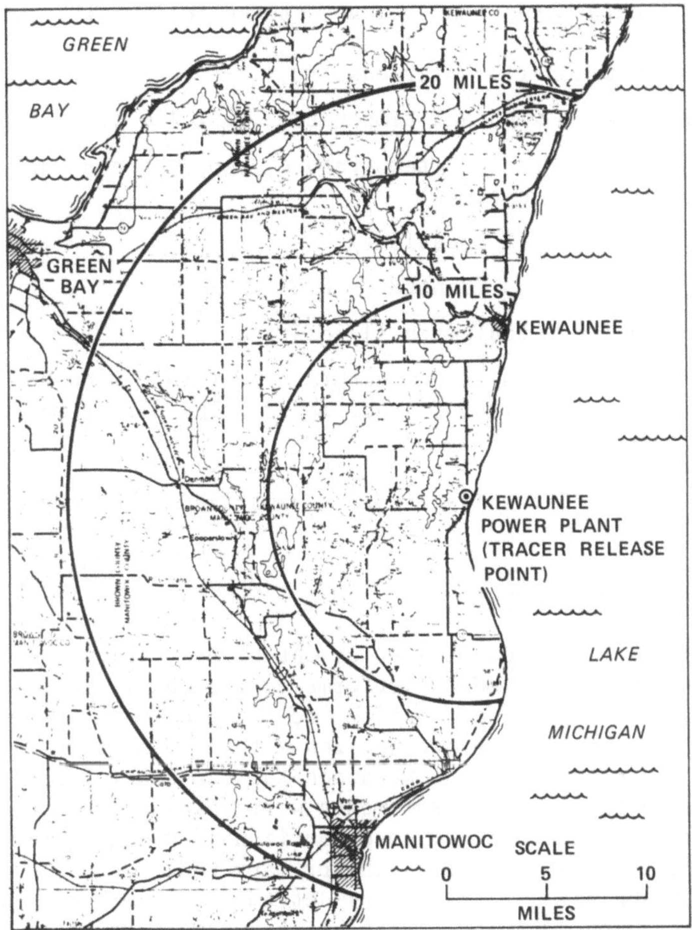

Fig. 1. Location of SEADEX-I field study area in Northeastern Wisconsin (USA).

Oil fog measurements were made by the downlooking ALPHA-1 (Airborne Lidar Plume and Haze Analyzer) system mounted in the SRI Queen Air aircraft. This system made crosswind traverses of the tracer plume at two or more downwind distances to map the oil fog concentration in the y-z plane for later use in calculating horizontal and vertical dispersion. In addition, the SRI Mark 9 ground-mobile lidar system was used along selected roadways to map the oil-fog plume and ambient boundary-layer aerosol structure. This combination of downward-looking ALPHA-1 and upward-looking Mark 9 measurements gave good plume coverage even during periods of moderate cloud cover.

For meteorology (boundary layer) characterization, the field test design incorporated measurements of the following parameters :

- Surface winds (13 locations)
- Near-surface wind and temperature profiles (2 locactions)
- Boundary-layer wind, temperature, dewpoint and turbulence profiles (5 locations)
- Mixing depths (2 locations)
- Air trajectories (constant-level tetroons)
- Solar and net radiation (1 location)
- Land and water surface temperatures (along-wind horizontal profiles from aircraft radiometer, plus direct measurements from boat).

Details of the field-test design are summarized more completely in Table 1.

Instrumentation Deployment

Figure 2 illustrates the deployment of the various instrumentation systems and measurement platforms during the field tests. The fixed air sampler array was confined to the area of primary interest for the field study, which was the area within 10 mi (16 km) of the tracer release point near the Kewaunee Nuclear Power Plant. However, extensive meteorological measurements were obtained well beyond this area to aid in the determination of horizontal variations of the meteorological variables across the primary test area. In addition, tetroon tracking and air trajectory determination extended out to distances of 30 to 40 mi (50 to 66 km) from the tetroon release point near the power plant.

The existing 72-m tower at the power plant was used for two purposes. The tracer gas and oil fog were released from the same point on the tower, at an elevation of 41 m. In addition, the tower was instrumented at two levels for obtaining wind and ΔT measurements. Similar meteorological instrumentation was installed and operated on a existing 91-m radio tower approximately 21 mi (34 km) west of the tracer release point.

As shown in Figure 2, the research vessel (R/V) EKOS was operated by ANL on Lake Michigan approximately 8 mi (13 km) offshore. At this distance offshore, the lake depth is approximately half of its maximum depth of 200 m. Lake water temperatures and tethersonde-based wind and temperature profiles were obtained from this research boat. To document the evolution of the internal boundary layer (IBL), a Doppler acoustic wind sounder (DAWS) was operated approximately 2 mi (3.2 km) inland from the shoreline, and a monostatic acoustic

Table 1. SEADEX-1 Design Parameters

Element/Activity	No. Units/Rate/Frequency
Number of tests	10 (design goal)
Test duration	5 hr nominal/2 hr minimum
Gas tracer release and analysis	80 lb/hr, 5 hr/test, 8 GC units for analysis
Fixed tracer sampling	130 samplers plus spares
Mobile tracer sampling	C-337 airplane with BNL continuous SF_6 analyzer
Oil-fog tracer release	100 gal/hr, 4 hr/test
Airborne lidar	ALPHA-1, 4 hr/test
Ground-mobile lidar	Mark 9, 4 hr/test
Surface wind field	11 (v,Θ) on 10-m towers
Near-surface wind and temperature profiles	2 (v,Θ) + ΔT on 91-m tower 2 (v,Θ) + ΔT on 72-m tower
Boundary-layer (BL) wind/temp. profiles	Rabals (two/hr) plus rawinsondes (one/3 hr)
Air trajectories	Radar-tracked tetroons (one/hr, 6 hr/test)
BL vertical profiles	C-337 aircraft (T, DP, turb.)
BL wind/temperature profiles over land	Doppler acoustic sounder (continuous 30-min averages) Tethersonde (1 profile/hr)
BL wind/temperature Profiles over lake	Tethersonde on R/V EKOS (1 profile/hr)
Lake surface water temp. (horiz. profiles)	From R/V EKOS (1 profile/test)
Solar and net radiation	Continuous
Time-lapse photos (oil fog)	Two cameras (4 hr/test)

sounder (sodar) was located at each of two sites : one near the tracer release point at the shoreline, and the other 3 mi (5 km) inland.

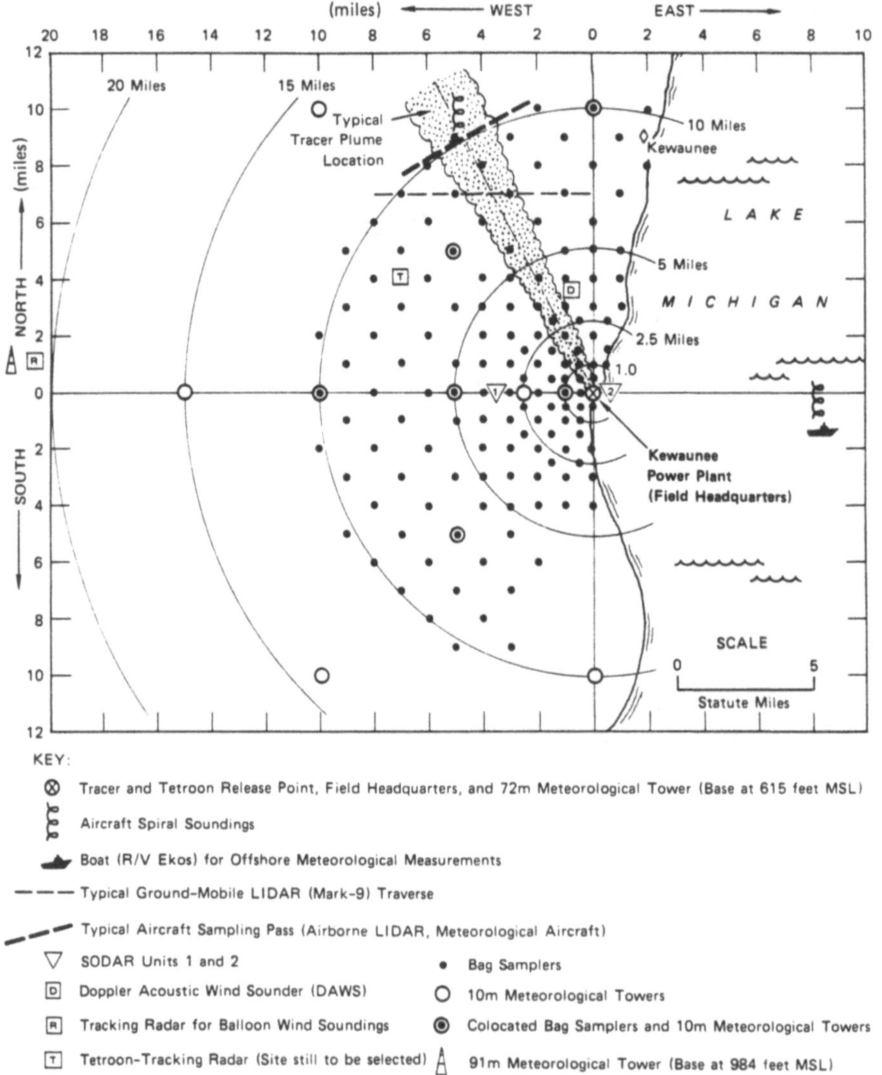

Fig. 2. Instrumentation deployment used in SEADEX-I.

DATA BASE OBTAINED DURING THE FIELD STUDY

As indicated in Table 2, a total of nine tracer tests covering a total period of 50 hours were conducted during SEADEX-I. Since lake-breeze (on-shore flow) conditions were of primary interest in this study, all of the tests were conducted during daylight hours. In general, favorable lake-breeze conditions prevailed during the experimental period because of the relatively large land-water surface temperature differences (typically 10 to 15°C) that occurred during that time.

EXAMPLE OF DATA OBTAINED DURING LAKE BREEZE RECIRCULATION CONDITIONS

As mentioned earlier, one of the primary instruments used for measuring horizontal and vertical dispersion during SEADEX-I was the ALPHA-I airborne lidar. This system was used in the manner illustrated in Figure 3. The aircraft flew over the oil-fog tracer plume, usually in a crosswind direction, and the downlooking lidar obtained a series of vertical particulate backscatter profiles, at the rate of 5 profiles/sec, during each traverse. At typical aircraft speeds of 65 m/sec, the resulting profile spacing was 13 m in the horizontal. These high-resolution data have been processed and displayed in the form of vertical cross sections of relative concentrations of the oil-fog tracer.

During the afternoon of 8 June 1982 (Test No. 9), a well-defined lake-breeze recirculation pattern developed that probably could not have been detected or confirmed without the vertical cross sections obtained by the ALPHA-1. As shown by the wind-direction profiles on the right side of Figure 4, the tethersondes at both the land and the lake stations measured strong vertical shears in wind direction at approximately 300 m above the surface during the early afternoon of 8 June. The wind direction in the lowest 300 m approximated 090°, while above 300 m it reversed and became roughly 270°.

The ALPHA-1 obtained a number of cross sections in the crosswind direction over the oil-fog tracer plume during this period. These data are presented in Figure 5. The aircraft flew a race-track pattern, which resulted in a sequence of cross sections alternately at 9 km and 28 km downwind of the tracer release point. The dark band at the bottom of each cross section in Figure 5 is the measured terrain profile. The gray shading above the terrain represents relative particulate concentrations in the atmosphere, with the darker shading corresponding to higher concentrations.

The ambient boundary-layer haze is apparent in each cross section, extending up to approximately 1100 m above the surface. In addition, the oil-fog tracer plume is clearly evident in each cross section. At the 9 km distance, the lower part of the tracer plume is about 1 to 2 km in width, while at 28 km downwind it is considerably wider (3 to 7 km).

Table 2. SEADEX-I Data Base

Test No.	Date (1982)	SF_6 Release (CDT)		Hours of Data
		Start	Stop	
1	28 May	1300	1800	5
2	29 May	1300	1800	5
3	30 May	1200	1700	5
4	2 June	1100	1600	5
5	3 June	1000	1500	5
6	4 June	0900	1400	5
7	5 June	0800	1200	4
8	6 June	0800	1500	7
9	8 June	0900	1800	9
Total Hours of Tracer Data				50

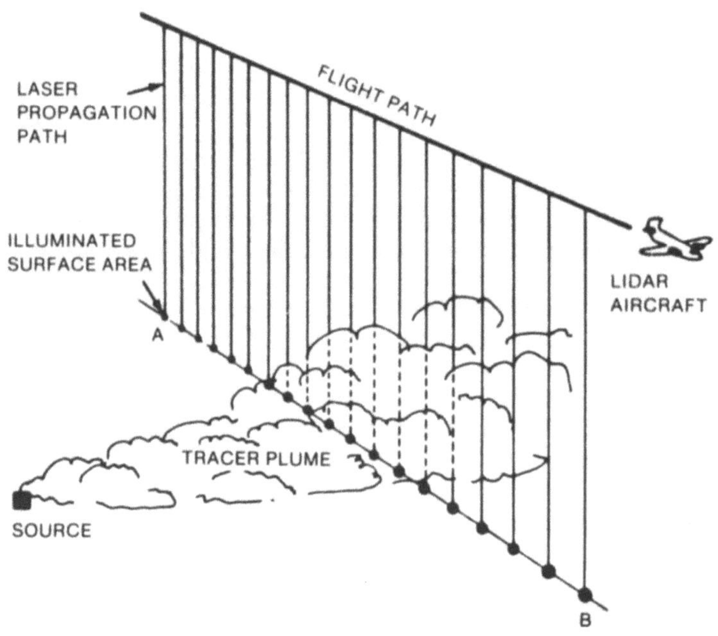

Fig. 3. Airborne lidar (ALPHA-1) technique for measuring tracer plume dispersion.

(a) Land station (3.5 km inland)

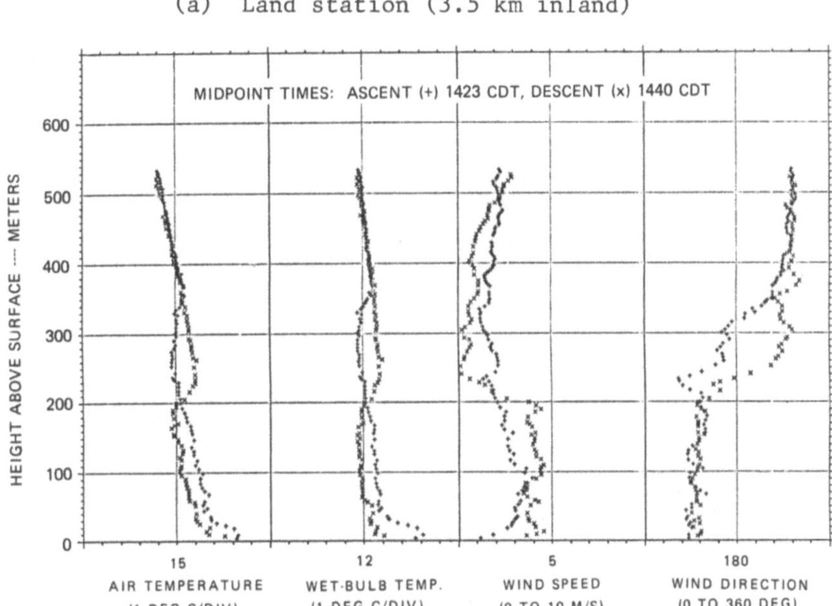

(b) Lake station (10 km offshore)

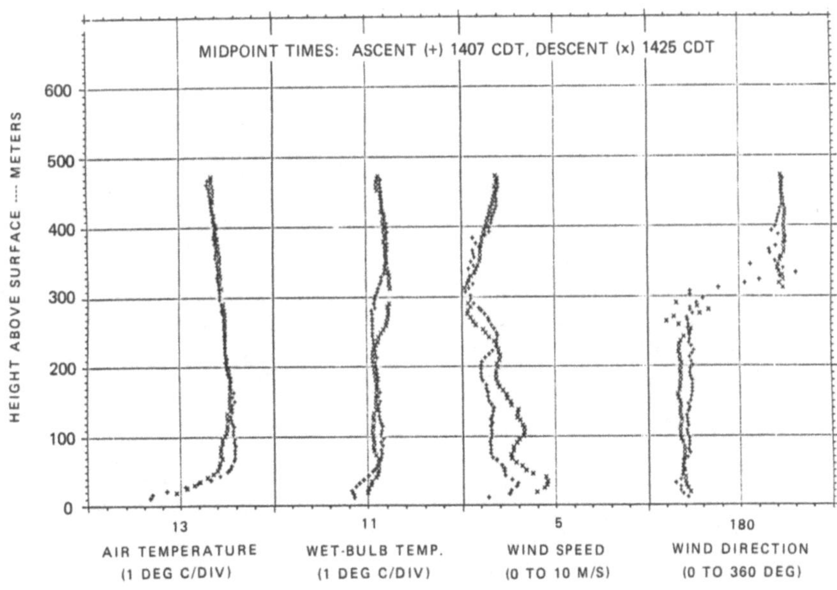

Fig. 4. Vertical profiles obtained by tethersonde at the (a) land and (b) lake stations during the early afternoon of 8 June 1982.

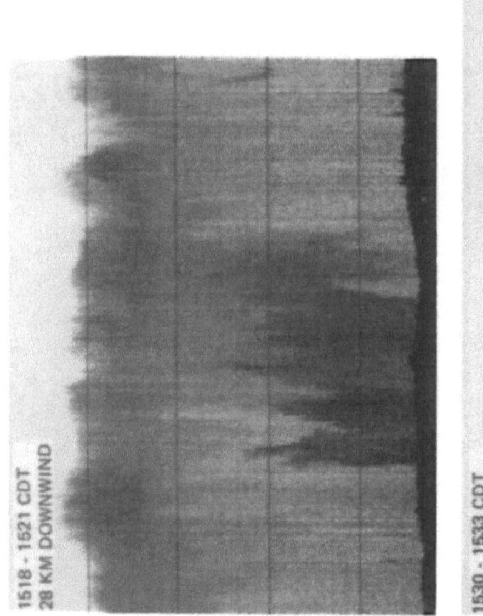

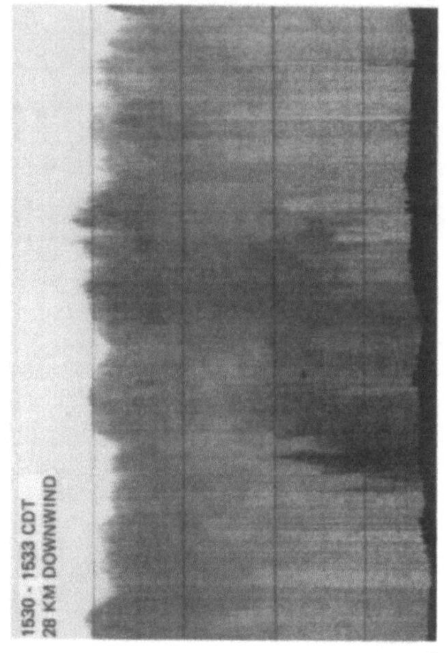

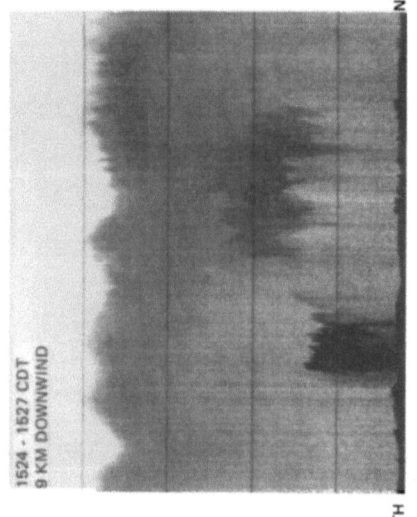

Fig. 5. Vertical cross sections of the oil fog plume obtained by the ALPHA-1 airborne lidar at 9 km and 28 km downwind of the release point during 1512-1612 CDT on 8 June 1982 (vertical scale is 300 m/division; horizontal scale shown applies throughout).

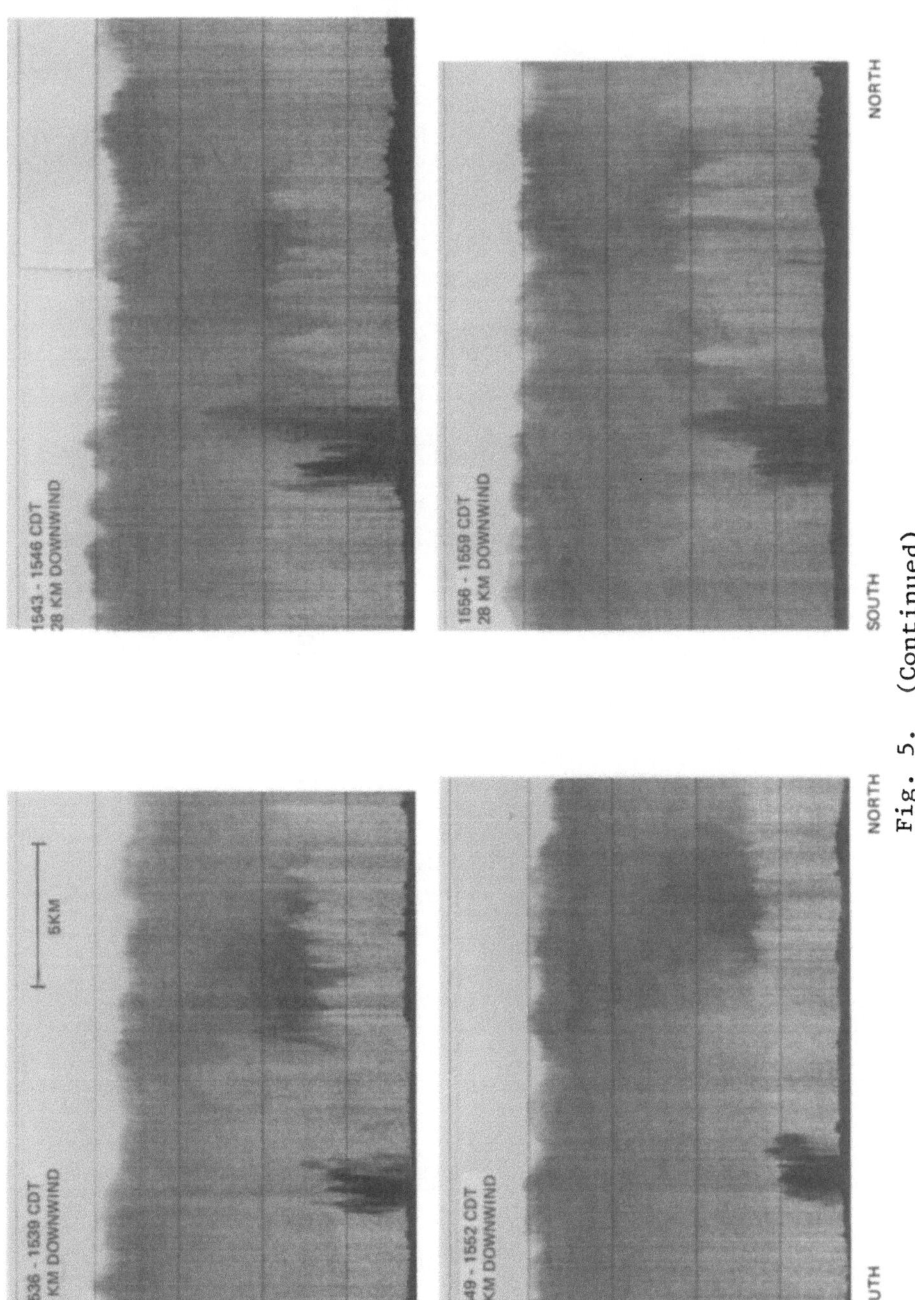

Fig. 5. (Continued)

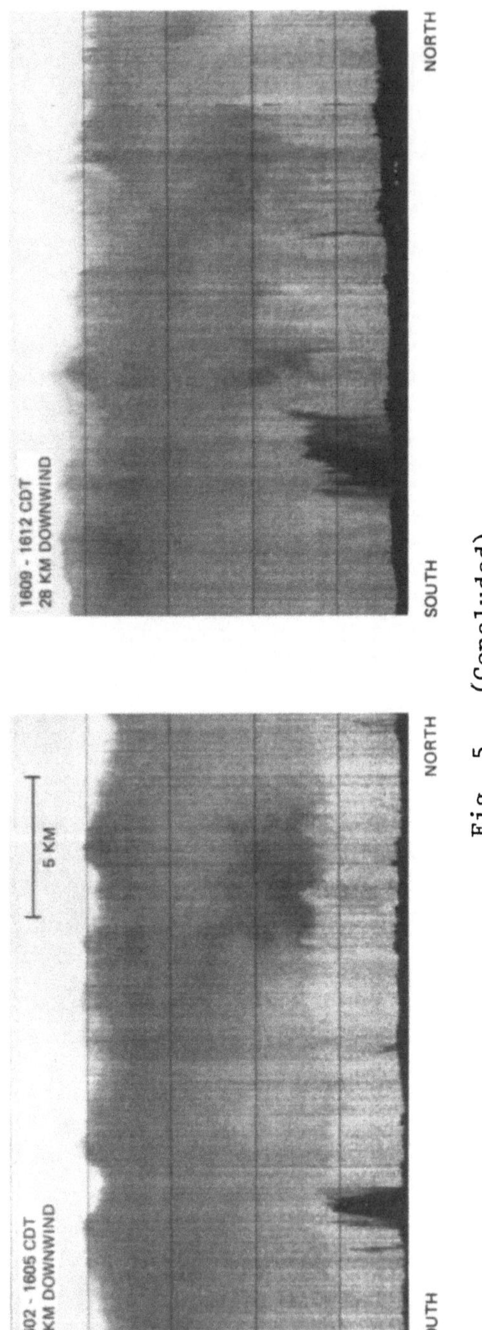

Fig. 5. (Concluded)

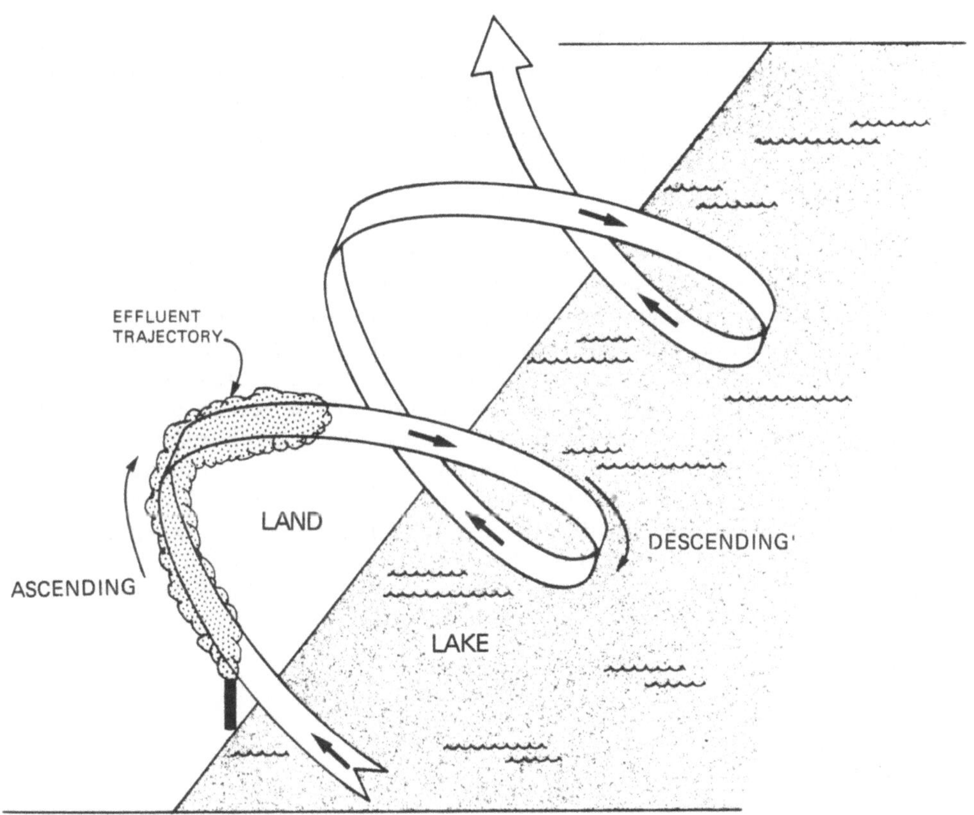

Fig. 6. Schematic of three-dimensional helical lake-breeze circulation pattern.

The most interesting feature of these data is the structure of the upper parts of the tracer plume at each distance. At the 9 km distance, two distinct plumes are evident, with the separation occurring at a height of approximately 300 m above the surface. This is the height of the wind reversal in the tethersonde data, and thus must correspond to the height of the internal boundary layer (IBL) encompassing the onshore flow. At 28 km, the upper portion of the tracer plume is still connected to the lower portion, but is sheared off toward the north.

Various features of the convective cells occurring within the IBL are revealed by the structure of the top of the lower plume and the bottom of the upper plume at the 9-km distance. It is clear that both sections of the plume at 9 km originated from the tracer release, because the second aircraft found significantly high SF_6 tracer concentrations in both areas.

When these lidar data are considered in conjunction with the wind-profile data, it is apparent that the lower plume at 9 km is moving to the west with the lake breeze, while the upper plume is moving to the east with the return flow aloft. The cross sections at 28 km were evidently obtained at a transition distance where the marine air was being forced aloft and to the north by horizontal convergence. This three-dimensinal circulation pattern may be visualized by referring to the sketch in Figure 6. Although only the initial portion of the pattern sketched in Figure 6 was actually observed, it seems reasonable to hypothesize that the extended trajectory of the air containing the tracer material would describe a helix as it descends and becomes re-entrained in the lake breeze.

CONCLUDING REMARKS

The foregoing data example illustrates the unique nature of the extensive data base acquired during the SEADEX-I field study. It is expected that the results from this and the other field studies to be conducted during the SEADEX program will significantly contribute to our understanding of the physical processes influencing the dispersion of air-borne materials in shoreline environments.

ACKNOWLEDGMENT

This work was supported by the Office of Nuclear Regulatory Research of the U.S. Regulatory Commission (NRC) under Contract No. NRC-04-81-211. Technical direction was provided by Robert A. Kornasiewicz of the NRC.

DISCUSSION

S. E. GRYNING From the set-up of the SF_6 tracer sampling units it appears that they are spread out over a rather large area, consequently only a few of the sampling-units will be in the plume during the individual experiments. Why did you spread the sampling-unit that much in the area ?

W.B. JOHNSON In all tracer studies where a finite number of samplers are available, a compromise has to be made between measurement <u>density</u> (sampler spacing) and measurement <u>coverage</u> (angular extent of the sampler network). In the SEADEX field study, we had to be prepared for a rather wide range of wind directions during the tracer tests and could not easily move the samplers at short notice because of the large distances involved. This required that the angular extent or coverage of the sampler network be large, with a corresponding decrease in sampler density. However, this was not too serious of a problem for two reasons :

(1) Much of the data will be used to evaluate models based upon four-hour averages. Over this averaging time, the tracer plume typically covered a fairly large angular segment, and was reasonably well measured by the sampler network.

(2) We had several other mobile measuring systems to help fill in the gaps between the fixed SF_6 sampling stations. These systems included an instrumented aircraft to measure horizontal SF_6 profiles, and groundbased and airborne lidars to measure vertical cross sections of the oil-fog tracer.

B. VANDERBORGHT You have a very high oil fog aerosol concentration at the emission site. Couldn't there be adsorption or absorption of SF_6 on the oil fog particles resulting in a systematic underestimation by the SF_6 gas concentration measurements ?

W.B. JOHNSON These is no indication that this occurred during the field study, nor is these any evidence from other field and laboratory studies that SF_6 absorption on oil fog droplets is a potential problem.

R.M. VAN AALST What was the size and size distribution of the oil droplets ?

W.B. JOHNSON I do not have this information since we did not directly measure the size distribution of the oil-fog tracer droplets during the SEADEX field study. However, the droplet sizes were probably relatively small, since downwind deposition did not appear to be significant. In addition, the oil-fog tracer plume was used only to obtain spatially relative, rather than absolute, concentrations. For this application, the data are not adversely affected by any droplet deposition that might occur.

DISPERSION CONDITIONS OVER LAND AND WATER IN A COASTAL ZONE REVEALED BY MEASUREMENTS AT TWO METEOROLOGICAL MASTS

S.E. Larsen and S.-E. Gryning

Meteorology Section
Risø National Laboratory
DK-4000 Roskilde, Denmark

ABSTRACT

In connection with a nuclear site evaluation program two meteorological masts have been operated for 3 years, one placed directly on a shoreline, the other 1 km inland. Substantial differences in the distribution of wind speed, wind direction and stability have been found between the two masts, reflecting the differences in surface temperature and roughness of the water and land surfaces. This paper solely deals with the distribution of dispersion categories at the two stations, seen in relation to the water - land differences.

The meteorological parameters measured are related to dispersion conditions by use of z_0, the roughness length, and L, the Monin-Obukhov length. This is done by relating Pasquill's σ_z-curves to the corresponding values obtained by use of a K model for a continuous ground-level source. Hereby the difficulties, associated with the different roughness and thermal properties of land and water, is avoided.

INTRODUCTION

Plans for future nuclear power plants in Denmark include a site on the southern tip of the Stevns peninsula, roughly 50 km south of Copenhagen, see Fig. 1. From the figure is seen that a plume from the power plant towards Copenhagen would pass over both land and water before reaching the city. Therefore, to establish suitable dispersion meteorological statistics, two meteorological masts were erected, one roughly 1 km inland north of the powerplant site and one on the southern coast of Copenhagen. From these masts 3 years of measurements of velocity and temperature profiles and wind direction have been recorded.

In the present paper we shall mostly be concerned with the difference in dispersion condition over land and water as measured when the wind comes through the two sectors I and II on Fig. 1. For both sectors the conditions at the Stevns mast are supposed to be characteristic for over land conditions owing to the roughly 1 km distance to the coast. For the mast on the southcoast of Copenhagen, sector I is characterized by a water fetch of the order of 100 km (down to the southcoast of the Baltic Sea). The water fetch in sector II is only of the order of 20 km. Previous to the water fetch the air has passed over 50-100 km of land.

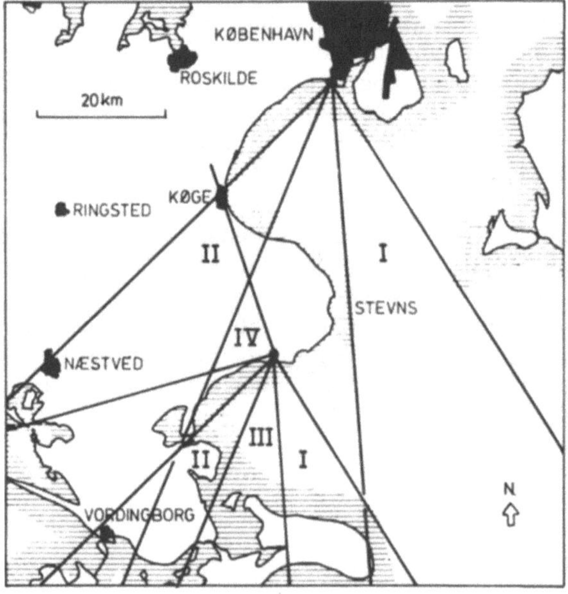

Fig. 1. Map of the region of Zealand, where the two masts, described in the text, are situated. The wind direction sectors shown are used in connection with the data analysis. The southern mast is referred to as the Stevns mast in the text, while the northern is denoted the Copenhagen mast.

DISPERSION PARAMETERS

Description of atmospheric dispersion necessitates estimates of wind direction, wind speed and at least one parameter describing the atmosphere's ability to disperse material by turbulent diffusion. Here we shall concentrate on establishing a measure of the turbulent diffusion, which allows us to compare conditions over land and water.

An often used system for categorizing the turbulent diffusion is the socalled Pasquill scheme that in reality consists of two systems, one scheme to categorize the atmospheric conditions in a number of classes and another scheme, that associates values for σ_z and σ_y, the dispersion parameters in a Gaussian plume model to each of the classes in the first scheme. The Pasquill classification applies a simplified Bulk-Richardson number. This has by several authors been used to describe the Pasquill classes in terms of more conventional micrometeorological parameters. With the measuring height for the velocity normally fixed to 10 m, the two characteristic parameters become z_0 and L, the roughness and the Monin-Obukhov length, respectively. Also the height of the boundary layer is believed to be important. Here, however, we shall neglect this parameter and stay solely within the frame work of the surface layer scaling laws. For a large number of sites Golder (1972) has correlated L and z_0 values with Pasquill classes, Figure 2a. The resulting Pasquill classes in terms of z_0 and L are given in Fig. 3b.

a

Surface Wind Speed (at 10 m), m sec^{-1}	Day: Incoming Solar Radiation			Night	
	Strong	Moderate	Slight	Thinly Overcast or $\geq$4/8 Low Cloud	$\leq$3/8 Cloud
< 2	A	A-B	B		
2-3	A-B	B	C	E	F
3-5	B	B-C	C	D	E
5-6	C	C-D	D	D	D
> 6	C	D	D	D	D

The neutral class, D, should be assumed for overcast conditions during day or night.

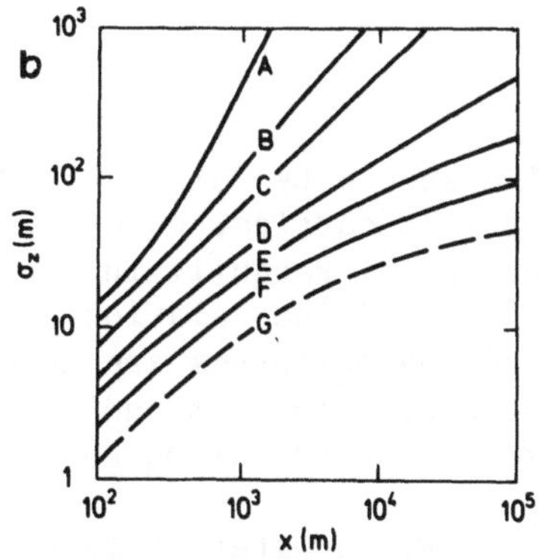

Fig. 2. a) Pasquill stability classification scheme (Turner, 1970).
b) The σ_z-curves corresponding to Turner (1970); σ_z for class G has been included, taken from USNRC (1974).

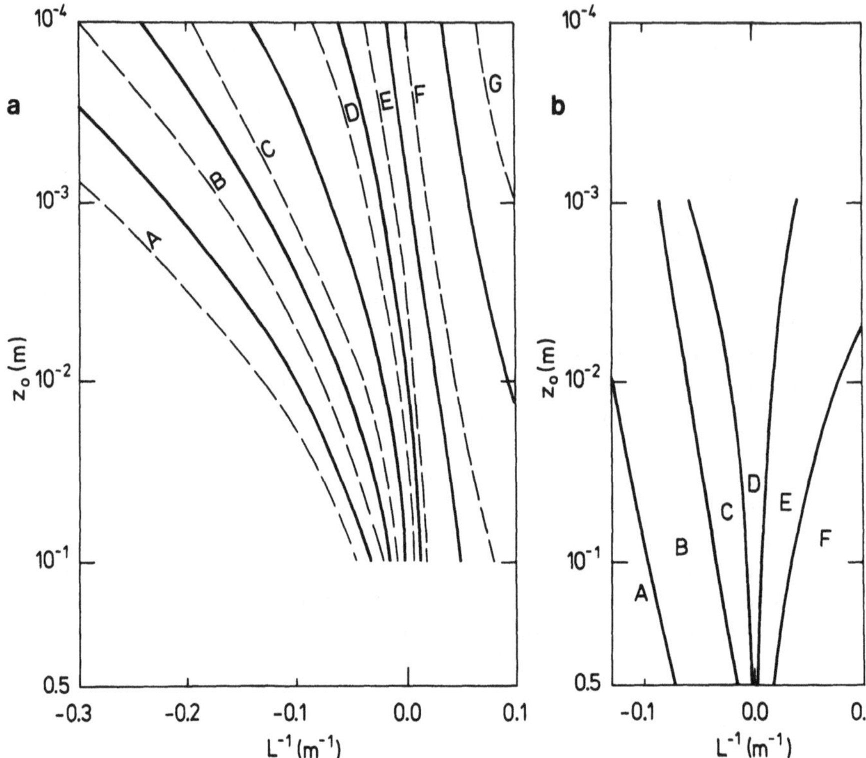

Fig. 3. a) shows the regions in the (z_0,L^{-1})-plane for which the different σ_z-curves in Fig. 2b can be used, b) shows the regions in the (z_0,L^{-1})-plane suggested by Golder (1972) corresponding to the stability classes given by Fig. 2a.

The scheme by Golder (1972) in Fig. 3b has found quite an extensive use. It carries, however, the same basic limitation as the original scheme by Pasquill in Fig. 2, of which we will mention (Pasquill, 1961): a) The relation between the classification scheme (Fig. 2a) and the σ_z-curves are essentially confirmed only for grassland with $z_0 \simeq$ 1-10 cm and associated thermal properties of the surface, and b) the σ_z-curves are for ground releases only.

For a water surface, we expect limitation a) to be serious owing to the large differences in roughness and thermal properties between water and land surfaces. Therefore we have tried to avoid this limitation, while we have used that the σ_z-curves in Fig. 2b are based on surface releases. Several authors have shown that a properly designed K-model can be used to describe the vertical diffusion from a continuous surface source. Recently Gryning et al. (1983) found excellent agreement between a K-model simulation and the Prairie Grass data (Barad, 1958). In the present discussion we have used the model of Gryning et al.:

The basic equation is:

$$u(z)\frac{\partial\chi(x,z)}{\partial z} = \frac{\partial}{\partial z}\left(K(z)\frac{\partial\chi(x,z)}{\partial z}\right), \tag{1}$$

where χ is the crosswind-integrated concentration, u the horizontal windspeed, K the eddy diffusivity of matter in the vertical direction, x the downwind distance from the source and z the height above the ground. For the eddy diffusivity, K, Gryning et al. found the following expression to give the best description

$$K = \kappa u_* z/\phi_H(z/L), \tag{2}$$

where κ is the von Kármán constant, $\phi_H(z/L)$ is the surface layer similarity function for the vertical temperature gradient, while L is the Monin-Obukhov length, u_* is the friction velocity. For ϕ_H we have used the expression given by Dyer (1974). In the surface boundary layer the velocity u(z) can be written as a function of z/z_o and z/L, that is

$$u(z) = \frac{u_*}{\kappa}\, f(z/z_o,\ z/L) \quad . \tag{3}$$

Through substitution of (3) and (2), Eq. (1) can be written

$$\frac{1}{\kappa} f(z/zo,\ z/L)\frac{\partial}{\partial(x/z_o)}\frac{\chi u_* z_o}{Q} = \tag{4}$$

$$\frac{\partial}{\partial(z/z_o)}\frac{\kappa}{\phi_H(z/L)}\frac{z}{z_o}\frac{\partial}{\partial(z/z_o)}\frac{(\chi u_* z_o)}{Q}$$

where Q is the source strength. Euqation (4) shows that

$$\chi u_* z_o/Q = g(x/z_o,\ z/z_o,\ z/L), \tag{5}$$

and therefore that any moment of the distribution along the z-axis will be a function of x/z_o, z_o and L.

In accordance with Pasquill (1961) the vertical spread h is defined as the height, where the concentration is reduced to one tenth of the ground level concentration. That is h can be found from (5)

$$0.1 = g(x/z_o,\ h/z_o,\ h/L)/g(x/z_o,\ o,\ o). \qquad (6)$$

Again following Pasquill (1961) h is related to σ_z in Fig. 2b through (assuming a Gaussian distribution)

$$h = 2.15 \cdot \sigma_z\ . \qquad (7)$$

Equations (6) and (7) are used to trace the σ_z-curves on Fig. 2b on a (z_o, L^{-1}) plane for fixed x-values less than 1 km.

The x-dependence of the σ_z-curves on Fig. 2b were in accordance with our estimates for z_o between 1 and 10 cm, while some deviations were found for a combination of small z_o-values and small negative L-values and for large z_o-values and small positive L-values. Essentially our σ_z-values grew somewhat slower for the latter situation and somewhat faster in the former situation relative to the curves in Fig. 2b. To give an overall impression of the vertical diffusion within the first kilometer from the source, we found $x \simeq 250$ m to be the best choice. The integration of (1) to obtain (6) was performed numerically, as described in Gryning et al. (1983). In the integration, we have neglected the effects of deposition, since it is not entirely clear how it should be included in order to describe the data on which Fig. 2b are based. In any case the influence of deposition will probably be small for $x \lesssim 250$ m, Gryning et al. (1983).

The results of the above described procedure is shown on Fig. 3a, where the broken curves show the (z_o,L)-values for which the σ_z-values at x = 250 m given by (6, 7) equals the values given in Fig. 2a. To separate the different classes handdrawn dividing lines are shown as unbroken curves. The two schemes on Fig. 3 show both similarities and differences. Most striking are the different slopes of the dividing lines of classes D,E, F between the two figures. This difference can be understood form the different origins of the two schemes.

Fig. 3b describes in terms of z_o and L^{-1} the boundaries between classes in Fig. 2a characterized with a given velocity and a given incoming/outgoing radiation. The definition of L^{-1} can be written

$$L^{-1} = -\frac{\kappa g}{T}\,\frac{\overline{w'\theta'}}{u_*^3}, \tag{8}$$

where g is the acceleration due to gravity and $\overline{w'\theta'}$ the sensible heat flux.

For a given radiation balance and a given wind velocity decreasing z_0 will reduce u_*^3 much more than $w'\theta'$, therefore the width of each class in the (z_0,L^{-1}) plot will increase for decreasing z_0. Furthermore, since the sign of $\overline{w'\theta'}$ will not change with z_0 for a given specified (in- or outgoing) radiation, an unstable category in Fig. 2a will remain unstable for all z_0's and correspondingly for a stable category. Fig. 3a on the other hand describes boundaries in the (z_0,L^{-1})-plane within which a given σ_z-curve from Fig. 2b can be used. From the K-model, described above, one finds that the growth rate of the plume will increase increase with increasing z_0 and decreasing L^{-1} (Gryning et al., 1983). This means that we to obtain a narrow plume, $\sigma_z(F)$ say, will need to increase L^{-1} less and less the smaller is z_0. Conversely to obtain a deep plume $\sigma_z(A)$ say, we must decrease L^{-1} more and more for decreasing z_0. It is noteworthy that for z_0 = 10^{-4} m the boundary between a $\sigma_z(E)$ and a $\sigma_z(F)$-plume is $L^{-1} \simeq - 0.2$, that is an unstable situation. Indeed, our reason for including a G-class in Fig. 2b and Fig. 3a is to be able to discriminate between different stable situations at all.

With these differences between the two sets of curves in Fig. 3 it is noteworthy that the L^{-1}-intervals for the different Pasquill-classes coincide very closely in the z_0-interval between 1 cm and 10 cm, i.e. in the z_0-interval where the matching between the classification scheme (Fig. 2a) and the σ_z-curves (Fig. 2b) was originally performed (Pasquill, 1961). We consider this a comforting argument for the concistency of the whole theoretical and experimental framework laying behind Figures 2 and 3.

MEASUREMENTS AND DATA ANALYSIS

The measurements were performed in the period 1978-81. The mast on Stevns was 18 m with velocity measurements at 3 levels and temperature measurements at 2 levels. The mast at the shoreline south of Copenhagen was 32 m with 8 levels of velocity measurements and 6 levels of temperature measurements. The greater height and denser instrumentation of the Copenhagen mast was to ensure that we obtained measurement above the shallow internal boundary layer originating from the coast-line 12 meter south of the mast. Data were recorded every 10 minutes, velocity as 10 min averages and temperature as instantaneous values. The data were averaged to yield hourly average profiles. From these profiles gradient Richardson numbers were derived and the corresponding L-values were determined from

$$z/L \simeq Ri \text{ for } z/L < 0$$
$$z/L \simeq Ri/(1-5Ri) \text{ for } z/L > 0, \quad (9)$$

where z is the height to which Ri pertains. Since here we are mainly concerned with wind from sectors I and II in Fig. 1, the z_0-values to be used in the schemes of Fig. 3 were taken as the average values for these two sectors. For Stevns we found $z_0 \simeq 1$ cm and for the Copenhagen mast $z_0 \simeq 0.01$ cm, that is the water roughness.

PASQUILL CLASS STATISTICS

From the data many aspects of flow conditions in a coastal zone can be studied. Here, however, we shall limit ourselves to such information as can be derived from simple distributions of Pasquill-classes, based on Fig. 3a.

Table 1 shows the yearly frequency of occurance of the different classes at the two stations for sectors I and II combined, see Fig. 1.

Table 1. Yearly distribution of Pasquill-classes at the Stevns and the Copenhagen station (St. and Co.) in percent of time in sectors I and II (Fig. 1). The z_0-values used are indicated, and the classification scheme employed is referred to by the figure number.

Station/Scheme	surface	z_0	Pasquill stability class						
			A	B	C	D	E	F	G
St./Fig. 3b	land	10^{-2} m	7	4	6	62	13	8	
St./Fig. 3a	land	10^{-2} m	8	4	5	43	27	7	6
Co./Fig. 3a	water	10^{-4} m	8	2	2	4	7	57	20
Co./Fig. 3a	water	10^{-2} m	13	3	5	24	31	19	5

In Table 1, rows 2 and 3 describe the yearly Pasquill class distributions after the schemes which are used in the rest of the paper. For comparison we have determined the distribution at Stevns according to the scheme by Golder (1972) from Fig. 3b as well. In connection with the discussion above about the similarities between the two schemes, it is illustrated that z_0 should be larger than 1 cm for coincidence between the two systems. Finally row 4 shows the distribution one obtains if the over water data are organized in the same L^{-1}-classes as the over land data, that is with the same assumed z_0-value.

From rows 2 and 3 is seen that the over water yearly distribution of Pasquill classes are displaced considerably toward classes F and G compared to the over land data. Owing to the difference between the heat capacity of water and land, the diurnal and annual temperature variations of the two surfaces are different. For Denmark the following features are typical (Larsen and Jensen, 1983). The diurnal variation of the water temperature is small, while the amplitude of the annual periods are about the same for land and water with a tendency to a slight delay of the water temperature relative to the land temperature.

The influence of this on the stability distributions over land and water is studied in Table 2, which shows the relative frequency of Pasquill classes for the year as a whole, for summer-days, where the land is warmest, and for winter nights, where the land is coldest. The distribution is shown for sectors I and II together.

Table 2. Distribution of Pasquill classes (in percent of time) for direction sectors I and II for the Stevns (St.) and the Copenhagen (Co.) position. Summer days correspond to the months of June, July and August between 11 a.m. and 4 p.m., while winter nights are specified as the months of December, January and February between 10 p.m. and 3 a.m.

Time	Surface/Pasq.	A	B	C	D	E	F	G
year	land, St.	8	4	5	43	27	7	6
	water, Co.	8	2	2	2	7	57	20
summer days	land, St.	17	11	17	53	2	0	0
	water, Co.	10	2	3	5	10	46	24
winter nights	land, St.	0	0	1	43	47	5	4
	water, Co.	10	2	3	5	9	70	1

Table 2 shows how the Pasquill class distribution over land changes with the sampling time according to the expectation that classes F F and G disappear during the summer days, while A and B disappear in winter nights. Over water, however, the frequency of classes A-E is seen to be the same for all three sampling periods employed. Furthermore, the most stable situations, class G, is all but eluminated during the winter nights over the water.

In Table 2 we have considered the periods when it is warmest (summer days) and when it is coldest (winter nights). In the following tables we will discuss the distribution of Pasquill classes for the periods March 21 - May 21 and July 23 - September 23. That is a two months period after the spring equinox and a two months period before the fall equinox.

Climatologically these two periods have the following characteristics (Larsen and Jensen, 1982): The sun goes through the same position on the sky, but due to differences in cloudcover the total incoming radiation is 10-20% larger in the fall period than in the spring period, the difference being largest in coastal regions. The average water temperature is about 1 ^{o}C colder than the daily average land temperature in the spring period and correspondingly 1 ^{o}C warmer in the fall period. The average air velocity is about the same for the two periods.

To reduce further the influence of airspeed differences in the two seasons the statistics presented in the following tables have been limited to situations, where the airspeed at 18 m is between 4 and 7 m/s.

Table 3 shows the distributions for sectors I and II (Fig. 1) of Pasquill classes for night and day data during the two seasons. A number of conclusions can be made: a) the diurnal changes of the over land data are obvious, b) daytime conditions are more unstable at the Stevns station during the spring period than during the fall period in spite of the larger insolation during the fall period, c) for the over water data the diurnal changes are less obvious, although present, d) the most obvious seasonal change is the reduction of the frequency of class G during the fall season, where the water is warm, relative to the frequency at spring. This tendency is strongest for nighttime data and corresponds to what was found for winter night data in Table 2.

The difference in the stability distributions between daytime spring and fall data from Stevns is considered further in Table 4. Here the distributions are presented for 4 different direction sectors (see Fig. 1). From the top to the bottom of the table the air blowing towards the station has been less and less influenced by the water temperature. In sector I there is previous to 1 km of

Table 3. Pasquill class distribution (percent of time) over land and water for sectors I and II and for the velocity at 18 m between 4 and 7 m/s. The period between 11 a.m. and 4 p.m. is called day, while night refers to the period between 10 p.m. and 3 a.m. Spring refers to the period March 21 - May 21, while fall refers to the period July 23 - September 23. St. means that data are from the Stevns mast, while Co. refers to the Copenhagen mast.

Time	surface/Pasq.	A	B	C	D	E	F	G
Spring	land, St.	53	18	13	12	4	0	0
days	water, Co.	6	1	1	2	4	35	51
Fall	land, St.	13	17	25	43	2	0	0
days	water, Co.	6	1	1	4	7	50	31
Spring	land, St.	0	0	0	20	53	23	3
nights	water, Co.	6	1	0	0	0	38	54
Fall	land, St.	0	0	0	18	58	14	11
nights	water, Co.	1	0	0	1	3	74	21

land south of the station about 100 km of water down to the south coast of the Baltic Sea. In sectors II and III the air trajectories pass an increasing amount of land surfaces until in sector IV the nearest 100 km to the mast is all land.

Due to the slightly larger insolation in the fall period one would expect the daytime fall stability distribution to show more strongly unstable situations than the spring distribution. Therefore, Table 4 illustrates the importance of nearby waters for stability distributions from meteorological stations close to the coast. This in turns illustrates the difficulties in applying schemes based on radiation considerations, like Fig. 2a, in the coastal zone.

Finally, Table 5 illustrates the difference in stability distribution at the Copenhagen station between sector I and II for spring days and fall nights. In sector II during spring days warm air from the land blows across a relative cool water surface before reaching the Copenhagen station, see Fig. 1. Hence we would expect the air to be more stable at this station, than during the fall nights, where cool land air blows across relatively warm sea. This

Table 4. Distribution of Pasquill-classes at the Stevns station for daytime data only (between 11 a.m. and 4 p.m.) and for the velocity at 18 m between 4 and 7 m/s. Spring refers to March 21 - May 21, while fall refers to July 23 - September 23. The sectors are shown on Fig. 1. From top to bottom the influence of the water temperature on the data decreases.

Sector	Period/Pasq.	A	B + C	D + E	F + G
I	Spring	59	29	11	0
	Fall	17	46	37	0
III	Spring	49	22	26	0
	Fall	19	41	37	0
II	Spring	37	37	25	0
	Fall	8	36	56	0
IV	Spring	2	14	84	0
	Fall	2	12	86	0

is in accordance with the table. In sector I the over water trajectories are much longer than in sector II, meaning that the air has adapted more to the water temperature. Hence, we expect the stability distributions in sector I to be displaced towards neutral relative to the distributions in sector II. The table shows this to be the case, when it is remembered that class F in the table actually corresponds to near neutral conditions in terms of the Monin-Obukhov length, L, compare Fig. 3a.

Table 5. Distribution of Pasquill-classes over water according to Fig. 3a, for the indicated sectors at the Copenhagen station and for the velocity at 18 m between 4 and 7 m/s. Spring days and fall nights are defined as in Tables 3 and 4.

Time	Sector/Pasq.	A	B	C	D	E	F	G
Spring	I	5	1	1	1	4	40	48
days	II	7	0	3	4	6	18	62
Fall	I	0	1	0	0	3	82	14
nights	II	2	0	0	1	1	68	26

DISCUSSION

The present paper discusses and illustrates a number of the problems encountered when the dispersion meteorology of a coastal zone has to be established. We do not believe that any of the phenomena, we have discussed, are unknown to airpollution meteorologists. However, quantitative discussions appear sufficiently rarely to justify the present description.

We want to emphasize that the dispersion condition classification system, described in Fig. 3a, is far from being a well developed scheme. It has the one advantage relative to the Pasquill class systems of Figs. 2a and 3b, that it allows one to estimate directly the relevant σ_z-curves from estimates for the Monin-Obukhov length and the roughness length for a very broad range of parameter values. However, it is limited by being based on a surface layer theory for a surface release with a somewhat uncertain relevance for elevated releases, a disadvantage it shares with Pasquill's original system of Fig. 2a.

Finally it shares the disadvantage with the scheme of Golder (1972), that it necessitates estimates of the roughness length, z_0. This gives rise to a good deal of uncertainty, since z_0 is the most elusive parameter in boundary layer meteorology. Over water it is known to vary with wind velocity. Indeed, Schacker et al. (1982) use this in their application of Golder's (1972) scheme. However, for a confined water area, as shown on Fig. 1, no satisfactory theory exists for the roughness length. Furthermore the profile data from which the used z_0-value was obtained as a central estimate yield values between 10^{-6} and 10^{-3} m, and as seen from Fig. 3a this scatter will give quite significant variations in the L^{-1}-limits between the different classes. Also the land data gave variations in z_0. The average z_0 = 1 cm for the two sectors I and II is an average of z_0-values of less than 1 mm during the winter and early spring, while more than 10 cm for the fully growth crops of the fields during the summer season, that is more than 2 decades of variation in the roughness length for the periods considered. Here we have decided to neglect the problems associated with the proper z_0-value(s) and used the chosen estimates as characteristic values.

Finally we should like to point out the difficulties associated with the use of climatological data to illustrate simple cause-effect relationships. The difficulty stems from that climatological data necessarily must include all the mechanisms and processes, that affect the weather in a given region. Therefore climatic data may often show a considerable more ambiguous response to a simple forcing, than one would imagine from an idealized model. Nevertheless, we believe that we have illustrated the necessity to include the conditions over both the water and the land when the dispersion conditions over any of these surfaces are to be described in coastal zone.

REFERENCES

Barad, M.L., 1958, Prairie Grass, a field program in diffusion, Report AFCRL-TR-235.

Dyer, A.J., 1974, A review of flux profile relationships. Boundary-Layer Meteorol., 7:363.

Golder, D., 1972, Relation among stability parameters in the surface layer. Boundary-Layer Meteorol., 3:47.

Gryning, S.-E., van Ulden, A.P., and Larsen, S.E., 1983, Dispersion from a continuous ground-level source investigated by a K-model. Quart. J.R. Met. Soc., 109:355.

Larsen, S.E., and Jensen, N.O., 1983, Summary and interpretation of some Danish climate statistics. Risø-Report No. 399.

Pasquill, F., 1961, The estimation of the dispersion of windborne material. The Meteorological Magazine, 90:33.

Schacher, G.E., Fairall, C.W., and Zanetti, P., 1982, Comparison of stability classification methods for parameterizing coastal over water dispersion. Proceedings of the first international conference on meteorology and air/sea interaction of the coastal zone, Am. Met. Soc. and KNMI, The Hague, Netherlands, 1982, 77-82.

Turner, D.B., 1970, Workbook of atmospheric turbulence. Public Health Service Publication No. 999-AP-26.

USNRC, 1974, Regulatory guide 1.111. Methods for estimating atmospheric transport and dispersion of gaseous effluents in routine releases from light-water-cooled reactors. U.S. Nuclear Regulatory Commssion, Wash. DC., USA.

DISCUSSION

A. VENKATRAM

Why did you go to such extraordinary lengths to fit your work into the framework of the Pasquill stability classification scheme ?

S.E. GRYNING

This system allows one to estimate directly the relevant σ_z-curve from estimates of the Monin-Obukhov length and the roughness length for a very broad range of parameter values; the system allows dispersion calculations over a water surface for example by use of Pasquilles σ_z-curves, and thus we avoid to introduce a new set of σ_z-curves.

A. GHOBADIAN F.B. Smith has proposed a classification scheme based on sensible heat flux and wind speed at 10 m with a set of σ_z curves which include the effect of roughness. The sigma curves are calculated by solution of the diffusion equation. The diffusivities employed are based on measurements. In view of this, I think your work and Smith work are similar.

S.E. GRYNING I agree that the two works are rather similar. However this system is much simpler to use than the Smith system (but not necessarily less accurate) because it simply gives you the relevant σ_z-curve for the conditions in question.

NUMERICAL SIMULATION OF COASTAL INTERNAL BOUNDARY LAYER DEVELOPMENTS AND A COMPARISON WITH SIMPLE MODELS

A. Ghobadian, A.J.H. Goddard and A.D. Gosman

Mechanical Engineering Department
Imperial College
London SW7 2BX

W. Nixon

United Kingdom Atomic Energy Authority
Safety and Reliability Directorate
Wigshaw Lane
Culcheth, Warrington, U.K.

INTRODUCTION

It is important, due to the frequent siting of industrial plant in such locations, to be able to predict diffusive conditions near shore-lines. An important feature of coastal meteorology is the presence of an internal boundary layer whenever an air flow crosses the coast line. The first quantitative attempt to describe internal boundary layer growth was given in the form of a nomogram by Prophet (1961) and reproduced by Van der Hoven (1967). Van der Hoven gives a graph based on the work of Prophet for determining the depth h of the mixing layer as a function of the initial over-water stability and over-land travel distance. An equation which represents these results is, (Meroney et al., 1975) :

$$h = 8.8\sqrt{\frac{x_1}{u_a\Delta\Theta}} \qquad (1)$$

where

X_1 = distance over land (m)
h = height of the mixed layer (m)
U_a = mean velocity $m.s.^{-1}$
$\Delta\Theta$ = over-water vertical difference in potential temperature within the inversion layer (°C.)

Raynor et al. (1975) proposed, based on their measurements, the following alternative formula for h :

$$h = \frac{u_*}{u_a} \left\{ \frac{x_1 \, | \Theta_1 - \Theta_2 |}{\left| \frac{\Delta\Theta}{\Delta x_3} \right|} \right\}^{\frac{1}{2}} \qquad (2)$$

where

u_* = friction velocity over down-wind surface
Θ_1 = low level potential air temperature over source region (°K)
Θ_2 = temperature of down-wind surface (°K)
$\left| \frac{\Delta\Theta}{\Delta x_3} \right|$ = absolute value of lapse rate over the source region.

Venkatram (1977) derived a slab model of the boundary layer and solved numerically the governing equations, which has been integrated over the boundary layer depth, in a Lagrangian framework. These equations were closed by employing an additional equation proposed by Zilitinkevitch (1975) and Tennekes (1975) describing the budget of turbulent kinetic energy near the inversion base. Venkatram also simplified the equation set and obtained an analytical solution, which compared favourably with the results of his numerical model. Furthermore, the analytical expression he obtained for the mixed layer height was very similar to that of Raynor et al. Kerman et al. (1982), who proposed, by separating the effect of land-water temperature difference and heat flux over land, an internal boundary layer growth formula based on a physical and a dimensional analysis and on the results of their experiments.

In the present paper, the APT computer code has been used to predict the effect of changes in surface conditions during onshore flows on the velocity, temperature and turbulence fields for conditions corresponding to the Nanticoke shore line diffusion experiment described by Kerman et al. A reasonable agreement has been obtained between predicted and measured internal boundary layer heights. The formulae presented by Raynor et al. and Van der Hoven have also been compared with the results of the Nanticoke experiment and found to underpredict internal boundary layer heights.

OUTLINE OF PREDICTION METHOD

In this section a mathematical model embodied in the APT method and the numerical method of the solution are firstly outlined; a more detailed account is available in the thesis of El Tahry (1979).

Mathematical Model

The assumption of stationary, two-dimensional, fully-turbulent boundary layer behaviour, which excludes consideration of strongly-unstable conditions and Coriolis effects along with other constraints, gives rise to the following equations for the fluid dynamics,in Cartesian tensor notation :

$$\frac{\partial\,(\rho U_1)}{\partial X_1} + \frac{\partial\,(\rho U_3)}{\partial X_3} = 0 \tag{3}$$

$$\frac{\partial\,(\rho U_1{}^2)}{\partial X_1} + \frac{\partial\,(\rho U_1 U_3)}{\partial X_3} = \frac{\partial P}{\partial X_1} - \frac{\rho\,\partial\overline{u_1 u_3}}{\partial X_3} \tag{4}$$

Here U and u represent the ensemble-average and fluctuating velocities respectively, subscripts 1 and 3 refer to the longitudinal and vertical directions, ρ and P are respectively density and pressure and the overbar denotes ensemble averaging. The distribution of potential temperature in the boundary layer, which is also assumed to be two-dimensional, is governed by :

$$\frac{\partial\,(\rho U_1 T)}{\partial X_1} + \frac{\partial(\rho U_3 T)}{\partial X_3} = - \frac{\rho\partial\overline{u_3 T'}}{\partial X_3} \tag{5}$$

where T' is the fluctuating component. Finally the concentration C of plume material is governed by the three dimensional transport equation :

$$\frac{\partial\,(\rho U_1 C)}{\partial X_1} + \frac{\partial\,(\rho U_3 C)}{\partial X_3} = - \frac{\partial}{\partial X_2}\left[\rho\overline{u_2 c}\right] - \frac{\partial}{\partial X_3}\left[\rho\overline{u_3 c}\right] \tag{6}$$

where c is the fluctuating component.

The turbulent fluxes in the above equations, which have the general form $\rho\overline{u_i\phi}$, where ϕ may be u, T' or c, are obtained from second-order transport models having their origins in the work of Launder et al. (1975, 1976) and Gibson and Launder (1978). The details are too lengthy to be presented here but may be found in El Tahry (1979). Briefly, each flux component has its own transport equation of the kind :

D $(\rho\overline{u_i\phi})$/Dt = Diffusion + (Production by Mean Gradients) + (Production/Destruction by Buoyancy) + Redistribution - (Destruction by Molecular Action)

where D/Dt is the substantive derivative and the labels on the right hand side denote groups of terms representing the physical processes in question. These differential equations are reduced to algebraic ones on the assumption that the transport of $\rho\overline{u_i\phi}$ by convection and diffusion is proportional to the transport of turbulence energy per unit mass $k = \overline{u_i u_i}/2$ by the same process. It is therefore only necessary to solve a differential transport equation for k and an additional one for the turbulence dissipation rate ε, which effectively provides the length scale for the transport processes : both equations are, within the boundary layer context, of the general form of (5), but with additional source and sink terms. The solution of these equations for the major part of the boundary layer is matched to a simplified representation of the flow near the ground based on one-dimensional equilibrium flow theory, so as to avoid problems of applicability and numerical resolution.

Numerical Solution Procedure

The foregoing equations are solved by an implicit, non-iterative, forward-marching finite-difference procedure which is conventional in many respects, the most noteworthy features being the use of computational grids which are; (i) different for the plume and boundary layer, in order to allow for the usually disparate scales of the two; and (ii) self-adjusting in a way which confines the calculations to regions of significant variations in the dependent variables.

These features, along with the other effort-saving facets of the modelling described earlier, give rise to a particularly economical prediction method.

APPLICATION TO NANTICOKE EXPERIMENT

In order to simulate the Nanticoke experiment with APT, initial conditions and boundary conditions must be specified : these conditions were obtained principally from Kerman et al (1982), while the experimental observations cited in Portelli (1982) and Anlauf (1982) were used to check the assumed boundary conditions. Observations on the conditions over Lake Erie were limited to temperatures at the shore line. The temperature values used in the present study were derived from the values, tabulated as a function of height, of the effective Brunt-Vaisala frequency N_e. The expression used by Kerman et al to evaluate N_e is

$$N_e h^2 = \int_0^h N^2 \, dx_3{}^2 \qquad (8)$$

If it is assumed that potential temperature gradient is constant in the height interval h_{i-1} to h_i, equation (1) can be expanded to :

$$N_e{}^2 h^2 = \int_0^{h_1} N_1{}^2 \, dx_3{}^2 + \ldots\ldots \int_{h_{i-1}}^{h_i} N_i \, dx_3{}^2 \qquad (9)$$

where $h = h_1 + \ldots . h_i$ and N_i is the Brunt-Vaisala frequency in the height interval $h_{i-1} < X_3 < h_i$.

Equation 2 was employed to estimate the initial temperature profile. Having started the APT calculation with a universal neutral profile, the new temperature profile was imposed and calculations were carried out over water for a distance $x_1 = 10\delta$ where δ is the prescribed height of the mixed layer over water. Trial and error was used to assess this height, which depends on the wind-speed over water, water surface roughness and the stable temperature profile imposed. If the imposed initial mixed layer height was too high, the predicted turbulence level near the top of the vertical domain of the calculations would fall to zero within $X_1 < 10$, necessitating the choice of a lower value of δ .

Analysis of the results indicated that the rate of decrease of turbulence level with X_1 is a decreasing function of X_1, as would be expected, initial values being those of neutral conditions ; the rate of decrease is small at $X_1 = 10\ \delta$, hence, varying this value makes negligible difference to conditions at the shore line.

The roughness length z_o was taken to be 0.005 m ; adjustment of this to 0.002 m resulted in an insignificant change in the internal boundary layer heights. The velocity at the top of the mixed layer was taken to be the tabulated mean velocity. The justification for this is as follows; (i) in an unstable boundary layer above the surface layer (which is a small fraction of the boundary layer height) the gradient of velocity is small. This is observed in the predictions and in the experimental results where, for example, at 11.00 am on the 1st of June the same mean velocity is quoted for a height of 200 m and a height of 500 m. (ii) The maximum predicted mixed layer height is 500 m and the velocities reported at plume level (see Anlauf et al (1982)), usually between 300 and 400 m, are close to the mean velocities.

As regards land surface roughness, Portelli (1982) reported that the surface flux measuring systems were located at a site with flat surrounding terrain and a long fetch ($\sim$ 1.5 km) free of obstructions (trees, buildings, etc.). The area was covered with winter wheat; Businger et al (1971) postulated a hydrodynamic roughness of 0.024 m for such a crop. In the present calculations a uniform land-surface roughness of 0.03 m and a progressively varying roughness between 0.03 to 0.3 m overland were both considered. It was found that : (i) roughness length within the above limits did not significantly effect the height of the calculated internal boundary layer under the relatively high land-surface heat fluxes that prevailed in the experiment, especially at 11.00 am 12.00 noon and 2.00 pm. (ii) Varying roughness length has significant effect on the turbulence level and hence the diffusive conditions within the internal boundary layer.

Kerman et al reported the wind speed at a height of 10 m at a location about 10 to 15 km from the shore line; this location was not within the direct line of measurements. It was found in the calculations that a progressive increase in z_o from 0.03 to 0.3 m with distance from the shore line resulted in velocities at 10 km from the shore line and 10 m above the surface which were very close to the measured velocities. Hence this varying roughness was specified in the main series of calculations. However it should be emphasised that the velocity at a height of 10 m depends strongly on the local roughness and agreement with the measurements at one point in the fetch does not imply a correct choice of roughness for the whole calculation.

But, as already indicated, roughness length has an insignificant effect on the calculated boundary layer height with the high heat fluxes that existed. In APT, the thermal boundary condition at the ground was taken to be the heat flux measured at the single measurement point in the middle of the fetch. Although a uniform ground heat flux based on a single measurement point is unsatisfactory, Kerman et al imply in their analysis that heat flux varies only with time, not with X_1.

APPLICATION OF ANALYTICAL FORMULAE

The over-water vertical difference in potential temperature, $\Delta\theta$, required in the Van der Hoven formula was taken to be the difference between the potential temperature at h (in equation (9)) and the water temperature. The value of u required in the Raynor et al formula was not reported by Kerman et al, hence the APT predicted value was used. (The APT calculation of u is based on the formulation of Businger et al (1971); predicted and measured velocities at a height of 10 m are very close and many APT simulations of wind tunnel and atmospheric measurements have demonstrated the correct level of surface shear.) Temperature differences between water and land required by the formula were tabulated by Kerman et al, although, again, it must be emphasised that the single land temperature was not taken on the line of the main measurements.

As already indicated, heat flux was prescribed as the ground thermal boundary condition rather than land surface temperature. Calculations suggested that the land temperature which could support the prescribed heat flux was generally higher than the reported land temperature but it must be emphasised that with the use of a constant overland heat flux and a constant water temperature, the calculated land temperature depends strongly on the intensity of turbulence in the surface layer and hence depends among other factors on the assumed roughness length. In their discussion, Kerman et al reported the land/water surface temperature difference as varying between 3 and 10 °C during the day on 1st June 1978. APT calculations indicated the same range of temperature differences during this day. However tabulated values by Kerman et al indicate a smaller range of temperature variation. In view of this inconsistency we have evaluated the parameter h from the Raynor et al formula using both the reported land temperature and the predicted land temperature. The effect of variation of lapserate with X_3 was incorporated by :

(i) applying the assumed constant value of the lapse rate to the down-stream distances for which the calculated h is within the height interval corresponding to the lapse rate. (See equation (9)).

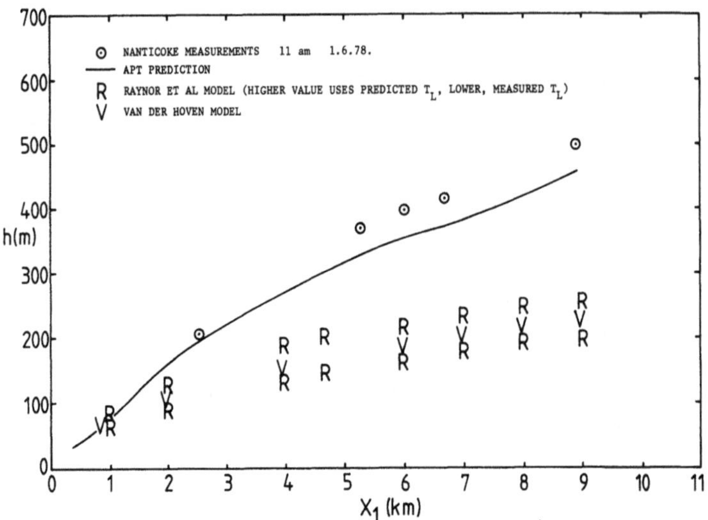

Figure 1. Comparison of APT predicted values, measured values and values from the analytical formulae of Raynor et al and Van der Hoven for the growth of the internal boundary layer in the vicinity of Lake Erie at 11.00 am on 1st June, 1978.

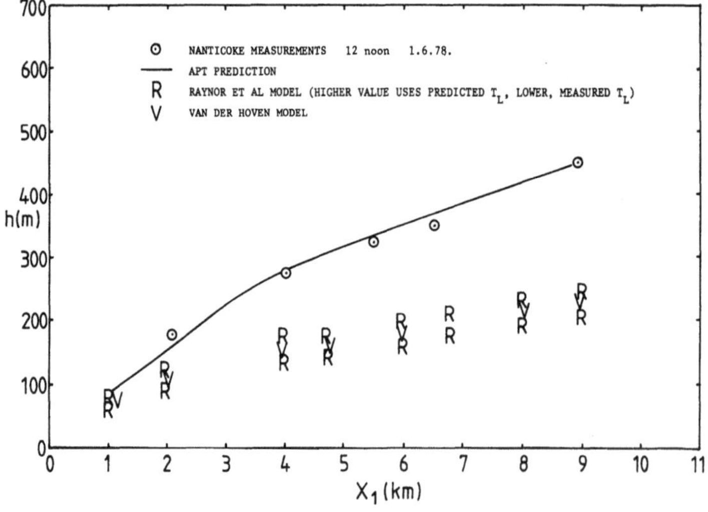

Figure 2. Comparison of the predicted and measured internal boundary layer growth, as in fig. 1, but for 12 noon, 1st June, 1978.

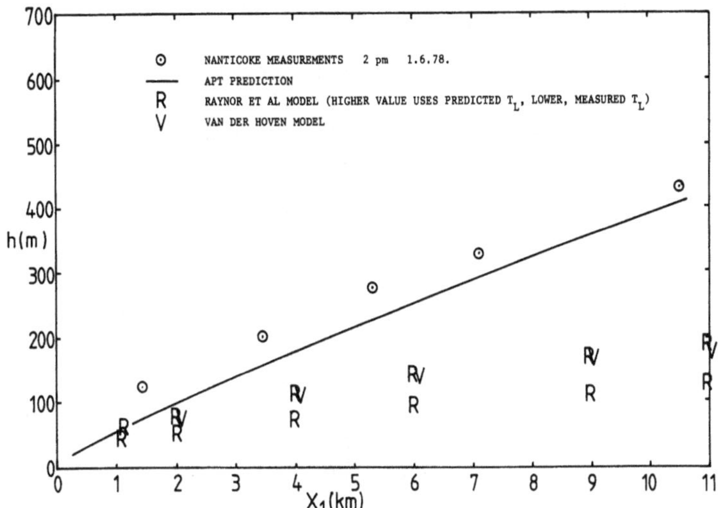

Figure 3. Comparison of the predicted and measured internal boundary layer growth, as in fig. 1, but for 2 pm, 1st June, 1978.

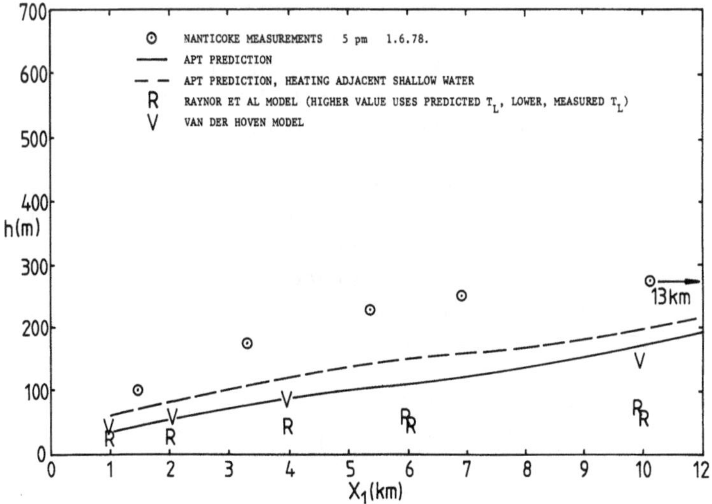

Figure 4. Comparison of the predicted and measured internal boundary layer growth, as in fig. 1, but for 5 pm, 1st June, 1978.

(ii) Evaluating the rate of increase of h with X_1 beyond the distance range in (i) above and, at each point, assuming a constant rate of increase for a further 1 km. As this rate of increase is a decreasing function of X_1, the calculated values of h are slightly overestimated.
In most cases considered the lapse rate decreased with height ; allowing this variation in evaluating equation (2) resulted in a slight overestimate in h compared with the case when variation of lapse rate with height was not considered.

DISCUSSION OF RESULTS

Figures 1 to 4 show the predicted and measured boundary layer heights and also those inferred from the Raynor et al and Van der Hoven formulae. The APT predictions agree well with the data in the morning, bearing in mind that the minisonde used for measuring the temperature has a spatial resolution of 10 m and an accuracy of $\pm$ 0.2°C, whereas the tethersondes, also used for measuring temperature, had an accuracy of $\pm$ 0.1°C. The analytical formulae both underpredict h.

In the afternoon, particularly at 5.00 pm, the APT predicted h is somewhat lower than the measured values. Kerman et al reported that a super-adiabatic layer was formed at the shore line by 1.00 pm indicating a non-zero (say 20 to 30 m) unstable internal boundary layer depth at this point. This is probably due to the very shallow lake waters extending some distance from the shore line which are heated up sufficiently to cause a boundary layer change off-shore. In APT a gradual increase in the heat flux to its land value was imposed for a distance of 2 km upstream of the shore line. As shown in figure 4, this increased the predicted h although the predicted values still remain lower than measurement. Possible reasons for the difference are :
(i) the predicted rate of erosion of very stable layers is too low ;
(ii) due to the super-adiabatic layer at the beach line, the turbulence levels near the surface were very high ;
(iii) there could be a region of enhanced roughness between the first measuring point and the shore line in the direction of the wind at 5.00 pm on 1st June 1978.

The assumed potential temperature profile $\Theta(X_3)$ at $X_1 = 0$ and the predicted $\Theta(X_3)$ and $H(X_3)$ at $X_1 = 6$ km and $X_1 = 8$ km at 11.00 am are shown in figure 5. The profiles have the shape observed in the atmosphere when an unstable mixing layer erodes an inversion layer above it (see, e.g. Zeman (1975)). The turbulent heat flux H has a constant value in the surface layer, decreases to zero and becomes negative before becoming zero again at the top of the mixing layer. This behaviour is observed in the atmosphere (Zeman (1975), Zeman and Lumley (1976)).

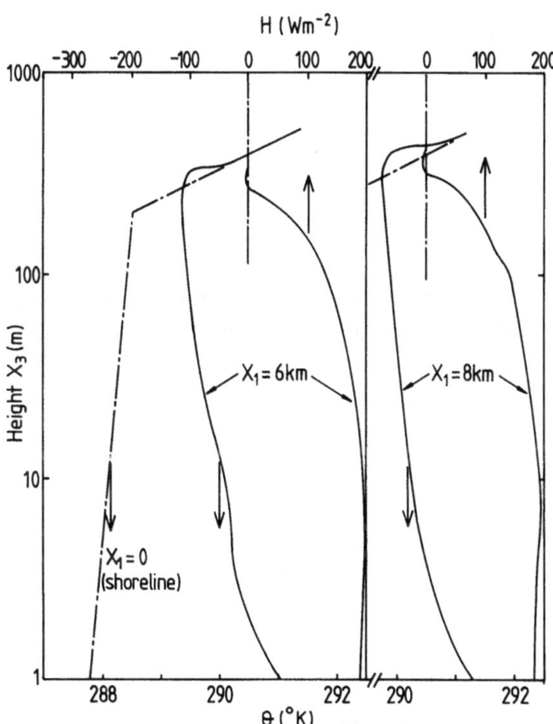

Figure 5. Potential temperature and heat flux profiles for two down-stream locations at 11.00 am, 1st June, 1978.

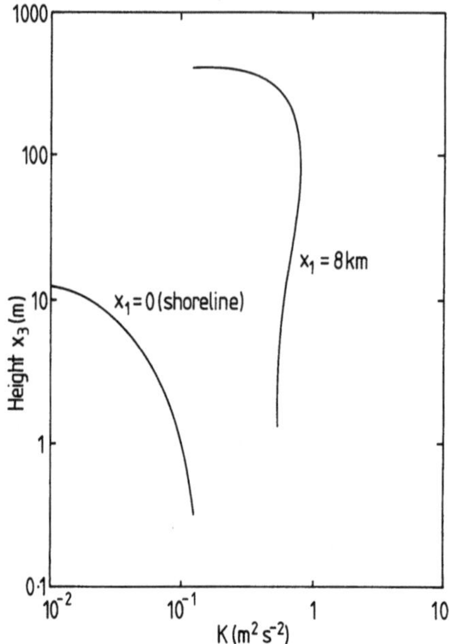

Figure 6. Predicted turbulent kinetic energy profiles at the shore-line and 8 km inland at 11 am, 1st June, 1978.

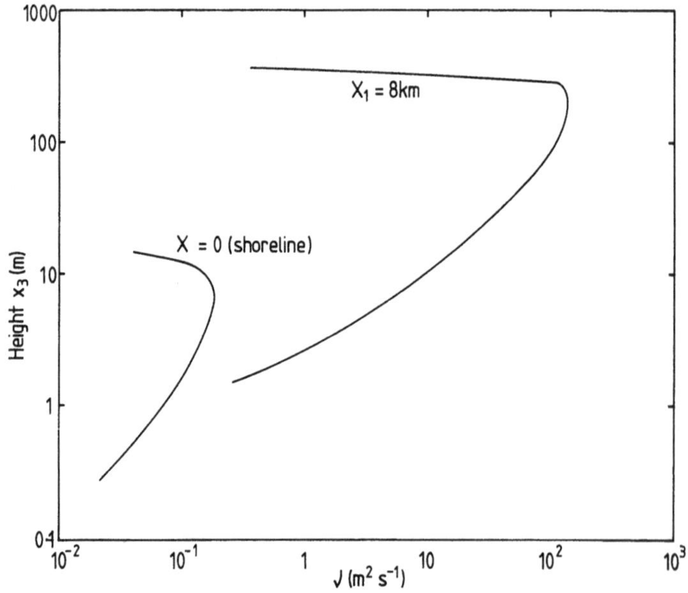

Figure 7. Predicted momentum diffusivity profiles at the shore-line and 8 km inland at 11 am, 1st June, 1978.

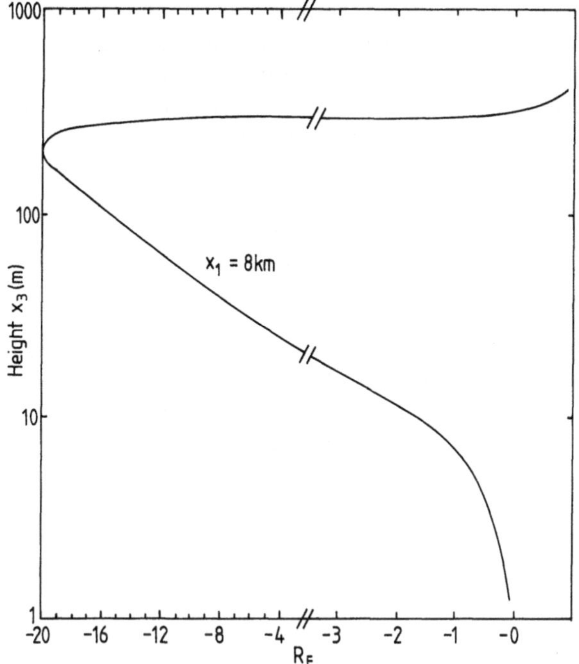

Figure 8. Predicted variation of flux Richardson number with height, 8 km down-stream of the shore-line at 11 am, 1st June, 1978.

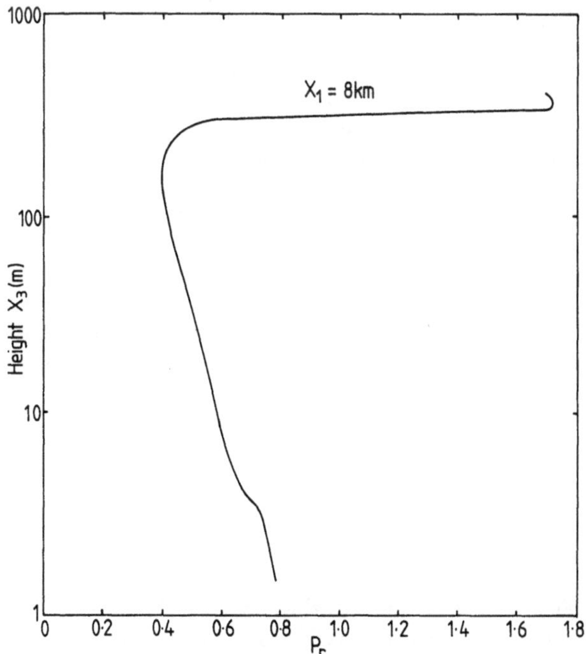

Figure 9. Predicted variation of Prandtl number with height, 8 km down-stream of the shore-line at 11 am, 1st June, 1978.

The predicted turbulence energy (k) profiles at $x_1 = 0$ and $X_1 = 8$ km are shown in figure 6. The increase in k due to the growth of the internal boundary layer is clearly demonstrated ; its uniformity and its increase to a maximum away from the surface in the unstable layer can also be seen. This increase in k away from the surface is observed in very unstable layers (Zeman (1975), Lumley et al (1978)). Vertical momentum diffusivity profiles are shown in figure 7. The large increase due to growth of the internal boundary layer is seen clearly. Figures 8 and 9 show the predicted Flux Richardson number R_F and Turbulent Prandtl number P_r at 11.00 am on 1st June at 8 km from the shore line. Comparison of these figures clearly shows that P_r decreases with instability and increases with stability. This trend is observed in both atmospheric measurements (Businger et al (1971), Arya, (1972)) and laboratory measurements (Ellison and Turner (1960) : and see also Turner (1973)).

CONCLUSIONS

In this paper it has been shown that the APT model predicts the growth of internal boundary layers in the Nanticoke experiments reasonably well, whereas the formulae proposed by Raynor et al and Van der Hoven both under-predict the growth. The height of the internal boundary layer is however an overall parameter and its accurate prediction by APT does not necessarily imply that the calculated velocity, temperature and turbulence fields are also accurate. It is to be hoped that these fields will be measured extensively in further experiments, thus contributing to the understanding of this transitory state and to the testing of theoretical models. The study has also shown that the APT model has correctly reproduced trends as to the effect of stratification on velocity, temperature and turbulence statistics in the atmospheric surface layer and has reproduced the observed effect of stratification on vertical and lateral dispersion of passive plumes (unpublished results). However, more experiments are needed to test its ability to predict processes in the atmospheric boundary layer under transitory conditions.

REFERENCES

Anlauf, K.G., Fellin, P., and Wiebe, H.A., 1982,
The Nanticoke shore-line diffusion experiment, June 1978 - IV, Atmospheric Environment, 16 : 455

Arya, S.P.S., 1972, The critical condition for Maintenance of turbulence in stratified flows, Quart.J.R.Met Soc., 28 : 264
Businger, J.A., Wynggard, J.C., Isumi, Y., and Bradley, E.F., 1971, Flux profile relationship in the atmospheric surface layer, J. Atmos. Sci., 28 : 181
Ellison, T.H., and Turner, J.S., 1960, Mixing of dense fluid in a turbulent pipe flow, J. Fluid Mech., 8 : 514
El Tahry, S., 1979, "Turbulent plume dispersal", PhD thesis, University of London
Gibson, M.M., and Launder, B.E., 1978, Ground effects upon pressure fluctuations in the atmospheric boundary layer, J. Fluid Mech., 86 : 491
Kerman, B.R., Mickle, R.E., Portelli, R.V. and Trivett, N.B. 1982, The Nanticoke shore-line diffusion experiment June 1978 - II, Atmospheric Environment, 16 : 423
Launder, B.E., 1976, Heat and mass transport in : - Topics in applied physics", (P. Bradshaw, Ed.) Vol 12
Launder, B.E., Reece, G.J., and Rodi, W., 1975, Progress in the development of a Reynolds stress turbulence closure, J. Fluid Mech., 68 : 537
Lumley, J.L., Zeman, O., and Siess, J., 1978, The influence of buoyancy on turbulent transport, J. Fluid Mech., 84 : 581
Meroney, R.N., Cermak, J.E., and Yang, B.T., 1975, Modelling of atmospheric transport and fumigation at shore-line sites, Boundary Layer Meteorol., 9 : 69
Portelli, R.V., 1982, The Nanticoke shore line diffusion experiment, June 1978-I, Atmospheric Environment, 16 : 413
Raynor, G.S., Michael, P., Brown, R.M. and Sethu Raman, 1975, Studies of atmospheric diffusion from a near shore oceanic site, J. Appl. Meteor., 14 : 1080
Tennekes, H. 1975, Reply to Zilitinkevich, 1975, J. Atmos. Sci., 32 : 992
Turner, J.S., 1979, "Buoyancy effects in fluids", Cambridge University Press, Cambridge.
Van der Hoven, I., 1967, Atmospheric transport and diffusion at coastal sites, Nucl. Saf., 8 : 490
Venkatram, A., 1977, A model of internal boundary layer development, Boundary Layer Meteorol., 11 : 419
Zeman, O., 1975, The dynamics of entrainment in the planetary boundary layer : A study in turbulence modelling and parameterization, PhD thesis, Pennsylvania State University.
Zeman, O. and Lumley, J.L., 1976, Modelling buoyancy driven mixed layers, J. Atmos. Sci., 3 : 1976
Zilitinkevich, S.S., 1975, Comments on "A model for dynamics of the inversion above a convective boundary layer", J. Atmos. Sci., 32 : 991

APPLICATION TO THE BELGIAN COAST OF A 2-DIMENSIONAL PRIMITIVE EQUATION MODEL USING σ COORDINATE

Hubert Gallée

Institute of Astronomy and Geophysics
Catholic University of Louvain
B-1348 Louvain-la-Neuve, Belgium

SUMMARY

This model is an improved and adapted version of the mesometeorological model described by Alpert P., Cohen A. and Neumann J., Hebrew University of Jerusalem (Alpert, 1980). Several simulations relative to a typical case of thermal breeze along the Belgian coast are presented, the results of which can be used as an input for the characteristic pollutant dispersion models of that region.

Alpert's model is a primitive equations bidimensional model using σ coordinate system.

First, the hight computational costs required by it were lowered by employing a split explicit integration scheme for the advection and diffusion term. A comparison between the results obtained from the versions for a typical case of thermal breeze at the Belgian coast (8/8/1976), demonstrates the accuracy of this method. The computational costs decrease by about 30%.

The infrared radiation term was then taken into account in the thermal energy equation, after introducing a diffusion equation for water vapor and the concept of virtual temperature. A simulation of the situation of 8/8/76 with these improvements is finally described and it is shown that the vertical temperature gradient is better simulated than without these last improvements.

DESCRIPTION OF THE MODEL

Basic Equations of the model

It is a 2-D model (in the x,z plane) assuming compressibility, variation of topography and hydrostaticity (Alpert, 1980).

The set of equations solved for the mean variables is shown below

a) The horizontal momentum equations

$$\frac{\partial u}{\partial t} = -u\frac{\partial u}{\partial x} - \dot{\sigma}\frac{\partial u}{\partial \sigma} + f(v-v_g) - \frac{RT}{p_* + \frac{p_t}{\sigma}}\frac{\partial p_*}{\partial x} - \left.\frac{\partial \phi}{\partial x}\right|_\sigma + F_x$$

$$\frac{\partial v}{\partial t} = -u\frac{\partial v}{\partial x} - \dot{\sigma}\frac{\partial v}{\partial \sigma} - f(u-u_g) + F_y$$

b) The hydrostatic equation

$$\frac{\partial \phi}{\partial (\frac{T}{\theta})} = - C_p \theta$$

c) The continuity equation in the σ system

$$\frac{\partial p_*}{\partial t} = - \int_0^1 \frac{\partial (p_* u)}{\partial x} \, d\sigma$$

d) The equation of state

$$p = \rho RT$$

e) The definition of potential temperature

$$\theta = T \left(\frac{p_0}{p}\right)^\kappa$$

f) The thermodynamic energy equation

$$\frac{\partial \theta}{\partial t} = -u\frac{\partial \theta}{\partial x} - \dot{\sigma}\frac{\partial \theta}{\partial \sigma} + F_\theta$$

g) The definitions

$$\sigma = \frac{p-p_t}{p_*}$$

$$p = p_s - p_t$$

$$\phi = gz$$

where we have

x,y : horizontal coordinates (x is parallel to the vertical simulation plane)

t : time coordinate

σ : vertical coordinate

p_t : top pressure of the model (which corresponds approximatively to the height of the P.B.L.)

p : pressure at level σ

$p_* = p_s - p_t$

p_s : soil pressure

$\phi = gz$: geopotential

$\left.\frac{\partial\phi}{\partial x}\right|_\sigma$: x component of the geopotential gradient on a σ surface

$g = 9{,}81 \text{ m sec}^{-2}$: gravity

z : altitude

(u_g, v_g) : geostrophic wind components

(u,v) : horizontal mean wind components

f : Coriolis parameter : $f = 1{,}114\ 10^{-4}\ \text{sec}^{-1}$

$\dot{\sigma} = \frac{d\sigma}{dt}$

T : temperature

R : perfect gas law constant for dry air : $R = 287$ J. $\text{kg}^{-1}\ \text{K}^{-1}$

ρ : specific mass of the air

$\kappa = R/C_p$

$C_p = 1004$ J. $\text{kg}^{-1}\ \text{K}^{-1}$: specific heat at constant pressure

$F_{x,y}$: diffusion contribution to the horizontal momentum equations

F_θ : diffusion contribution to the thermodynamic energy equation

<u>The diffusion forces</u>

The parameterization of diffusion term follows a first order closure scheme. For example for x momentum equation

$$\overline{u'w'} = -K_z \frac{\partial \bar{u}}{\partial z}$$

and

$$F_x \simeq \frac{g^2}{p_*^2} \frac{\partial}{\partial \sigma} (\rho^2 K_z \frac{\partial u}{\partial \sigma})$$

taking only into account the vertical turbulent fluxes. For K_z in the surface layer, using the logarithmic law, one has

$$K_z = \frac{k^2 z \bar{u}_{10m}}{\log \frac{z+z_0}{z_0}}$$

where z_0 is the roughness length
k is the von Karman constant
$\bar{u}_{10m}$: horizontal component of wind at anemometer level.

Above the surface layer, one has

$$K_z = \ell^2 S f(R_i)$$

where

$$\ell = \frac{kz}{1 + \frac{kz}{\lambda}} \quad \text{is the mixing length}$$

$$= 0.00027 V_g/f \quad \text{(Blackadar, 1962)}$$

V_g is the norm of the geostrophic wind vector

$$S = \sqrt{(\frac{\partial \bar{u}}{\partial z})^2 + (\frac{\partial \bar{v}}{\partial z})^2} \quad \text{is the wind shear norm}$$

$$R_i = \frac{\frac{\partial \phi}{\partial \sigma} \frac{\partial \theta}{\partial \sigma}}{\bar{\theta}_\sigma |\frac{\partial \bar{V}}{\partial \sigma}|^2} \quad \text{is the local Richardson number}$$

admit $\bar{\theta}_\sigma$ is mean of θ between 2 σ levels.

The stability functions used are :

$$f(R_i) = (1 - 18 R_i)^{1/2} \qquad R_i < 0 \quad \text{(unstable)}$$

$$f(R_i) = (1 + \alpha R_i)^2 \qquad 0 < R_i < -\frac{1}{\alpha} \quad \text{(stable)}$$

α =-0.003

for very stable conditions, Alpert imposes a lower limit of $K_z = 2.10^4 \text{ cm}^2 \text{ sec}^{-1}$

For y momentum equation, one has :

$$F_y \simeq \frac{g^2}{p_*^2} \frac{\partial}{\partial \sigma} (\rho^2 K_z \frac{\partial v}{\partial \sigma})$$

For thermodynamic energy equation

$$F_\theta \simeq \frac{g^2}{p_*^2} \frac{\partial}{\partial \sigma} (\rho^2 K_z \frac{\partial \theta}{\partial \sigma})$$

Specific humidity

We have also introduced the specific humidity conservation law:

$$\frac{\partial q}{\partial t} = -u \frac{\partial q}{\partial x} - \dot{\sigma} \frac{\partial q}{\partial \sigma} + F_q$$

where q is the specific humidity [kg/kg] and F_q is the diffusion term:

$$F_q \simeq \frac{g^2}{p_*^2} \frac{\partial}{\partial \sigma} (\rho^2 K_z \frac{\partial q}{\partial \sigma})$$

Specific humidity is also introduced in the perfect gas law

$$p = \rho R T_v$$

where $T_v = (1 + 0.608\ q)T$ is the virtual temperature.

Hydrostatic equation must also be modified :

$$\frac{\partial \theta}{\partial (\frac{T}{\theta})} = -C_p\ \theta(1 + 0.608\ q)$$

Infrared radiation contribution

Finaly in the thermodynamic energy equation we also take into account vertical radiative infrared fluxes $R_z$after (Sasamori, 1968) interacting with water vapor and carbon dioxide : so we get

$$\frac{\partial \theta}{\partial t} = -u \frac{\partial \theta}{\partial x} - \dot{\sigma} \frac{\partial \theta}{\partial \sigma} - \frac{\partial R_z}{\partial \sigma}$$

Initial conditions

The simulation begin at the time of the day when it can be reasonably assumed that there is no horizontal gradients of any importance (beginning of day time, roughly 6h GMT). Initialization of temperature and humidity vertical profile is made with the sounding of Uccle at 00 t.u. These profiles are assumed to be uniform along the whole integration domain.

Initialization of wind is given by the mean geostrophic wind during the simulated day.

Boundary conditions specified are :

Lateral boundaries

$$\frac{\partial u}{\partial x} = \frac{\partial v}{\partial x} = 0$$

$$\frac{\partial \theta}{\partial x} = 0$$

$$\frac{\partial q}{\partial x} = 0$$

$$\frac{\partial p_*}{\partial x} = 0$$

Top boundary (σ=0)

$$\dot{\sigma} = 0$$

$$\frac{\partial u}{\partial \sigma} = \frac{\partial v}{\partial \sigma} = 0$$

$$\frac{\partial \theta}{\partial \sigma} = 0$$

$$\frac{\partial q}{\partial \sigma} = 0$$

Bottom boundary

$$\dot{\sigma} = 0$$

$$u = v = 0$$

temperature is a forcing : sea surface temperature is assumed to be constant and soil surface temperature is calculated knowing under shelter temperature data, taking into account the diffusion effects in the surface layer.
Specific humidity is also fixed at the bottom boundary with the help of synoptic data's.

Numerical scheme

The integration domain is a vertical plane perpendicular to the coast. It is divided into 53 horizontal grid points with 4 km grid spacing. There are 8 vertical levels + the top and bottom boundary levels. The spacing between each level follows a logarith-

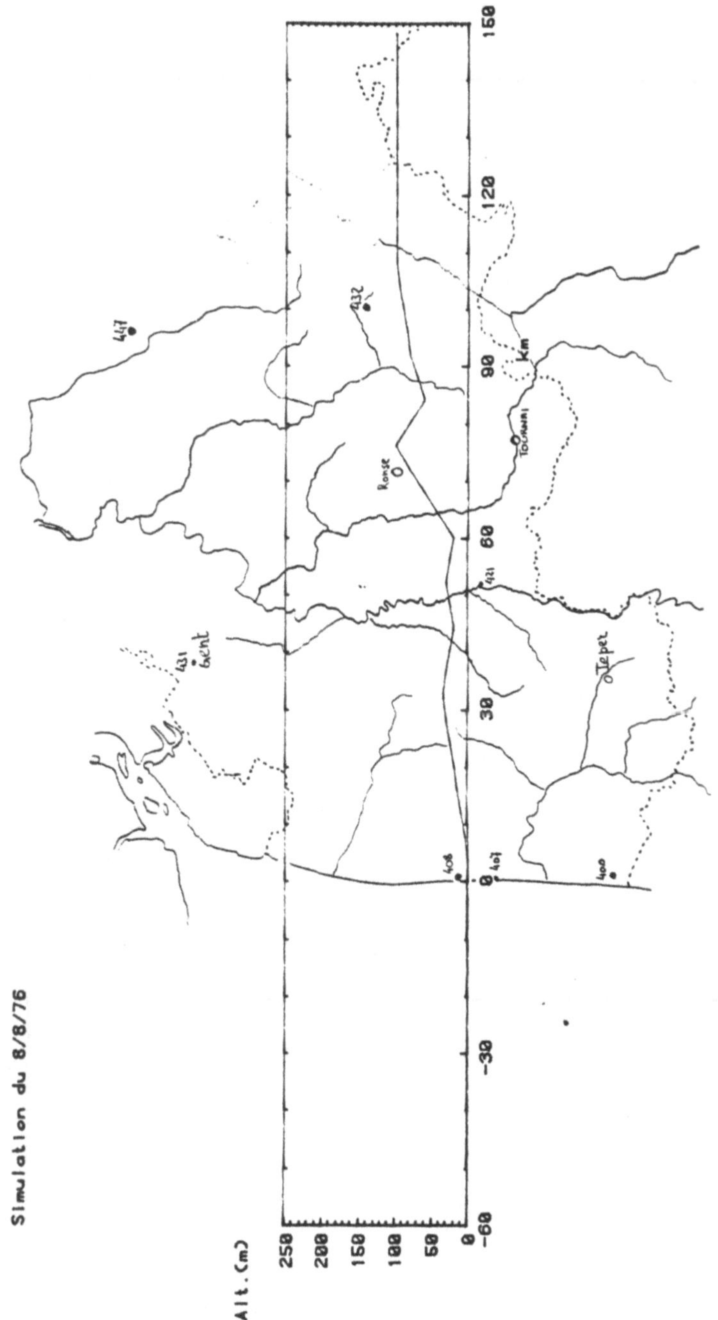

Figure 1 The initial conditions and topography of Belgium.

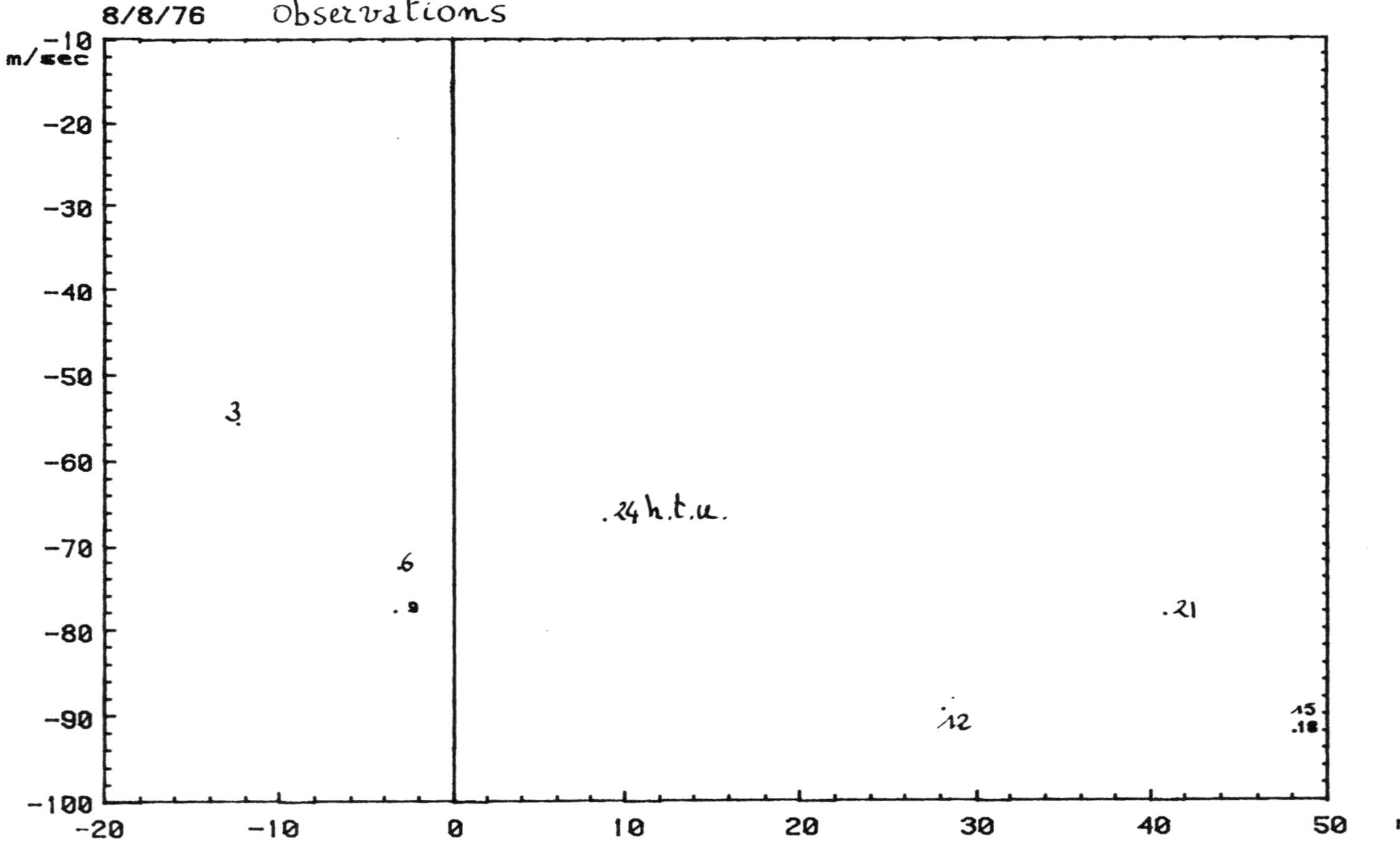

Figure 2.a. Observed hodograph at Belgian coast.

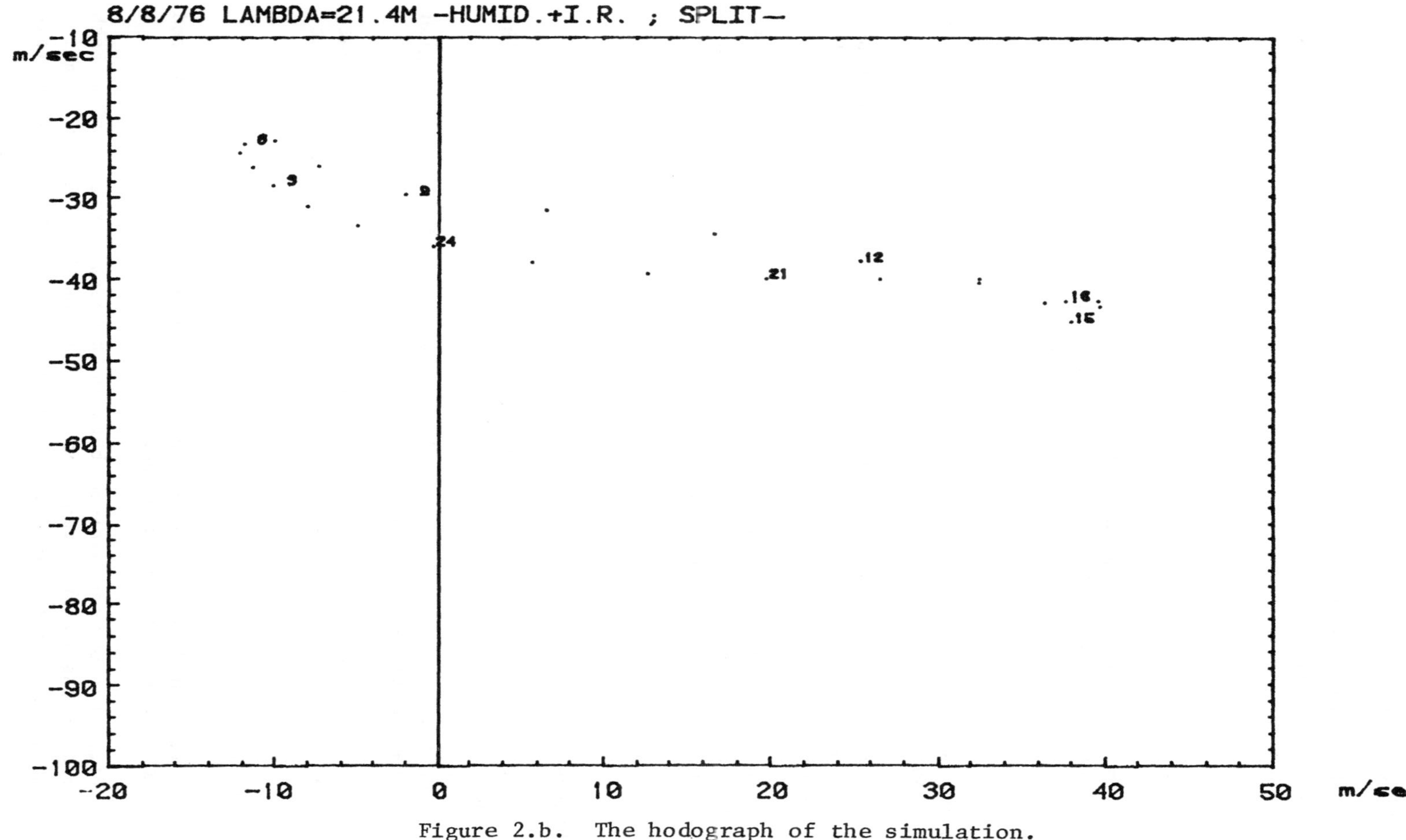

Figure 2.b. The hodograph of the simulation.

mic growth with the first intermediair level as the 10m anemometer level and the last level chosen above the P.B.L.

The time step is chosen so we avoid unstable growth of the fastest waves which can propagate into the model. They are 2-dimensional accoustic lamb waves of 340 m/sec wave speed. So we choose

$$\Delta t < \frac{\Delta x}{c}$$

$$\Delta t = 8 \text{ sec}$$

Integration of pressure gradient term and Coriolis term is performed with a leepfrog scheme. Integration of diffusion term requires an implicit scheme and integration of advection term is achieved with a semi-Lagrangian technique with help of cubic spline interpolation of horizontal velocity between grid points. But diffusion term as well advection term do not carry fast waves at all so we calculate their contribution only respectively 3 and 5 time step because they need hight computational time.

Finaly to avoid computational instability at the lateral boundaries we end the scheme with a filter, in order to elimite the waves of length greater or equal to $2\Delta x$.

The simulation is done on a mini-computer PDP11 (only 56 kbytes). The introduction of a split time differencing technique does not introduce any modifications in the results of simulation.

THE THERMAL BREEZE OF 8/8/76 AT BELGIAN COAST

Observations

On Fig. 1 we have plotted the initial conditions, and the topography (continuous line) on a map of Belgium.

On Fig. 2.a. we have plotted the observed hodograph at Belgian coast (Middelkerke, synoptic station n°407).

The geostrophic wind was (-2.7 m/sec, -8.4 m/sec) (coordinate x directed on shore) so this situation leads to a sea breeze front.

Simulation

Fig. 2.b shows the hodograph of the simulation and we see a good agreement in the amplitude of breeze.

Fig. 3 shows comparison between temperature evolution both observed (crosses) and simulated (continuous line) for a location for inland (Chièvres, synoptic station nr 432) from which we take

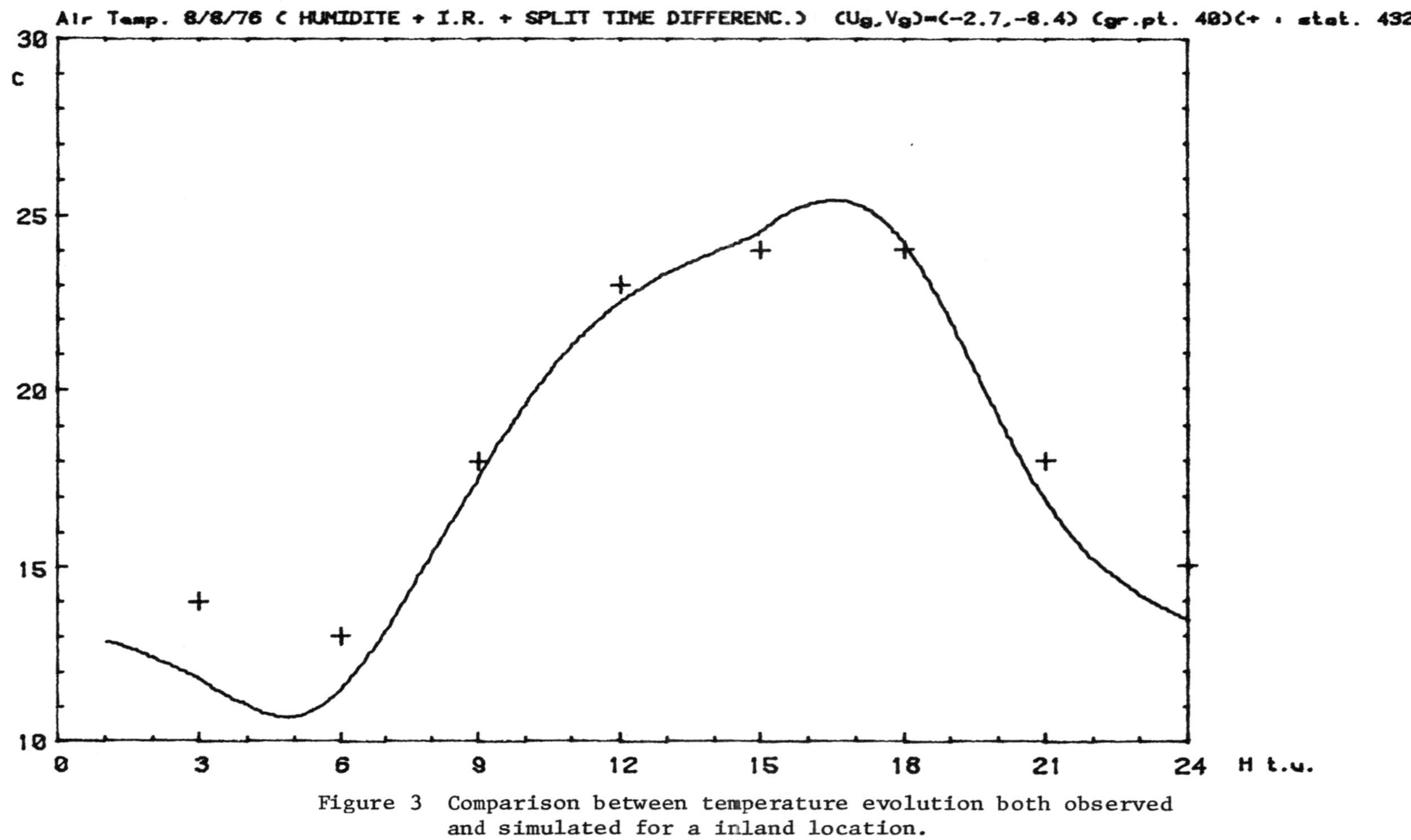

Figure 3 Comparison between temperature evolution both observed and simulated for a inland location.

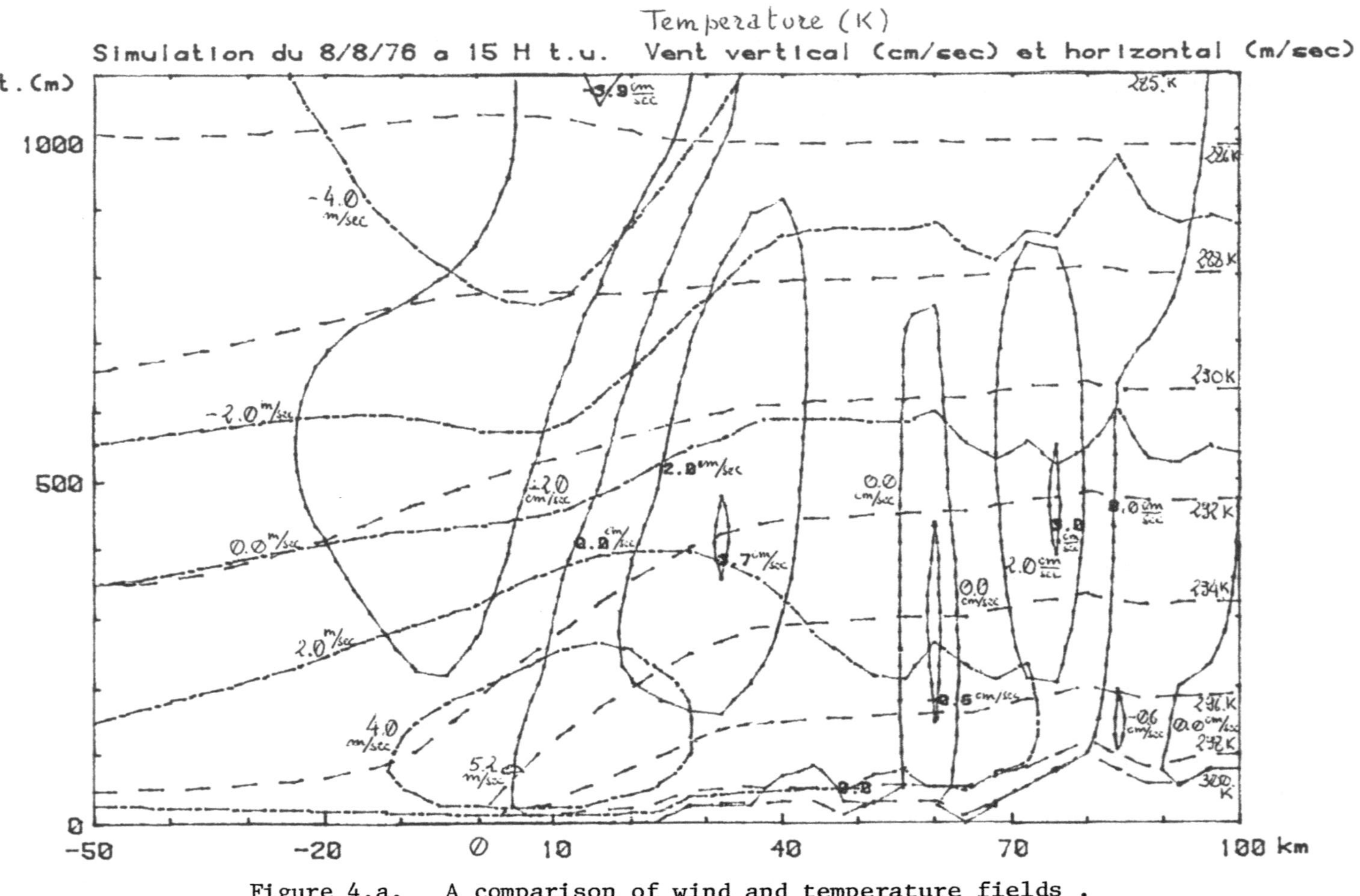

Figure 4.a. A comparison of wind and temperature fields .

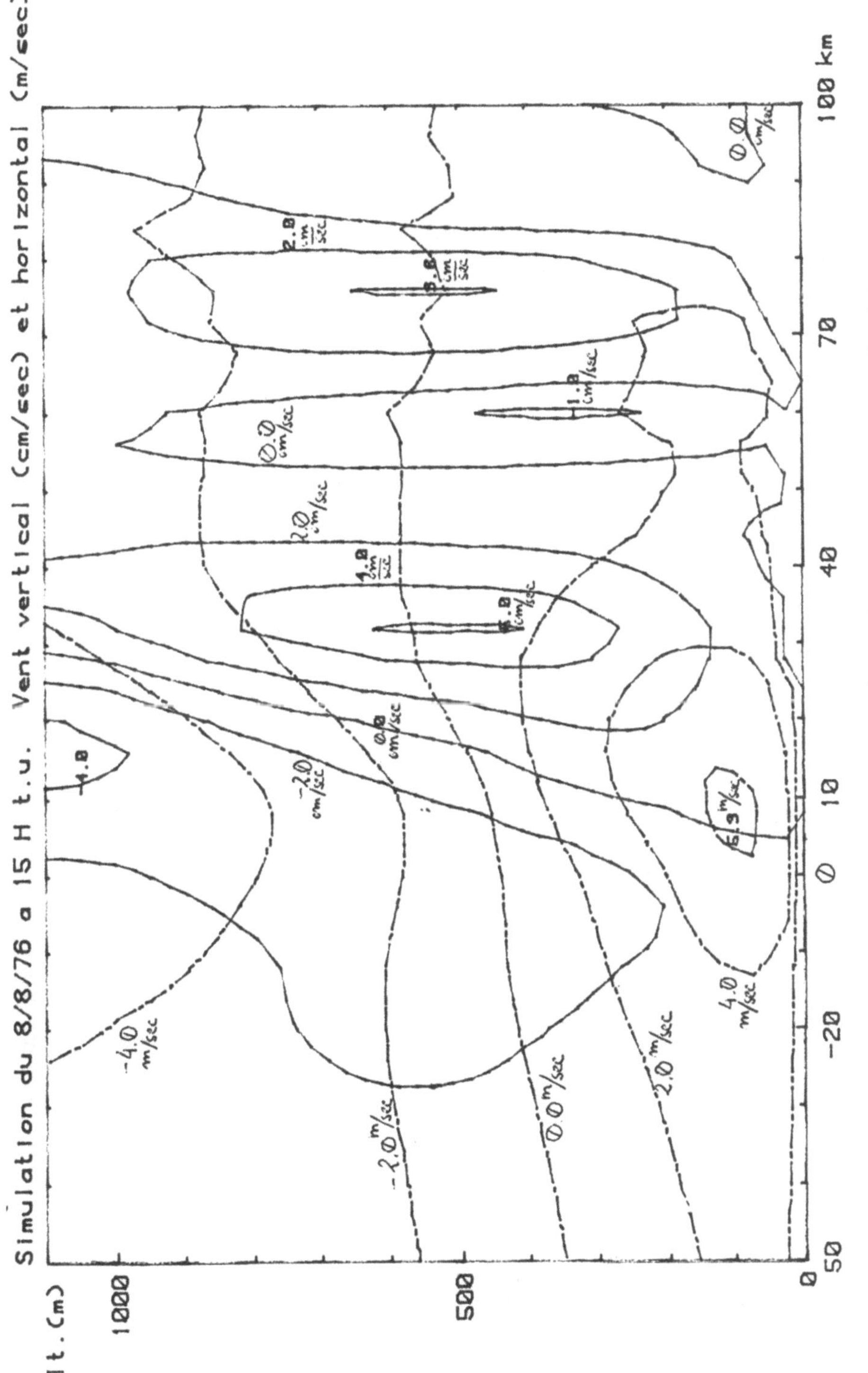

Figure 4.b.1. A comparison of wind and temperature fields.

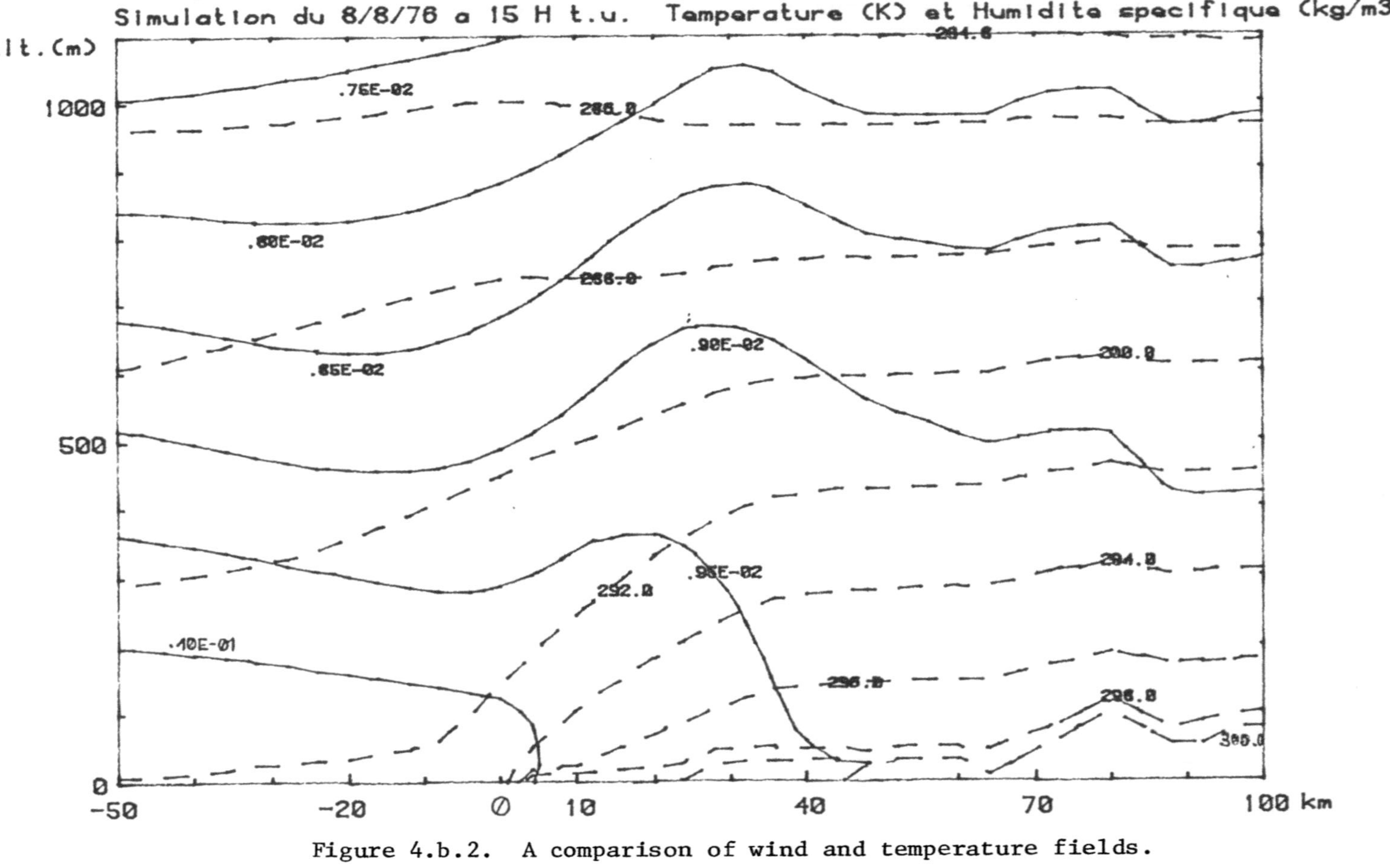

Figure 4.b.2. A comparison of wind and temperature fields.

into account under shelter temperature to calculate soil forcing temperature.

Fig. 4.a., 4.b.1 & 4.b.2 show a comparison of wind, temperature (and humidity) fields between computation with and without specific humidity conservation and infrared radiation contribution (versions a & b respectively) and we see that the radiational warming of the onshore lower levels during the day leads to an enhancement of sea breeze circulation. On the other hand, neglecting the variations of synoptic conditions during the simulation, we get a better evolution of the high levels temperature as we take into account I.R. contribution.

Both figures show the effect of topography on sea breeze circulation : the ascending cell is divided into 2 cells : one 32 km onshore, in front of the first hillocks and one 75 km onshore, in front of the "vlaamse hoofden", which is a old volcanoes range parallel to the coast.

CONCLUSIONS

The numerical model of Alpert and Neumann J., with the implementation of the specific humidity conservation law and infrared radiation contribution, and the introduction of a split time differencing technique, was applied to the simulation of a sea breeze event at Belgian coast.

The amplitude of simulated sea breeze was in good agreement with the observations and its propagation was also good simulated. The introduction of soil elevation in the model by σ coordinate permits a good comprehension of its effect on the sea breeze circulation. The introduction of humidity conservation law and infrared radiation contribution does not lead to a great difference in the sea breeze simulation except an improvement in the temperature profiles simulation.

REFERENCES

Ahlberg, J.H., Nilson, E.N., and Walsh, J.L., 1967, "The theory of splines and their applications", Academic Press, New York, 284pp.

Alpert, P., 1980, A mesometeorological model with topography, The Hebrew University of Jerusalem.

Alpert, P., Cohen, A., Neumann, J., and Doron, E., 1982, A model simulation of the summer circulation from the eastern Mediterranean past Lake Kimeret in the Jordan Valley, Mon. W. Review, 110 : pp. 994-1006.

Atkinson, B.W., 1981, "Mesoscale atmospheric circulations", Academic Press, New York, pp. 125-214.

Blackadar, A.K., 1979, Hight resolution models of the Planetary Boundary Layer, Advances in Environmental Science and Engineering, vol.1, J.R. Pfafflin and E.N. Ziegler, Eds, Gordon and Breach, 256pp.

Coantic, M.F., 1978, An introduction to turbulence in geophysics and air-sea interactions, Advisory Group for Aerospace research and Development, AG-232, NATO.

Defant, F., 1951, Local winds, Compendium of Meteorology, Americ. Met. Soc., pp. 655-672.

De Moor, G., 1978, Les théories de la turbulence dans la couche limite atmosphérique, Secrétariat d'Etat auprès du Ministère de l'Equipement et de l'Aménagement du territoire, République Française.

Gadd, A.J., 1978, A split explicit integration scheme for numerical weather prediction, Quart. J.R. Met. Soc., 104 : pp. 569-582.

Gallée, H., Schayes, G., Berger, A., 1981, Contribution à un modèle bidimensionnel de la couche limite planétaire avec relief : étude dimensionnelle, application à la côte belge et programmation sur mini-ordinateur, UCLLN, Progress report 1981/8, Institut d'Astronomie et de Géophysique.

Holton, J.R., 1979, "An Introduction to Dynamic Meteorology", Academic Press, New York, 319pp.

Longhetto, A., 1980, "Atmospheric Planetary Boundary Layer Physics", Elsevier.

Meaden, G.T., 1981, Whirlwind formation at a sea-breeze front, Weather, 36 : pp. 47-48.

Neumann, J., Mahrer, Y., 1971, A theoretical study of the land and sea-breeze, J. Atm. Sc., 28 : pp. 532-542.

Ozoe, H., Shibata, T., Sayama, H. and Ueda, H., 1983, Characteristics of air pollution in the presence of land and sea-breeze - A numerical simulation, Atm. Environ., 17 : pp. 35-42.

Queney, P., 1974, "Eléments de Météorologie", Masson & Cie, 300pp.

Quinet, A., 1975, Ondes atmosphériques aux latitudes moyennes, Inst. Roy. meteor. de Belgique, Miscellanea Série C - n°11, 31pp.

Richmeyer, R.D., and Morton, K.W., 1967, "Difference Methods for Initial Value Problems", J. Wiley, New York, 405pp.

Sasamori, T., 1968, The radiative cooling calculation for application to general circulation experiments, J. Appl. Met., 7, pp. 721-729.

Seaman, N.L., and Anthès, R.A., 1981, A mesoscale semi-implicit numerical model, Quart. J.R. Met. Soc., 107 : pp. 167-190.

Schayes, G., and Cravatte, M., 1981, Diffusivity profiles deduced from synoptic data, "Air Pollution Modeling and Its Applications", J.C. De Wispelaere, Ed., NATO, Plenum Press.

Triplet, J.P., and Roche, G., 1977, "Météorologie Générale", Ecole Nationale de Météorologie, 317pp.

ANALYSIS OF TETROON FLIGHTS PERFORMED DURING THE PUKK MESO-SCALE EXPERIMENT

Siegfried Vogt, and Peter Thomas

Kernforschungszentrum Karlsruhe GmbH
Hauptabteilung Sicherheit/Programm Klimaforschung
D-7500 Karlsruhe 1

1. THE FRAMEWORK OF THE PUKK EXPERIMENT

PUKK is an acronym for Project zur Untersuchung des Küstenklimas = Project for Investigation of the Coastal Climate. Therefore, PUKK was planned to investigate the different structures of Atmospheric Boundary Layers (ABL) above sea and land surfaces and to study the transition in the ABL behavior perpendicular to the coast /KR82/.

PUKK was a joint effort of 17 institutions located in the Fed. Rep. of Germany and in the Netherlands. During the field phase September 25 until October 9, 1981 140 scientists and coworkers carried out measurements between Cuxhaven and Sprakensehl, see Fig. 1.

Most of the PUKK stations were situated on the meso-scale line. This line extended from the North Sea to southwest and intersected at rightangle the coastline between Bremerhaven and Cuxhaven. Besides these 20 stations a synoptic-scale radiosonde network was available over land and sea. Additionally, a net of 9 high masts supplied wind data in the lowest ABL. The network was supplemented by 4 aircrafts, namely three motor-gliders ASK-16 and one Cessna 310 well equipped with meteorological sensors.

To identify special areas of research and to clearly specify the objectives, several subprograms have been defined.

- To measure the internal boundary layer the very dense network near the coast was set up. The investigation of the wind and temperature fields in cases of coast-parallel

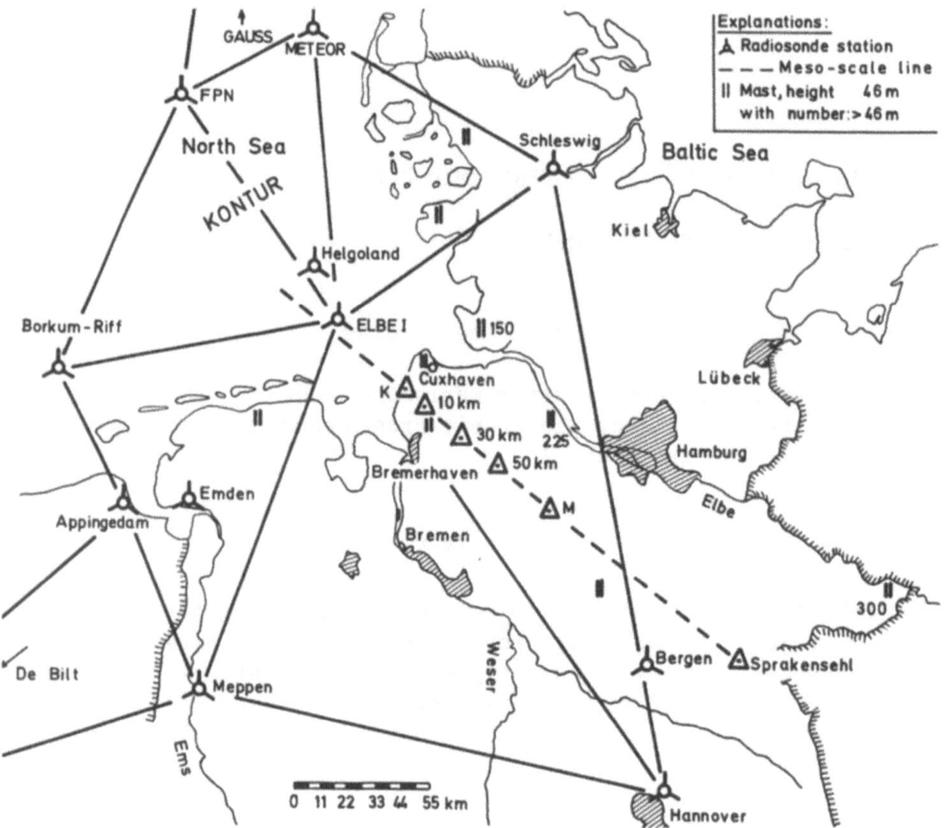

Fig. 1. PUKK stations, mainly those on the meso-scale line (▲) and those of radiosonde (⅄) launching.

flow is of similar importance.

- Also the thermally induced secondary circulation is most effective close to the coast.

- The main tool for studying the structure of organized convection (in its difference over land and sea) were the motor-gliders.

- The radiation fields (again different over land and sea) were investigated in the Cessna flights.

- The remote sounding program was performed mainly by sodar (Doppler and monostatic) and lidar measurements and by radar tracked tetroons.

- The variability of height of turbulent fluxes is studied for one station located well over the land and one over the North Sea. Station M was especially equipped with a triangle (side length 20 km) of tethered balloons.

- The boundary layer jet is investigated in its horizontal distribution. Structure-sondes and mainly the tethered sondes served for this study.

All data of the field campaign are now sorted and stored by the PUKK-Data-Center at Sweetteramt in Hamburg. They are available on request for manifold investigations of the ABL.

2. TETROON FLIGHTS PERFORMED UNDER PUKK

The balloons used in the remote sounding subprogram are tetrahedral in shape and commonly referred to as tetroons. Once a tetroon has been inflated and released, it will fly at an isopycnic surface. The floating level of a tetroon is adjusted by ballast depending on the atmospheric temperature and pressure at ground and at the desired floating level. The inflation procedure and balancing technique are described in /VO83/. To facilitate tracking of the tetroon, it is equipped with a corner reflector made of wooden bars and aluminium coated paper. The radar used during the PUKK experiment and tracking is described in /VO82/.

On six days 29 flights in total were performed. The most important data about the flights have been compiled in Table 1. More than half of all flights could be tracked over 30 km. All tetroons were launched near the coast or about 12 km into the interior. Depending on the wind direction, 14 tetroons flew past the tideland and to the open sea. The projection of all tetroon trajectories into the x, y-plane is shown in Fig. 2. The mean flight levels varied between 120 m and 1250 m although all tetroons had been set to a specified flight level of 400 m to 500 m above ground. The differences between the desired and the actual flight levels were probably due to inadequate ballasting, which had been rather difficult on account of the high wind velocities near ground.

3. EVALUATION OF TETROON FLIGHTS

3.1 LAND-SEA TRANSITION OF TETROONS

Kraus /KR82/ describes in general terms how the inhomogenity of the coast affects the structure of the ABL over the sea and land. Three main influences are to be expected:

- influences of dynamics (roughness change and orographical effects),

Table 1. Tetroon flight information
*Time in GMT

No. of Tetroon	Release Date	Release Time*	Operating base of radar (BR)	Release point	Mean wind direction	Mean transport speed	Mean flight level	Range: Distance tracked	Range: Duration of flight
PK8101	29.09.	11.00	Highway entrance Dorum-Neuenwalde	1500 m west of BR	294°	6.8 m/s	500 m	12 km	25 min
PK8102	29.09.	13.00	" " " "	"	295°	8.8 "	1250 m	41 "	82 "
PK8103	29.09.	16.00	" " " "	"	292°	9.2 "	800 m	33 "	61 "
PK8104	30.09.	8.05	Highway entrance Dorum-Neuenwalde	2000 m south of BR	204°	11.0 m/s	1200 m	47 km	74 min
PK8105	30.09.	9.47	" " " "	"	203°	12.1 "	750 m	29 "	43 "
PK8106	30.09.	11.01	" " " "	"	193°	9.8 "	850 m	45 "	81 "
PK8107	30.09.	15.30	300 m north of swimming pool Dorumer Neufeld	100 m northwest of BR	-	-	-	-	-
PK8108	30.09.	17.05	" " "	"	183°	12.2 "	300 m	12 "	17 "
PK8109	01.10.	8.05	300 m north of swimming pool Dorumer Neufeld	Campingplace Dorumer-Neufeld	164°	9.0 m/s	120 m	11 km	19 min
PK8110	01.10.	9.32	" " "	"	175°	11.8 "	900 m	34 "	49 "
PK8111	01.10.	11.00	" " "	"	166°	9.4 "	700 m	24 "	43 "
PK8112	01.10.	12.33	" " "	"	151°	10.5 "	800 m	29 "	48 "
PK8113	01.10.	14.00	" " "	"	126°	9.1 "	400 m	25 "	46 "
PK8114	01.10.	15.30	" " "	"	128°	11.6 "	400 m	41 "	60 "
PK8115	02.10.	8.00	200 m south of swimming pool Dorumer Neufeld	near Radar	152°	12.4 m/s	850 m	37 km	51 min
PK8116	02.10.	9.30	" " "	"	159°	10.1 "	1250 m	38 "	64 "
PK8117	02.10.	11.00	" " "	"	155°	7.0 "	500 m	26 "	62 "
PK8118	02.10.	12.30	" " "	"	147°	7.4 "	550 m	28 "	64 "
PK8119	02.10.	14.00	" " "	"	138°	7.2 "	400 m	20 "	46 "
PK8120	02.10.	15.30	" " "	"	141°	6.4 "	400 m	16 "	41 "
PK8121	05.10.	8.15	Highway entrance Dorum-Neuenwalde	1500 m west of BR	236°	10.5 m/s	500 m	34 km	54 min
PK8122	05.10.	11.05	" " " "	"	241°	12.1 "	1250 m	51 "	73 "
PK8123	05.10.	12.52	" " " "	"	230°	9.1 "	250 m	32 "	60 "
PK8124	05.10.	14.37	" " " "	"	-	- "	-	-	- "
PK8125	05.10.	15.53	" " " "	"	233°	15.4 "	700 m	35 "	41 "
PK8126	05.10.	17.00	" " " "	"	250°	8.9 "	250 m	16 "	34 "
PK8127	06.10.	8.06	Highway entrance Dorum-Neuenwalde	1500 m west of BR	177°	9.0 m/s	550 m	46 km	88 min
PK8128	06.10.	9.45	" " " "	"	180°	8.2 "	950 m	40 "	82 "
PK8129	06.10.	11.17	" " " "	"	179°	5.6 "	700 m	46 "	140 "

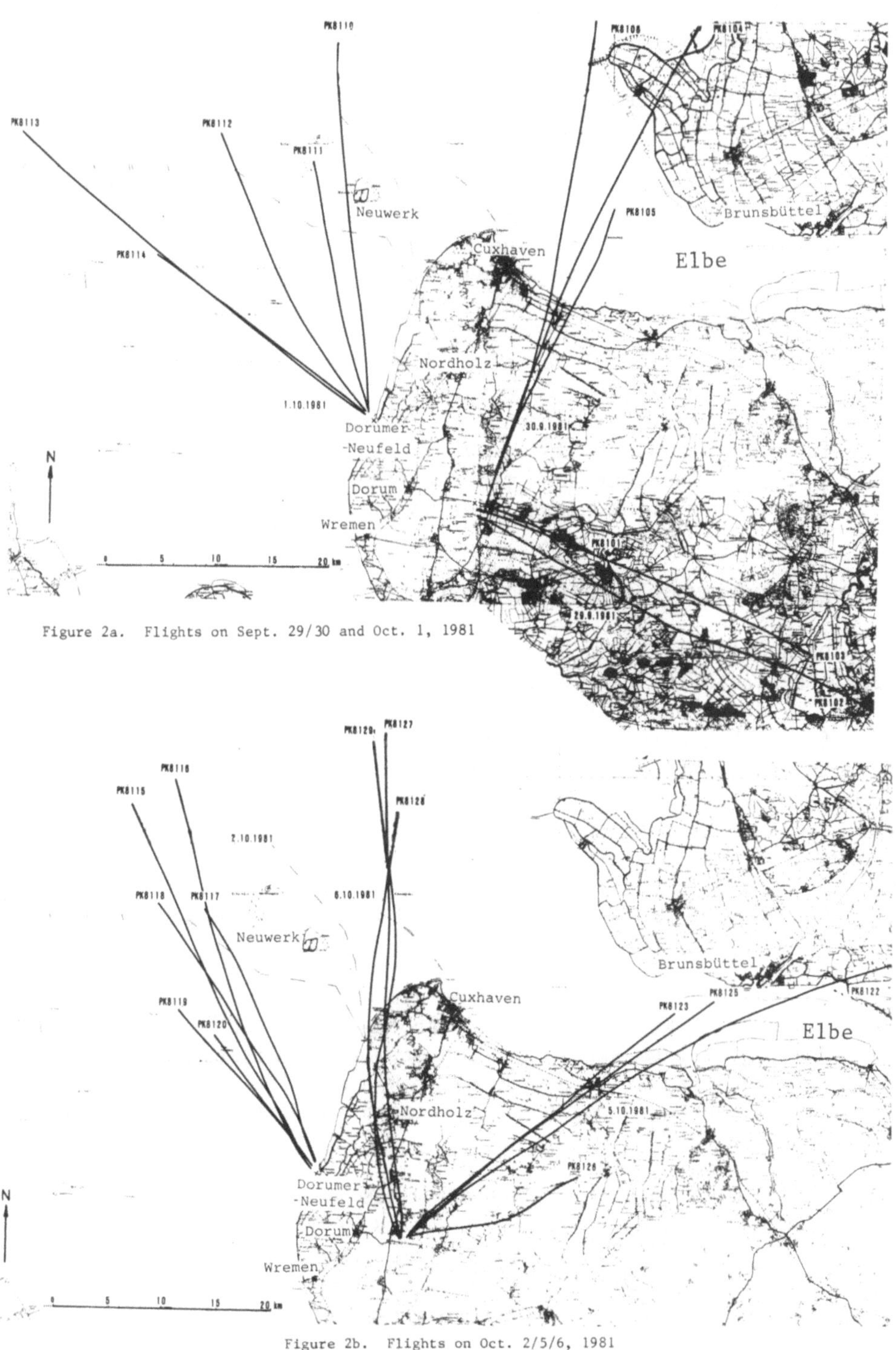

Figure 2a. Flights on Sept. 29/30 and Oct. 1, 1981

Figure 2b. Flights on Oct. 2/5/6, 1981

Fig. 2. Trajektories of tetroon flights in PUKK

- influence of temperature (change of turbulent flux of sensible heat),

- influence of moisture (most effective with very dry land surfaces).

We expected, to be able to quantify some of these influendes by tetroon flights, especially the vertical wind speed and the horizontal wind direction. The flights on Oct. 6, 1981 seem to be well suited. Having traveled about 20 km over land all three tetroons flew to the open sea, (Fig. 2b). A detailed investigation showed different tetroon behaviour over land and sea:
- First : the horizontal velocity of the tetroons perpendicular to the mean flight direction is higher over land
- Second : the vertical velocity of the tetroons is lower over land
- Third : the turbulence intensity in the x and z-directions is lower over land, whereas the turbulence intensity in the y-direction is higher over land.

3.2 VERIFICATION OF CALCULATED TRAJECTORIES

Data with which a comparison of calculated trajectories with 'real' trajectories can be made are not numerous on the meso-scale. As described in Pack /PA78/, tetroon data are a very good tool for such a comparison. For example Druyan /DR68/, using tetroon data, obtained compared trajectories in coastal regimes. They were derived from a 30-60 stations network of surface winds and from a much less dense network of obsrvations by pilot balloon. Druyan found that the use of surface winds is inferior to the use of aloft winds. He also found that adjusting the surface data by means of an Ekman wind profile increased the error in about one third of the comparisons.

To see whether his conclusions are also valid for the Heligoland Bight we compared the tetroon data of Oct. 6, 1981 with data on surface winds. From a network of 21 stations comprising an area of 260 x 230 km² a windfield was calculated. The wind at each grid point is calculated and weighted according to Fig. 4 and /HE80/.

In Figs. 5a and b the observed winds and the wind field interpolated with the tetroon trajectory PK8127, respectively are plotted. The trajectory of PK8127 (launch time 8:06 GMT) differs by about 25° from the surface windfield (time of observation 9:00 GMT). Assuming surface winds the speed of the tetroon is underestimated by a factor of 3.4. In an additional calculation based on tower winds (46 m height) this speed is underestimated by a factor of 2.3, whereas the difference of directions of tetroon and tower windfield changes insignificantly to 23°.

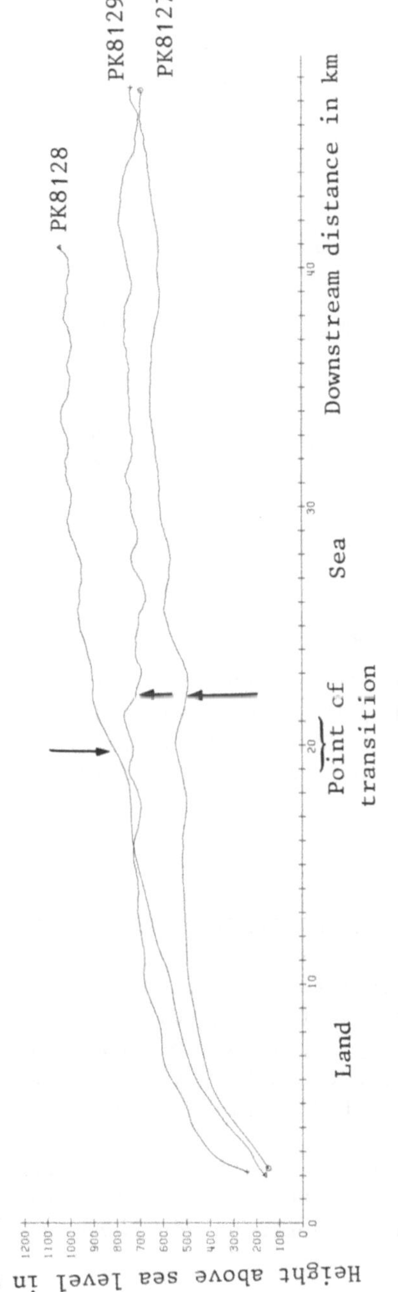

Fig. 3. Profile of tetroon flights on Oct. 6, 1981

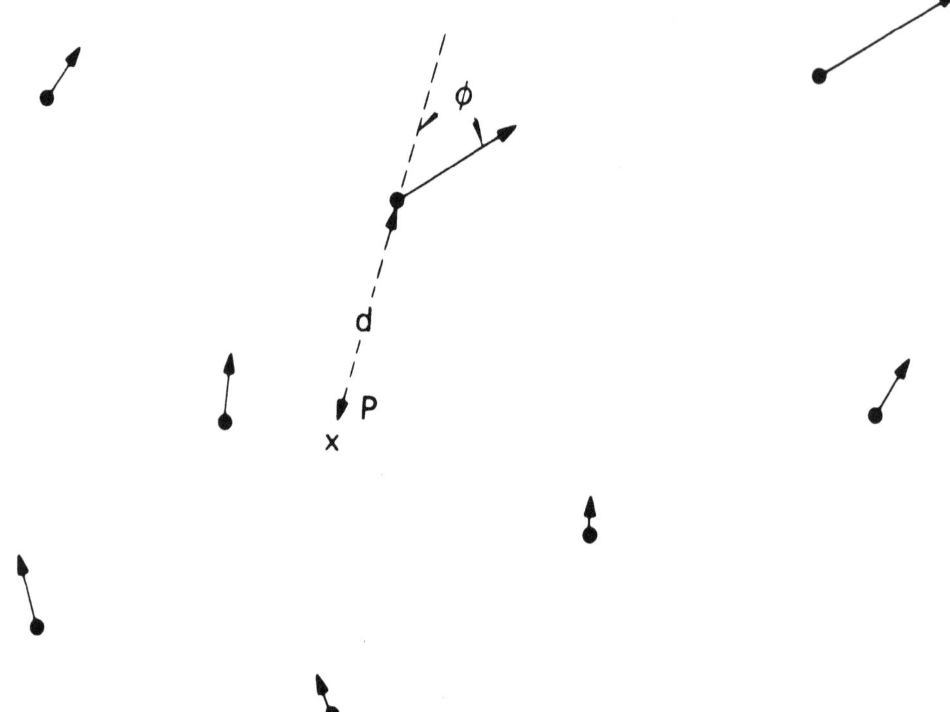

Fig. 4. Example of wind vectors, observed at eight surface stations, which must be used to estimate the wind vector at point P. Each observation is weighted by $1/d^2$ and by $1-0.5/\sin\phi$, /HE80/.

3.3 DETERMINATION OF THE HORIZONTAL DISPERSION PARAMETER

As shown in Slade /SL68/, it is possible to estimate lateral diffusion from successive tetroon releases via the horizontal dispersion parameter σ_y. For this purpose, the root-mean-square value of the spacings between the mean trajectory and individual trajectories are evaluated at given downstream distances.

We adopted this technique for several flight series performed at various sites /HU82/. The evaluations showed that σ_y is neither significantly dependent on the atmospheric stability nor on the sites.

In the present paper σ_y will be evaluated with respect to the time interval between the first and the last tetroon launched in a series. The time interval corresponds to the sampling time τ of airborne material.

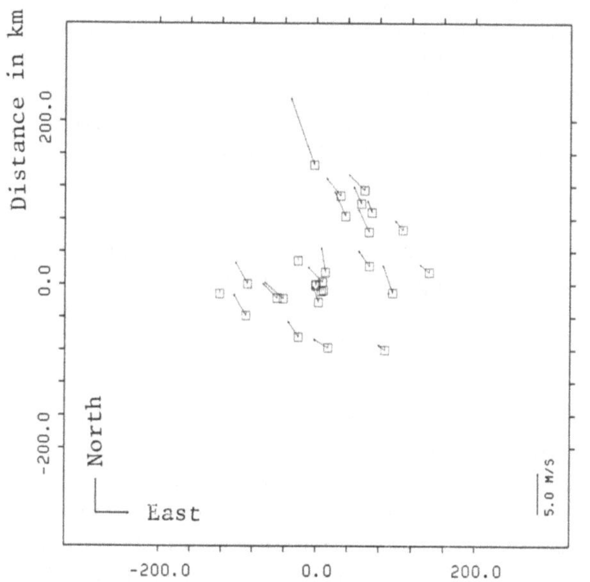

Fig. 5a. Wind data observed on Oct. 6, 1981, 9:00 GMT

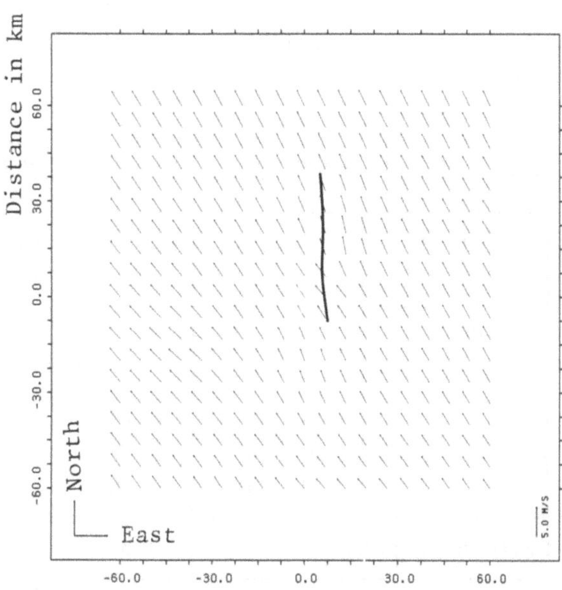

Fig. 5b. Windfield and tetroon trajectory of PK8127 (launch time 8:06 GMT)

With respect to this sampling time τ Table 2 shows the possible series of tetroon flights suitable for evaluating σ_y. In case of τ = 2 h only 2 trajectories are used per flight series, whereas in case of τ = 6 h σ_y can be evaluated from up to 5 successive trajectories within one series. By way of example, the σ_y-curves of τ = 6 h of 4 series are plotted in Fig. 6.

For the values σ_y obtained in this way the dependence on the distance (x-axis) is supposed to take the form

$$\sigma_y = \sigma_o x^p$$

There are two reasons for choosing such a function:
Each σ_y-curve for each of the 30 single series suggests such a function. (Mean correlation coefficient better than 0.97). Moreover, this functional dependence on the downstream distance is frequently used in the evaluation of tracer experiments /TH81/, /VO78/,/MC69/. By means of a least squares fit σ_o and p are evaluated seperately for each sampling time τ. The results are listed in Table 2.

The slopes p of the σ_y curves on a double-logarithmic scale are close to the values found in the literature. Summarizing tetroon flights launched at six sites with $\tau \leq 6$ h, Slade /SL68/ reported a p-value of 0.84. Tracer experiments carried out at the Nuclear Research Centers of Karlsruhe and Jülich with a sampling time of 0.5 h showed a dependence on atmospheric stability with the p-values lying in a range similar to that of our tetroon flights, see Table 2.

Table 2. Series of tetroon trajectories for σ_y-evaluation

Sampling time τ	Number of series	Number of trajectories within one series	σ_o	p	References
2 h	12	2	0.345	0.882	
3 h	8	2÷3	0.407	0.867	
5 h	6	3÷4	0.576	0.868	
6 h	4	4÷5	0.478	0.905	
≤6 h	other tetroon flights			0.84	/SL68/
0.5÷1.0 h	tracer experiments		0.170÷0.671	0.818÷1.296	/TH81/,/VO78/ /GE81/

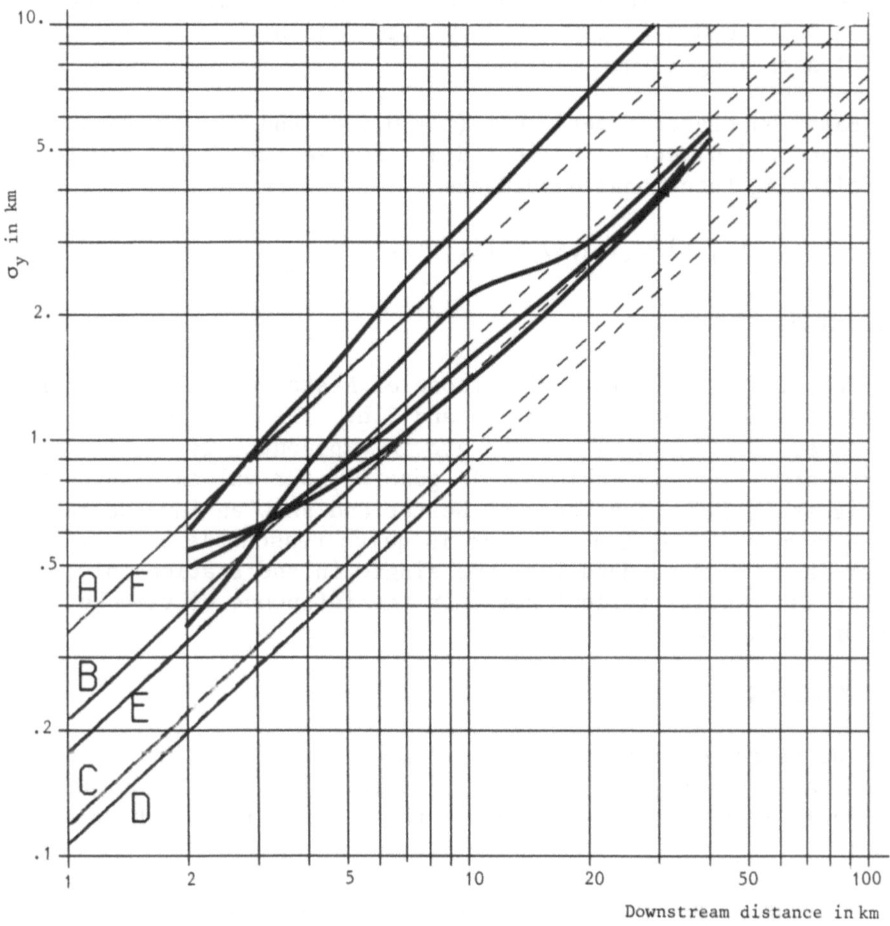

Figure 6. Horizontal dispersion parameter σ_y
▬▬▬ : individual flight series, sampling time $\tau = 6$ h
——— : tracer experiments
-------- : extrapolated from tracer experiments

On the average, our σ_0 increases with the sampling time whereas p is independent of τ. To investigate the dependence on sampling time of the horizontal dispersion parameters σy is calculated using the coefficients σ_0 and p of Table for different downstream distances and for all sampling times seperately. According to Angel /AN62/ a power function is chosen for the dependence of σ_y on sampling time:

$$\sigma_y(\tau)/\sigma_y(2\ h) = (\tau/2\ h)^{\alpha}$$

Table 3 shows a small increase of the exponent α with increasing downwind distance. These values correspond to growth exponents between 0.24 to 0.5 as indicated by Slade and Barry /SL68/,/BA75/ for releases longer than one hour.

Based on tracer experiments (τ = 0.5 h) and on 1975 data of wind speed and direction collected at KfK, Nester /NE79/ developed the following functional dependence

$$\sigma_y(\tau_1, x)/\sigma_y(\tau_2, x) = f(\tau_1, \tau_2, x)$$

up to times of τ = 3 h.

To compare our values with those of Nester we will look for example to the growth between 2 h and 3 h. The function f(2 h, 3 h, x) equals $(3\ h/2\ h)^{\alpha}$ if the growth exponent α is used. Both figures are listed in Table 3. In general, our values of growth are higher (∿ 10 %) than those of Nester and show an inverse ratio to the downstream distance. This is quite reasonable because Nester is using wind data from one single inland station only, whereas our results reflect not only the influence of time but also of surface area.

4. CONCLUSION

The tetroon flight series performed during the PUKK experiment demonstrated clearly the possible advantages of this technique for meso-scale modeling. In this paper three possibilities have been shown:

- Study of a single tetroon trajectory with a view to the behavior of the ABL at the point of transition from land to sea.

Table 3. Growth exponent α as a function of downstream distance and sampling time τ

$\sigma_{y1}/\sigma_{y2} = (\tau_1/\tau_2)^{\alpha}$ compared to

$\sigma_{y1}/\sigma_{y2} = f(\tau_1, \tau_2, x)$ /NE79/

Downstream distance	Growth exponent α	$(\frac{3\ h}{2\ h})^{\alpha}$	$\frac{f(3\ h,x)}{f(2\ h,x)}$ /NE79/
10 km	0.535	1.242	1.14
20 km	0.544	1.246	1.13
30 km	0.550	1.249	1.12
50 km	0.557	1.253	1.11

- Verification of calculated trajectories. The comparison of observed and constructed trajectories should not only be based on wind measurements as done here, but also on sophisticated meso-scale ABL models which include calculations of stream lines. As the data base of all PUKK measurements is nearly completed, we hope that such models can be applied and checked by our tetroon flights.

- Estimation of lateral diffusion from successive tetroon releases. Our evaluations yielded σ_y-curves as a function of the downstream distance and sampling time. The evaluations will be continued with respect to a dependence on travel time rather than on downstream distance.

REFERENCES

/AN62/ Angell, J. K.; On the Use of Tetroons for Estimating of Atmospheric Dispersion on the Mesoscale. Monthly Weather Rev. 90 (7) pp 263-270 (1962).

/BA75/ Barry, P. J.; Stochastic Properties of Atmospheric Diffusivity, Atomic Energy of Canada Limited, AECL5012 (1975).

/DR68/ Druyan, L. M.; A Comparison of Low-level Trajectories in an Urban Atmosphere. J. appl. Met. 7, pp 583-590 (1968).

/GE81/ Geiß, H., Nester, K., Thomas, P., Vogt, K. J.; In der Bundesrepublik Deutschland experimentell ermittelte Ausbreitungsparameter für 100 m Emissionshöhe, JÜL 1707, KfK 3095 (1981).

/HA67/ Hass, W. A., Hoecker, W. H., Pack, D. H. and Angell, J. K.; Analysis of Low-level Constant Volume Balloon (Tetroon) Flights Over New York City, Q. J. R. Meteor. Soc. 93, 483-493 (1967).

/HE80/ Heftter, J. L.; Air Resources Laboratories Atmospheric Transport and Dispersion Model (ARL-ATAD), Report ERLTM-ARL-81. NOAA, Air Resources Lab. (1980).

/HU82/ Hübschmann, W. G., Thomas, P., Vogt, S.; Tetroon Flights as a Tool in Atmospheric Meso-scale Transport Investigations. 13th Int. Techn. Meeting on Air Pollution Modelling and its Application. Ile des Embiez 14-17 Sept. (1982).

/KR82/ Kraus, H.; PUKK: A Meso-scale Experiment at the German North Sea Coast, Contr. to Atmosph. Physics Vol. 55 No. 4, pp 370-382 (1982).

/MC69/ Mc Elroy, J.; A Comparative Study of Urban and Rural Dispersion, J. of Appl. Meteorol. 8, No. 1 pp 19-31 (1969).

/NE79/ Nester, K.; Einfluß der Mittelungszeit auf σ_y, in Annual Report 1978 of the Safety Department. pp 195-196, KfK 2775 (1977).

/PA78/ Pack, D. H., et al.; Meteorology of Long-range Transport. Atm. Environm. Vol. 12 pp 425-455 (1978).

/SL68/ Slade, D. H.; (Editor); Meteorology and Atomic Energy, TID-24190, pp 180 (1968).

/TH81/ Thomas, P., Nester, K.; Experimental Determination of the Atmospheric Dispersion Parameters at the Karlsruhe Nuclear Research Center for 60 m and 100 m Emission Heights, KfK 3091 (1981)

/VO78/ Vogt, K. J., Geiß, H., Polster, G.; New Sets of Diffusion Parameters Resulting from Tracer Experiments in 50 m and 100 m Source Height, 9th Int. Techn. Meeting on Air Pollution Modelling and its Application, Toronto (1978).

/VO82/ Vogt, S., Thomas, P.; Investigation of Meso-scale Atmospheric Transport by Means of Radar Tracked Tetroons During PUKK. Contr. to Atmosph. Physics Vol. 55 No. 4 pp. 409-416 (1982).

/VO83/ Vogt, S., Thomas, P.; Untersuchung mesoskaliger Luftströmungen in der Umgebung des Kernforschungszentrums Karlsruhe KfK 3565 (1983).

DISCUSSION

S.E. GRYNING

When σ_y is characteristic for a 2 hour sampling period, the estimates of σ_y are based on typically the track from 3 tetroons flights. What is the uncertainty on the estimates of σ_y, when they are based on three tetroon flights only.

S. VOGT

The formula used in my paper for estimating σ_y is based on the assumption that the trajectories are randomly based around the mean value. The distribution is a Gaussian one. So one can calculate the uncertainty or the error with error formula (N = 3). Of course if N(= number) of trajectories) is bigger, the uncertainty is decreasing. Comparing the uncertainties for σ_y-calculations associated with tetroon trajectories with the uncertainties involved with tracer experiments the first uncertainties are much less.

3: LAGRANGIAN MODELING FOR SYNOPTIC RANGE TRANSPORT

Chairman: A. Venkatram
Rapporteur: R. Stern

LONG RANGE TRANSPORT OF AIR POLLUTANTS ON THE SYNOPTIC SCALE

Joseph B. Knox

Lawrence Livermore National Laboratory
Atmospheric and Geophysical Sciences Division
Livermore, CA 94550

ABSTRACT

Long range transport (LRT) of air pollutants has in recent years become increasingly more important, in part, because of the technical challenge of the current applications (e.g. acidic deposition, visibility impairment, and source attribution issues), and in part, because of the growing awareness of the technical difficulties of providing definitive models. This overview paper reviews the evidence for an accelerated-diffusion regime for puffs and plumes; some of the current LRT diffusion formulations in view of the summary information; the sensitivity of LRT diffusion solutions to the potential range of conditions; the potential for inherent errors in current simplified, but commonly used, trajectory estimates; some avenues emerging from research for avoiding these accumulating errors in trajectory estimates. In addition, the most recently developed techniques from applied mathematics hold promising potential for use in global pollutant budgets. Some of the salient dilemmas of the long range transport problem are highlighted.

INTRODUCTION

For more than one decade, serious scientific attention has been given to the subject of transport of primary and secondary

---

This work was performed under the auspices of the U. S. Department of Energy by the Lawrence Livermore National Laboratory under contract No. W-7405-Eng-48.

air pollutants on spatial scales of one thousand to perhaps several thousand kilometers. As is well known, the motivation for subject-studies lies in the acidic deposition, visibility impairment and source attribution issues. The present paper is restricted to considerations of synoptic-scale transport, whether by (a) the hydrodynamic field of motion or by (b) processes of diffusion in the boundary layer or in the troposphere. In other words, we do not consider the additional processes of dry or wet deposition on the underlying surfaces, the treatment of new sources enroute, or chemical transformation. In a sense, there is sufficient technical challenge in discussing the subject of transport and diffusion of identifiable, conservative substances over extensive spatial paths.

It is precisely this subject, that has received considerable attention in review articles during the past year, including a review of accelerated diffusion by Dr. Gifford in Boston in 1983, the appearance of an articulate Chapter in Engineering Meteorology (1982) by Steve Hanna, and the recent comprehensive catalogue of most of the pertinent models of LRT by Dr. Warren Johnson in June (1983). The recent appearance of these contributions makes the current overview either easy or difficult, depending on one's point of view.

As Dr. Gifford stated in Boston, the meteorologist when confronted with long range transport is being challenged to extrapolate what is known about these processes on very much shorter ranges to ranges very much longer; this is in fact a very difficult thing to do in an intellectually satisfying way. The reasons for this will become apparent.

The problem of treating combined transport and diffusion of puffs and plumes over long distances or large travel times can be stated reasonably tersely as a multi-particle problem. Imagine a puff composed of a large number of particles placed into a turbulent daytime boundary layer; the mass centroid of the particles is transported down-wind while the remainder move hydrodynamically in response to time averaged winds that are locally different due to shear within an evolving boundary layer. Shear alone increases the surface within which the particles reside, and the wind shear modifies the particle gradients applicable to diffusive processes that are determined by every increasing scales of motion as the puff expands. Depending on the initial size of the puff, the boundary layer can gradually become filled with particles through its full depth. The puff may then spatially fractionate with the diurnal evolution of the boundary layer, such that during the night the upper portion of the now modified puff moves with the winds above the new nocturnal boundary layer, and the lower decoupled puff fragment moves in some other manner with the winds in the surface layer. By the end of the first night, these puff fragments can be widely separated, as the next cycle of boundary layer evolution begins. In addition, the exact sequence in which all of this occurs depends on the time of creation of our imagined puff in the boundary layer. One additional complication should be mentioned, that is several diurnal cycles might well occur between a given source and a proposed receptor. In the real world, however, a few other complications arise; the nature of the boundary layer evolution depends on geography, season, and interactions with the synoptic-scale disturbances.

The most obvious reaction to this statement of the problem is to suggest that we examine the data and ascertain if there are any suggestions from nature as to how to proceed. This would, of course, require that suitable experiments to have been performed with puff or plume tracking for up to a few days travel time for appropriate conservative pollutants or tracers. Such data are meager and come from those who study the evolution of volcanic clouds, from those who observe large emissions from stacks in isolated regions (e.g. Mt. Isa), or from those who comb the archives regarding the fate of nuclear debris clouds in the lower atmosphere. Volcanic clouds, as usually referred to, will not be applicable to our posed problem for the boundary layer; but historically, the volcanic clouds gave early evidence of accelerated diffusion in the atmosphere. So we are left with isolated sources like Mt. Isa and some archival data from nuclear debris clouds. The really well designed tracer experiments involving multiple arc sampling data and dense meteorological data including upper air observations at frequent intervals probably are still in the future for potential execution. There has been a recent and renewed interest in the use of pollen to verify calculated trajectories (Raynor et al., 1983), but the area sources of pollen appear, at least to this author, as sufficiently imprecise in regard to source location so as to be relatively uninteresting. In other words, such data is probably unsuitable for discriminating between proposed model improvements, for example, in trajectories originating from two different methods of calculation.

The cross wind puff or plume widths (or standard deviation, σ_y) has been studied for puffs by Crawford (1966), and for plumes by Gifford (1983). Before reviewing the data or empirically based models from these investigators, we need to recall the relevant diffusion regimes that arise as a function of time t. These regimes are essentially those discussed by Batchelor (1950) for short-ranges, and extended to much larger ranges by Crawford (1966).

1. Initial dilution. Regime 1: Dominated by the inertial sub range (puffs), or dominated by self-induced mixing (plumes): $\sigma_y \propto t$.

2. Accelerated Diffusion. Regime 2: $\sigma_y \propto t^{3/2}$.

3. Late time diffusion, Regime 3 Parabolic dependency, $t^{1/2}$.

Gifford has analyzed the data regarding $\sigma_y(t)$ for numerous observations of the Mt. Isa plume; these data are summarized in Figure 1, wherein the exponent p, in $\sigma_y(t) = \alpha t^p$, is plotted as a function of time. These data suggest three regimes: (a) the time period less than 2 hours wherein the plume width is proportional to t, (b) the period of 2 to about 20 hours in which σ_y is dependent on t^p, where $p > 1$, and (c) a final regime not well established, where σ_y is proportional to $t^{1/2}$. Crawford's (1966) large-cloud Lagrangian diffusion model, reported by Knox (1974), has been verified against a reasonable number of nuclear debris clouds in the troposphere and the boundary layer. This latter empirically based model predicts $\sigma_y(t)$ through the three above regimes using a scale dependent diffusion coefficient. By selecting the parameters of Crawford's 2BPUFF model to represent the subtropical boundary layer over a heated continent, the resulting behavior of p as a function of time is shown in Figure 2a and 2b, superimposed on Gifford's results. The 2BPUFF predictions for puff-diffusion in the Mt. Isa diffusion-environment are

Regime 1	$p = 1$	$t < 1$ hour,
Regime 2	$p = 3/2$	1 to 24 hours,
Regime 3	$p = 1/2$	$t > 24$ hours,

It would appear that there is a set of investigators, including Richardson in 1926, Crawford in the 1960's, and Gifford (1983) whose studies support a regime of accelerated diffusion in the atmosphere. But noteably to find it, they analyzed clouds, puffs, or plumes on an individual basis, looking for the time-dependency of σ_y carefully, on a case-by-case basis. If the data is composited into a common pool prior to analysis for time dependency, then the evidence for accelerated diffusion and the three regimes is easily lost. Under these latter conditions other conclusions can be readily drawn from

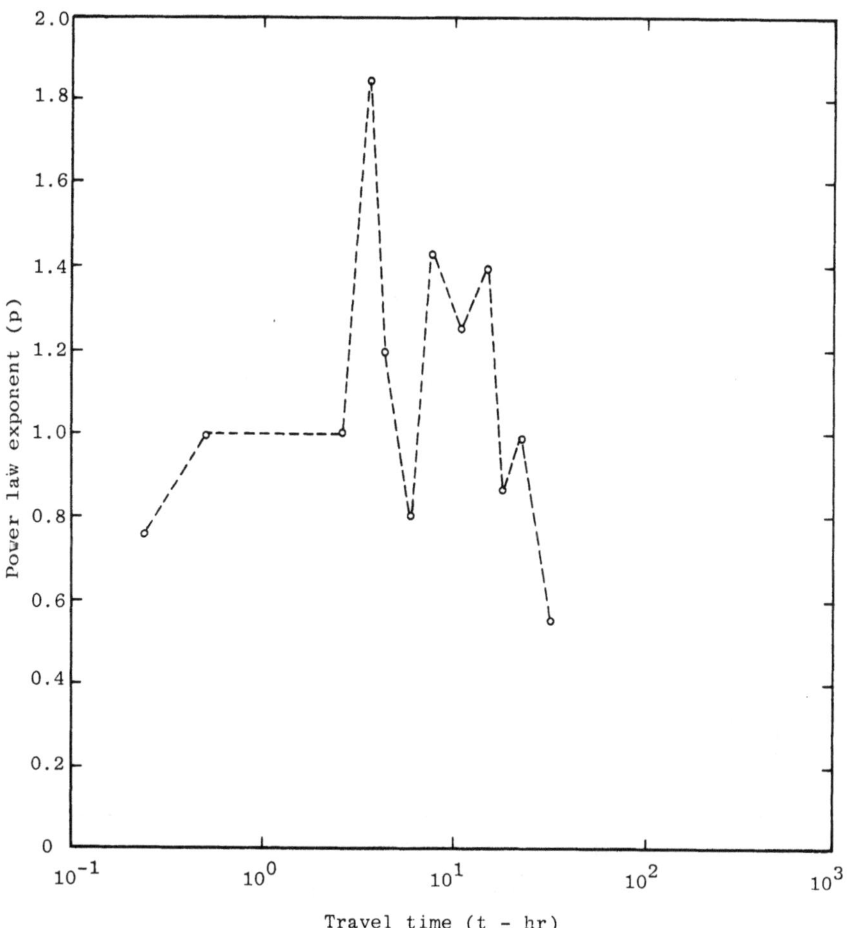

Fig. 1. Plot of the power law exponent, p, vs. travel time, t, for Gifford's Mt. Isa data (Gifford, 1983).

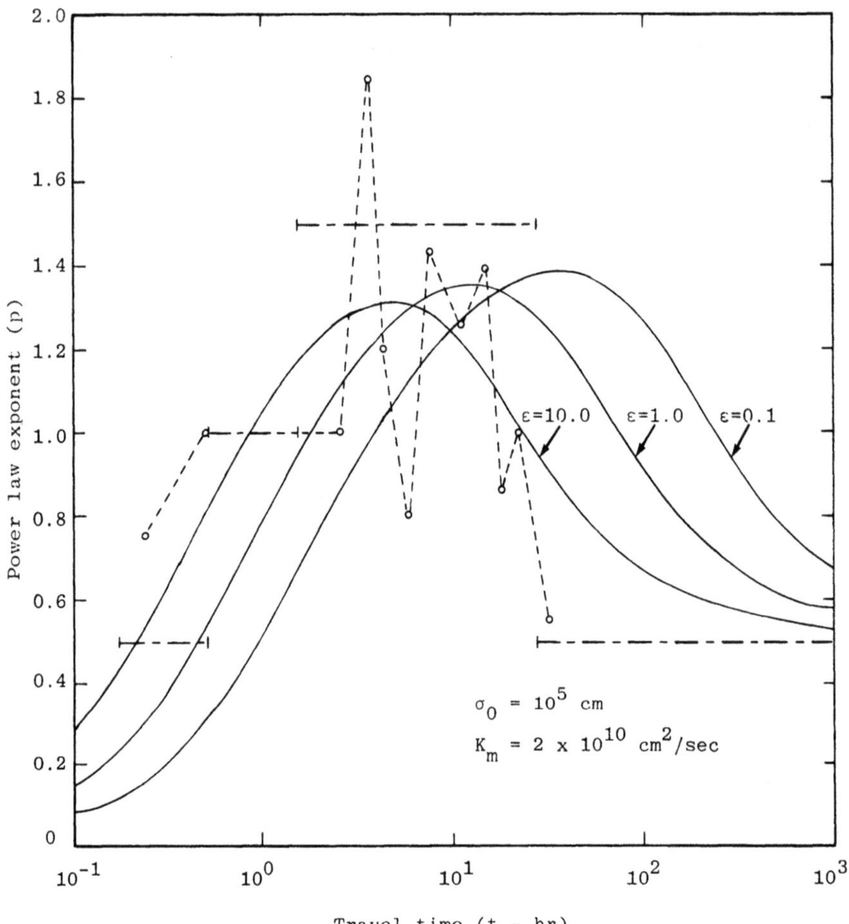

Fig. 2a. Plots of the power law exponent, p, vs. travel time, t, for Gifford (1983); Crawford 2BPUFF (1966) and Walton (1972) with Km = 2 x 10^{10} x cm^2/sec.

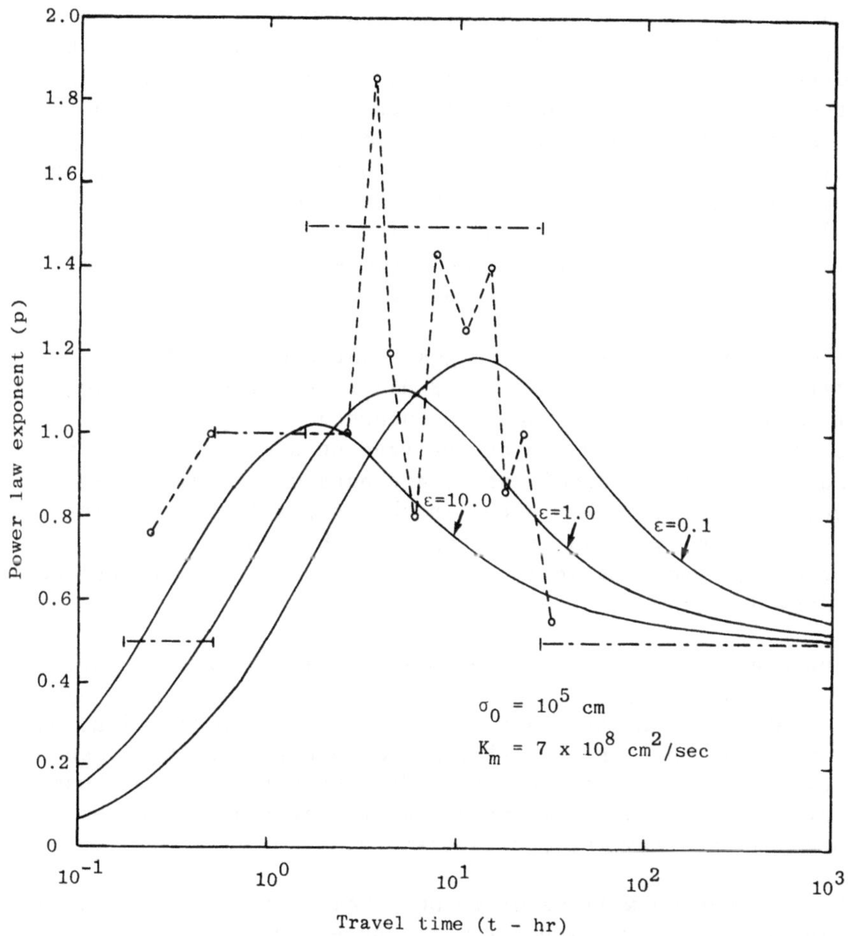

Fig. 2b. Plots of the power law exponent, p, vs. travel time, t, from Gifford (1983); Crawford 2BPUFF (1966) and Walton (1972) with Km = 7 x $10^8 cm^2/sec$.

the same total data set, for instance $\sigma_y = .5\ t$ (where t is the time in seconds) is widely used in the US perhaps for this reason.

The diffusion modelers who have observed this evolution of interpretation of nature remark that the $\sigma_y(t)$ data is influenced in the real world by wind shear. Further, the analysis of the data has not separated the growth in width due to wind shear from the growth due to random motions in the atmosphere. To compound the problem, if the modeler calculates the effect of wind shear on the cloud (or puff) very carefully and then uses the diffusion coefficients construed from the accelerated diffusion regime, he may very well get the wrong answer with the estimated growth in width being much too fast. The experience of these modelers (Luther and Lange, 1983) is signaling us that a significant portion of the observed accelerated diffusion may well be associated with wind shear from the mesoscale and synoptic scales of motion.

The author would like to suggest that there are two important themes contained in this historical overview. The first theme is that if we want to understand long range transport data, it needs to be analyzed on a case-by-case basis so that the initial, accelerated, and parabolic regimes can emerge. Secondly, if we invest in the computational physics necessary to do hydrodynamical transport by mesoscale or synoptic scale motions well, then we must select the turbulence submodels or parameterizations with care.

The organizers of the conference when inviting this paper requested that the author not repeat the material covered by Prof. Arnt Eliassen last year, nor, perhaps by implication, on that which has been prolifically produced this year by those already cited. This can be easily done by expanding on these last two points within the framework of long range transport research old and new, at our Laboratory.

Our discussion will be drawn on three different portions of experience; the 1960's when we were concerned with transport-diffusion of large clouds to long distances and used the empirically based, scale-dependent diffusion (Crawford, 1966 and Walton, 1973); the 1970's when we were involved in particle-in-cell model developments on the regional scale (Lange, 1978) and on the global scale in 1981 by Lange and Walton (reported by Knox, 1982) ; and the latest developments of computational techniques for equilibrium global budgets performed-in-simulation by a new kind of particle, a Lagrangian sampling parcel (Walton et al., 1983). Our task here is not to review the foundation of all these models, but through the window of our research and experience recount some of the most important features or themes. The author offers the following as some of the salient features/themes of this experience.

1. The analyses of several large, uniquely-tagged clouds of the 1960's strongly suggests the existence of a regime of accelerated diffusion.

2. An ad hoc method of extending this knowledge of accelerated diffusion to different diffusion scenarios and portions of the lower atmosphere, namely scale-dependent diffusion, is a useful concept.

3. The analysis of long range transport data must be done by a case-by-case basis that takes into account differing hydrodynamic or synoptic scale fields of motion and various possible states of the PBL.

4. The application of particle-in-cell techniques to tens of thousands of kilometers has appeared to be intractable until recently. Cloud evolutions over such distances were numerically simulated with two innovations. The first involved applying the constraint of mass consistency to the time sequence of analyzed fields of motion acquired from the AFGWC, so that in each Eulerian cell on the globe the accepted flow field was 3D non-divergent. The other improvement was that for such particle-in-cell methods to be successful, one needed a very precise method of calculating each particle trajectory through a sequence of prescribed wind fields. Precisions surpassing those associated with the common NOAA/ARL methods of using the first term in a Taylor expansion ($S = v\ \Delta t$) were clearly needed. Recent research by Warner, et al. (1983) substantiates this thrust. Such techniques may now be applied to individual plumes or puffs.

5. Advances in computational physics have occurred (Walton, 1983) so that new particle-in-cell techniques can be applied on the scale of many thousands of kilometers for multiple area sources (or sinks). The assumption of mass consistency within each zone of a global flow field permits the analytic (rather than series expansion) calculation of the particle trajectory within or through an Eulerian cell with rigorous accuracy. The continuous piecing together of rigorously accurate trajectories permits very extended integrations. Accleration or deceleration along the trajectory of the "particle" is included in the solution of future "particle" positions. The trajectory approach in TRANSZAM (Walton et al., 1983) relies upon a unique characteristic of

grided wind fields, i.e., the meridional velocities in a zone do not vary between the zones pressure boundaries and the vertical velocities do not vary between the latitude bounds. So this approach would not be generally applicable. From this work, it would appear that applied mathematics is emerging with alternative methods to simple trajectory models, based on $S = v \Delta t$.

6. For a few years we have known that normal particle-in-cell techniques can not be applied to global equilibrium budgets because of the requirement for too many particles (many tens of thousands) or too many Eulerian zones. A new type of particle, a Lagrangian sampling parcel (LSP), was invented, first by the Soviet, Seidov (1976) and independently by Walton et al. (1983), whose preliminary use indicates that we are on the verge of being able to calculate global equilibrium pollutant budgets, for instance time-dependent CO_2 concentrations, with a detail not achieved before. The key to this development lies in improved applied mathematical techniques; in summary, the Lagrangian sampling parcels are moved through Eulerian cells within which the flow is mass-consistent, and rigorously accurate trajectories are calculated in analytic form for each marching step in time. Supplemental processes of diffusion and chemical transformation can be included in the theory readily, if prescribed and known.

Illustrations of these salient characteristics and features of our experience are now given.

Illustration 1: Intercomparison of Diffusion Formulations.

It is pertinent to very briefly bring together and intercompare the various formulations of accelerated diffusion that have emerged over the past few decades. Figure 2a shows (a) the experience of Gifford in analyzing the plume width data from Mt. Isa overlain by (b) the predictions of Batchelor (1950) and those inherent in the formulation of Crawford's Model (1966), and (c) the more modern predictions from scale dependent diffusion (Walton, 1973) for three different values of ε, namely $\varepsilon = 10$, 1, and 0.1 cm^2/sec^3. The predictions of Crawford have used the value of $\varepsilon = 10$ which we believe to be consistent with subtropical continental PBL properties. The predictions of Crawford (1966) and Walton (1973) using $\varepsilon = 10$ are really in quite good agreement with Gifford's findings. The projection of $\sigma_y(t)$ with $\varepsilon = 10$ does not achieve the value of 3/2 during the peak of the accelerated regime, but merely ~ 1.3. As epsilon is decreased the time period of the accelerated regime increases and the power approaches 3/2 a bit better. The intercomparison, however, suggests some agreement and some important differences. The differences in regard to duration of regime 2, the exponent p as a function of time, and the decay into the parabolic regime all suggest that plume width behavior is a scenario dependent matter. But even the predictions for the plume growth in scenario dependent manner need to be considered with some humility in that the prediction probably apply to an average plume with the scenario specified. Since the value of epsilon can vary by five or more orders of magnitude for different parts of the atmosphere and/or diffusion scenario the predictions of plume width and corresponding surface air concentration vary widely also in response to this parameter selection.

Historically, the limiting value of K in Walton's treating of scaled-dependent diffusion has been selected as 2×10^{10} cm^2/sec or 7×10^8 cm^2/sec, Walton (1973). Figure 2b corresponds to 2a, with the exception of the selection of the limit of $7 \times 10^8 cm^2/sec$. The solutions for $\sigma_y(t)$ are reasonably sensitive to this bounding K-value in that the value of p in the accelerated regime is suppressed, the transition to regime 3 occurs earlier at the K-limit as decreased. Several extended range experiments would be required in boundary layers characterized by various values of ε to directly determine the K-limit.

Illustration 2: Scenario Dependent LRT solutions.

The air pollution community is quite accustomed to Pasquill-Gifford categories and the corresponding data regarding σ_y and σ_z for ranges of 1 to 10 or 20 km; some references have extended these widely used curves to ranges as large as 100 km. However, analogous classification schemes or categorizations of long range transport experience is generally lacking. A sensitivity study by Knox (1981) illustrates the wide variety of anticipated LRT experience when the parameters in Crawford's Model with scale-dependent-diffusion are diffusion-scenario dependent. The scenarios represent LRT conditions from subtropical environs to sub-arctic. Table 1 shows the parameters selected for each of the scenarios. Table 2 compares the predicted surface air concentration for a unit (1 CI) release of noble gas as calculated by the ARL model (which has p = 1 for all times and geograhic locations), and as calculated by Crawford's large cloud diffusion code (2BPUFF) using diffusion-scenario dependent inputs. Figure 3a shows the predicted values of $\sigma_y(t)$ for the scenarios as prescribed in the 2BPUFF code by scale dependent diffusion (Walton, 1973).

Table 1. Input parameters used in the 2BPUFF sensitivity study

Parameter	Subtropics	U.S.A. Winter	U.S.A. Summer	Mid-Latitude Ocean Winter	Mid-Latitude Ocean Summer	Subarctic Snow	Cabriolet	Buggy	Schooner
1. Sfc boundary layer hgt (m)	200	150-300	200-400	200	200	30-100	100-300	100-400	100-300
2. Top of mixed layer (m)	1500	500-1000	800-1800	1000	2000	100-500	1400-4000	1000 4400	1600-3800
3. K_z at 1 metre (m^2/s)	0.2	0.04-0.15	0.1-0.2	0.4	0.3	0.02-0.03	0.04-0.7	0.02-0.7	0.004-0.3
4. K_z in mixed layer (m^2/s)	10	0.5-2	2-5	1	3	0.05	0.1-20	0.5-10	0.5-5.0
5. ε at puff center (joule/kg s)	$5x10^{-4}$	$4x10^{-6}$-$1x10^{-4}$	$6x10^{-5}$-$3x10^{-4}$	$5x10^{-5}$	$2x10^{-4}$	$6x10^{-8}$-$1x10^{-6}$	$3x10^{-4}$-$1x10^{-2}$	$5x10^{-5}$-$2x10^{-3}$	$4x10^{-6}$ $2x10^{-3}$
6. Puff center hgt AGL (m)	100	100	100	100	100	30	210	350	400
7. Ave wind along traj (m/s)	6.5	6	5	11	7	0.5	14	14	12
8. Tropopause hgt (km)	15.0	11.0	14.0	11.0	13.0	9.0	11.5	9.0	10.0

Note: 1 joule/kg-s = 10^4 erg/g-s
$1\ m^2/s^3 = 10^4\ cm^2/s^3$

Table 2. Surface air concentration estimates for a 1 Ci release at three times downwind for all scenarios. Calculations are from the ARL and 2BPUFF models; the ratio is for the ARL/2BPUFF estimates

	DOWNWIND TRAVEL TIMES (HOURS)								
	10			40			100		
	ARL	2BPUFF	RATIO	ARL	2BPUFF	RATIO	ARL	2BPUFF	RATIO
Subartic snow cover	$6.5x10^{-13}$	$9.5x10^{-11}$	0.0068	$2.1x10^{-14}$	$7.0x10^{-13}$	0.030	$3.3x10^{-15}$	$3.1x10^{-14}$	0.11
USA-summer	"	$5.8x10^{-14}$	11.	"	$5.1x10^{-16}$	41.	"	$7.6x10^{-17}$	43.
-winter	"	$3.0x10^{-13}$	2.2	"	$2.3x10^{-15}$	9.1	"	$1.6x10^{-16}$	21.
Mid-Latitude Ocean-summer	"	$1.1x10^{-13}$	5.9	"	$7.8x10^{-16}$	27.	"	$8.2x10^{-17}$	40.
-winter	"	$3.5x10^{-13}$	1.9	"	$2.1x10^{-15}$	10.	"	$1.6x10^{-16}$	21.
Subtropics-summer	"	$1.8x10^{-14}$	36.	"	$2.0x10^{-16}$	105.	"	$5.6x10^{-17}$	59.
Cabriolet (V_d=0)	"	$8.0x10^{-14}$	8.1	"	$5.3x10^{-16}$	40.	"	$8.8x10^{-17}$	38.
Buggy (V_d=0)	"	$1.1x10^{-13}$	5.9	"	$9.0x10^{-16}$	23.	"	$1.0x10^{-16}$	33.
Schooner (V_d=0)	"	$1.0x10^{-12}$	0.65	"	$1.4x10^{-14}$	1.5	"	$1.7x10^{-16}$	19.

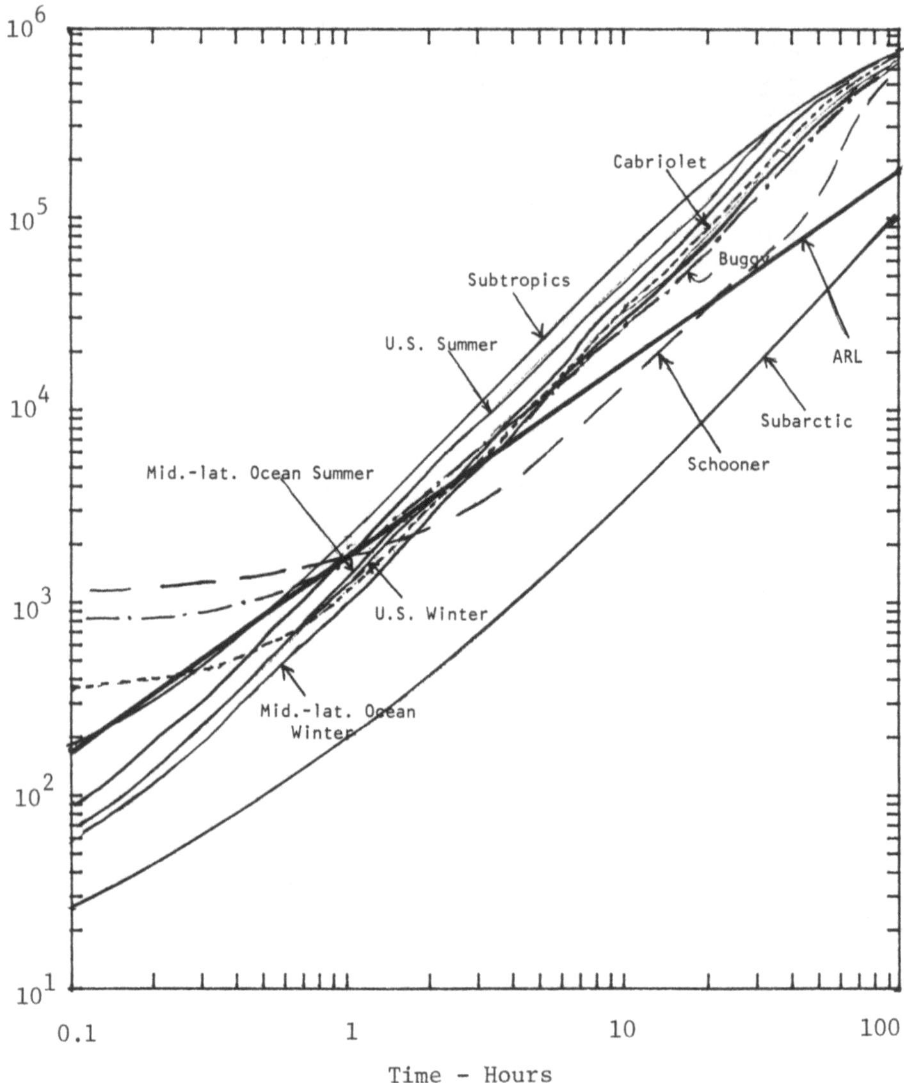

Fig. 3a. Horizontal standard deviation vs. time for ARL and 2BPUFF models. The heavy solid line is for the ARL model; the other curves are for the nine scenarios as calculated by 2BPUFF (K-limit = 2×10^{10} cm^2/sec).

Illustration 3: Long Range Transport of the Schooner, NTS, Debris Cloud.

These first two illustrations contain a number of potentially useful predictions, however, the aggregated LRT experimental data base is still too meager, as previously mentioned, to validate the 2BPUFF model and its techniques of input parameter selection. This model is extensively reported by Reiter (1978) and by Knox (1974). The latter reference, however, does not contain the current inclusion of scale dependent diffusion in the model formulation wherein,

$$\sigma^2(t) = (\sigma^{2/3}(0) + 2/3\ \varepsilon^{1/3} t)^3$$

is used until $K = 2 \times 10^{10}$ cm^2/second, the limiting value of the horizontal eddy diffusion coefficient for the pollutant puff. This latter restriction forces Eq 3-1 to approach $t^{0.5}$ at late times as previously discussed. The predicted isotopic depositions and those observed along the path of the Schooner cloud are shown in Figure 3b. This comparison, not untypical, suggests that the depositions estimated by the model verify within a factor of 2 or 3 for a period of calculation that is in the range of 40 to 50 hours. This comparison has been made without particular reference as to whether the actual cloud center was in fact coincident in space with that projected. Rather without regard to trajectory errors, the depositions under the moving cloud center appear reasonable when compared to depositions under actual cloud center. It is pertinent now to consider the question of hydrodynamic transport and some of the current trajectory practices and attendent problems.

Illustration #4a: The rotating-particle problem.

Consider a 2D steady-state flow field of unbound-solid-rotation about point 0 (see Figure 4). The velocity vector at four grid points of an Eulerian cell are depicted; velocity any point is $\vec{V} = \vec{\Omega} \times \vec{r}$. (It is easily shown that (a) on cell face $\underline{i+1}$ that $v = \Omega \frac{\Delta x}{2}$, (b) on cell face j+1, $u = -\Omega \frac{\Delta y}{2}$, (c) on cell face i, $v = -\Omega \frac{\Delta x}{2}$, and (d) on cell face j, $u = \Omega \frac{\Delta y}{2}$. The particle (x) has the velocity $v = .8\ \Omega\ \Delta x/2$ initial). In this field, the particle (x) has the velocity $v = \Omega \times R_j$, the centripetal acceleration $\frac{|V|^2}{R}$ inwards, so that x

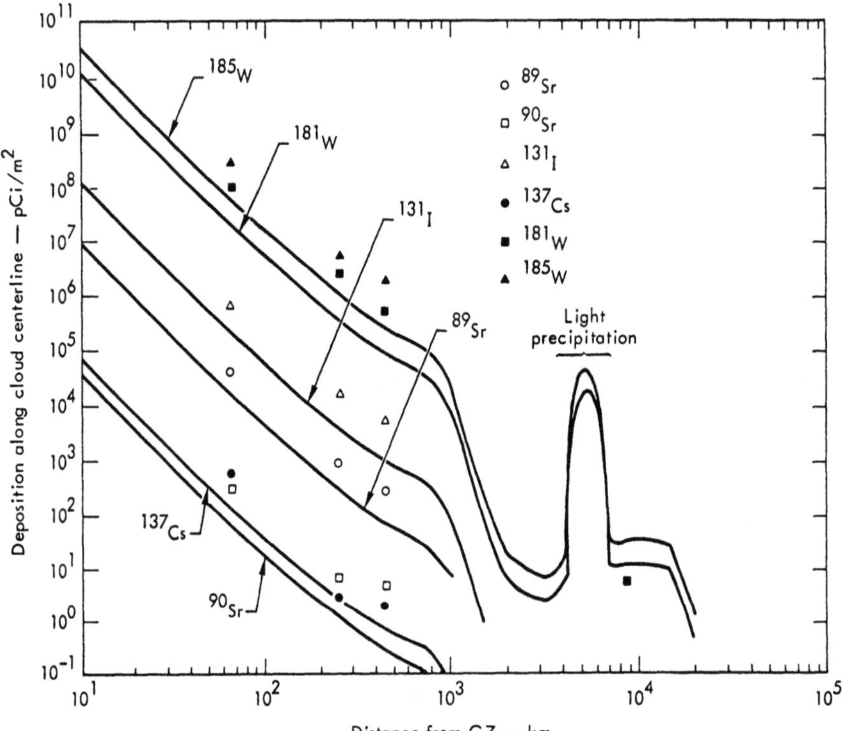

Fig. 3b. Schooner main cloud: deposition under the cloud center as a function of distance from ground zero (GZ) for several isotopes.

moves around the circular trajectory at constant speed. Typical trajectory models, widely used in the US, would project the future movement of particle x as follows: initially (x) would be moved for a time Δt (3 hours) due north, leaving the true circular path of (x). A displacement error of $\frac{V^2}{R_j} \frac{\Delta t^2}{2}$ would have occurred from the neglect of higher order terms in Taylor expansion. The serial accumulation of this error in the simple trajectory model, leads to the particle gradually spiraling outwards away from its true circular trajectory and speeding-up for each Δt. The fractional error ε is of the order of

$$\varepsilon \simeq \frac{V^2}{R_j} \frac{\Delta t^2/2}{V \Delta t} = 1/2 \frac{V\Delta t}{R_j} \simeq 1/2 \frac{S}{R_j} \simeq \frac{2\pi Rj}{2R} \frac{\Delta t}{T}$$

where T is the orbital cycle time.

In one cylce time for particle (x), the accumulated fractional error exceeds π , which means the estimated position is at least in error π times the initial (3 hr) displacement of particle (x). It appears that commonly used Heffter trajectory model (Heffter, 1980) fails the rotating particle trajectory test. Extended trajectories prepared by such models appear highly questionable. The accumulated error illustrated above would occur in practical applications independent of and in addition to analysis errors attributable to the meteorological data itself.

Illustration 4b: Accurate Particle Trajectories on the Spherical Earth (GRANTOUR).

Illustration 4b (Figs. 5 and 6) presents an independent estimate of this accumulated error in comparison to trajectories possible from new improved approaches inherent in GRANTOUR (Walton, private communication, 1983).

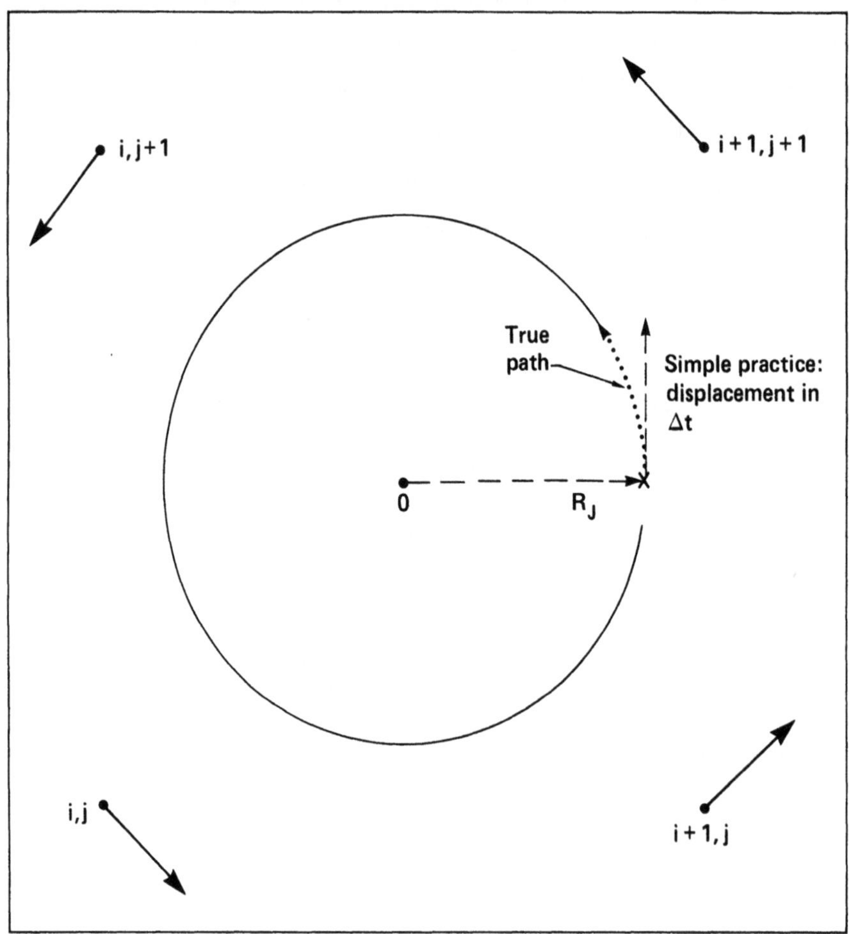

Fig. 4. Idealized Problem - the rotating particle (x) in a flow field of unbounded-solid-rotation (about 0).

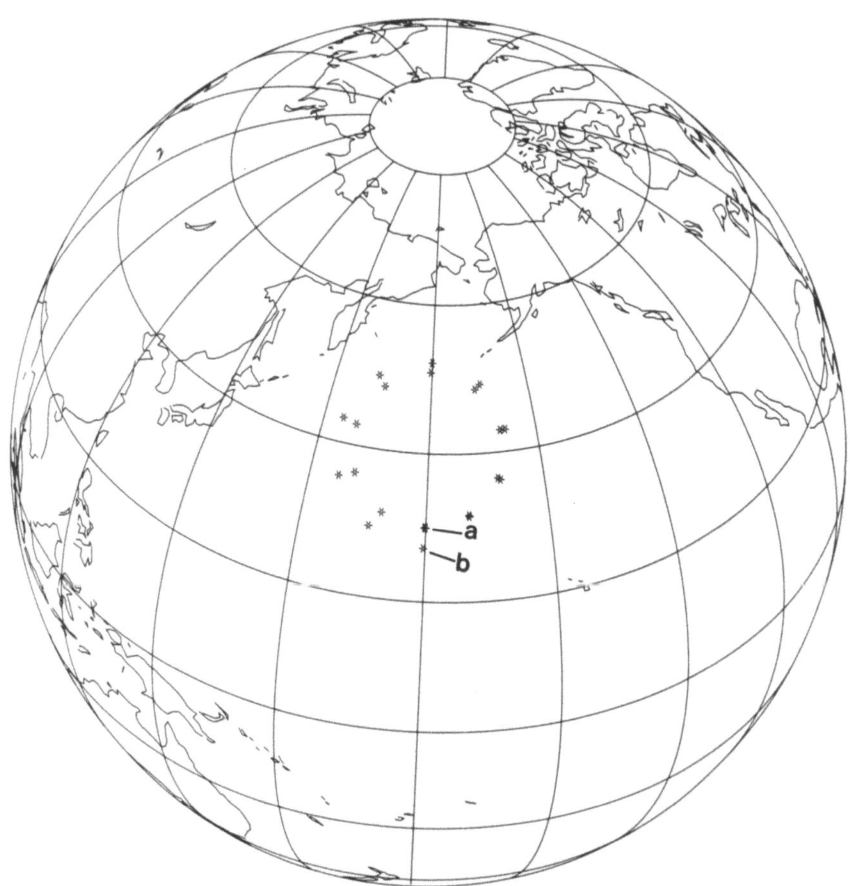

Fig. 5 Trajectories of two parcels subject to zonal and meridional winds resulting from solid body rotation (ω = 1.5°/hr) about an axis to the center of the earth and 180°E, 40°N. A fixed time-step at 3 hours was used. The motion of parcel 'a' had the accelerational correction that of parcel 'b' did not.

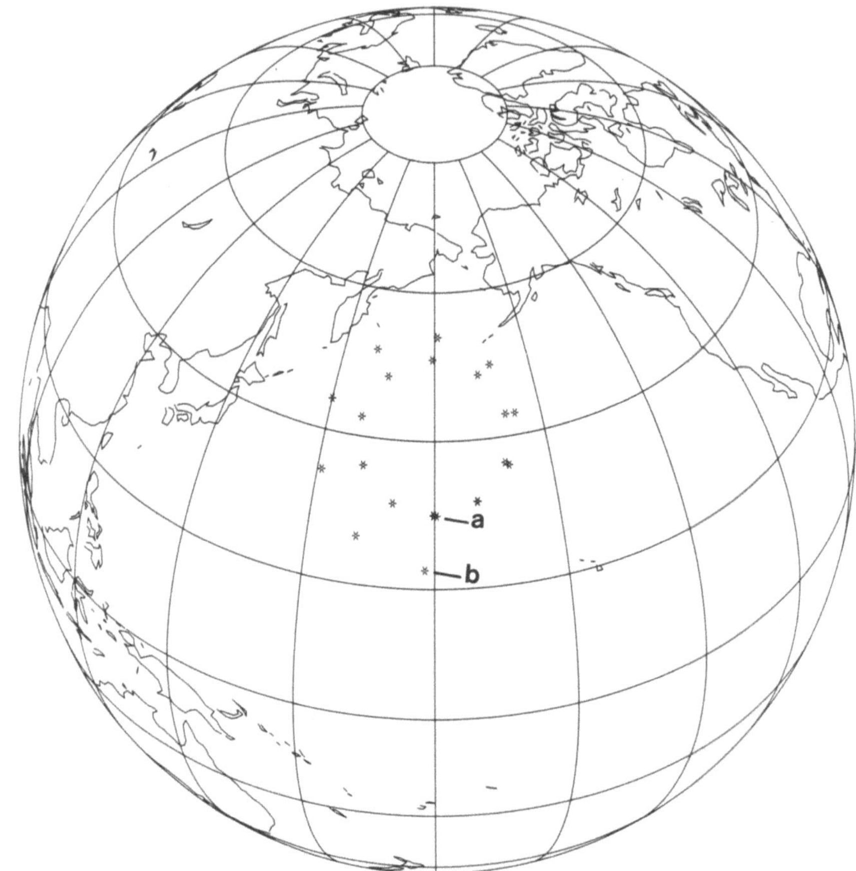

Fig. 6. Trajectories of two parcels subject to zonal and meridional winds resulting from solid body rotation (π = 1.5°/hr) about an axis through the center of the earth and 180°E, 40°N. Available time-step was used so that the maximum parcel displacement is less then 700 km. The motion of parcel 'a' had the accelerational correction, that at parcel 'b' did not.

Illustration 5: Global PATRIC

The adaptation of particle-in-cell techniques to be applicable to global assessment problems began in 1977, and the first successful results of applying the new capability GLOBAL PATRIC to a Chinese nuclear test cloud were reported by Knox et al. (1982). These integrations spanned at time interval of 144 hours, or transport roughly one-half way around the earth. The techniques development involved the establishment of channel for receipt of AFGWC analyzed winds for 300, 200, 150, 100, 70, and 50 millibar surfaces at twelve hourly intervals over the northern hemisphere, the testing of methods for making each analysed wind set mass-consistent, and devising and testing accurate techniques for calculating trajectories of particles within zones of mass-consistent winds at a single time. Such improved methods involved solution of particle trajectories using second order-terms in Taylor series for the velocity of the particles (Walton, 1979). The later development was, as previously mentioned, key to making particle-in-cell techniques applicable on the global scale for extended times, for a single puff. The treatment of the source term, initial puff size, cloud center (or height) and the selection of the appropriately somewhat reduced (1/10th) diffusion parameters is given in Table 3. The latter is required in that the particle displacement, and the cloud deformation by hydrodynamic flow fields is performed in some detail.

The relevant Figures 7 to 12 show sample analyzed wind fields for the 300 mb surface and the 50 mb surface (October 16, 1980), the puff particle footprint at 45 hours with the upper, middle, and lower third of the cloud at that time distinctively designated, the vertical profile of the puff at 54 hours, and at 138 hours when the cloud arrived in the US the final figures show the vertical profile of the particles, and the puff particle footprint. The deformed and shear cloud form at 138 hours is mainly accomplished by differential displacement of the particles throughout the cloud, and distortion of the cloud. The wide "apparent" breadth of the cloud in the western US is primarily accomplished by wind shear. At any particular level, the puff growth is depicted as being much, much smaller than that achieved by shear. The verification of these above calculations against data is still in progress, and can not unfortunately be included in this report at this date.

Table 3.

SOURCE TERM:

- ACTUAL CLOUD GEOMETRY AT STABILIZATION TIME
- NORMALIZED YIELD TO 1 KT AT H+1
- CLOUD CENTER HEIGHT 14 Km

DIFFUSION PARAMETERS:

- VERTICAL DIFFUSION

  $K_z = 0.1\ m^2/s$

- HORIZONTAL DIFFUSION

  SCALE DEPENDENT (Walton: $K_h = \frac{d\sigma^2}{dt} = \varepsilon^{1/3}\sigma^{4/3}$)

  $\sigma^2 = [\sigma_o^{2/3} + 2/3\ \varepsilon^{1/3}\ t]^3$

  WITH $\varepsilon = 10^{-6}\ m^2/s^3$

OUTPUT:

- INSTANTANEOUS CONCENTRATION CONTOURS AT SELECTED HEIGHTS WITH SAMPLE GRID FOLLOWING DEBRIS CLOUD CENTER
- TIME INTEGRATED CONCENTRATION CONTOURS ON FIXED, PRESELECTED SAMPLING GRID.

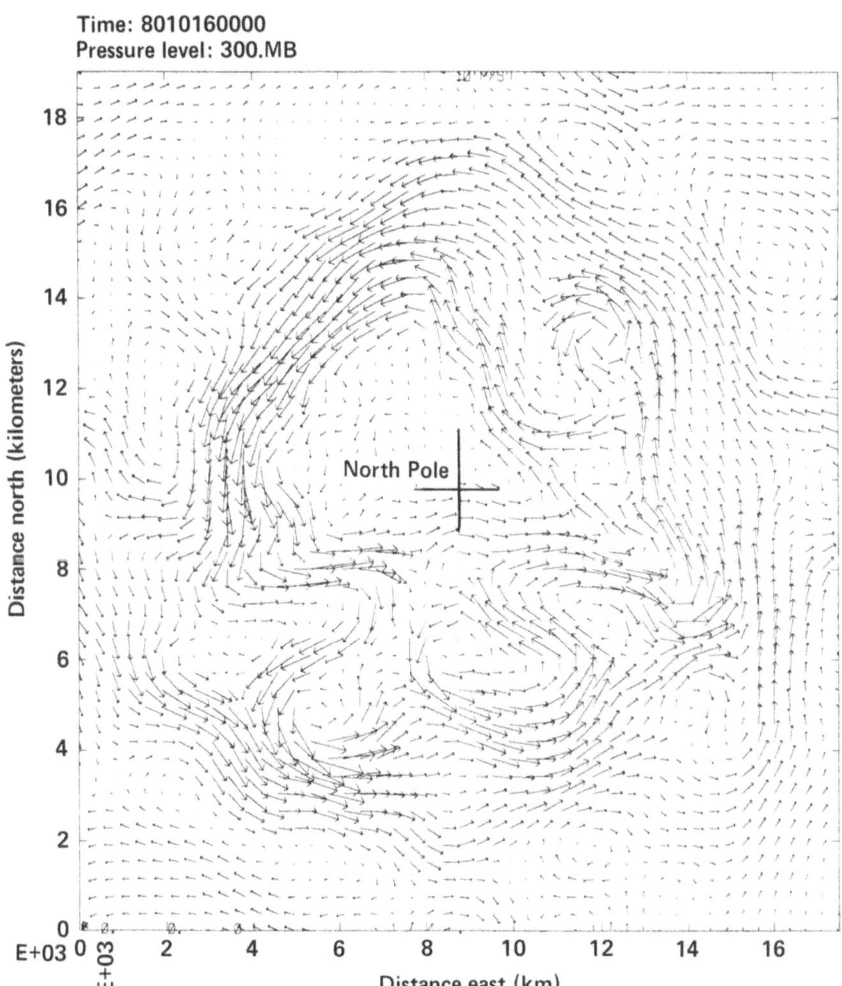

Fig. 7. AFGWC gridded analysis of the 300 mb winds over the Northern Hemisphere in Polar Stereographic projection.

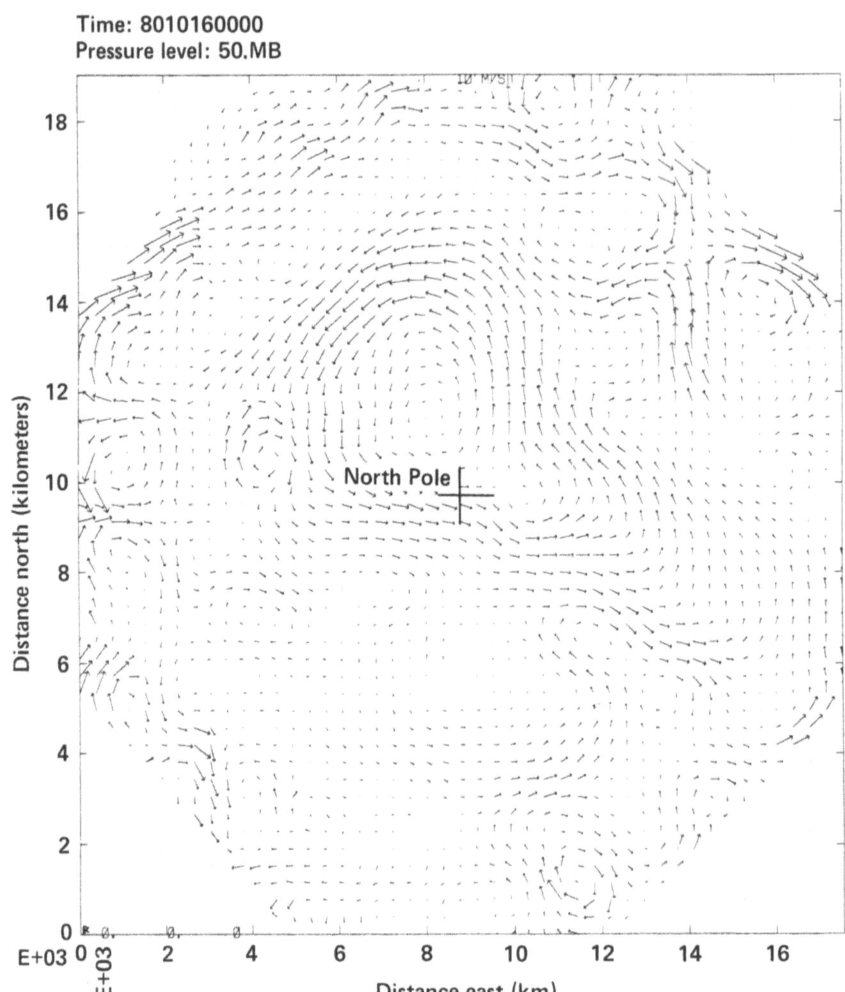

Fig. 8. AFGWC gridded analysis of the 50 mb winds over the Northern Hemisphere in Polar Stereographic projection.

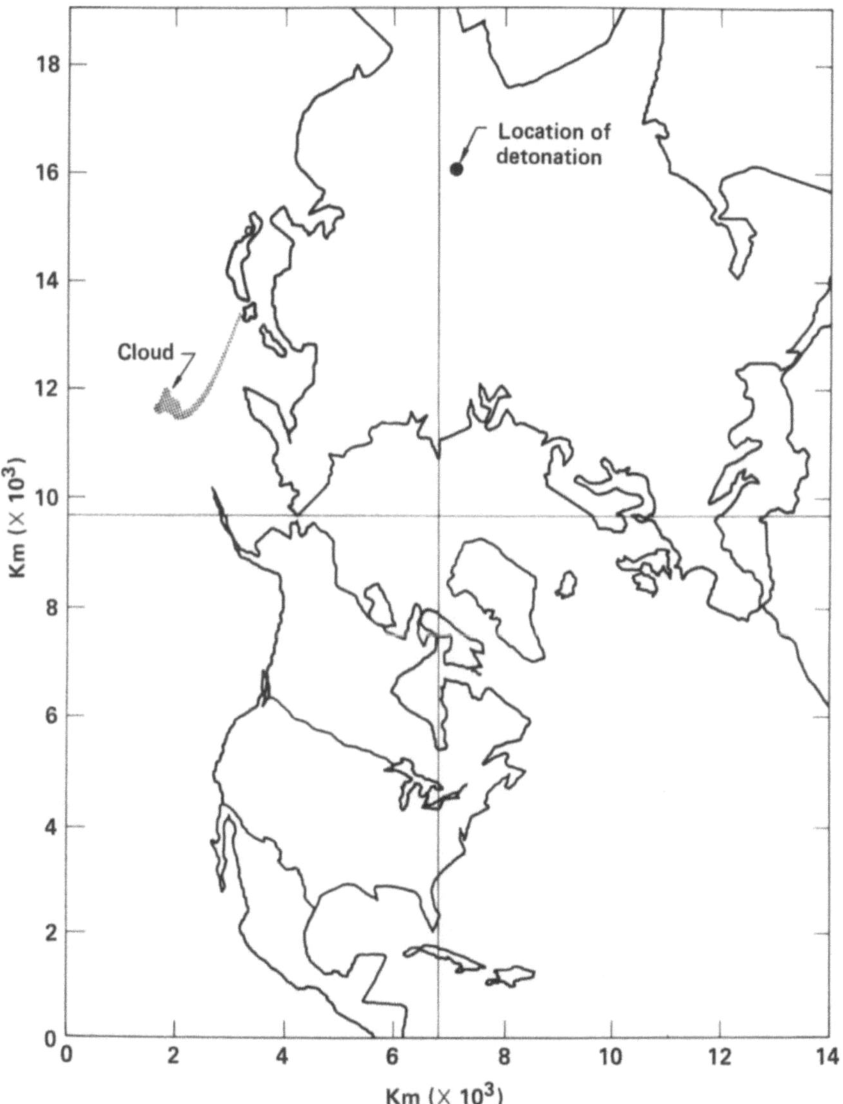

Fig. 9. Projected view of the cloud just west of Japan 45 hrs after stabilization time. The distribution of the cloud is due to wind shear.

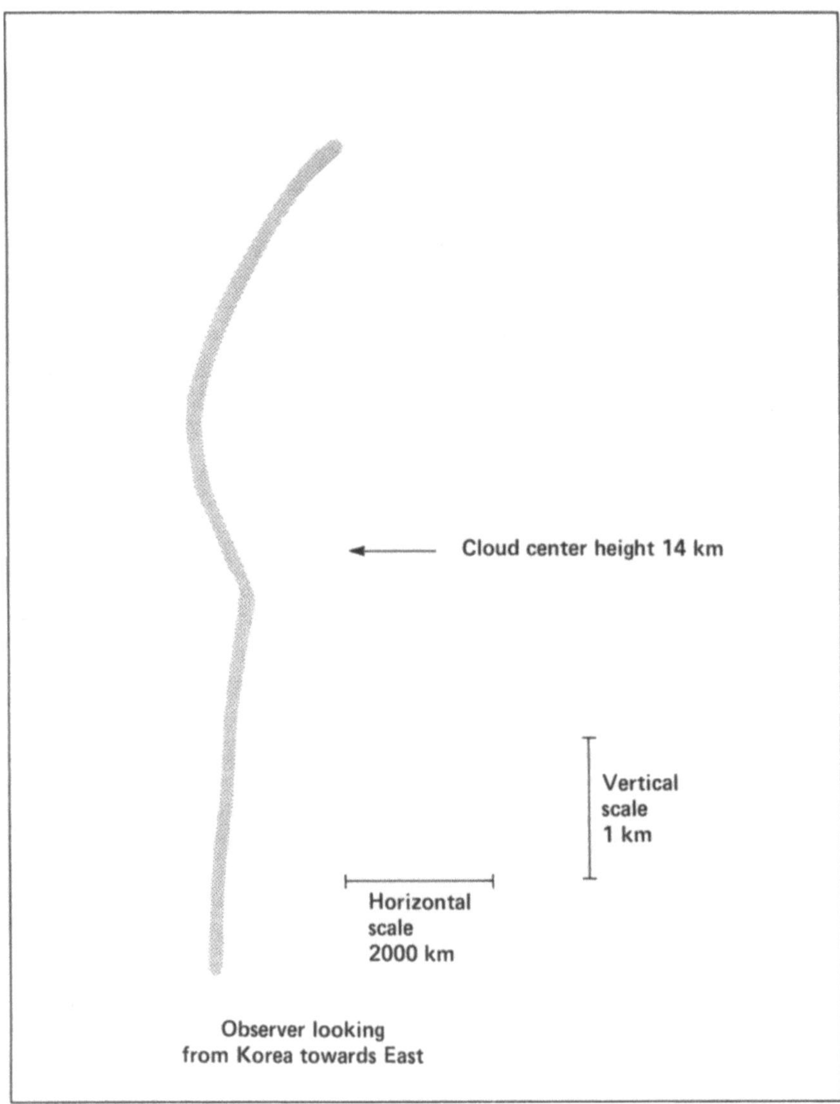

Fig. 10. Vertical profile of the cloud 54 hours after stabilization time. Horizontal scale is 2000 times vertical scale.

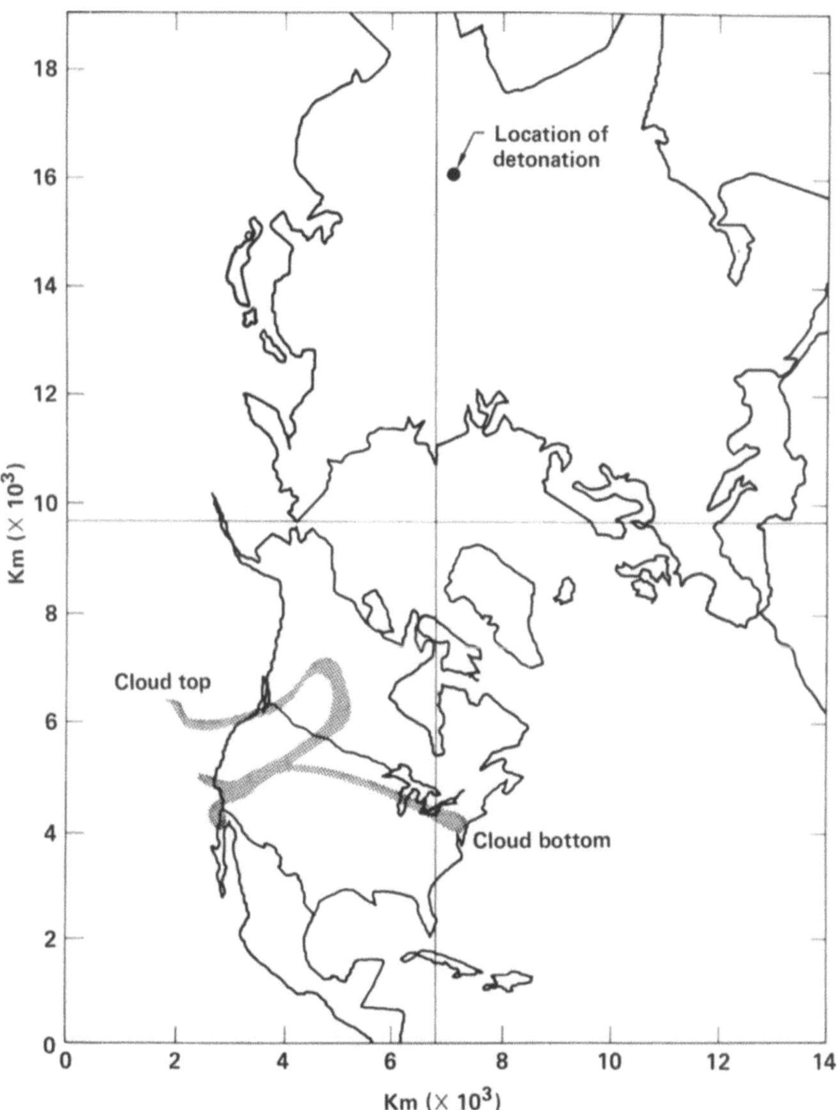

Fig. 11. Projected view of the cloud over the United States 138 hours after stabilization time. The enormous distortion of the cloud is due to wind shear.

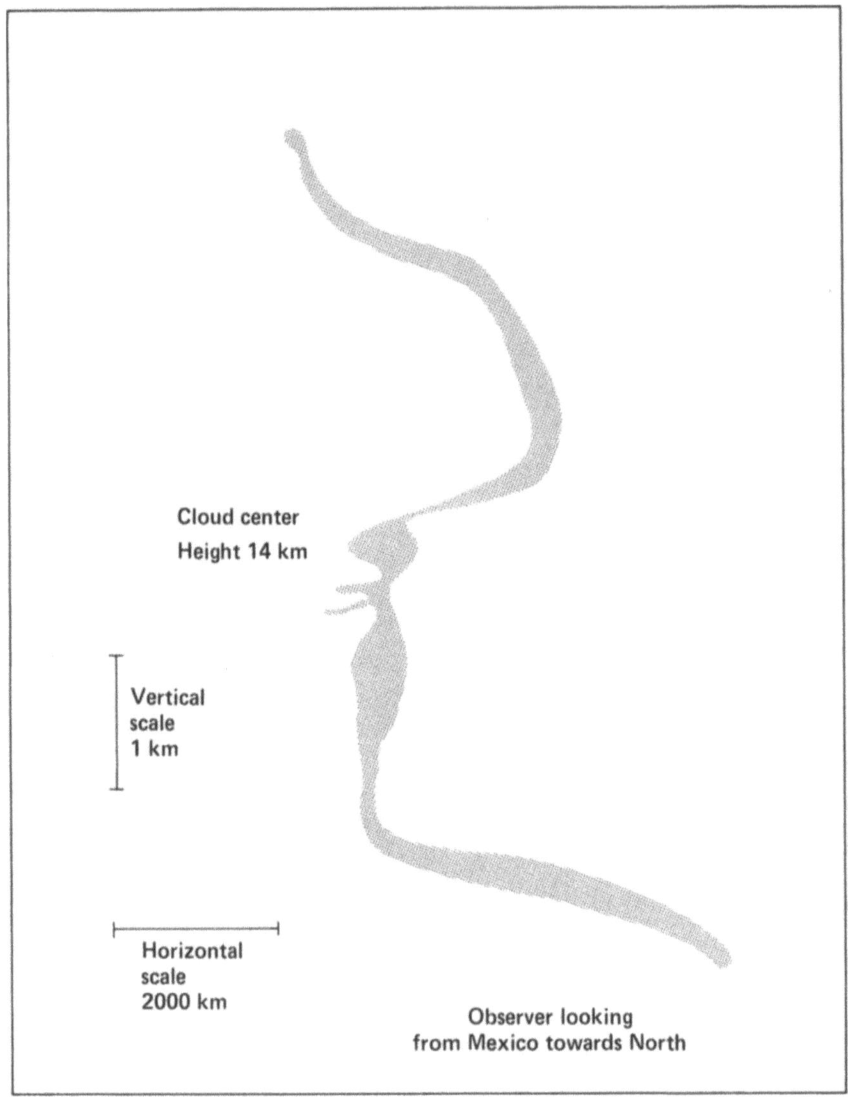

Fig. 12. Vertical profile of the cloud 138 hours after stabilization time. Horizontal scale is 2000 times vertical scale.

Illustration 6: Simulation of the CO_2 Global Budget as a Function of time by TRANSZAM.

During the past year, a new modeling capability has been developed for the purpose of simulating global budgets of gases like carbon dioxide and its time dependent distribution in response to anthropogenic and biogenic sources and sinks. In order to perform multi-year integrations, a number of innovations are required as reported by Walton et al., (1983). These innovations are briefly enumerated below.

1. The concept of a Lagrangian sampling parcel (LSP) is introduced; a LSP is assumed to be tagged by a minute, conservative, and known mass for its lifetime, but in addition the parcel carries a mass of one or more trace gases. The parcel as it moves in prescribed wind fields is subjected to sources, sinks, chemical transformations, and diffusion as appropriate for the trace species. In the initial development and testing the wind fields have been derived from the Livermore Statistical Dynamical Model (LSDM), MacCracken et al., (1981) whereby the wind fields are prescribed in the meridional plane from the north pole to the south pole.

2. The time-dependent wind fields provided by LSDM to the Global LSP model, called TRANSZAM, are mass-consistent and have the properties, discussed in Illustration #4, that permit the solution of the LSP trajectory within a cell in analytical form. Tests of the type presented in Illustration #4 have been performed over very extensive periods to convince the developers of the model that the trajectory methodology is precise enough for extended integrations.

3. The mixing ratio of a trace gas at any point in space and at a given time is obtained by interpolating from the Lagrangian sampling parcels that are in the vicinity of the grid point.

4. The simulation of the diffusion processes is performed from these interpolated mixing ratios at grid points for two different scales, one for the scale of the fixed grid cell, and a second for a larger scale background field. The changes in trace gas mixing ratio from these two diffusive processes are then interpolated back to the LSP.

5. Due to displacement and deformation it is quite possible for an atmosphere with only a few thousand LSP's to generate a distribution in which a fixed Eulerian cell might have no LSP within it at sometime during the integration. A

rational data filling process has been devised so that a new parcel is created as needed. Since the new parcel carries mixing ratio information, the trace gas budget is uneffected by the creation of the new LSP.

6. Initialization is performed by seeding the atmosphere randomly with a few thousand LSP and prescribing reasonable parcel mass and mixing ratio statistics; in time the distributed sources and sinks force the trace-gas distribution to evolve. To date too few extended simulations have been performed to determine the sensitivity of long term solutions to the current process of initialization.

The preliminary or exploratory calculation that we now present is for carbon dioxide; hence, the attribute of the LSP that permits the inclusion of chemical transformation is not exercised.

The sources and sinks for carbon dioxide assumed in this illustrative simulation are contained in Table 4.

Table 4. Surface sources of CO_2 used to drive the TRANZAM model. Values were chosen based on the work of Pearman and Hyson (1980) and Hyson and Pearman (1980). Time t is in hours since January 1; a 360 day year is assumed.

Source	Latitudinal distribution	Net source strength $[10^{12}$ Kg/yr$]$
Fossil fuel	33° N - 50° N	5
Biosphere	33° N - 58° N	$3 \sin [2\pi(t-6480)/8640]$
Preindustrial ocean (net zero)	72° S - 33° S	-0.875
	25° S - 25° N	1.75
	33° N - 72° N	-0.875
Present ocean (taking up half of fossil fuel source)	72° S - 33° S	-1.5
	25° S - 25° N	0.5
	33° N - 72° N	-1.5

Over the twelve month period the evolution of the CO_2 distribution seems reasonable, and the calculated annual global increase in background CO_2 corresponds closely to that observed. We have, however, output the time series of calculated CO_2 concentration at 600 mb at the geographical locations of the south pole and Mauna Loa Observatory in Hawaii. The remarkable agreement of the amplitude of the annual cycle between the observed record and the simulated time series for these two locations is quite promising. (See Figure 13.) The concepts and the relatively precise applied mathematical techniques contained in TRANSZAM may make the numerical simulation of global budgets of trace gases a tractable problem. In principle, such techniques offer the hope of being able to estimate the impact of new sources on the distribution of background-trace gas on very large geometrical scales.

CONCLUDING REMARKS

The present overview of the long range transport of air pollutants has focused on insights, concepts, and modeling approaches over the past ten to twenty years of our studies. We have reviewed some of the evidence for the existence of a regime of accelerated diffusion in the atmosphere. The puff

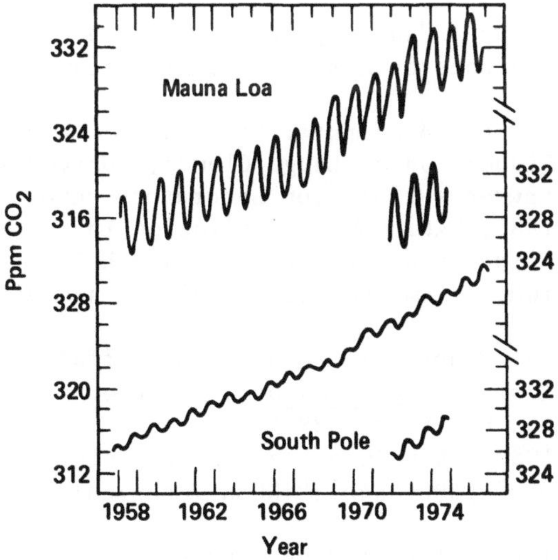

Fig. 13. Comparison of computed seasonal cycles at 600 mb at latitudes of Mauna Loa and the South Pole with the observed records from these stations as given by Bacastow and Keeling (1979).

(or plume) growth data is influenced by the intertwined effects of shear and diffusion; the examples presented indicate that fairly precise calculational techniques for shear effects may well be necessary before these simultaneously operating processes are separable on a case-by-case basis. The escape from the inherent errors of oversimplified trajectory models (e.g. $S = v\Delta t$) appears plausible with the emergence of accurate applied mathematical techniques. Further, the invention of Lagrangian sampling parcels and the creative methods introduced by Walton et al. (1983) opens up new avenues for investigating global budgets. It is not inconceivable that these concepts discussed herein will play a role in improved source attribution studies on the mesoscale to synoptic scale distances during the next several years.

Acknowledgments

The author is indebted to Mr. Rolf Lange for providing the calculations and illustrations regarding the GLOBAL PATRIC simulations which have only been reported in the most brief manner in the past. The author also expresses his gratitude to Dr. John J. Walton for providing several figures regarding the TRANSZAM simulations of the carbon dioxide global budget that have not been previously available.

References

Batchelor, G. K., 1950, "The Application of the Similarity Theory of Turbulence to Atmospheric Diffusion, QJRMS, 76:133.

Crawford, T. V., 1966, "A Computer Program for Calculating the Atmospheric Dispersion of Large Clouds," University of California, Lawrence Radiation Laboratory report UCRL-50179.

Gifford, F. A., 1983, "Atmospheric Diffusion in the Mesoscale Range: The Evidence of Recent Plume Width Observations," Sixth Symposium on Atmospheric Turbulence and Diffusion, 300-303.

Hanna, S. R., 1982, Chapter 10: Turbulent Diffusion: Chimneys and Cooling Towers, Engr. Meteor., Elsevier Scientific Publishing Co., New York, 429-474.

Heffter, J. L., 1980, "Air Resources Laboratories Transport and Dispersion Model (ARL-ARAD)," NOAA Technical Memorandum ERL ARL-81

Johnson, W. B., 1983, "Interregional Exchanges of Air Pollution: Model Types and Applications, JAPCA, 33:6:563.

Knox, J. B., 1974, "Numerical Modeling of the Transport Diffusion and Deposition of Pollutants for Regions and Extended Scales," JAPCA, 24:7:660.

Knox, J. B., 1981, "Theory and Domain of Validation: Long Range Transport Model, Lawrence Livermore National Laboratory Internal Memorandum UASG 81-35.

Knox, J. B., M. C. MacCracken, M. H. Dickerson, P. M. Gresho, F. M. Luther and R. C. Orphan, 1982, "Program Report for FY 1981 Atmospheric and Geophysical Sciences Division of the Physics Department," Lawrence Livermore National Laboratory report UCRL-51444-81.

Lange, R., 1978, "ADPIC-A Three Dimensional Particle-in-Cell Model for the Dispersal of Atmospheric Pollutants and its Comparison to Regional Tracer Studies," J. Appl. Met., 17:320-329.

Luther, F. M. and R. Lange, 1983, private communication.

MacCracken, M. C., J. S. Ellis, H. W. Ellsaesser, F. M. Luther and G. L. Potter, 1981, "The Livermore Statistical Dynamical Climate Model, Lawrence Livermore Laboratory report UCID-19060.

Raynor, G. S., J. V. Hayes, and D. M. Lewis, 1983, "Testing of the Air Resources Laboratories Trajectory Model on Cases of Pollen Wet Deposition after Long-Distance Transport from Known Sources Regions," Atmospheric Environment, 17:2:213-220.

Reiter, E. R., 1978, "Atmospheric Transport Processes: Part 4 Radioactive Tracers," U.S. Department of Energy.

Richardson, L. F., 1926, "Atmospheric Diffusion Shown on a Distance-Neighbor Graph," Proc. of Royal Society (London) A110: 709.

Seidov, D. G., 1976, "A Numerical Euler- LaGrange Model of the Currents in a Inhomogeneous Ocean, Izv. Atmos. and Oceanic Physics, 12:686.

Walton, J. J., 1973, "Scale Dependent Diffusion," J. Appl. Met., 12:547-549.

Walton, J. J., 1979, "Development of a Lagrangian Treatment of Advection in a Moving Fluid," Lawrence Livermore National Laboratory Internal Memorandum UASG 79-2.

Walton, J. J., M. C. MacCracken, and H. W. Ellsaesser, 1983, "Preliminary Report on the LSDM Transport Sub-Model TRANSZAM," Lawrence Livermore National Laboratory report UCID-19819.

Warner, T. T., R. R. Fizz, and N. L. Seaman, 1983, "A Comparison of Two Types of Atmospheric Models - Use of Observed Versus Dynamically Predicted Winds," J. Cli. and Appl. Met., 22:394.

DISCUSSION

A. VENKATRAM In modeling long-range transport, one runs into the problem of defining turbulence. Do you have any comments on the solution of this problem ?

J.B. KNOX Turbulence is indeed defined differently depending on scale : for instance, in the LSDC model the horizontal eddy transport is parameterized and it represents primarily the transport by the synoptic disturbulences. Whereas for the transport of the nuclear debris cloud from China to U.S., illustrated in the paper, the transport and determination was accomplished by average fields of motion from the synoptic scale. Indeed, one man's turbulence is another's time averaged flow on transport field.

SENSITIVITY STUDIES WITH A LRT MODEL

Daniel Lavenu, Serge Legouis, and Pierre Bessemoulin

Direction de la Météorologie
Etablissement d'Etudes et de Recherches Météorologiques
77, Rue de Sèvres, 92106 Boulogne Billancourt, France

INTRODUCTION

In recent years, several long range transport (LRT) models have been developed (see Eliassen, 1980, for a review). The work discussed in this paper results from a request of the French "Ministère de l'Environnement" who wanted to have at his disposal an interpretation tool in case of pollution episodes due to LRT. The complexity of the model is a compromise between the necessary realism in simulating the most important physico-chemical processes involved, and a relative simplicity which might allow a routine use. It is to be noted that the work described here reflects more the development of a methodology than that of an operational model.

The conclusion of a bibliographical study was that for that purpose, such a model like EURMAP 1 (Bhumralkar et al, 1979) could work, provided some improvements were included for the calculation of concentrations values for specific episodes. The way to improve the model consists mainly in a sensitivity study described further. TABLE 1 is a summary of the main features of the two models.

DESIGN OF THE MODEL

The model has two modules corresponding respectively to the estimates of trajectories, diffusion and dry deposition.

Trajectories computations

Trajectories are computed from the 850 mb level date. It is well known that the geostrophic approximation fails when the curvature of isohypses (contour lines) is important. So we have made

Table 1 : Respective treatment of physico-chemical elements in the two models.

	EURMAP 1	This model
Emission data	SO_2 amount constant and based on annual total	SO_2 amount consists in constant and variable (seasonal) components
Mixing height	Constant (1 km)	Varies diurnally according to TRAPPES radio sounding
Transport wind	$V_T = 0.75\ V_{850}$ $\theta = \theta_{850} - 15°$ (index 850 refers to the geostrophic wind at the 850 mb level)	$V_T = \alpha\ U_{850}$ $\phi = \phi_{850} - \phi_o$ (index 850 refers to the gradient wind at the 850 mb level)
Horizontal diffusion	Fickian	Fickian
Vertical diffusion	Immediate uniform mixing up to the mixing height	Immediate uniform mixing up to the mixing height
Decay rates	Dry deposition : 0.8 $cm.s^{-1}$ for SO_2 0.2 $cm.s^{-1}$ for SO_4^{--} Transformation : 0.01. h^{-1} for $SO_2 \rightarrow SO_4$	Dry deposition : 0.8 $cm.s^{-1}$ for SO_2 0.2 $cm.s^{-1}$ for SO_4^{--} Transformation : 0.014 to 0.036 h for $SO_2 \rightarrow SO_4^{--}$

use of the gradient wind which takes into account such curvatures. The wind shear induced by the frictional effects in the boundary layer is considered according to the following formulation :

$$|\vec{U}| = \alpha\ |\vec{U}_{850}| \qquad \theta = \theta_{850} - \theta_o$$

where α is a proportionality coefficient, and θ_o a systematic anticlockwise rotation which can be shown to be dependent on the location and on meteorological conditions (stability). For the purpose of the model, different constant values have been tested.

Diffusion and deposition estimates

The diffusion model is a receptor-oriented puff-model. Each of the SO_2 puffs is released at 12- hourly intervals for each of the emission cells (according to the availability of upper-air meteorological charts), and followed during 120 hours.

The time step used is 15 minutes, for :

- computation of the wind field by linear interpolation between two successive altitude charts
- and, if necessary, linear interpolation of the mixing height estimated from the TRAPPES Radio Sounding.

OECD emission data used were available for an array of cells 0.5° in latitude by 1.0° in longitude (approximately 55 km N-S by 70 km E-W), covering the area from 5°W to 20°E, and from 41°N to 56.5°N.

The familiar gaussian formula is used to describe the pollutant concentration behaviour in the puff :

$$C = \frac{2Q}{(2\pi)^{3/2}\,\sigma_H^2\,\sigma_Z} \exp - \frac{R^2}{2\sigma_H^2}$$

where Q = amount of pollutant emitted during 12 hours
R = distance from the sampling point to the center of the puff
σ_H = horizontal diffusion coefficient = f (travel time)
σ_Z = vertical diffusion coefficient = g (mixing height).

Physico-chemical processes considered here are mainly dry deposition for sulphur dioxide and sulphates, and the transformation SO_2 to SO_4^{--}. It is assumed that the rates of change of pollutant mass in a puff due to deposition and chemical transformation are proportional to the respective masses :

$$\frac{dm}{dt} = - k\,m \quad , \; k = k_t + \frac{V_d}{H}$$

where m = mass of SO_2 in the puff at time t
k_t = $SO_2 \rightarrow SO_4^{--}$ transformation coefficient (h^{-1})
V_d = dry deposition velocity ($m.s^{-1}$)
H = mixing height (m)

From this hypothesis, the evolution of the SO_2 mass content of the puff is described by :

$$m = m_o .\ \exp - (k_t + \frac{V_d}{H})\ t \qquad (1)$$

Likewise, the SO_4^{--} content variation is proportional to the total mass M of SO_4^{--} :

$$\frac{dM}{dt} = - \frac{g}{H} M + k_t m$$

where g = deposition velocity for sulphates ($m.s^{-1}$).

Integration between o and t leads to the following expression

$$M = m_o \exp -\frac{g}{H} t . \frac{k_tH}{k_tH = V_d - g} . (1 - \exp - (\frac{V_d - g + k_tH}{H})t) \quad (2)$$

Equations (1) and (2) are instructive (Fig.1) to get some idea about the residence time of different species :

- the SO_2 residence time falls between 40 and 80 hours, depending on the values of k_t, V_d and H.

- the sulphates'one is much longer; it can be noted that there is always a significant residue of SO_4^{--} after 120 hours.

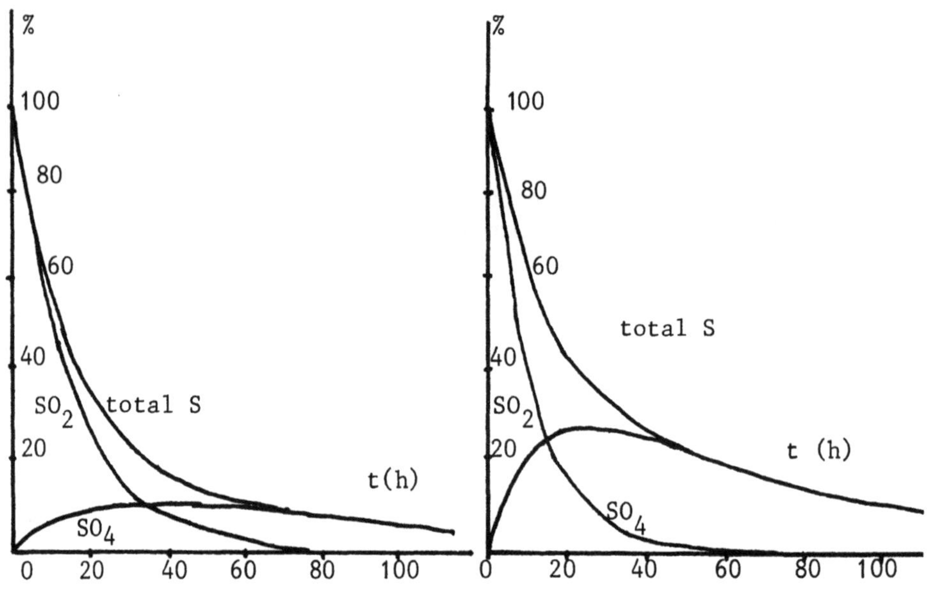

Fig.1. 1-a Mixing height = 250 m k_t = 0.025 h^{-1} 1-b Mixing height = 1000 m k_t = 0.025 h^{-1}

SENSITIVITY ANALYSIS

The period covered by the study is from the 1st to the 19th February, 1975. It is characterized by the persistance of an anti-cyclonic situation over Western Europe, which gives rise on some days to high levels of pollution. The performances of the model have been checked against the La Hague station data.

Sensitivity to the backing angle of transport wind relative to the 850 mb wind, and to speed reduction

Fig.2 to 5 show correlation coefficients between calculated and observed values of sulphur dioxide and sulphates as a fuction of the backing angle θ_0 (for a constant stretching coefficient Fig.2-3), or as a function of the stretching coefficient α (for a constant θ_0, Fig.4-5). It is reminded that for a 17 values sample, significant correlation coefficients are 0.41 at the 90 % confidence level, and 0.61 at 99 %.

Obviously, results for SO_4^{--} are less correlated than for SO_2 : 0.6, against 0.8 on the mean. This is logical if one considers the respective life time of SO_2 and SO_4^{--}, and the fact that for SO_4^{--} remote emissions which don't appear in the inventory may contribute significantly (Eastern countries in particular).

It can be seen that curves are rather flat, and that the location of the peak value is not inevitably the same for SO_2 and SO_4^{--}.

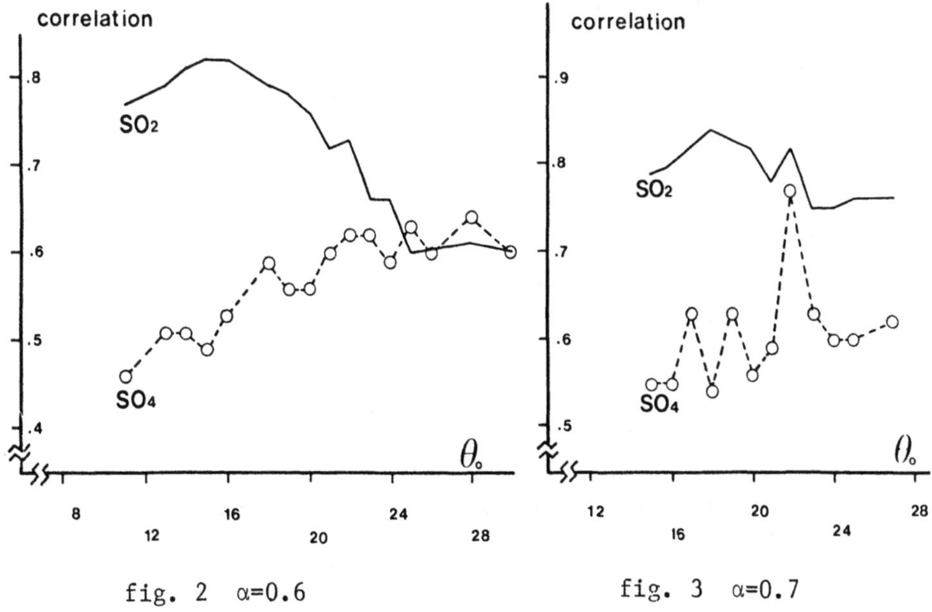

fig. 2 α=0.6 fig. 3 α=0.7

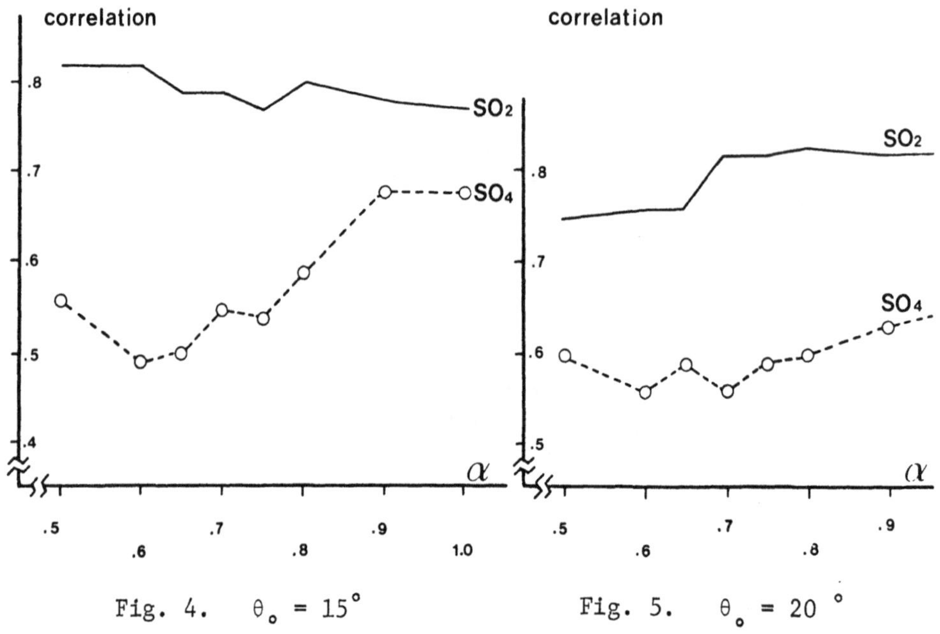

Fig. 4. $\theta_o = 15°$

Fig. 5. $\theta_o = 20°$

Curvature correction

Fig.6 indicates that for this case study, the curvature correction is not a significant improvement. However, greater differences would likely occur in the case of meteorological situations associated with ridges and throughs.

Fig.7 gives an example of the curvature correction effect.

Variable mixing height

The use of a variable mixing height, as measured at Trappes, instead of a constant one (1000 m) results in a significant improvement (Fig.8).

Decrease of dry deposition in the case of ground level inversion

It is well known that vertical fluxes inside a stable layer are very weak. It is the reason why we set the deposition velocity to zero when the inversion is at ground level. When this feature is not introduced in the model, the correlation coefficient falls from 0.83 to 0.71.

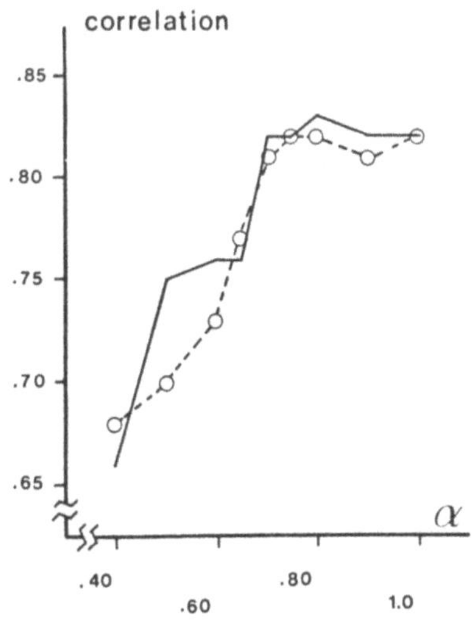

Fig.6. $\theta_o = 20°$
o without curvature correction
- with curvature correction

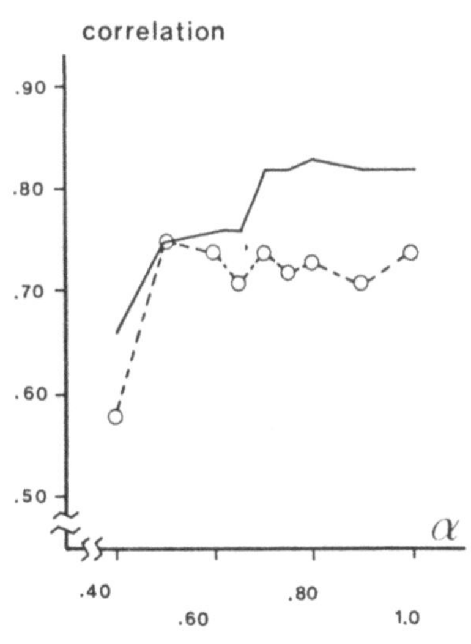

Fig.8. $\theta_o = 20°$
o H = 1000 m
- H from RS

Fig.7. Example of the curvature correction influence (x : with, o : without)

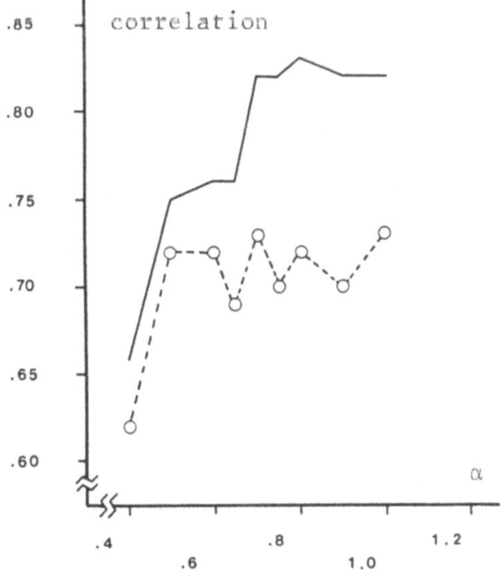

Fig.9. $\theta_o = 20°$
o without reduction of dry deposition
- with reduction of dry deposition

This hypothesis can be also justified in a simple way : near the ground, the vertical flux is :

$$K_Z \frac{\partial C}{\partial z} = V_d \cdot C \text{ , so } \frac{\partial C}{\partial z} = \frac{V_d}{K_Z} \cdot C \quad (3)$$

From the vertical diffusion equation, it comes :

$$\frac{dC}{dt} = \frac{\partial}{\partial z} (K_z \frac{\partial C}{\partial z}) - R$$

where R = removal term considered here only as dry deposition = Λ C

Λ = capture rate by the ground

K_Z = turbulent vertical diffusivity

For a stationnary process, $\frac{\partial}{\partial z} (K_Z \frac{\partial C}{\partial z}) = \Lambda C = \frac{\partial}{\partial z} (V_d \cdot C)$

$$\text{so} \quad \frac{\partial C}{\partial z} = \frac{\Lambda . C}{V_d} \quad (4)$$

Comparing (3) and (4) gives $V_d = (K_Z \Lambda)^{1/2}$

Due to low K_z in stable condition V_d must decrease in stable condition (see also GARLAND, 1979).

Furthermore, nocturnal inversion reduces concentrations at the ground and limits the amount of material available for deposition.

SOME MODEL RESULTS AND CONCLUSIONS

Fig.10 and 11 give an idea of the model results, according the two different choices for θ_o and α, namely (-20°, 0.8) and (-25°, 0.6).

The two correlation coefficients are 0.83 and 0.60.
It can be seen that :

- the peak value, occuring on the 6/2/1975 is correctly simulated in both cases (pollution coming from the Rhine Valley).
- trends between 3rd and 5th, February, are opposite and on the 8th the SO_2 estimates are very different according to the different hypothesis.

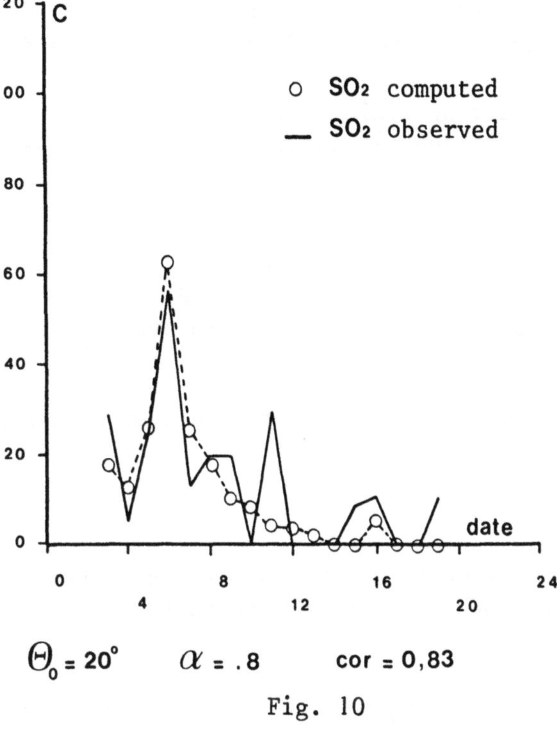

$\Theta_0 = 20°$ $\alpha = .8$ cor = 0,83

Fig. 10

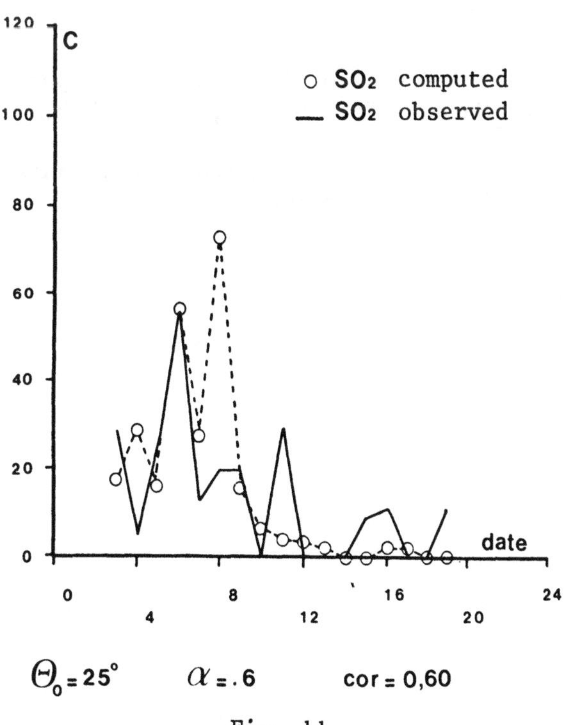

$\Theta_0 = 25°$ $\alpha = .6$ cor = 0,60

Fig. 11

This study suggests that a model such as EURMAP 1 can be significantly improved if some simple modifications are incorporated namely :

- using daily varying mixing heights and
- reducing the dry deposition in stable conditions

Further refinements should be incorporated :

- greater emission coverage
- analysed winds every 6 hours by ECWMF
- objection analysis of mixing heights over all Europe
- wet deposition

The model also would have to be tested over longer periods to be fully validated.

Acknowledgements. This work has been sponsored by the French "Ministère de l'Environnement" under Contract N° A-12/1982.

REFERENCES

BHUMRALKAR C.M., JOHNSON W.B, MANCUSO R.L, WOLF D.E (1979) : Regional patterns and transfrontier exchanges of airborne sulfur pollution in Europe.
SRI International - Report N° 4797

ELIASSEN A. (1980) : A Review of long-range transport modeling. Journal of Applied Meteor. Vol 19, N°3, pp 231-240.

GARLAND (1979) : Dry deposition of gaseous pollutants. WMO Technical Note N°538, pp 95-104.

THE GREEN RIVER AMBIENT MODEL ASSESSMENT PROGRAM

Francis A. Schiermeier* and Alan H. Huber*

U.S. Environmental Protection Agency
Research Triangle Park, North Carolina 27711

C. David Whiteman and K. Jerry Allwine

Pacific Northwest Laboratory
Richland, Washington 99352

INTRODUCTION

The Clean Air Act Amendment of 1977 requires the U.S. Environmental Protection Agency (EPA) to promulgate regulations that specify the use of atmospheric dispersion models to achieve air quality limits pertinent to Prevention of Significant Deterioration (PSD) and attainment of National Ambient Air Quality Standards (NAAQS). A major need exists for predictive models of pollutant transport and diffusion in complex terrain, especially in the pristine areas of western United States where energy development and utilization are rapidly expanding.

Due to the proposed development of the Green River Oil Shale Formation encompassing the areas of southwestern Wyoming, northeastern Utah and northwestern Colorado (Figure 1), the EPA Region VIII Headquarters in Denver has a specific need for the development of site-specific ambient dispersion models. To meet this requirement, the Environmental Sciences Research Laboratory has inititated the Green River Ambient Model Assessment (GRAMA) program. The main objective of this program is to develop improved air quality models for analyzing the impacts of the oil shale industry with respect to PSD increments over pristine wilderness areas and to ambient air quality standards for criteria pollutants. The GRAMA program is being conducted under an Interagency Agreement by the Pacific Northwest Laboratory (PNL), operated for the U.S. Department of Energy by Battelle Memorial Institute.

*On assignment from National Oceanic and Atmospheric Administration

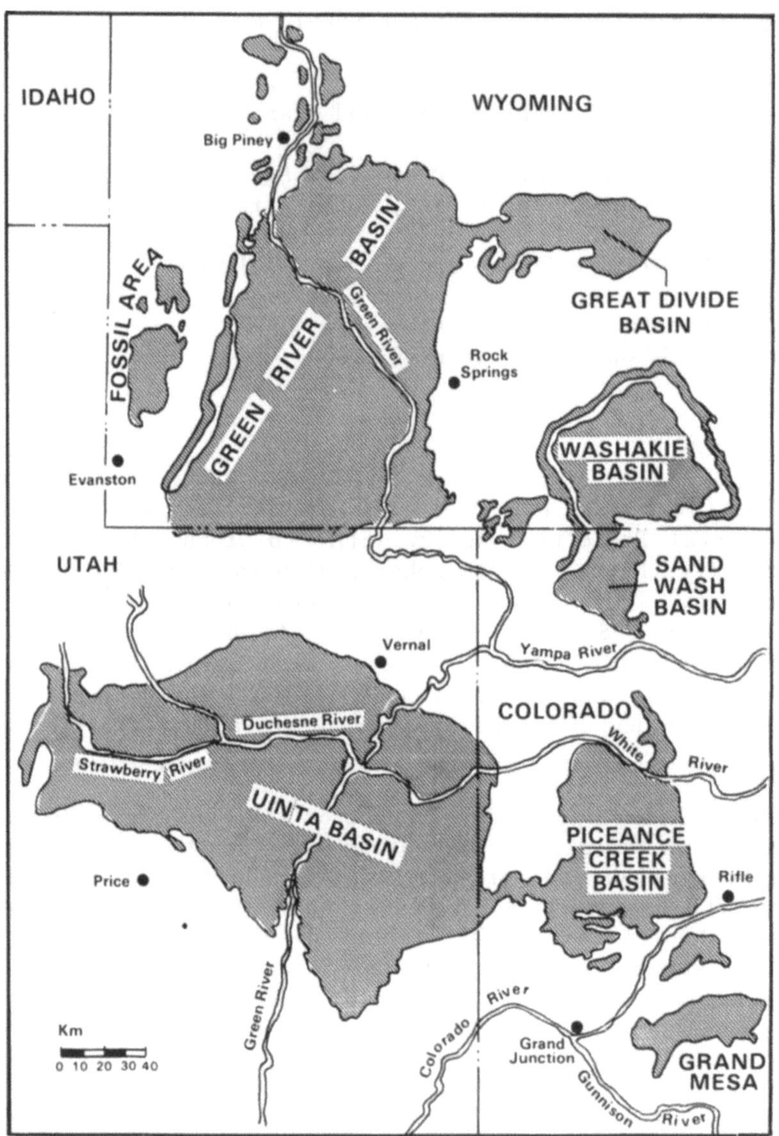

Figure 1. Green River Oil Shale Formation areas (shaded) of western United States.

Two air quality simulation models are currently being developed by PNL for this topographically and meteorologically complex region. Both are designed in a modular structure so that additional modules may subsequently be incorporated without significantly affecting the overall code framework. One model, VALMET, is a local-scale model for predicting concentrations of nonreactive pollutants within a well-defined mountain valley configuration. The current version of VALMET has not yet undergone sensitivity tests, has not been evaluated with respect to its performance in simulating air pollution concentrations in actual valleys, and contains modules that are not yet fully developed. A tracer experiment was conducted recently in a deep mountain valley in Colorado to provide data for preliminary evaluation of VALMET.

The second model, MELSAR, is a mesoscale Lagrangian puff model for predicting pollutant concentrations within a 500 km by 450 km region. This model is currently in the developmental stage, although a preliminary version is operational which computes short-term (up to 24-hour) pollutant concentrations at specified receptors. The computer code has not been fully tested and several program modules are incomplete.

This paper gives a description of the current VALMET model followed by brief explanations of the MELSAR model and the GRAMA field experiments. The paper concludes with a summary of the GRAMA program.

VALMET MODEL DEVELOPMENT

The VALMET (VALley METeorology) air quality model was developed to predict concentrations of a nonreactive pollutant arising from an elevated continuous point source located within a well-defined mountain valley. The pollutant concentrations are computed at the valley floor and sidewalls on a valley cross section a selected distance downwind from the pollutant source. The model is intended to simulate the effects on pollutant transport and diffusion of various meteorological processes that are thought to result in worst-case pollutant concentrations. The model is designed for situations in which pollutants are carried in locally developed circulations within a valley when these circulations are decoupled from prevailing regional flow above the valley. Detailed descriptions of the meteorological hypotheses on which VALMET is based and of the model itself are given by Whiteman and Allwine (1983a).

The VALMET model, while including a variety of meteorological processes, is highly parameterized so it is simple in concept and easy to run. VALMET is composed of 13 modules, or subroutines, arranged so that an improved understanding of individual valley meteorological phenomena can be easily incorporated into future versions. The modules can be replaced by measurement data if such are available. Thus, the model can be used in either of two modes. It can be used in a screening mode to calculate pollutant concentrations within a valley when little site-specific data are available or it can be used as a site-specific model by calibrating it with actual measurement data.

VALMET was developed primarily to predict pollutant concentrations during the post-sunrise temperature inversion breakup period. However, the air pollution concentrations within the valley cross section at sunrise must be known for the post-sunrise simulation. VALMET is therefore comprised of two parts--a nighttime part to predict concentrations on the valley cross section at sunrise, and a daytime part to predict concentrations on the valley floor and sidewalls during the post-sunrise temperature inversion breakup period.

VALMET Nocturnal Component

Steady-state nocturnal pollutant concentrations within a valley cross section are determined by a Gaussian plume model that is modified to account for channeling and reflections off the sidewalls within the valley, for dilution due to clean air volume flux convergence within the valley, and for diffusion of pollutants out the top of the valley inversion. The model begins with an elevated pollutant source located anywhere on the valley floor. (The present version of the model cannot handle sources on the valley sidewalls.) The source is assumed to emit a continuous elevated pollutant plume into a nocturnal valley temperature inversion. A plume rise formulation simulates the initial rise of the pollutant plume at the stack due to momentum and buoyancy of the effluent.

Nocturnal dispersion coefficients for the Gaussian plume equations are determined from the empirically derived Pasquill-Gifford curves, which were developed from atmospheric tracer experiments conducted over homogeneous terrain for various downwind distances and stability categories. Because of the enhanced nocturnal diffusion in the rough valley terrain, the valley dispersion coefficients are derived from the original Pasquill-Gifford curves by simply adjusting the stability categories. From existing data, it appears that the enhancement of horizontal dispersion can be approximated by dropping the stability two categories (e.g., F becomes D), and the enhancement of vertical dispersion can be estimated by dropping the stability one category (e.g., F becomes E). These simple adjustments to the stability categories are a temporary measure until more appropriate means are developed for handling nocturnal dispersion in deep valley terrain.

Dispersion of pollutants during the plume's nocturnal transport down the valley will cause a spreading of the plume in both the vertical and horizontal, resulting in the plume's eventual contact with the valley floor and sidewalls. Additional spreading of the plume, using the basic Gaussian equations, would result in physically unrealistic concentrations occurring "beyond" the valley sidewalls or "below" the valley floor. This was handled by superimposing the valley cross section on the theoretical plume cross section and, through integrations, determining the fraction of pollutant mass within the valley below the inversion top, as well as the mass that has diffused above the inversion top and beyond the valley sidewalls and valley floor. To conserve pollutant mass within the valley, the mass fraction that has diffused "beyond" the valley sidewalls and "below" the valley floor is folded back into the valley cross section, assuming uniform mixing within the cross section.

Calculations with the Gaussian plume model are made assuming the Gaussian plume is channeled down the valley axis, despite meanders in the valley's course (Figure 2). The down-valley distance of a given cross section must therefore be determined from topographic charts to account for such valley axis meander.

An important factor affecting plume concentrations is the dilution of the plume during its travel down a valley due to tributary flows that bring clean air into the plume. Other processes occurring in valleys that affect plume dilution include converging downslope drainage flows and entrainment of clean air into a valley from above. Any process capable of producing volume flux convergence between two vertical valley cross sections is capable of diluting a plume. At present, the physical understanding of these processes is insufficient to incorporate them explicitly into an air quality model. The extent to which volume flux convergence is a feature of a particular valley's meteorology can be determined, however, by wind observations.

In the present version of VALMET, the Gaussian plume equation is modified initially to correct for the dilution of a pollutant plume during its down-valley travel. The correction is applied to the plume-transporting wind speed at the pollution source cross section to obtain a modified or virtual wind

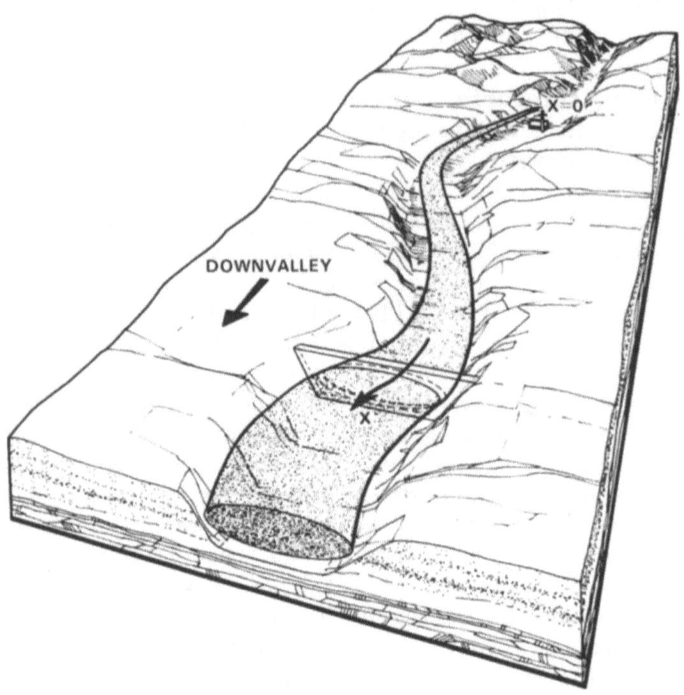

Figure 2. Illustration of the nocturnal down-valley transport and diffusion of a pollutant plume as related to terrain-following coordinate system.

speed u_v. The virtual wind speed is a function of the ratio of the volume fluxes across the source and receptor cross sections. The value of u_v is substituted for u_s (actual wind speed) in the denominator of the Gaussian plume equation to calculate pollutant concentrations at the receptor cross section. To avoid physically unrealistic results, u_v must always be equal to or greater than u_s.

VALMET Daytime Component

The daytime simulation uses numerical techniques that simulate the fumigation of the nocturnal plume onto the valley floor and sidewalls as a convective boundary layer grows upward from the heated valley surfaces and as subsiding motions occur over the valley center after sunrise. This daytime portion of VALMET begins with solar radiation calculations, parameterization of the surface energy budget, and estimation of the temporal variation of sensible heat flux which destroys the temperature inversion after sunrise. The valley energy budget model of Whiteman and McKee (1982) is invoked to simulate convective boundary layer (CBL) growth and inversion descent. The nighttime Gaussian plume on a valley cross section is used as the initial condition from which the numerical model calculates concentrations in grid elements that are fixed to the valley floor and sidewalls (Figure 3). The depth of these vertically growing grid elements is identified with the CBL height. Calculations are made of concentrations within the grid elements taking account of fumigation of the Gaussian plume into the elements and upslope advection from grid element to grid element. Pollutants are mixed uniformly within each grid element, roughly simulating atmospheric diffusion processes in the convective boundary layer.

This daytime model is driven by sensible heat flux, estimated as a fraction of the solar radiation using a highly parameterized surface energy budget. Equations were derived to determine extraterrestrial solar flux at any site and for any given time. The sensible heat flux was parameterized as a constant fraction of the time-dependent extraterrestrial solar flux. The effects of such factors as snow cover, soil moisture, cloud cover, or surface albedo are not explicitly included in the daytime model, but can be incorporated through an expanded energy budget module, using as a basis the extensive work already conducted on these topics by other investigators.

The daytime component of VALMET uses a numerical technique to simulate time-varying pollutant concentrations in the grid elements fixed to the valley floor and sidewalls. Grid elements are arrayed from the valley center across the valley floor and up the sidewall, with grid element height representing the height of the CBL, assumed not to vary from grid element to grid element. At sunrise, the model is initiated with a shallow CBL height and an inversion top height obtained from observations in the modeled valley.

The atmospheric energy budget approach used by Whiteman and McKee (1982) is capable of partitioning energy between these two different processes to produce inversion destruction solely by CBL growth, solely by inversion descent (assuming a non-growing CBL is present initially in a simulation), or by a combination of the two processes. The partitioning is controlled by a single parameter defined as the fraction of sensible heat flux going to CBL growth. The remaining fraction is assumed to be responsible for mass transport up the sidewalls, which results in inversion descent. The proper partitioning of energy in a particular valley can be determined by fitting the thermodynamic model to observations from the valley.

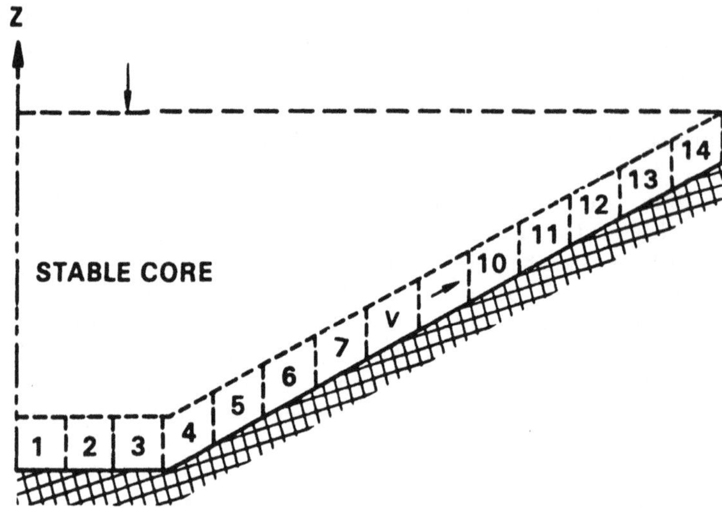

Figure 3. Cross section of the valley floor and sidewalls illustrating the grid elements whose heights correspond to the CBL height. The rate of sinking of the inversion top (vertical arrow) is related to the upslope transport of mass (diagonal arrow) by the equation of mass continuity.

Daytime pollutant concentrations are computed by treating the Gaussian plume within the cross section as being "frozen" within the inversion. After sunrise, concentrations within individual grid elements change as the grid elements grow upwards into the frozen plume, as the inversion sinks causing pollutants to sink into the tops of the grid elements, and as upslope flows develop within the CBL. These slope flows arise as air parcels are heated by sensible heat flux over the sidewalls and rise up the slope. The speeds of the slope flows are calculated under a continuity of mass constraint within the cross section below the (sinking) inversion top. That is, sinking of the top of the inversion with time implies that mass is removed from the cross section below the inversion top level.

Pollution concentrations are calculated for each time step for each grid element following the general outlines of a numerical method described by Whiteman and McKee (1977). Calculations at each time step are performed sequentially for all of the grid elements, beginning with the grid element at the valley center and progressing up the valley sidewall. Calculations within each grid element are made assuming pollutant mass conservation. The pollutant mass sinking into the top of each growing grid element and the pollutant mass advected into and out of each grid element is computed using the velocity calculations at grid element septa as determined using a total air mass conservation constraint.

VALMET uses a continuity equation for air of constant density to calculate wind velocities at the boundaries or septa of the grid elements in the slope flow layer. Assuming the wind velocity on the left septum at the valley center is zero (i.e., there is no mass transferred across the valley center within the CBL), a sinking inversion will produce a wind speed increase from the left septum to the right septum of the grid element (Figure 3). The velocity normal to a given septum of any grid element can thus be determined by summing the velocity changes calculated across each grid element beginning at the valley center and ending at the septum of interest.

A modification to the pollutant calculation procedure is necessary for individual grid elements on the sidewalls when the top of the inversion descends below the level of the grid element. The grid element depth, defined as the depth of the CBL, suddenly increases when this occurs, necessitating a rapid decrease in grid concentrations. This is handled in VALMET by simply allowing the last concentration calculated in the grid element to decay exponentially with time after the inversion top descends below the grid element. Because observational evidence is lacking, the time constant of the exponential decay is chosen arbitrarily so that the concentration decreases by 2% at each subsequent time step.

The criteria for maintaining numerical stability constrain the model user to maintain a certain relationship between the model time step and the grid element crosswind length, depending on the upslope wind speeds encountered in the simulation. The maximum upslope wind speed simulation in the model, when multiplied by the time step length, must be less than the grid element length to maintain computational stability. These constraints are performed automatically within VALMET.

Pollutant concentrations may be calculated for every time step for every grid element. Rather than storing each of these concentrations, which would require a large amount of memory, 5-min average concentrations are stored and constitute one of the major VALMET outputs. Additional outputs are the maximum 1- and 3-hour average concentrations and their times of occurrence in each of the grid elements. These calculations are made from the basic 5-min average concentration array for each grid element by means of a moving average method.

The results of a model run are written to two output disk files. The first output file is a formatted file which contains the results of the model run. After the model is run, the contents of this file may be printed to obtain a summary record of the model run, including model input parameters and outputs. A second file is also automatically generated with every computer run. This formatted file contains time series of pollutant concentrations, temperature inversion depths, and convective boundary layer depths, as well as many of the basic parameters used in the model run. This file is created for the user who wishes to develop his own programs to plot the results of a model run.

Following is a sample simulation by Whiteman and Allwine (1983b) for a 600 m wide valley with 15 degree sidewalls. The valley inversion at sunrise is assumed to be 500 m deep with an average potential temperature gradient of 25°K per km. The pollution source emits a non-reactive pollutant at the rate of 1 g/s at a height of 250 m after plume rise into a 4 m/s down-valley wind on September 21 at latitude 40°N and longitude 105°W. The model time step is 10 s, and the grid elements are 100 m long. The sensible heat flux is 24% of the solar heat flux, and 15% of the sensible heat flux is used to increase CBL height. The remaining 85% is used to transport mass up the sidewalls. The simulation for a downvalley distance of 10 km from the source is shown in Figure 4 as concentration plots for grid elements 03, 06, 09, 12, and 15 between 0500 and 1100 LST.

Low and steady nocturnal concentrations are initially observed at the 10-km down-valley distance in all grid elements since diffusion from the centerline is not yet sufficient to produce large concentrations on the valley floor or sidewalls. The plume centerline concentration 250 m above the valley floor is 1.19 $\mu g/m^3$. Only 0.1% of the plume has diffused out the top of the valley. After sunrise (just before 0600), fumigation and upslope advection begin, raising concentrations on the sidewalls and valley floor. The highest concentrations due to fumigation occur over the center of the valley floor just before 0900. Here, the highest one-hour average concentration of 0.70 $\mu g/m^3$ occurs. Concentrations in the highest grid elements on the sidewalls are relatively low, although upslope flows carry up the higher concentrations observed first on the lower slopes. Diffusion in the growing CBL occurs during this transport. Concentrations drop to zero rapidly in the sidewall grid elements as the top of the temperature inversion sinks below their elevations

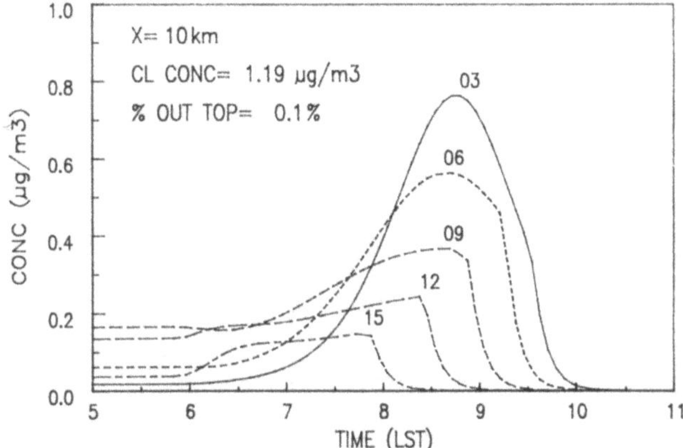

Figure 4. VALMET pollutant concentration versus time for selected grid elements. In the legend, X is down-valley distance, CL CONC is centerline concentration, and % OUT TOP is the fraction of mass diffusing out the top of the valley inversion during the plume's nocturnal transport.

and the mixing layer suddenly becomes unlimited. The concentration in Box 03, the last valley floor grid element, drops to zero at the time of temperature inversion destruction (ca. 0940).

MELSAR MODEL DEVELOPMENT

A preliminary version of a puff trajectory air quality model has been developed by PNL to simulate air pollutant transport and diffusion over a specific region 500 km in the east-west direction by 450 km in the north-south direction. This region covers the mountainous areas of western Colorado, eastern Utah and southern Wyoming as shown in Figure 5. The modeling system called MELSAR (MEsoscale Location Specific Air Resources model) currently consists of four programming components: a terrain processor; a meteorological data processor; a pollutant transport, diffusion and concentration computing module; and a post-processor. MELSAR is to be used principally for estimating short-term (3-hour and 24-hour) SO_2 and Total Suspended

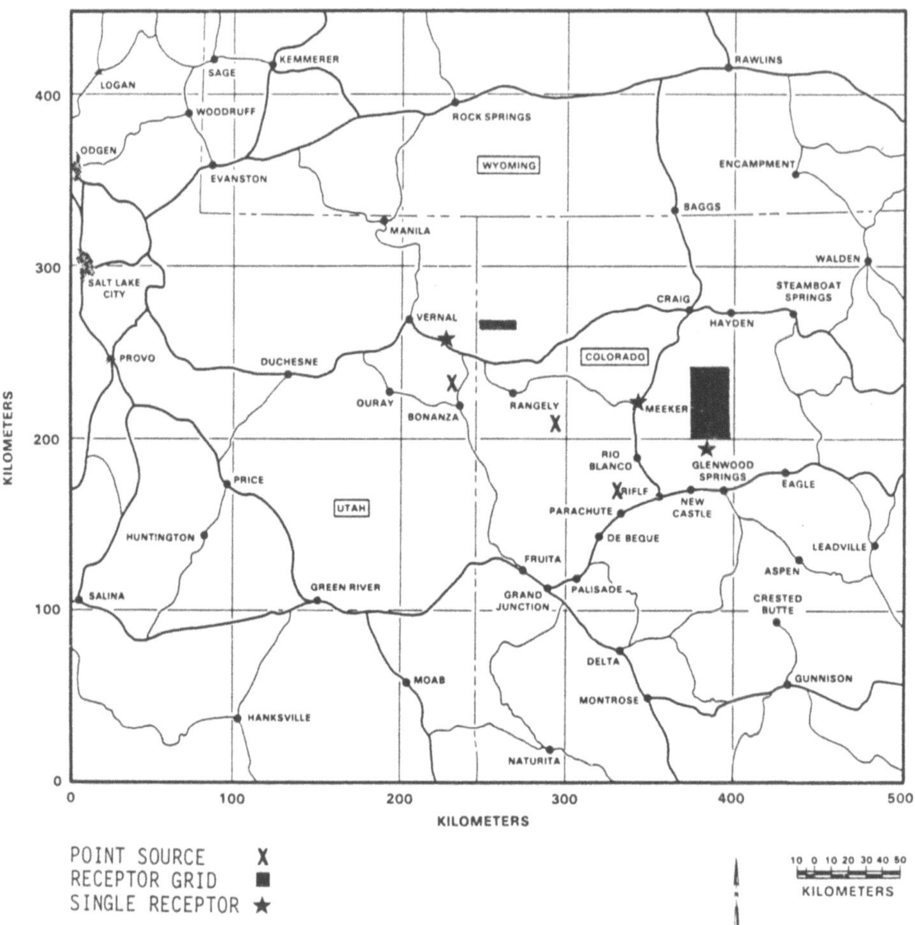

Figure 5. MELSAR air quality model domain with hypothetical locations of point sources and receptors for illustration.

Particulate (TSP) average air concentrations in determining compliance with PSD regulations.

The model is currently configured to compute pollutant concentrations from up to five point sources on up to three user-specified receptor grids and at up to ten user-specified individual receptor points. Each receptor grid can contain a maximum of 25 by 25 (625) receptors. Highest and second highest average pollutant concentrations are computed at each receptor for the sum of all sources and the contribution of each source to the sum. The averages are moving averages and up to three averaging periods can be computed ranging from one hour to 24 hours. The model operates on the whole 500 km by 450 km region or on any sub-region within the domain. The user must specify the sub-region desired at the start of the model run. The decision will be based on the desired resolution of the concentration fields and the amount of input data available.

Hourly, mass-consistent, three-dimensional wind fields, gridded mixing heights, and gridded stabilities are used to drive the model. The flow module currently being used in MELSAR is a three-dimensional, mass-consistent flow model using a terrain-following coordinate system. It was developed by Drake and Huang (1980) and has been described by Drake et al. (1981). Flow channeling around major terrain obstacles under stable atmospheric conditions is treated through consideration of a spatially and temporally dependent Froude number, as described by Whiteman and Allwine (1982). Gridded mixing heights are computed from upper air and surface wind and temperature observations following the technique described by Benkley and Schulman (1979). The gridded stabilities are computed using the method in the model MESOPUFF (Benkley and Bass, 1979). An important feature of the meteorological data processor of MELSAR is that it is configured to receive input data from any location in the domain and the input data do not have to be uniformly spaced in time. The data can come in at any frequency and for any time period of the simulation.

MELSAR is a puff trajectory model in which pollutant concentrations are described in a Gaussian fashion about the puff center of mass, although the distribution is modified by reflections from the ground and inversion surface. Current techniques in puff modeling are used. These include the method proposed by Zannetti (1981) for treating highly non-stationary, inhomogeneous flow conditions such as one would encounter in complex terrain. This method also provides the flexibility for choosing low puff release and sampling rates while simulating a continuous plume. The method is currently configured to operate at a release and sampling rate varying from six puffs per hour to one puff per hour. A release rate of one puff per hour allows an efficient computation for long distance transport while continuing to give adequate plume coverage near the source (less than 10 km, depending on the wind speed).

The dispersion coefficients can be computed in the model using any of three different methods: (1) the formulation in MESOPUFF (Benkley and Bass, 1979); (2) the open country formulation by Briggs (1973); and (3) the method of accounting for terrain enhanced diffusion (MaCready et al., 1974). Spatial and temporal variations in pollutant diffusion as well as the effects of wind shear on diffusion are treated in MELSAR.

This preliminary version of MELSAR consists of four modules named TERRAIN, MET, POLUT, and POLPRC. TERRAIN uses the base terrain file for the region of interest and produces disk files of spatially averaged terrain statistics (e.g., heights, slopes, roughness) for use in the MET and POLUT programs. MET produces hourly winds, gridded mixing heights, and gridded stabilities using upper air and surface winds and temperature observations. POLUT uses the output files from MET and produces pollutant concentrations on a user-specified receptor network for each time step of a simulation. The concentration fields are output to a disk file that is then used by POLPRC for computing moving-average pollutant concentrations for up to three averaging times. The outputs from POLPRC are tables of highest and second highest pollutant concentrations at each receptor. Tables are output for the sum of all sources along with the resulting contribution of each source to the sum. Also, highest and second highest concentration tables are output for each receptor grid and set of individual receptors, considering each source individually.

The principal module yet to be included in MELSAR is the treatment of pollutant transport and diffusion during the period of coupling of valley winds with above-ridgetop winds. The approach will be to release pollution trapped in valley inversions to above ridgetops using the formulation for valley inversion breakup contained in VALMET.

GRAMA FIELD EXPERIMENTS

Two GRAMA field experiments have been conducted in the Green River oil shale region of western Colorado. The first was conducted in a shallow gulch in the Piceance Basin during a ten-day period in August 1980. The data and analyses from this experiment were described by Whiteman et al. (1981) and Laulainen et al. (1981). The second experiment was conducted in cooperation with the U.S. Department of Energy's Atmospheric Studies in Complex Terrain (ASCOT) program during a two-week period in July and August 1982. The GRAMA component of the experiment involved multiple tracer experiments in the deep, well-defined Brush Creek Valley. SF_6 was used to generate an elevated continuous plume, and experiments were designed to determine how SF_6 concentrations would change on the valley floor and sidewalls down-valley from the release site during the post-sunrise temperature inversion breakup period. Analysis of the data from these experiments is continuing and will be reported in future papers. These data will be used to conduct initial evaluations of the VALMET and MELSAR models.

SUMMARY

The model development performed to date in the GRAMA program is leading toward a modeling system with two major components for analyzing the impacts of the oil shale industry with respect to PSD increments over pristine wilderness areas and ambient air quality standards for criteria pollutants. Recognizing that many aspects of complex terrain meteorology and air pollution transport are not well understood, the GRAMA models are being composed in a highly modular fashion so that future upgrading will be simplified. Several of the modules are designed to serve only when site-specific data are not available for direct input.

The first component, VALMET, is an air quality simulation model that predicts concentrations of a nonreactive pollutant arising from an elevated continuous point source located within a well-defined mountain valley. VALMET simulates conditions in which pollutants are carried in locally developed valley circulations that are decoupled from the prevailing regional flow.

The second component, MELSAR, is a mesoscale Lagrangian puff model for predicting pollutant concentrations within a 500 km by 450 km domain encompassing the oil shale development area. MELSAR is to be used principally for estimating short-term SO_2 and TSP average concentrations in determining compliance with PSD regulations. Although a preliminary version of VALMET exists for evaluation, MELSAR is still in the developmental stage. The principal module yet to be included in MELSAR is the treatment of pollutant transport and diffusion during periods of coupling of the regional flow with the valley winds, using the VALMET formulation for valley inversion breakup.

REFERENCES

Benkley, C. W. and A. Bass, 1979: "Development of Mesoscale Air Quality Simulation Models. Volume 3 - User's Guide to MESOPUFF Model". Environmental Research & Technology, Inc., Concord, MA, 124 pp.

Benkley, C. W. and L. L. Schulman, 1979: Estimating Hourly Mixing Depths from Historical Meteorological Data. J. Appl. Meteor., 18, 772-780.

Briggs, G. A., 1973: Diffusion Estimates for Small Emissions. ATDL Contribution File No. 79, Atmospheric Turbulence and Diffusion Laboratory, Oak Ridge, TN.

Drake, R. L. and C. H. Huang, 1980: Mass-Consistent Interpolated Wind Fields for Complex Terrain. Presented at Symposium on Intermediate Range Atmospheric Transport Processes and Technology Assessment, Gatlinburg, TN.

Drake, R. L., C. H. Huang, and W. E. Davis, 1981: "Green River Ambient Model Assessment Program - Progress Report for the Regional and Mesoscale Flow Modeling Components". PNL-3988. Pacific Northwest Laboratory, Richland, WA, 123 pp.

Laulainen, N. S., C. D. Whiteman, W. E. Davis, and J. M. Thorp, 1981: Mixing Layer Growth and Background Air Quality Measurements Over the Colorado Oil Shale Area. Preprints, Second Conference on Mountain Meteorology, American Meteorological Society, Boston, MA, 165-172.

MacCready, P. B., Jr., L. B. Baboolal, and P. B. S. Lissaman, 1974: Diffusion and Turbulence Aloft over Complex Terrain. Presented at AMS Symposium on Atmospheric Diffusion and Air Pollution, Santa Barbara, CA.

Whiteman, C. D. and K. J. Allwine, 1982: "Green River Ambient Model Assessment Program - FY-1982 Progress Report". PNL-4250. Pacific Northwest Laboratory, Richland, WA, 187 pp.

Whiteman, C. D. and K. J. Allwine, 1983a: "VALMET-A Valley Air Pollution Model". PNL-4728. Pacific Northwest Laboratory, Richland, WA, 163 pp.

Whiteman, C. D. and K. J. Allwine, 1983b: Time-Dependent Model of Pollutant Transport and Diffusion in Mountain Valleys. Extended Abstracts, Sixth Symposium on Turbulence and Diffusion, American Meteorological Society, Boston, MA, 173-176.

Whiteman, C. D., N. S. Laulainen, G. A. Sehmel, and J. M. Thorp, 1981: "Green River Air Quality Model Development: Meteorological Data - August 1980 Field Study in the Piceance Creek Basin Oil Shale Resources Area". EPA-600/S7-82-047. EPA, Office of Research and Development, Research Triangle Park, NC. 172 pp.

Whiteman, C. D. and T. B. McKee, 1977: Air Pollution Implications of Inversion Descent in Mountain Valleys. Atmos. Environ., 12, 2151-2158.

Whiteman, C. D. and T. B. McKee, 1982: Breakup of Temperature Inversions in Deep Mountain Valleys: Part II - Thermodynamic Model. J. Appl. Meteor., 21, 290-302.

Zannetti, P., 1981: An Improved Puff Algorithm for Plume Dispersion Simulation. J. Appl. Meteor., 20, 1203-1211.

DISCUSSION

H. HASENJAGER

You explained well the decay of the stable flow in that valley. Could you also give some explanation on the build-up of that flow, that is what is happening during the early night hours ?

C.D. WHITEMAN

The temperature inversion builds up rapidly in the late afternoon and evening after valley surfaces become shaded. Downslope flows are the first to form. Down-valley flows begin at the valley floor and rapidly increase in depth and strength. A large number of observations of the inversion buildup were collected by the U.S. DOE ASCOT participants. Analyses of these data should be available soon.

A. VENKATRAM

Why doesn't the temperature gradient intensify above the mixed layer as the stable layer sinks ? The vertical velocity cannot be constant through the stable layer.

C.D. WHITEMAN

Stretching or shrinking of a column of air will change the potential temperature gradient in the column. In actual soundings of the stable core we see intermittent zones in which the temperature gradients vary from sounding to sounding. Some of these variations are representative of vertical velocity oscillations. Others may be due to horizontal turbulent transport of heat or other processes. In fact, observations show a modest net destabilization of the stable core during the inversion breakup period. Consistent with the observations, it is a useful first approximation for modeling to consider the stable core temperature gradient as remaining constant during the inversion break up period.

A. VENKATRAM

Why don't you have a temperature jump across the top of the mixed layer ?

C.D. WHITEMAN

The valley atmosphere can respond to temperature differences that develop between the mixed layer and the overlying stable core in a way that is not possible over flat terrain. Namely, the slope flows are available and are in fact, driven by such temperature contrasts.

APPLICATION OF A MESOSCALE LAGRANGIAN PUFF-MODEL TO THE MEASUREMENTS OF SO2-POLLUTION TRANSPORTS OVER BELGIUM

G. Dumont, F. Vervliet, E. De Saeger and G. Verduyn

Institute for Hygiene and Epidemiology
J. Wytsmanstraat 14
B-1050 Brussels (Belgium)

INTRODUCTION

Within the framework of the BeNeLux cooperation[x], it was decided to use a same puff-model for interpretation of data from stationary air quality networks as well as for the data from dynamic remote sensing measurements. The model calculates the effects of local and mesoscale SO2-transports on the air quality in the BeNeLux countries. Because same measuring techniques and a same interpretation scheme are now used within the BeNeLux, the analysis of individual country contribution to the general SO2-patterns may be done in a uniform way.

The puff-model was developed at the R.I.V. (Bilthoven, The Netherlands) and presented earlier by van Egmond and Kesseboom (1983a). The measured mesoscale SO2-patterns in the BeNeLux during 1980 were analysed by the model in a series of case-studies (van Egmond et al.,1983b). Recently, routine Belgian meteorological measurements have been included, the SO2-emission inventory of the model area was revised in detail and the earth surface radiation budget was refined.

In this paper, these recent developments are emphasized. The SO2-transports over Belgium by easterly winds are studied by comparing airplane measurements to model calculations. An example of a total deposition budget is discussed. This study is a part of the general study of 'SO2-transports over the Flanders by remote sensing' which was ordered by the Flemish Ministery of Environment.

*The BeNeLux workgroup 'Lucht-netten' subcommission 'Laboratoria'

MODEL DESCRIPTION

The model area has been fixed as to incorporate the BeNelux-countries and major SO2-emitting surrounding areas (figure 1). The model is based on the transport of SO2-emissions from inside this 480 x 540 km2 area.

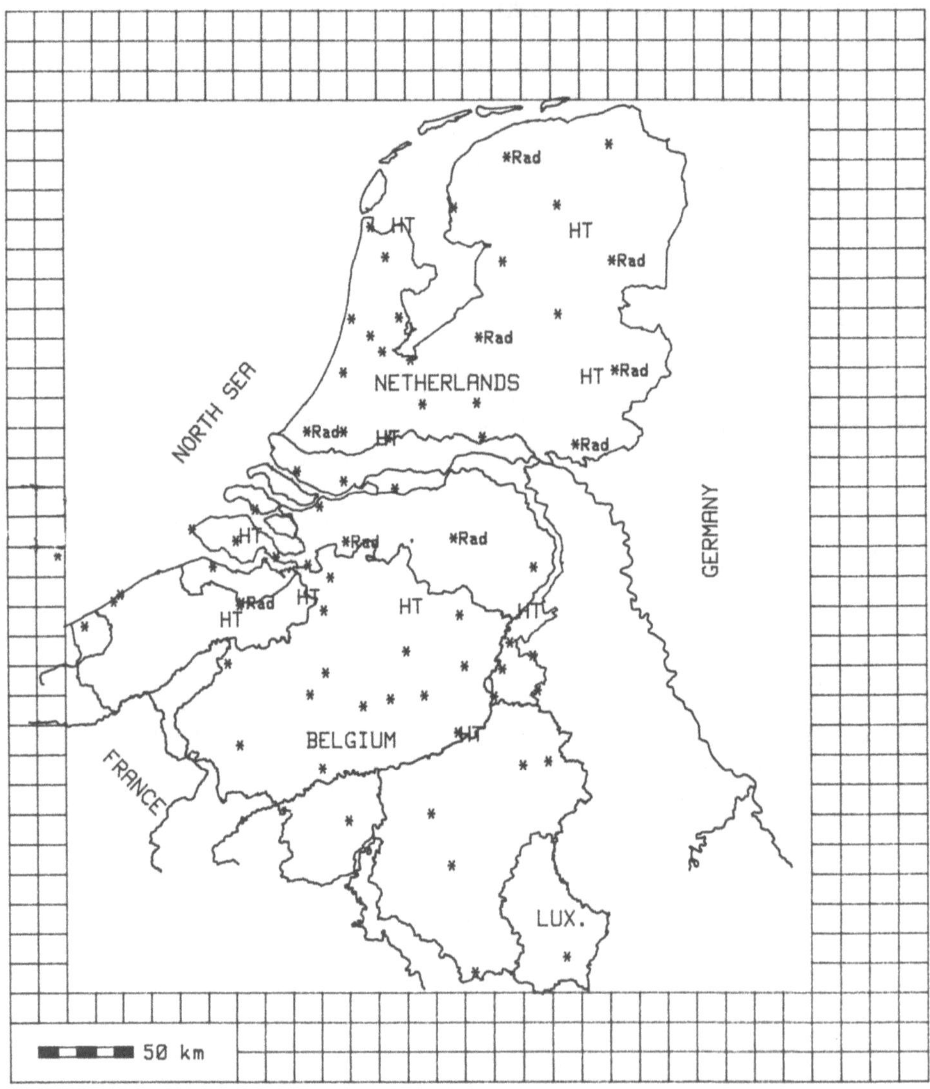

Figure 1 : Model area and locations of meteorological measurements. * shows location of 10 m-level windvector measurements.HT indicates high tower windvector measurements.Rad locates incoming solar radiation measurements.

Model outlines

The emissions are represented as an ensemble of Gaussian puffs with concentration distribution given by

$$C(r,z) = f(z) \frac{M}{2\pi\sigma_r^2} \exp(-r^2/2\sigma_r^2), \quad (1)$$

where r is the distance to the puff center
σ_r the horizontal extend of the puff
f(z) the vertical distribution function (see below)
M the pollution mass stored in the puff. $M = Q.\Delta t$ with Q the emission rate of the source and Δt the model time increment.

Advection of the puffs is achieved according to two basic windfields which are constructed every hour from 10m level and high tower windvector measurements. The lowest windfield is located at 70m altitude considered as a main transport height for pollution in the mixing layer. The second one is at the actual center height of the reservoir layer (see vertical stratification). A 15 by 15 km grid is used for windfield construction. The model time increment Δ t is chosen in a way to obtain 10 km interpuff distance.

Horizontal diffusion is modelled in a Lagrangian way by increasing σ_r according to

$$\sigma^2_{r,t+\Delta t} = \sigma^2_{r,t} + 2K_H\Delta t + 2\,K_s\Delta t$$

The time dependent horizontal diffusivity K_H is given by

$$K_H(t) = v_m^2 \int_o^t R_L(t')dt'$$

where v_m is the hourly mean crosswind component derived from windspeed and winddirection standard deviation data. $R_L(t')$ is the Lagrangian correlation function assumed to be a negative exponential $\exp(-t'/t_L)$ with time scale $t_L = 40$ minutes. The effect of windshear is described as a supplementary horizontal diffusivity K_S approximated by

$$K_S(t) = 0.5\sigma_z^2\, S^2 t,$$

where σ_z is the vertical extend of the puff and S the measured windshear in rad/m multiplied by the windspeed.

Vertical stratification is obtained by considering three layers : a pollution reservoir layer on top of a mixing layer of which the lowest part is treated as the surface layer. Mixing depth is derived from acoustic soundings (acdars). The depth is assumed to be constant at night and to increase linearly during daytime untill acdar measurements indicate the reset of the night value (in the late afternoon). At the time of its emission, the puff mass is partitioned into the reservoir and the mixing layer according to the effective puff height and the actual value of the mixing depth. When mixing depth increases pollution mass from the reservoir layer is transferred to the mixing layer (fumigation).

As the part of the puff mass stored in the reservoir layer does not contribute to the ground level concentrations, the vertical distribution in that layer is taken constant within certain limits. This means that f(z) from (1) for the reservoir layer is given by

$$f(z)_{res} = 1/h \quad \text{for } 0.5\,h < z < 1.5\,h \text{ (h:effective puff height)}$$

$$f(z)_{res} = 0 \quad \text{else}$$

In the mixing layer and close to the source the vertical distribution is Gaussian. Hence, ground level concentration is obtained by introducing f (z=0) in (1)

$$f(z=0) = \frac{2}{\sqrt{2\pi}\sigma_z}\{\exp(-\frac{1}{2}\frac{h^2}{\sigma_z^2})+\exp(-\frac{1}{2}\frac{(2H-h)^2}{\sigma_z^2})+\exp(-\frac{1}{2}\frac{(2H+h)^2}{\sigma_z^2})\}$$

with σ_z the vertical dispersion coefficient determined according to the Pasquill-scheme (Pasquill-class derived from surface layer parameters)
H the mixing height
h the puff effective height.

At larger distances, when $2\sigma_z$ becomes comparable to mixing height, the Gaussian vertical distribution smoothly turns to a homogeneous one : $f(z)_{mix} = 1/H$.

Dry deposition at the earth surface causes a z-independent flux F trough the surface layer which is described by

$$F = v_g(z).C(z)$$

with $v_g(z)$ the deposition velocity at depth z
C(z) the concentration at z

Because the concentration at the top of the surface layer is known (i.e. the mixing layer concentration), the concentration at 4m can be found when deposition velocities at both depths are known. These deposition velocities are calculated from the surface layer parameters L and u_x (Obukhov length and friction velocity) (see below).

Chemical transformation of SO_2 during transport is splitted into a radiation independent part estimated as a 1 % decay per hour and a radiation dependent part (Gillani, 1978) :

$$\frac{dC}{dt} = -(c_2\ K^+)C$$

with $c_2 = 0.032\ h^{-1}\ m^2\ kW^{-1}$

K^+ the incoming solar radiation (kWm^{-2})

Input data

It was agreed to use only routinely available input data, not only to avoid long pre-run time for gathering the necessary information but also in view of a future on-line application of the model.

At 42 10m-level stations of the Dutch National Air Pollution Monitoring Network (see fig. 1) horizontal windspeed, winddirection and standard deviation of the winddirection are registered on a hourly base. At 26 locations in Belgium and Luxemburg synoptic measurements of windspeed and winddirection are performed every hour. The synoptic stations belong to the Royal Meteorological Institute network (KMI) and to the airport network of the 'Regie der Luchtwegen'. These synoptic measurements were preferred to the

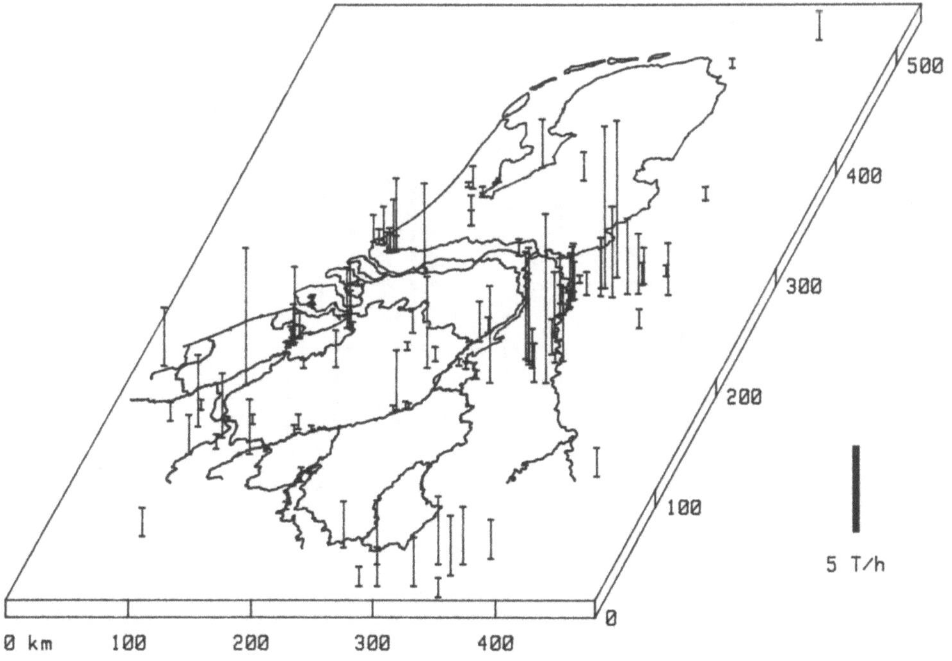

Figure 2 : SO_2-point sources (255 ton/hour).

meteorological measurements of the Belgian Automatic Air Monitoring Network (concentrated in five agglomerations) because the latter are influenced by city turbulences and because their spatial pile-up is not suited for representattive construction of mesoscale windfields. At 6 TV-towers in the Netherlands at levels between 150 and 300 m, windspeed, winddirection and standard deviation are measured. Four Belgian high towers yield windspeed and winddirection at levels ranging from 100 to 153 m. Total incoming solar radiation is measured at 9 Dutch locations. Acoustic soundings are performed continuously at Bilthoven (The Netherlands) and at Ukkel (Belgium). Temperature measurements and rainfall data are available in the synoptic networks.

In cooperation with the EIVR (Emission Inventory of the Flemish Region,Belgium) and the TNO (Apeldoorn, The Netherlands) 112 major point sources of SO2 over the model area have been selected (see fig 2). They are continuous industrial sources such as power plants, refineries, etc. Physical stack heights, volume flow rates and exhaust gas temperatures enter into the model. The SO2-emission rates of the BeNeLux point sources are the most recently updated annual emission figures of the Dutch and the Flemish emission inventories (no data are older than 1980). Emission rates of the German and French point sources are derived from data in the open li-

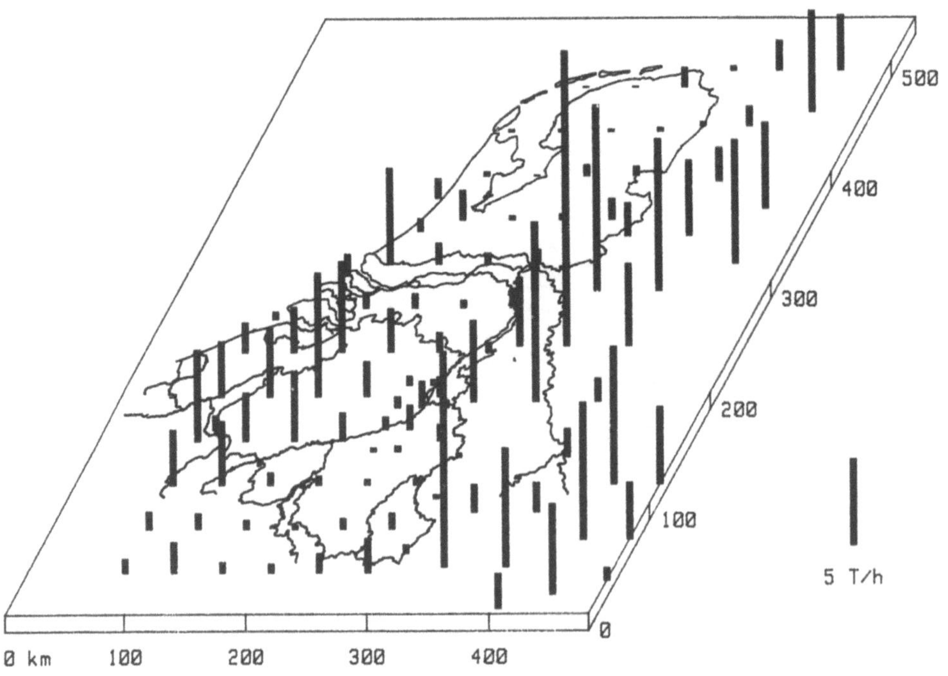

Figure 3 : SO_2-surface sources including emissions due to building heating for the month of february (290 ton/hour).

terature on energy production and industrial activities in those countries. Remaining industrial $SO2$-emissions are collected into 118 large circular surface sources with a diameter varying from 20 to 50 km. Their emission height was fixed at 30 m. Emissions from building heating were also collected in the same surface sources according to population density and national energy consumption data. Total annual emission figures due to the building heating were time-modulated over the months september till april. The sum of industrial surface sources and emissions due to building heating during the february month are presented in figure 3.

Surface layer parameters and dry deposition

The way the model calculates the dry deposition will be presented here in detail because in the following section of this paper, some budgets of total dry deposition rates over Belgium will be estimated.

The depostion velocity $v_g(z)$ at height z in the surface layer is calculated by

$$v_g(z) = (r_a(z) + r_s + r_c)^{-1}.$$

In the model, the canopy resistance r_c is taken space and time independent and estimated at 70 s/m (Fowler, 1978). The surface resistance r_s and the aerodynamic resistance $r_a(z)$ are calculated using the Monin-Obukhov similarity theory (Businger, 1973 and Weseley and Hicks,1977). The surface resistance is given by $2.6/(k.u_x)$ with k=0.35 the von Karman constant and u_x the friction velocity. The aerodynamic resistance $r_a(z)$ over a depth z is calculated by

$$r_a(z) = (\ln z/z_0 - \Phi_c)/(k.u_x)$$

with z_o being the roughness length, taken as a constant over the model area ($z_o = 0.25$ m) which is the roughness length for open fields with scattered trees and hedges (Wieringa,1980). The stability correction function Φ_c is given by

$$\Phi_c = \exp\{0.598 + 0.39 \ln(-z/L) - 0.09(\ln(-z/L))^2\} \quad \text{for } L<0 \text{(unstable)}$$

$$\Phi_c = -5z/L \quad \text{for } L>0 \text{(stable)}$$

where L is the Obukhov length.

The Obukhov length L and the friction velocity u_x are derived for every simulation hour from the spatial 10m windspeed average and from the averaged measured incoming solar radiation K^+. For low or zero incoming radiation levels (night time) the empirical relation of Venkatram (1980) is applied :

$$L = 1100\, u_x^2$$

and u_x is computed directly from the 10m windspeed, because for stable conditions the surface layer windprofile is represented by

$$u(z)= \frac{u_x}{k} (\ln(z/z_o) + 5z/L)$$

During day time the values of L and u_x are computed iteratively from 10m windspeed and the sensible heat flux H_o according to :

$$L = -91\ u_x^3/H_o \quad \text{(from the definition of the Obukhov length)}$$

and the surface layer wind profile :

$$u(z)= \frac{u_x}{k} (\ln(z/z_o) - \psi(z/L))$$

with the stability correction ψ (z/L) :

$$\psi(z/L) = 2 \ln((1+x)/2)+\ln((1+x^2)/2)-2\text{arctg } x + \pi/2$$

with $x = (1-15z/L)^{1/4}$ for unstable and neutral conditions, or

$$\psi(z/L) = -5 z/L \text{ for } L > 0 \text{ (stable conditions).}$$

The only parameter still to be derived is the sensible heat flux H_0. Following De Bruin and Holtslag (1982) it can be approximated by

$$H_o = \frac{(1-\alpha) + \gamma/s}{1 + \gamma/s} (Q^*-G)-\beta.\alpha$$

with α a surface moisture parameter (=0.65 for dry periods ; 0.95 normally)
- s the slope of the saturation-vapor-pressure temperature curve
- γ the psychometric constant
- Q^* the net radiation
- G the soil heat flux density (normally taken as 0.1 Q^*)
- β a constant (20 Wm^{-2})

In this expression, repartition of the net radiation over latent heat flux density and sensible density is accounted for. $\int$/s-values are tabulated versus temperature in Holtslag and van Ulden (1983). They vary from 2.01 at -5C° to 0.27 at 30 C°. Finally, the net radiation Q^* is derived from the measured incoming solar radiation K^+ as follows :

$$Q^* = \frac{(1-r)K^+ + c_1T^6 - \sigma T^4}{1 + c_3} \quad \text{(for open sky)} \qquad (2)$$

where K^+ is the measured incoming radiation (Wm^{-2})
- r is the albedo reflection coefficient (normally : 0.2)

c_1 5.31 10^{-13} $Wm^{-2}K^{-6}$
σ the Boltzmann constant : 5.67 10^{-8} $Wm^{-2}K^{-4}$
T the absolute temperature at 2 m
c_3 a surface heating coefficient (=0.12)

In (2) the terms c_1T^6 and $\sigma T4$ account for the incoming and outgoing infra-red radiation.

STUDY OF TRANSPORTS BY EASTERLY WINDS

As the model already satisfactorily described immission patterns in The Netherlands (spatial correlation in the order of 0.7, van Egmond and Kesseboom, 1983a), more attention is paid in this study to the model accuracy in simulating dynamic profiles obtained by mobile measurements.

A Dornier Skyservant airplane has been equipped with two Cospec V correlation spectrometers (Barringer, Canada) to measure SO2 and NO2 gasburden spectra (vertical integrated concentration). A Picoflux monitor (Hartmann & Braun,Germany) measures the SO2 concentration, a Nitrogen Oxydes Analyser (Monitor Labs,U.S.A.) registers the NO and NO2 concentrations and finally a nephelometer (MRI,U.S.A.) measures the dust concentration. The fleight altitude for all scans presented in this section was 500 ft. Average fleight speed during the scans was 230 km per hour. Only SO2-related measurements will be treated here.

For testing the model performance it was considered necessarily to collect a set of measurements under almost indentical atmospheric and emission conditons. Only in this way some statistical representativity can be attributed to the measurements. From 17th till 23rd of february 1983, a high pressure zone moved slowly from Denmark over Scotland to northern Germany, advecting air masses from ENE, ESE and SE directions over Belgium. During these period there was an open sky (0 oktas) and a windspeed at 10m level ranging from 2.2m/s to 7.1m/s. Acoustic soundings observed nightly mixing depths from 150m up to 250m with a fumigation start time at about 8.30h UT, a linear ascension of the mixing depth of 70 m/hour till 12.30 UT. Between 17h and 19h UT the nightly mixing layer was restored. A campaign of consecutive daily transport measurements over the Belgian territory was scheduled. A selection of those measurements is simulated by the puff-model.

Comparison procedure

For each modelled day a standard comparison drawing is given. In the upper frames, measured and calculated SO2-concentration le-

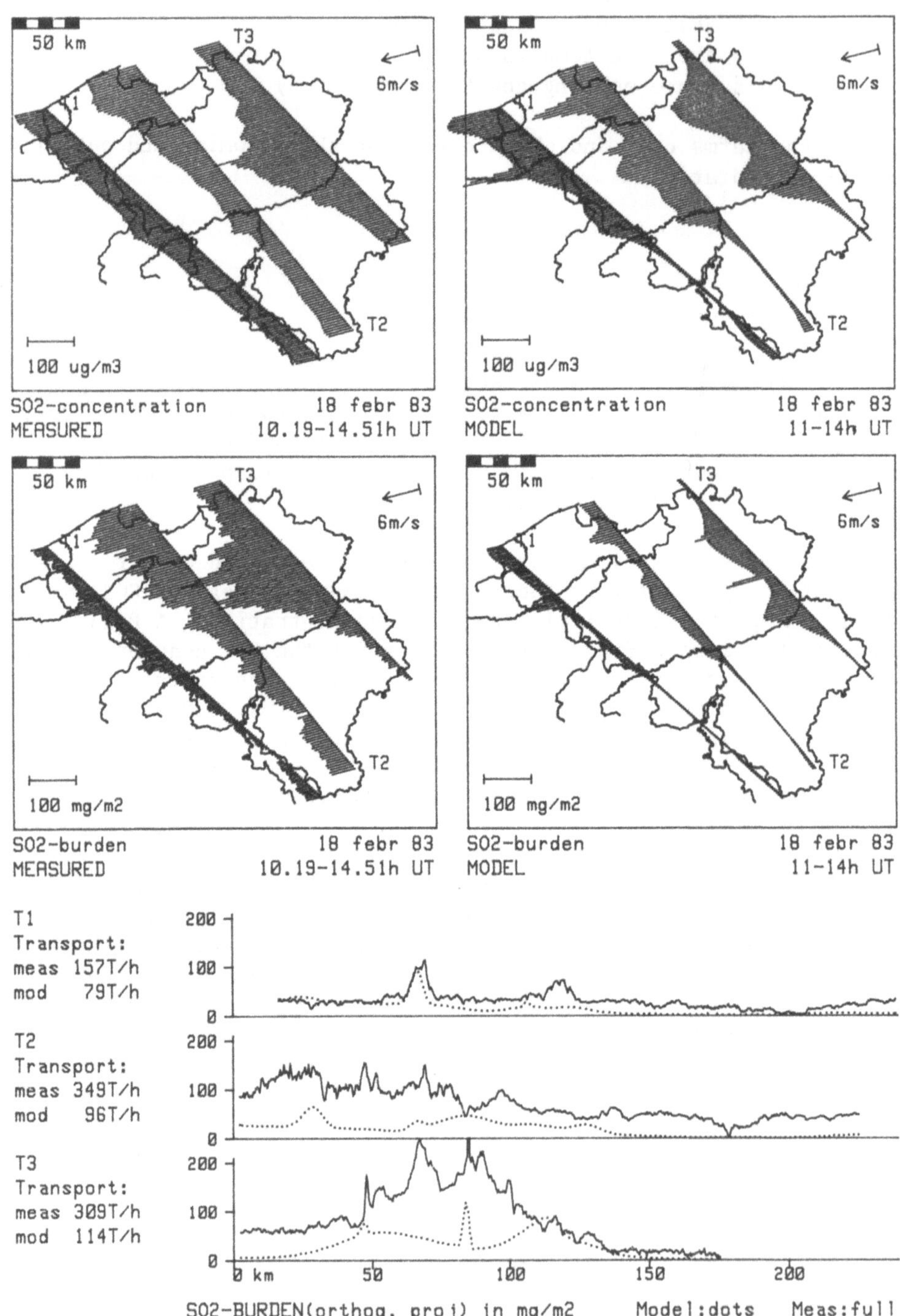

Fig. 4. Survey of measurements and model results for 18 febr 83

vels along the trajectories are represented by the length of bars which are tilted according to the winddirection indicated in the upper right corner. In the frames at the middle of the survey drawings, the same is done for SO2-gasburden levels. For each trajectory, the projected modelled and measured spectra are printed over each other. Projection is performed orthogonally to winddirection. Indicated transport figures are calculated as the road integral of the gasburden times the windspeed component normal to the trajectory. The trajectory averaged windspeed component applied to the measurements already results from the model. For a specific itinerary, the model indeed calculates an average windvector according to the local windfield vectors and weighted over the measuring trajectory according to the amount of pollution mass crossing this itinerary.

18 february 1983 (figure 4)

The mean levels of measured and modelled concentration profiles agree better than do the gasburden profile levels: total transports across the trajectories are undermodelled by a factor 2 to 3.5. Belgian power plants emissions are rather well simulated in the trajectories labelled T1 and T3. Industrial area emissions, such as the Antwerp area in T2 at km 30, are more pronounced in the model than they are in the measurements although its magnitude seems to be underestimated by the model. In T3, closest to the eastern border, a broad transport between km 50 and km 100 is not reproduced in the model.

However, it should be emphasized that in correlation spectrometry no absolute zero level can be assigned. A common practice is to assign the burden zero level to the lowest value measured over the trajectory. This may be justified when the trajectory exceeds by far the polluted air mass, but this almost never is the case for mesoscale profiles. By just omitting the doubtfull lowest level at km 175 of T2, the zero level of the whole spectrum would rise by 25 μg/m2 which represents a reduction of the measured transport by 120 Ton/hour.

22 february 1983 (figure 5)

Agreement between measured and modelled concentration and gasburden profiles is encouraging. The winddirection changed to ESE which made the transport patterns dominated mainly by Belgian emissions which are more accurately inventorised in the model than are the emissions beyond the eastern border. Again the Antwerp emissions look like underestimated by the model (upper side of T2). Differences in measured and calculated transports across the itineraries are smaller than in the cases of 18 february. The incoherency of these differences illustrates again the difficulty to assign zero levels in the measured profiles of gasburden.

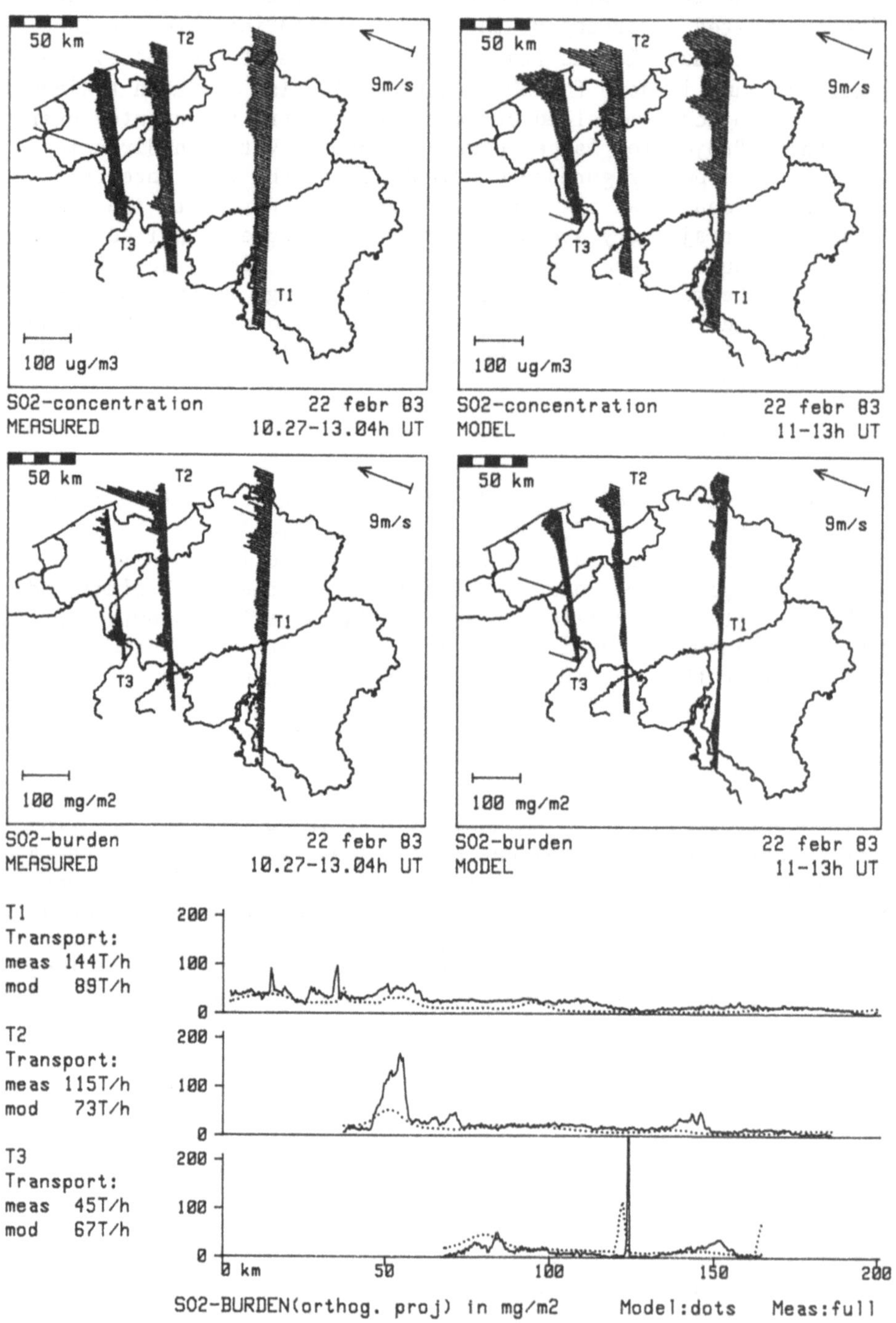

fig. 5. Survey of measurements and model results for 22 febr 83

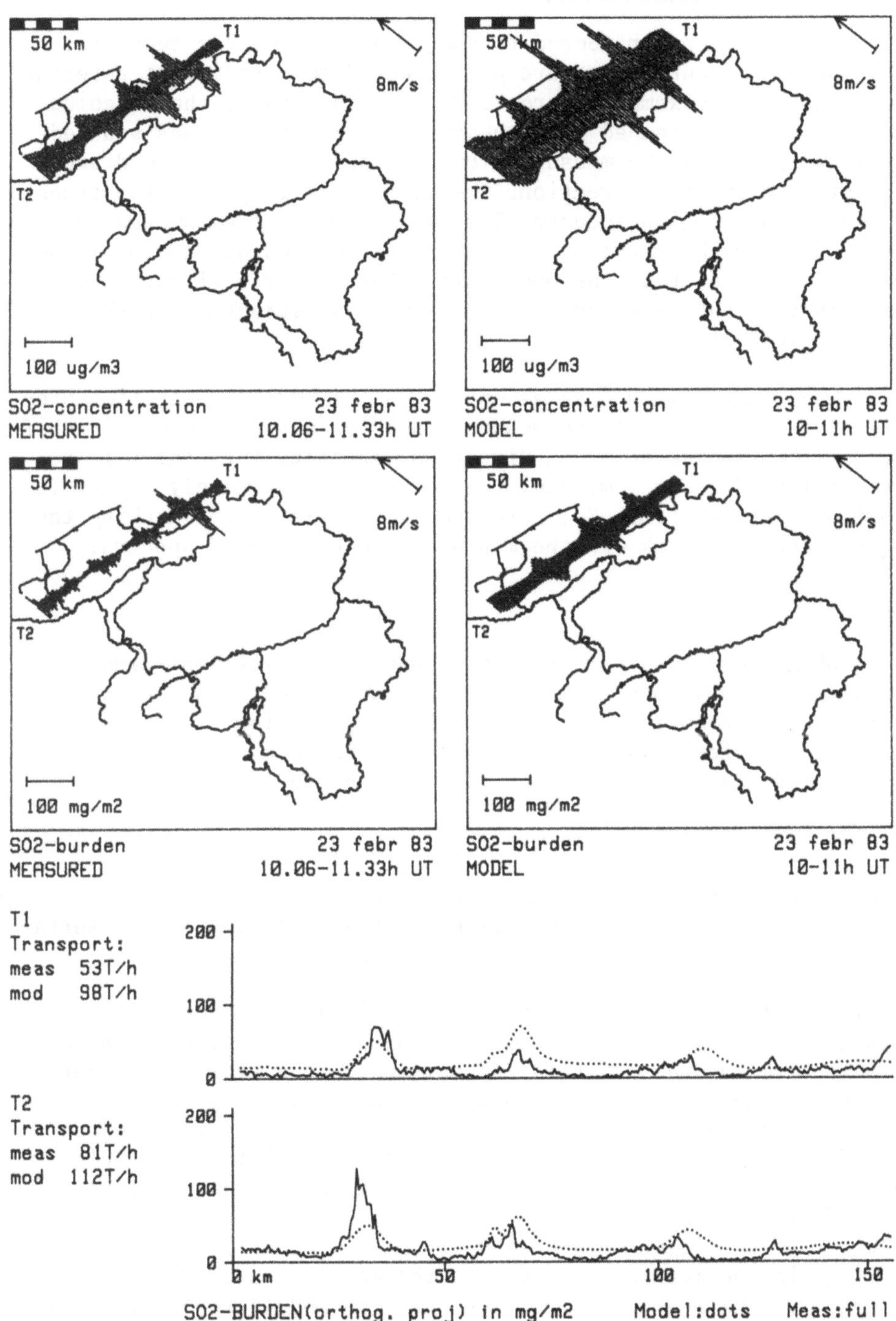

fig. 6. survey of measurements and model results for 23 febr 83

23 february 1983 (figure 6)

The wind further veered to the SE direction. For technical reasons, no measurements were performed of the transports entering Belgium at the Luxemburg border. On the other hand, the outgoing transports were scanned in both ways : trajectory T2 was scanned from SW to NE and the measured levels along this scan are drawn opposite to the winddirection. The profiles mainly show the transports of two Belgian industrial areas (Antwerp at km 30 and Ghent at km 65) and the plume of a power plant located at some 50 km upwind. It is seen that the model simulates fairly well both types of emissions and even accounts for some fine structure in the Ghent transport.

When paying some more attention to the SO2-concentration profiles, it is seen that the measured spectra show a systematic shift of some 4km in the scanning direction. In fact, this shift is a time delay of about 1 minute. Part of this delay surely is due to monitor response time. So, care should be taken when checking the exact overly of peaks in the measured and calculated spectra.

Deposition budget

When there is a good agreement between measurements and model results, a deposition analysis may be carried out. As the model constructs for every hour a 4m-level concentration field, an analogue deposition flux field can be established because for each model-gridpoint the deposition flux F is given by

$$F = v_g(4).C(4),$$

where C(4) is the concentration at 4m-level
$v_g(4)$ is the deposition velocity at 4m calculated from surface layer parameters as derived above.

By integrating this flux F over a specific area or country and by tracing the origin of the SO2-emissions during model calculation, a selective analysis of total deposition in the area can be performed. As an example, the deposition on 22 february 1983 over the Belgian territory is analysed. In table 1, a survey is given of the deposition related parameters and the deposition over Belgium :

- H_{mix} is the mixing height as derived from acdar measurements
- L is the Obukhov length
- u_x is the friction velocity
- ff_{70} is the mean windspeed over the model area at 70m height
- dd_{70} is the mean winddirection over the model area at 70m height
- $v_g(4)$ is the depostion velocity at 4m
- $D_{Belgium}$ gives the total deposition over Belgium.
- D_{import} gives the imported part of the deposition.

TABLE 1 : Deposition analysis and related model parameters for 22 february 1983

h (UT)	H_{mix} (m)	L (m)	u_x (m/s)	dd_{70} deg	ff_{70} m/s	v_g(4) cm/s	$D_{Belgium}$ Ton/h	D_{import} Ton/h
01	150	90	0.29	93	7.1	0.84	39	20
02	150	95	0.29	95	6.9	0.85	41	22
03	150	96	0.30	98	7.5	0.85	42	23
04	150	94	0.29	98	8.0	0.85	42	23
05	150	102	0.30	100	8.3	0.86	44	25
06	150	92	0.29	100	8.1	0.85	42	24
07	150	138	0.30	101	8.0	0.87	45	25
08	150	-262	0.37	101	7.5	0.94	53	29
09	200	-118	0.46	105	7.0	1.02	52	27
10	250	-116	0.53	111	6.8	1.06	49	25
11	300	-149	0.62	112	7.5	1.10	46	23
12	350	-172	0.67	110	8.0	1.12	42	21
13	400	-202	0.70	109	8.5	1.13	40	21
14	450	-230	0.70	107	8.5	1.12	38	21
15	500	-315	0.70	107	8.7	1.12	36	21
16	550	-500	0.64	106	8.4	1.10	35	20
17	600	∞	0.55	106	8.5	1.06	32	18
18	650	229	0.46	104	8.7	1.00	31	17
19	150	249	0.48	101	9.7	1.01	27	15
20	150	263	0.49	106	9.8	1.02	31	14
21	150	295	0.52	109	9.5	1.04	34	15
22	150	261	0.49	112	9.7	1.02	37	16
23	150	248	0.47	113	9.0	1.01	38	17
24	150	256	0.48	113	9.1	1.02	41	18

The daily averaged deposition for this day in Belgium is 40 ton/hour from which 21 ton/hour has been imported. The daily averaged deposition over the whole model area is 180 ton/hour. In figure 7, the deposition at 12h UT on 22 february 83 is shown. The height of the bars represents the total deposition in a 15 by 15 km square. For each square, the Belgian contribution is distinguished from the foreign-generated part. Pollution export figures may be calculated in the same way. It was not done for this particular day as the Belgian export exceeded the model area limits.

CONCLUSIONS

The introduction of the Belgian meteorological measurements enhanced the accuracy of the model results in the southern part of the model area when compared to mobile remote sensing measurements. Also because of the improved emission inventory, the SO2-transport patterns measured in Belgium are well reproduced specially when they

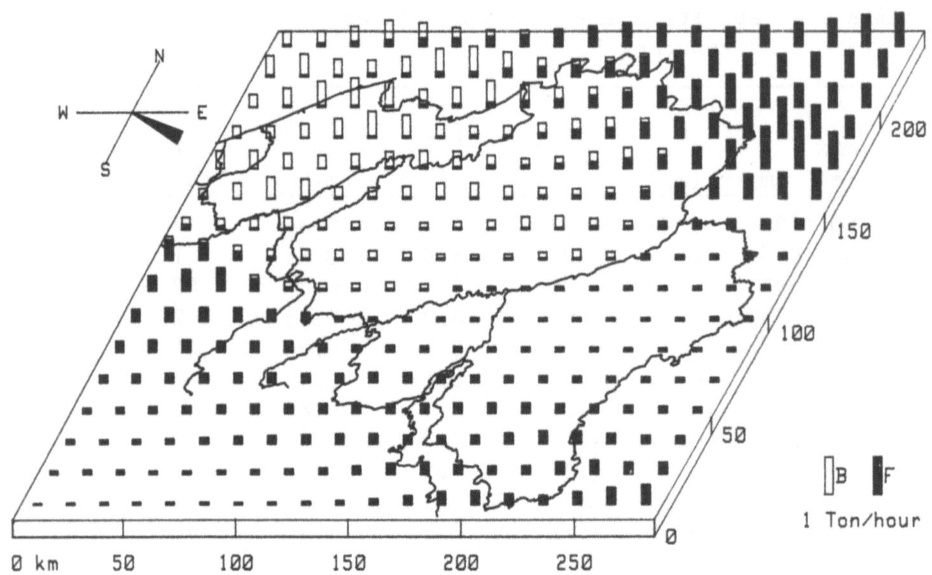

Figure 7 : SO2-DRY DEPOSITION at 12h UT on 22 february 1983. White bars represent deposition due to Belgian (B) sources. Dark bars indicate deposition from foreign (F) emissions.

are dominated by BeNeLux emissions. The refined description of the surface radiation budget allows the establishment of total deposition rates on the Belgian territory, inclusive an origin analysis.

It is seen that in mesoscale transport studies both modelling and measuring are indispensable because complementary tools. In this study the model learns from measurements that an important pollution flow entering Belgium by ENE winddirection is missed or badly located. It is also clear on several occasions that the Antwerp emissions are underestimed by the present inventory. On the other hand, the model is needed as a kind of a referee when zero levels in measured gasburden profiles are to be assigned or judged. It also provides some usefull parameters such as a mean windvector over a certain trajectory, weighted by the pollution mass traversing this trajectory.

As far as the further model developments are concerned, priority still should be given to studies of emissions outside the BeNeLux. Introducing more time modulation of the averaged emission figures may ameliorate the simulation of mobile measured profiles. Only then, some refinements such as spatial diversification of surface layer parameters and more sophistication of the vertical structure of pollution distribution will appear to full advantage.

The authors acknowledge the BeNeLux workgroup, the R.I.V.,TNO and EIVR for their cooperation to study the mesoscale transports over the BeNeLux countries.

REFERENCES

Businger, J.A.,1973, in "Workshop on Micrometeorology", D.A. Haugen, ed.,Atm. Met. Soc.

De Bruin, H.A.R., and Holtslag, A.A.M.,1982, A simple parameterization of the surface fluxes of sensible and latent heat during daytime compared with the Penman-Monteith concept, Journ. Appl. Meteor., 21:1610.

van Egmond, N.D., and Kesseboom, H.,1983, Mesoscale air pollution dispersion models-I and -II, Atm. Env., 17:257.

van Egmond, N.D., and van Jaarsveld, J.A., De Saeger, E., and Vervliet, F., Weber, W.T., 1983, Comparison of measured and modelled mesoscale SO2-patterns in the BeNeLux, Science Tot. Env., 27:1.

Fowler, D.,1978, Dry deposition on agricultural crops, Atm. Env., 12:369.

Gillani, N.V.,1978, Mesoscale plume modelling of the dispersion, transformation and ground removal of SO2, Atm. Env., 12:569.

Holtslag, A.A.M., and van Ulden, A.P.,1983, De meteorologische aspecten van luchtverontreinigingsmodellen, K.N.M.I. W.R. 83-4, De Bilt, The Netherlands.

Venkatram, A.,1980, Estimating the Monin-Obukhov length in the stable boundary layer for dispersion calculations, Boundary-layer Met., 19:481.

Weseley, M.L., and Hicks, B.B.,1977, Some factors that effect the deposition rates of sulphur dioxide and similar gases on vegetation, Journ. Air Pollut. Control Ass., 27:1110.

Wieringa, J.,1980, Estimation of mesoscale and local-scale roughness for atmospheric transport modelling, in "Proc. 11th Int. Techn. Meeting on Air Pollut. Mod. and its Appl.", NATO-CCMS, Amsterdam.

TESTING A STATISTICAL LONG-RANGE TRANSPORT MODEL ON EUROPEAN AND NORTH AMERICAN OBSERVATIONS

B.E.A. Fisher and P.A. Clark

Central Electricity Research Laboratories
Leatherhead
Surrey KT22 7SE. U.K.

A statistical long-range transport model of sulphur oxides was applied to Europe and compared with observations of SO_2 and sulphate in air and the wet deposition of sulphate. A sensitivity analysis indicated that SO_2 wet removal is an efficient process and SO_2 oxidation in dry conditions is generally slow. The model was then applied to North America and the same choice of parameter values describing the main removal processes gives acceptable agreement in spite of the markedly different source density. This is evidence that the total sulphur system contains two components one of which depends roughly linearly on source strength and one related to a background concentration of sulphate in precipitation which is not directly related to sources within the regions modelled. Source-receptor matrices for both regions are presented and show that the background can be important at receptors remote from major emissions in areas of high precipitation.

1. INTRODUCTION

In this paper a statistical model of the long-range transport of sulphur is applied to eastern North America and to north west Europe and compared with annual average observed levels of SO_2 and sulphate in air, and the annual rate of wet deposition of sulphur. The primary advantage of a statistical model is that it depends on just a few main parameters, namely the removal rates of SO_2 and sulphate in dry and wet conditions and the rates of transformation of SO_2 to sulphate in dry and wet conditions. Thus it is a relatively simple matter to undertake a sensitivity analysis to determine what ranges of parameters give acceptable agreement with measurements. Though results are somewhat insensitive to the choice of parameter values, some restrictions are necessary.

These are found to be the following (i) the average rate of oxidation of SO_2 to sulphate is not much greater than 1% h^{-1}, or predicted airborne sulphate levels would be too high (2) SO_2 removal in rain has to be moderately efficient to produce sufficient sulphur wet deposition. The sensitivity analysis is described in Section 2 and was performed on a European database. Its conclusions are in line with earlier modelling exercises.

These results imply that in the overall sulphur cycle the rate of transformation of SO_2 to sulphate in dry conditions is of relatively low importance, the main removal rates being dry and wet deposition of SO_2. There has been much discussion recently over the question of "linearity", that is whether a reduction in emission will produce a proportional reduction in deposition at a fixed distance. Clearly calculated dry deposition, because of the way it is parameterised, is proportional to emission strength in all long-range transport models. Wet deposition need not be, even though it appears that at current emission levels in Europe SO_2 removal is fairly efficient in rain. This efficient removal may arise because there is sufficient oxidising and/or neutralising material entering rain systems with sulphur oxides to ensure efficient removal. The chemical mechanisms have not been established for this, but clearly one requires oxidising material, such as ozone, or a neutralising species, such as ammonia, in fairly abundant quantities to ensure SO_2 removal. To model the situation properly the cycles of all the relevant species would need to be included, as well as mixing processes on small and large scales in three dimensions within the rain system. This is beyond present day atmospheric physics. As an alternative indicator of the degree of coupling between removal and transformation rates and source density a long range transport model applied to Europe has been transferred to an equivalent area of eastern North America, to see whether the same values of the main removal coefficients give acceptable agreement with measurements.

The thinking behind this exercise is as follows. Sulphur emissions over eastern North America are about half those in an equivalent area of Europe (see Table 1), while nitrogen oxide and hydrocarbon emissions are proportionally higher. With its lower latitude one may reasonably expect the oxidising potential, due to photochemically related oxidants, of the eastern North American atmosphere to be higher than that in Europe. It therefore seemed a good idea to test a European model on North American observations, with appropriate allowance for differences in emissions and meteorology, to see whether the implicit dependence of parameters on other atmospheric chemical concentrations was strong enough to be noticeable.

Some results of this exercise are given in Section 3. Two points must be recognised initially. The model is not a perfect description of the real world and therefore it is impossible to achieve an exact correspondence between observed and measured values. Discrepancies between observed and predicted values can arise from many causes, only some of which may be due to uncertainty over the rates at which processes occur. We are therefore looking for some large systematic effect produced by an almost 50% change in source strength.

Secondly it will be apparent from Sections 2 and 3 that even with the assumption of efficient removal of SO_2 in rain systems there appears to be a tendency to underpredict wet deposition particularly at sites more remote from the major emissions. This shows itself as an intercept in the regression between observations and predictions. This extra sulphur in wet deposition is often referred to as the "background" wet deposition, but it may also be described as sulphur not directly attributable to sources within the model. Such sulphur may be of natural origin, from man-made sources outside the study area or from sources within the area, the sulphur having left the area along a curved trajectory and later returning. Since in wet deposition virtually the whole depth of the troposphere is scavenged, it is not surprising that the extra sulphur should appear as a component of wet deposition, especially in regions of high rainfall.

If there exists a "background" contribution to wet deposition then this has implications for the question of linearity. Our general conclusion from the European and North American comparison is that the same choice of parameter values works reasonably well, but in both cases we need to introduce an extra term in the sulphur wet deposition. One cannot determine the effect of a reduction in emissions if one does not know the effect of such a reduction on this extra term. The magnitude of the extra term can be very large over 20% in parts of Canada and of order 40% in some sensitive areas with high rainfall, such as southern Norway. Thus the reality and origin of the sulphur "background" remains a primary non-linearity in the system, even if other non-linearities are only of secondary importance.

2. SENSITIVITY ANALYSIS ON MODEL PERFORMANCE OVER EUROPE

2.1 Application to All Monitoring Sites

The model used is a statistical model from which annual average concentrations and annual deposition rates are derived. There are descriptions of the model in the literature (Fisher, 1975, 1978) but the main features are the following. Straight trajectories are assumed with the mean directional dependence representative of transport on a European scale. Transport occurs

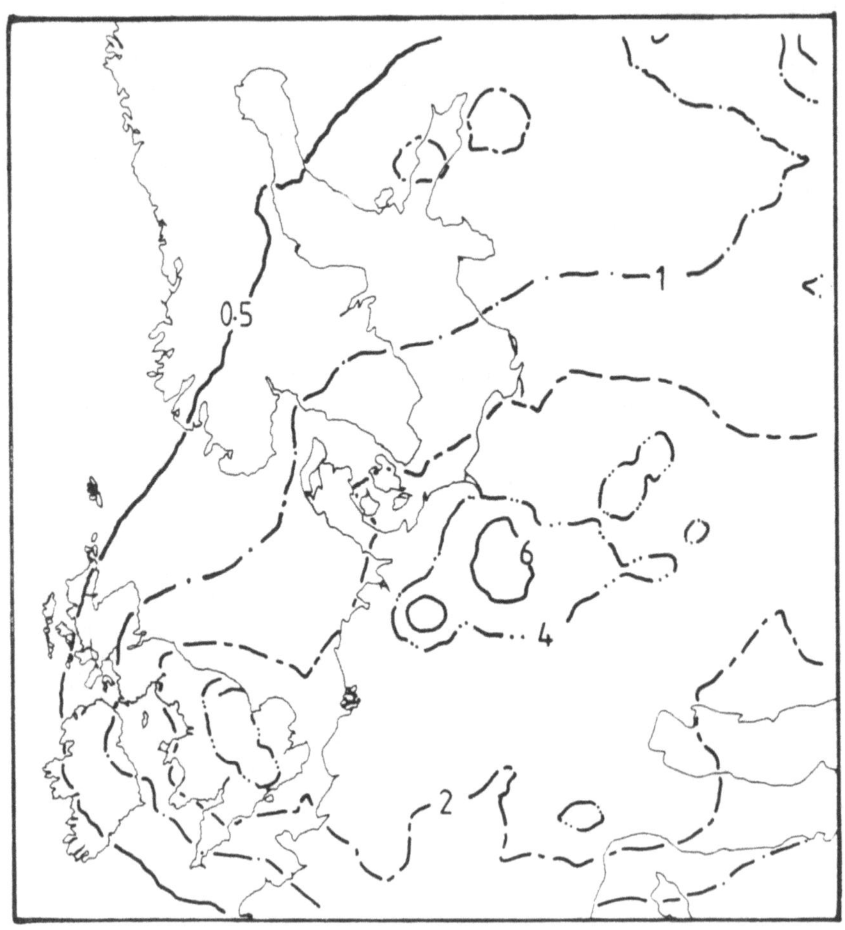

Fig. 1 Total Sulphur Deposition over Europe (g $S/m^2/a$)

within a mixing layer with vertical diffusion treated using the diffusion approximation. Five categories of windspeed, mixing depth and vertical eddy diffusivity are allowed representing a range of atmospheric behaviour characteristic of each region. Each has a given probability of occurrence. Rainfall is treated statistically, so that at any given instant in the history of an air mass there is a fixed probability per unit time of rainfall starting during a dry period, and similarly there is a fixed probability per unit time that rain will stop during a wet period. No special treatment of southern Norway, or other regions of high precipitation, is adopted in this simple scheme. Since the purpose of this analysis is to obtain some broad indications of the relative importance of the main removal processes, this is not thought to be too important. In particular, conclusions concerning the "background" contribution are little affected by omitting sites in these regions.

The source strengths are taken from the EMEP source inventory and refer to 1978. The model is designed to look at the contributions to deposition at long distances. The near-field deposition is estimated by assuming that within each grid square emissions have a uniform distribution. The contribution to deposition in a grid square from sources within the same grid square can then be estimated approximately. Each grid square has dimensions of 127 km by 127 km.

Dry deposition of SO_2 is calculated by assuming a constant deposition of 5 mm s^{-1} everywhere. The deposition velocity of sulphate particles is taken to be zero. Though of prime interest in the sensitivity analysis, the deposition velocity of SO_2 has been kept fixed as it is fairly well known. The primary effect of small changes in deposition velocity is to change airborne SO_2 concentrations and dry deposition, with only a small effect on wet deposition. Instead other parameters have been varied. The other parameters are:

(1) oxidation of SO_2 to sulphate in dry conditions (k_d) assumed to be a first order rate constant (s^{-1})

(2) oxidation of SO_2 to sulphate when it is raining (k_w (s^{-1}))

(3) the removal rate of SO_2 during rain (Λ_2 (s^{-1}))

(4) the removal of airborne sulphate during rain (Λ_4 (s^{-1})).

The results of the model have been compared with observations from the EMEP network for the period 1978-9. Observations of annual average SO_2 in air, sulphate in air and the annual rate of deposition of sulphate in precipitation from about 30 sites in north west Europe are used. The observed rate of deposition of sulphate in precipitation is assumed to equal the sum of the rate of removal by precipitation of SO_2 and sulphate in air.

There is insufficient room in this paper to present details of the sensitivity analysis, but its results will be described qualitatively. A range of values of the four removal rates are assumed and the degree of agreement between observations and calculations for each selection is assessed. In the initial sensitivity analysis the following 36 combinations of the four main parameter values were tested:

$$k_d = 3 \times 10^{-6},\ 1.5 \times 10^{-5},\ 3 \times 10^{-5}\ s^{-1}\quad (1,5,10\%\ h^{-1})$$

$$k_w = 3 \times 10^{-6},\quad 3 \times 10^{-5},\ 9 \times 10^{-5}\ s^{-1}\quad (1,10,30\%\ h^{-1})$$

$$\Lambda_2 = 10^{-5},\ 10^{-4}\ s^{-1}$$

$$\Lambda_4 = 10^{-5},\ 10^{-4}\ s^{-1}$$

These values cover and sometimes exceed the likely range of these parameters. It was found that the runs which gave the best overall agreement (lowest residual) have in common an oxidation rate of 1% h^{-1} and an oxidation rate much faster than this is not consistent with measurements of SO_4 in air. A linear regression between measurements and calculations suggests that the runs with the more rapid rate of removal of SO_2 in precipitation (10^{-4} s^{-1}) give the best fit. Results are insensitive to the rate of oxidation of SO_2 in wet conditions (k_w). Comparisons of wet deposition consistently indicate the presence of a "background" concentration of sulphate in precipitation, of magnitude 1 g SO_4 m^{-2} $year^{-1}$, from sources not directly included in the model. This occurs despite the "high" value of the removal rate of SO_2 in precipitation, and has been identified in earlier work (Eliassen 1978).

A finer sensitivity analysis was tried to obtain closer bounds on the four main parameters. This time the parameter values were:

$k_d = 2 \times 10^{-6}, 3 \times 10^{-6}, 6 \times 10^{-6}\ s^{-1}$ (.7,1,2% h^{-1})

$k_w = 3 \times 10^{-6}, 3 \times 10^{-5}, 9 \times 10^{-5}\ s^{-1}$ (1,10,30% h^{-1})

$\Lambda_2 = 10^{-5}, 5 \times 10^{-5}, 10^{-4}\ s^{-1}$

$\Lambda_4 = 10^{-5}, 5 \times 10^{-5}, 10^{-4}\ s^{-1}$

The results indicate that an oxidation rate of 2% h^{-1} in dry conditions is inconsistent with the measurements of airborne sulphate. Otherwise the runs are rather insensitive to alterations in parameter values. Best agreement comes from runs with the fastest rates of removal of SO_2 and sulphate in precipitation.

The comparison of measured and calculated SO_2 values show remarkable consistency, indicating that source location is the main factor determining SO_2 concentrations. SO_2 is overpredicted for all runs. This may be because sites are deliberately chosen to be remote from sources and the local correction is too large. It could also arise from the choice of deposition velocity which may be too low at 5 mm s^{-1}. When repeated with a higher deposition velocity of 8 mm s^{-1} the runs showed a distinct improvement in the agreement between calculated and observed SO_2 concentrations, but conclusions regarding the other parameters were little affected.

2.2 Sensitivity Analysis Applied to Data from Three Sites

Further clarification of the sensitivity of calculated concentrations to changes in parameter values can be obtained by considering the predictions for three sites in detail. Three sites in the EMEP network were chosen. One, N1, a southern Norwegian site is remote from major sources and is in an area of high rainfall. NL5 is a Dutch site which is close to major source areas and has rainfall characteristic of coastal, lowland regions of northwest Europe. The British site UK2, at Eskdalemuir, is a long running site influenced by industrial regions in Britain. A sensitivity analysis for these sites was made using the finer tuning of variables and a deposition velocity of 5 mm s^{-1}.

SO_2 is overpredicted at all three sites for all choices of parameters. The large overprediction at the Norwegian site may not be significant as the measured values are very low indeed and are therefore subject to considerable measurement error. The use of a higher deposition velocity of 8 mm s^{-1} gives better agreement at all three sites especially UK2. The agreement at UK2 and NL5 is well within the factor of 2 expected in this type of modelling.

Agreement between measured and calculated sulphate concentrations in air is poor at N1 and UK2 for the highest dry oxidation rate (2% h^{-1}) suggesting that generally the oxidation rate is closer to 1% h^{-1} At NL5 this effect is not so marked, an oxidation rate of about 1.5% h^{-1} looking sensible, and this may reflect greater photochemically induced oxidation, because of higher nitrogen oxide and hydrocarbon emissions in the Netherlands.

The wet deposition is underpredicted at all sites. At the Norwegian site predictions are far too low (1/5 measured), but at the other sites the runs giving best agreement in the previous sensitivity analysis using data from all sites give agreement to within a factor of 2. For most of these runs the fraction of the total wet deposition from SO_2 scavenging is 50% or more suggesting that agreement would be much poorer without reasonably efficient SO_2 removal in rain.

Even with this assumption the wet deposition is underpredicted. The inclusion of a "background" concentration in precipitation of 1 mg l^{-1} SO_4 would produce satisfactory agreement at NL5 and UK2, but is insufficient to explain the high wet deposition at N1. Qualitatively the explanation is thought to lie in the orographic precipitation in southern Norway, which is likely to be polluted. In some regions, such as western Scotland and even at Eskdalemuir, the airstreams bringing the highest precipitation are unlikely to be the most polluted. In the former situation it has been shown (Fisher, 1978) that an adjustment in the incidence of precipitation can make differences of up to a factor of 3 in the wet deposition. This would go a long way towards explaining the differences between measured and calculated wet deposition at N1.

The conclusions of this analysis entirely support the sensitivity analysis performed on all the EMEP monitoring sites.

3. PATTERNS OF SULPHUR DEPOSITION IN EUROPE AND EASTERN NORTH AMERICA

The statistical model applied in Section 2 can also be used to calculate the pattern of sulphur deposition over Europe and with some adjustments in meteorological conditions and the appropriate emissions, the deposition pattern over eastern North America. Using the parameter values $k_d = 3 \times 10^{-6}$ s^{-1}, $k_w = 3 \times 10^{-6}$ s^{-1}, $\Lambda_2 = 10^{-4}$ s^{-1} and $\Lambda_4 = 10^{-4}$ s^{-1}, Figs 1 and 2 show the pattern of total sulphur deposition, on the same scale, over the two regions. Total deposition includes the contribution from wet and dry deposition, but does not include the "background" deposition, which increases deposition by about 0.3 gS m^{-2} $year^{-1}$.

Fig.2 Tatal Sulphur Deposition over North America (g $S/m^2/a$)

The maximum deposition occurs over central Europe reflecting the greater source strength in Europe. However it is not proportionally greater than the maximum deposition in eastern North America according to source strength and this reflects differences in emission pattern. Table 1 summarises the sulphur budget for the two grid areas and shows that a greater proportion leaves the European grid area, mainly towards the east.
A sulphur budget is always sensitive to the size of the grid area under consideration.

Table 1. Sulphur Budget for Grid Areas

Region	Europe	N. America
Total man-made emissions for grid area (kTS $year^{-1}$)	19300	11600
Percentage dry deposited over grid area	33.5%	46.3%
Percentage wet deposited over grid area	24.1%	28.0%
Fraction man-made emissions deposited over grid area	.58	.74

It is interesting to compare how well calculations fit measurements for both regions. Table 2 shows the comparison. a and b are the intercept and gradient of the regression:

$$\begin{bmatrix}\text{measured}\\ \text{value}\end{bmatrix} = a + b \begin{bmatrix}\text{calculated}\\ \text{value}\end{bmatrix}$$

The contribution from local sources was neglected on the grounds that sampling sites would generally be chosen to be remote from local sources and our usual treatment of averaging emissions over a grid square would lead to an overestimate. The North American data comes from the SURE and MAP3S networks (1978/9) supplemented by readings (1980/1) on Bermuda (Jickells, Knap, Church, Galloway and Miller, 1982) to include an easterly remote site, data from Canadian sites run by Ontario Hydro (1979-80) and the APN network operated by the Atmospheric Environment Service (1979).

Table 2. Comparison of Calculated and Measured Data

Region	Europe	N. America
Airborne SO_2		
correlation coefficient	·84	·63
regression coefficients		
(a [μg m^{-3} SO_2], b)	(1.5, ·78)	(-4.6, 1.6)
Number of sampling sites	28	18
Airborne SO_4		
correlation coefficient	·69	·75
regression coefficients		
(a [μg m^{-3} SO_4], b)	(-1.3, ·99)	(-1.1, 1.0)
Number of sampling sites	29	18
Wet deposition (zero background)		
correlation coefficient	·61	·74
regression coefficients		
(a, [g SO_4 m^{-2} $year^{-1}$], b)	(1.4, 1.1)	(1.2, ·95)
Number of sampling sites	31	28
Wet deposition (background = 0.8 mg l^{-1} SO_4)		
correlation coefficient	·74	·72
regression coefficients		
(a, [g SO_4 m^{-2} $year^{-1}$], b)	(·17, 1.3)	(64, ·89)
Number of sampling sites	31	28

Acceptable agreement for all three sulphur components can be obtained in both Europe and North America using the same values of the main parameters. A "background" wet deposition of the same magnitude 0.3 g S m^{-2} $year^{-1}$) is indicated in both regions although SO_2 agreement seems worse over N. America. Annual precipitation amount varies considerable between European sites and the addition of background concentration of 0.8 mg SO_4 l^{-1} x precipitation amount is a site dependent correction, which improves agreement markedly in Europe, but not in North America.

The importance of this background is highlighted if we consider source-receptor matrices for the two regions. Sensitive or potentially interesting receptor areas were chosen in the two regions and the relative contribution of major source areas estimated (Tables 3 and 4). With the inclusion of a background wet deposition of 0.3 g S m^{-2} $year^{-1}$ two points emerge. The background can make a sizeable contribution, especially at receptor sites more remote from major sources. Secondly sources close to a receptor area make a larger contribution than the equivalent source further away. Hence deposition in the Muskoka region of Ontario is mainly from sources in Ontario and deposition in the Solling region of West Germany is mainly from west continental sources, predominantly West Germany itself.

4. CONCLUSIONS

It has been found when determining the long-term average deposition of sulphur emitted in Europe and North America that the largest fraction is dry deposited, a significant fraction escapes from the area (42% for Europe and 26% for North America) and the fraction oxidised to sulphate under dry conditions, that subsequently appears in rain, makes only a small contribution to the total wet deposition of sulphur. To explain the observed wet deposition patterns requires in both cases either a very efficient wet removal process or invoking contributions from sources outside the regions.

Even though eastern North America has half the sulphur emissions of Europe, a different spatial distribution of sources, and greater oxidising potential, similar parameter values apply as in Europe. This strongly suggest that, while numerous non-linear processes may be contributing to the scatter between measurements and calculations, in addition to the natural variability and measurement error, the sulphur system as a whole is not strongly non-linear. The underlying reason for this is that the rate removal of sulphur dioxide is found to be largely determined by dry deposition velocity and rainfall frequency, neither of which is (strongly) coupled to emissions, although more surprisingly airborne sulphate concentrations are also well explained by similar parameters.

To account for the wet deposition of sulphate, particularly in areas of high deposition remote from sources, a "background" concentration of sulphate similar to that suggested by the OECD study (Eliassen, 1978) needs to be incorporated. The physical and chemical origins of this background need to be explained before the effects of emission reductions on sulphur depositions can be realistically assessed.

ACKNOWLEDGEMENT

This work was carried out at the Central Electricity Research Laboratories and is published by permission of the Central Electricity Generating Board.

Table 3. Total Sulphur Deposition (%)

Source Region (emission $kTSO_2$ $year^{-1}$)	Receptors					
	Holland	S. Norway	W. Sweden	Central Sweden	Cairngorms, Scotland	Solling, W. Germany
British Isles 5300	18	16	9	8	47	4
West Continent 12200	40	19	29	23	12	59
East Continent 14800	11	16	21	25	7	28
Norway 154	0.1	11	2	1	.3	0.1
Sweden 551	0.3	4	16	16	.5	.2
Finland/Russia 5100	1	2	2	4	1	.5
Holland 480	19.5	2	1	1	1	2
Background	11	29	19	22	31	7
Total Deposition g S m^{-2} $year^{-1}$	2.8	1.0	1.6	1.4	1.0	4.6

Table 4. Total Sulphur Deposition (%)

Source Region (emission kTSO_2 year^{-1})	Receptors						
	Algoma	Muskoka	Quebec City	S. Nova Scotia	Adirondacks	Pennsylvania	S. Appalachians
Michigan 940	18	5	2	1	2	1	1
Illinois + Indiana 3200	7	4	2	1	3	4	11
Ohio 2900	9	12	5	3	11	24	5
Pennsylvania 1600	2	4.5	3	4	10	32.5	2
New York-Maine 1800	2	4	6	16	19	7	1
Kent-Tennessee 2200	2	1.5	1	1	1.5	2	35
W.Virg-N.Carolina 2400	2	3	2	3	5	10	6
Rest of USA 4700	11	3	2	1	2	2	23
Ontario 1900	20	41	10	4	20	6	1
Quebec 1100	5	8	44	7	8	1	0
Atlantic Provinces 230	0	.1	1	21	0	.1	0
Background	22	15	23	37	19	10	16
Total Deposition g S m^{-2} year^{-1}	1.4	2.0	1.3	0.3	1.6	3.0	1.8

REFERENCES

Eliassen, A., 1978, The OECD study of long range transport of air pollutants: long range transport modelling. Atmos. Environment, 12, 479-487

Fisher, B.E.A., 1975, The long-range transport of sulphur dioxide. Atmos. Environment, 9, 1063-1070

Fisher, B.E.A., 1978, The calculation of long term sulphur deposition in Europe. Atmos. Environment, 12, 489-501

Jickells, T., Knap, A., Church, T., Galloway, J. and Miller, J., 1982, Acid rain on Bermuda, Nature, 297, 55-57

DISCUSSION

W. ASMAN How large is the background concentration in moles l^{-1} ?
Remark : on an ocean weathership 800 km west from Scotland we find concentrations of about 5 micromoles of SO_4 in rainwater.

P. CLARK The answer to this depends on whether one ascribes this background to a concentration in rain water or a uniform background deposition (in which case it clearly depends on rainfall amount). If one assumes the former, a level of about 8 μMl^{-1} is indicated, and in this case the background takes in even greater importance in high rainfall areas.

R.M. VAN AALST Could you comment on the possible contribution of dry deposition of sulfur in rainwater gauges ?

P. CLARK Although I would refer you to those involved in making the measurements, this certainly can happen and may significantly increase the apparent wet deposition. At present this is difficult to quantify and requires further research.

ANALYSIS OF EPISODES WITH HIGH POLLUTANT CONCENTRATIONS IN BERLIN (WEST) USING BACKWARD TRAJECTORIES

Felicitas Wilcke and Franz J. Ossing

Institut für geophysikalische Wissenschaften
Freie Universität Berlin
D 1000 Berlin 33

INTRODUCTION

A system of numerical models is described used to trace back trajectories of air parcels from Berlin (W) at a level of 100 m. Data is based on hourly wind observations of the WMO-net. These were interpolated onto a 300 x 400 km^2 grid with a 10 km mesh size. With some simplifying assumptions on the vertical windprofile and the traveling height of the air parcels, the model system was developed for the application to specific pollution episodes during which the contribution of remote external pollutant sources to the SO_2-concentration in Berlin (W) is relevant and implemented on the CYBER 835 computer of the Free University Berlin.

For a test, the models were applied to some cases marked by SO_2-concentrations at the upwind boundary stations of the Berlin urban SO_2 monitoring network (BLUME) which were significantly higher than the simultaneous half-hourly network average, suggesting advection of highly concentrated pollutants from remote external sources. Two examples of episodes (23/24 Feb. '78, 5/6 June '78) have been examinated more closely to discuss the capabilities and limits of the method. As the results show the computed trajectories gave the opportunity, together with some more meteorological variables, to explain quite a number of details of the pollution situations analyzed. Certain SO_2 events could be traced back to individual main centres within the industrial belt of the GDR southeast, south and southwest of Berlin. Uncertainties of the interpretation of the trajectories emerge from the actual hourly vertical change of the wind, and in case of stronger convective processes, besides the fact that an emission inventory of the area concerned isn't accessible. These, and other, impacts on the representativity of the computed trajectories need some more systematic research (sensitivity analysis).

OBJECTIVES

Our basic aim was to contribute to the practise of air pollution controll in Berlin (W), especially to surveying the import of atmospheric pollutants from the neighbouring regions by means of a numerical instrumentarium. It was to be applicable in practise by the local authorities, (e.g. Senator für Stadtentwicklung und Umweltschutz), as easily as possible. This was attempted by numerical methods which allow to analyze, in concrete cases, the import of highly concentrated pollutants by tracing back trajectories to the regional neighbouring industrial centres, regarding the track of dispersion, the localisation of the source areas at the time, and dependencies of the dispersion processes on the weather situation.

If developed adequately, the model system should be able to be applied in a numerous tasks of scientific, administrative and political kind, among others e.g.

- identification and selection of the special pollution episodes referred to above on the basis of the Berlin SO_2 measuring array
- research and display of the meteorological conditions responsible for the interregional transport of highly concentrated pollutant plumes
- recording and description of the frequency of interference of the Berlin city area by the neighbouring industrial areas
- review and forecast of situations of dangerous pollutant concentrations (e.g. referring to smog alarms)
- as a basis of a qualitative analysis and daily forecast of the current pollutant import situation
- putting up the basical material for planning and, if necessary, controlling measures
- in the formulation of the means to reduce transborder dispersion of pollutants
- in proving and estimating the mutual interference of regionaly neighbouring areas of conurbation (e.g. as a basis for international negotiations).

These aims are attainable only step by step, viz. by formulating, programming, implementing and testing of the numerical models, including exemplary case studies, comprehensive sensitivity analyses, possibly necessary improvements and a subsequent operationalisation depending on the different aims of application, and, finally, a comprehensive, including statistical, evaluation by analyzing a sufficiently large number of cases.

In this study the first objective was reached to full extent : an operational numerical model was developed and applied to some cases, described exemplarily in the following. Further steps, from sensitivity analysis to operational applicability, are reserved for continuing work on this project.

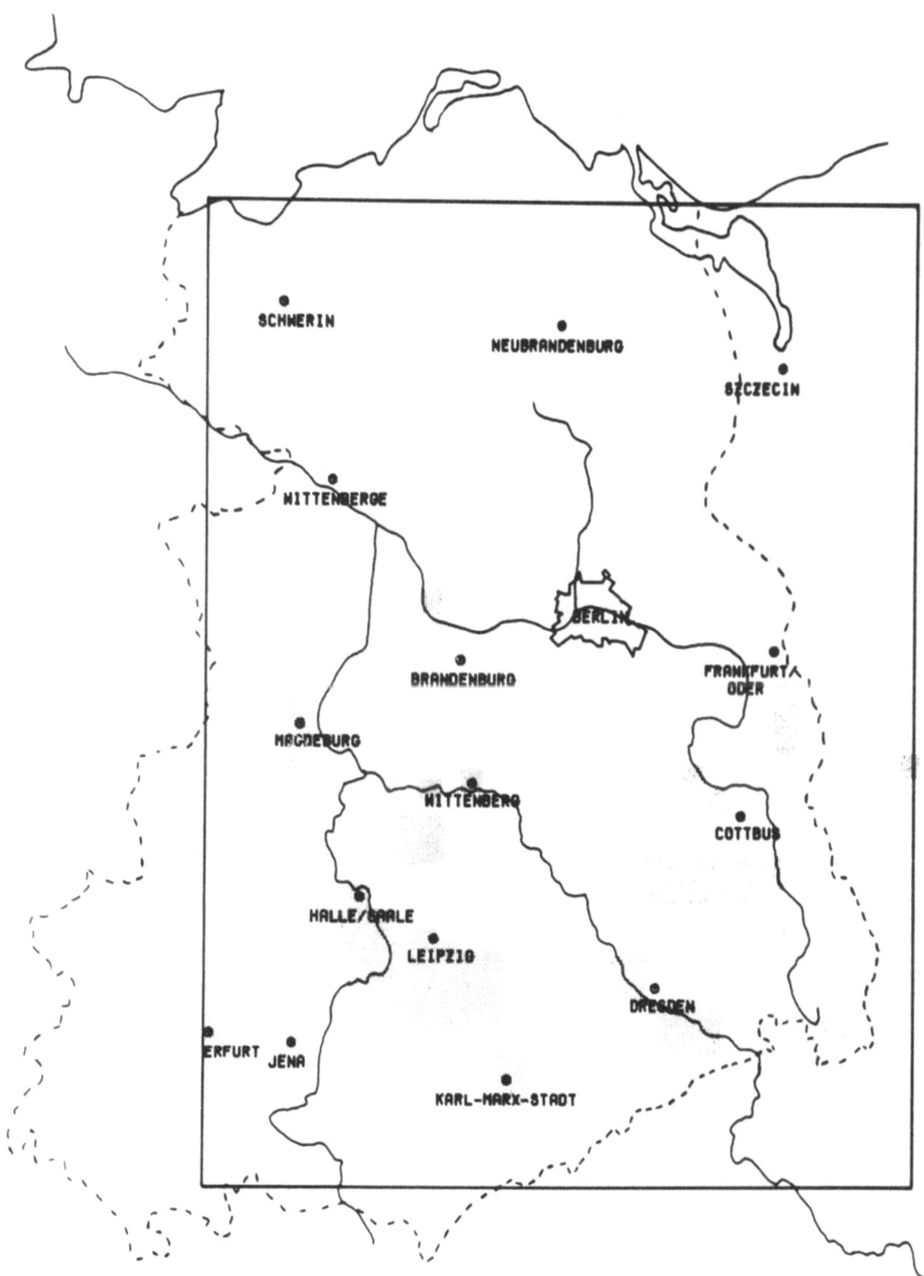

Fig. 1. Geographical position of the area concerned

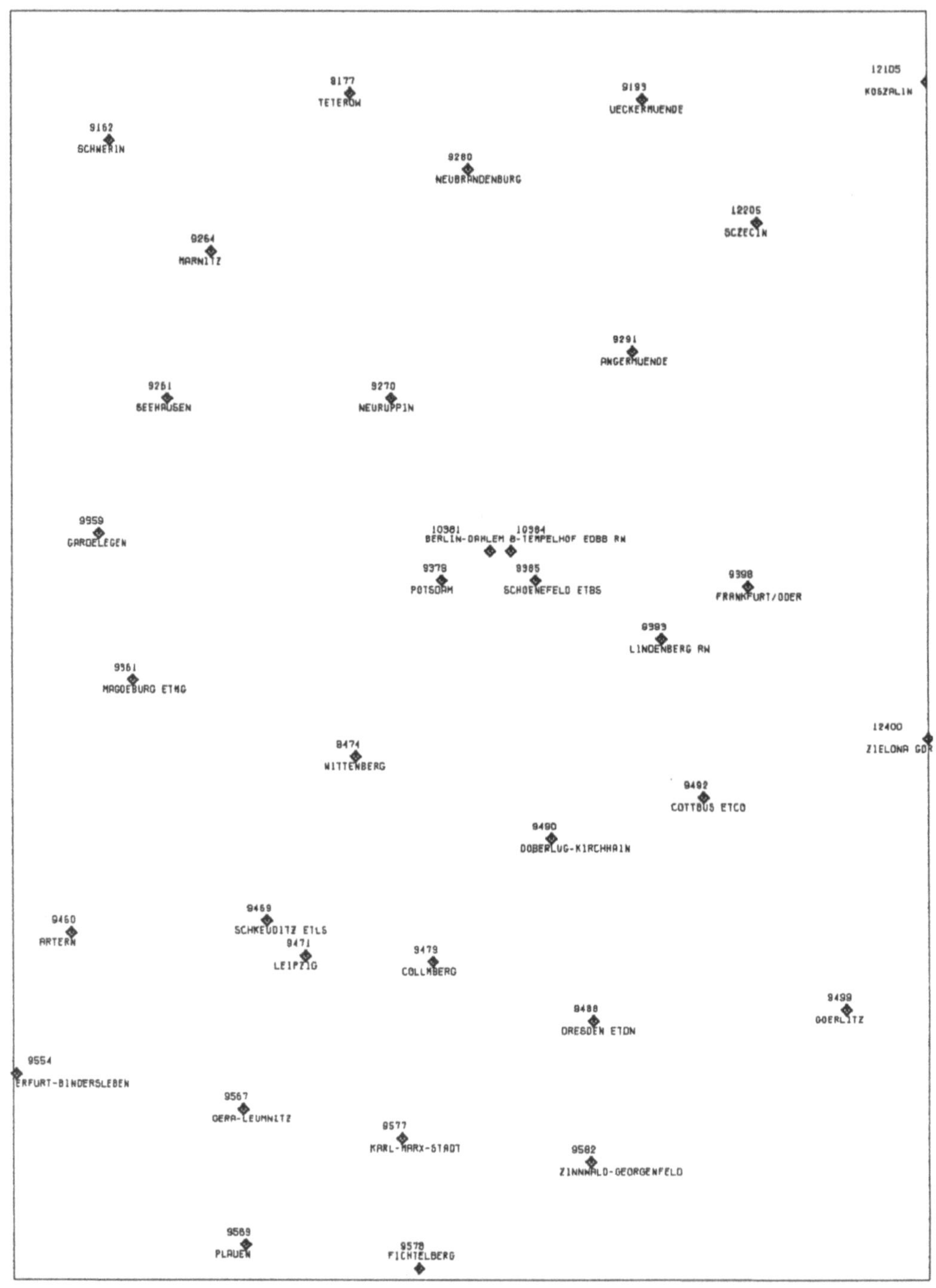

Fig. 2. Position of weather stations

AREA OF RESEARCH AND DATA BASIS

For the simulation of trajectories of air parcels a grid was constructed with Berlin in its center, having a 10 x 10 km mesh size (fig. 1,2) and covering an area of 120.000 km², including the major part of the GDR and parts of western Poland. It comprises 30 grid squares in x-direction opposed to 40 in the y-direction (i.e. 300 km by 400 km) and is fixed on its north-western corner with the geographical coordinates 54°N, 11° E (which is in grid coordinates (1,1)).

Data basis for the required wind field was supplied by the hourly weather reports of the Berlin stations Tempelhof Central Airport (DWD) and Dahlem (FU) and of the weather service of the GDR, and of three-hourly reports of the polish meteorological service as they are available by broadcasting through the WMO-net.

The data used in the selection of cases come from the SO_2 monitoring network of Berlin (Berliner Luftgüte-Meßnetz BLUME) which is in operation since 1975. BLUME consists out of 31 SO_2-monitoring stations distributed in a nearly regular 4 x 4 km network over the city area of Berlin (W). (The construction of the network is displayed in the film by C. Zick / B. Carus : "Das Berliner SO_2-Meßnetz", part 1, HFR 1976). This distribution can be descripted in a 7 x 8 matrix (fig. 3).

THE MODEL SYSTEM

The model system developed here consists essentially of three parts (fig. 4) :

i) Precursor-program for the selection (by measured SO_2 distributions) and suitability analysis (by the state of atmospheric mixing) of the episode to be analysed, this is made by two sub-programs CHECK and CUNIMB.

ii) Main program which prepares the measured wind data and computes the trajectories according to Peterssen (1956). The subprograms UMPOL, POLIER, TRACK belong to this section.

iii) Plot programs to display the trajectories (TRAPLO) and wind fields (REWIPT).

By a system of control cards the programs UMPOL, POLIER, REWIPT, TRACK, TRAPLO are linked, so that after preparing the meteorological data accordingly the program system runs automatically.

EXAMPLE 1 : THE WINTERLY SMOG-EPISODE OF 23/24 FEB. 1978

This episode was chosen because of the results of an examination by Heimann et al. (1980) which showed that reasonable transport of SO_2 into Berlin must have taken place on these days. While on other

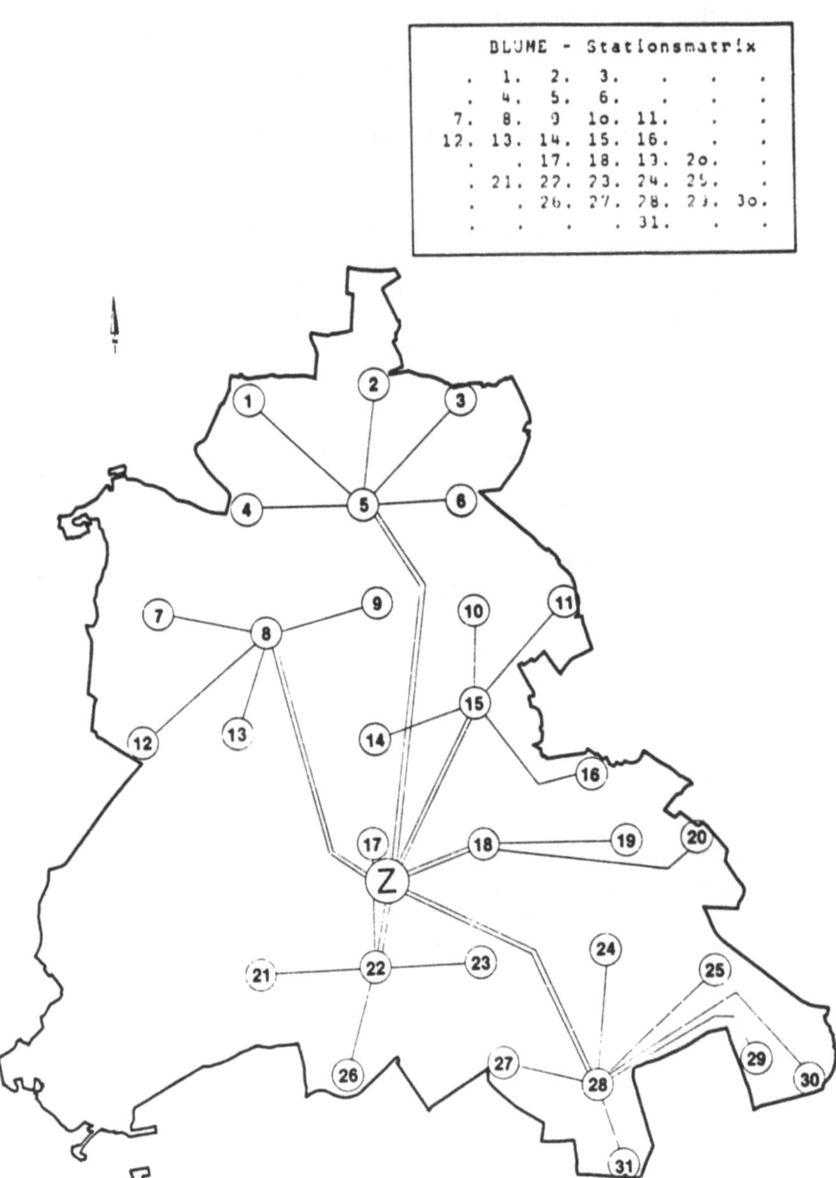

Fig. 3. Berlin SO_2 monitoring network

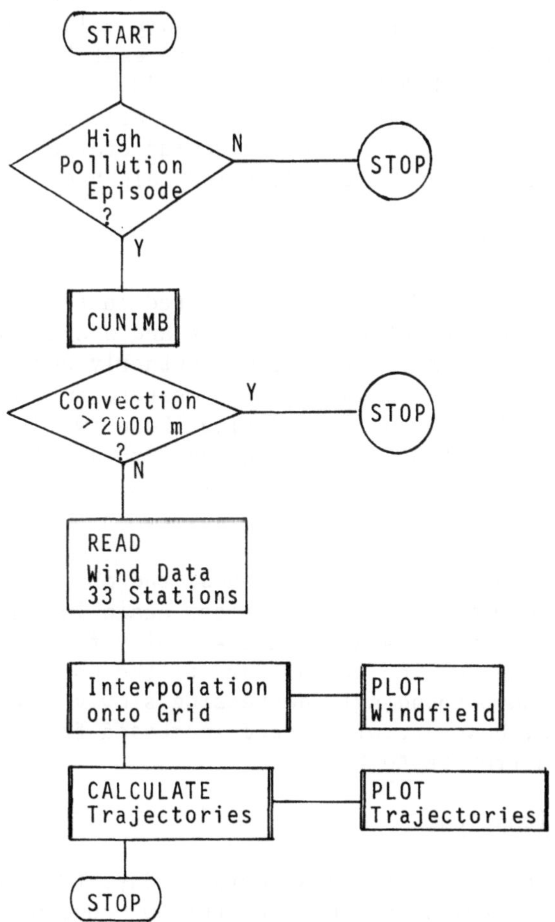

Fig. 4 Program System; for details see Ossing et al. 1983

days the SO_2 concentration patterns in Berlin could be simulated rather well with a diffusion model (Heimann et al. (1981), on this day (24th Feb.) the simulation values showed were too low, the difference between measured and computed values being largest at the southern edge of town, with winds from the south. Therefore external sources were made responsible for the height of the observed SO_2-concentrations without an exact proof.

This is what we tried to substanciate. As for the selection criteria in general, the upwind city boundary values had to exceed the city average value concentration by at least 50µg/m³. This took place on Friday the 24th Feb. 1978, from 5.30 to 10.30 MEZ on 9 of 11 half-hourly measuring times. Besides, nearly all the half-hourly averages of SO_2 concentrations at the upwind stations at the city boundary were higher than at the downwind neighbouring stations.

The Weather Situation

On Feb. 23/24 1978 Germany was situated in the transition zone of an east-european zone of high pressure and a chain of lows off the european coasts. In the resulting southerly current a surge of air too warm for the season was lead to western Europe that made the snow cover in the flat areas melt during these two days, reaching maximum temperatures of 3°C to 7°C on the 23rd and 10°C over large areas the following day. In Berlin-Dahlem the temperature record of this day was beat by pre-spring-like 12.1°C.

After the night's overcast vanished, the morning was sunny till shortly afternoon when it became overcast again leading to rain in the evening. The surface wind turned steadily from east to southerly directions from the morning of the 23rd to the morning of the 24th, with wind speed correspondingly decreasing slowly from 5 m/s to 2.5 m/s in 500 m, however, wind speed increased from 4.5 m/s to 7.5 m/s, turning from SE to SW.

The Pollution Situation

Though the two days are usual working days (Thursday, Friday), the hight and temporal development of the SO_2-concentration of the 23rd and 24th are fundamentally different. Fig. 7 shows the development of the half-hourly means averaged over the whole set of BLUME-stations from OOCET to 12CET. Except for the time from 8h to 8.30h a.m. the values of the 24th throughout were higher than the day before, at noon even more than twice the time. The 23rd from midnight to 6 a.m. showed a nearly constant SO_2-concentration, followed by an increase of more than two times the city area average within 2.5 hours and an correspondingly quick decrease to the levels of the former values. Contrarily, the next day showed a nearly steady increase of the pollutant concentration by factor 2.6 within 12 hours. Both temporal developments do not accord to the usual daily course ;

Fig. 5. Weather situation 23.2.1978 7 CET (above)

Fig. 6. Weather situation 24.2.1978 7 CET (below)

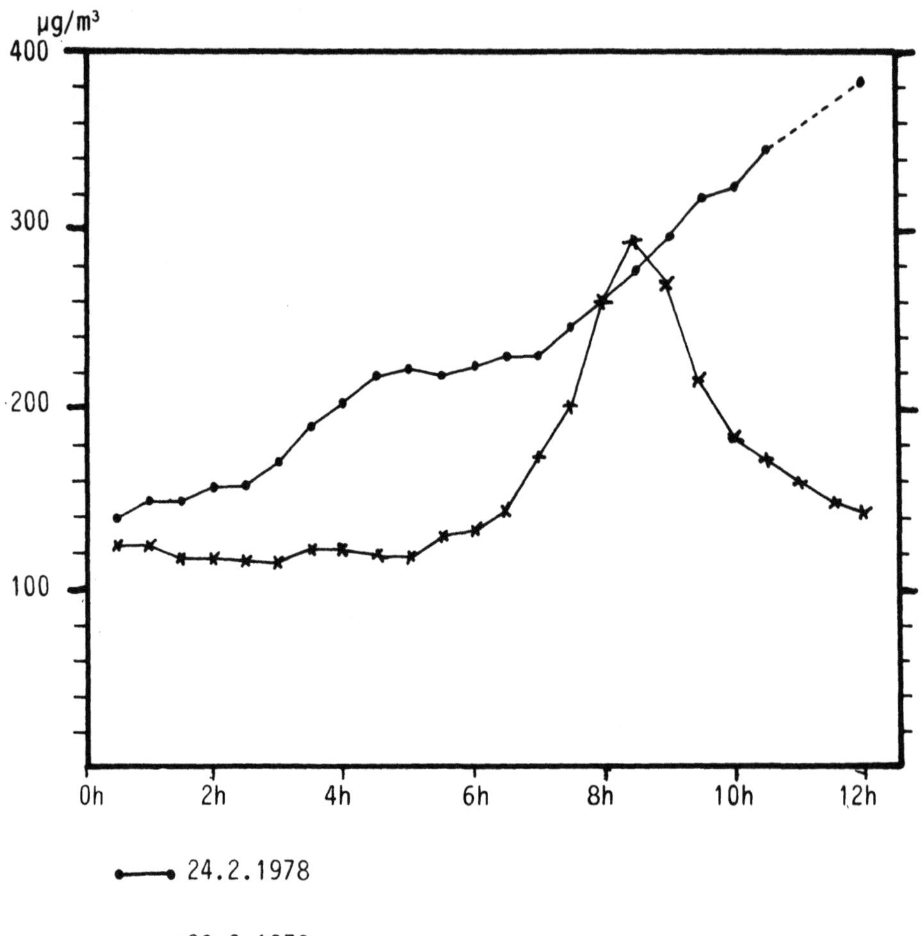

Fig. 7. Course of the area-average of half-hourly-average of SO_2-concentration

on the 24th even more surprisingly the increase corresponds to pre-spring-like temperature increases, which should have lead to a concentration decrease due to decrease of heating power.

Interpretation of the Pollution Episodes by Means Of Computed Trajectories

The analysis shown in this paper uses data from the smog episode of 24th Feb. 1978. The data available allowed to construct sufficiently long back reaching trajectories.

The computed trajectories (fig. 8-11) in the course of the first half of the day cover the whole south-east sector, turning slowly from east to south, so that they lead the path of the air crossing the following industrial areas and power plants coming into consideration as SO_2-sources : Frankfurt/Oder, Finkenherd power plant, Eisenhüttenstadt, Zielona Gora, Cottbus, Lübbenau power plant, Vetschau power plant, Boxberg power plant, Schwarze Pumpe, Senftenberg, Lauchhammer, Dresden. From Tabel 1 can be deduced that the concentration values measured at BLUME-station 26, situated at the southern edge of Zehlendorf, did not exceed the steadily increasing city average value. But from the time the bundle of trajectories that had passed Cottbus and the neighbouring chain of big power plants, reached Zehlendorf, the values of SO_2-concentration began to increase more strongly than the city average, exceeding the latter by 40 % under south-eastly winds. This situation continued until at least 10.30 MEZ. By 12 MEZ the trajectories came from the "industrial gap" between Dresden and Karl-Marx-Stadt, leading to Zehlendorf values that were no longer higher than the city average.

The plausibility of the temporal and spatial correlations supports the idea of sufficient exactness of the computed trajectories, at least in this case of a slow to moderate, but steady drift (Beaufort scale 2-3). The assumption made by Heimann et al. (1980), that in this case import of SO_2 from external sources, can be certified by the correlations found.

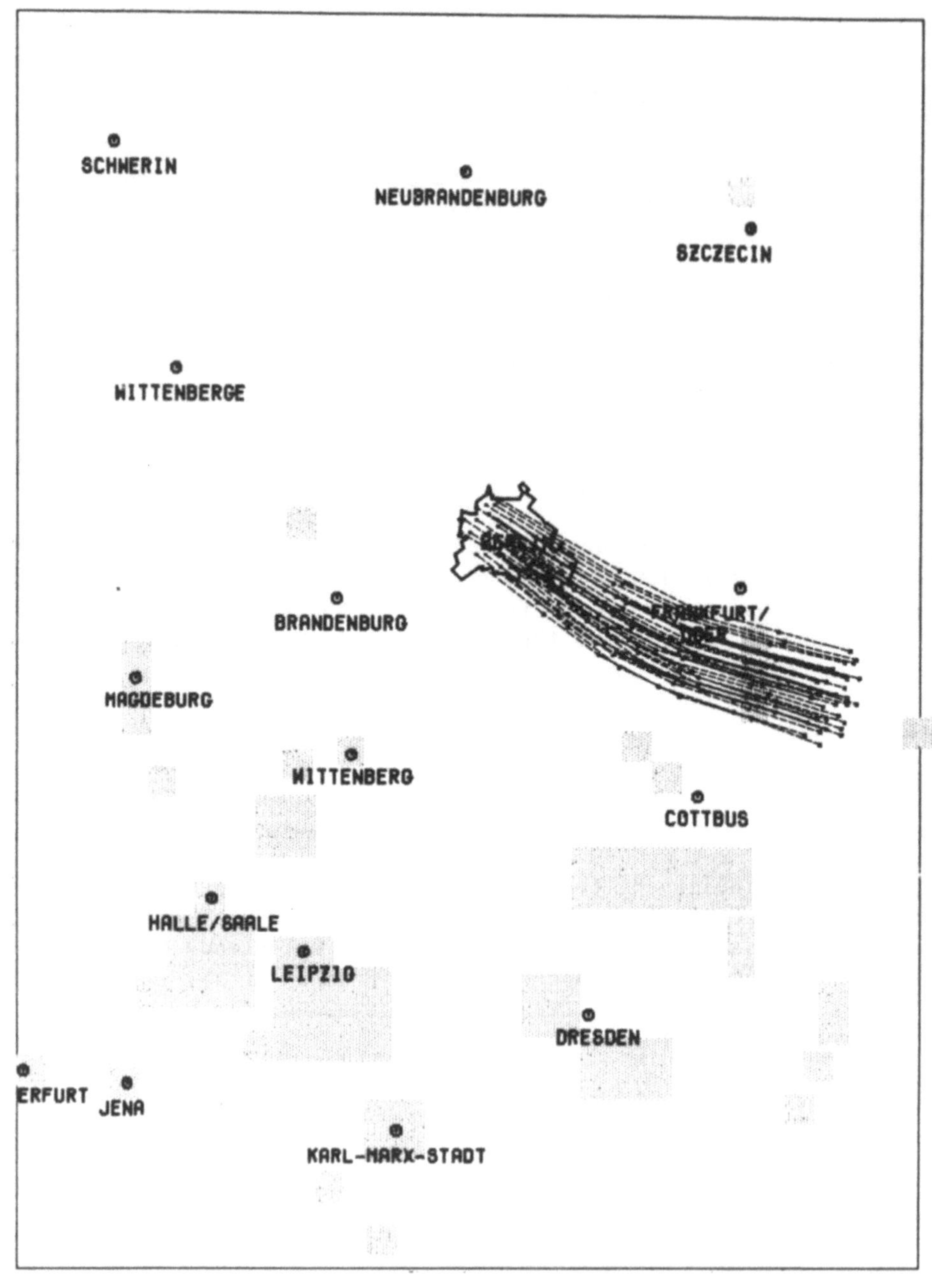

Fig. 8. Trajectories for the 24/ 2/ 78 3 Cet

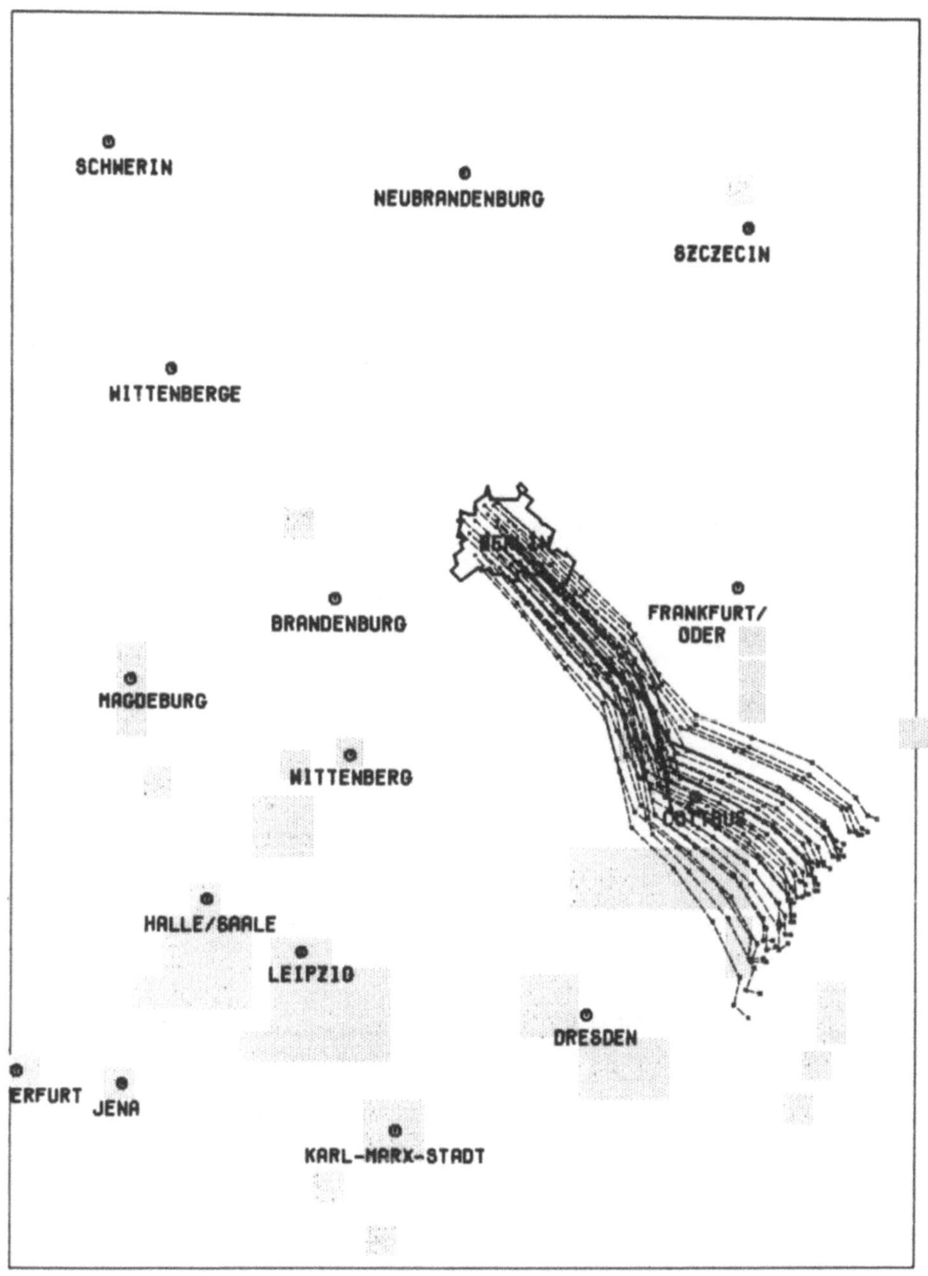

Fig. 9. Trajectories for the 24/ 2/ 78 6 Cet

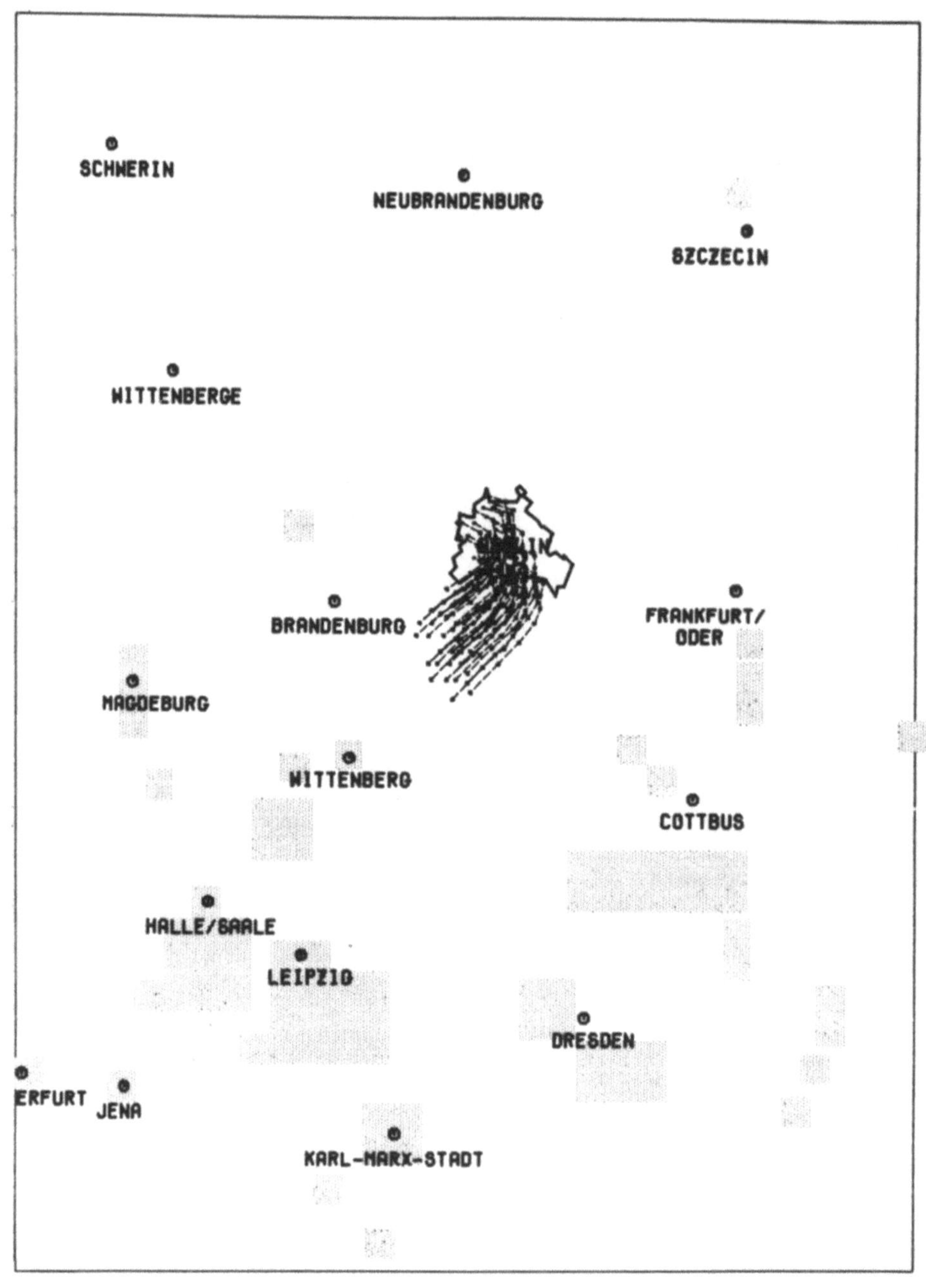

Fig. 10. Trajectories for the 23/ 2/ 78 9 Cet

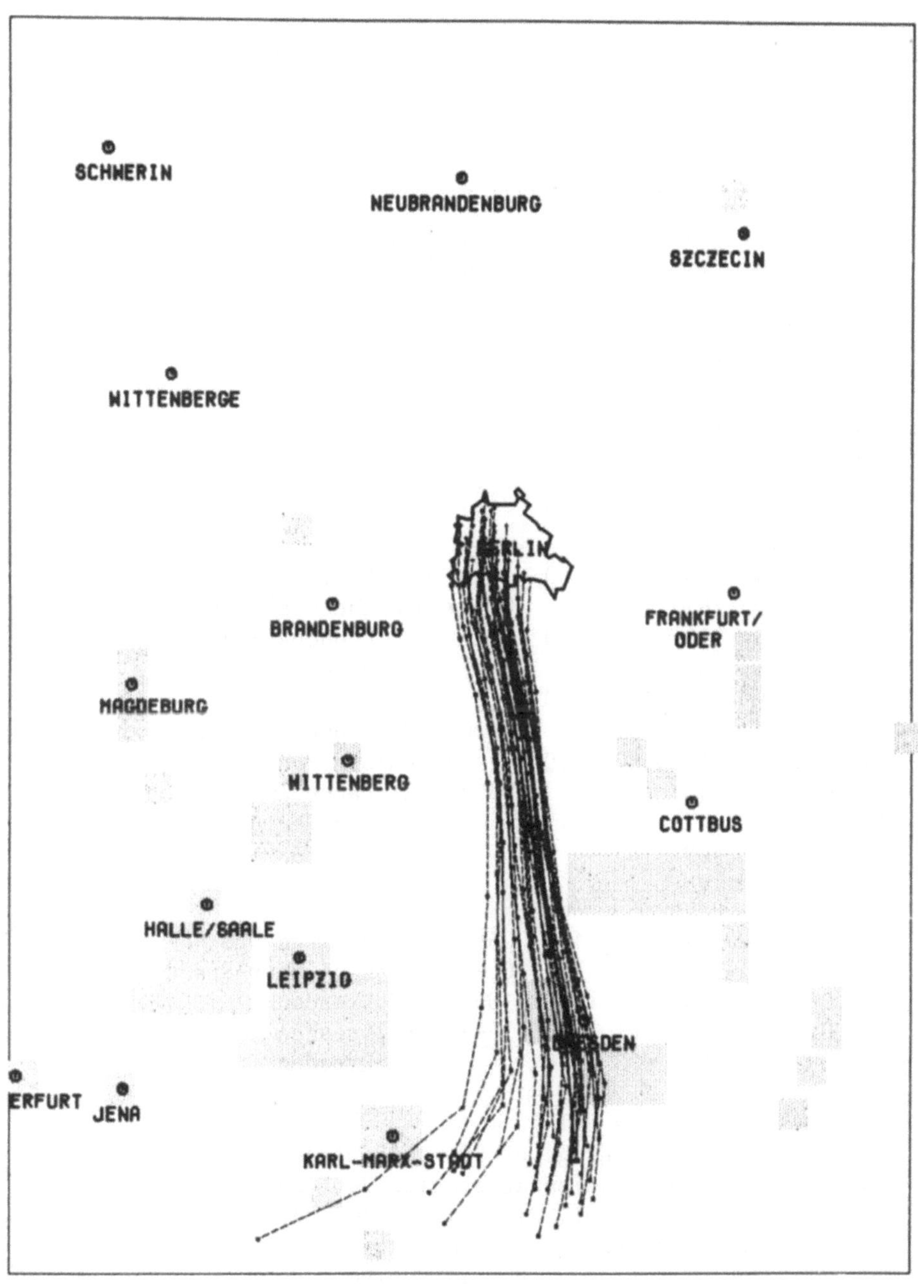

Fig. 11. Trajectories for the 24/ 2/ 78 12 Cet

Table 1. Relations between SO_2-concentration and trajectories on 24th Feb. 1978

1	2	3	4	5	6
0.30	139	145	6		20-21 MEZ:Frankfurt/Oder
1.00	150	154	4	120	20-02 MEZ:Finkenheerd
1.30	151	153	2		
2.00	158	161	3	120	
2.30	154	157	3		23 MEZ:Finkenheerd
3.00	171	184	13	130	Eisenhüttenstadt
3.30	190	197	7		21 MEZ:Zielona Gora
4.00	202	206	4	135	
4.30	217	207	-10		
5.00	222	219	- 3	140	
5.30	218	236	18		
6.00	224	244	20	160	01 MEZ:Cottbus as well as
6.30	229	261	32		02-04 MEZ:the chain of power plants
7.00	231	293	62	150	Lübbenau-Vetschau-
7.30	246	331	85		Boxberg
8.00	259	360	101	160	
8.30	276	344	63		02 MEZ:Schwarze Pumpe
9.00	296	346	60	160	03 MEZ:Senftenberg
9.30	319	418	99		03.30 MEZ:Lauchhammer
10.00	324	411	87	175	04 MEZ:Dresden
10.30	346	409	63		
.	.	.	.		
.	.	.	.		
12.00	384	317	-67	190	05-08 MEZ:between Dresden and
15.00	329	259	-70	165	04-07 MEZ:Karl-Marx-Stadt
18.00	292	194	-98	180	10-11 MEZ:Dresden
21.00	273	169	-104	160	15-16 MEZ:Dresden
23.00	232	139	-93	160	Dresden

col. 1: time (MEZ = CET)
col. 2: city area average of SO_2-concentration (µg/m³)
col. 3: SO_2-concentration at station 26 (µg/m³)
col. 4: (col. 3) - (col. 2) : difference between stat. 26 and city area average (µg/m³)
col. 5: wind direction at Berlin-Dahlem (degrees)
col. 6: emission areas touched by the trajectories

CONCLUDING REMARKS

In the course of the project described, a system of models was developed and made operationable. With the examples presented its efficiency could be demonstrated. Refering to the further objectives of the project, a first essential part was concluded, viz. the development of an instrumentarium capable of the tasks mentioned above. In the following, a systematic experimental tests of the model system with respect to exactness and susceptability to interference (sensitivity analysis) has to be made ; furthermore it is necessary to apply the models to a greater number of smog episodes to be sure that, analyzing the multiple basical meteorological conditions for the LRT of highly concentrated pollutant plumes, the whole spectrum of variable conditions is convered (statistics of cases), and to make possible the weighting of the importance of the single components. To realize this, viz. the treatment of real episodes in retrospect, the vertical change of wind referring to actual measurements should be approximated instead of the probably too much generalizing theoretical access. Furthermore, the inclusion of air-chemical measurements in analyzing the causes on the one side, and in checking the exactness of the trajectory processing on the other, should be a further aim.

REFERENCES

Heimann, D., Stern, R., Strobel, B., Timm, B., 1980, Modelling of SO_2 Concentration Patterns in Berlin (West) Using Linked PBL and Dispersion Patterns, in : Proc. of the 2nd Joint Conference on Applications of Air Pollution Meteorology and 2nd Conference on Industrial Meteorology, Amer. Met. Soc., Boston, Mass.

Heimann, D., Ossing, F.J., Strobel, B., Timm, B., 1981, Entwicklung eines mathematisch-meteorologischen Modells, in : "Untersuchung der Schadstoffausbreitung als Grundlage für Ausbreitungsrechnungen bei windschwachen Wetterlagen", Umlandverband Frankfurt, ed., D6000 Frankfurt/Main.

Ossing, F.J., Tepper H., Warnecke, G., Wilcke, F., 1983, "Erstellung und Erprobung eines numerischen Trajektorienmodells zur Rückverfolgung von Berlin erreichenden grenzüberschreitenden Schadstoffwolken", Publication by the Free University Berlin.

Pettersen, S., 1956, "Weather Analysis and Forecasting", McGraw-Hill New York.

4: ATMOSPHERIC EXPERIMENTS PERTINENT TO AIR QUALITY MODELING

Chairmen: R. Berckowicz
F. Schiermeier

Rapporteurs: S. E. Gryning
S. Sandroni

ATMOSPHERIC FIELD EXPERIMENTS FOR EVALUATING POLLUTANT TRANSPORT AND DISPERSION IN COMPLEX TERRAIN*

P. H. Gudiksen, M. H. Dickerson, R. Lange, and J. B. Knox

Atmospheric and Geophysical Sciences Division
Lawrence Livermore National Laboratory
Livermore, CA 94550

INTRODUCTION

The Department of Energy is currently sponsoring a program of Atmospheric Studies in Complex Terrain (ASCOT) to improve the technology needed to assess the air quality impacts of developing energy resources in areas of complex terrain. The program uses field experiments, theoretical atmospheric physics research, and mathematical models to develop a measurements and modeling methodology that can be used to provide the air quality assessments in these areas. The ASCOT team is composed of scientists from several DOE-supported research laboratories and university programs.

With the program's initial focus being on the study of transport and dispersion of materials injected in or near nocturnal drainage flows, three series of field experiments have been carried out in The Geysers geothermal area in northern California. The initial series, conducted during July 1979, in the Anderson Creek valley of The Geysers area were exploratory in nature and of limited scope to gain initial insight into the structure of the drainage flows. As a result of these experiments, a second series of experiments were conducted during September 1980 in the same valley to acquire more detailed

* This work was performed under the auspices of the U. S. Department of Energy by the Lawrence Livermore National Laboratory under contract No. W-7405-Eng-48.

information about the temporal and spatial characteristics of drainage flows and their interaction with the regional and synoptic scale flows. In contrast to these generic studies of the transport and dispersion of pollutants entrained in drainage flows, the third series of experiments, conducted during August 1981 in the Big Sulfur Creek valley of The Geysers, was designed to (1) acquire data needed to predict the impact of the hydrogen sulfide emissions from future geothermal power plant cooling tower plumes during nocturnal drainage flows, and (2) to perform nocturnal drainage flow studies in a new environmental setting and to test the general applicability of the methodologies developed from the experimental studies in the Anderson Creek valley. This report provides an overview of the experimental designs, highlights some of the results, and discusses the application of the results for improving a three-dimensional mass-consistent wind field model and particle-in-cell transport and diffusion model; one of several modeling capabilities being developed under the auspices of the ASCOT program.

EXPERIMENTAL DESIGNS

With the program goal being to study the transport and dispersion of materials injected in or near drainage flows, the design of the three series of experiments mainly reflected the following general objectives:

- to evaluate the entire nocturnal drainage cycle to include initiation, perpetuation, and breakdown.
- to evaluate the temporal and spatial characteristics of the drainage flows within the valley, including the evolution of pooling of drainage flows within the Anderson Creek valley and the subsequent outflow of this air out of the valley.
- to evaluate the influence and the extent of mixing between the external flows and the drainage flows within the valley, and
- to evaluate the effect of changing surface roughness due to forest canopies on the drainage flows.

In addition, one of the objectives of the August 1981 experiments in the Big Sulfur Creek valley was to evaluate the interactions of the geothermal power plant cooling-tower plumes, containing H_2S emissions, with the nocturnal drainage flows.

To illustrate how these objectives were addressed, we have chosen an overview of the design of the September 1980 series of

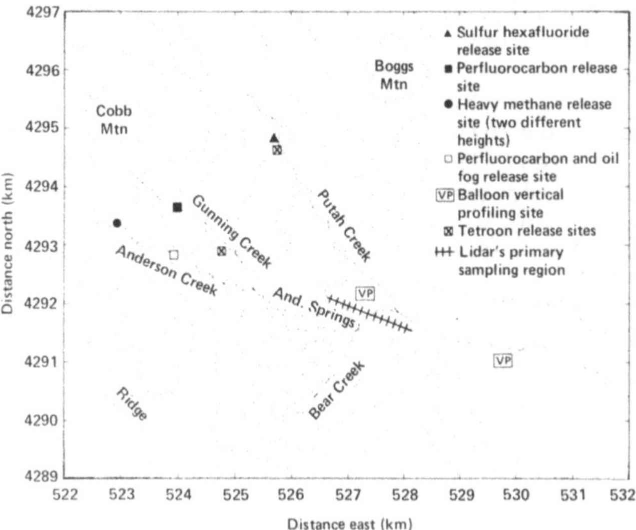

Fig. 1. Layout of the September 1980 tracer experiments within the Anderson Creek valley.

experiments in the Anderson Creek valley. This series consisted of five identical experiments. Each experiment consisted of a series of tracer studies that were coordinated with a host of meteorological studies. The scope and complexity of these experiments required personnel and equipment resources from 17 organizations.

The Anderson Creek valley has the characteristics of a basin. Its topographic features and the layout of the tracer studies are shown in Fig. 1. The valley is bounded by Cobb Mountain on the north, by a ridge on the west and south, and by Boggs Mountain on the east. The Anderson, Gunning, and Putah Creeks, which form the principal drainage areas, merge near Anderson Springs with outflow toward the southeast. The tracer studies included the use of two perfluorocarbons, two heavy methanes, and sulfur hexafluoride gases that were measured using conventional sampling techniques, as well as oil fog tracked by lidar and tetroons tracked by radar. One of the perfluorocarbon tracers (PMCH; C_7F_{14}) was released into the nocturnal drainage flows from an open, but very sheltered area in Anderson Creek; while the other perfluorocarbon tracer (PDCH; C_8F_{16}) was released within a forest canopy in Gunning Creek. These sites are roughly halfway up the slopes. These experiments were carried out as a cooperative effort involving the NOAA Air

Resources Laboratories, the Department of Energy Environmental Measurements Laboratory, and the Brookhaven National Laboratory. Two heavy methane tracers, methane-20 ($^{12}CD_4$) and methane-21 ($^{13}CD_4$) were released by investigators from the Los Alamos National Laboratory within the upper reaches of the Anderson Creek drainage area. The methane-21 was released at the surface directly into the drainage flows; while methane-20 was released simultaneously into the transition layer flows at a height of 60 to 75 m above the surface to investigate the extent of mixing between the transition layer flows and the drainage flows. The sulfur hexafluoride was released in the upper part of the Putah Creek drainage area by researchers from Meteorology Research Inc., to evaluate the merging of the flows from Putah Creek with those from the Anderson and Gunning Creek drainage areas. All of the releases were of one hour duration. The downwind surface concentrations were sampled at 30 to 50 locations depending upon the tracer. The sampling periods ranged up to eight hours with time integration periods varying from a few minutes to several hours. In addition two vertical profiling systems were used to define the temporal variations in the vertical distributions of the tracers within the valley basin and outflow region. These consisted of balloon borne sampling systems developed and operated by the Brookhaven National Laboratory and the Sandia National Laboratories. The sampling and analytical techniques for the perfluorocarbon tracers have been reported by Ferber et al. (1981) and by Lovelock and Ferber (1982); while those for the heavy methane tracers have been reported by Cowan et al. (1976) and Fowler (1979).

In order to acquire more detailed structural information about the three-dimensional evolution of these tracers, oil fog was released at the same site and simultaneously with the PMCH perfluorocarbon tracer and tracked by a lidar operated by the NOAA Wave Propagation Laboratory. For each release the lidar, which was situated near the valley outflow region, performed a series of scans in various vertical planes to observe the evolution of the plume. The region of most frequent sampling by the lidar and, hence, the most detailed analysis is shown in Fig. 1. The lidar used in this work has been described by Eberhard (1981). The remaining studies included the release of tetroons that were tracked by radar within the Anderson and Putah Creek drainage areas by researchers from the U. S. Forest Service (Riverside). These were released individually as well as in clusters of three at a height of 100 m from the two sites shown in Fig. 1. Thus, the tetroons were flown in the transition layer overlying the drainage flows within the two valleys, and provided direct measurements of individual air parcel trajectories and the dispersion characteristics of these air parcels. A description of the radar and the data analysis techniques are reported by Fosberg and Lanham (1983).

The meteorological measurement systems dedicated to these experiments included an array of nine acoustic sounders, seven tethersondes, eight optical anemometers, two rawinsondes, one minisonde, 27 surface meteorological stations with real-time telemetry, as well as ten 10 m and one 60 m meteorological towers. Most of the acoustic sounders, tethersondes, and optical anemometers were dedicated to evaluating the characteristics of the drainage flows within the three major drainage areas and to evaluate the interactions of these flows with the transition layer flows and the pooling of the flows within the valley basin. The surface stations, on the other hand, provided a comprehensive view of the surface flow characteristics over the entire valley. Finally, the rawinsondes and the minisonde, which were located outside the valley, were dedicated to defining the regional scale flow situation within which the Anderson Creek valley was imbedded. These meteorological studies were conducted by investigators from the Argonne National Laboratory, Battelle-Pacific Northwest Laboratory, Los Alamos National Laboratory, Lawrence Livermore National Laboratory, National Center for Atmospheric Research, NOAA Atmospheric and Turbulence Diffusion Laboratory, NOAA Wave Propagation Laboratory, and the Savannah River Laboratory.

The conduct of the August 1981 experiments in the Big Sulfur Creek valley was similar to that described above for the September 1980 experiments in the Anderson Creek valley except less resources were available. The Big Sulfur Creek valley, situated immediately west of the Anderson Creek valley, resembles a standard V-shaped valley aligned along a northwest-southeast direction with outflow toward the northwest. The major difference between the Anderson Creek valley and the Big Sulfur Creek valley experiments was associated with definition of the cooling tower plume dimensions and the interactions of these plumes with the nocturnal drainage flows. Using the plume from the Pacific Gas and Electric Company geothermal power plant No. 14, situated in the bottom of the Big Sulfur Creek valley, the plume dimensions were defined by an airborne lidar and the visible portion of the plume was catalogued by surface-based time lapse photography. The lidar was developed and operated by researchers from SRI International, and has been described by Uthe, et al. (1980). The photographic documentation was performed by participants from the Battelle-Pacific Northwest Laboratory. The perfluorocarbon tracers were again used in these experiments. The PMCH tracer was injected into the cooling tower plume, while the PDCH tracer was released simultaneously on the surface (directly into the drainage flows) near the power plant to test the transferability of the methodologies, developed in the Anderson Creek valley, to the Big Sulfur Creek valley.

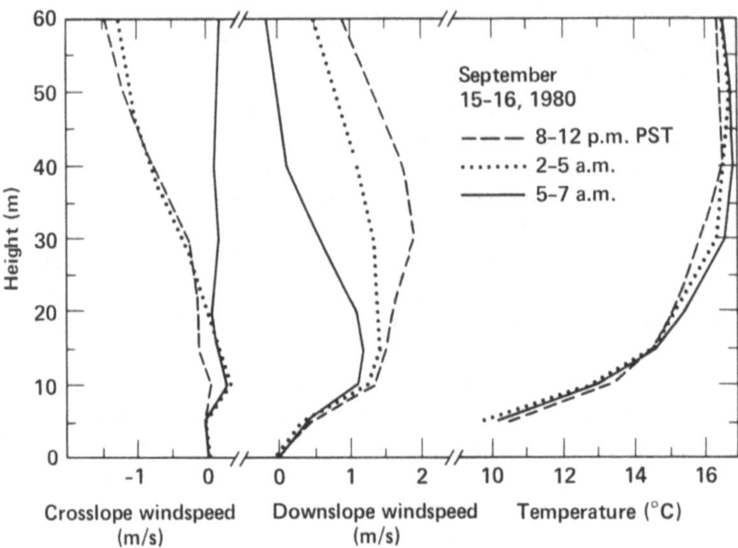

Fig. 2. Typical wind and temperatue profiles from tower on south slope of Cobb Mountain.

EXPERIMENTAL RESULTS

The following provides a description of the drainage flow characteristics observed within the Anderson Creek valley. This includes a description of the vertical and horizontal temperature and wind structure, the interactions of the drainage flows with the external environment, and the tracer transport and dispersion characteristics of the drainage flows. Also included are some of the results of the airborne lidar observations of the cooling tower plumes and the tracer distributions resulting from the cooling tower plume and surface releases in the Big Sulfur Creek valley.

Meteorological Observations in the Anderson Creek Valley

Typical observations of the temperature and wind structure, reported by Horst and Doran (1981), for good drainage flow situations within an area exhibiting relatively simple slope flows are shown in Fig. 2. These data were acquired from a 60 m meteorological tower situated on the south slope of Cobb Mountain between the Anderson and Gunning Creeks. The temperature profile is characterized by a surface based inversion of 40 to 50 m depth produced by the radiational cooling effects. Generally, the depths varied between 30 m and above 60 m during good drainage

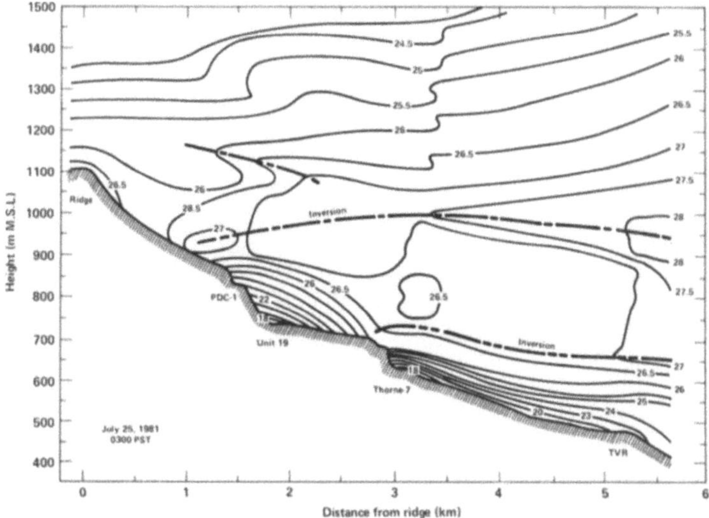

Fig. 3. Temperature distribution along a northwest-southeast cross-section of the Anderson Creek valley (Analysis by Orgill).

flow situations. Below the temperature inversion, the winds in Fig. 2 are essentially downslope with a low level maximum of 1 to 2 m/s at a height that is roughly half the depth of the inversion. In the vicinity of the inversion and above it, the cross-slope component may increase with height depending on the direction and speed of the regional scale winds.

As flows from the higher elevations merge with those from other drainage areas and encounter irregular topographic features and forest canopies, very complicated flow structures may appear within the valley. This is illustrated by Orgill (1982) in Fig. 3, which shows the temperature distribution along a northwest-southeast cross-section of the Anderson Creek valley observed during the July 1979 field experiments. Near the ridge only very shallow drainage currents are formed. As the flows from adjacent slopes merge, they may flow over one another depending on their relative densities, and may collect within minor topographic depressions as indicated at the Unit 19 site in Fig. 3. The flows deepen upon descent into the basin due to the continuous entrainment of air into the drainage layer and finally become involved in a basin wide circulation that appears to have the characteristics of a pool of cold air that slowly exits the valley toward the southeast. As indicated by the temperature

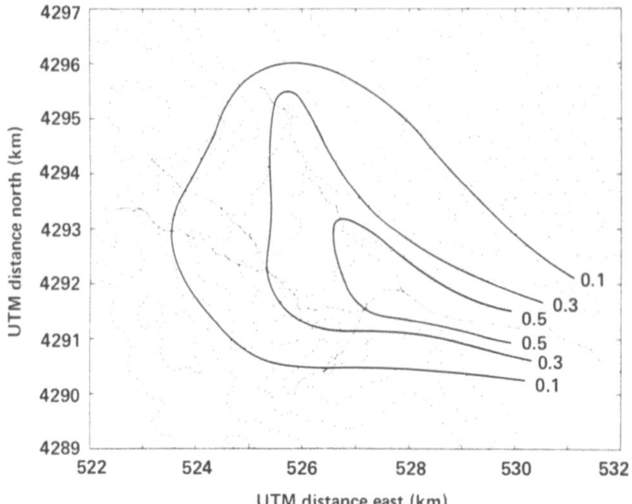

Fig. 4. Cooling rates (°C/hr) measured two meters above the surface on September 19-20, 1980 (Experiment 4) when drainage flows were well established.

inversion in Fig. 3, the pool of cold air is generally about 200 to 300 m deep over the valley basin and decreases in depth as one moves up the slope. One may also note a higher level inversion at about 950 m above the valley basin delineating the upper boundary of the transition layer flows overlying the drainage flows.

The drainage flows observed within the Anderson Creek valley are generally influenced to some extent by the larger scale flows. Besides the synoptic scale migratory systems, the principal influences are due to westerly sea breezes and descending upper-level easterly flows. The intrusion of off-shore marine air during the summer has been previously reported by Fosberg and Schroeder (1966). Sea-breeze intrusions into the Anderson Creek valley are not always easy to detect because the wind directions within the valley are still downslope. However, the rate of cooling of surface air and the resulting temperature distributions along the valley slopes are rather sensitive indicators of external flow intrusions. Fig. 4 depicts the cooling rates observed during Experiment 4 conducted September 19-20, 1980, when well established drainage flows existed throughout the valley. Note the cooling rates are highest within the valley basin leading to the lowest temperatures within the bottom of the

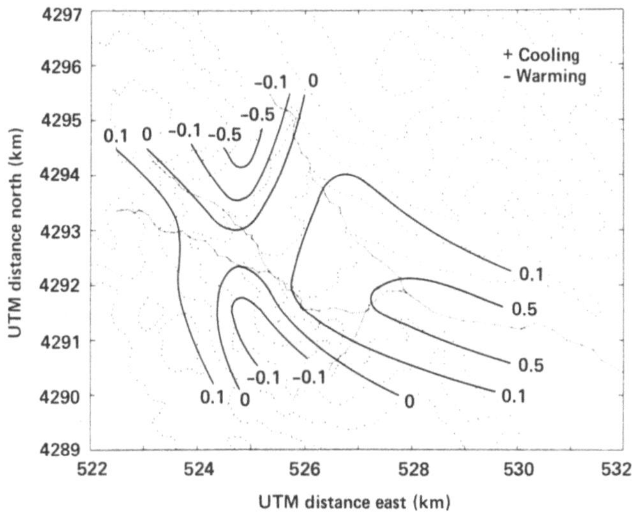

Fig. 5. Cooling rates (°C/hr) measured two meters above the surface on September 23-24, 1980 during period of descending northeasterly flows into the Anderson Creek valley.

valley; in agreement with the observations of Nappo (1983) and Gudiksen and Walton (1981). This contrasts significantly with the essentially uniform cooling rates observed during periods of strong marine intrusions when radiational cooling is not able to maintain the surfaced based inversion needed to develop the drainage flows due to enhanced vertical mixing processes except in relatively protected areas.

Quite frequently, easterly flows over the ridges were noted to protrude into the Anderson Creek valley. These almost invariably produced a warming phenomenon within the middle slopes of the valley as depicted by the cooling rates observed on September 23-24, 1980 and shown in Fig. 5. Hence, these easterlies are capable of preventing the development of the surface based inversion that is needed to maintain the drainage flows at the middle elevations; while leaving the flows at the lower elevations relatively unaffected.

Tracer Studies Within the Anderson Creek Valley

A considerable amount of tracer data were acquired during the September 1980 studies in the Anderson Creek valley. For the sake of brevity we will only summarize some of the highlights of

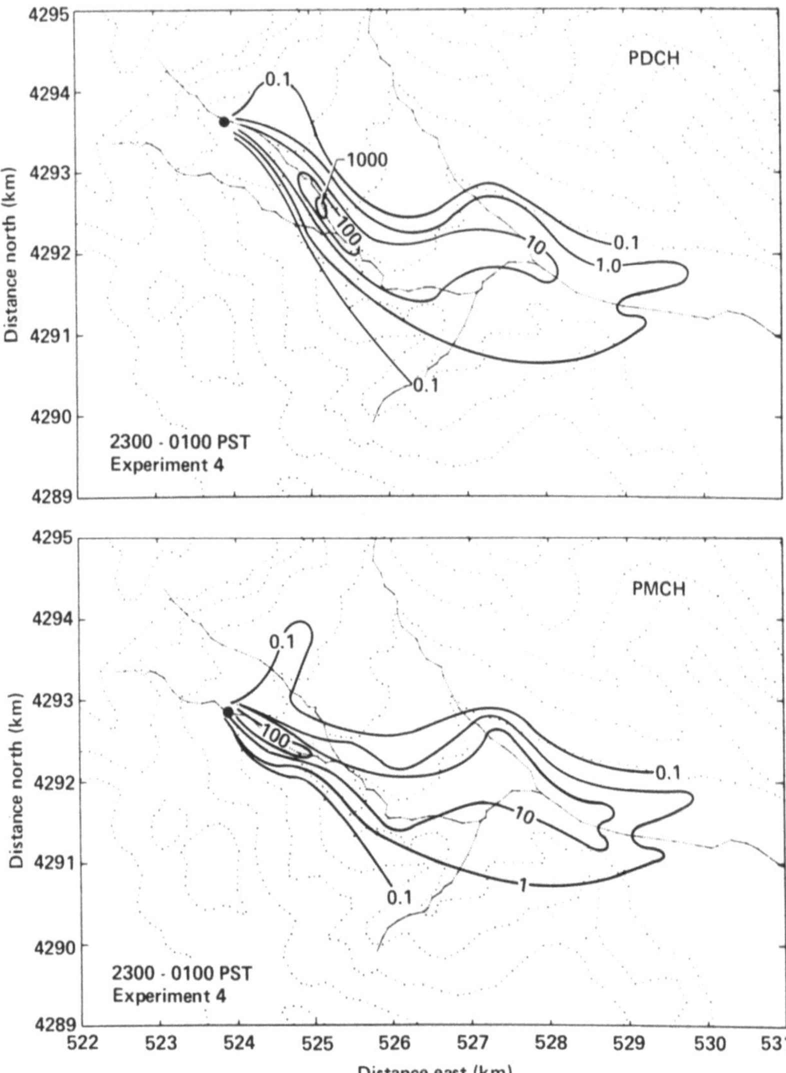

Fig. 6. Surface concentration patterns of the PDCH released from Gunning Creek and the PMCH released from Anderson Creek on September 19-20, 1980 (Experiment 4). The concentrations are in units of ppt and are averaged over the first two hours (2300-0100 PST) after the initiation of the releases. A total of 471 g of PDCH and 416 g of PMCH were released during the one hour release period.

Experiment 4 of the September 1980 series. A complete summary of the experiments have been reported by Gudiksen, et al. (1983). The surface distributions of the PMCH and PDCH tracers, released over a one-hour period within the Anderson Creek and Gunning Creek drainage areas, are shown in Fig. 6 for the first two hour period after the initiation of the release.

The two tracers showed very similar concentration patterns with the plume centerlines following the Creeks rather closely. The two plumes merge near the confluence of the Anderson and Gunning Creeks, then proceed in a southeasterly direction toward the Anderson Springs area. However, a significant northward transport may be noted by the bending of the plumes up into the Putah Creek drainage area. Note also that the plumes are relatively narrow within the first two kilometres of the source and then appear to spread out horizontally at an accelerated rate with an attendent decrease in the concentrations. A comparison of the relative PDCH and PMCH concentration isopleths reveal that the PDCH plume exhibits centerline concentrations within the thickly forested Gunning creek drainage area that are about 3 to 10 times higher than the corresponding PMCH concentrations in the Anderson Creek drainage areas for the first 1 to 2 km from the release sites; thereafter, the concentrations of the two tracers are essentially identical. Thus, in this particular case, it appears that the principal effect of the forest canopy on the transport and dispersion of the tracer was to inhibit mixing within the canopy to produce the more concentrated plume. These surface concentration patterns appear surprisingly similar from one experiment to another, both spatially and temporally as well as in magnitude in spite of rather wide variations in the regional flows. Thus, the drainage flows responsible for the transport and dispersion of the perfluorocarbon tracers appeared to be fairly well decoupled from the external environment when strong drainage flows have developed. This may not always be the case, however, depending upon the physical exposure of the release site and the transport path to the larger scale flows. The sequential samplers, situated at selected downslope distances from the release sites, indicate an average transport speed of about 1 ms^{-1} to the valley floor. The plume passage times within the valley basin and the outflow region were about 5 to 6 hours for the one hour releases, which supports the concept, proposed by Barr (1983) of a slowly drifting and meandering tracer plume within the valley basin prior to flowing out of the basin toward the southeast.

The vertical distribution of the tracers over the valley basin were noted to be strongly governed by the widely varying external flow situations. Very complex circulation systems can be generated aloft due to the influence of the regional scale

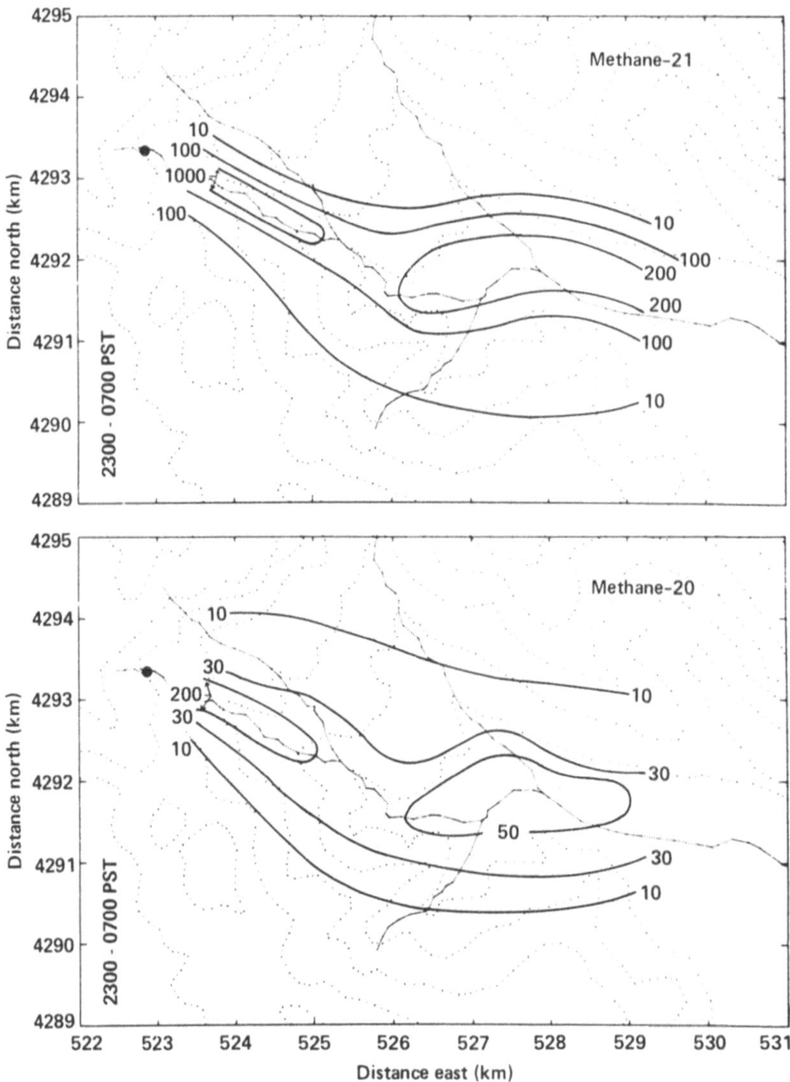

Fig. 7. Eight-hour averaged surface concentration patterns for methane-21 and methane-20 observed from 2300-0700 PST on September 19-20, 1980 (Experiment 4). The units are in ppt based on 1 kg releases. The actual amounts released were 1.27 g of methane-21 and 9.24 g of methane-20.

flows, which may serve to markedly alter the vertical concentration gradients. In spite of these influences, the tracer plume fronts persistently seemed to be transported within elevated layers over the valley basin with the bulk of the tracer arriving at the surface of the valley basin within 1 to 2 hours.

With the exception of being somewhat broader, the surface concentration distributions produced by the two heavy methanes were quite similar to those for the perfluorocarbons as shown in Fig. 7. By comparing the relative concentrations of the methane-21, released at the surface within the upper reaches of the Anderson Creek drainage area with the methane-20, released at a height of 60 to 75 m within the lower levels of the transition layer, one finds a pattern that is typical of all experiments. The methane-21 to methane-20 ratios are mostly near unity, except near the centerline of the plumes where the ratios are typically within the 3 to 5 range. This indicates that considerable mixing did occur between the transition layer flows and the underlying drainage flows; possibly, because the tracers were released near the ridge top where the drainage flows may be more exposed to the external conditions. Ratios less than unity, observed on the fringes of the plumes, reveal enhanced horizontal dispersion of the methane-20 tracer at the elevated heights.

The tetroons, which were released at a height of 100 m within the Putah and Anderson Creek drainage areas, were flown within the transition layer flows. The individual tetroon trajectories, acquired by the radar during several of the September 1980 experiments, showed persistent down canyon flows within the steep-walled Putah Creek drainage area while those within the much wider and broader area showed considerable variability. The dispersion coefficients derived from these trajectories, when compared with those acquired by Pasquill, Gifford, and Turner over flat terrain, fall near the A and B stability curves. Additional values, determined from the perfluorocarbon surface tracer distributions, within the first 1.5 km from the release point before the plumes become involved in complex circulation systems over the valley basin, appeared to be somewhat lower; possibly reflecting the constraints of topography and canopy influences. Thus, the dispersion coefficients obtained from the PMCH patterns fall between the B and C stability categories; while those derived from the PDCH distributions lie between the C and F curves.

Drainage Flow Studies in the Big Sulfur Creek Valley

One notable difference between the nocturnal flows generated within the Big Sulfur Creek valley with those in the Anderson Creek valley is that the drainage flows generated along the steep

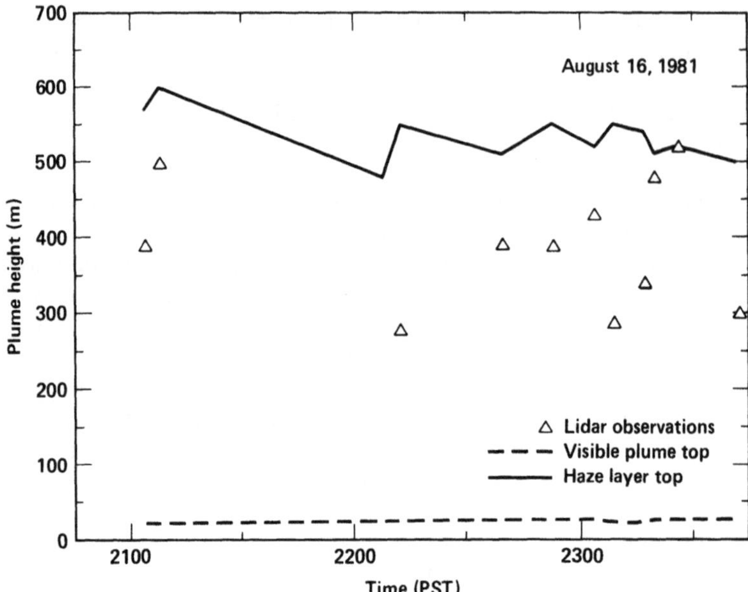

Fig. 8. Cooling tower plume heights determined by airborne lidar and surface based photography.

sidewalls in the Big Sulfur Creek valley drain into the valley to form a roughly 200 m deep valley flow regime which outflows down-valley toward the northwest - in direct opposition to the north-westerly sea breezes; whereas, in the Anderson Creek valley the drainage flows and the seabreezes are generally aligned with one another.

The cooling tower plume heights determined by the airborne lidar and the surfaced based photography during Experiment 2, conducted on August 16-17, 1981, are shown in Fig. 8. The heights of the visible plume appears to be less than 50 m while the lidar was able to detect the subvisible portions of the plume to heights generally ranging between 300 to 400 m. Both perfluorocarbon tracers were used in these experiments. The PMCH was released over a one hour period into the cooling tower plume after the drainage flows were well established; while the PDCH was released simultaneously at a surface location adjacent to the cooling tower. The hourly-averaged surface concentrations of these tracers two hours after the initiation of the releases are shown in Fig. 9. Note that the PMCH concentrations are generally an order of magnitude less than the corresponding PDCH concentrations due to the enhanced dilution within the cooling tower plume. Also note the generally downvalley transport of both

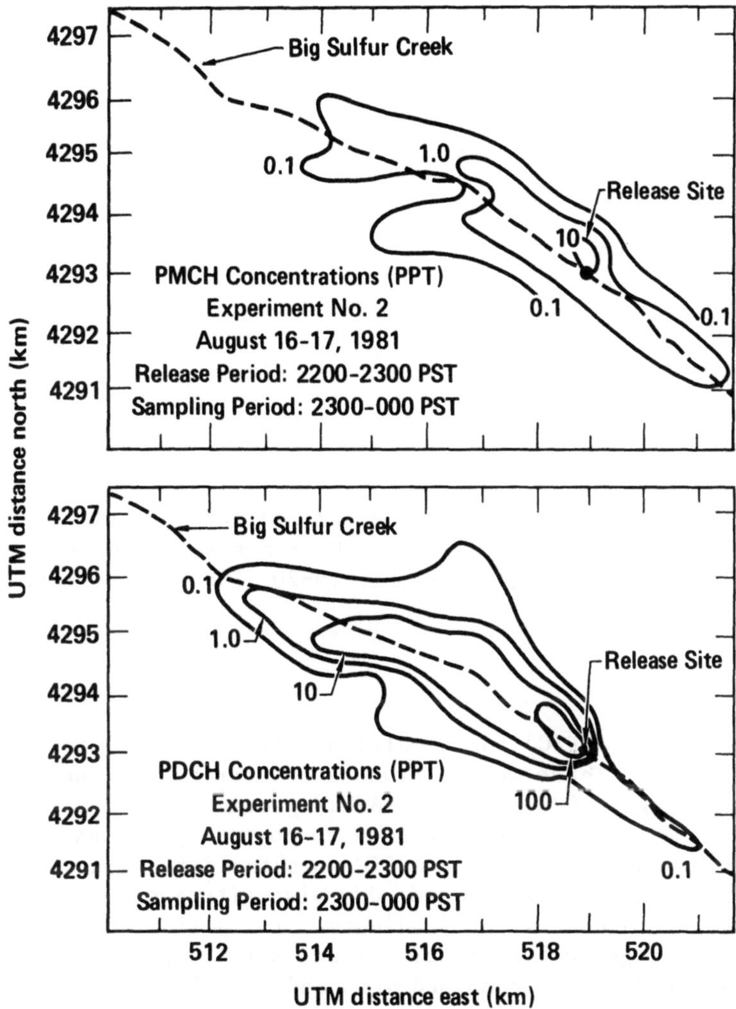

Fig. 9. Surface distributions of PDCH and PMCH tracers.

tracers, except a noticeable fraction of the PMCH was transported up-valley from the power plant since the cooling tower plume was sufficiently elevated to become at least partially entrained in the northwesterly sea breezes above the down-valley drainage flows.

MODELING STUDIES

The results from these experiments have been utilized to evaluate a three-dimensional mass-consistent diagnostic wind field model (MATHEW) and a particle-in-cell transport and

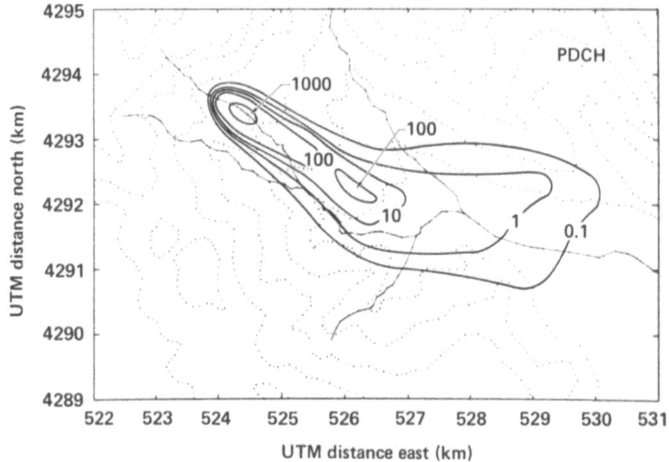

Fig. 10. MATHEW/ADPIC computed surface concentration distribution of the PDCH tracer released in Gunning Creek during September 19-20, 1980 (Experiment 4).

diffusion model (ADPIC) for simulating pollutant behavior in drainage flows. These models have been described by Sherman (1978) and Lange (1973, 1978). MATHEW provides ADPIC with hourly averaged wind fields which are used to compute pollutant transport by the mean winds. Then ADPIC computes diffusion velocities based on empirically-determined eddy viscosity coefficients and the pollutant concentration gradients. The pollutant is represented by a large number of Lagrangian "marker" particles which are transported by the combined mean and turbulent velocities. The pollutant concentrations are obtained by counting the particles within each grid cell.

To date, models were run using the data from Experiment 4 of the September 1980 series in the Anderson Creek valley and Experiment 2 of the August 1981 series in the Big Sulfur Creek valley. For the Anderson Creek valley study, the calculational domain was 7 by 10 km extending 1100 m above the lowest point in the topography with the individual cells being 250 by 250 by 50 m in the x, y and z directions, respectively. Using the hourly averaged meteorological data from the surface stations, tethersondes, and acoustic sounders along with pertinent tracer release information, the models computed the concentration distributions of all the gaseous tracers utilized in this experiment. As an example of the results, Fig. 10 shows the computed surface concentration pattern of the PDCH tracer released within the Gunning Creek drainage area. This pattern may be compared directly with

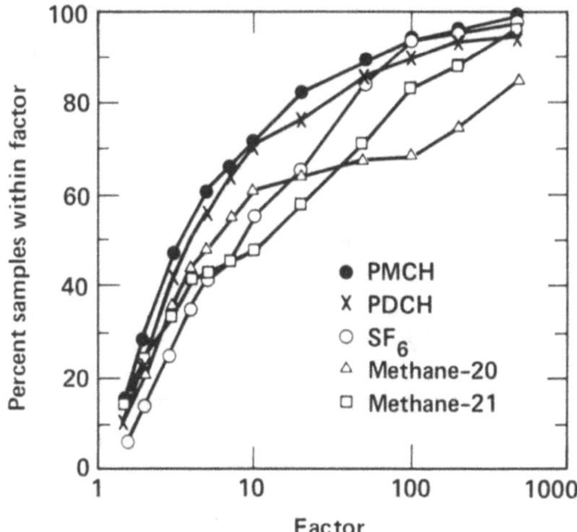

Fig. 11. Percent of observed-to-computed concentration ratios that are within a specified factor.

the measured distribution shown in Fig. 6. Generally the results of the calculations agree reasonably well with the observed concentration patterns. The ratios of the measured to the computed concentrations for the two perfluorocarbon, the two heavy methane, and sulfur hexafluoride tracers released in Experiment 4 were within roughly a factor of 5 for 50% of all samples as shown in Fig. 11. However, agreement within a factor of two occurred only in 25% of the samples. Some of the discrepancies between the calculations and the observations are due to point measurements being compared with grid-volume averaged values in the model. This problem becomes most critical close-in to the sources and in areas strongly influenced by local topography. In consonance with the observations, the calculations indicated that the diffusivity parameters over the study area are larger by a factor of 2 to 5 than the corresponding values over flat terrain.

Similar calculations were undertaken to estimate the surface concentrations of the PMCH tracer injected into the cooling tower plume in the Big Sulfur creek valley as well as the PDCH tracer released at the surface directly into the drainage flows during Experiment 2 of the August 1981 studies. Using the lidar observations of the cooling tower plume to define the initial plume dimensions, and the hourly averaged surface and upper air meteorological observations, a series of calculated surface concentration patterns were derived and compared to the hourly-averaged tracer concentrations acquired during the first five hours after the release at approximately 50 sites. The ratios of

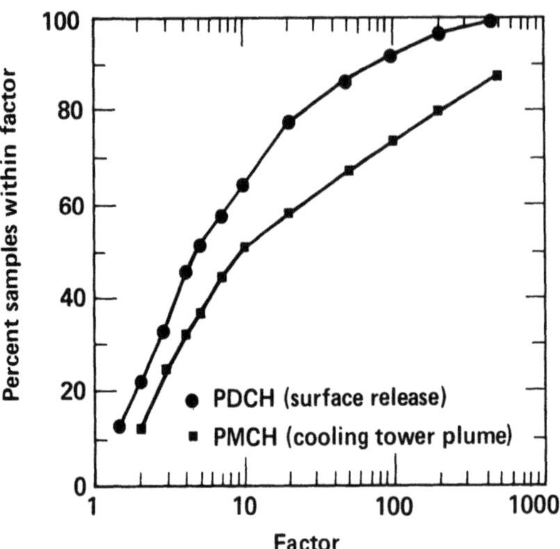

Fig. 12. Percent of observed-to-computed concentrations that are within a specified factor.

the observed to the computed surface concentrations for the two tracer plumes are shown in Fig. 12. Note the ratios for the PDCH plume, which are within a factor of 5 for 50% of the samples, agree well with similar surface releases within the Anderson Creek valley; while the PMCH plume, which is within a factor of 10 for 50% of the samples, agrees fairly closely with the elevated methane-20 release within the Anderson Creek valley. Thus, similar results were obtained within the two study areas indicating that the modeling capabilities are expected to be transferable to other complex terrain sites.

SUMMARY

The results of these experiments indicate that the drainage flows generated at the higher elevations within the Anderson Creek valley are generally a few tens of meters in depth. As the flows from the different slopes merge and encounter irregular terrain and forest canopies, very complicated flow structures are developed. The depths of these flows generally increase upon descent due to continuous entrainment of air from the overlying transition layer to produce a pool of cold air within the valley basin. This pool, typically 200 to 300 m in depth, is often involved in valley-wide circulations. The drainage flows along the upper elevations as well as this pool of cold air in the basin may be influenced to varying degrees by the larger scale flows; principally, the westerly sea breezes and descending upper-level easterly flows. These influences may prevent the development of the drainage flows or significantly alter their characteristics even though strong radiational cooling is occurring.

A series of tracer experiments were carried out in the Anderson Creek valley. One series, which involved the release of tracers near the surface at mid-slope elevations, indicated that during good drainage flow conditions the surface tracer distributions were similar from one drainage flow situation to another. Thus, these tracers followed the creeks down the slopes and then subsequently became involved in the valley wide circulations. Another series of tracer releases revealed that considerable mixing does occur between the transition layer flows observed high up on the slopes near the ridge top. Thus, the surface concentrations within the valley basin resulting from a tracer injected into the lower levels of the transition layer at such a site may only be a factor of 3 to 5 less than those produced by a tracer injected directly into the drainage flows at the same site.

The dispersion rates obtained from tetroon flights in the transition layer and from the surface tracer distributions reflect the constraints of topography and forest canopies. Typically, the dispersion coefficients within the transition layer fall near the Pasquill-Gifford-Turner A and B stability categories for flat terrain, while those obtained from the surface tracer releases lie between the B and F stability curves.

Studies of the interactions of a geothermal power plant cooling tower plume indicate that the plumes, rising 300 to 500 m above the tower, do become involved with drainage flows as well as the overlying larger scale flow regime. Thus, a tracer released in the cooling tower plume produces surface concentrations that are generally an order of magnitude less than those produced by a surface release.

Some of the results of these field experiments have at this time been used to improve and evaluate three-dimensional, mass-consistent wind field and particle-in-cell transport and diffusion models. It appears that the models' performance, measured by the ratios of the computed to the observed tracer concentrations acquired during one series of tracer releases in the Anderson Creek valley, were within a factor of 5 for 50% of the comparisons. Similar results were obtained when using the data from one of the surface tracer releases in the Big Sulfur Creek valley, implying that these modeling capabilities are expected to be transferable from one area to another.

REFERENCES

Barr, S, 1983, A comparison of lateral and vertical diffusion in several valleys. Presented at the Fourth Conference on the Meteorology of the Upper Atmosphere, American Meteorological Society, March 22-25, 1983, Boston, MA.

Cowan, G. A., Ott, D. G., Turkevich, A., Machta, L. Ferber, G. J., and Daly, N. R., 1976, Heavy methanes as atmospheric tracers. Science, 191, 1048-1050.

Ferber, G. J. Telegadas, K., Heffter, J. L., Dickson, C. R., Dietz, R. N., and Krey, P. W., 1981, Demonstration of a long range atmospheric tracer system using perfluorocarbons. NOAA Tech. Memo. ERL ARL-101. NOAA Air Resources Laboratories, Silver Spring, MD.

Fosberg, M. A., Lanham, L. M., 1983, Above-canopy dispersion in nighttime downslope flow. Presented at Seventh Conference on Fire and Forest Meteorology, American Meteorological Society, April 25-29, 1983, Ft. Collins, CO.

Fosberg, M. A., and Schroeder, M. J., 1966, Marine air penetration in central California, J. Appl. Meteor., 5, 573-589.

Fowler, M., 1979, Methane tracer system development. Proceedings of the Atmospheric Tracers and Tracer Application Workshop. LA-8144-C. Los Alamos Scientific Laboratory, Los Alamos, New Mexico.

Gudiksen, P. H., and Walton, J. J., 1981, Categorization of nocturnal drainage flows in the Anderson Creek valley. Presented at Second Conference on Mountain Meteorology, November 9-12, 1981, Steamboat Springs, CO.

Gudiksen, P. H., Ferber, G. J., Fowler, M. M., Eberhard, W. L., Fosbert, M. A., and Knuth, W. R., 1983, Field studies of transport and dispersion of atmospheric tracers in nocturnal drainage flows. University of California Lawrence Livermore National Laboratory, UCRL-88931.

Horst, T. W., and Doran, J. C., 1981, Observations of the structure and development of nocturnal slope winds. Presented at Second Conference on mountain Meteorology, November 9-12, 1981, Steamboat Springs, CO.

Lange, R., 1978, ADPIC - A three-dimensional particle-in-cell model for the dispersal of atmospheric pollutants and its comparison to regional tracer studies, J. Appl. Meteor., 17, 320-329.

Lovelock, J. E., and Ferber, G. J., 1982, Exotic tracers for atmospheric studies, Atmospheric Environment, 16, 1467-1471.

Nappo, C., 1983, private communication.

Orgill, M. M., 1982, Wind and temperature discontinuities from three-dimensional slope effects. Presented at ASCOT meeting, May 10-14, Gettysburg, PA.

Sherman, C. A., 1978, A mass-consistent model for wind fields over complex terrain, J. Appl. Meteor., 17, 312-319.

Uthe, E. E., Nielsen, N. B., and Jimison, W. L., 1980, Airborne lidar plume and haze analyzer, Bull. Amer. Meteor. Soc., 61, 1035-1043.

DISCUSSION

A. VENKATRAM — Does your flow field model use the momentum or energy conservation equations ?

P. GUDIKSEN — The MATHEW/ADPIC models do not include the momentum or energy conservation equations; however, other models within the ASCOT program do.

S.E. GRYNING — Did you observe oscillations in the strength of the drainage winds; were you able to model these.

P. GUDIKSEN — Yes, we did observe oscillations in the strength of the drainage flows. Many of these variations are due to interactions of the meso-scale flows over the ridges with the drainage flows. Variations that have time and spatial dimension comparable to or greater than those in the models may be resolved by the models.

REGIONAL-SCALE POLLUTANT TRANSPORT

STUDIES IN THE NORTHEASTERN UNITED STATES

J.F. Clarke*, J.K.S. Ching*, T.L. Clark*
and N.C. Possiel*

Meteorology and Assessment Division
Environmental Sciences Research Laboratory
U.S. Environmental Protection Agency
Research Triangle Park, NC 27711 USA

INTRODUCTION

There has been an effort in recent years to assess the importance to local regulatory strategies of regional-scale pollutant transport, particularly for secondary pollutants such as oxidants, nitrate, and sulfate. The U.S. Environmental Protection Agency (EPA) has established modeling and field programs to address regional-scale transport of both ambient oxidant and acidic materials.

The EPA program to address regional-scale transport of oxidants is summarized here. The focal point of the program is the development and application of the Regional Oxidant Model (ROM) as part of the Northeast Corridor Regional Modeling Project (NECRMP). The ROM (Lamb, 1983) is a regional-scale Eulerian transport and photochemistry simulation model that will be used to provide inflow boundary conditions of ozone and precursors for application of urban scale air quality simulation models to cities in the northeastern U.S. The ROM will also provide a regional perspective for assessing the impact of oxidant control strategies of individual cities. The model is formulated to treat those chemical and physical processes that are presently thought to affect photochemical oxidant concentrations over diurnal time scales (e.g., nocturnal wind shear, nighttime chemistry, and effects of clouds on perturbing

*On assignment from the National Oceanic and Atmospheric Administration, U.S. Department of Commerce

the photochemical reactions and venting of pollutants from the boundary layer). The area being modeled is truly regional in scope, extending from western Ohio to the East Coast and from northern Virginia across southern Maine and Ontario (see Fig. 1).

Several field studies were conducted in support of NECRMP. These were the Northeast Regional Oxidant Study (NEROS) field programs and the Urban Field Studies. The NEROS field programs, conducted during August 1979 and from mid-July to mid-August 1980, consisted of three basic sampling scenarios: 1) Regional Air Mass Characterization (RAMC) sampling designed to obtain model initialization and evaluation data over the extent of the ROM grid; 2) Regional Atmospheric Lagrangian (REAL) experiments involving Lagrangian characterization of the photochemistry within an air mass segment marked by a tetroon; and 3) Regional Urban Plume Characterization (RUPC) studies designed to obtain data to characterize the evolution of the urban plume from its source area to the point where it blends with the regional background ozone burden. In addition, Urban Field Studies were conducted in four major Northeast Corridor cities during the summer of 1980 to obtain data for application of urban scale models.

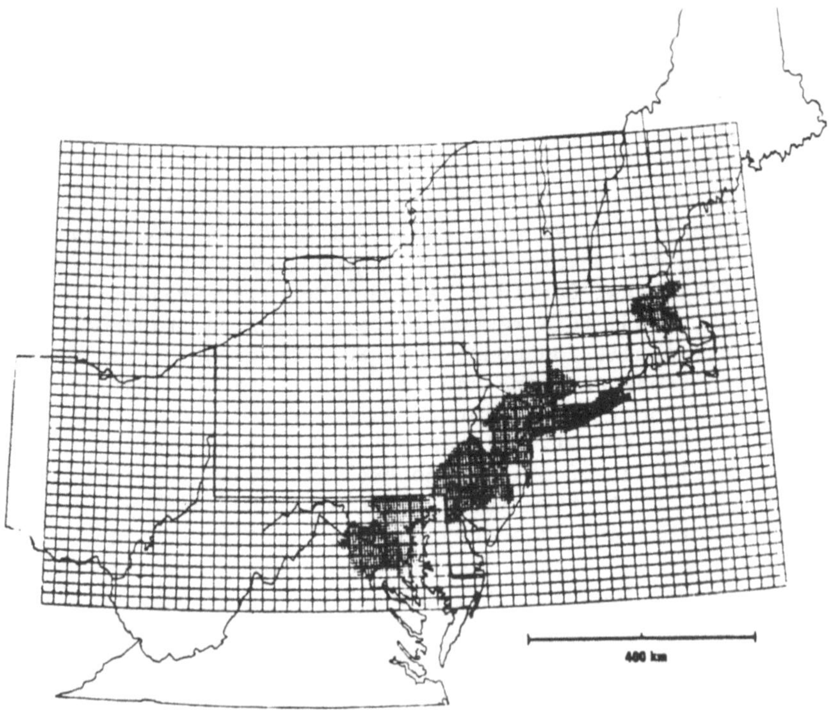

Fig. 1. Boundaries of regional (large grids) and urban (small grids) modeling domain.

This paper provides a brief overview of the various field studies and resulting data bases. The primary sampling scenarios are described and some results that were analyzed for the purpose of data base preparation and evaluation for modeling exercises are summarized.

DESCRIPTION OF FIELD PROGRAM

Aircraft were the basic sampling platforms for all the field programs. Most aircraft were instrumented for *in situ* continuous sampling of nitric oxide (NO), total oxides of nitrogen (NO_x), ozone (O_3), nephelometer scattering coefficient, ambient temperature, relative humidity or dew point, and for grab sampling for later analysis of hydrocarbon species. Five aircraft, based at three locations in the Northeast, were used in the 1979 program, which emphasized RAMC and REAL experiments coordinated from Raleigh, North Carolina. Thirteen aircraft were used in various components of the 1980 program. REAL experiments were coordinated from Columbus, Ohio; REAL and RUPC experiments were conducted from Columbus, Ohio, and from Baltimore, Maryland. Urban Field Studies were conducted in Washington, D.C.; Baltimore, Maryland; New York, New York; and Boston, Massachusetts. A number of special experiments, including studies on ozone deposition and cloud venting of ozone from the boundary layer, were conducted near Lancaster, Pennsylvania, in 1979 and Columbus, Ohio, in 1980. The NECRMP field program and scenarios are diagrammed in Fig. 2.

NEROS RAMC Scenario

The Regional Air Mass Characterization (RAMC) scenario was designed to obtain data over the three-dimensional domain of the NEROS grid for evaluating the overall performance of the ROM. Aircraft were deployed to obtain air quality data over 24- to 48-h periods. Sampling was conducted throughout the diurnal cycle and included flights within the residual mixed layer at night and in the cloud layer above the mixed layer. Seven RAMC scenarios were conducted in 1979 and two in 1980 with varying degrees of success. Most RAMC aircraft sampling was conducted in a forecast Lagrangian mode, i.e., an initial volume of air was sampled near midday along the upwind edge of the ROM grid at different altitudes within and above the mixed layer. Subsequent sampling traverses were made at approximately 6-h intervals at the predicted downwind Lagrangian location of the air mass sampled on the original traverse. Thus each aircraft traverse provided evaluation data based on the same initial boundary conditions.

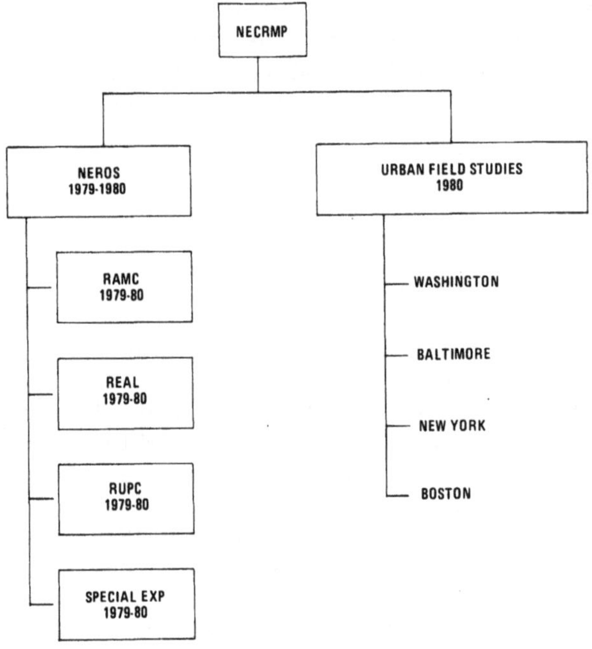

Fig. 2. Diagram of NECRMP field programs and scenarios.

The frequency of National Weather Service upper-air soundings at stations in the vicinity of the NEROS grid was increased to 6-h intervals and one additional site was added during both 1979 and 1980 programs (see Fig. 3). The expanded network provided a meteorological data base for the execution of the regional model and for sensitivity assessment of various diagnostic trajectory models to density of observations (see Clarke et al., 1983).

NEROS REAL/RUPC Scenarios

Regional Atmospheric Lagrangian (REAL) experiments were also conducted during 1979 and 1980. The scenario for these experiments involved the release and tracking of tetroons and subsequent Lagrangian pollutant sampling at the location of the tetroon. The tetroons were used to mark specific segments of polluted air masses, power plant plumes, and urban plumes. Each tetroon carried an instrument package weighing approximately 2 kg and consisting of an aircraft transponder and pressure encoder. The location and altitude of the tetroon was identified by Federal Aviation Administration regional flight service centers. The spatial and temporal coordinate information were relayed to the NEROS field headquarters from which the sampling aircraft were deployed. Twelve of the tetroons were released and tracked: two in 1979 and ten in 1980.

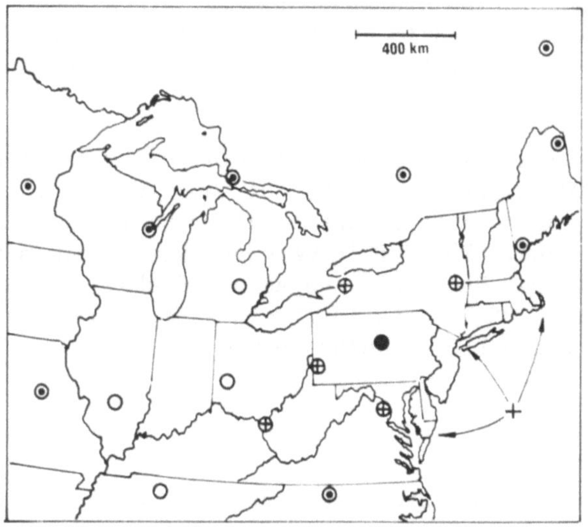

Fig. 3. Radiosonde stations near modeling area. The nine stations indicated by open circles and circles with plus signs took radiosondes at 6-h intervals during the 1980 NEROS; the eight stations indicated by a plus sign took radiosondes at 6-h intervals during the 1979 NEROS; the site in central Pennsylvania indicated by a solid circle was a special 6-h station established for the 1979-1980 NEROS programs.

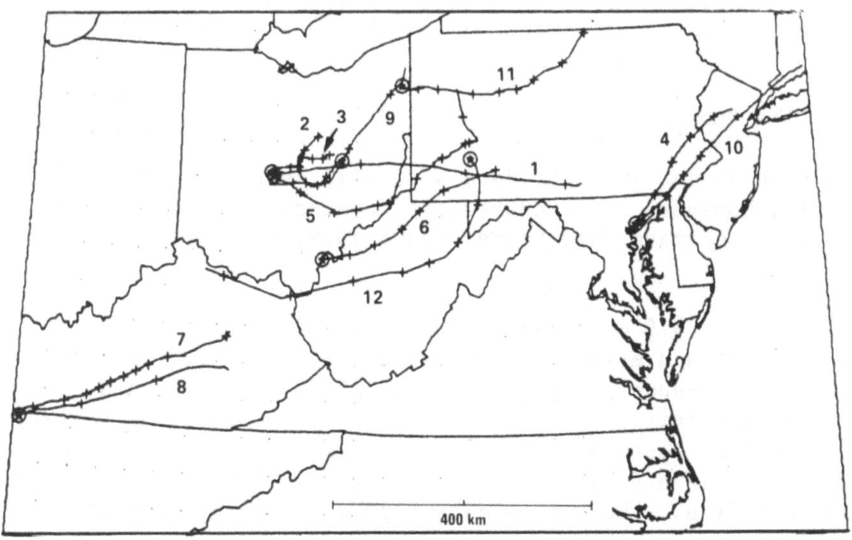

Fig. 4. Flight tracks of tetroons with positions every three hours marked by (+).

The tetroon trajectories are shown in Fig. 4. A case study of a REAL experiment conducted in conjunction with a Regional Urban Plume Characterization (RUPC) experiment is given by Clark and Clarke (1984).

RUPC experiments were conducted only during the 1980 field study. The RUPC scenario consisted of sampling within an urban plume from its source area to the point where it blended into the regional background. The RUPC's were conducted from both Columbus, Ohio, and Baltimore, Maryland, and as noted above, sometimes in conjunction with a REAL scenario, i.e., a tetroon was sometimes used to mark a segment of an urban plume.

Urban Field Studies

Intensive field measurements were obtained during the summer of 1980 in Washington, Baltimore, New York City, and Boston to acquire an ambient air quality and meteorological data base for application of the Airshed urban-scale model (Reynolds and Reid, 1978) to these cities.

The Airshed model requires that boundary and initial conditions of pollutant concentration, wind, and temperature be specified in three dimensions for each day modeled. Verification of the model requires specification of pollutant concentrations for each predictive time step. These requirements led to the development of a monitoring program consisting of continuous pollutant and meteorological measurements at the surface, pollutant measurements aloft (via aircraft), and upper-air meteorological measurements. An overview of these studies is provided by Possiel and Freas (1982).

NECRMP DATA BASE

A comprehensive NECRMP data base will eventually be compiled. Presently, however, it is convenient to discuss the individual elements as separate data bases: NEROS 1979, NEROS 1980, and Urban Field Studies.

The NEROS 1979 data base consists of RAMC measurements from three aircraft and enhanced radiosonde observations. In addition, there were site-specific data from Lancaster, Pennsylvania, consisting of aircraft profiles of pollutant concentrations (primarily ozone), surface ozone deposition measurements, and aircraft measurements of surface and cloud fluxes of ozone. These data are archived by platform, i.e., there is a data tape for each platform.

The 1980 NEROS data base is organized separately for two cities: Columbus, Ohio, the headquarters for the field program; and Baltimore, Maryland. The Columbus data base consists of pollutant and meteorological measurements from eight aircraft and many surface sites for RAMC, REAL, and RUPC experiments. A special feature of this data base is indirect measurements of ozone from a NASA aircraft equipped with a dual-wavelength lidar. Aircraft ozone flux measurements, both at the surface and in the vicinity of clouds, data from several ancillary platforms, and the enhanced radiosonde network (Fig.3) are also a part of this data base. An overview of the Columbus program with a listing of platforms is given by Vaughan et al. (1982). This data base is not complete but is to be archived both by platform and chronologically, i.e., the data will also be organized by date.

The 1980 NEROS Baltimore data base consists of *in situ* pollutant and meteorological measurements from three aircraft and a mobile van. There are also pollutant data from an enhanced surface network and a tethered balloon system, and meteorological data from pilot balloons, radiosondes, and acoustic sounders (Possiel and Freas, 1982). This data base is archived by platform and chronologically for 22 experimental days.

The Urban Field Studies data base consists of pollutant and meteorological measurements via aircraft for 13 days near Washington, 19 days near New York City, and 15 days near Boston. There are also pollutant and meteorological data from an enhanced surface network within 100 km of each of the cities and upper-air meteorological measurements from special radiosondes, pibals, and acoustic sounder in and near the cities.

A major effort to sample hydrocarbon species was incorporated into the overall field program. Over 550 grab samples (one to three minutes in duration) were collected from aircraft, and 790 one-hour integrated samples were collected at fixed urban surface sites in Columbus, Baltimore, Washington, New York City, and Boston. The samples were analyzed by gas chromatography for concentrations of 113 species.

The hydrocarbon measurements for both years and all cities have been collated into a comprehensive data base, which includes supplemental information on the ambient concentrations of other pollutants measured by the *in situ* aircraft and at the surface monitoring sites.

NECRMP DATA ANALYSES

Application of the data for modeling requires some preliminary preparation and evaluation regarding the quality and availability of data from the various platforms. Mission summaries were prepared describing the experimental scenarios, meteorological synopses, and platform deployment of the various field programs. More detailed presentations and evaluations were carried out for potential candidate days (based on quantity and quality of data) for model evaluation runs. This often led to interesting analyses and conclusions, some of which are summarized here.

Regional-Scale Transport

The following discussion summarizes a case study of the RAMC scenario for 3-4 August 1979. The analysis, which is based primarily on aircraft measurements, pertains to an air mass followed and sampled in a Lagrangian mode through a diurnal cycle as it was transported from Ohio to the East Coast. Details are given by Clarke and Ching (1983).

On 2 August a weak cold front passed through the Ohio Valley and moved to the East Coast where it became stationary for about 48 h. Boundary layer transport winds behind the cold front were generally westerly and initial conditions for Lagrangian air mass sampling (and upwind boundary conditions for the ROM) were established by aerial sampling along track A-B in Fig. 5 from 1151 to 1359 EST on 3 August. Diagnostic trajectories initialized at 1300 EST on 3 August at three points along track A-B are also shown in Fig. 5. Based on meteorological data and forecast trajectories obtained over the next 24 h, four subsequent aircraft traverses were made along tracks C-D, E-F, G-H, and I-J at the times indicated. The tracks were specified to approximate Lagrangian sampling of the air mass as it was transported to the East Coast.

The air quality and meteorological data for each aircraft transect were subjectively analyzed to describe the large-scale features along the transect. The analyzed ozone cross-section for the second transect (track C-D) is shown in Fig. 6. The track was the predicted 6-h displacement of the air mass sampled earlier along track A-B. The analysis depicts a plume with an ozone concentration near 160 ppb along the northern part of the transect and ozone concentrations near 80 ppb above the mixed layer in both the northern and southern parts of the track.

Isopleths of ozone concentration near the center of the mixed layer (or its nocturnal residual), extracted from cross sections of all the transects, are shown in Fig. 5. The analysis represents a

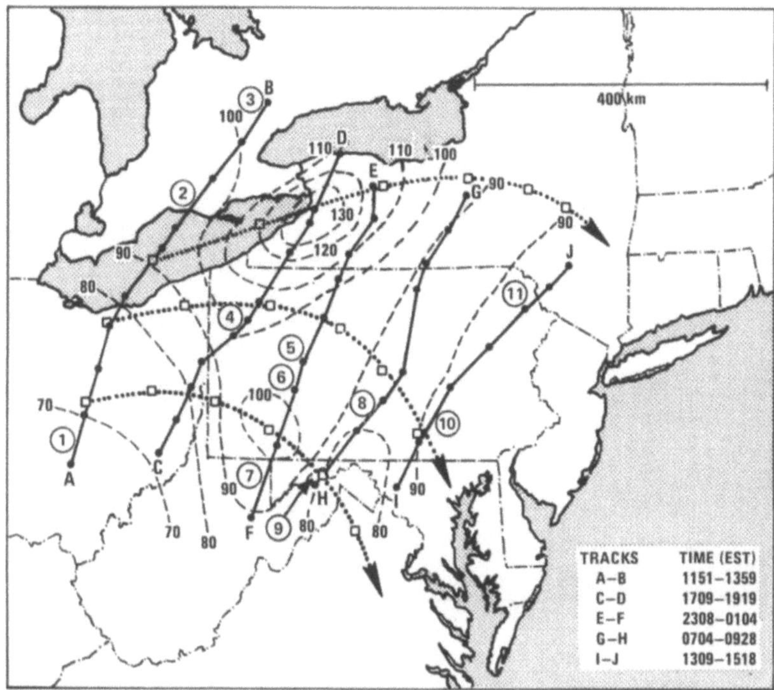

Fig. 5. Aircraft flight tracks, ozone concentrations (ppb) near the center of the mixed layer, and boundary layer trajectories initialized along track A-B at 1300 EST on 3 August 1979 (squares indicate 6-h positions). The circled numbers indicate the location of hydrocarbon samples.

quasi-Lagrangian field in that changes in concentrations along a trajectory roughly represent the Lagrangian derivative rather than a spatial change. Ozone concentrations were generally high over the grid and exceeded the primary 1-h average National Ambient Air Quality Standard for ozone (0.12 ppm) over a large area of western New York. A backward trajectory analysis suggested that the air mass over western New York at 1900 EST 04 August was over Chicago, Illinois, 36 h earlier and was transported over urban source areas in northern Indiana and northern Ohio.

The temporal gradients of ozone concentrations were consistent with those expected from production and depletion processes. Between 1300 and 1800 EST on 3 August (tracks A-B and C-D respectively in Fig. 5), ozone concentrations increased as would be expected from chemical production processes. The increase was largest in the northern portion of the grid, the region of highest precursor

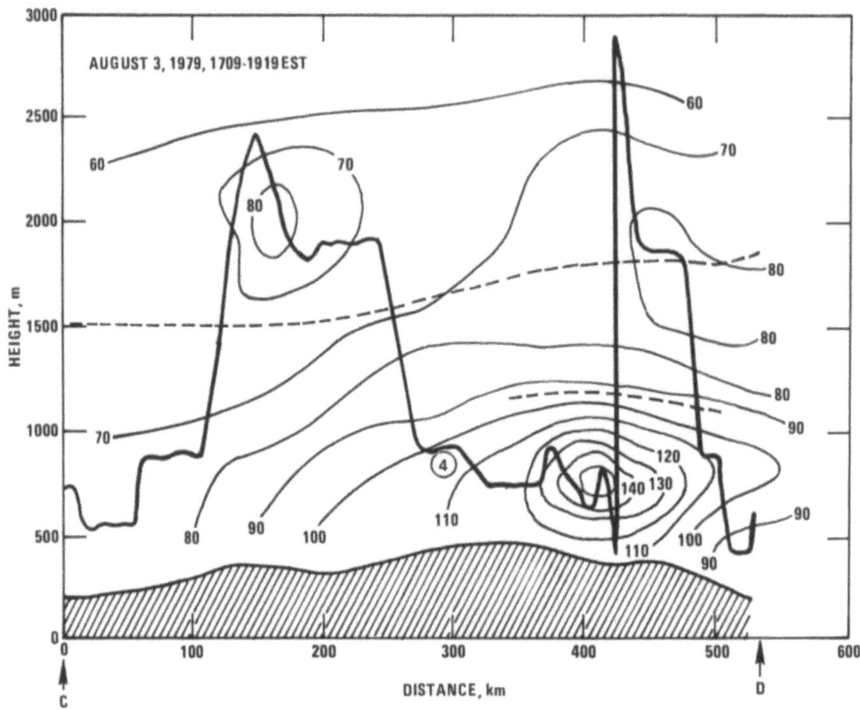

Fig. 6. Subjective analysis of ozone concentrations (ppb) measured along track C-D (see Fig. 5). The heavy solid lines indicate the aircraft flight path; the dashed lines indicate the top of the afternoon mixed layer; the circled number the location of a hydrocarbon sample.

emissions (Clark, 1980). Concentrations generally decreased between 1800 and 0800 EST the following morning and then increased on the afternoon of 4 August (track I-J). Concentrations were generally lowest during the early morning transect (track G-H). During the late evening through early morning hours, surface concentrations of ozone were between 0 to 30 ppb over the extent of the grid. These data support the concept of ozone being transported aloft at night while being depleted in the surface layer.

Transport of Ozone in Urban Plumes

On 14 August 1980, a segment of the Baltimore-Washington composite plume, marked by a tetroon, was tracked and sampled by aircraft for 400 km along the Northeast Corridor (Clark and Clarke, 1984). The tetroon trajectory and aircraft sampling tracks are shown in Fig. 7. The tetroon was released immediately downwind from Baltimore at 0945 EST and was transported about 11 h along the Northeast Corridor into southern Connecticut. Aircraft sampling was conducted upwind from Baltimore prior to the release of the tetroon and in the vicinity of the tetroon (Lagrangian sampling) at locations immediately downwind from Baltimore, 70 km downwind from Baltimore, immediately downwind from Philadelphia, and in southwestern Connecticut.

Aircraft sampling upwind of Baltimore established the existence of an air mass ozone concentration of about 70 ppb. This concentration was fairly extensive both within the Washington plume, in the mixed layer (about 600 m at that time), and in the residual mixed layer from the previous afternoon which extended to about 1800 m. Background ozone concentrations measured above the residual mixed layer were about 40 ppb.

Fig. 8 illustrates the ozone and NO_x traces at three flight levels immediately downwind from Baltimore between 0955 and 1041 EST. The mixing height was about 700 m. Below the mixing height, a broad NO_x plume from Baltimore sources apparently scavenged much of the developing Washington ozone plume. Ozone concentrations outside the NO_x plumes increased from 70 ppb upwind to between 80 and 90 ppb downwind from Baltimore. Ozone continued to develop in the composite Washington-Baltimore plume and reached 170 ppb northwest of Philadelphia by late afternoon.

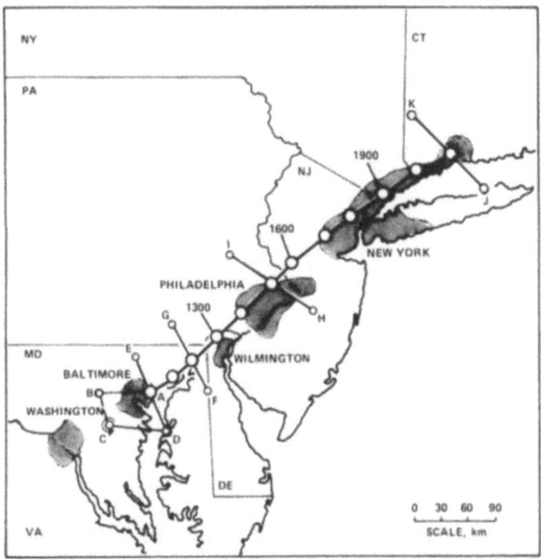

Fig. 7. Hourly location of tetroon (larger circles) and aircraft flight tracks on 14 August 1980.

Ozone concentrations on the last set of transects in southwestern Connecticut were significantly reduced from those measured northwest of Philadelphia. Near the surface, ozone was severely scavenged by NO emissions from New York which produced NO_x concentrations approaching 90 ppb. A sampling traverse at 1100 m showed nearly uniform ozone concentrations of about 100 ppb--a reservoir transported throughout the night that could potentially be mixed to the surface the following day.

Simulation of Regional Transport

The tetroon data from the REAL experiments combined with the radiosonde data provided a unique resource to evaluate trajectory models and to assess the effects of increasing the temporal density of the radiosonde observations on model performance. The tetroon trajectories shown in Fig. 4 and those for two other tetroon data sets were simulated by three different diagnostic trajectory models which differed significantly in sophistication and input data requirements (Clarke et al., 1983). The trajectory models evaluated

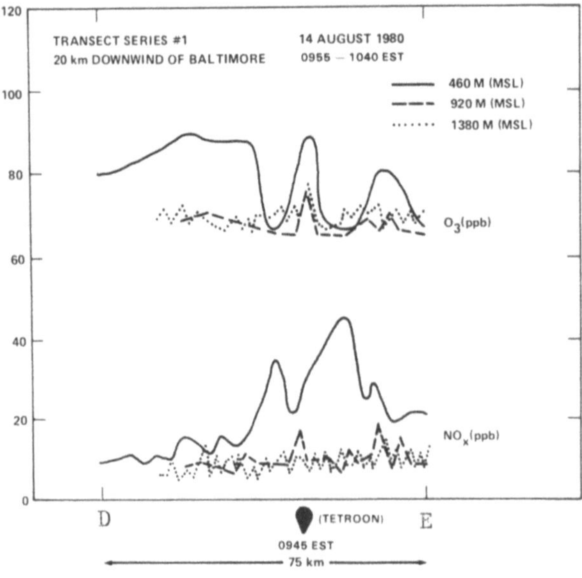

Fig. 8. Ozone and NO_x concentrations along Lagrangian traverses D-E (see Fig. 7) between 0955 and 1041 EST on 14 August 1980. The location of the tetroon at 1000 EST is indicated at the bottom of the figure.

were: 1) the National Oceanic and Atmospheric Administration, Air Resources Laboratory Atmospheric Transport and Dispersion (ATAD) model (Heffter, 1980), 2) the National Center for Atmospheric Research (NCAR) Isentropic model (Haggenson and Shapiro, 1979), and 3) the Center for Air Pollution Impact and Trend Analysis (CAPITA) Monte Carlo model (Patterson et al., 1981). The ATAD model was evaluated for both 6- and 12-h radiosonde data. The NCAR model used only the standard 12-h radiosonde data, and the CAPITA model used only midday surface wind speed and direction measurements.

Comparison of the results generated by these techniques did not reveal a clearly superior model. There was a tendency in all the models for the tetroon to be to the left of its diagnostic position by about 10% of the trajectory length with a standard deviation of about 25% of the trajectory length. Apparently the midday surface winds used in the CAPITA model described the large-scale flow features about as well as the ATAD model using four

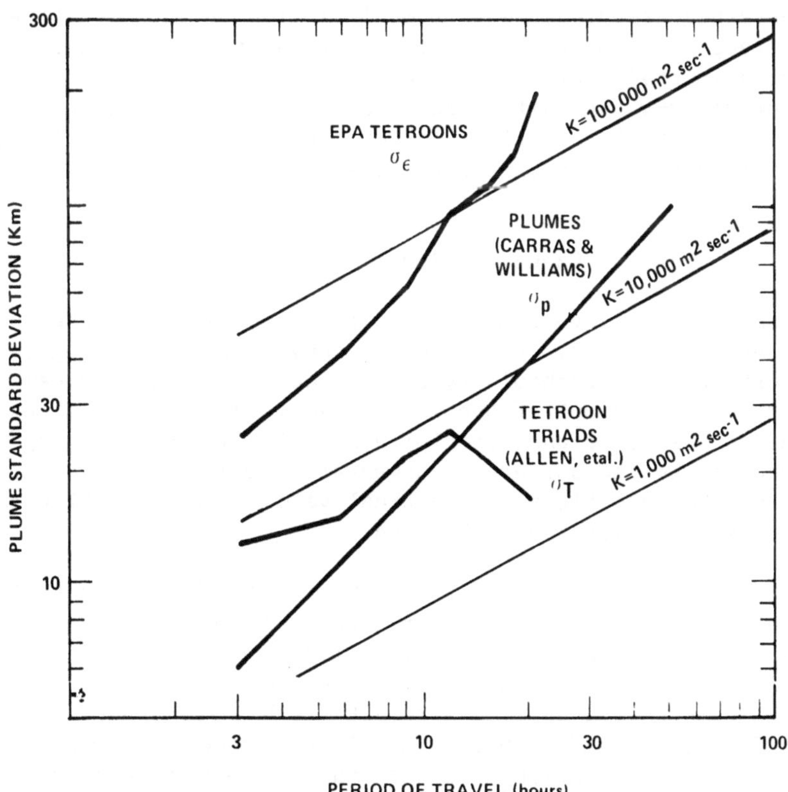

Fig. 9. Standard deviations of the differences between tetroon and diagnostic trajectories σ_ε, tetroon triad separation σ_T, and plume data σ_P.

radiosondes per day, suggesting that a significant portion of the random features of the flow must be associated with space and time scales smaller than the 6-h radiosonde network.

The standard deviation of the differences as a function of travel time between the REAL tetroons (Fig. 4) and diagnostic positions by the ATAD model using 6-h data are shown as σ_ε in Fig. 9. Also shown are the standard deviations of the separation of tetroon triads, σ_T, (Allen et al., 1967) and plume standard deviations, σ_p, based on data presented by Carras and Williams (1981).

The curves representing the tetroon triad data and plume data are not greatly inconsistent and are suggested to represent plume spreading due to turbulence, wind shear, and mesoscale atmospheric processes. Diagnostic trajectories are idealized simulations of the mean flow. The variances of the difference between the diagnostic and tetroon trajectories are due primarily to the inability of the wind fields and models to adequately simulate the motion of the air parcel associated with the tetroon. Thus, the differences between the σ_ε curve and the σ_P and σ_T curves are an indication of the uncertainty associated with the simulation of the mean flow. Had the transport component of the wind been precisely simulated by the diagnostic trajectories, σ_ε should have been similar in magnitude to σ_P and σ_T. The differences increase with increasing travel time and σ_ε is about five times larger than σ_P after about 12 h, a result similar to that obtained by Allen et al. (1967).

Vertical Transport of Ozone by Clouds

Clouds play an important role in transport and photochemistry of ozone and precursors. For example, they reflect incoming radiation important in photochemistry of the lower layers; they can transport pollutants out of the mixed layer; they provide a medium for liquid phase reactions; precipitating clouds can return upper-level material to the surface in downdrafts; and they can alter the strength of the synoptic-scale vertical velocity field, which in turn affects mixed layer growth rates.

An important component of the NEROS field program was to investigate the effect of vertical transport by clouds, hereafter referred to as venting, on the variability of ozone concentrations in the mixed layer. In order to study the net transport of ozone via the cumulus route, special aircraft flights were made during both NEROS 1979 and NEROS 1980. The aircraft was instrumented with a gust probe system complemented with an on-board inertial navigation system, to yield data on both the mean and turbulent component of temperature, moisture, 3-D wind, and ozone. Flights were made both within and above the mixed layer and included penetration of convective cloud elements. Preliminary results and further details regarding the deployment of this aircraft have been reported by Ching et al. (1983).

Fig. 10 shows plots of the departures from the mean of a few of the variety of parameters measured during a cloud layer traverse in the 1979 NEROS study. The actual cloud penetrations are indicated at the top of the figure and are associated with positive anomalies in all of the parameters. The ozone anomaly is about 5 ppb and ozone concentrations within the clouds were similar to those within the mixed layer.

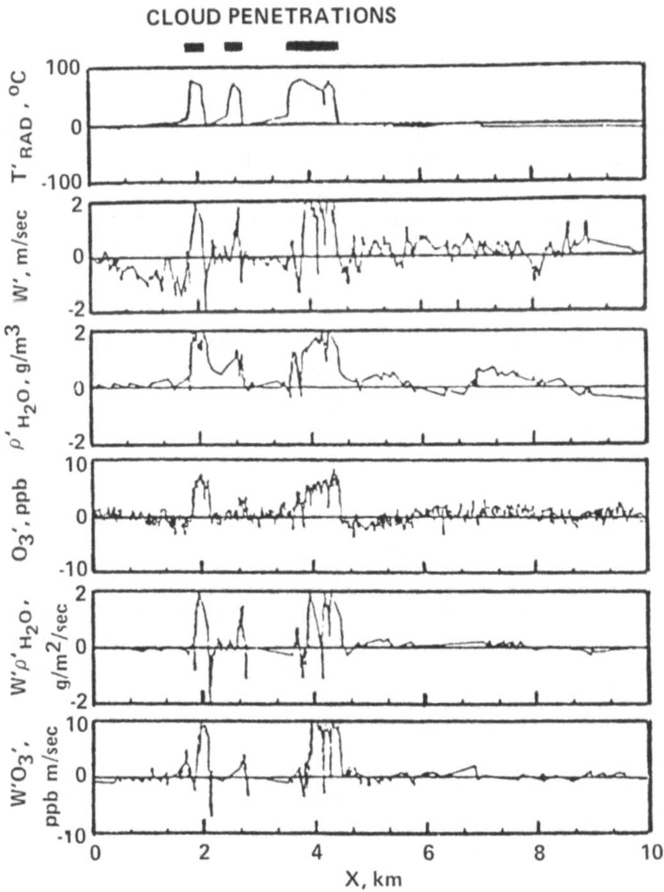

Fig. 10. Fluctuation data during cloud penetration flight in 1979 NEROS.

Urban areas and their plumes may exhibit large ozone concentrations and therefore large differences between mixed and cloud layer values. Consequently, venting is expected to be a significant feature of the ozone budget of urban plumes. For example, a cloud layer flight over Philadelphia, Pennsylvania, on 27 August 1979 showed ozone during cloud penetrations of greater than 100 ppb, 50 ppb above ambient levels. Ozone concentrations in the mixed layer below the flight traverse were about 90 to 150 ppb.

NEROS field programs have confirmed cloud venting as an important process in the mixed layer ozone budget. Pollutants vented above the mixed layer may be transported without significant dilution or chemical transformations and eventually entrained back into the mixed layer to impact surface concentrations.

Characterization of Ozone Episodes

As part of the process of selecting days for application of the ROM and urban-scale models, each high ozone day in 1980 is being evaluated in terms of 1) the magnitude and spatial gradient of ozone transported across the modeling domain and within urban plumes and 2) the occurrence and complexity of mesoscale meteorological phenomena within the Northeast Corridor. The following is a synopsis of this type of episode characterization analysis for 5 August 1980.

A high pressure system was centered over the southeastern U.S. and a warm front extended east-west across Massachusetts and northern New York. In general, clear skies persisted from morning throughout mid-afternoon and maximum surface temperatures were between 32 and 34°C.

Obvious ozone plumes extended downwind of the major Corridor cities as evidenced from the isopleths of maximum 1-h surface ozone concentrations (see Fig. 11). Areas of high ozone concentrations were observed to the northeast of Washington, Baltimore, Philadelphia, and New York City urban areas consistent with the boundary layer wind flow. Near Boston, the peak surface ozone concentration, 159 ppb, was measured to the west of the city in an area of convergent low-level flow associated with the slow northward advancement of a warm front across southern New England. Ozone concentrations and the configuration of the urban plumes aloft (measured by aircraft) were similar, except near Boston. A peak ozone concentration of about 200 pbb was measured aloft 50 km to the northeast of Boston in a layer of southeasterly flow.

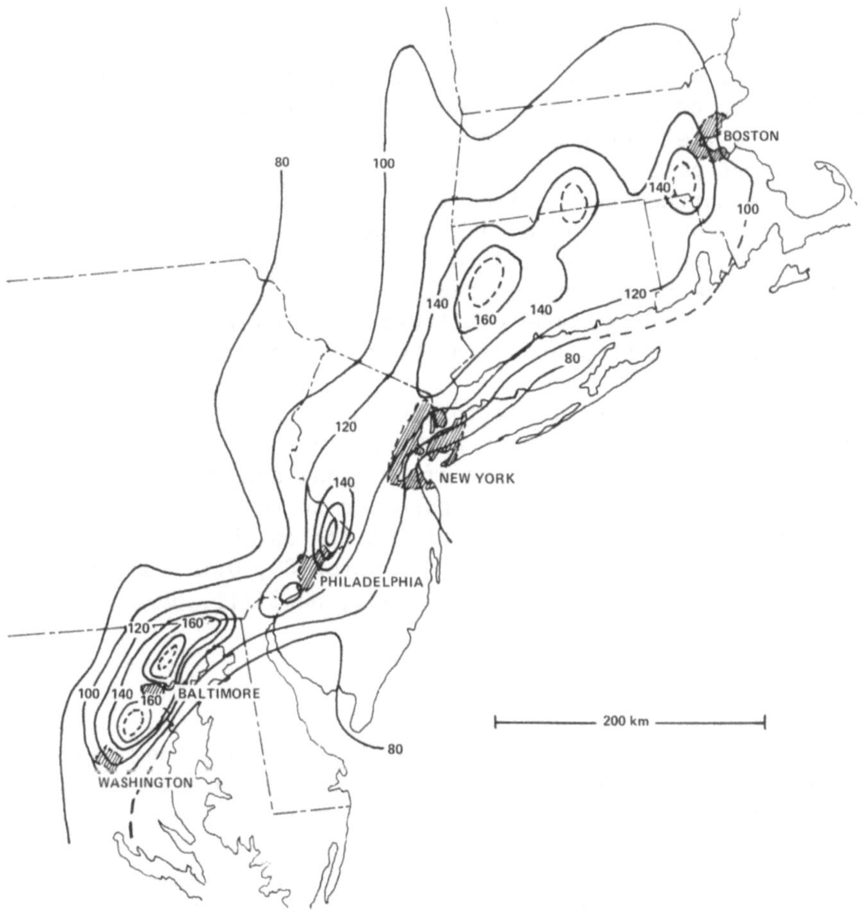

Fig. 11. Maximum 1-h ozone surface concentration isopleths (ppb) for 5 August 1980.

Ozone concentrations outside the area of direct impact of Corridor emissions were fairly homogeneous at about 80 ppb. Within the Corridor, ozone concentrations were generally high and exhibited sharp gradients. The National Air Quality Standard (0.12 ppm) was exceeded over a large area and even upwind of most major urban areas. Thus it appears that a relatively high background ozone burden is supplementing strong ozone production downwind of the major urban areas and these plumes subsequently impact other urban areas.

CONCLUDING REMARKS

Field programs were conducted during the summers of 1979 and 1980 in support of the development and evaluation of the ROM and for application of the Airshed model to cities in the Northeast Corridor. The field programs have resulted in an extensive data base of ambient pollutant concentrations and meteorological measurements for the northeastern United States. Some preliminary analyses for the purpose of preparation and evaluation of the data base for ROM modeling exercises have led to some interesting insights on horizontal and vertical transport processes of importance to regional-scale transport, e.g., the confirmation of convective cloud venting as an important process in the mixed layer ozone budget.

The data base is available for further analyses and model evaluation programs. It is most appropriate for research programs leading to enhanced understanding of physical and chemical processes by which ozone is produced and transported to regions both near and far from its source. Additional information on the data base can be obtained from the Chief, DMSAB, Mail Drop 80, U.S. Environmental Protection Agency, Research Triangle Park, NC 27711, USA.

REFERENCES

Allen, P. W., Jessup, E. A., and White, R. E., 1967, Long range trajectories, Proc. of U.S.A.E.C. Symposium Meteorological Information Meeting, Chalk River, Ontario, Canada, Sept. 11-14, 1967, AEC L-2787, 176-190.

Carras, J. N., and Williams, D. J., 1981, The long-range dispersion of a plume from an isolated point source , Atmos. Environ., 15:2205.

Ching, J. K. S., Clarke, J. F., Irwin, J. S., and Godowitch, J. M., 1983, Relevance of mixed layer scaling for daytime dispersion based on RAPS and other field programs, Atmos. Environ., 17:859.

Clark, T. L., 1980, Annual anthropogenic pollutant emissions in the United States and southern Canada east of the Rocky Mountains, Atmos. Environ., 14:961.

Clark, T. L., and Clarke, J. F., 1984, A Lagrangian study of boundary layer transport of pollutants in the northeastern United States, Atmos. Environ., 17:1703.

Clarke, J. F., and Ching, J. K. S., 1983, Aircraft observations of regional transport of oxidant in the northeastern United States, Atmos. Environ., 18:287.

Clarke, J. F., Ching, J. K. S., Clark, T. L., Haagenson, P. L., Husar, R. B., and Patterson, D. E., 1983, Assessment of model simulation of long-distance transport, Atmos. Environ., 17:2449

Haagenson, P. L., and Shapiro, M. A., 1979, Isentropic trajectories for derivation of objectively analyzed meteorological parameters, NCAR Tech. Note TN-149 STR, Boulder, CO.

Heffter, J. L., 1980, Air Resources Laboratories atmospheric transport and dispersion model, NOAA Tech. Memo. ERL ARL-81.

Lamb, R. G., 1983, A regional scale (1000 km) model of photochemical air pollution - Part I: theoretical formulation, U.S. Environmental Protection Agency Technical Report EPA-600/3-83-035, Research Triangle Park, NC.

Patterson, D. E., Husar, R. B., Wilson, W. E., and Smith, L. F., 1981, Monte Carlo simulation of daily regional sulfur distribution: comparison with SURE sulfate data and visual range observations during August 1977, J. Appl. Meteorol., 20:402.

Possiel, N. C., and Freas, W. P., 1982, Northeast Corridor Regional Modeling Project - description of the 1980 urban field studies., U.S. EPA Tech. Report, EPA 450/4-82-018, Research Triangle Park, NC 27711.

Reynolds, S. D., and Reid, L. E., 1978, An Introduction to the SAI Airshed Model and its Usage, System Applications, Inc., Report EF 78-53R.

Vaughan, W. M., Chan, M., Cantrell, B., and Pooler, F., 1982, A study of persistent elevated pollution episodes in the northeastern United States, Bull. Amer. Meteor. Soc., 63:258.

DISCUSSION

J.B. KNOX — If as reported the standard deviation of predicted trajectory end points error is 5 times the pollutant standard deviation, then in absence of the tetroon as a marker, there would be little chance of vectoring an aircraft to the same "parcel" on successive days. In absence of marking the parcel with a tetroon the experiment would most likely fail.

J.F. CLARKE — I agree, particularly for the very complex meteorological regimes under which these experiments were conducted. The design of the program was biased to low wind speed, near stagnation conditions.

S. VOGT — What's about the mean flight heights of the tetroons and the plume heights?

J.F. CLARKE — The plume being sampled was assumed to be uniformally mixed in the vertical through the mixed layer. The tetroons were normally set to float at 0.8 Z_i (where Z_i is the height of the mixed layer).

F.L. LUDWIG Are the data bases readily available to outside organizations ?

J.F. CLARKE Data tapes are available to outside organizations now. Full documentation of these tapes is available for the Baltimore data (urban plume data) and the Urban Field Studies data (New York, Boston, Washington).
Full documentation will be available by March 84 for the Columbus data (Urban plume and regional sampling), the hydrocarbon data base and the 1979 data base (regional sampling).

T. LAVERY The 6-hour sounding data did not improve the estimates of the trajectories. Why ?

J.F. CLARKE I think this indicates that atmospheric processes on time scales less than 6-hour have a significant cumulative effect on long-range transport.

T. LAVERY The implication for regional modeling is that you need better than 6-hour resolution wind data for the simulations ?

J.F. CLARKE The results indicate this, particularly for the complex meteorological regimes occurring during many of the tetroon releases.

INSTANTANEOUS OBSERVATIONS OF PLUME DISPERSION IN THE SURFACE LAYER

Torben Mikkelsen and Richard Eckman

Physics Department
Risø National Laboratory
DK-4000 Roskilde Denmark

ABSTRACT

In many transformation and removal processes encountered in the atmosphere, instantaneous rather than time averaged concentrations are of interest. Determination of the instantaneous concentration requires knowledge of the dispersion of the cloud about its center of mass rather than with respect to a fixed coordinate system.

During a recent series of experiments held over homogeneous terrain in Denmark, the relative dispersion of surface released smoke plumes was calculated from aerial photographs. Simultaneously, wind data were obtained from a horizontal array of tower mounted sonic anemometers. This provided information about the spatial and the temporal variability of the dispersing wind field.

The wind data were used to compare commonly used relative dispersion models with the experimental results. A recently proposed statistical model based on the space-time variability of the turbulence field was found to agree with the experimental data over the limited scale considered.

INTRODUCTION

During the summers of 1980 and 1981, the Risø meteorology section conducted a series of surface-layer smoke diffusion experiments over homogeneous terrain in Jutland, Denmark. The purpose of the experiments was to provide relative diffusion data for comparison with different diffusion models. In this paper we describe the experimental procedure used and discuss the results in relation to a statistical model based on the space-time variability of the turbulence field.

THE EXPERIMENT

During two experimental campaigns carried out in August 1980 and August 1981, studies of ground-level smoke releases took place just south of the Danish town Borris in Western Jutland. The experiment site was a 1 km × 1 km area in the middle of the Borris moors, which are part of a Danish army exercise area. The largest differences in elevation over the area are of the order of 1 meter. The vegetation during August consists of naturally growing grasses and blossoming heather plants, which produce a homogeneously distributed roughness over the terrain of approximately 1 cm. Due to this flatness and homogeneity, the site is ideal for meteorological experiments.

Meteorological Instrumentation

Four 10 m high meteorology masts were placed in a north-south oriented array as shown in Fig. 1 (the prevailing wind direction during August is from the west). A 3-axis sonic anemometer/thermometer (Kaijo Denki type DAT 310) was mounted on each of the masts. The positioning of the masts made it possible to obtain estimates of the cross covariances of all three velocity components for six, approximately logarithmically equidistant, lateral displacements: 4.5 m, 7.5 m, 12 m, 18 m, 25.5 m and 30 m. In addition, we measured the wind speed and direction with a sensitive cup anemometer and wind vane which were mounted on one of the masts.

Release and Detection of Smoke

Four smoke-pots, each containing 10 kilos of the powder Hexit*

*Hexit (HC) is a mixture of zinc- and chlorine enriched powders. It reacts strongly with the moisture in the atmosphere and forms a thick, white smoke.

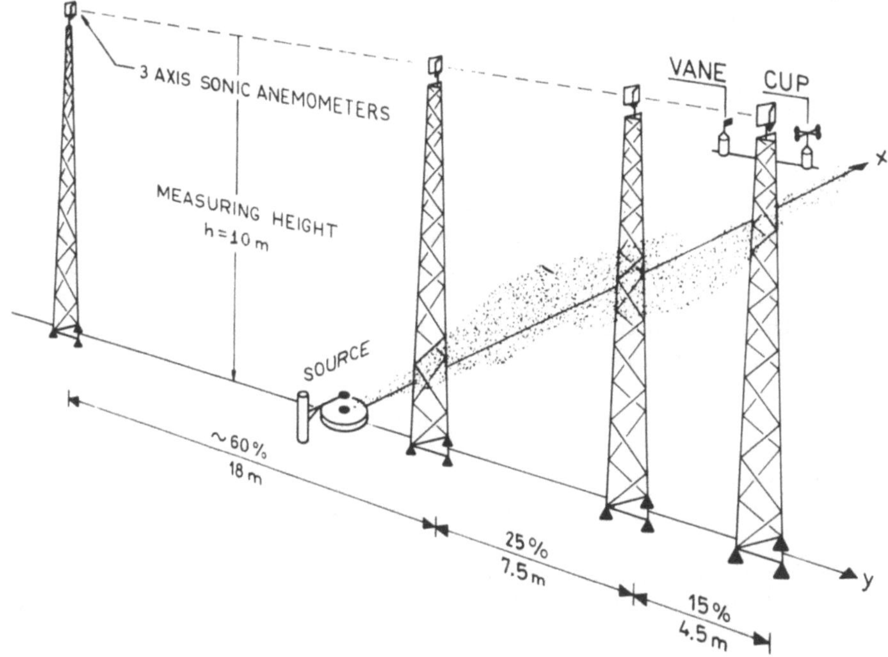

Fig. 1. The meteorological instrumentation. With the individual positioning of the four masts as shown, velocity co-variances could be obtained for the following lateral displacements: 4.5, 7.5, 12, 18, 25.5 and 30 meters. The smoke was released at the position between the first two masts as indicated.

Each of the masts was mounted with a Kaijo Denki DAT 310 3-axis sonic anemometer/thermometer.

were placed in a stack along the baseline between the masts. When ignited, the smoke-pots produced a visible plume that could be followed from a small aircraft as far as ~ 1 km downwind under light wind conditions. An electrical scale equipped with a strip chart measured the mass of the smoke pots continuously and therefore also the smoke release rate. In addition, a thermometer mounted 1 meter downwind from the release point measured the exhaust temperature of the smoke. At this short distance, the smoke temperature was detected to be surprisingly low, on the average less than ~ 50°C above the ambient air temperature. The low excess temperature results from the high affinity of the hexit gas for the moisture in the atmosphere and a corresponding increased initial entrainment rate. Taking into account the relatively small volume flux from the source, the buoyancy and subsequent plume rise (approx. 10 meter at x = 1000 m) was deemed to be insignificant relative to the vertical mixing of the smoke in the neutral to unstable atmosphere encountered during the measurements.

The visible contour of the white smoke plume was registered on photographs taken from an airplane circling above the release point. Crosswind arrays of white contrast plates were placed on the ground at downwind distances of 31.25 m, 62.5 m, 125 m, 250 m, and 500 m. The spacing between the contrast plates was 5 meters at the first three downwind distances, 10 meters at the 250 m line, and 20 meters at the 500-m downwind distance.

Figure 2 is an example of a smoke plume photograph which was taken at an altitude of approximately 1000 feet. The visible contour of the smoke plume is seen to be easily detectable, and its position can be determined from the white contrast plates.

The 35-mm camera used in the experiments was equipped with either an 18-mm wide-angle (100°) objective or a 50-mm (47°) objective, depending on the altitude of the aircraft. In addition, the camera had a timer which marked the pictures with the time and date of the exposure. A flying technique was developed in order to compensate for the drift of the plane with the mean wind, and in this way a picture of the visible smoke plume could be taken from above the source point approximately every other minute. A series of approximately 20 pictures could be obtained in this way during each 40-min smoke release.

Radio-sondes were launched during each experiment, from which information was gained about the wet and dry bulb temperature profiles of the atmosphere. This in turn gave information about the stability and mixing height of the boundary layer.

DATA REGISTRATION AND PROCESSING

The signals from the four sonic anemometers, cupanemometer, and wind vane were all sampled with a 20-Hz sampling frequency and subsequently stored on magnetic tapes in a digitized form. Including the calibration and test-voltages before and after each recording, each tape could hold 3 hours of data.

The velocity components stored on the magnetic tape were necessarily those in the sonic anemometers' reference system, since prior to the analysis, the mean wind direction was of course unknown. The first stage in the statistical analysis was therefore to calculate the direction of the mean wind vector for each run.

Fig. 2. Air photo of an instantaneous smoke plume taken from approximately 1000 feet above the source point. The white contrast plates constitute a network of lines normal to the x-axis at the following downwind distances: 31.25 m, 62,5 m, 125 m, 250 m and 500 m. Their lateral spacing is 5 m in the first three rows, 10 m in the 250 m row, and 20 m in the 500 m row.

These directions were then used in the subsequent analysis to rotate (first about the vertical, then about the lateral anemometer axis) the measured components into a reference frame with the x-axis along the mean wind direction and the y-axis in the crosswind direction. The resulting longitudinal (u), lateral (v) and vertical (w) velocity components were thereby oriented so that $\bar{u} = U$ and $\bar{v} = \bar{w} = 0$.

The sonic-alignment program transformed the sonic anemometer data from arbitrarily chosen 54.6 min periods (2^{16} data points each) into calibrated, internally and externally aligned time series. The time series were then plotted and any linear trends were removed. Subsequently, we calculated the power spectra and the coherences.

Smoke Data Collection and Analysis

During each of the BORris EXperiments (BOREX), approximately 20 pictures were taken over the source point from a small airplane at 2 minute intervals. The cruising altitude of the airplane varied from experiment to experiment between 1000 and 6000 feet, depending on the height of the cloud base and on the overall visibility of of the atmosphere. Higher-altitude pictures produced less distortion of the experiment site on the film, but they also resulted in reduced contrast due to moisture and dust in the air. The aircraft height above the source point was measured with the plane's altimeter. However, this could also be inferred by measuring the distance δ between the marker plates on the film and by use of the formula $h = f_o(A/\delta)$, where f_o is the focal length of the objective and A the distance between the marker plates on the ground.

By comparing the outline of the plume in the picture with the marker plates it was possible to determine the visible contour of the smoke plume as function of the distance from the source point. By subsequent use of an opacity method (Gifford, 1980), the corresponding lateral standard deviation σ_r of the instantaneous plume could be inferred by assuming that the lateral concentration distribution for the instantaneous plume is Gaussian in form. With a constant release rate, the relation between the visible half-width y_c and the instantaneous standard deviation σ_r of the smoke cloud is (Gifford, 1980)

$$\sigma_r = y_c / \{\ln(e y_{c,max}^2 / \sigma_r^2)\}^{1/2} \qquad (1)$$

where the quantity $2y_{c,max}$ denotes the maximum visible width of the smoke plume and e is the base to the natural logarithm. When Eq. (1) is used in the limit for $y_c \ll y_{c,max}$, the relation between σ_r and y_c is rather insensitive to changes in $y_{c,max}$, and therefore to fluctuations in the smoke release rate. The maximum visible plume width, $2y_{c,max}$ was typically between 50 and 100 meters, and it was observed at downwind distances between 500 and 1000 meters.

The data processing of the large amount of smoke data was facilitated by digitizing the visible contour positions on a desk top computer. The digitizing program automatically calculated y_c after the user moved a light pen to the correct positions on the visible plume edge. Both the plume's instantaneous lateral spread σ_r and its instantaneous lateral position $y_{cm} = (y_c^+ - y_c^-)/2$ (where y_c^+ and y_c^- are the upper and lower contour positions at fixed x, respectively) could in this way automatically be registered at logarithmically increasing downwind distances. Average values of the relative standard deviation $\langle\sigma_r\rangle$, and the center of mass variance, $\langle y^2_{cm}\rangle$, were obtained by averaging all the pictures in a run together. With the Gaussian particle distribution assumed in Eq. (1), the quantity $\langle y^2_{cm}\rangle$ can be interpreted as the spreading connected with the center-of-mass movement of the instantaneous plume. The single-particle dispersion Σ referred to in the fixed coordinate system $\underline{y}$, has in this way been estimated by the relation $\Sigma^2 = \langle\sigma_r\rangle^2 + \langle y^2_{cm}\rangle$.

RESULTS AND DISCUSSION

In order to calibrate the four sonic anemometers which were to be used at Borris, we conducted a series of test runs over water near Risø National Laboratory. Several 2^{16} data point runs (corresponding to a 56 min sampling period) were analyzed from these test experiments. For each of the four sonic anemometers used in one of these test runs, Fig. 3 shows the velocity spectra for all three wind components u, v and w. An inertial subrange with a $k^{-5/3}$ dependency is evident in all the spectra.

In the high wavenumber end, the 4/3 ratio between the longitudinal and two transverse (v, w) spectral amplitudes indicate that local isotropy is present in this region. The almost identical shape and amplitude of the spectra of the four different anemometers seem to justify the calibration and alignment procedures used. From the sonic anemometer/thermometers, we were able to obtain an average value of the Monin-Obukhov length, L. In terms of the total heat flux $\langle\theta'\omega'\rangle$ * and the friction velocity u_*, L is

$$L = \frac{T}{\kappa g}\, u_*^3/\overline{\theta' w'}. \tag{2}$$

Here, T is the average surface layer temperature, g the gravitational acceleration, and κ the von Kármán constant.

* The quantity θ' denotes the moisture-corrected virtual temperature, and therefore $\langle\theta'\omega'\rangle$ represents the sum of the latent and sensible heat flux. Sonic anemometers measure the sound virtual temperature, but for all practical purposes this equals θ', so that an estimate of the total heat flux could be obtained from the sonic anemometer/thermometers (see also Schotanus et al., 1983).

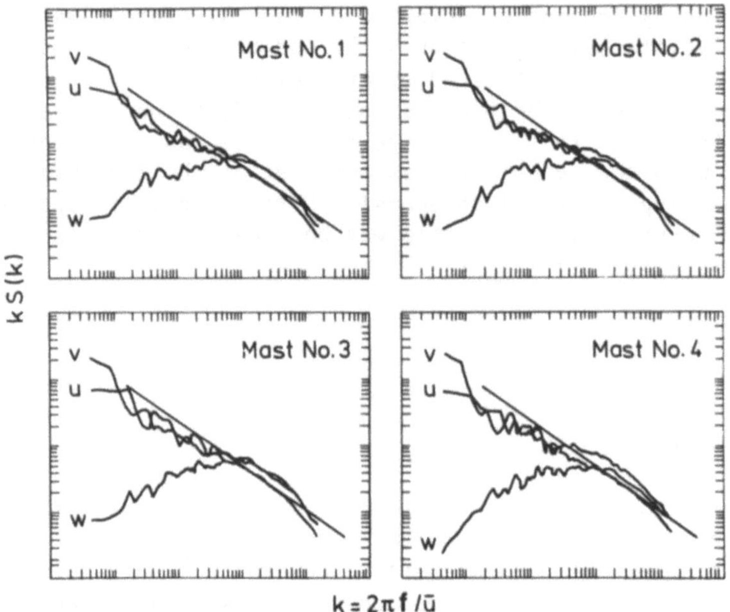

Fig. 3. Simultaneously measured power spectra as function of wavenumber, kS(k), of the three velocity components u, v, and w for each of the four sonic anemometers in Fig. 1. The wavenumber k is related to the frequency f by: $k = 2\pi f/\bar{u}$. The full line shown has slope equal to -2/3. Range of k: 10^{-3} - $10 m^{-1}$. Range of ordinate: 10^{-4} - $1\ m^2/s^2$. Relative bandwidth: 0.20.

Smoke release experiment, BOREX 80 Run 6

In the middle of the afternoon on Thursday, August 28, the sky was almost totally clear and light winds of the order of 4-5 m/s indicated rather unstable atmospheric conditions. At 3.20 p.m., the wind shifted to a direction almost perpendicular to the mast array, so we began to release smoke at this time.

Over the following 56 min sampling period, the mean wind was determined to be 4.72 m/s and the longitudinal (u) and the lateral (v) variances were $0.62\ m^2/s^2$ and $0.98\ m^2/s^2$, respectively. At the experiment site, the surface temperature (z = 1 meter) reached 20.4 ^{o}C at 3 p.m. and the relative humidity at that time ~ 57%. Between 3.20 and 4.00 p.m., a series of 22 plume photographs were taken at approximately 2 min intervals from an altitude of 1000 feet. In this experiment, the visible smoke plume could be detected in several pictures as far as ~ 1 km downwind from the source. The smoke data obtained at downwind distances greater than 500 meters are, however, somewhat uncertain. By referring to the ground-placed contrast plates, it was later possible to calculate the instantaneous lateral visible width of the smoke plume, and the corresponding instantaneous (lateral) standard deviations, σ_r, by use of the opacity method described earlier.

Figure 4 represents the ensemble of instantaneous standard deviations σ_r from BOREX 80, Run 6 as function of the downwind distance. A considerable scatter in the data at a fixed downwind position is to be expected in connection with the turbulent diffusion process considered and this should not be regarded as measurement error. By averaging the observed instantaneous σ_r values at each downwind spread $\langle\sigma_r\rangle$ was obtained as function of downwind distance x. Figure 5 shows this averaged standard deviation $\langle\sigma_r\rangle$ together with the centroid dispersion $\langle y_{cm}^2\rangle^{1/2}$. The latter is measured with respect to a coordinate axis aligned with the mean wind direction over the 40-min period. Also shown in Fig. 5 is the total lateral dispersion corresponding to Taylor's (1921) theoretical analysis of single-particle diffusion $\Sigma = (\langle y_{cm}^2\rangle + \langle\sigma_r\rangle^2)^{1/2}$. This is seen to be dominated entirely by the center of mass dispersion at the downwind positions considered in the experiment.

The average standard deviation of the instantaneous spread is, to a first approximation, found to grow linearly with downwind distance. However, some scatter in $\langle\sigma_r\rangle$ between the logarithmically equidistant (increment factor = $2^{1/3}$) downwind distances is present. This is probably a consequence of statistical uncertainty due to the relatively small ensemble of 22 pictures. It is reasonable to assume that smoke plumes observed at two-minute intervals represent independent outcomes of the stochastic diffusion process that is responsible for the instantaneous width of the plume. At least this must be true for the downwind distances considered here, since most of the smoke particles observed in the individual pictures were advected out of the experiment site during the time period between two photographs. The statistical uncertainty, $\Delta\langle\sigma_r\rangle$, of the estimated average spread of $\langle\sigma_r\rangle$ is therefore of the order $\sim\sqrt{22}/22 = 21\%$. In Fig. 5, the shaded bar shows the 68% confidence level $(2\Delta\langle\sigma_r\rangle)$ of $\langle\sigma_r\rangle$, and it is seen that most of the observed scatter in $\langle\sigma_r\rangle$ can be explained in terms of this.

At distances shorter than 500 meters, the observed values of $(\langle y_{cm}^2\rangle)^{1/2}$, and thereby also of the absolute or fixed frame diffusion Σ, are also found to increase linearly with downwind distance. Here, however, the scatter about a straight line fitted

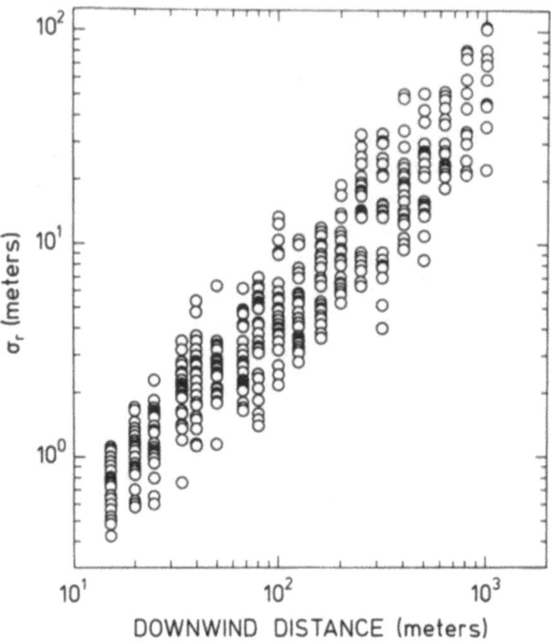

Fig. 4. Observed instantaneous standard deviation σ_r during BOREX 80 RUN 6, as function of the downwind distance x.

to the observations seems to be more correlated from point to point. But the stochastic process responsible for the advection and positioning of the plume over the experiment site is also of a different character than that responsible for the instantaneous width of the plume. The plume advection has to do with the more energetic turbulent eddies that exist on the scale comparable with the experiment site. The positioning of the smoke plume in two consecutive observations will therefore not be independent to the same extent as the observations of the instantaneous width,

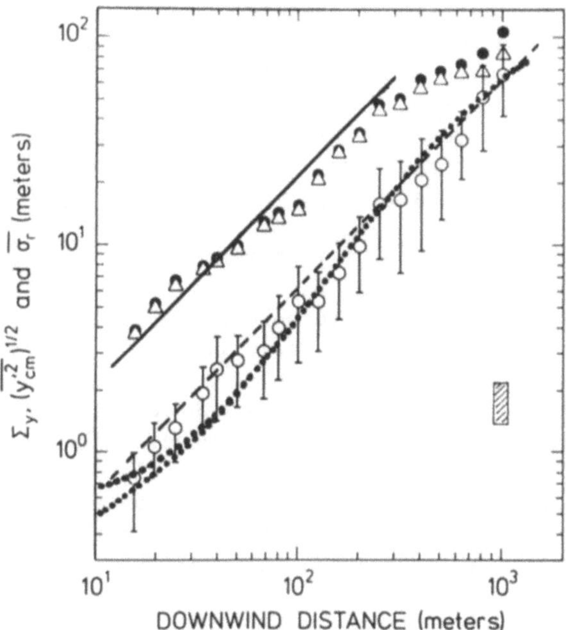

Fig. 5. Ensemble-averaged standard deviation $\langle\sigma_r\rangle$ of the instantaneous observed spread in Fig. 4 versus the downwind distance x. The bars through the datapoints represent the standard deviation of the ensemble of instantaneously observed spreads at a particular downwind distance. Dispersion associated with the movement of the centerline of the instantaneous plume, $(\langle y^2_{cm}\rangle)^{1/2}$, has been indicated by Δ. The total (single-particle) dispersion Σ which is referred to a fixed coordinate at the ground, is indicated by •. The upper solid line was calculated on the basis of Taylor's (1921) absolute diffusion theory: $\Sigma = i_v x$, whereas the stipled line is from the formula given by Smith and Hay (1961) for the expansion of a three-dimensional isotropic Gaussian puff: $\sigma = 0.22\ ix$. The shaded vertical bar, which is inserted in the lower right corner of the figure represents the statistical uncertainty (± one standard deviation) on the estimate $\langle\sigma_r\rangle$. The dotted lines show numerical solutions of σ(x) obtained from the statistical diffusion theory Eq. 3-5, with two different initial valves: σ(o) = 0.5 meter (upper curve) and σ(o) = 0.25 meter (lower curve).

and this may explain the more correlated departures in the scatter observed in Σ. At distances greater than 500 meters, the observed values of the absolute and instantaneous spread seem to converge, but both the plume location technique, as well as the opacity method used, are rather uncertain at these distances, and less confidence should be allocated to these measurements. The solid curve in Fig. 5, which falls close to the observation of the absolute diffusion, was calculated by means of G.I. Taylor's (1921) absolute diffusion theory. For travel times that are short relative to the Lagrangian time scale t_L of the turbulence, this theory yields $\Sigma = i_v x$. The lateral turbulence intensity, i_v, is defined by $i_v^2 = \langle v^2 \rangle / \bar{u}^2$.

The dashed line in Fig. 5 represents a relative diffusion formula suggested by Smith and Hay (1961): $\sigma \;\; = 0.22\, i \cdot x$, where $i^2 = (\langle u^2 + v^2 + ^2 \rangle)/\bar{u}^2$ is the turbulence intensity based on the total energy of the turbulence. Pasquill (1974) recommends Smith and Hay's formula for the expansion of a Gaussian cloud in a field of isotropic turbulence. This formula has successfully been used in the Risø small-scale puff diffusion model (Mikkelsen et al., 1980 and Mikkelsen et al., 1983b). For calculating the dashed curve in Fig. 5, however, only the two horizontal variances were used in calculating i. Therefore, the dashed curve is a lower limit of Smith and Hay's formula. It is slightly high compared to the data.

Smoke Release Experiment, BOREX 81, Run 1B

During the period from August 23 to 29, 1981 the entire experiment set-up was reestablished at the site in the Borris Sönderland moors. During these experiments the weather conditions were dominated by somewhat stronger winds, and the atmospheric stability was close to neutral. Detailed data analyses of all (10) of the runs has not yet ended. Here, we will discuss Run 1B which was successful both with respect to the smoke and wind registration.

In the morning of Monday the 24, clouds at approximately 3000 feet were present at the experiment site, and the wind came rather strongly out of the W-NW direction at 5-8 m/s. The 2-m temperature was 15.0 °C and the relative humidity 40%. The atmospheric stability was judged to be close to neutral, in agreement with the Monin-Obukhov length later obtained from the sonic anemometers (L = -100 m). A radio-sonde launched at 11.00 a.m. indicated that the inversion height was ~ 950 m.

At 11.30 a.m., the wind turned almost perpendicular to the mast array, and over the subsequent 40-min smoke release period, the mean wind direction was found to differ by less than 20 degrees from the predefined x-axis of the set-up. This time, 21 useful

Table 1. Smoke release experiment: BOREX 81 RUN 1B

MAST NO.	U m/s	$\langle u^2\rangle$ $(m/s)^2$	$\langle v^2\rangle$ $(m/s)^2$	$\langle w^2\rangle$ $(m/s)^2$	$\langle \theta^2\rangle$ K^2	u_* m/s	L m
1	5.88	1.03	1.43	0.359	0.099	0.447	- 69
2	5.72	0.942	1.49	0.363	0.068	0.425	- 98
3	5.93	0.974	1.39	0.361	-	0.443	
4	6.20	1.09	1.45	0.321	0.052	0.426	- 99
MEAN	5.93	1.01	1.44	0.351	0.073	0.435	- 90

$\overline{\theta} = 20.9^o$ $z_i = 950$ m

$i_v = \sqrt{1.44}/5.93 = \underline{0.202}$

$i = (\sqrt{1.01 + 1.44 + 0.351})/5.93 = \underline{0.280}$

pictures were taken from approximately 2000 feet above the release point. As a consequence of the stronger wind conditions during this experiment, the smoke plume was only visible down to about 250 - 500 meters from the source. The various statistical parameters calculated from the sonic anemometers are shown in Table 1. Figure 6 shows $\langle\sigma_r\rangle$ (circles), $\langle y_{cm}^2\rangle^{1/2}$ (stars), and Σ (crosses) for BOREX 81, Run 1B. The vertical lines represent the rms spread of σ_r about the mean. The solid line is Taylor's near-field absolute diffusion formula and the dashed line is Smith and Hay's relative diffusion formula (i and i_v were obtained from Table 1). In this experiment, a power law $\langle\sigma_r\rangle \propto x^{0.8}$ seems to fit the observations better than the linear relation suggested by Smith and Hay.

By assuming that Taylor's frozen eddy hypothesis is valid, we can compute velocity correlations for the 3 velocity components as a function of both downwind and crosswind directions (Figure 7). The turbulence is neither isotropic nor even axis-symmetric on the scale considered. The correlations, however, drop off faster with distance here than they did in BOREX 80, Run 6. This is a result of the fact that Run 6 was more unstable, and therefore there were more larger-scale correlated eddies present.

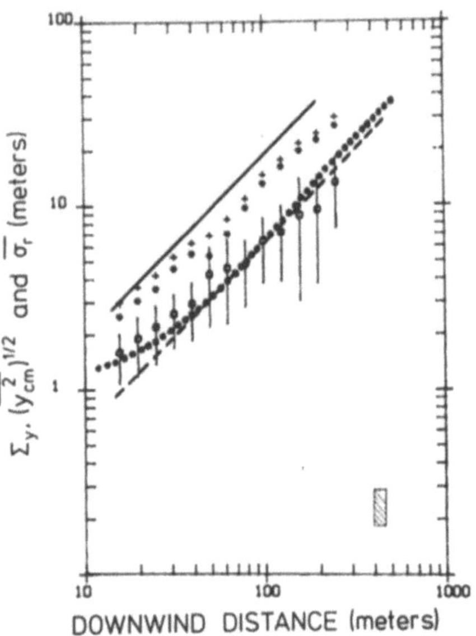

Fig. 6. Ensemble averaged standard deviation, $\langle\sigma_r\rangle$ of the instantaneous spreads during the BOREX 81 experiment. As in Fig. 5 the crossbars represent the standard deviation of the ensemble at a particular downwind distance. The centroid dispersion $(\langle y_{cm}^2\rangle)^{1/2}$ is here idndicated by * and the total or single-particle dispersion by +. The solid line $\Sigma = i_v x$, represents the near-field limit of Taylor's (1921) (single particle) diffusion theory whereas the dotted line is Smith and Hay's (1961) formula $\sigma = 0.22\ ix$. As in Fig. 5 the shaded vertical bar represents the statistical uncertainty (+/- one standard deviation) of $\langle\sigma_r\rangle$. The dotted line shows the corresponding numerical solution of Eqs. 3-5 with $\sigma(o) = 1$ meter.

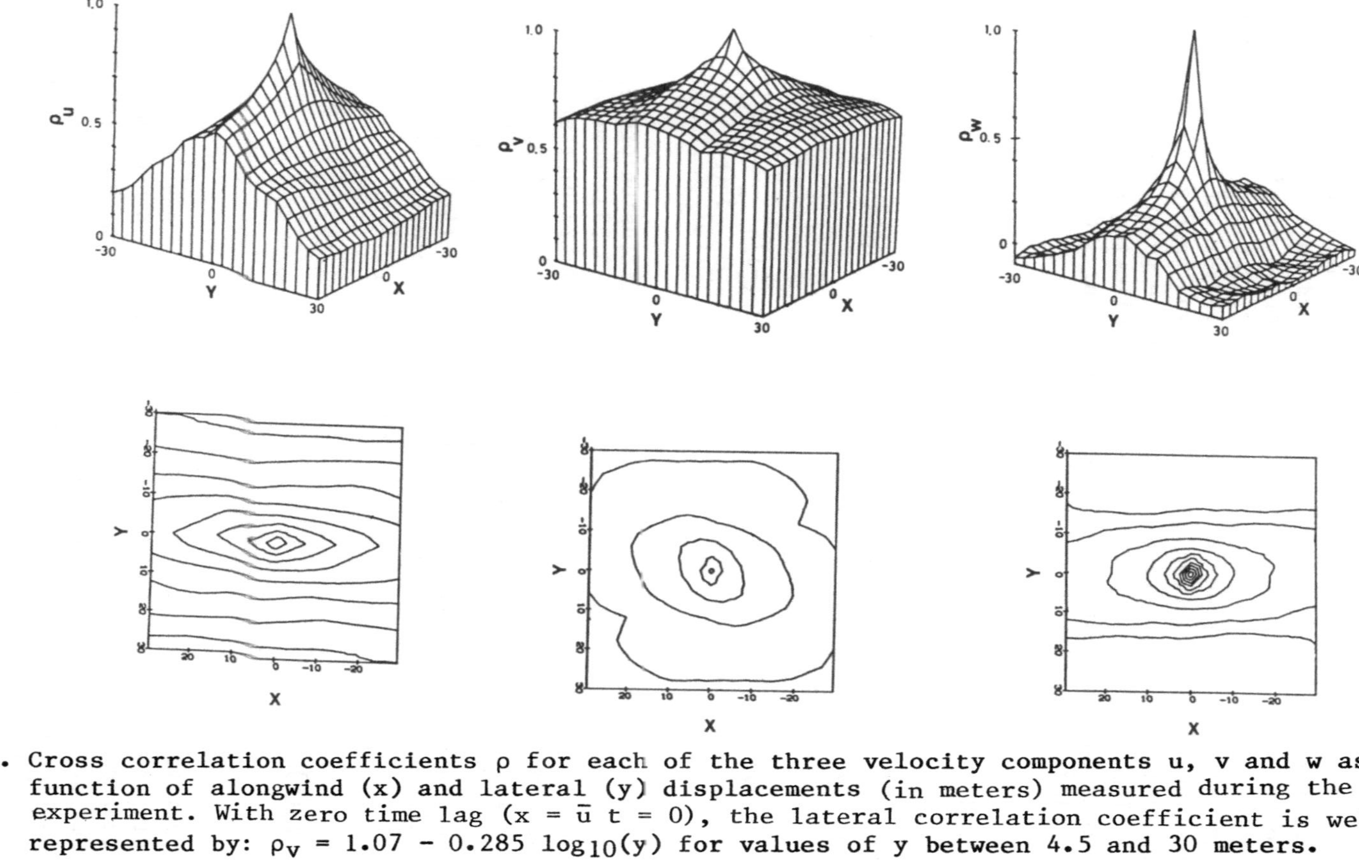

Fig. 7. Cross correlation coefficients ρ for each of the three velocity components u, v and w as function of alongwind (x) and lateral (y) displacements (in meters) measured during the 1981 experiment. With zero time lag ($x = \bar{u}\, t = 0$), the lateral correlation coefficient is well represented by: $\rho_v = 1.07 - 0.285 \log_{10}(y)$ for values of y between 4.5 and 30 meters.

COMPARISON OF EXPERIMENTAL RESULTS WITH A GAUSSIAN PUFF MODEL

The observations of instantaneous plume dispersion have already in Figs. 5 and 6 been compared with Smith and Hay's (1961) formula for the expansion of a Gaussian cloud in a field of isotropic turbulence: $\sigma = 0.22$ ix. In this section we will interprete the BOREX data in light of a recently proposed equation for the growth of a Gaussian puff (Mikkelsen, 1982). Let σ denote the lateral standard deviation(in the y-direction in Fig. 1) of a Gaussian cloud or puff released at the source point $x = 0$ at time $t = 0$. The statistical theory yields the following set of equations for determining $\sigma(t)$ (cf. Mikkelsen, 1982). Eqs. (3.37) and (3.38).

$$1/2 \frac{d\,\sigma^2}{dt} = \langle v_p^2\rangle(t)\cdot t_p(t) \tag{3}$$

where the velocity scale appropriate for the relative diffusion process is

$$\langle v_p^2\rangle(t) = \langle v^2\rangle\{1 - \int_{-\infty}^{\infty} \rho_v(\xi) \frac{1}{2\sqrt{\pi}\sigma(t)} \exp(-1/4 \frac{\xi^2}{\sigma^2(t)})d\xi\} \tag{4}$$

The time scale appropriate for the relative diffusion process, $t_p(t)$, is also given by Mikkelsen (1982) (eq. 3.39). However, in the small time limit considered here, the puff travel time t can be assumed to be small relative to the Lagrangian integral time scale of turbulence, t_L. In this case, it can be shown that $t_p(t) \simeq t$ (cf. Mikkelsen, 1982, eqs. (3.63) and 3.64)). A simple model for t_p, which is consistent with the near-field limit $t_p(t) \simeq t$ and which in addition provides a simple first-order correction when t becomes of a magnitude comparable to t_L, is

$$t_p(t) = \frac{t_L}{1+t_L/t}. \tag{5}$$

By specifying the lateral velocity variance $\langle v^2\rangle$ and the corresponding correlation function for the lateral displacements ρ_v in accordance with the quantities obtained during the experiments, the set of equations (3) to (5) has been solved numerically for σ as function of the diffusion time t, and by use of the relation $x = \bar{u}t$, the predicted puff size σ could then be compared with the experimental values of the ensemble mean value of the instantaneous plume width $\langle\sigma_r\rangle$, as function of downwind distance.

Fig. 5 shows the numerical solution for $\sigma(x)$ obtained from Eqs. (3), (4) and (5), with the variance and lateral cross correlation coefficient ρ_v measured during the BOREX 80 Run 6 experiment. Two different initial puff sizes, $\sigma(0) = 0.5$ m and $\sigma(o) = 0.25$ m have been used here and the Lagrangian time scale t_L has in accordance with the measurements been set equal to 100 sec. (using $t_E/t_L = 4$, where t_E is the measured Eulerian timescale). Since the measured correlation coefficient $\sigma_v = 1.05 - 0.20 \log_{10}(D)$ applies only for values of the separation D between 4.5 and 30 meters, it has been necessary to consider the contribution from the integral in Eq. (3) for values of $|\xi| < 4.5$ meters as an undetermined constant. In Fig. 5 the value of the part of the integral for which $|\xi| < 4.5$ meter has consequently been chosen in order to give best agreement between the theory and the observations at the shortest distances considered. At longer distances, where the cloud had become large relative to the smallest mast separation, the solutions for $\sigma(x)$ were found to be insensitive to the initial value chosen for this integral.

Fig. 6 shows in a similar way the calculated curve of $\sigma(t)$ from the more neutral experiment BOREX Run 1B. Here $\rho_v = 1.07 - 0.285 \log_{10}(D)$ for values of D between 4.5 and 30 meters, $\langle v^2 \rangle = 1.44\ m^2/s^2$ and t_L is approximately 100 sec. The inital size $\sigma(0)$ is set equal to 1 meter. As before the prediction falls inside the scatter in the observations.

Eckman (1983) has been able to solve Eqs. (3) and (4) analytically in the near-field limit (when $t \ll t_L$) assuming a simple form for the correlation ρ_v. If the longitudinal integral length scale ℓ is assumed to be constant, the analytical solution for σ is proportional to $t^{3/2}$. This is in agreement with Batchelor's (1950) dimensional arguments for "intermediate" diffusion times (times where $\sigma/\ell \ll 1$ and the initial puff size is no longer an important parameter).

If one assumes that ℓ is directly proportional to height and that the puff height is directly proportional to t (which is reasonable for a surface source in neutral conditions) the solution for σ is linear with time. Eckman shows that a linear growth with time is in good agreement with the BOREX 81, Run 1B data if the proper relationship between x and t is used. (Instead of the simple equation $t = x/\bar{u}$, Eckman uses a more complicated equation that involves the logarithmic wind profile. This equation is more appropriate near the surface than the simple linear relation.) The increase of the BOREX 80, Run 6 data is somewhat greater than t to the first power. This is probably a result of the fact that BOREX 80 Run 6 was more unstable than BOREX 81 Run 1B. Further details on the statistical puff diffusion model are described in Mikkelsen (1982a), and Eckman (1983), and further details on the BOREX experiments are described in Mikkelsen (1983a).

CONCLUSIONS

The BOREX experiments are unusual in the sense that both velocity correlations (as function of crosswind separation and time lag) and the relative diffusion of a surface released smoke plume were measured simultaneously. This allows us to compare the data with some of the more advanced theoretical diffusion models.

Except for small time lags and lateral separations, the BOREX cross correlations do not show the characteristics associated with axisymmetric turbulence. This is not unexpected at a measuring height of 10 meters. Axisymmetry may be an acceptable approximation for small diffusion times, however, since in this limit, only the smallest eddies (which are nearly axisymmetric) diffuse the cloud.

Smith and Hay's (1961) relative diffusion formula seems to agree with the more unstable BOREX 80, Run 6 data, but it has a steeper slope in comparison to the more neutral BOREX 81, Run 1B data. Its main advantage is that it is easy to use. Mikkelsen's (1982) equation is more complicated, but it has the advantage of allowing one to view the relative diffusion in terms of variations in the turbulence structure. For example, assuming that the integral length scale of the turbulence, ℓ, varies linearly with height, as it would in neutral conditions, Eckman (1983) shows that Mikkelsen's equation predicts a linear puff growth rate for a surface source during "intermediate" diffusion times. (It is commonly assumed that the puff size is proportional to $t^{3/2}$ in "intermediate" times. This relationship only holds, however, when ℓ is a constant). This is in agreement with the near-neutral BOREX 81, Run 1B data.

REFEERENCES

Batchelor, G.K., 1950, The Application of the Similarity Theory of Turbulence to Atmospheric Diffusion. Quart. J.R. Met. Soc., 76:133.

Eckman, R.M., 1983, Relative Diffusion and the Underlying Turbulence Structure, Master's Thesis, The Pennsylvania State University.

Gifford, F.A., 1980, Smoke as a Quantitative Atmospheric Diffusion Tracer, Atmospheric Environment, 14:1119.

Mikkelsen, T., Larsen, S.E., and Troen, I, 1980, Use of a Puff Model to calculate Dispersion from a Strongly Time-dependent Source. In "Proceeding of the European Seminar on Radioactive Releases and their Dispersion in the Atmosphere following a Hypothetical Reactor Accident". 2:575, Risø, 22 - 25 April.

Mikkelsen, T., 1982, A Statistical Theory on the Turbulent Diffusion of Gaussian Puffs, 106 pp. Risø-R-475. Available from: Risø Library, Risø National Laboratory, P.O. Box 49, DK-4000 Roskilde, Denmark.

Mikkelsen, T., 1983a, The Borris Field Experiment: Observations of Smoke Diffusion in the Surface Layer over Homogeneous Terrain. 86 pp. Risø-R-478. Available from: Risø National Laboratory, P.O. Box 49, DK-4000, Roskilde, Denmark.

Mikkelsen, T., 1983b, Larsen, S.E., and Thykier-Nielsen, S., 1983, Description of the Risø Puff Diffusion Model. Submitted to: Nuclear Technology. Available from: Risø Library, Risø National Laboratory, P.O. Box 49, DK-4000 Roskilde, Denmark.

Pasquill, F., 1974. Atmospheric Diffusion, 2nd Ed., John Wiley & Sons, New York, xi + 429 pp.

Schotanus, P., Nieuwstadt, F.T.M., and De Bruin, H.A.R., 1983: The Use of a Sonic Anemometer for Measurements of Heat and Moisture Fluxes. Boundary Layer Meteorology, 26:81.

Smith, F.B., and Hay, J.S., 1961, The Expansion of Clusters of Particles in the Atmosphere, Quart. I.R. Met. Soc., 87:82.

Taylor, G.I., 1921, Diffusion by Continous Movements, Proc. London Math. Soc. A 20:196.

DISCUSSION

A. VENKATRAM

Why did you not calculate the ensemble averaged σ_y by superimposing the centroids of the instantaneous plumes ? Averaging the σ_y's of the instantaneous plumes is not correct in principle because the distribution for each realization is definitely not Gaussian.

T. MIKKELSEN

I see both advantages and difficulties by, as you suggest, first calculating an ensemble visible plume width and subsequently transforming this into a corresponding Gaussian standard deviation. In the first place, how would I know the centroid position in the instantaneous plume, if I relax the assumption of a normal (or any other symmetric) particle distribution. And even if I assume the centroid to be in the middle of the visible plume, I would have to draw constant-contrast contour lines on a film-badge constituted by superimposing approximately 20 films. This is possible in principle, however. An advantage would be that these superimposed puff-contours probably would be more Gaussian than the particle distribution in individual puffs, due to the central limit theorem. But I also think, that the opacity method used here on individual plumes as described by F. GIFFORD is able to give an quantitative estimate of the instantaneous relative diffusion coefficient σ_r, when it is remembered that we have assumed a Gaussian particle distribution in the cloud.

C.D. JONES

One of the difficulties with photograpic techniques is obtaining information on the vertical dispersion. I would be interested to hear your comments on this.

T. MIKKELSEN

Estimating vertical surface released dispersion by means of the here used photographic technique is to my opinion more difficult. First, you would have to release real separated puffs, and not a continuously released plume as was done here. Second, it may be more difficult to obtain high contrast pictures of the individual puffs against a diffuse bright horizon, than the case was here with a homogeneous dark background. Nevertheless, such an experiment would be interesting.

S. VOGT

Your smoke experiments extended up to at least 1 km. For practical application in dispersion-modeling this short range is not of big interest. Do you intent to extend your experiment into bigger distances (~ 10 - 15 km) ?

T. MIKKELSEN

Yes, we indeed would like to extend the experiment to larger downwind distances, but with the visible smoke technique used here, this does not immediately seem to be possible. Therefore, we would like to consider other techniques such as SF_6-tracer releases, but in order to obtain data on the instantaneous rather than on the time averaged concentration distribution function, it will be necessary to use mobile flame photometers. This is beyond us at the moment.

J.B. KNOX

In regard to inclusion of K_z, one should note that Weinstein (1978) showed that in stably stratified flows $K_z = -81 \frac{\varepsilon}{N^2}$, when H is the Brunt-Väisäla frequency.
With this equation, and scale dependent diffusion σ_y could be related to K_z in consistent manner in stable structural flows.

T. MIKKELSEN Thank you for the comment. At Risø we also treat vertical diffusion of puffs by means of a K-diffusivity model, and we agree with a constant diffusivity for very stable stratified flows.

A LITERATURE STUDY ON TRACER EXPERIMENTS FOR ATMOSPHERIC DISPERSION STUDY

B. Vanderborght and J.G. Kretzschmar

Studiecentrum voor Kernenergie
SCK/CEN
B-2400 Mol, Belgium

INTRODUCTION

Most mathematical models for the simulation of air pollution dispersion are semi-empirical approaches in which the value of some model-parameters must be experimentally determined. In the widely used bi-gaussian dispersion approach, the horizontal and vertical dispersion parameters (σ_y and σ_z) are based on a number of dispersion measurements, and for application they are determined in a turbulence typing scheme as a function of meteorological conditions and of distance. In fact, they are only valid for terrain conditions similar to these of the original tests. The dispersion parameters of some schemes are entirely based on meteorological turbulence measurements, others are based on tracer dispersion experiments. These are experiments in which an artificial tracer gas or aerosol is released under controlled conditions and the resulting ambient concentrations are measured. Validation of a model can be done using the emissions of a single or a multiple industrial source configuration and the ambient concentration measurements in the vicinities over extended time periods or - usually with less experimental uncertainties like emission amount and plume rise - by using tracer dispersion experiments.

Considering the importance of tracer experiments for the determination of semi-empirical parameters and for model validation and considering the increasing interest for these field experiments during recent years, a literature study has been undertaken.

Due to the limited space available for this article and because the reports on many tests are not readily available in the open literature, this review must be restricted to the most important trends. A more elaborated work is in preparation.

The available references can be classified under the following headings :

- experimental : papers describing the possibilities of different atmospheric tracers
- qualitative tracer experiments : a tracer is released in the atmosphere and measured downwind, however without attempt for a quantitative description of the dispersion
- determination of dispersion parameters : tracer releases are performed for the determination of dispersion parameters to be used in further model calculations.
- validation of mathematical models : the results of tracer experiments are compared with model calculations.

EXPERIMENTAL

Earlier tracer experiments used aerosols such as fluorescein acid (Bultynck et al., 1972), inorganic fluorescent particles, e.g. zinc sulfides and zinc cadmium sulfide (Eggleton et al., 1961 ; Mulholland, 1980), uranine dye (Singer et al., 1966), oil fog smoke (Raynor et al., 1975), spores (Hay et al., 1957) and metal oxides (Shum et al., 1975).

Different physical phenomena such as difficulties with aerosol generation, sedimentation, capture by the soil, impact on tree leaves and pine-needles, dissolution in water and loss of fluorescence due to sunlight action caused major problems when using these aerosol tracers. Most of these products have been given up by now as more sensitive tracers and analytical techniques became available.

In a few experiments radioactive gases like ^{85}Kr (Draxler, 1979), ^{41}Ar (Singer, 1966), ^{133}Xe (Eggleton, 1961) have been used as a tracer. In most cases the emissions were routine discharges from nuclear installations. The public objections against artificial radioactive releases for dispersion studies are straightforward.

Most of the recent work has been done with inert gases containing several halogene elements, by preference fluorine, chlorine or bromine, which can be determined with high sensitivity by gas chromatography with electron capture detection. The most popular tracer gas is sulfurhexafluoride (SF_6). It has almost all the properties of an excellent tracer : non-toxic (even at the 80 % by volume concentration level (Vershueren, 1978), odourless, colourless, gaseous at temperatures well below ambient, insoluble in water, chemically stable towards hydrolysis, oxidation, photolysis and heat, low ambient background levels, detectable in the ppt range in air, easy sampling, easy to emit at a constant rate and readily commercially available. Emission in the field is usually done with moderate pressure cylinders placed in a trailer and connected to a flow restrictor and rotameter. Depending on the emission rate the cylinders must eventually be heated. Typical emission rates are a few kg SF_6 per hour for measurements up to 10 km, and 50 to 100 kg per hour for up to 700 km (Ferber et al., 1981 ; Kallend et al., 1981). Sampling can be done in PVC bags (Van Duuren et al., 1975 ; Vanderborght, 1981), Saran bags (Clemons et al., 1968 ; Gryning et al., 1978), polyethylene (Collins et al., 1965), steel (Dietz et al., 1973 ; Collins et al., 1968) or glass bottles (Nieuwstadt et al., 1979), also plastic syringes have been used (Lamb et al., 1978). Saltzman et al.,(1966) have investigated the stability of SF_6 mixtures in different containers and under different conditions in the presence of other pollutants. Re-use of sampling containers is dissuaded because of possible memory effects (especially in glass containers (Turk et al., 1968)) and increased risk for microscopic cracks in plastic bags (Clemons et al., 1968).

In the first chromatographic separations, the SF_6 eluted from the separation column after the O_2 so that the sensitivity of the SF_6 determination was hampered because a small SF_6 peak was superimposed on the tail of a large O_2 peak (Clemons et al., 1968 ; Collins et al., 1965 ; Dietz et al., 1973 ; Saltzman et al., 1966 ; Turk et al., 1968). With Molecular Sieve 5 A as column packing the SF_6 comes out before O_2 improving the detection limit and the precision (Gryning et al., 1978 ; Lamb et al., 1978 ; Nieuwstadt et al., 1979 ; Van Duuren et al., 1975).

The detection limit of the SF_6 determination by direct injection in a gas chromatograph is typically about 2 to 5 ppt. Activated carbon has been used for preconcentration (Clemons et al., 1968 ; Van Duuren et al., 1975) in order to reduce the minimum detectable amount by a factor 100. Recovery was however often much less than 100 % (Clemons et al., 1968).

The very high stability and the increased emission (for dispersion experiments and by leaks in e.g. high tension switches) of SF_6 risk to increase the actual global background. It is now about 0.3 ppt in the northern and southern hemisphere (Singh et al., 1979 ; De Bortoli et al., 1976). For this reason other halogenated compounds have been proposed, like e.g. $CBrF_3$ (Lamb et al., 1978) ; CCl_4, CF_2Br_2 and $CFCl_3$ (Nester et al., 1979) ; CCl_2F_2, $CFCl_3$ and CF_2Br_2 (Vogt, 1977). The handling of all these products is comparable to the one for SF_6 but for some of them the background is considerably higher (Singh et al., 1979 ; De Bortoli et al., 1976).

Another possibility is the use of perfluorocarbons (C_8F_{16}, C_6F_{12} and C_7F_{14}) which have a very low background and, thanks to the many F atoms in the molecule, very high sensitivity. Since they are liquid at ambient temperature, their emission is more difficult. Analysis by GC is usually more complicated and time consuming. Lovelock (1982) has calculated the cost per experiment (relative to SF_6) taking into account the atmospheric background, cost per mol and detection limit. Perfluorocarbon experiments are usually cheeper than SF_6. Up to now the perfluorocarbons have only been used for long (up to 100 km) (Draxler, 1979) and for very long range dispersion experiments (up to 500 - 1000 km) (Kallend et al., 1981 ; Ferber et al., 1981).

Finally some "laboratory made heavy methanes ($^{12}CD_4$, $^{13}CD_4$)" have been proposed for long range experiments (Draxler, 1979 ; Ferber et al., 1981 ; Lovelock et al., 1982).

QUALITATIVE TRACER EXPERIMENTS

Several papers (some of them already mentioned in what preceeds) report on the methodology of tracer release experiments and qualitative results of field experiments are given as an illustration of the possibilities. Turk et al. (1968) have used SF_6 to track the plume of an industrial stack, rising from a 18 m high roof to a height of 41 m above ground-level. Samples were taken in 60° segments 650 and 1000 m from the source. No model calculations were tried. Dietz and Cote (1973) measured SF_6 on-line in an airplane up to 15 km downwind a power plant to determine the SO_2 to SO_4 conversion rate. Clearly defined crosswind concentration profiles were obtained but analytical errors hindered further in-depth analysis of the results. Collins et al. (1965) have used SF_6 and CF_2Cl_2 as indicators of air trajectories and for determining actual diffusion rates in complex terrains. Comparison with an oversimplified model was not successful. Clemons et al. (1968) have used SF_6 and preconcentration on activated carbon to track air movements over distances of 120 km by release of about 500 g per minute for 1 hour. SF_6 was measured as a function of time. Drivas and Shair (1974) performed a fairly similar experiment by releasing 750 g SF_6 per minute for 3/4 hour for a large-scale test over the complex Los Angeles area. SF_6 transport was measured as a function of time at different communities up to 124 km from the point of release. Attempts to explain the results gave rather poor results. Emberlin (1981) has used SF_6 to track in a coastal area with complex topography a plume up to 40 km downwind the 200 m high power plant stack. The experiment illustrated the use of SF_6 to trace the dispersal of a plume and to estimate the plume contribution to the actual total SO_2 ground-level concentration. Partly due to the poor meteorological input data the proposed empirical regression model did not explain very well the observations.

A considerable amount of experimental work has been done by Le Quinio (1970)(Doury, 1980) in France. Le Quinio performed some 100 tracer dispersion experiments using uranine emissions during 1 h periods at 5 to 50 m altitude, and sampling during 40 minutes up to 15 km from the source. The data were not treated to obtain a mathematical formalism, but he drew practical graphs which can be directly used to estimate the concentration to emission ratio under the plume axis and the plume width as a function of distance.

The CEA systems discern only two types of dispersion conditions : normal and bad diffusion. With such a rough classification, many dispersion situations can occur within the same dispersion class. Therefore Le Quinio (1973) introduced a probabilistic dimension in his graphs by observing from his measurements that a log normal distribution of the dispersion parameters within one class is a good approximation.

Ferber et al. (1981) have demonstrated the possibilities of perfluorocarbons in long range dispersion experiments going as far as 600 km downwind. During 3 hours five tracers were simultaneously released approximately 1 m above ground-level. SF_6 emission rate was 91 kg/h, enough to be detectable at 100 km, the other emission rates produced concentrations well above the detection limit at 600 km : 64 kg C_7F_{14}/h, 62 kg C_8F_{16}/h, 51 g $^{12}CD_4$/h and 28 g $^{13}CD_4$/h. Sampling time on the 100 km arc was 45 minutes while on the 600 km arc 22 three hour samples were taken. The five tracers behaved the same in the atmosphere and this over the entire distance of 600 km. The tests reported in this article are a preparation for a large campaign to be held in Augustus - September '83 and intended to be going as far as 1200 km from the source.
A research team of the Central Electricity Research Laboratories (UK) (Kallend et al., 1981) has measured SF_6 tracer up to 700 km distance from the emission source (i.e. the world record for SF_6). The experiments were set up to investigate the possible sources of acid rain and surface water acidification in Southern Scandinavia. 50 kg SF_6/h was injected continuously into a stack of the Eggborough power plant and the plume was time-marked by injecting 25 kg of C_7H_{14} during a 30 min period repeated at 6 hourly intervals. SF_6, SO_2, NO, NO_2 and SO_4 were continuously measured in an airplane flying from UK to Denmark. At the latter distance the SF_6 profile had a maximum of some 40 ppt and some 15 km plume width. The SO_2 to SO_4 oxidation rate was calculated. Results of model calculations are not reported.

DETERMINATION OF DISPERSION PARAMETERS

The widely used bi-gaussian distribution theorem is a semi-empirical model in which the horizontal (σ_y) and vertical (σ_z) dispersion parameters are experimentally determined and linked to meteorologically determined stability classes. Different turbulence typing schemes have been reviewed and their results have been compared by Kretzschmar et al. (1980 and 1982). Some dispersion parameters have been determined by means of direct turbulence measurements while others are based on the results of tracer experiments. The next paragraphs review the experiments on which the dispersion parameters of the turbulence typing schemes discussed in (Kretzschmar et al., 1980 and 1982) are based.

SCK/CEN System

The dispersion parameters (Bultynck et al., 1972) were obtained from wind vector fluctuations measured by means of a tridimensional anemometer, and not by tracer experiments. Fluorescein acid tracer experiments were used to check the results obtained by the wind vector method. An aerosol generator at 69 m altitude released 5 to 6 kg fluoresceine (particle size < 4 μm) during 50 minutes. 50 to 80 collectors were placed along arcs 100 to 15.000 m downwind the source. Fifteen experiments were completed. The $\sigma_y(x)$ measured were neither significantly nor systematically different from the computed ones. Nevertheless the computed concentrations were larger than the measured ones by a factor 1.5 to 3. The reason may be systematic experimental errors due to the use of unstable aerosols.

The Pasquill Systems

The Pasquill diffusion parameters are based not only on tracer experiments (primarily the Prairie Grass Experiments), but also on measurements of the wind direction fluctuations. The original paper from Pasquill (1961) does not give much experimental details. In the Prairie Grass Experiments the SO_2 ground-level emission took place over very flat and smooth (roughness length 1 cm) terrain, sampling time was 10 minutes and the downwind distances were less than 1 km. Dispersion parameter values for longer distances, other emission heights or terrain roughnesses are merely extrapolations. Despite these restrictions already pointed out by Pasquill himself, his parameters are still widely used.
Other authors (Martin and Tikvart, Nieuwstadt, Hosker, Gifford, Briggs, Turner, Klug (Kretzschmar e.a., 1982)) have determined other turbulence typing schemes and dispersion parameters starting from the same experimental data.

The Karlsruhe System

The original Karlsruhe parameters, also used in (Kretzschmar et al., 1980 and 1982) were obtained by Klug by reevaluating the data material available at that time from major US tracer test series already used by Pasquill (1961). From 1969 to 1979 the Karlsruhe Nuclear Research Center has performed 62 tracer experiments (Nester et al., 1979 ; Thomas, 1980) using tritiated water (HTO) (sampling on cooled Al plates), CCl_4, CF_2Br_2 and $CFCl_3$ as tracer. Emission heights were 60, 100, 160 and 195 m. The stability categories were determined by the fluctuations of the vertical wind direction at 100 m. The new dispersion parameters σ_y and σ_z were obtained by least square fitting the many ground-level concentration profiles. Different parameters have been calculated for emissions below 100 m and above 100 m. Since the tracer releases have not been done simultaneously at different heights, the conclusions are not yet definitive, but it seems that σ_y decreases with altitude, while the dependence of σ_z with emission height is not clear.

The Jülich System

Another set of diffusion parameters has been experimentally determined in Germany by Vogt (1974 and 1977) at the Jülich Nuclear Research Center. For that purpose, Vogt released $CuSO_4$ and $HoSO_4$ aerosols radioactively labeled with ^{64}Cu and ^{166}Ho, for 1 h periods from 50 m and 100 m altitude and measured the ground level concentration up to 11 km from the source during 65 tests. The dispersion parameters σ_y and σ_z were simultaneously obtained by means of a double least square fitting of the many observed glc profiles. They were derived as a function of distance and meteorological conditions, in a scheme similar to the Pasquill turbulence typing scheme. It was observed that the dispersion parameters decreased with height but that they are nevertheless systematically larger than the Pasquill parameters for ground level release. This is attributed to a much larger surface roughness. This has to be seen not as the local roughness conditions, but merely as the mean roughness lengths over extended entrance regions and diffusion distances.

The Brookhaven System

The dispersion parameters of the Singer - Smith system (1966) are based on many tracer dispersion experiments conducted over a 15 years period (from 1950 to 1965). The height and nature of the sources and samplers have varied considerably (oil fog, uranine dye and ^{41}Ar) as a function of time. The final result was a large accumulation of concentration data observed a few meters as well as many kilometers away from the release points.

MODEL VALIDATION

Finally some tracer experiments have been performed specifically for validation of mathematical models.

Gaussian Models up to 5 km

Van Duuren et al. (1975) have done SF_6 tracer releases under different conditions. SF_6 was emitted from cylinders at ground level, through a stack with cold emissions and through a high stack with hot gases. Sampling was done by adsorption on activated carbon in the field. 4 ground-level releases were done with sampling at 100 m distance. Not the measured and calculated concentrations but the measured and calculated (Pasquill-Gifford) σ_y and σ_z were compared, they agreed within 50 %. For the emissions through a 110 m high cold chimney SF_6 was measured up to 2000 m distance. Taking into account a 20 % uncertainty on the σ_y and σ_z values of Singer and Smith, they calculated the concentration range for the plume maximum. Generally the measured maxima fell within the calculated limits. This was however not the case for the experiments (up to 4 km) with the 210 m high hot chimney were the model overpredicted the measurements.

Nieuwstadt and Van Duuren (1979) measured SF_6 between 3 and 7 km from a 213 m high mast. Samples were taken with water filled glass vessels. Validation was done by fitting a gaussian curve through the measurements and comparison of the measured σ_y with the ones of Turner and Singer-Smith. With a mixing height between 200 and 600 m vertical distribution was already homogeneous at 3 km.

De Bortoli et al. (1981) have traced the emissions of the Turbigo thermal power station in Italy to study the dispersion in the Po valley. They discharged 10 g $SF_6.s^{-1}$ during 2 - 3 hours from 48 and 96 m stacks. Samples were taken on 2 arcs between 2 and 10 km from the stack and sampler spacing of 2 to 10 degrees. Sampling time was 8 or 16 min. Using the bi-gaussian distribution algorithm they determined σ_y and σ_z from the observed SF_6 glc-profiles. A comparison with the Pasquill-Gifford system showed that the obtained dispersion parameters are systematically larger than the P-G ones. Especially under low wind speed conditions can plume meandering induce increase of σ_y. The angular difference between the mean wind direction and the plume axis can amount to 10 degrees. In a later stage also remote sensing techniques were used to track the aerosol and SO_2 plume which was labeled with SF_6 (Cerutti et al., 1983).

A clear illustration of the uncertainties of the results of field experiments is given in a Joint Research Centre (JRC) Ispra - Karlsruhe Nuclear Research Centre (KNRC) joint exercise (Schuettelkopf et al., 1981). Different tracers $CFCl_3$, CF_2Br_2 and SF_6 were released simultaneously from the 160 m and 195 m platforms of the KNRC tower. Near ground sampling was performed at up to 67 positions at downwind distances between 500 m and 8.5 km during 2 periods of 30 min. each. The tracer $CFCl_3$ was analysed both by JRC and KNRC. The average ratio obtained by dividing the KNRC by the JRC data is 1.79 for the $CFCl_3$ tracer and 1.07 for the other two tracers. They also demonstrated that the determination of σ_y is very sensitive to the distribution and positioning of the samplers. Determination of σ_z is less sensitive in this respect. The σ_y values of the experiment are smaller than obtained from the Karlsruhe system due to the higher wind speeds and smaller direction fluctuations.

Gryning et al. (1978 ; 1980) have performed several SF_6 experiments over rural area. SF_6 was released at a height of 60 m and was sampled with automatic bag samplers at 2 m height in the streets of Risø up to 5 km from the source. They also validated their model merely by comparing measured and calculated σ_y values rather than concentrations. Because σ_z could only be determined indirectly, it was not compared with theoretical values. Measured σ_y values agreed well with the methods of Hay-Pasquill, Draxler and Taylor. Excellent agreement was also obtained when the SCK/CEN parameters for open area were multiplied with a factor 0.6. This indicates that the effect of surface roughness can be separated from the effect of stability.

Numerical Models

Numerical models have been validated with the results of tracer experiments in complex terrain and under non-steady meteorological conditions by Mulholland (1980) and Taylor e.a. (1982). Since the dispersion in these situations could not be described by a simple bi-gaussian distribution, the validation could not be done through the comparison of measured and calculated dispersion parameters but was done by comparison of concentrations.

Mulholland used ZnS-CdS aerosol emissions (25 m high) and 10 sampling points (1 m ; t_{av} = 1 h) to measure the distribution of a tracer up to 15 km under stable winter conditions in a complex coastal area. Tracer dissemination was begun before midnight during stable weather, and continued for 6 to 16 h through the fumigation period after sunrise. Ten meteorological masts with wind speed direction and temperature at 2 and 11 m were used to obtain enough data on the wind field. The aerosol concentration in the 10 sampling points was calculated as a function of time by a Dynamic Puff Model and an equivalent Gaussian Puff Model including Stokes sedimentation. The Gaussian Puff Model failed to simulate the effects of vertical wind shear and diffusion and generally overestimated the measured values. The Dynamic Puff Model showed no tendency for systematic deviations from the measurements but the scatter of the calculated around the measured concentrations still amounted to a factor 10.

Taylor et al. (1982) released SF_6 at 150 to 175 m from a tethered balloon and took 1 h samples at 12 locations up to 40 km from the release site, to validate also a three-dimensional numerical model in a complex terrain under stable and fumigating conditions. An extensive array of meteorological data systems was operated during the program, including a Doppler Sodar, two tethersondes, two pilot balloon systems and five mechanical ground-level weather stations. In general model predictions correlated very closely with observed tracer gas concentrations. Overall correlation coefficients for 6 tracer experiments ranged from 0.73 to 0.92. The overall average of predicted-to-observed ratios for the peak concentrations was 0.87.

Long range models

Lamb e.a. (1978) have measured 1 h average SF_6 and $CBrF_3$ concentrations at 28 locations up to 100 km from a 5 m high tracer emission, which lasted some 5 h, under complex, coastal meteorological conditions. Although the flow patterns were very complex they were reproducible and in some cases steady. Ground-level centerline concentrations were calculated using the bi-gaussian dispersion model. The dispersion parameters σ_y and σ_z - to be used in the calculations - were obtained in two ways :
- from the Pasquill-Gifford-Turner turbulence typing scheme
- from best-fit lines through the measured profiles for each test

The wind speed was obtained by simply averaging all available data. Despite the fact that most basic assumptions of the bi-gaussian theorem were not fulfilled (complex topography, non-steady wind conditions, Pasquill parameters are determined for 10 minutes and up to 10 km but used for 1 h averaged up to 100 km) the authors claim a good comparison between measured and calculated concentrations. Calculated trajectories and arrival times of air parcels based upon a numerical two-dimensional model agreed well with tracer data.

Draxler (1979) developed a mesoscale trajectory model with gaussian vertical distribution and compared the results with two tracer experiments up to 100 km from the release point. In a first experiment SF_6, $^{13}CD_4$ tracers and ^{85}Kr from 62 m high routine emissions were measured on an arc 100 km from the source. In the second SF_6, $^{12}CD_4$, $^{13}CD_4$, C_8F_{16}, C_6F_{12} and C_7C_{14} were released at ground-level and sampled on arcs 50 and 90 km downwind. Four meteorological data bases were used (10 m high masts network with and without correction for surface winds in the model, 60 m high mast, hourly pilot balloons) and the influence of different meteorological input data on the results is illustrated. Travel times vary as much as 50 %. The experiments demonstrate realism in the mechanism of the computational methods.

Some preliminary results on the validation of a long range model with a SF_6 tracer exercise have been reported by Doury e.a.,

(1975). In a first approximation the measurements agree with the model predictions but more experimental work is in progress.

CONCLUSIONS

Tracer experiments have been widely recommended as an excellent tool for the study of the dispersion of atmospheric pollutants. SF_6 is the most popular tracer and experimental details are well documented in the literature. Perfluorocarbons are gaining interest for long range dispersion experiments. Many papers describe tracer experiments without attempt of quantitative description of the dispersion. Only the Karlsruhe and Jülich turbulence typing schemes for the bi-gaussian model are strongly based on systematic tracer experiments. Validation of bi-gaussian models with tracer experiments is often realized by comparing measured and calculated dispersion parameters rather than concentrations. It is clear that turbulence classification and dispersion parameters have to be reinvestigated in order to correlate σ_y with wind direction fluctuations, σ_z with temperature gradient and wind speed and to take site specificity into account. Numerical and long range models have been validated through the concentrations, although in some cases serious doubts on the validity of the comparison exist. The quality and quantity of meteorological observations during the tracer experiments is extremely important for a successful model calculation. Considering the importance of mathematical modeling in the nuclear and the non-nuclear air pollution control policy the number of model validation experiments is extremely small.

REFERENCES

Bultynck, H., and Malet, L., M., 1972, Evaluation of atmospheric dilution factors for effluents diffused from an elevated continuous point source, Tellus 24:455.

Cerutti, C., et al., 1983, Characterization of plume dispersion in breeze regime by remote sensing and tracer techniques, Report EUR8545EN

Clemons, C., A., Coleman, A., I., and Saltzman, B., E., 1968, Concentration and ultrasensitive chromatographic determination of sulfur hexafluoride for application to meteorological tracing, Env. Sci. Techn. 2,551.

Collins, F., Bartlett, F., E., Turk, A., Edmonds, S., M., and Mark, H., L., 1965, A preliminary evaluation of gas air tracers, JAPCA 15, 109.

De Bortoli, M., and Pecchio, E., 1976, Measurements of some halogenated compounds in air over Europe, Atm. Env. 10,921.

De Bortoli, M., et al., 1981, SF_6 dispersion experiments at a power plant, Eur. appl. Res. Rept., Environ. and Nat. Res. Sect., 3:287.

Dietz, R., N., and Cote, E., A., 1973, Tracing atmospheric pollutants by gas chromatographic determination of sulfur hexafluoride, Env. Sci. Techn. 7:338.

Draxler, R., R., 1979, Modeling the results of two recent mesoscale dispersion experiments, Atm. Env. 13:1523.
Drivas, P., J., and Shair, F., H., 1974, A tracer study of pollutant transport and dispersion in the Los Angeles Area, Atm.Env. 8: 1155.
Doury, A., et al., 1975, "Simulation de Transferts Atmosphériques à Grande Distance", Rapport SESR-R-33.
Doury, A., 1980, Pratiques françaises en matière de prévision quantitative de la pollution atmosphérique potentielle liée aux activités nucléaires, in : "CEC seminar on radioactive releases and their dispersion in the atmosphere following a hypothetical reactor accident", CEC, Risø.
Eggleton, A., and Thompson, N., 1961, Loss of fluorescent particles in atmospheric diffusion experiments by comparison with radioxenon tracer, Nature, 192:935.
Emberlin, J., C., 1981, A sulphur hexafluoride tracer experiment from a tall stack over complex topography in a coastal area of southern England, Atm. Env. 15:1523.
Ferber, G., J., et al., 1981, Demonstration of a long-range atmospheric tracer system using perfluorocarbons, NOAA Technical Memorandum ERL ARL-101.
Gryning, S., E., Lyck, E., and Hedegaard, K., 1978, Short-range diffusion experiments in unstable conditions over inhomogeneous terrain, Tellus, 30:392.
Gryning, S., E., and Lyck, E., 1980, Medium-range dispersion experiments downwind from a shoreline in near neutral conditions, Atm. Env., 14:923.
Gryning, S., and Lyck, E., 1980, Elevated source SF_6 tracer dispersion experiments in the Copenhagen area, in : "CEC seminar on radioactive releases and their dispersion in the atmosphere following a hypothetical reactor accident", CEC Risø.
Hay, J., S., and Pasquill, F., 1957, Diffusion from a fixed source at a height of a few hundred feet in the atmosphere, Journal of Fluid Mechanics 2:299
Heffter, J., Ferber, G., & Krey, P., 1980, Atmospheric tracer experiments for regional dispersion studies, in : "CEC seminar on radioactive releases and their dispersion in the atmosphere following a hypothetical reactor accident", CEC Risø.
Kallend, A., S., et al., 1981, Studies of the fate of atmospheric emissions in power plant plumes over the North Sea, in : "CEC symposium on physico-chemical behaviour of atmospheric pollutants", Varese, Italy.
Kretzschmar, J., and Mertens, I., 1980, Influence of the turbulence typing schemes upon the yearly average ground-level concentrations calculated by means of a mean wind direction model, Atm. Env., 14:947.
Kretzschmar, J., G., Mertens, I., De Baere, G., and Vandervee, J., 1982, Influence of the turbulence typing scheme upon the cumulative frequency distributions of the calculated relative concentrations for different averaging times, CEC Project 10 C : Contract SR-028-B. Final Report task 1.

Lamb, B., K., Shair, F., H., and Smith, T., B., 1978, Atmospheric dispersion and transport within coastal regions, Part II, Atm. Env., 12:2101.

Le Quinio, R., 1970, Evaluation de la diffusion d'effluents gazeux en atmosphère libre à partir d'une source ponctuelle continue, Rapport CEA-R-3945.

Le Quinio, R., 1973, Concentrations sur une heure de polluants, dues à des émissions ponctuelles près du sol. Présentation probabiliste, IAEA-SM-169/14, Vienne, 215-222.

Lovelock, J., E., and Ferber, G., J., 1982, Exotic tracers for atmospheric studies, Atm. Env., 16:1467.

Mulholland, M., 1980, Simulation of tracer experiments using a numerical model for point sources in a sheared atmosphere, Atm. Env. 14:1347.

Nester, K., and Thomas, P., 1979, Im Kernforschungszentrum Karlsruhe experimentell ermittelte Ausbreitungsparameter für Emissionshöhen bis 195 m, Staub, 39:291.

Nieuwstadt, F., T., M., and Van Duuren, H., 1979, Dispersion experiments with SF_6 from the 213 m high meteorological mast at Cabauw in The Netherlands, Fourth Symposium on turbulence, diffusion, and air pollution, Boston, Mass., 34-40.

Pasquill, F., 1961, The estimation of the dispersion of windborne material, The meteorological magazine, 90:33.

Raynor, G., S., Michael, P., Brown, R., M., and Sethuraman, S., 1975, Studies of atmospheric diffusion from a nearshore oceanic site, J. Appl. Met. 14:1080.

Saltzman, B., E., Coleman, A., I., and Clemons, C., A., 1966, Halogenated compounds as gaseous meteorological tracers, Stability and ultrasensitive analysis by gas chromatography, Anal. Chem., 38:753.

Schuettelkopf, H., Thomas, P., Vogt, S., De Bortoli, M., and Gaglione, P., 1981, Experimental determination of the atmospheric dispersion parameters, Report EUR7577EN.

Singh, H., Salas, L., Shigeishi, H., & Scribner, E., 1979, Atmospheric halocarbons,hydrocarbons, and sulfur hexafluoride : global distributions, sources and sinks, Science 203:899.

Shum, Y., S., Loveland, W., D., and Hewson, E., W., 1975, The use of artificial activable trace elements to monitor pollutant source strengths and dispersal patterns, JAPCA 25:1123.

Taylor, G., H., Schanot, A., J., Tran, K., T., and Marsh, S., L., 1982, A tracer study and model validation program for the proposed Lucerne valley generating station, 3rd joint conference on applications of air pollution meteorology, 190-193.

Thomas, P., 1980, "Atmosphärische Ausbreitungsversuche am Kernforschungszentrum Karlsruhe", in : "CEC seminar on radioactive releases and their dispersion in the atmosphere following a hypothetical reactor accident", CEC Risø.

Turk, A., Edmonds, S., Mark, H., L., and Collins, G., F., 1968, Sulfur hexafluoride as a gas-air tracer, Env. Sci.Techn. 2:44.

Vanderborght, B., Kretzschmar, J., Rymen, T., Candreva, F., and Dams, R., 1981, On the use of SF_6 tracer releases for the determination of fugitive emissions, 2nd European symposium on "Physico-chemical behaviour of atmospheric pollutants", Varese.

Van Duuren, H., Krijt, G., D., and Elshout, A., J., 1975, Zwavelhexafluoride als tracer voor de verspreiding van luchtverontreinigende componenten vanuit puntbronnen, Electrotechniek 53: 135.

Verschueren, K., 1978, "Handbook of environmental data on organic chemicals", Van Nostrand Reinhold Company, New York.

Vogt, K., and Geiss, H., 1974, Tracer experiments on the dispersion of plumes over terrain of major surface roughness, Report Jul-1131-ST.

Vogt, K., J., 1977, Empirical investigations of the diffusion of waste air plumes in the atmosphere, Nuclear Technology, 34:43.

Singer, I.A., and Smith, M.E., 1966, Atmospheric dispersion at Brookhaven National Laboratory, Air and Wat. Pollut. Int. J. 10:125.

DISCUSSION

S.E. GRYNING — What do you think the effect is on the ground-level concentrations due to deposition of the tracer, when particles or SO_2 are used as tracer, as was done in e.g. the Porton Downs experiments, the Prairie Grass experiments and the experiments of Le Quinio. Can these experiments be used to obtain reliable estimates of σ_z without correcting for the effect of deposition?

B. VANDERBORGHT — As far as I know, gravitational settling or deposition of tracer material has not yet been measured directly.
Some model calculations on tracer experimental results indicate however that deposition can be important, especially for ground level release experiments. Work by Bultynck-Malet (1972) and Little e.a. (Atm. Env. 43, 1982, 355) shows that the mean ratio of predicted/observed glc improves from a value between 2 an 3 without correction for deposition to a value between 1 and 2, when deposition is taken into account. Values of σ_z, estimated from SO_2 or particulate tracer experiments, may be subjected to a presently unknown bias.

DOWNWIND HAZARD DISTANCES FOR POLLUTANTS

OVER LAND AND SEA

A. Groll, W. aufm. Kampe, and H. Weber

German Military Geophysical Office
Mont Royal
D-5580 Traben-Trarbach, FRG

INTRODUCTION

The German Military Geophysical Office, with the support of several other military and civilian organisations, has carried out several diffusion experiments both over land and sea areas between 1979 and 1982.

The sea trials were part of the GEOMAR Experiment /4/ which took place in the North Sea about 80 km NW of Helgoland near the research platform "Nordsee". During a total of 15 weeks measurements of the crosswind concentration profiles of plumes were made at different distances up to 35 km downwind of a continuous point source located at about 5 m above the sea surface on board a military research vessel. Three different tracers were used : HC smoke /3/ as a wet aerosol, copper doped zinc sulfide as a dry aerosol, and sulfurhexafluoride (SF_6) as a gas tracer. Only the last will be discussed here.

Two land trials were carried out as part of the MESOKLIP and MERKUR experiments which took place in the upper Rhine Valley near Speyer and the Alpine foothills near Rosenheim, respectively, giving a total of 3 weeks of measurements up to distances of 4 km downwind of a continuous SF_6 point source at 2 m above the ground.

The experiment design was principally the same over land and sea, the aim being to determine the diffusion parameters σ_y σ_z near the surface for a biaxial Gaussian plume model as a function of atmospheric stability described by some asequate classification scheme.

In all cases vehicles equipped with infrared analysers or flame photometers made passes through the plume perpendicular to the downwind direction at different distances from the source. On land the source was positioned in such a way to allow traverses of the plume at angles as close to 90° as possible while making use of local roads.

BASIC THEORY

Using the Gaussian Plume Model the crosswind concentration (C) of the plume is described by :

$$C(x, y, z) = \frac{Q}{2\pi\bar{u}\sigma_y\sigma_z} \exp\left(-\frac{(y-\bar{y})^2}{2\sigma_y^2}\right) \left(\exp\left(-\frac{(z-h)^2}{2\sigma_z^2}\right) + \exp\left(-\frac{(z+h)^2}{2\sigma_z^2}\right)\right) \quad (1)$$

in a coordinate system oriented so that the x-axis coincides with the downwind direction, y is the crosswind, and z the vertical coordinate.

h = source height, z = sampling height,
Q = source strength, $\bar{u}$ = mean wind speed.

It is further assumed that source and samplers are both at the surface :

$$h = z = 0 \quad (2)$$

The error made by assuming (2) instead of the actual values h = 6.5 m and z = 5 m for the sea and h = z = 2 m for the land trials is negligible at those distances from the source at which the measurements took place. This results in :

$$C = \frac{Q}{\pi\bar{u}\sigma_y\sigma_z} \exp\left(-\frac{(y-\bar{y})^2}{2\sigma_y^2}\right) \quad (3)$$

The procedure used to determine $\sigma_y(x)$ and $\sigma_z(x)$ from the data is described in detail below.

In evaluating experimental data one must distinguish between the instantaneous spreading of the plume relative to its centroid (relative diffusion) and the lateral spreading of concentration over a larger area due to the meandering of the plume's axis in the course of time. The latter is referred to as Taylor or single particle diffusion /1/, /2/, /5/, /6/.

Thus σ_y and σ_z are functions of the sampling time.

In the experiments described here it was possible to distinguish between relative and single particle diffusion. The concentration profiles obtained on a single pass (1 - 3 minutes) are taken to be representative for relative diffusion and the ensemble of profiles measured during 1.5 - 2-hour periods is taken to be representative for single particle diffusion /7/.

The determination of the diffusion category for each profile poses somewhat of a problem. In the case of the land trials a scheme developed by Wamser et.al. /8/ was used, which relates routine synoptic data such as cloud cover and height of ceiling as well as wind speed and surface roughness to σ_w (vertical wind variance) and finally σ_y and σ_z values.

The corresponding σ_w values were determined for each profile measured and the measurements then grouped to form the following diffusion categories.

Table 1. Diffusion Categories over Land

Category		σ_w [m/s]
1	unstable	[.84, 1)
2		[.63, .84)
3		[.45, .63)
4		[.26, .45)
5		[.16, .26)
6	stable	[0, .16)

Table 2. Diffusion Categories over Sea

Category		$\Delta T/u^2$ [Ks^2/m^2]
1	unstable	($-\infty$, -.3)
2		(-.3, -.15)
3		(-.15, -.01)
4		(-.01, .01)
5		(.01, .15)
6		(.15, .3)
7	stable	(.3, ∞)

At sea, however, this scheme was not directly applicable. The quantity $\Delta T/u^2$ was used as a measure for stability, ΔT being the temperature difference between air temperature at 3 m and the surface water temperature and u being the mean surface wind speed (10 m).

The category boundaries were determined empirically to show some significant differences in the measured maximum concentrations as a function of downwind distance from the source between categories. Then equal-sized intervals were formed with the exceptions that the lowest and highest categories are open intervals toward $\pm\infty$ and the neutral case ($\Delta T/u^2 = 0$) is singled out as a finite interval $[-0.01, 0.01)$, thus resulting in table 2.

RESULTS

Determination of σ_y and σ_z for Relative Diffusion

The strip chart recordings of each measured crosswind concentration profile were digitized to obtain the data points C_{mi} which are then fitted by a Gaussian curve (3) under variation of only σ_y and $\bar{y}$ using the least squares method. This gives the values for σ_y as a function of downwind distance from the source. The corresponding σ_z is determined beforehand using the integral relationship :

$$A = \int_{-\infty}^{+\infty} C dy = \frac{Q}{\sqrt{\frac{\pi}{2}}\,\bar{u}\sigma_z} \tag{4}$$

A may be approximated as the sum of the trapezoid areas defined by the data points and the Y-Axis as follows :

$$A = \frac{1}{2}\sum_{i=1}^{N} (y_{i+1} - y_i) \cdot (C_{mi+1} + C_{mi}) \tag{5}$$

N = number of digitized data points.

Using this value σ_z can be calculated from (4)

The functional relationship between σ_y or σ_z and the distance from the source (x) is described in the common way as :

$$\sigma_y = F\, x^{\,f} \qquad\qquad \sigma_z = G\, x^{\,g} \tag{6}$$

for each diffusion category.

The free variables F, f, G, and g are determined by a least squares fit. Due to the fact that each curve has two free variables which may compensate each other during the fitting procedure, the problem arises that the curves for the different diffusion categories may intersect and lead to incorrect results when extrapolating beyond the range of the measurements, which is inevitable for practical applications. In order to avoid this the exponents f and g are fitted linearly as a function of the diffusion category. In doing this, however, only those categories that are well represented by measurements are used for the fit.

For the sea trials no significant linear dependence of f on the diffusion categories could be found, so it was replaced by the average of f over all categories. The data was then again fitted with fixed exponents taken from the linear curves described above for each category allowing only the factors F, G as free variables.

These factors in turn were fitted with an exponential function to insure a smooth transition from one category to the next while at the same time ruling out values $\leqslant 0$.

Fig. 1 shows the results of the different steps described above using as an example the σ_z for diffusion category 4 (neutral).

Fig. 1a shows the data points for σ_z at different distances from the source together with the original fit. For better clarity Fig. 1b shows only the mean values of σ_z averaged over 1 km intervals together with the error bars for the standard deviation. The dashed line is the original fit (as in Fig. 1a) while the dotted line is the fit resulting from fixing the exponent f and only letting the factor F vary, and the solid line is arrived at when fixing both F and f in accordance with the fits of both variables as a function of the diffusion category.

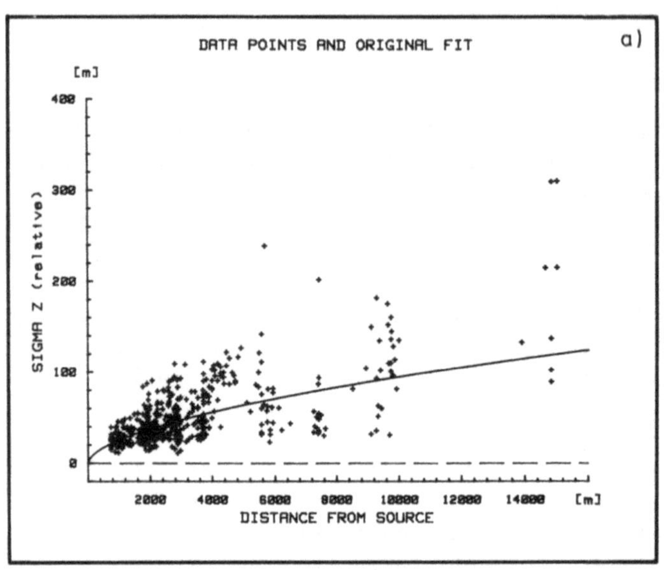

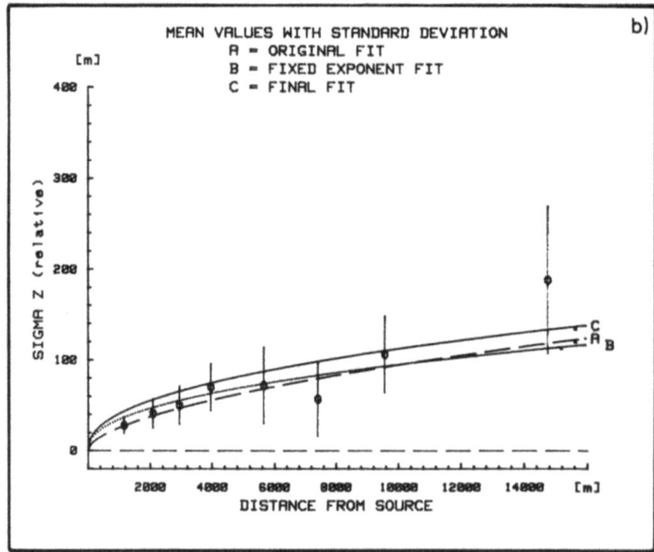

Fig. 1. Example of fit procedure for σ_z (relative diffusion) for neutral stability over sea.

The procedure described before was repeated analogously for the land trials. The resulting set of curves for both land and sea are shown in Fig. 2a - d. The corresponding functions for the parameters F, f, G, and g are given in Table 3.

Table 3. Factors and Exponents for σ_y and σ_z as a Function of the Diffusion Category I

Land	Sea
F = 0.2997· exp (0.3175·I)	0.4570· exp (-0.0863·I)
f = 0.89 - 0.08· I	0.7
G = 0.1229· exp (0.3844· I)	0.9740. exp (0.1750· I)
g = 0.97 - 0.10· I	0.68 - 0.06· I

Determination of σ_y for Single Particle Diffusion

Land trials. While passing through the plume significant points (road crossings, major trees, milestones, etc.) were marked on the strip chart recorder making it possible to determine the UTM-Coordinates of the point with maximum concentration with a resolution of better than 10 m using detailed maps of the area.

For each separate 2-hour period the positions of maximum concentration were plotted to scale and the set of points rotated as a whole to minimize the deviation of the crosswind components (y-coordinates) from the x-axis of the coordinate system, thus forcing the actual mean wind direction for the 2-hour period to coincide with the x-axis.

The data points for all 2-hour periods within the same diffusion category are then superimposed to give plots as shown in Fig. 3 for category 4. For equidistant intervals on the x-axis the standard deviation of the y-coordinates relative to the x-axis (mean downwind direction) is determined for each interval.

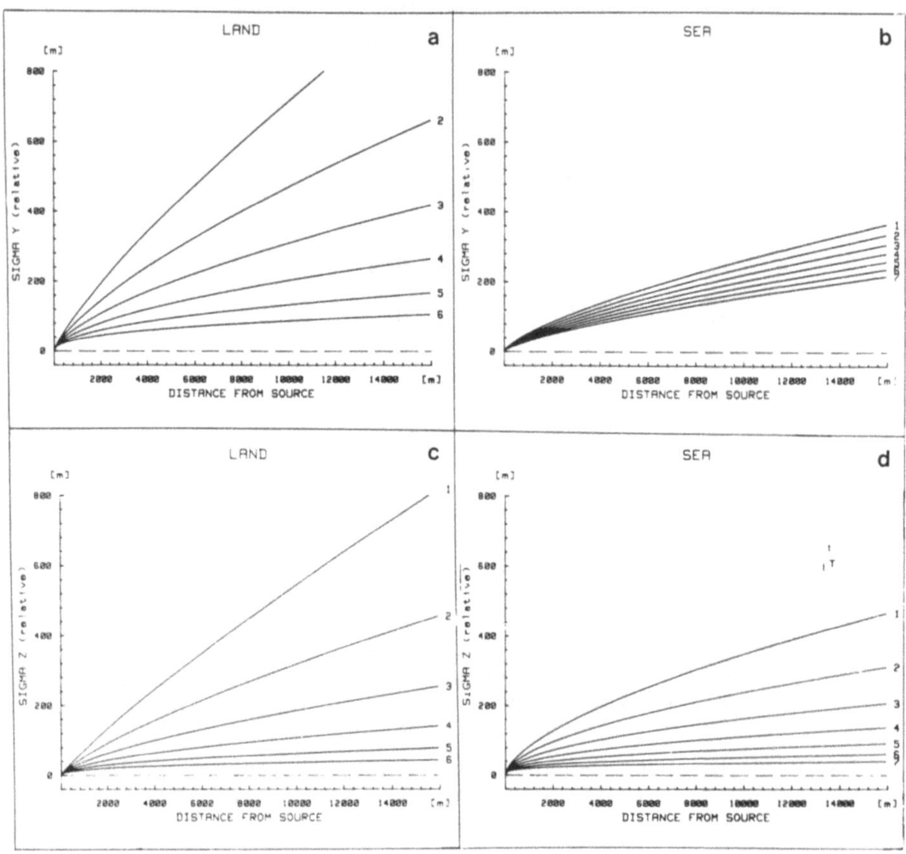

Fig. 2. Final σ_y and σ_z Fits for Relative Diffusion over Land and Sea for the Different Diffusion Categories.

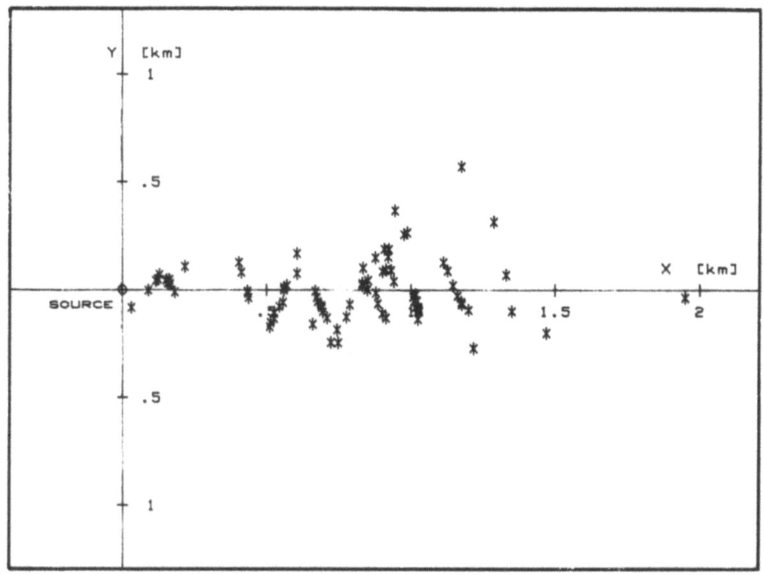

Fig. 3. Positions of Maximum Concentration for 2-Hour Periode under Neutral Conditions over Land.

The standard deviation is taken to be the σ_y for single particle diffusion and is fitted by a least squares method as a function of the distance from the source as follows :

$$\sigma_y = F\ x^{\ f}$$

The diffusion category for each 2-hour period was determined using the σ_w values derived from synoptic data after Wamser and classifying them as described above.

No significant dependence of σ_y on the diffusion category could be found so a second attempt was made to classify the measured data using only the wind speed as a parameter. Fig. 4 shows the resulting data points together with the curves for u < 10 kn and u ⩾ 10 kn. The fits were obtained by fixing the exponent to 0.7, which was uniformly done for all σ_y fits on land and sea. The resulting functions are

$$\text{Land}: \quad \sigma_y\ (x) = \begin{cases} 1.577\ .\ x^{0.7} & u < 10\ \text{kn} \\ 1.130\ .\ x^{0.7} & u \geqslant 10\ \text{kn} \end{cases}$$

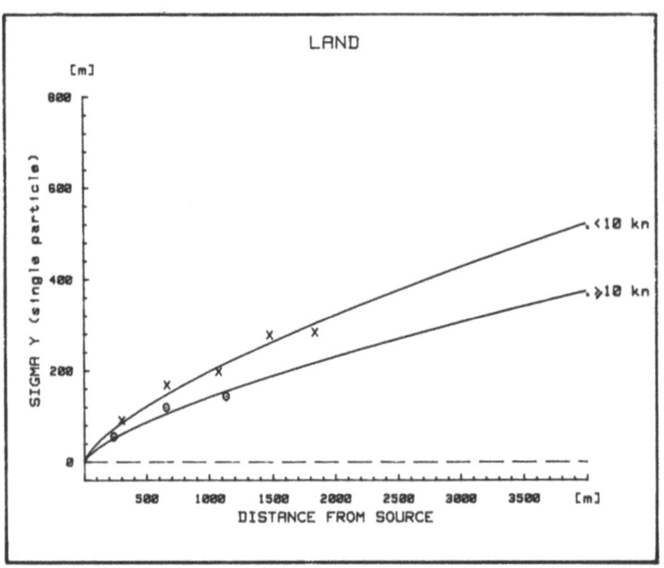

Fig. 4. σ_y for Single Particle Diffusion over Land Dependent on Wind Speed Categories.

Sea trials. The ships' positions were tracked using the DECCA HIFIX navigation system, which has a resolution of 1 to 10 m in the area of operation. On each crossing of the SF_6 plume the position of the maximum concentration was registered by this method. The daily measurements were subdivided into four or five 2-hour periods during which the ships continuously traversed back and forth through the plume at different distances from the source.

The parameter used to characterise the diffusion category ($\Delta T/u^2$) is averaged over each 2-hour period and used for classification as shown in Table 2.

As for the land trials no significant dependence on the diffusion categories could be detected.

Using only the wind speed as a classification paramater and repeating the procedure described above gives the following results :

$$\text{Sea :} \quad \sigma_y = \begin{cases} 1.538.x^{0.7} < 10 \text{ kn} \\ 1.038.x^{0.7} \geqslant 10 \text{ kn} \end{cases} \tag{8}$$

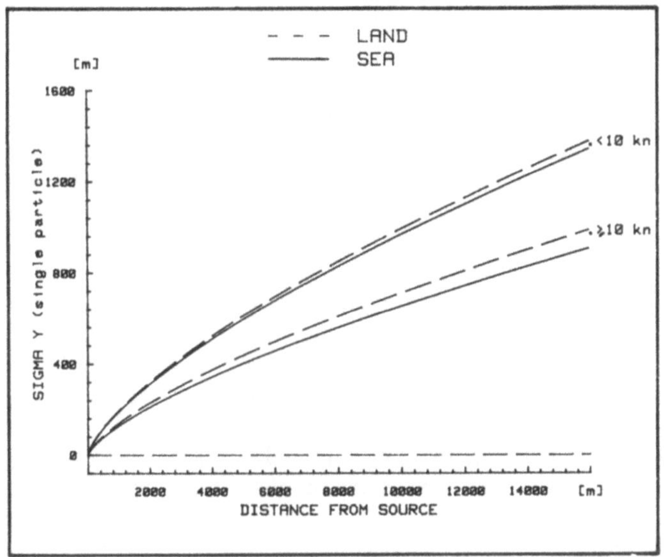

Fig. 5. Comparison of σ_y for Single Particle Diffusion over Land and Sea.

COMPARISON OF POLLUTANT HAZARD DISTANCES

In order to determine hazard distances the "hazard" must be defined for each different pollutant in terms of a certain concentration which would lead to a hazard. Depending on the type of pollutant and the type of emission, one may be interested in the peak concentrations encountered for short time periods or the longer-term, average concentrations. Tables 4 - 7 are compiled using a continuous point source at the surface with a source rate of Q = 1 kg/s and hazard concentrations (C) of 10, 5, and 1 mg/m^3, and a deposition velocity v_d of 1 mm/s.

The concentration is then determined by :

$$C = \frac{Q}{2\pi\bar{u}\sigma_y\sigma_z} \left(\exp\left(- \frac{(\frac{v_d}{\bar{u}}\cdot x)^2}{2\sigma_z^2}\right) + \exp\left(\frac{(\frac{v_d}{\bar{u}}\cdot x)^2}{2\sigma_z^2}\right)\right)$$

The hazard distance is determined iteratively by solving for x given fixed C's Tables 4 and 5 are calculated using only the relative diffusion giving the distances for short term peak concentrations while Tables 6 and 7 also account for single-particle diffusion with

$$\sigma_y = \sqrt{\sigma^2_{y_{relative}} + \sigma^2_{y_{sing.\ part.}}}$$

σ_z is taken from the relative diffusion in all cases since σ_z for single particle diffusion could not be derived from the measurements. It is assumed the vertical meandering of a plume from a ground source may be neglected when considering the concentration at the surface.

It is important to note that the diffusion categories can not be compared one to one over land and sea. Besides the fact that different parameters must be used to define a diffusion category and the fact of having 6 categories over land and 7 over the sea, there are the questions of the possible ranges of stability over land and sea and the range encountered during the experiments. A situation described by the very stable category 6 over land may not be possible over sea and if it were, may not have been encountered during the 15 weeks of GEOMAR, thus limiting the range of the 7 diffusion categories for sea areas described here.

The tables show all calculated values regardless of the fact, whether the combination of wind speed and diffusion category is at all climatologically or physically possible, or whether extrapolations to the distances derived are acceptable from experiments which covered only distances from the source not exceeding 4 km over land and 20 km over sea areas.

In viewing the tables one should also bear in mind that, for example, hazard distances of over 100 km with wind speed of about 5 knots require travel times of 10 hours during which the meteorological situation must remain unchanged. This is not generally the case in reality and a change in the meteorological situation will tend to reduce the hazard distances in these cases.

Table 4 Land

Category :	1	2	3	4	5	6	
Wind kn							
	1	2	3	6	15	55	- 10 mg/m^3
0 - 4	2	3	5	11	30	134	- 5 mg/m^3
	5	9	18	45	163	1054	- 1 mg/m^3
	1	1	2	3	7	23	- 10 mg/m^3
5 - 9	1	2	3	6	15	55	- 5 mg/m^3
	3	6	11	25	79	434	- 1 mg/m^3
	1	1	1	2	5	13	- 10 mg/m^3
10 - 14	1	1	2	4	10	33	- 5 mg/m^3
	3	4	8	17	52	258	- 1 mg/m^3
	1	1	1	2	3	9	- 10 mg/m^3
15 - 19	1	1	2	3	7	23	- 5 mg/m^3
	2	3	6	13	38	178	- 1 mg/m^3

Table 5 Sea

Category :	1	2	3	4	5	6	7	
Wind kn								
	2	3	4	6	9	14	22	- 10 mg/m^3
0 - 4	4	5	7	11	16	27	46	- 5 mg/m^3
	13	18	28	44	73	130	247	- 1 mg/m^3
	1	2	2	3	5	7	11	- 10 mg/m^3
5 - 9	2	3	4	6	9	14	22	- 5 mg/m^3
	7	11	16	24	39	66	120	- 1 mg/m^3
	1	1	2	2	3	5	7	- 10 mg/m^3
10 - 14	2	2	3	4	6	9	15	- 5 mg/m^3
	5	8	11	17	27	44	79	- 1 mg/m^3
	1	1	1	2	2	3	5	- 10 mg/m^3
15 - 19	1	2	2	3	5	7	11	- 5 mg/m^3
	4	6	9	13	20	33	58	- 1 mg/m^3

Hazard Distances in km for Relative Diffusion

Table 6 Land

Category :	1	2	3	4	5	6	
Wind kn							
	1	1	1	2	2	4	- 10 mg/m^3
0 - 4	1	2	2	3	4	7	- 5 mg/m^3
	4	5	7	11	18	31	- 1 mg/m^3
	1	1	1	1	2	3	- 10 mg/m^3
5 - 9	1	1	2	2	3	5	- 5 mg/m^3
	3	4	5	8	13	22	- 1 mg/m^3
	-	1	1	1	1	2	- 10 mg/m^3
10 - 14	1	1	1	2	2	3	- 5 mg/m^3
	2	3	4	6	9	15	- 1 mg/m^3
	-	-	1	1	1	1	- 10 mg/m^3
15 - 19	1	1	1	1	2	3	- 5 mg/m^3
	2	2	3	5	7	11	- 1 mg/m^3

Table 7 Sea

Caterogy :	1	2	3	4	5	6	7	
Wind kn								
	1	1	1	1	2	2	3	- 10 mg/m^3
0 - 4	1	2	2	3	4	5	7	- 5 mg/m^3
	5	6	8	11	16	23	37	- 1 mg/m^3
	1	1	1	1	1	2	2	- 10 mg/m^3
5 - 9	1	1	2	2	3	4	5	- 5 mg/m^3
	4	5	6	8	12	17	26	- 1 mg/m^3
	-	1	1	1	1	1	2	- 10 mg/m^3
10 - 14	1	1	1	1	2	2	3	- 5 mg/m^3
	3	3	4	6	8	12	17	- 1 mg/m^3
	-	-	-	1	1	1	1	- 10 mg/m^3
15 - 19	1	1	1	1	1	2	2	- 5 mg/m^3
	2	3	3	4	6	9	13	- 1 mg/m^3

Hazard Distances in km for Single Particle Diffusion.

SUMMARY

The experimental results show that the Gaussian plume model gives a good approximation of the cross wind concentration profiles especially for sea areas, even though there is considerable scatter in the σ -values.

There is a significant difference both over land and sea between the relative diffusion and the single particle diffusion which must be considered in pollutant hazard calculations.

Both over land and sea the σ_z for relative diffusion is strongly dependent on the diffusion category, and both are of the same order of magnitude (Fig. 2c, d). For the corresponding σ_y's, however, we find that they are strongly dependent on the diffusion category only for land areas (Fig. 2c), while there is virtually no dependence on the category at sea (Fig. 2d). This may be attributed to the fact that terrain and vegetation produce horizontal inhomogeneities over land which in turn increase the lateral diffusion much more than the relatively homogeneous sea surface will.

The single particle diffusion showed no systematic dependence on the diffusion category at all while it appears to depend on the wind speed, as shown in Fig. 5, with a tendency to reduce σ_y as a function of the distance from the source with an increase in wind speed. However, the experiments only took place up to wind speeds of about 30 knots. The values for land areas are slightly higher than for sea areas.

As a result, the hazard distances over land and sea are virtually the same for neutral conditions, while they vary considerably both in the unstable and stable situations. The range of stability encountered over land may be considerably larger than over the sea, so that a direct comparison may not be possible.

REFERENCES

/1/ Gifford, F.A., Horizontal Diffusion in the Atmosphere : A Lagrangian Dynamical Theory, Report No. LA-8667-MS, Los Alamos Scientific Lab., (1981).

/2/ Gifford, F.A., Peak to Average Concentration Ratios According to a Fluctuating Plume Dispersion Model, Int. J. Air Poll., Vol 3, No. 4, 253-260, (1960).

/3/ Groll, A., George, R., Ritter, B., Luftbildaufnahmen kuenstlicher Nebelfahnen, Meteorol. Rdsch. 34, 162-166, (1981).

/4/ Groll, A., aufm Kampe, W., Single Particle and Relative Diffusion over Sea Areas, Sixth Symposium on Turbulence and Diffusion, Am. Meteor. Soc., 5-9, (1983).

/5/ Kristensen, L., Jensen, N.O., Petersen, E.L., Lateral Dispersion of Pollutants in a Very Stable Atmosphere - the Effect of Meandering, Atmos. Environ., Vol 15, No. 5, 837-844, (1981).

/6/ Nappo, C.J. Jr., Atmospheric Turbulence and Diffusion Estimates Derived from Observations of a Smoke Plume, Atmos. Environ., Vol 15, 541-547, (1981).

/7/ Sheih, C.M., On Lateral Dispersion Coefficients as a Function of Averaging Time, J. Appl. Met., Vol 19, 557-561, (1980).

/8/ Wamser, C., Schröter, J., Hinrichsen, K., Darstellung und Anwendung eines verbesserten universell gültigen Ausbreitungskriteriums, Staub-Reinhalt. Luft 40, 253-257, (1980).

ACKNOWLEDGEMENTS

The authors are specially indebted to the captain and crew of the "WFS Planet", the captains and crews of the research vessels of the Erprobungsstelle 71 in Eckernfoerde, the technicians of the Wehrwissenschaftliche Dienststelle in Munster and of the DFVLR in Oberpfaffenhofen among others not specifically mentioned here - who helped with great effort to make these experiments a success.

DISCUSSION

S. VOGT

I am wondering about the fact that you didn't find a dependency of σ_y (time averaged) over land on the stability class. From the previous paper of Vanderborght we learned that a lot of experimentators in those experiments over land had found this behaviour.

W. AUFM. KAMPE

We found a strong dependance of σ_y over land on the stability for the relative diffusion, but not for the 2hour averages. This may be due to the fact that we do not have enough measurements on land yet, but maybe also to the length of the averaging time. We don't have a sound reason for this, but we simply couldn't find a significant dependance.

REMOTE SENSING OF STABILITY CONDITIONS DURING SEVERE FOG EPISODES

G. Bonino, D. Anfossi, P. Bacci (*), and A. Longhetto

Istituto di Cosmogeofisica del CNR - C.so Fiume, 4 - 10133 Torino, Italy
Istituto di Fisica Generale dell'Università - Torino

INTRODUCTION

It is a widely recognized fact that the occurrence of most severe air pollution episodes is connected to deep fog layers. This happens not only as a mere consequence of that particular thermodynamic stratification of the lower troposphere wich establishes under fog conditions and reduces vertical dispersion, but also because fog takes usually place during the cold season, when a substantial contribution from low level domestic, uncontrolled sources is added to the industrial ones. The increased pollution background due to the former makes it necessary to exert stricter controls on the latter, through strategies of emission reduction, desulphuration, temperature increase of flue gas, burning of low sulphur content fuel, and so on.

The effectiveness of each above mentioned action is different according to the different thermodynamic vertical structures of fog. In order to choose the most reliable one, it is necessary to know the vertical structure of the whole fog layer, but this doesn't suffice; also one has to know the inversion strength of the layer capping the fog, whose vertical extent can reach as far as 500 m or more. This information is a crucial one, on account of the height of the tallest modern chimneys and, provided it's available, it allows an actual possibility to be seen to carry out cost effective, real time controls and management of air quality.

(*) Centro di Ricerca Termica e Nucleare dell'ENEL - Milano, Italy

This explains the interest involved in studying fog formation and evolution, in connection with air pollution problems, mainly when heavily industrialized and densely built-up areas, with humid, unstable and weak atmospheric circulation, like the Po Valley, are concerned.

Not only are fogs very frequent on the Po Valley (Pagliari and Persano, 1969), but also they appear as long lasting episodes, which make air pollution conditions critical, in that diffusion processes are superimposed upon an increasing background due to pollutant accumulation.

Deep (from 200 m to 600 m) and long lasting (as far as 10 days) fog episodes are observed at least once a year, twice or three times a year not being a rare phenomenon in winter.

Due to the important problems (like safety of surface and air transportation, insulation of high voltage electric lines, bronchial morbidity and diseases, and so on) which, in addition to those of air pollution, are related to the fog, this latter has been the object of a lot of studies in the past few years. Owing to difficulty, if not impossibility, in getting information on dynamic and thermal vertical structure of deep fog layers (it must be recalled that conventional sounding facilities cannot operate during fog conditions and, in any case, their information should be of little use in real time air quality control, due to long time-intervals elapsing among subsequent measurements), most investigators preferred to carefully measure as many near-ground fog parameters as they could, including thermal gradients, wind shear, drop size spectra, turbulence intensity, and so on in the first few meters of atmosphere (Choularton et al, 1981 - Roach, 1976).

The hope to be able to find any relationship, or at least a correlation, between the above parameters and higher up structures of fog turned out to be, in most cases, misleading. In fact, the only models which succeeded in describing radiation and water balance in the fog to a satisfactory degree of completeness and consistence referred to shallow fog layers (H_i < 200 m) (Choularton et al, 1981).

The recent development of modern and powerful remote sensing techniques of lower troposphere opened new prospects of success in studying both vertical structure of atmospheric boundary layer and fog formation, allowing observations, with high space and time resolution, of vertical profiles of air temperature and wind speed and direction as far as 500 m and, often, higher up.

The above techniques have been employed by us, in winter 1981-1982, at the Turbigo Remote Sensing Facility, described in the next section, located in the premises of a large power plant in the western region of the Po Valley (∿ 30 km West of Milan), which already hosted a lot of experimental studies of lower atmosphere and air pollution, among which an important field experiment organized by the European Community Commission (Longhetto et al, 1982).

The present report shows some preliminary results of a case study of a deep and thick fog episode, in which measurement of vertical distribution of stability of the layer prevailed over other more conventional observation near the ground.

Time trends of the main fog structure parameters in the first 600 meters (heights of fog and upper inversion top, "fine structure" of inner inhomogeneities connected to peculiar shapes of temperature and wind shear profiles, and so on) have then been correlated with the corresponding time trends of ground level SO_2 concentration.

Particular attention has been payed to the possibility of qualifying remote sensing techniques as powerful tools of emission planning from important sources (like power plants), with a view to reaching the best trade-off between meteorological conditions and power demand.

THE TURBIGO REMOTE SENSING STATION

In recent years a station has been developed at Turbigo for remote sensing of atmospheric physical parameters. There, are operating convenctional systems like anemometers, thermometers, hygrometers, solarimeters together with advanced systems to measure wind and thermal vertical profiles, worked out in the last few years and described below. The sulphur dioxide concentration at ground level is measured by five monitoring stations N-S oriented, with respect to the Power Plant, where wind direction prevails, and extending as far as 6 km.

Wind profile

A Doppler SODAR is used for vertical profile measurements of horizontal and vertical wind components together with standard deviation of the vertical velocity. The principle of operation is known (see e.g. Beran and Clifford, 1972). The system is formed by three acoustic antennas : one vertical oriented and two 30° sloping respect to zenith.

The main parameters are : acoustic frequency 1600 Hz : burst lenght 50 - 200 ms; repetition rate $\sim 10^{-1}$ s^{-1}; altitude measurement range 0 - 200 m minimum, 0 - 1000 m maximum, depending on atmospheric conditions with one average datum every 50 m; power 150 - 200 W; data elaborated and recorded on magnetic tape as half an hour average values. Validation of the system was satisfactorily performed by comparison with data obtained with pilot balloons and meteorological tower (see e.g. Gland 1982).

Thermal profiles

A metric Radio Acoustic Sounding System (RASS) was experimentally assembled and tested, in the last few years, at the remote sensing station of Turbigo in order to measure thermal profiles in the PBL (Bonino et al, 1979, 1980).

The RASS consists in its essence of a powerful acoustic source beaming a short burst of sinusoidal waves toward the zenith. The upward speed of this pulse is proportional at every height to the square root of the local temperature. The pulse speed is continuously measured from the ground by means of a Doppler radar. The radar echo is due to the change in the refractive index of air compressed by the acoustic waves. The faint echo is maximized by choosing an acoustic wave in Bragg resonance with the radiowave. The measured sound speed as a function of the delay leads to the acquisition of the thermal profile. So the temperature, relative to every height considered, is derived using the equation :

$$T = \left[\frac{4.7\ 10^{-2}\ f_D}{(1 + 4.714\ 10^{-5}\ (H - 30))\ \cos\ (\mathrm{arctg}\ (d/z))} \right]^2$$

where f_D is the Doppler frequency, H the relative humidity in %, d the half distance between the two radar antennas and z the height of observation. In the above expression, the term cos (arctg (d/z)) is used to convert radial velocities into vertical velocities. The numerical constants are peculiar to the Turbigo geometry and assume a speed of sound linearly related to H within the relative humidity range of 30 - 100 %. The radial echo is automatically processed and the output thermal profile is the average of a preset number of soundings (generally 10-20). RASS soundings could be repeated at a rate of one every 2-3 minutes.

Comparison of RASS measurements with other tropospheric soundings (radiosonde on tethered ballons and on disposable balloons, radiosonde on model aircraft and aircraft-borne thermosonde) performed at Turbigo in particular during the Fourth ECC Atmospheric Pollution Survey Campaign (Longhetto et al, 1982) and held in Turbigo during September, 1979, has demostrated :
a) the RASS ability to produce vertical thermal profiles in the range of altitudes 100 - 1000 m with a temperature accuracy and a height discrimination comparable with conventional soundings;
b) the RASS performances typical of a remote sensing device, i.e. short time of measurement, minimum cost of operation (Bonino et al, 1980).

Limitations associated with this technique are essentially due to strong horizontal winds and large convective vertical motions. Wind shears associated to strong wind distorting the acoustic burst worsen the coherent matching between radio and acoustic waves and cause a lowering of the sounding range. A test on the thermal profile measured during June-December 1982 has demostrated that in about 95 % of cases, the maximum sounding range is larger than 600 m. The temperature error Δ t (°C), arising from the effect of the vertical speed component of the wind on the radio-acoustic measurements, is given by Δ t = - 1.7 w. This error could cause trouble when evaluating thermal profiles during diurnal convective conditions, but it can be eliminated using SODAR technique to measure w or it can be reduced by averaging temperature measurements over a suitable period of time. During fog conditions, however, these limitations are negligible.

The performance of the RASS during fog conditions was discussed elsewhere (Bonino et al, 1981). In that paper it was demostrated that the RASS thermal profile has a good temperature accuracy and vertical resolution. In particular it's possible to know : a) the thickness of fog layer; b) the intensity of thermal inversion at the top of the fog; c) the atmospheric stability parameter above the fog-capping inversion.

FOG EPISODE

Synoptic conditions

The fog episode started on January 16, 1982 and ended on the 26th of the same month.

A few days previous to onset of fog, a wide area of levelled high pressure (around 1030 mb) was established over the Balkans at the surface level, extending a ridge towards Central Europe at 700 mb.

The Po Valley was then affected by a weak South-Easterly circulation near the ground and Easterly higher up; the surface pressure pattern was therefore, effective in advecting, over the region under consideration, cold air from the Balkans and warm and moist air from the Mediterranean basin, through the Adriatic Sea. The main frontal systems were embedded in the high latitude (> 50° N) zonal circulation and could hardly reach the Alps. Air was first quite dry below 500 m, very dry between 500 and 1000 m and moist in the upper layer, as far as 5000 m.

During the following days, also the lower atmosphere got wet, mainly in the daytime hours, and, as from the 15th and during the fog development, the anticyclonic system tilted its axis from the initial direction WNW-ESE towards a more meridional one (N-S), supporting flow of warm air from the Mediterranean basin towards higher latitudes, West of the Alps; air resulted saturated as far as 500 m and got progressively drier in the upper layers. In the last phase of the episode, the axis of the Balkan anticyclone took again its previous zonal orientation, joining the Azores high, which extended a ridge over Central Europe.

In the boundary region between the two anticyclones, some frontal perturbations began entering the western part of the Po Valley, bringing cold air descending from the Alps; as a consequence, pressure dropped towards 1020 mb and snowfalls appeared here and there in the Po Valley and fog dispersed.

Synoptic thermodynamic structure

Thermodynamic soundings, carried out at the Linate airport (Milano), showed the development of a subsidence phenomenon, (see fig. 1), occuring from the 17th to the 26th, whose maximum was observed on the 20th; in fact, above the top fog, air got very dry and the fog top itself was progressively lowering, from 500 m to 150 m (see next section). As from the 21th till the 26th, the subsidence weakened, weather got perturbed and the fog vanished.

In addition to the lowering of the height of the fog top, fig. 1 shows the time trends of other physical quantities showing the occurence and the development of a subsidence process. In fact, during the subsidence period, the heightof the lower thermal inversion (observed, on average, at nearly 850 mb) decreased from 800 to 900 mb, its temperature increased and so did the geopotential at 700 and 850 mb.

Local thermodynamic structure

As it will be shown in more detail in the following section, other observations, carried out at the site or in its neighbourhood, confirmed the occurence of a subsidence development from the 17th to the 23th, with a maximum between the 18th and 21st.

With reference to fig. 2, the heights H_i of the fog top and H_u of the upper inversion top (see for reference figs. 4-6) decreased noticeably from the 19th to the 21st, while the corresponding thermal difference $(T_u - T_i)$ (see for reference fig. 3b) between the two levels strengthened. In fig. 3a are shown the time evolution of the thicknesses of the layers $(H_u - H_b)$, $(H_u - H_i)$ and $(H_i - H_b)$; while $(H_u - H_i)$ is nearly constant, the other two differences attain their maximum value on between the 18th and 19th. Moreover, wind speed at various levels (fig. 3c), along with optical visibility (evaluated at the nearby airport, 10 km apart) and cooling and heating rates, attain their minimum values, the latters fluctuating around zero.

DATA

The vertical soundings of RASS and Doppler SODAR, mentioned in the previous section, made it possible to reconstruct a few of the main characteristics of the dynamic and thermodynamic structure of deep fogs.

In the figs. 4, 5, 6 are reported, as examples, some thermal profiles measured by the RASS. They show the fine structure of the Planetary Boundary Layer as far as 1000 m. The profiles are characterized by three separation layers whose heights are indicated respectively by H_b, H_i, H_u. Right of the thermal profiles, in the tables, are reported : σ_w measured by SODAR as an average value in correspondence of the inversion layer (central value) and in the layer adjacent to H_u and H_i (respectively upper and lower value); the stability parameter s in correspondence of the inversion layer; the wind shear in direction, $\Delta\alpha$, and intensity Δu in correspondence of the layer adjacent H_u and H_i. It may be noticed that a good correspondence exists between thermal stability measured by RASS and σ_w measured by SODAR. The wind shears are very variable from one sounding to another because of the low wind condition characterizing the fog episode : $\Delta\alpha$ ranging from 0° to 170° and Δu from 0 ms^{-1} to 2 - 3 $m\ s^{-1}$.

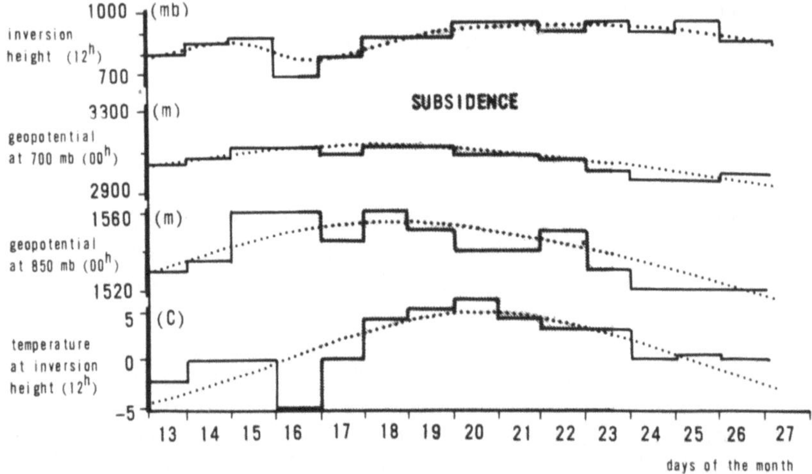

Fig. 1 Time trends of thermodynamic parameters measured by vertical soundings at Milan airport (January 1982)

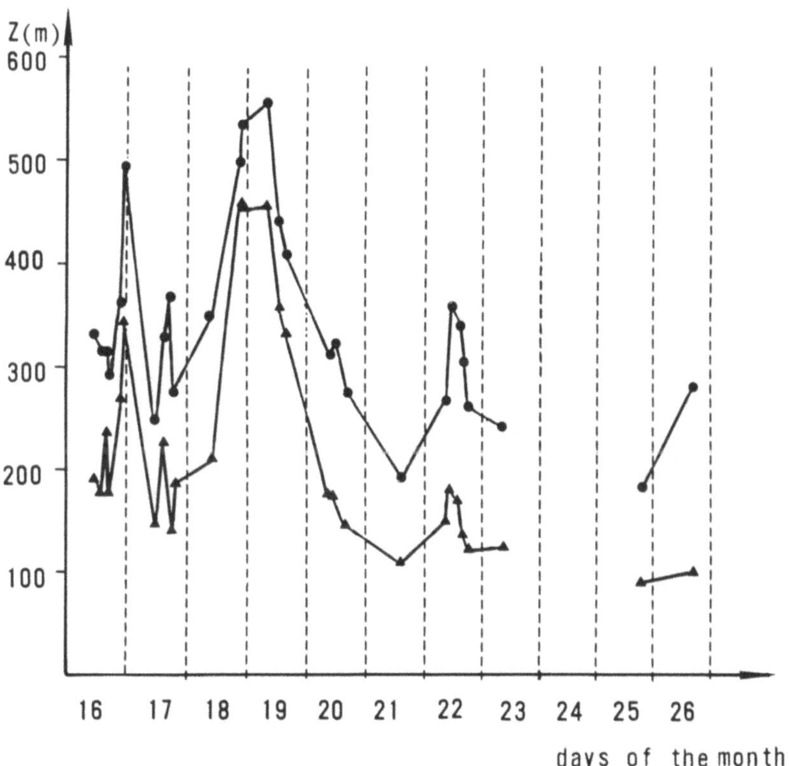

Fig. 2 Time evolution of H_u (●) and H_i (▲) (see for reference figs. 4-6) as measured by RASS (January 1982).

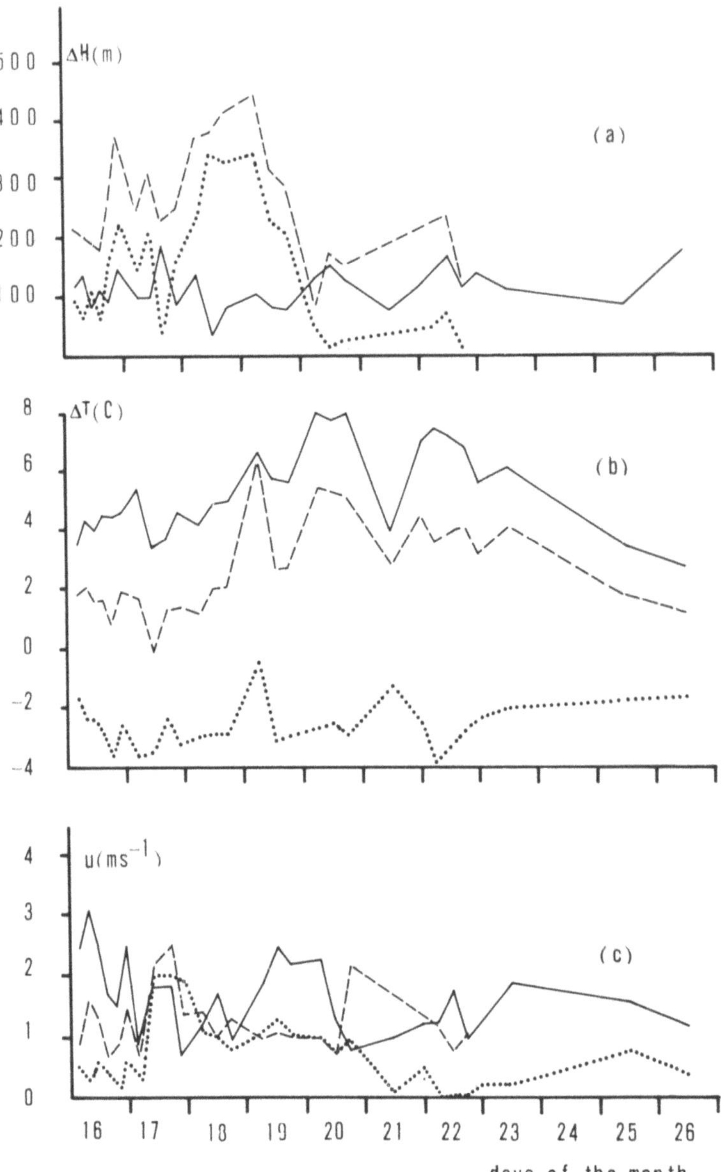

Fig. 3 Time evolution of : a) the thickness of the layers ($H_u - H_b$), (---), ($H_u - H_i$), (——), ($H_i - H_b$), (........) - see for reference figs. 4-6; b) the temperature differences ($T_u - T_b$), (——), ($T_u - T_g$), (---), ($T_i - T_g$), (.......) - see for references figs. 4-6; c) wind speed at 30 m (.......), at the inversion height (——) and in the isothermal layer (---).

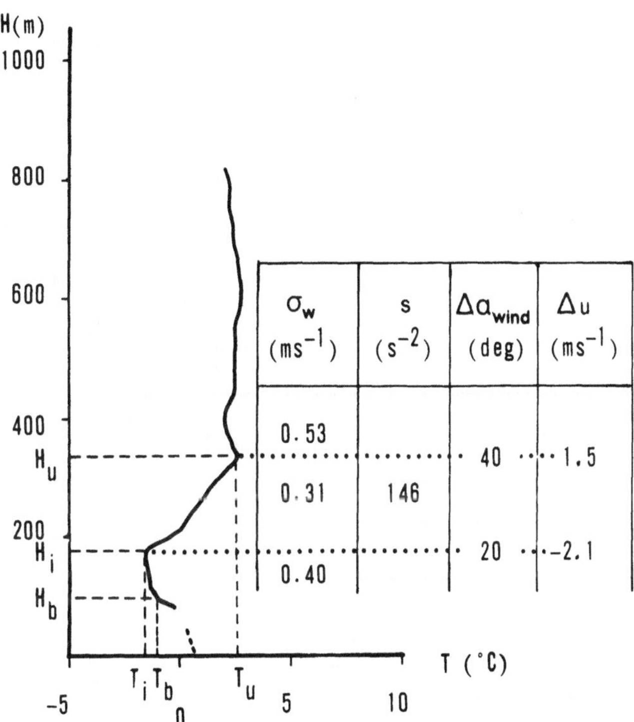

Fig. 4 Radio acoustic temperature profile (16-01-1982, 13.50 L.T.). σ_W : standard deviation of vertical component of wind speed measured by Doppler SODAR;

$s = (g/t)\ \frac{\partial\Theta}{\partial z}$: stability parameter; $\Delta\alpha$: wind direction shear; Δu : wind speed shear.

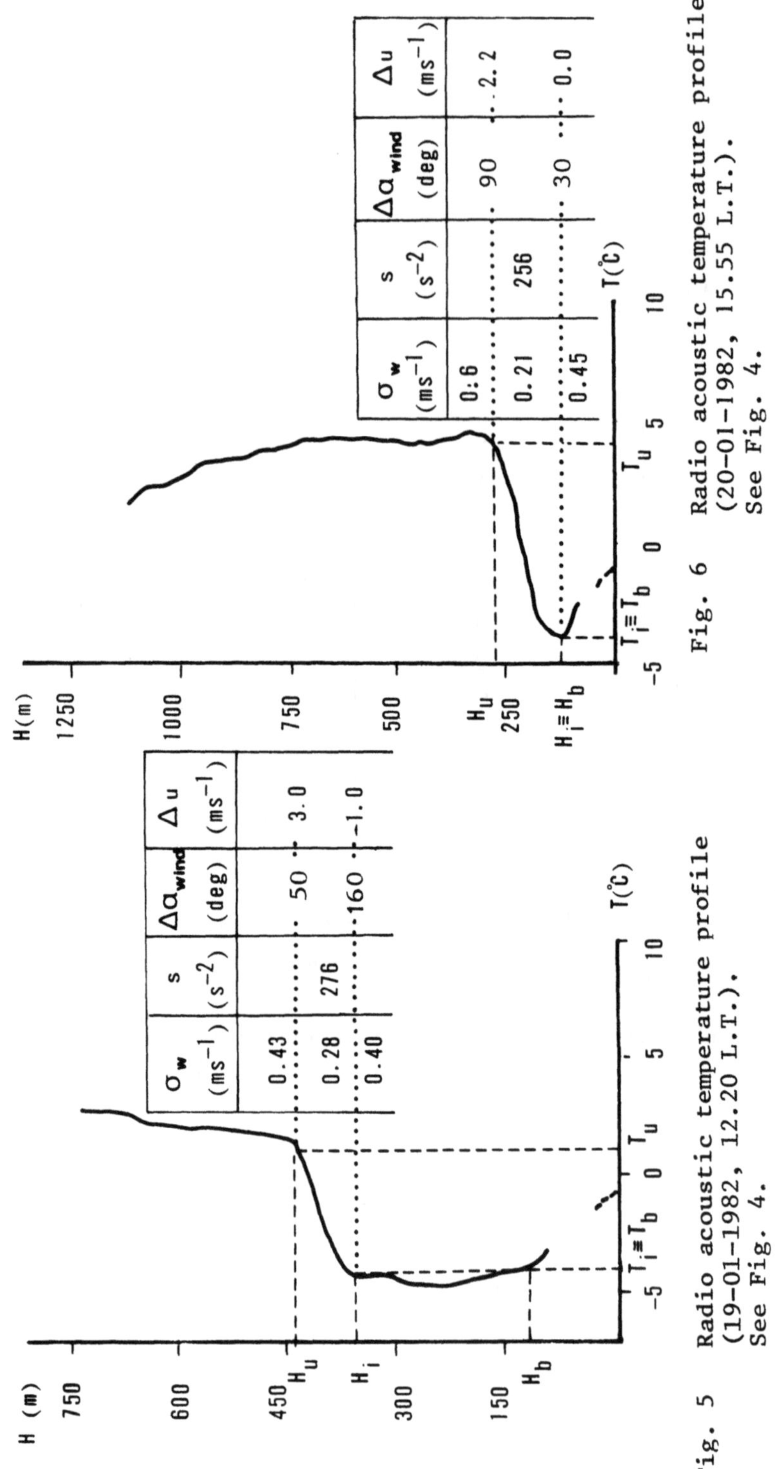

Fig. 5 Radio acoustic temperature profile (19-01-1982, 12.20 L.T.). See Fig. 4.

Fig. 6 Radio acoustic temperature profile (20-01-1982, 15.55 L.T.). See Fig. 4.

Among the lot of analysed data, we chose those which seemed particularly suitable to give as synthetic and consistent as possible a picture of the episode. First, all thermal RASS soundings, when grouped in homogeneous classes, allowed one to recognize three distinct kinds of fog :

a) shallow fog (∿ 150 m deep), at its first stage of development, marked by moderate lapses in the thermal gradients, which where unstable in the interior of the fog and stable in a layer 150 m deep above the fog top. Along with the fog of type c), this was the one more looking like the fogs described by others authors (like Roach, 1976), with a convective structure inside the fog layer and the basis of the capping inversion lifted at its top.

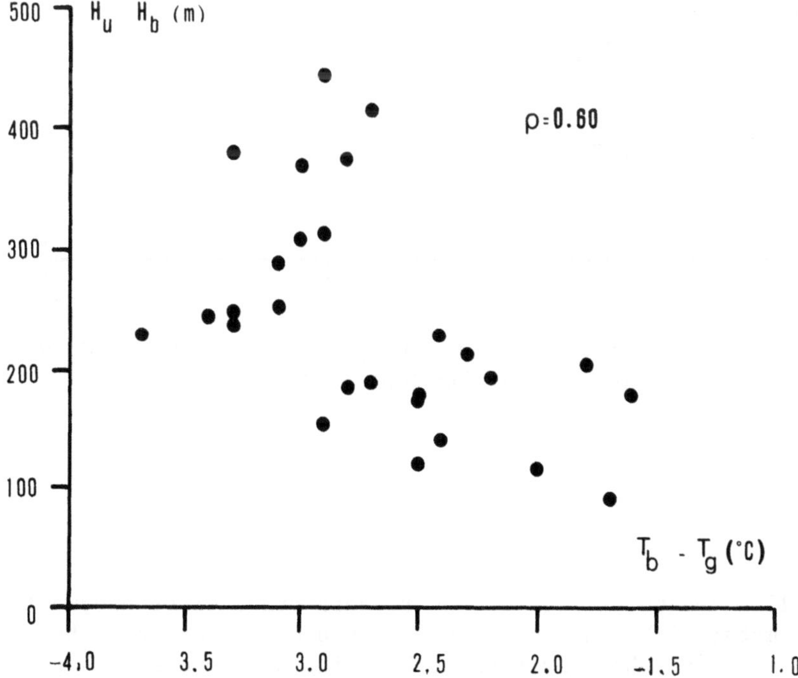

Fig. 7 Scatter diagram of the thickness of the stable layer (H_u - H_b) as a function of the lower layer thermal gradient (T_b - T_g).

b) deep and thick fog, 300 - 500 m deep, at the intermediate stage of development, capped by a thermal inversion based at the top of the fog layer. Thermal gradient was unstable in the lower 100 m, isothermal from 100 m till the top of the fog and very stable above this level, as far as 600 - 700 m.

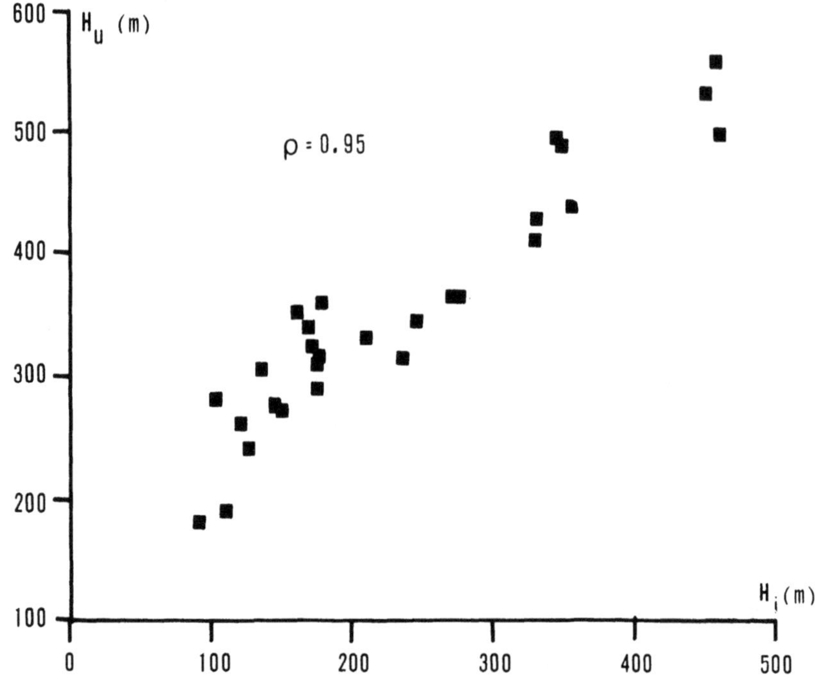

Fig. 8 Scatter diagram of the upper inversion height as a function of the fog top layer.

c) shallow and thick fog, at the ripness stage of development, 150 meter high; this kind of fog occured during the subsidence period when lowering of both fog and inversion tops were observed. Thermal gradients were shaped as in the kind a) fogs, but they were sharper. During this stage, temperatures of fog top reached their lowest values (-6°C), showing strong radiative cooling.

A fairly good correlation was observed between the strenght $-(T_b - T_g)$ (see fig. 7), of the convective lapse rate in the first 100 m and the thickness $(H_u - H_b)$ of the whole stable layer (fog plus upper inversion).

As H_b was almost constant during the episode, we can conclude that H_u showed to be an increasing function of $-(T_b - T_g)$.

As far as the top of the fog H_i is concerned, it follows a similar time trend as H_u; a regression analysis of data showed (fig. 8) that the increase of H_i is linearly correlated to that of H_u through the following equation.

$$H_u = 0.83\ H_i + 160 \quad \text{(m)}$$

This equation indicates that the higher H_u becomes, the shallower $(H_u - H_i)$ results. Not any similar conclusion seemed to be valid between the thickness $(H_u - H_i)$ and its correspondent thermal gradient $(T_u - T_i)$, what doesn't allow one to establish any relationships between the lower convective thermal gradient and stability parameters of the capping, upper inversion, which can become very effective in forbidding vertical dispersion of tall emissions.

DISCUSSION OF FOG STRUCTURE WITH REFERENCE TO POLLUTION DATA

During the fog episode, both SO_2 emissions from each of the 4 sections of the power plant and SO_2 concentrations at five ground level stations were recorded every 30 min. Also recorded were surface values of wind speed and direction, air temperature and moisture, global solar radiation. As the Turbigo power plant consists of two groups of sections, the first one with low chimneys (48 m) and small load capacity (140 MW_e) and the second with 3 high stacks (95, 150 and 150 m respectively), connected to 4 sections of 1200 MW_e full load capacity, the plant emissions have been accordingly subdivided, as shown in fig. 9, on account of the different contribution they could give to SO_2 pollution during fog periods.

As fig. 9 shows, a severe pollution episode occured during the period of fog, on between 18th and 19th of January. The only sources we considered here are plant emissions, in that for the domestic ones we can just have rough estimates based on population density, while other industrial emissions can be hardly taken into account, so that they both are taken in this analysis as contributing to the winter pollution background, typical of this region.

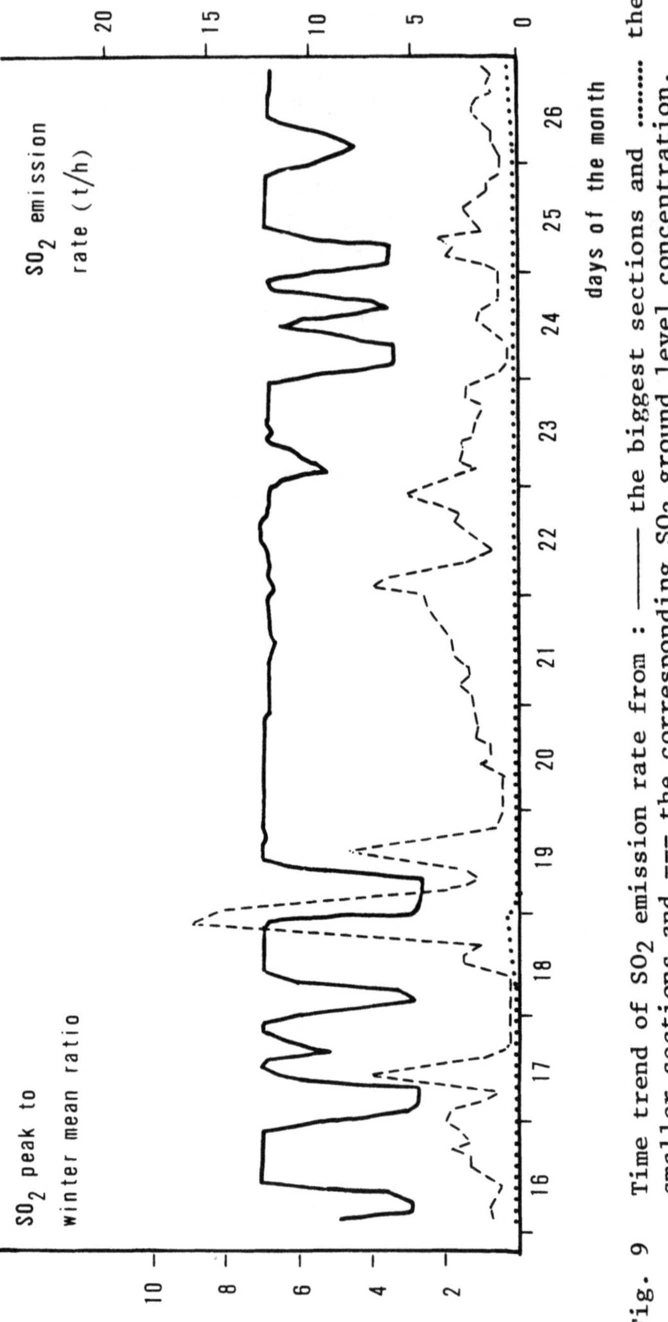

Fig. 9 Time trend of SO_2 emission rate from : ——— the biggest sections and .... the smaller sections and --- the corresponding SO_2 ground level concentration.

If we compare simultaneous trend of plant emissions and SO_2 ground level concentration (we mean here the average value of concentrations measured by the five recorders), we can attempt some considerations regarding effluent dispersion of the plant in connection to vertical structure of fog layer. Our reasoning is mainly oriented towards emissions from the largest group, in that the other ones were generally absent, excepting on day 18th.

It appears at once a quite similar role the emission rate and the fog depth played on SO_2 concentration at the ground. When the emission rate was kept at its maximum value, SO_2 concentration appeared well modulated accordingly to the trend of H_i (and also of H_u) height. The same was apparent when H_i and H_u attained their maximum heights and the power station emissions were changed; this was dramatic on between the 18th and 19th days, when the pollution episode occurred. It is worth stressing that when emissions of both groups of the power plant were lowered to their minimum values, the same did the SO_2 concentration and when the emissions of the biggest group were again raised to their full load, on 19th morning, SO_2 concentrations increased some 30 minutes after.

What happened since that moment is very interesting from the point of view of a real time air pollution control strategy. If the plant management had to take the decision to whether increase or decrease emissions, on the sole basis of readings given by the SO_2 recorders, the answer should have been in favour of a charge lowering. On the contrary, the RASS sounding was showing a steady decrease of top fog and inversion heights and, in fact, when they crossed respectively the 350 and 200 meter levels, the SO_2 average concentration started decreasing, keeping low or normal values during the remnant of the episode, in spite of the maximum values of the plant charge.

This kind of remote sensing, which allows real time control of dispersion properties of deep fog layer, proved to be more efficient and safer than the classic method. For calculation of inversion overpassing by plume to be reliable, one must have got good estimates of the stability parameters along the vertical, what can only be attained if temperature and wind vertical profiles, as far as 500 m, are available.

CONCLUSIONS

The analysis of the fog episode, object of the case study described in the previous sections, has demonstrated the feasibility of employing advanced remote sensing techniques for the management of industrial emissions, in order to keep air quality standards within acceptable levels even during severe dispersion conditions. In particular, two recently developed instruments, the RASS and the Doppler Sodar, proved to be particularly effective in giving essential information on the vertical distribution of stability of lower troposphere through a deep and thick fog layer, so realizing a very satisfactory matching with readings of conventional networks of SO_2 ground level detectors. This matching allowed rational and aware choices to be made during a pollution episode, which otherwise could hardly have been taken.

Of course, further examinations of other fog situations are needed, for a full understanding of performances of these new techniques to be reached. This is, in fact, the objective of the continuation of our research.

REFERENCES

Beran, D. W., and Clifford, S. F., 1972, Acoustic Doppler measurements of the total wind vector, Proc. 2th Symposium on Meteorological Observations and Instrumentation, Amer. Meteorol. Soc., San Diego, Calif., March 27-30, pag. 100.

Bonino, G., Lombardini, P. P., and Trivero, P., 1979, A metric wave radio-acoustic tropospheric sounder, IEEE Trans. on Geos. Electr., GE-17, 4 : 179.

Bonino, G., Lombardini, P. P., and Trivero, P., 1980, Comparison of RASS temperature profiles with other tropospheric soundings, Nuovo Cimento, 1C, 3 : 217.

Bonino, G., Lombardini, P. P., Longhetto, A., and Trivero, P., 1981, Radio acoustic measurements of fog-capping thermal inversions, Nature, 290, 5802 : 121.

Choularton, T. W., Fullarton, G., Latham, J., Mill, C. S., Smith, M. H., and Stromberg, I.M., 1981, A field study of radiation fog in Meppen, West Germany, Quart.J. R. Met. Soc., 107 : 381.

Gland, H., 1982, Acoustic sounder data as meteorological input in dispersion estimates, Preprints NATO/CCMS 13th Int. Tech. Meeting on Air Pollution Modelling and its Application, Ile des Embiez, September 14-16.

Longhetto, A., Guillot, P., Anfossi, D., Bacci, P., Elisei, G., Frego, G., Sandroni, S., and Varey, R., 1982, Atmospheric dispersion experiments in the near and medium field (IV CEC campaign, Turbigo, Italy, Sept. 1979), Il Nuovo Cimento, 5C, 3 : 299.

Pagliari, M., and Persano, A., 1969, Nota sulla distribuzione della frequenza della nebbia in stazioni della Valle Padana, Proceedings XVIII Meeting of It. Geoph. Ass. (AGI), Naples, Italy, 1 - 4 October.

Roach, W.T., Brown, R., Caughey, S.J., Garland, J. A., and Readings, C. J., 1976, The physics of radiation fog : I - a field study, Quart. J. R. Met. Soc., 102 : 313.

SOME PRELIMINARY RESULTS OBTAINED USING A RECENTLY DEVELOPED CONDITIONAL TRACER RELEASE SYSTEM FOR STUDYING ATMOSPHERIC DISPERSION

C.D. Jones

CDE Porton Down
Salisbury
Wilts, United Kingdom

INTRODUCTION

For many years conventional dispersion modelling has relied, often rather slavishly, on the use of Gaussian profiles - however it is clear that if more realistic estimates of hazards arising from highly toxic, inflammable or even just nuisance from malodourous materials are to be made then a radically different approach to the problem is required. Indeed this has been recognised by a number of workers in the field but notably by Chatwin[1] who advocates the use of ensemble-type statistical approaches to the development of models. A major difficulty, from the experimental viewpoint, in implementing such methodology is the non-stationary (in the statistical sense) nature of nearly all atmospheric flows. Hence, for example, one may be presented with a slowly, but systematically, varying wind direction perhaps ahead of a frontal system and thus the ensuing have an unwanted bias impressed upon them. One could of course counter this argument by pointing out that such synoptic scale wind variations form part of a larger "climatic" ensemble and thus should not therefore be excluded from the sample. Whilst in some respects this may certainly be an intellectually cogent point it does not provide a very helpful way forward for the experimenter who generally has to operate on a more pragmatic basis.

Although the main concern of this paper is the presentation of results obtained with a novel conditional tracer release system employed as an adjunct to improving the statistical quality of the concentration data obtained it is instructive to briefly review the sequence of events which precipitated its development. Recently a series of field experiments[2] has been carried out to investigate the transient aspects of atmospheric transport and dispersion in the vicinity of a small, flat roofed and regularly shaped building. In these experiments a tracer, negatively ionized air, was released into the recirculating wake region, ie that volume on the leeward face containing a slack and reversed flow regime, and the behaviour of the tracer examined with a number of fast response ($\sim$ 0.01 s) detectors. The initial experiments were carried out using a continuous release of tracer but the results obtained were rather difficult to interpret largely because of the constantly and unpredictably varying nature of the oncoming wind - both in terms of speed and direction. Consequently a second series of tests was carried out in which the tracer was injected into the wake in the form of regularly spaced short bursts - the idea being that, using this technique, it should be feasible to examine the motion and trajectory of individual 'packets' of ions and so form a much better appreciation of the physical processes occurring. This approach proved quite successful and enabled a number of interesting features of dispersion within wakes to be studied and elucidated.

However it was very evident from both the continuous and pulsed tracer release experiments that a major difficulty from the analytical, and thus model development, points of view arose from the non-stationary nature of the wind flow at the upwind edge of the building. (This is particularly so when an obstruction to the flow is involved because it is possible for relatively small changes in upwind flow conditions to provoke substantial alterations in the flow regime around the sides and in the lee of the building, see for example Hussain and Lee[3].) One, and perhaps rather over-emphasised, method of circumventing this particular problem is to abandon full scale experiments entirely and resort to modelling physically the structures and flows of interest in a wind tunnel. Undoubtedly this technique has much to commend it in terms of experimental convenience and rates of data acquisition as compared with field tests but it must be noted that our present knowledge of the effects of scale is not yet adequate to predict the behaviour of the full size flow and dispersion patterns, particularly their transient aspects, with the requisite accuracy and confidence. Thus there is still a definite need for full scale dispersion experiments particularly in situations where the obstructions or structures present are sufficiently substantial to be likely to cause significant deviations to the 'normal' flow patterns. In view of the above and taking into consideration the necessity to obtain data which can be analysed on a statistically rational and physically meaningful basis it was felt appropriate to explore the possibilities of conditional tracer release so that, in effect, a specially selected, but hopefully representative, data set could be assembled.

Note that due to the prevailing wind conditions it was not possible to obtain as high a quality set of data as was originally anticipated in the experiments with the building. However in a sequence of parallel experiments in open terrain conditions were rather more favourable and it is therefore these results which are given more emphasis here. Nonetheless the experiments provided a very useful basis for assessment of the potential of this new technique and suggest various ways in which it can be exploited to advantage in the future.

2. EXPERIMENTAL

The experimental arrangements are depicted in plan in Fig 1. The bivane*, which had a distance constant of $\sim$ 2.2 m, was placed at a point 5 m upwind of the ion generator and at a height of 2 m above a closely mown grass surface. The ion generator has been fully described elsewhere in the literature[4,5,6] and is not dealt with in detail here. Suffice it to say that the device functions by utilizing the properties of a corona discharge in such a way as to provide a convenient, controllable and reliable source of unipolar ions. In these particular experiments the generator was operated at a potential of - 3kV and, from previous calibrations, this was known to produce an output in the region of - 20nA when the device was activated.

Four detectors, ion collectors as they are normally referred to, were located in the nominal (but not always actual!) downwind direction from the ion generator. Distances were intentionally kept short (see Fig 1) since the experiments were essentially exploratory in nature and this conferred the additional benefits of convenience and flexibility in the operating procedures used. Again detailed descriptions of the ion collectors are available in the literature already cited and are thus not repeated in this paper. However it is perhaps worth pointing out that their main advantage, from the experimental standpoint, is that of very fast response to changes in ion concentration (typically 0.01 s) making them almost ideal for the purposes of studying the transient concentrations which occur in wakes particularly but also of course in the vicinity of sources in other types of terrain. Furthermore these devices and their associated electronics are relatively cheap to produce and do not possess certain irritating deficiencies such as non-linearity, drift and hysterisis so often apparently inherent in chemical tracer detectors.

The arrangements for providing triggered release of the ionized air tracer and recording the subsequent concentration and other information obtained are indicated schematically in Fig 2. In brief the wind vector (both its azimuthal and vertical components) is sensed by the bivane and after electronic processing two

* Model PBV1 Vector Instruments.

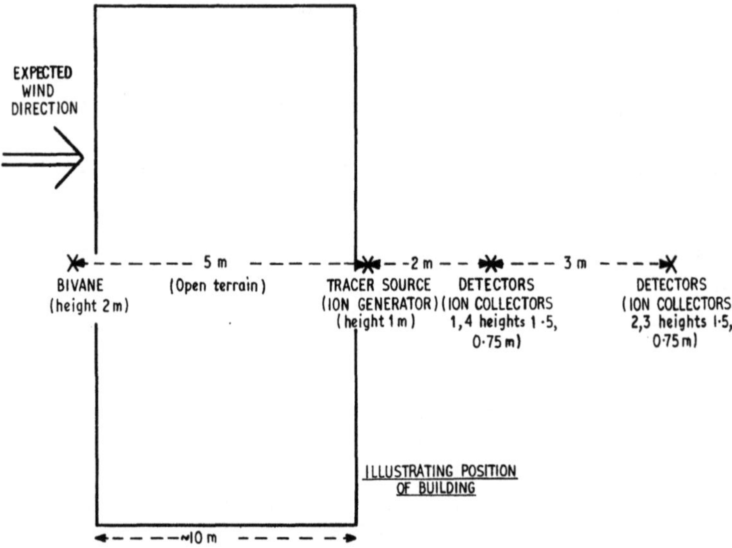

Fig. 1. Arrangements for the conditional release experiments

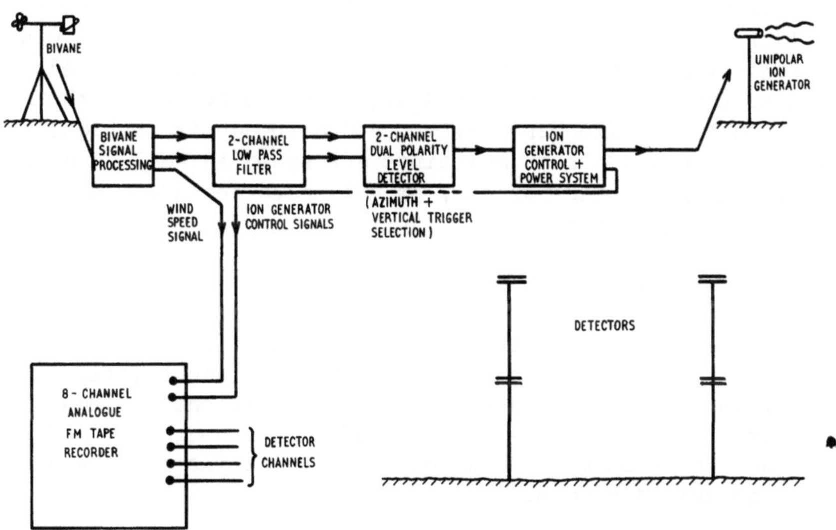

Fig, 2. The instrumentation system - schematic diagram

corresponding voltage signals are presented to a two channel low pass filter and thereafter to an adjustable bi-polar two channel threshold sensing device. The filter was included because it was considered undesirable (this may have been a misjudgement in retrospect) for the ion generator to be triggered on and off in response to every minute and rapid fluctuation in wind vector which happened to pass through the acceptance band of the threshold system. Indeed also from the dispersion and transport standpoint this course of action seemed justufiable in that, particularly in the vicinity of a small building, one was mainly concerned with the effects of flow regimes of at least a few seconds duration - hence it was desirable to effectively average the upwind flow variations over this sort of time scale.

The threshold sensing system consisted essentially of an arrangement of four independently adjustable comparators (one for each component of the wind and one for the upper and the lower thresholds) and could be set to trigger the source when the wind vector fell within the vertical and azimuthal angles required. In addition to the above a wind speed signal was provided by the bivane, derived from a propeller anemometer, and although it is highly likely that many flow regimes near buildings are speed as well as direction sensitive this parameter has not, as yet, been included as an element in the triggering system.

3. RESULTS AND DISCUSSION

As mentioned in the introductory section of this paper meteorological conditions did not permit the detailed investigation of wake properties that had been planned with the conditional release system. Nevertheless the results that were obtained, although the majority were in fact in unobstructed terrain, are quite instructive and are therefore described herein.

Turning to the open terrain experiments first Fig 1 shows the instrumental layout employed. The terrain was flat in all directions for at least several km and the ground surface in the immediate vicinity of the apparatus was closely mown grass - there was however a small number of fairly squat trees and shrubs quite close to the experimental site the nearest of these being approximately 60 m away. Table I shows the parameter selection (ie the horizontal and vertical angle trigger levels together with the filter time constant) for the three runs comprising the experiment plus the mean wind speed obtained from a simple propeller type run of wind meter. The weather conditions were rather convective initially with high surface temperatures, strong insolation and quite light (1-3 ms^{-1}) and variable winds. However the wind direction was, at least at times, sufficiently well defined and consistent to enable meaningful experiments to be carried out.

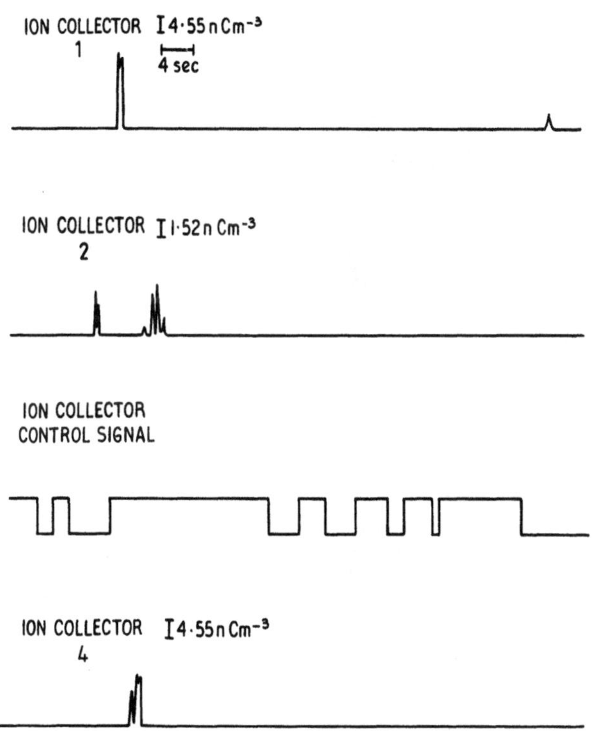

Fig. 3. Typical detector behavior - wide ($\pm 30^{\circ}$) horizontal acceptance angle

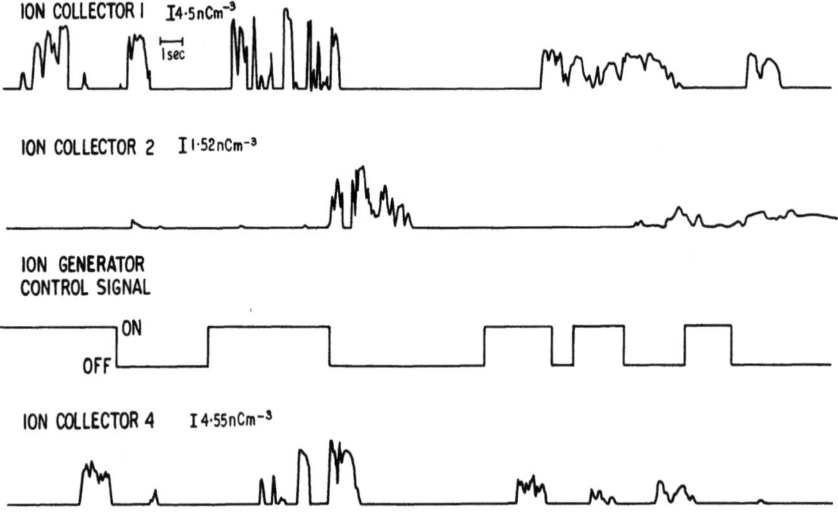

Fig. 4. Typical behavior - narrow (±10°) horizontal acceptance angel

TABLE I

DATA FOR THE OPEN TERRAIN EXPERIMENTS

Run No	Duration	Mean Wind Speed at 2 m Height	Ion Generator Triggering Angle: Horizontal	Vertical	Low Pass Filter Setting (Horizontal and Vertical)
1	110 mins	∿ 1.3 ms^{-1}	± 30°	± 10°	0.3 Hz
2	110 mins	1.49 ms^{-1}	± 20°	± "	"
3*	59 mins	1.75 ms^{-1}	± 10°	± "	"

* This run had to be curtailed because of imminent thundery activity!

Whilst the wind speeds are rather below average, certainly for the UK, this probably does not detract too seriously from the quality of the results since it is probably the wind direction that is more relevant here. If one considers the mechanism of the dispersion process, more particularly at distances close to the source, it appears that elements of fluid are likely to travel in approximately straight lines from source to detector - this hypothesis can be demonstrated for example by observing the trajectories of soap bubbles or smoke released from a small source. It must be noted however that most instruments, and this includes the bivane, sample the oncoming flow at a fixed point ie in an Eulerian frame whereas the transport and dispersion of tracer takes place in a frame attached to individual fluid elements - the so-called Lagrangian frame of reference. As Pasquill[7] has pointed out the main distinction, insofar as the practical implications for dispersion are concerned, is the not unexpected property that the Lagrangian turbulence spectrum resembles the Eulerian but with the variations shifted to lower frequencies. This factor thus reinforces the kinematically based hypothesis concerning quasi-linear motion of tracer and suggests that the vertical and horizontal triggering angles selected for the release of tracer must have a critical effect on whether or not it is subsequently detected by the ion collectors.

Figs 3 and 4 are typical examples of chart records obtained from Runs 1 and 3 respectively and reveal convincingly the effect of triggering angle on detection of tracer. In the first run, where this angle was ± 30° in the horizontal, the ion generator is switched on for substantial periods of time but detection of ions is an infrequent occurrence because obviously the ion plume, which is quite narrow, often completely misses all four collectors. However when this angle is reduced to ± 10° the ion collectors are activated for a greater proportion of the time that the ion generator is switched on and, based on an analysis of plume width according to Jones[8], this could probably be increased still further by decreasing the acceptance angles (both horizontal and vertical) to as little as ± 3° - at which point the 'success rate' should begin to approach 100%.

A more quantitative presentation of these effects is contained in Table II which shows the results obtained using a threshold sensor in conjunction with a totalising counter system to determine the total time over an entire run for which the signal voltage remains above a particular selected value (a fuller description of this system can be found in Jones (loc. cit.)).

The threshold values selected to produce the results in the Table were such that all reasonably sized ion collector signals would have been included in the totals but it is not impossible that a few of the very smallest concentration fluctuations may have been excluded. Consequently the T_{IC2} values tabulated are probably slight underestimates. Although there is certainly a trend for an increase in the value of the ratio T_{IC1}/T_{ig}, which is indicative of the success rate for detection, with reducing horizontal acceptance angle it is not particularly marked and this is largely a result of the 3 angles selected all being rather bigger than the optimum. Furthermore the choice of filter time constant (0.3 Hz) may not have been appropriate despite the fact that its selection was made to correspond approximately with the expected time of travel between the various instruments concerned (additional experiments specifically addressed to study the effects of varying the filter time constants

TABLE II

THE EFFECT OF HORIZONTAL TRIGGERING ANGLE ON TRACER DETECTION

Run No	Parameter	Value (T_{ig})	Parameter	Value (T_{IC1})	Ratio T_{IC1}/T_{ig}
	Ion generator:		Ion collector 1:		
1	Percentage of total run time switched on	26.5%	Percentage of total run time switched on	4.10%	0.16
2		20.6%		3.35%	0.16
3		6.07%		1.39%	0.23

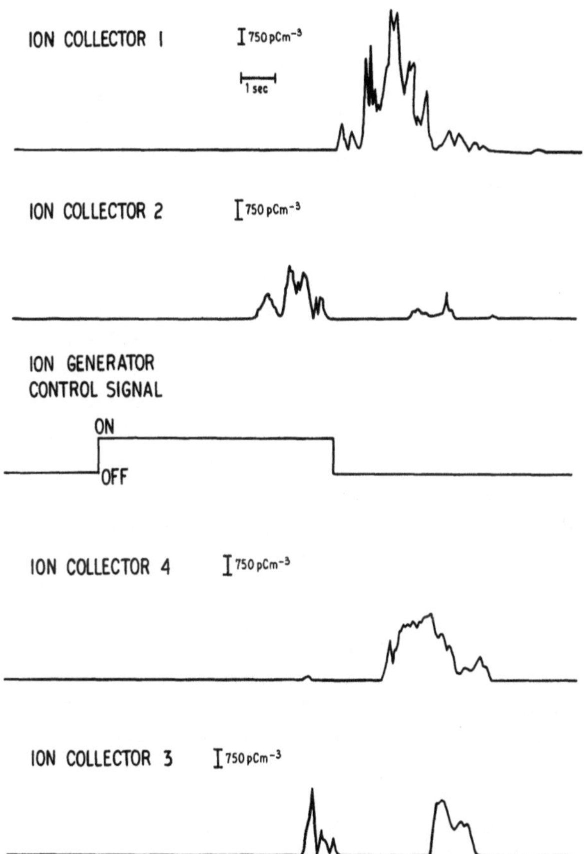

Fig. 5. Typical detector behavior in the recirculating wake region

were in fact planned but could not be executed due to the persistence of unsuitable meteorological conditions).

An interesting corollary of the above findings, if the effects observed scale-up in a straightforward manner, is that it would appear feasible to direct toxic waste into areas where risks were low by employing some type of conditional release system. Other problems may be created however, particularly one of storage, which may incur cost penalties sufficiently heavy to negate the environmental advantages.

A completely different type of processing was carried out with a signal analyser. The object here being to determine the distribution in temporal terms, of individual events eg the switching on and off of the ion generator or the length of time the ion concentration is above zero for a particular sequence. In the past this method of analysis has proved informative in studies of ion concentration data obtained from conventional atmospheric dispersion experiments - ie those with the tracer source operating continuously - see Jones (loc. cit.). The distribution of pulse or 'burst' lengths is determined automatically by the instrument (SHADA)* and the results are stored in 32 time 'bins' in semi-conductor memory for subsequent retrieval. The results obtained for the same three runs examined above for the ion generator (IG) trigger signal and ion collector 1 (IC1) are presented in Table III. In this case the figures quoted are the numbers of events (ie ion concentration bursts or ion generator 'triggerings') in relation to the sequence of listed time 'bins'. As previously these data are rather inconclusive although, as might be expected, the ion concentration fluctuations tend to occupty shorter time scales than the generator triggerings presumably owing to the comminuting effect of the turbulent field. There is perhaps also some indication that the shorter time bins have a proportionately higher occupancy for both the ion generator and collectors signals as one progresses to smaller bivane acceptance angles. Again this somewhat inconclusive behaviour is probably a result of the arbritarily too large choice for the instrumental settings on the triggering system. Note that the event totals, whilst not large for this type of experiment (cf the continuous source data described in Jones (1983) where totals are an order of magnitude greater) are acceptable and suggest that the results ought to be reasonably reliable statistically.

Fig 5 portrays the type of tracer concentration variations that were recorded when a similar arrangement of ion generator and collectors was situated in the expected position of the re-circulating wake of a small single-storey flat-roofed building (ref to Fig 1). In this case the bivane was sited on the upwind side of the building the idea being to sample only a properly developed wake rather than the miscellany of rather ill-defined flow regime encountered in previous field experiments. Unfortunately the

* Signal Height and Duration Analyser.

TABLE III

RESULTS OF SIGNAL DURATION ANALYSIS FOR THE OPEN TERRAIN EXPERIMENT

Time 'Bin' Limits	Run 1		Run 2		Run 3	
	IG	IC1	IG	IC1	IG	IC1
< 15 msec	-	-	-	13	-	10
15 - 21 msec	-	-	-	9	-	8
21 - 30	-	13	-	16	-	16
30 - 42	-	11	1	42	-	22
42 - 60	-	13	1	59	1	17
60 - 85	-	30	2	77	1	16
85 -120	1	50	1	77	1	16
120 -170	2	62	-	82	1	19
170 -240	4	47	4	77	1	16
240 -340	1	56	7	69	3	16
340 -480	5	69	13	65	13	23
480 -680	7	59	19	54	10	16
680 -960 msec	8	76	21	25	23	14
960 - 1.36 sec	13	59	37	21	19	4
1.36- 1.92 sec	16	45	40	16	20	1
1.92- 2.72	42	31	50	4	19	-
2.72- 3.84	36	15	56	3	22	-
3.84- 5.44	35	10	40	-	13	1
5.44- 7.68	39	7	39	-	5	-
7.68- 10.88	42	1	27	-	1	-
10.88- 15.36	27	-	13	-	-	-
15.36- 21.76	23	-	7	-	-	-
21.76- 30.72	14	-	-	-	-	-
30.72- 43.52 sec	7	-	-	-	-	-
43.52- 1.02 min	1	-	-	-	-	-
1.02- 1.45 min	-	-	-	-	-	-
1.45- 2.05	-	-	-	-	-	-
2.05- 2.90	-	-	-	-	-	-
2.90- 4.10	-	-	-	-	-	-
4.10- 5.79	-	-	-	-	-	-
5.79- 8.19	-	-	-	-	-	-
> 8.19 min						
TOTAL 'EVENTS'	323	654	378	712	153	216

meteorological conditions were less than propitious on this occasion - the winds being rather lighter than in the open terrain experiments discussed above. Consequently the development of a true recirculation zone under such circumstances is questionable and thus the results are included here mainly for reasons of completeness rather than to demonstrate the value of conditional release per se.

A feature which is of considerable interest in the figure is the arrival of ionized air at collectors 2 and 3, which were located 5 m from the ion generator, some time before detection occurred at the nearer collectors (1 and 4). One can only surmise that the tracer must have either taken a very tortuous route or have remained circulating around the wake for a considerable time (ie have been the result of an earlier release of ions) relatively speaking, to produce this effect. Either way it is certainly quite a surprising result and it is clear that our current knowledge of tracer movement and dispersion around and near to even relatively simple geometrical configurations is far from satisfactory.

Nonetheless despite the unfavourable conditions even a cursory glance at the chart record of the entire recirculating wake experiment, which was of nearly 6 hrs duration, shows a markedly improved tendency, when compared with the earlier continuous tracer experiments, to release tracer at times when it was far more likely to be detected.

4. CONCLUSIONS

Despite difficult experimental circumstances, mainly very light winds, the use of a conditional release technique based on the wind vector variation as a means of improving the statistical quality of results obtained in short range atmospheric dispersion experiments has been demonstrated. Indeed the method has already shown considerable promise as offering a powerful tool for providing a consistent basis for obtaining the ensemble type statistics of concentration fluctuations that will be required for the development of future dispersion models. Such models are essential if realistic assessments and predictions of hazards due to short acting toxic materials and flammability/explosion risks arising from, for example, LPG releases are to be made.

REFERENCES

1. P. C. Chatwin, "The use of statistics in describing and predicting the effect of dispersing gas clouds", J. Haz. Mat., 6, 213-230 (1982).

2. C. D. Jones and R. F. Griffiths, "Full scale experiments on dispersion around an isolated building" (to be published in Atmos. Env.).

3. M. Hussain and B. E. Lee, "A wind tunnel study of the mean pressure forces acting on large groups of low rise buildings", J. Wind Eng. and Aerodynamics, 6, 207-225 (1980).

4. C. D. Jones, "Ion concentration variations at short distances downwind of continuous and quasi-instantaneous point sources", Pestic. Sci. 8, 84-95 (1977).

5. C. D. Jones, "Ionized air as a wind tunnel tracer", J. Phys. E. Sci. Instr. 10, 1287-1291 (1977).

6. C. D. Jones and N. T. Gulliford, "Developments in the use of ionized air as a wind tunnel tracer", J. Phys. E. Sci. Instr. 12, 321-327 (1979).

7. F. Pasquill, "Atmospheric dispersion, 2nd edition", Ellis-Horwood Publ. Chichester, UK (1973).

8. C. D. Jones, "On the structure of instantaneous plumes in the atmosphere", J. Haz. Mat. 7, 87-112 (1983).

DISCUSSION

P. GUDIKSEN: How conservative is the tracer relative to the transport time between source and receptor ?

C.D. JONES: The tracer is reasonably good at this range. Detailed calculations have been made in the references given in the paper.

T. MIKKELSEN: How important do you think the effect of electrostatic forces between the equally charged particles may be for the short range diffusion experiment you have presented ?

C.D. JONES: The effects are not too serious at short range (i.e. up to 50 m or SO). It is possible to calculate (with reasonable accuracy) the effects these processes will have on plumes and puffs and thus make simple corrections. Further details of this procedure can be found in the references to the paper.

EPA MODEL DEVELOPMENT FOR STABLE PLUME IMPINGEMENT ON ELEVATED TERRAIN OBSTACLES

Francis A. Schiermeier*

U.S. Environmental Protection Agency
Research Triangle Park, North Carolina 27711

Thomas F. Lavery, David G. Strimaitis, Akula Venkatram, Benjamin R. Greene, and Bruce A. Egan

Environmental Research & Technology, Inc.
Concord, Massachusetts 01742

ABSTRACT

The U.S. Environmental Protection Agency's Complex Terrain Model Development program is designed as a series of progressively advanced model development efforts accompanied by requisite field studies to provide data for model evaluation. Plume impingement studies have been performed during 1980 at Cinder Cone Butte near Boise, Idaho, and during 1982 at Hogback Ridge near Farmington, New Mexico. Experimental protocol consisted of terrain-surface measurements of dual-tracer plumes emitted from mobile cranes during stable atmospheric conditions. Dimensions of simultaneously-released oil - fog plumes were obtained prior to impingment by lidar and by time exposure photographs. Accompanying meteorological sensors included instrumented towers, tethersondes, monostatic and Doppler acoustic sounders, and optical crosswind anemometers.

Selected data from the Cinder Cone Butte study were initially used to evaluate the performance of two new preliminary algorithms (Neutral and Impingement) and of four existing complex terrain plume dispersion models (Valley, Complex I, Complex II, and Potential Flow Model). After a case study analysis to

---

*On assignment from National Oceanic and Atmospheric Administration.

improve understanding of the physical processes involved, the Neutral and Impingement models were refined into the Lift and Wrap models, respectively. Currently, the Lift and Wrap components have been combined to form the basis of the Complex Terrain Dispersion Model which has now been evaluated on all of the Cinder Cone Butte data and on selected hours of Hogback Ridge data. Simultaneous scaled physical modeling is being performed to determine proper similarity criteria and limits of applicability for fluid modeling simulations of plume-terrain interactions.

INTRODUCTION

The U.S. Environmental Protection Agency (EPA) is currently sponsoring the Complex Terrain Model Development (CTMD) program, a multi-year integrated program to develop and evaluate practical plume dispersion models for calculating ground-level air pollutant concentrations that result from large emission sources located in mountainous terrain. The initial focus of the CTMD program is to develop models with known accuracy and limitations for simulating one-hour-average ground-level concentrations during stable atmospheric conditions.

The objectives of the CTMD program were described by Holzworth (1980) and are based on recommendations arising from an EPA-sponsored workshop to consider the issues and problems of simulating air pollutant dispersion in complex terrain (Hovind et al. 1979). In response to the workshop, the EPA initiated a program emphasizing the production of a useful model (or models) with demonstrated reliability and prescribed applicability. During the requisite field measurements and laboratory experiments, the observational needs of the modelers were to be foremost in importance. As described by Holzworth (1980), the CTMD program was perceived as an integrated and highly coordinated effort that involves:

- dispersion model development/evaluation/improvement,
- scaled physical modeling in a fluid modeling laboratory,
- field measurements/experiments centered on isolated terrain features, and
- field measurements/experiments centered on an existing power plant in terrain with opportunities for plume impingement and other types of plume-terrain interactions.

The CTMD program was begun in June 1980. The first major component was a field study conducted during the fall of 1980 at Cinder Cone Butte (CCB), a roughly axisymmetric, isolated 100-m tall hill located in the broad Snake River Basin near Boise, Idaho. The field study consisted of 10 flow visualization (oil-fog) experiments and 18 multi-hour dual-tracer gas (SF_6 and CF_3Br) and oil-fog experiments conducted during stable flow conditions with supporting meteorological, lidar, and photographic measurements.

The CCB data were used to develop and evaluate two new preliminary modeling algorithms (Impingement and Neutral) and to evaluate the performance of current EPA complex terrain plume dispersion models (Valley, Complex I, Complex II, and Potential Flow Model). The Impingement Model simulates dispersion during strongly stable flows in which plumes remain approximately horizontal and impinge on and travel around the sides of CCB. The Neutral Model simulates dispersion during slightly stable and neutral flows in which plumes rise over CCB. For the initial CCB model evaluations, 45 hours were selected from 153 case hours obtained during the 18 tracer experiments. The meteorological and tracer data were used in a "hands-off" mode; no modifications were made to the algorithms to improve their predictions.

The EPA CTMD First Milestone Report - 1981 (Lavery et al. 1981) describes the air quality models that were evaluated, the evaluation methods, and overall model performance. The report also provides comparative statistics of the model evaluations and summarizes the CCB field experiments and associated quality assurance activities. A series of experiments (Snyder and Lawson 1981) conducted in the EPA Fluid Modeling Facility water channel towing tank to quantitatively simulate the field studies are described in an appendix. The participation of the Fluid Modeling Facility is an integral part of the CTMD program, with the aim of determining proper similarity criteria and limits of applicability for fluid modeling simulations of plume-terrain interactions.

The CCB tracer gas source data (emission rates, locations and heights of tracer and oil-fog releases), concentration data (over 12,000 bag samples) and meteorological data (from six towers, a tethersonde and pilot balloons) were subsequently delivered to the EPA, and are available upon request.

The Second Milestone Report - 1982 (Strimaitis et al. 1982) documents subsequent work to improve the Impingement and Neutral modeling approaches, to refine the CCB meteorological and tracer gas data bases, and to further analyze the data to relate plume behavior and observed meteorological conditions. Unlike the previous hands-off evaluations, the approach taken in this report is a case study analysis. From the 45 hours used in the initial model evaluation, 14 hours were selected for detailed individual case study to illustrate and contrast various flow situations and ranges of model performance.

The Second Milestone Report also contains, as an appendix, a paper prepared by the EPA Fluid Modeling Facility (Snyder et al. 1982). The paper describes a series of wind tunnel and towing tank experiments designed to study various aspects of stratified, sheared flow over a variety of simple obstacles, as well as flow over a scale model of CCB. In particular, the dividing-streamline concept is evaluated for obstacles of differing aspect ratio embedded in stably stratified flows, both with and without wind shear; and the effects of hill slope and wind shear on the flow patterns are discussed. On the basis of these experiments, the dividing-streamline concept was found to be applicable for a wide variety of terrain obstacles and stratifications.

In January 1982, the second phase of the CTMD program was initiated. This component included continuing model development and evaluation with the CCB data and performance of a second plume impingement study at the Hogback Ridge (HBR) near Farmington, New Mexico. The information collected at the HBR has enlarged the modeling data base to include measurements of flows and tracer gas concentrations in the vicinity of a long ridge.

MODELING STABLE FLOW CONDITIONS

The modeling of tracer gas concentrations measured at CCB is based on the concept of two distinct regimes of flow encompassing the hill (Figure 1). In Region 1, which extends from the base of the hill to a height H_c, the flow does not have enough kinetic energy to go over the hill and remains roughly horizontal as it goes around the hill. In Region 2, which lies above the dividing streamline that separates the two regions, the flow has enough energy to pass up and over the hill. The concept of the critical dividing streamline height, $H_c \equiv H\,(1 - Fr_H)$, is related to the notion that a fluid parcel can rise only through a height U_o/N, where H is the hill height, $U_o(H_c)$ is the wind speed at $z = H_c$, N(z) is the local Brunt-Vaisala frequency defined by

$$N(z) = [g/\Theta(z)\ (\partial\Theta/\partial z)]^{\frac{1}{2}}, \qquad (1)$$

g is the acceleration due to gravity, Θ is the potential temperature (°K), and Fr_H (the Froude number) is defined as

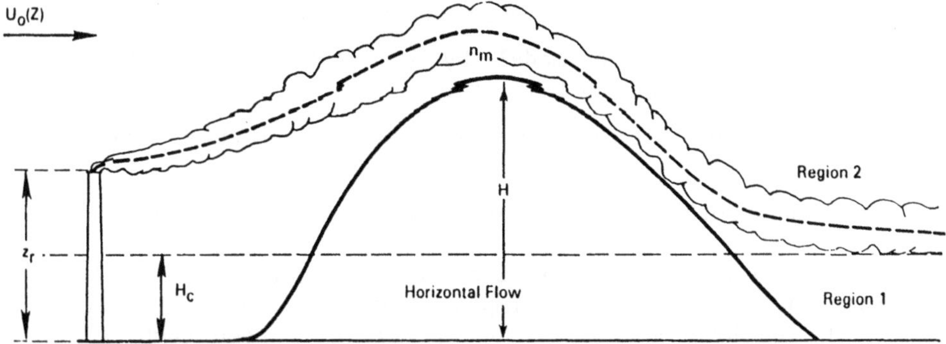

Figure 1. Schematic Diagram of Plume Behavior in Stable Flows Around Obstacles.

$$Fr_H \equiv U_o/NH. \tag{2}$$

If the wind speed and stratification are functions of height (as is generally the case), H_c is implicitly defined as follows

$$\frac{1}{2} U_o^2(H_c) = \int_{H_c}^{H} N^2(z)(H-z)dz. \tag{3}$$

The left-hand side of Equation 3 is the kinetic energy of the fluid at $z = H_c$, and the right-hand side is the potential energy gained by the fluid in rising through the height $H - H_c$.

The preliminary algorithms to simulate the flow in Region 1 and 2 were described in the First Milestone Report and by Venkatram et al. (1982). Model improvements and further model testing were discussed in the Second Milestone Report. The improvements to the Region 1 Impingement flow model (now termed the Wrap Model) included;

- use of an observed wind direction distribution function instead of an assumed Gaussian distribution,
- ability to calculate concentrations at all points on the hill instead of just at the stagnation point, and
- a new method for estimating σ_z (Venkatram et al. 1983).

The improvements to the Region 2 Neutral flow model (now termed the Lift model) included:

- modeling the air flow below H_c as a "dead" layer,
- generalizing the model by calculating the radius of curvature based on hill shape,
- use of an observed wind direction distribution function, and
- a new method for estimating σ_z (Venkatram et al. 1983).

As described by Strimaitis et al. (1983), the refined versions of the Neutral and Impingement models (referred to as the Lift and Wrap models, respectively) have been combined to form the basis of the Complex Terrain Dispersion Model (CTDM). The CTDM has been subsequently tested on all 153 hours of CCB tracer data and, for an initial assessment of the model's performance at a ridge, on 17 selected hours of HBR tracer data. The results of this latest model evaluation are contained in the draft Third Milestone Report - 1983 (Lavery et al. 1983). Also presented in this report are comparisons of laboratory simulations of neutral plume dispersion over CCB with field measurement data, as performed by the EPA Fluid Modeling Facility.

CCB CASE STUDY ANALYSIS

Fourteen case hours were selected from the 45 CCB hours discussed in the First Milestone Report. The concurrent SF_6, CF_3Br, meteorological, and photographic data, and various model sensitivity calculations were analyzed in detail to gain a better understanding of the relationships among emissions, observed meteorological conditions, and measured concentrations in order to improve the model formulations. In addition, the 45-hour subset of the CCB data base was analyzed to develop systematic relationships between the plume observations and other measurements to help in the modeling effort. The specific analyses included:

- a comparison of vertical plume growth (σ_z) estimated from photographs with σ_z estimated from turbulent intensity measurements,
- a comparison of σ_z estimated from photographs with σ_z estimated from lidar data,
- a comparison of σ_y and σ_z estimated from lidar data with σ_y and σ_z estimated from turbulent intensity measurements,
- time series and spectral analyses of wind speed data to determine time scales of turbulence, and

- analysis of 5-min temperature data to estimate the vertical distribution of isentropic surfaces as surrogates for the streamline patterns over CCB.

The principal conclusions from the CCB case study and data analysis are:

- A plume released above H_c tended to travel over the hill. The plume path appeared to be consistent with streamlines in weakly stratified flow, as described both in theory and laboratory experiments.
- A plume released well below H_c tended to impinge and flow around the hill, lacking sufficient kinetic energy to surmount the hill.
- A plume released very near H_c tended to travel directly towards the hill, with only slight-to-moderate vertical displacement.
- When the vertical velocity gradient near release height in the approach flow was large and positive, the maximum concentration on CCB was often measured at an elevation lower than the release height due to downward plume deflection.
- When Fr_L ($\equiv U_o/NL$) based on the hill half-length was close to unity, lee side plume depression was observed at CCB.
- $\sigma_y \cong i_y x$, when i_y is measured near the release elevation.
- $\sigma_z \cong \sigma_w t/(1+Nt/p)^{\frac{1}{2}}$, where $p = 0.5$ and $t = x/U_o$.
- U_o/N small is a necessary but not sufficient condition for large values of σ_y. (Hanna 1983)
- The isentropic analyses suggest that the air flow below H_c can be treated as a "dead" layer and that flows above H_c can be considered approximately neutral.
- The model improvements result in better agreement between model calculations and observations. The use of the observed wind distribution function is especially important for wind speeds less than 3 m/sec.
- Meteorological measurements need to be taken as close to the source as is practical because of spatial and temporal inhomogeneities produced by slow mesoscale meanders and vertical shear.

THE HOGBACK DISPERSION EXPERIMENT

A second plume impingement study was conducted at the Hogback Ridge (HBR) near Farmington, New Mexico during October 1982. The HBR is a long, roughly 90-m tall ridge with a gap at the San Juan River. Three flow visualization and 11 multi-hour, dual-tracer gas experiments were conducted. The experimental methods were similar to those used and tested at CCB:

- releases of two tracer gases (SF_6 and CF_3Br) and oil-fog using a mobile crane and a tower as source platforms,
- fixed meteorological measurements obtained by four instrumented towers, two monostatic acoustic sounders, one Doppler acoustic sounder, and three optical crosswind anemometers,
- one tethersonde each at the source location and on the upwind face of the HBR,
- ground-level tracer gas concentrations, and
- plume dispersion documentation by lidar measurements, videotapes, and time-exposure photographs.

The meteorological data were archived and displayed by an array of onsite minicomputers. The availability of real-time information on ambient meteorological conditions and the flexibility of releasing oil-fog and tracer gas at a variety of heights and locations enabled selective placement of source positions to obtain useful information for model development purposes. The real-time information was supplemented by near real-time lidar data and by prompt development of plume photographs. Approximately 86 hours of dual-tracer gas experiments were conducted during the HBR field experiments.

A preliminary examination of the unedited meteorological, tracer gas concentration, and photographic data indicates that the field study was successful. The new HBR experimental data base will provide a good basis for testing and extending the CTMD modeling approaches and the dividing streamline concept for two-dimensional ridges. Preliminary analyses and observations in the field suggest that plumes released above H_c tended to flow over the ridge and plumes below H_c tended to flow along the ridge. Occasionally at night, the elevated plume mixed to the ground upwind of and on the ridge. The mixing was apparently caused by turbulence produced during breakdown of the stable boundary layer.

CONCLUSIONS

Results from the analyses undertaken to date on the refined CCB data base are leading toward a complex terrain modeling system with two major flow components. The usefulness of the critical height (H_c) as a parameter to differentiate the two distinct flow regimes has been amply demonstrated at CCB. Plume trajectories well below and well above H_c are consistent with low and high Froude number flow theory, respectively. The behavior of trajectories near H_c is difficult to describe analytically, but in practice the plume behavior on the windward face of the hill appears amenable to low Froude number modeling.

Because H_c is a critical parameter in modeling stable flow in complex terrain, the use of this two-model approach in a regulatory setting will require truly representative meteorological information. On-site vertical profiles of wind and temperature are necessary because there is no substitute for H_c to differentiate between impingement cases and near-neutral flow cases.

The modifications made to the models facilitate their applicability to sites other than CCB, and increase their usefulness for estimating concentrations over several hours. The Lift Model now incorporates a measure of the crosswind aspect ratio of the terrain feature. The Wrap Model estimates concentrations at more than one point across the hill surface.

The Wrap Model impingement estimates (and, in fact, the observed concentrations) are very dependent on the actual horizontal wind direction distribution when the winds are light and variable. When the plume is released above H_c, concentrations over the top of the hill are extremely sensitive to the size of σ_z. Therefore, on-site meteorological input is required at levels representative of plume height.

Initial analysis of the HBR data base suggests that the concept of a critical dividing-streamline height is also appropriate for stable flows toward a two-dimensional ridge. Tracer gas released below H_c dispersed in a highly variable, "blocked" flow and produced relatively large ground-level concentrations on the windward face of the ridge. Tracer gas emitted above H_c dispersed in a flow that traveled over the ridge and produced peak ground-level concentrations near the ridge crest and on its lee side.

REFERENCES

Hanna, S. R., 1983: Lateral Diffusion Due to Mesoscale Eddies with Period One to Two Hours at Cinder Cone Butte. Extended Abstracts, Sixth Symposium on Turbulence and Diffusion, American Meteorological Society, Boston, MA, 162-165.

Holzworth, G. C., 1980: The EPA Program for Dispersion Model Development for Sources in Complex Terrain. Preprints, Second Joint Conference on Applications of Air Pollution Meteorology, American Meteorological Society, Boston, MA, 465-468.

Hovind, E. L., M. W. Edelstein, and V. C. Sutherland, 1979: "Workshop on Atmospheric Dispersion Models in Complex Terrain". EPA-600/9-79-041. EPA, Office of Research and Development, Research Triangle Park, NC, 213 pp.

Lavery, T. F., A. Bass, D. G. Strimaitis, A. Venkatram, B. R. Greene, P. J. Drivas, and B. A. Egan, 1981: "EPA Complex Terrain Model Development, First Milestone Report-1981". EPA-600/3-82-036. EPA, Office of Research and Development, Research Triangle Park, NC, 327 pp.

Lavery, T. F., D. G. Strimaitis, A. Venkatram, B. R. Greene, D. C. DiChristofaro, and B. A. Egan, 1983 : "EPA Complex Terrain Model Development, Third Milestone Report-1983" (in preparation). EPA, Office of Research and Development, Research Triangle Park, NC.

Snyder, W. H. and R. E. Lawson, 1981: Laboratory Simulation of Stable Plume Dispersion over Cinder Cone Butte. "Appendix B, EPA Complex Terrain Model Development, First Milestone Report-1981". EPA-600/3-82-036. EPA. Office of Research and Development, Research Triangle Park, NC, 52 pp.

Snyder, W. H., R. S. Thompson, R. E. Eskridge, R. E. Lawson, Jr., I. P. Castro, J. T. Lee, J. C. R. Hunt, and Y. Ogawa, 1982: The Structure of Strongly Stratified Flow Over Hills, Dividing-Streamline Concept. "Appendix A, EPA Complex Terrain Model Development, Second Milestone Report-1982". EPA-600/3-83-015. EPA, Office of Research and Development, Research Triangle Park, NC, 56 pp.

Strimaitis, D. G., A. Venkatram, B. R. Greene, S. Hanna, S. Heisler, T. F. Lavery, A. Bass, and B. A. Egan, 1982: "EPA Complex Terrain Model Development, Second Milestone Report-1982". EPA-600/3-83-015. EPA, Office of Research and Development, Research Triangle Park, NC, 402 pp.

Strimaitis, D. G., A. Venkatram, and T. F. Lavery, 1983: A Model to Estimate Concentrations During Plume Impingement. Extended Abstracts, Sixth Symposium on Turbulence and Diffusion, American Meteorological Society, Boston, MA, 28-31.

Venkatram, A., D. Strimaitis, D. DiChristofaro, J. Pleim, T. Lavery, A. Bass, and B. Egan, 1982: The Development and Evaluation of Advanced Mathematical Models to Simulate Dispersion in Complex Terrain. Preprints, Third Joint Conference on Applications of Air Pollution Meteorology, American Meteorological Society, Boston, MA, 167-170.

Venkatram, A., D. Strimaitis, and W. Eberhard, 1983: Dispersion of Elevated Releases in the Stable Boundary Layer. Extended Abstracts, Sixth Symposium on Turbulence and Diffusion, American Meterological Society, Boston, MA, 297-299.

5: PRACTICAL APPLICATIONS OF AIR QUALITY MODELING

Chairmen: L. Prahm
G. Omstedt
J. Kretzschmar

Rapporteurs: P. Bessemoulin
B. Van Der Borght

METHODOLOGIES TO VALIDATE MULTIPLE SOURCE MODELS : AN APPLICATION TO THE 5TH EUROPEAN CAMPAIGN ON REMOTE SENSING TECHMIQUES (GHENT, BELGIUM, 1981)

J.G. Kretzschmar, G. Cosemans and B. Vanderborght

Studiecentrum voor Kernenergie
SCK/CEN
B-2400 Mol, Belgium

INTRODUCTION

From 12 to 26 June 1981 the "5th CEC Campaign on Remote Sensing" took place in and around the industrial area located along the Ghent-Terneuzen canal, north of the city of Ghent. This campaign was jointly organized by the Commission of the European Communities and the Belgian Ministry for Science Policy within the framework of the National R & D Programme on the Environment.

During that fortnight's campaign several Belgian and foreign research teams combined their skills and technical means to obtain an as complete as possible picture of the local industrial SO_2 emissions, to follow the varying meteorological conditions, to study the transport and dispersion of the released SO_2, to monitor the resulting gas-burden and ground-level concentrations and to model the entire system : emission - transmission - emmission.

In what follows some of the relevant data obtained during the campaign will be used to illustrate and evaluate different procedures to validate or verify unsophisticated dispersion models, when used to simulate short-term SO_2-pollution levels in a multiple source region such as the Gent urban-industrial area.

MULTIPLE SOURCE MODEL VALIDATION PROCEDURES

Ambient air pollution concentration levels in a multiple source region are variable as a function of time and space. In order to be useful and acceptable from a practical point of view, a mathematical model must therefore be able to quantitatively simulate both variations in a realistic and consistent way as a function of time and space. This means that over a sufficiently long period of time, e.g. several weeks to one or more years, the successive spatial patterns of the concentration levels averaged over short time intervals, e.g. 30 min. to 1 day, must be reproduced in an acceptable way. From this point of view the ideal model validation or verification procedure has to be based on the comparison of matrices of calculated and simultaneously measured short-term concentration averages, with the columns of the matrices referring to different receptor (monitoring) points and their lines to the successive (short) time steps in the period taken into consideration.

It must be emphasized that the general philosophy outlined in what preceeds can be applied in two quite different ways. In the frequently used statistical model validation procedures, similar to the one used by Cosemans et al. (1983a) for the large scale validation of the simulation of the SO_2 situation in the Gent area, the corresponding statistical characteristics of large ensembles of simultaneously measured and calculated short-term averages in a limited number of receptor points (monitoring sites) are used. This means that the columns of the corresponding matrices are first of all summarized by means of the appropriate statistical parameters and that the rest of the verification is based on the intercomparison of corresponding statistical quantities (e.g. yearly averages, percentiles of cumulative frequency distribution, extreme values, variances ...). The second possible approach is based on a line by line comparison of the matrix values, implying that corresponding situations averaged over short periods of time (30 min., 1 h, 8 h, 24 h) are compared. It's obvious that in the first approach the global impact of a source configuration, in the sense of an impact assessment study, is the major aim of the analysis, while in the second approach much more details with respect to the spatial and temporal resolution of the simulation are emerging. Both ways of attacking the validation problem rely nevertheless basically on the same output data although the requirements for the input data are much stricter for validation exercises of the latter kind.

Due to practical hardware constraints and financial limitations, especially with respect to the rather limited number of monitoring sites wherefore reliable and simultaneously measured information on the short-term average concentrations can be made available, whatever validation is restricted as the matrix of the measured values quite often turns out to be a sparse one. More specifically with respect to the spatial average of the available information validation exercises turn out to fall into one of the following 4 classes: point validation, line validation, area validation and volumetric or 3D-validation.

"Point validation" is based on the available (measured) information in a limited number of fixed points in the region under investigation. These points are the monitoring sites of a permanent and continuously operating stationary network, sometimes set-up specifically for the validation study but most frequently designed and run for more general survey and control purposes, or the mobile or semi-mobile campaign's measuring points wherein, as a function of the objectives of the campaign, short-term averages are discontinuously determined within the framework of a specific sampling scheme (single measurements or limited series of successive ones). It's obvious that especially for short averaging times individual "point data" are very sensitive to the wind direction and its fluctuations and that their representativeness for the existing pollution-levels in the surrounding area is quite often doubtful or unknown. The great majority of the model validations are nevertheless "point validations" but fortunately most of the time of the statistical type.

A first attempt to validate the spatial consistency of simulated short-term air pollution levels is the "line validation" technique. A first possibility to collect measured profiles, of ground-level concentration (glc) or gas-burden, is in a pure mobile way by scanning with a mobile laboratory the same road stretch downwind the source(s) a sufficient number of times within the averaging time (t_{av} = 30 min to 2 h) used for the simultaneous calculations. The required number of successive scans to obtain a representative average profile over the given averaging time is of course a function of the temporal variability of the profile. The individual (instantaneous) profiles are not or hardly usable for model validations but they illustrate the temporal variability quite well (Cosemans et al., 1983b ; Vanderborght et al., 1983). A second semi-mobile approach, only suited for glc measurements, is based on portable sampling units (or portable monitors) set out on a road stretch, as a function of the wind direction and the source configuration, and sampling simultaneously the ambient air during the specified averaging time. The required number of sampling units or the mutual distance between them is of course a function of the variability of the pollution pattern along the road. A mobile laboratory with a continuous monitor is in this context a very useful complementary device to evaluate the spatial variability. Similar to the combination of several points in the "point validation" approach, different (less or more parallel) road stretches can be scanned at the same time or over the same time interval and the "line validation" technique becomes then very useful to determine the evolution of cross-wind profiles as a function of the downwind distance. This is a very well-known and frequently used technique in air pollution dispersion tracer experiments. The mobile "line validation" technique is the only possibility to verify gas-burden simulations and/or measurements.

For the "area validation" technique, aiming at the comparison of simultaneously calculated and measured two-dimensional glc-patterns it is necessary to determine the latter.

In a multiple source configuration it is almost excluded to try to do this with a stationary monitoring network as the required site density is too high. The only pragmatic approach is in the semi-mobile way by means of portable sampling units or (simple and cheap) monitors integrating over the choosen averaging time in a representative number of points of the area under investigation. It's obvious that lines or road stretches and individual points are combinable in this area approach.

The ultimate model validation is certainly the three dimensional one. The required technical means to collect representative data for "3D-validation" in a multiple source configuration of the type we are dealing with are not readily available. In long-range transport studies (the far-field zone) landborne and airborne mobile units can be combined to study the three-dimensional distribution of certain pollutants in a region far enough away from the main source areas (so that the spatial and temporal variability of the levels are not excessive). Also in the very near field zone, e.g. for the study of street canyon phenomena, building wake effects, automative emissions on a highway or specific fugitive emissions, three-dimensional (stationary) measurements are feasible and have been done.

Some of the data made available or obtained during the 5th CEC campaign in Gent were suited for further use in a "point" and/or "line validation" as illustrated in what follows. Although eventually possible no systematic attempts have been made yet to use the data within an "area validation" approach.

THE 5th CEC CAMPAIGN, GENT 1981

Since 1975 the Commission of the European Community periodically organizes intercomparison campaigns for the remote sensing of atmospheric pollution (Guillot et al., 1979). In these campaigns teams of different member countries simultaneously measure SO_2 ground-level concentrations and vertical SO_2 burdens with mobile laboratories. The main objective of the campaigns is the intercomparison of sampling strategies, data acquisition and treatment systems as well as measuring results.

The 5th CEC-campaign was organized in Gent from 12 to 26 June 1981. The main difference of this campaign compared to previous ones (Lacq and Cordemais in France, Drax in the UK and Turbigo in Italy) was that instead of a single source, a complex configuration of sources has been studied. Besides the already mentioned general aims of this type of CEC campaigns, the specific objectives of this campaign were :

- provide an overall SO_2 balance for the studied area : this balance consists of transport towards, generation inside and transport away from the area ;

- provide a set of input data for modeling purposes ;
- determine the relative influence of one important source of SO_2 within the complex source configuration of the canal zone.

Gent is a city with approximately 250.000 inhabitants situated in the flat northern part of Belgium. Downtown Gent is situated at the southern end of the Gent-Terneuzen Canal (figure 1). Along the last 15 km of the canal heavy industries have been set up including steel mills (Si on figure 1), a non-ferrous metal refinery (Sa), an oil refinery(T),two power plants (R and L) and various chemical plants (P,K,U and S). Emissions of these sources approximate 100.000 t SO_2/year. A local automatic monitoring network continuously follows (30 min averages) the SO_2-pollution levels in 11 sites (R7xx, figure 1) and the meteorological situation in 6 sites (four M7yy sites measuring at 30 m height, the JRC site measuring at 22 m height and the 108 m high meteorological MAST, fig. 1).

The different participating teams can be classified according to their respective task(s) and/or available instrumentation. Monitoring of glc or burden was done by :

- AES, Atmospheric Environment Service, Canada
- CEA, Commissariat à l'Energie Atomique, France
- CERL, Central Electricity Research Laboratories, UK
- IHE, Institute for Hygiene and Epidemiology, Belgium
- JRC, Joint Research Centre Ispra, Italy
- RIV, National Institute for Public Health, the Netherlands
- SCK/CEN, Nuclear Energy Research Centre, Belgium

Meteorological measurements were carried out by :

- CERL, Central Electricity Research Laboratories, UK
- ENEL, National Electrical Energy Laboratories, Italy
- IHE, Institute for Hygiene and Epidemiology, Belgium
- JRC, Joint Research Centre Ispra, Italy
- KMI/IRM, Royal Meteorological Institute, Belgium
- MIH, Meteorological Institute, University Hamburg, GDR
- SCK/CEN, Nuclear Energy Research Centre, Belgium
- UCL, Université Catholique de Louvain, Belgium

Emission measurements and inventory were under the responsibility of:

- INW, Institute for Nuclear Sciences, RUG, Belgium
- UELg, Université de Liège, Belgium

Modeling of glc and burden was done by :

- SCK/CEN, Nuclear Energy Research Centre, Belgium
- UCL, Université Catholique de Louvain, Belgium

During the Gent campaign, where the pollution downwind a large source area had to be measured, the different teams measured profiles along rather long road stretches downwind the sources, so that

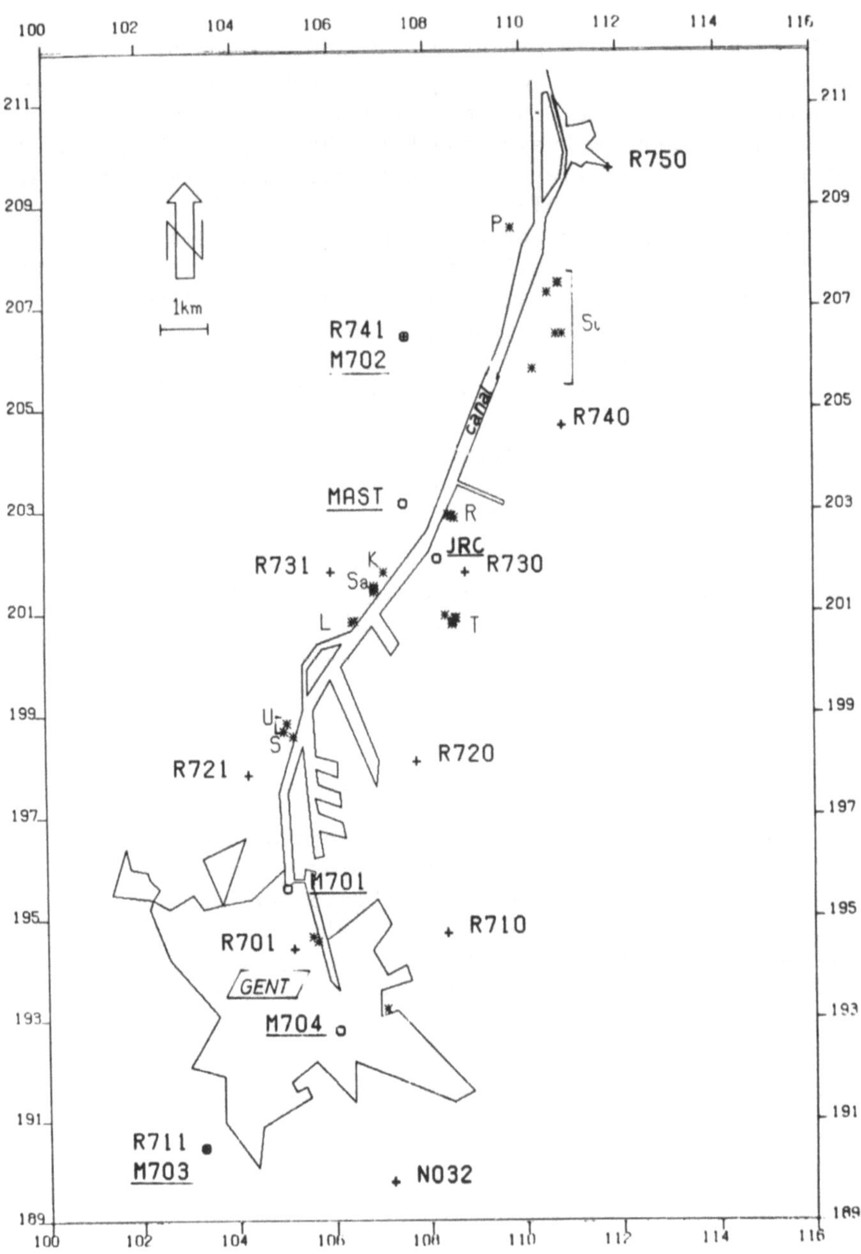

Fig. 1. The Gent test area with the main industrial sources and the sites (+ SO_2, o meteo, ⊕ SO_2 and meteo) of the Belgian Automatic Monitoring Network.

it was impossible to perform many traverses within 1 hour. Quite often, one traverse took 15 to 30 minutes to even one hour. The consequences of this sampling strategy is discussed in Vanderborght et al., 1983. Stationary 30 min. average glc-measurements were obtained in the sites of the local automatic monitoring network (AMG-IHE) and supplemented with semi-mobile measurements along road stretches by means of sample collection (t_{av} = 30 min) with 10 portable sampling units and subsequent analysis by means of the TCM or West-Gaeke method (SCK/CEN).

Detailed information on the raw data gathered during the campaign, as well as the up to now available interpretation of the campaign's results, can be found in the relevant papers or reports given in the list of references.

SHORT DESCRIPTION OF THE DISPERSION MODEL

Concentrations (glc) at ground-level C(x,y,o) and gas-burden B(x,y) are calculated by means of a straightforward bi-Gaussian dispersion code (IFDM, SCK/CEN, Mol) implying that :

$$C(x,y,o) = \frac{10^9 Q}{\pi . \bar{u}(h_e) . \sigma_y(x) . \sigma_z(x)} \cdot \exp\left(\frac{-y^2}{2\sigma_y^2(x)}\right) \cdot \exp\left(\frac{-h_e^2}{2\sigma_z^2(x)}\right) \quad (\mu g/m^3)$$

$$B(x,y) = \frac{10^6 Q}{\sqrt{2\pi} . \bar{u}(h_e) . \sigma_y(x)} \cdot \exp\left(\frac{-y^2}{2\sigma_y^2(x)}\right) \quad (mg/m^2)$$

with : Q : release rate in kg.SO_2/s

$\sigma_y(x)$, $\sigma_z(x)$: dispersion parameters according the Bultynck-Malet (1972) scheme (m)

$h_e = h_g + \Delta h$: effective plume height (m)

h_g : geometric stack height (m)

Δh : plume rise according the Stümke II (1963) formula

$\bar{u}(h_e)$: average wind speed at height h_e metres

$\bar{u}(h_e) = \bar{u}(h_o)(h_e/h_o)^{m_i}$

$\bar{u}(h_o)$: wind speed at measuring height h_o

m_i : derived from the observed wind speeds at different heights during the relevant period for the calculations.

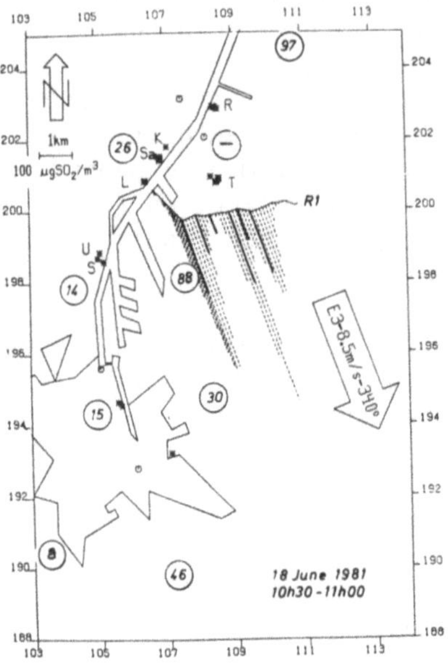

Fig. 2a. Measured and calculated glc-profiles along road R1.

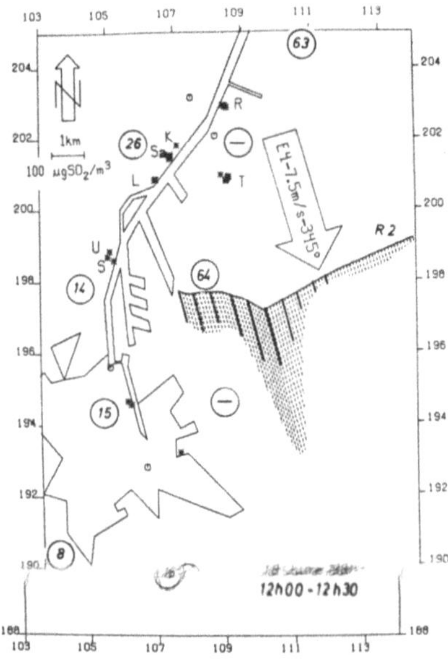

Fig. 2b. Measured and calculated glc-profiles along road R2.

It must be emphasized here that all subsequent model calculations will in essence be based on the routine meteorological observations at the 30 m height in the local automatic air pollution monitoring network (AMG -IHE) and along the 108 m high meteorological mast (AMG-SCK/CEN). When possible reference is also made to supplementary meteorological information obtained specifically during the 5th CEC campaign by means of the tethered balloon soundings.

POINT AND LINE VALIDATION ON 18 JUNE 1981

On 18 June, a day with rather steady strong winds from the NNW, semi-mobile TCM-measurements were carried out on two different road stretches, respectively R1, at a downwind distance of 1 to 3 km from the most important SO_2-sources and R2 at 4 to 6 km downwind the same sources (figure 2a and 2b). During each set of measurements ten portable sampling units are placed along the road side downwind the sources and sample simultaneously during 30 min the ambient air at $\approx$ 1.5 m above ground-level (Kretzschmar et al., 1981a). After transfer to the laboratory the samples are analysed by means of the reference TCM- or West-Gaeke method.

Measured concentration values (t_{av} = 30 min) in the discrete monitoring points, as well as the corresponding calculated glc-profiles along the road stretches are graphically represented on figure 2a for the first measuring period (10h30-11h00 LT, road R1) and on figure 2b for the second one (12h00-12h30 LT, road R2). The choosen meteorological input data, respectively neutral atmospheric stability (E3), 8.5 m/s by 340 degrees and slightly unstable (E4), 7.5 m/s and 345 degrees, are also represented on the figures. Wind direction and wind speed are based on the observed half-hourly averages summarized in table 1. The information on figure 2a and 2b is completed with the half-hourly SO_2-averages as measured in the AMG-sites (encircled values).

Table 1 : Half-hourly averages for wind speed and wind direction as measured in the AMG monitoring sites, on the mast and by the JRC team (Hasenjäger, 1981) on 18 June 1981.

		10h30-11h00		12h00-12h30	
site	height	u(m/s)	dd(degrees)	u(m/s)	dd(degrees)
M701	30 m	5.5	354	6.9	355
M702	30 m	5.6	341	6.9	349
M703	30 m	5.2	345	5.4	343
M704	30 m	4.4	340	5.7	351
JRC	22 m	8.5	325	7.5	345
mast	108 m	6.0	338	6.5	345

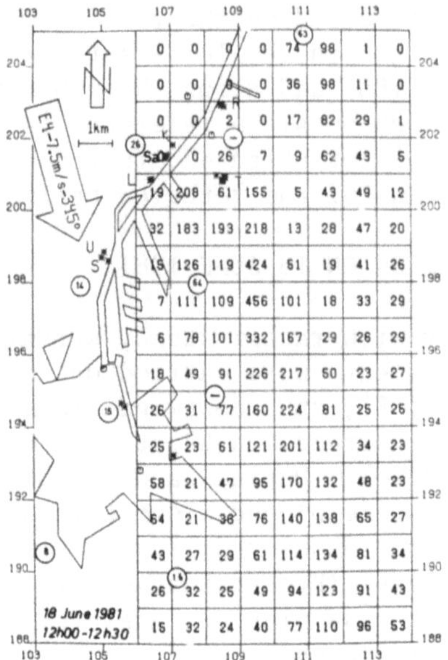

Fig. 3a. Calculated glc-spatial distribution.

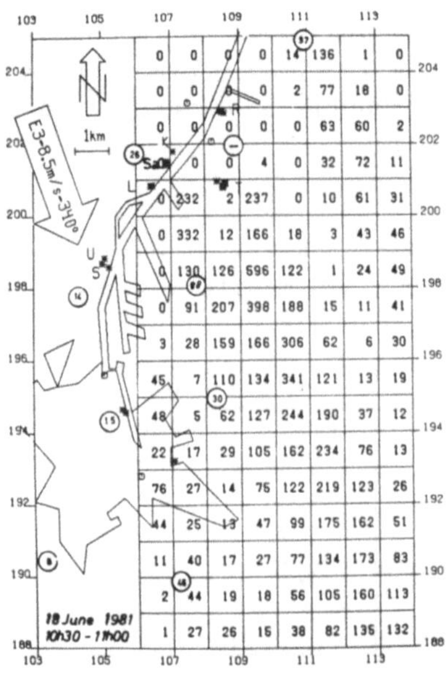

Fig. 3b. Calculated glc-spatial distribution.

Two facts are quite striking on figure 2a and 2b. First of all a very high spatial variability along the first traverse R1 is noted as well in the measured as in the calculated glc-profile. Further away on R2 plumes have obviously merged resulting in a much smoother profile. The high spatial variability, a combination of the specific source configuration and the occurring meteorological conditions on 18 June, is further highlighted on figure 3a and 3b by the spatial distribution of the glc-values calculated in the centre points of a 1 km by 1 km square grid (Lambert coordinates). On more than one occasion two orders of magnitude change in the glc-values are noted over 1 km distance. With that high degree of spatial variability in the glc-pattern it is no wonder that "point validation", e.g. by comparing calculated and measured glc-values in the very limited number of monitoring sites R7xx (figures 1, 2a, 2b, 3a and 3b), will give disappointing results in the situation at hand, as illustrated by some of UCL's results (Demuth et al., 1982) summarized in table 2.

Table 2 : Measured and computed glc's ($\mu g.SO_2/m^3$) in some receptors of the AMG on 18 June 1981 (Demuth et al., 1982)

period	site (AMG)	measured ($\mu g/m^3$)	computed Bi*	Ka*	Kn*	meteo
10.30-11.00	R710	30	26	31	34	$\bar{u}(30)=5.6m/s$
	R720	88	7	4	4	$dd(108)=325°$
	R740	97	6	2	3	
	N032	46	0	0	0	
12.00-12.30	R720	64	117	119	121	$\bar{u}(30)=6.9m/s$
	R740	63	59	40	40	$dd(108)=345°$
	N032	16	38	45	50	

* Bi : bi-Gaussian model (Pasquill-Briggs)
* Ka : analytical K-model
* Kn : semi-numerical K-model

A model "point-validation" limited to the seven points of table 2 would be misleading as figures 3a and 3b clearly illustrate that the models do not simulate all that bad when the measured (point) data are placed in their (proper) spatial context. Taking into account the drastic influence of the choice of the specific wind direction upon the glc-value in a given point the measured data do fit into the calculated two-dimensional glc-patterns now.

The second striking feature in figure 2a and 2b is the tendency of the model to overpredict the peak glc-values (up to a factor 2) along the traverses R1 and R2 where TCM-measurements were carried

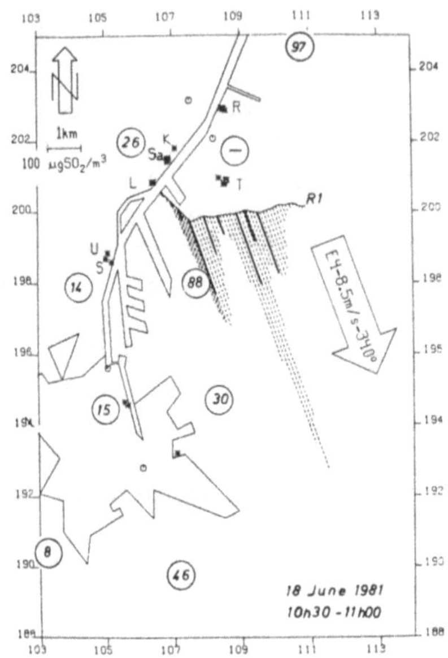

Fig. 4a. Same as figure 2a, but with stability class E_4 i.o. E_3

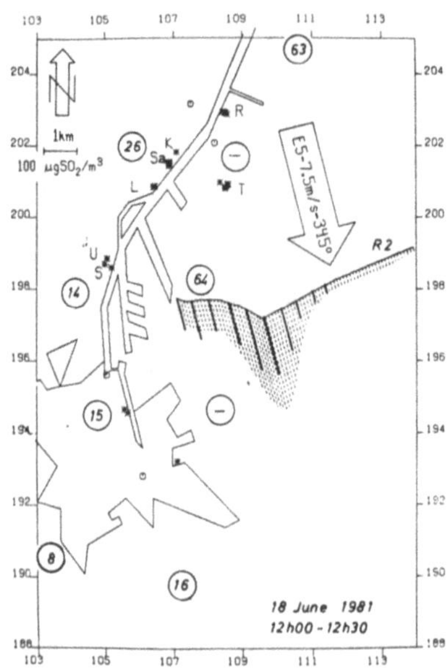

Fig. 4b. Same as figure 2b, but with stability class E_5 i.o. E_4

out. Although the previously discussed large spatial variability can play an important role here, as the peaks are possibly missed in the measurements, it is also possible that the overestimations are inherent to the used model and its (meteorological) input data. To illustrate this the R1 and R2-profiles were recalculated for a somewhat more unstable atmospheric stratification namely E4 instead of E3 and E5 instead of E4 for the respective half hours. The corresponding results given in figure 4a and 4b show net improvement for 3 of the 4 peaks and further degradation for the fourth one, due to the nearby sources of plant T and being better approximated in a neutral (E3) to slightly stable (E2) atmosphere as shown by G. Cosemans et al., 1983.

As on 18 June 1981 the mobile teams measuring simultaneously SO_2-glc and SO_2-gas-burden only started their measurements around 13h00 it was not possible to combine model calculations, semi-mobile TCM measurements and mobile glc- and gas-burden measurements in a single "line validation" example for that day. Measurements along R1 from 13h00 to 15h00 by JRC (Cerutti et al., 1981) and along R2 by CERL and IHE from 13h00 to 16h00 (De Saeger et al., 1982) confirmed nevertheless the earlier measured profiles (fig. 2a and 2b) and the already discussed spatial variability and patterns.

POINT AND LINE VALIDATION ON 19 JUNE 1981

In contrast to the previous day 19 June 1981 was a day with persistently light southwesterly winds. Semi-mobile TCM measurements, as well as mobile measurements by different teams, were carried out on a road some 7 to 12 km downwind the source area. The first profile, measured from 11h30 to 12h00, can be considered to be the combined result of all the industrial sources along the canal, as illustrated by the measured (full line) and the calculated (dashed lines) glc-values graphically represented on the map in figure 5a. As before the choice of the meteorological input data for the model were based on the local stationary half-hourly measurements, as summarized in table 3, and the resulting relative position of the profile's peaks with respect to the location of the respective industrial sources.

Figure 5a clearly shows that the calculated profile, apart from the overestimation in its central part, does fit the measured one very well.

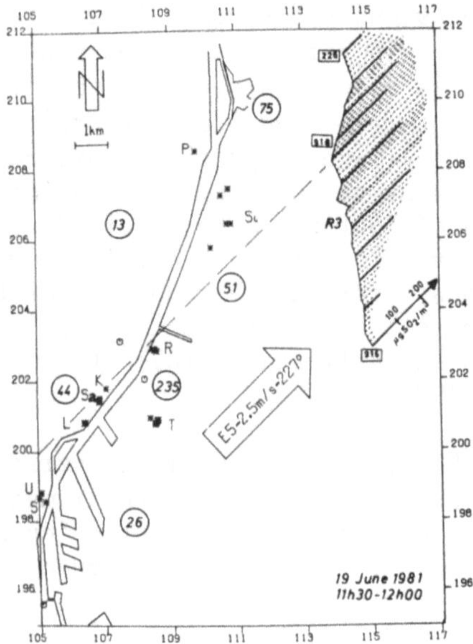

Fig. 5a. Measured (—) and calculated (---) glc-profile along R3.

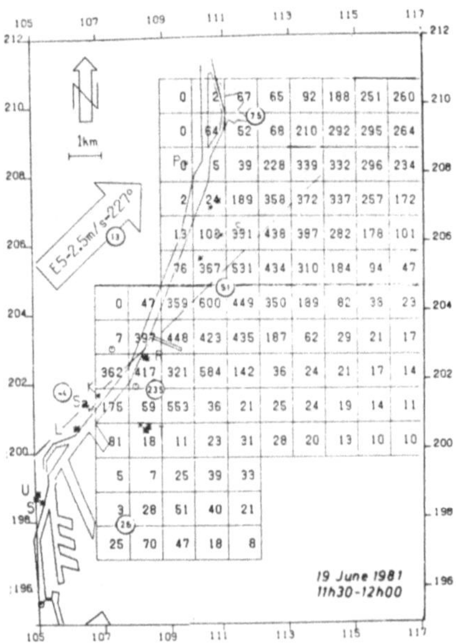

Fig. 5b. Measured glc-values in AMG (◯) and calculated ones in a 1 km x 1 km grid.

Table 3 : Half-hourly average wind speeds and wind directions as measured in the AMG, on the MAST and by the JRC team (Hasenjäger, 1981) at noon on 19 June 1981

	M701	M702	M703	M704	MAST	JRC
height (m)	30	30	30	30	108	22
$\bar{u}$ (m/s)	2.6	2.4	1.8	1.0	2.5	1.9
dd (degrees)	239	228	235	209	220	263

A seeming similar phenomenon of overestimation in the computed values is noted when comparing the individual half-hourly averages measured in the downwind monitoring sites R7xx of the automatic network with the calculated two-dimensional glc-pattern, as illustrated by figure 5b. The four measured values nicely fit into the general pattern except the value of monitoring site R740, with a measured value of 51 $\mu gSO_2/m^3$ to be compared with calculated values ranging from 367 to 600 $\mu gSO_2/m^3$. Note that this discrepancy occurs downwind the same source(s) as it was the case for the TCM-profile measured on road stretch R3 (figure 5a). On the other hand it must be pointed out that under the prevailing unstable atmospheric stability class (E5), the meandering plumes give rise to a very variable glc pattern along the canal and in the neighbourhood of the sources. This is first of all illustrated in table 4 by means of the successively measured glc half-hourly averages in the monitoring sites R720, R730, R740 and R750, where under pretty persistent meteorological conditions drastic concentration changes as a function of time were noted in the monitoring sites. A second illustration is given by two consecutive (instantaneous) glc-profiles obtained by the mobile team of JRC (Cerutti et al., 1981) on the Kennedylaan, a road passing close to the monitoring sites R720 to R750. These profiles given in figure 6 do illustrate the high spatial (and temporal) variability as well as the possibility for elevated SO_2 ground concentration levels especially in the vicinity of R730 and R740.

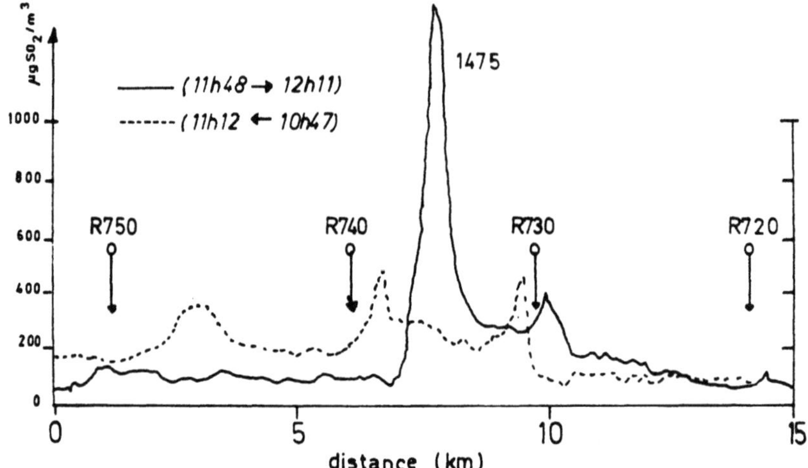

Fig. 6. Glc-profiles measured in a mobile way by JRC on 19 June 1981 on the Kennedylaan between R750 and R720 (Cerutti et al., 1981).

Table 4 ; Temporal variation of the glc-values in the downwind monitoring sites of the Gent automatic monitoring network (IHE, Brussels) and corresponding meteorological conditions.

** time	glc (μg/m³)				wind direction		wind speed		Stability
	R720	R730	R740	R750	M702	MAST	M702	MAST	
10.00	51	208	335	34	215	-	2.5	2.9	5
10.30	66	167	284	90	228	-	2.6	2.9	5
11.00	35	598	232	136	234	225	2.5	2.6	5
11.30	26	309	83	73	231	230	2.4	2.8	5
12.00*	26*	235*	51*	75*	228*	220*	2.4*	2.9*	5*
12.30	26	339	53	38	233	210	2.9	3.2	5
13.00	27	285	104	39	228	220	2.6	2.9	5
13.30	42	202	146	45	229	210	2.0	2.6	5

* corresponds with the simulations on figures 5a and 5b

** 10.00 means from 09h30 to 14h00 LT

Due to their instantaneous nature glc-profiles obtained while driving must be interpreted with great care especially in relating them to time-integrated profiles (e.g. semi-mobile TCM-measurements) or to time-integrated glc-values in individual points (e.g.stationary monitoring sites R7xx) on the same road stretch or in its immediate vicinity. Reliable comparisons can only be done on the basis of the average of a significant number of successive instantaneous profiles with "significant number" being a function of the observed temporal and spatial variability of the profile. In this context it is interesting to compare the measured half-hourly average TCM-profile on transect R3 (figure 5a) with the instantaneous profiles measured within the same half hour by respectively the SCK/CEN, the AES and the RIV mobile teams. Table 5 illustrates that in each of the measuring points, where a half-hourly average glc-value was obtained by means of the portable TCM sampling unit, the instantaneous concentration values obtained by the three mobile teams can be quite different, although their averages (and thus the profile) do correspond quite well with the corresponding TCM averages over 30 minutes. It can thus be concluded that under certain conditions the average result of successive mobile scans on the same road stretch can be used for "line validation" purposes (Cosemans et al., 1983b and Vanderborght et al., 1983)

Table 5 : Comparison between 30 min. averaged stationary measurements and mobile instantaneous ones on road stretch R3, 19 June 1981, 11h30-12h00 (Vanderborght et al., 1983)

	mobile measurements (μg/m^3)				stationary TCM(μg/m^3)
point	AES	RIV	SCK	average	
44	84	240	180	170	210
45	120	330	170	210	160
46	130	340	220	230	245
47	75	270	210	185	240
48	46	240	140	140	160
49	30	190	125	115	86
50	30	100	110	80	64

AES : from 11h49 to 11h56 at 60 km/h
RIV : from 11h33 to 11h40 at 56 km/h
SCK : from 11h30 to 12h00 at 28 km/h

Although not discussed in detail in this paper the gas burden profiles measured on transects can be verified in a "line validation" procedure similar to the one used for glc-profiles. It must nevertheless be pointed out that with respect to the use for validation purposes of the numerous gas burden profiles obtained during the 5th CEC campaign some major problems do exist. Road stretches covered by the different mobile teams were in general rather long so that the number of (instantaneous) profiles obtained during a given period of time, e.g. 2 hours, was very limited. As important changes in shape as well as in total gas burden were quite frequently noted in successive scans, despite a reported almost constant emission rate for the inventorized sources, the obtained average profiles have a high degree of uncertainty as well with respect to their shape as with respect to the derived total gas burden (De Saeger et al., 1982 and Cosemans et al., 1983b). For the time being it looks like some more (harmonization and interpretation) work will have to be done on the 5th CEC campaign's raw data before the comparison of (reliably) measured and modeled gas burden profiles could be undertaken in the usual straightforward way.

CONCLUSIONS

The verification of modeled short-term averages requires an insight in the spatial variability of the concentration and/or gas burden patterns, even if the emissions and the occurring meteorological conditions are relatively well known. Point validation e.g. by means of the short-term average glc-values measured in the sites of a local monitoring network set-up for other purposes than model validation, can be very disappointing and misleading as the spatial representativeness of these sites for short-term average profiles and/or patterns is quite often questionable. "Line validation", by combining in an appropriate way a representative number of (temporary) measuring points is a minimum minimorum requirement to tackle the problem. The "multiple line approach" or the "area approach", changing its configuration each time as a function of the (known) source configuration and the prevailing meteorological conditions, is the appropriate solution. This is not at all new as the methodology is commonly used in tracer release experiments, set-up for dispersion studies or model validation exercises. From the pragmatic point of view semi-mobile measurements based on simple (and cheap) portable and programmable sampling units, with subsequent analysis in the laboratory, are quite useful as illustrated in the 5th CEC campaign carried out in the complex source area of Gent in June 1981.

For model validation purposes results of mobile measurements must be interpreted with great care. When used in the appropriate way they can nevertheless give essential supplementary information especially with respect to the required number of stationary sampling points, as spatial and temporal variability of the phenomena under investigation can be scanned in a fast and reliable way by means of mobile monitors.

The 5th CEC campaign on remote sensing of pollutants gave the opportunity to different Belgian and foreign research teams to exchange under field conditions their know-how and experiences with less or more sophisticated methodologies and equipment. By concentrating the available technical means and the human resources during two weeks in the area under investigation a tremendous amount of raw data, hiding a lot of interesting information, was collected. Although by now remarkable attempts have been made by the different participants, some more coordinated interpretation of the raw data will be required if the optimum is to be reached.

REFERENCES

Bultynck, H., and Malet, L., 1972, Evaluation of atmospheric dilution factors for effluents diffused from an elevated continuous point source, Tellus, 24:455-472

Cerutti, C., De Groot, M., and Sandroni, S., 1981, "Sulphur Dioxide Mass Flow in the Gent Industrial Area Measured at the 5th CEC Campaign on Remote Sensing", EUR7788EN JRC-Ispra, Italy.

Cosemans, G., Kretzschmar, J., De Baere, G., and Vandervee, J., 1983a, Large scale validation of a bi-Gaussian dispersion model in a multiple source urban and industrial area, in "Air Pollution Modeling and its Application II", C. De Wispelaere, ed., Plenum Press, New-York and London.

Cosemans, G., and Kretzschmar, J.G., 1983b, On the simultaneous simulation of SO_2 ground-level concentration and corresponding gas burden profiles, The Science of the Total Environment, to be published.

Demuth, Cl., and Heck, P., 1982, Application of three stationary diffusion models for estimating the concentration field during the campaign, in : "Draft Report : 5th CEC Campaign on Remote Sensing of Atmospheric Pollutants,Ghent (Belgium), June 1981", DPWB, Brussels.

De Saeger, E., Dumont, G., Verduyn, G., and Vervliet, F., 1982, Study of a complex source area by remote sensing, in : "Draft Report, 5th CEC Campaign on Remote Sensing of Atmospheric Pollutants, Ghent (Belgium) June 1981", DPWB, Brussels.

Guillot,P.,et al.,1979,First European Community campaign for remote sensing of atmospheric pollution - Lacq 1975, Atm. Env., 13: 895-917.

Hasenjäger, H., 1981, "Atmospheric Parameters for Plume Propagation Studies Measured and Elaborated at the 5th CEC Campaign at Ghent," EUR 7772 EN - preprint, JRC-Ispra, Italy.

Hoff, R.M., and Gallant, A.J., 1982, "AES Participation in the 5th Remote Sensing Campaign of the Commission of the European Communities, Ghent, Belgium", AQRB Report-030/T, internal report AES, Canada.

Kretzschmar, J.G., and Cosemans, G., 1981a, Random- and minimax-campaigns for the determination of the actual air-pollution levels in an unknown region, Atm. Env., 15:1047-1058.

Kretzschmar, J.G., Cieslik, S., De Baere, G., Julien, J., and Van Tongerloo, J., 1981b, "Mesures Météorologiques dans la Basse Atmosphère, Gand, 13 - 24.06.81", Rapport Scientifique 1981/AL4.5/0.2, SCK/CEN, Mol, Belgium.

Kretzschmar, J., and Cosemans, G., 1982a, Simulation of the measured ground-level concentrations and the gas burden, in : "Draft Report, 5th CEC Campaign on Remote Sensing of Atmospheric Pollutants, Ghent (Belgium) June 1981", DPWB, Brussels.

Kretzschmar, J., and Vanderborght, B., 1982b, Analysis of the measured ground-level concentration profiles, in : "Draft Report, 5th CEC Campaign on Remote Sensing of Atmospheric Pollutants, Ghent (Belgium) June 1981", DPWB, Brussels.

Stümke, H., 1963, Vorschlag einer empirischen Formel für die Schornsteinüberhohung, Staub 23:549-556

Vanderborght, B.M., and Kretzschmar, J.G., 1983, On the comparison of mobile and stationary measurements of SO_2 ground-level concentrations, The Science of the Total Environment, to be published.

Van der Meulen, A., and van Jaarsveld, J.A., 1982, "1981 CEC Remote Sensing Campaign at Gent", Report 247604012, RIV, Bilthoven, the Netherlands.

Van Evercooren, J., 1982, Emission Data, in : "Draft Report, 5th CEC Campaign on Remote Sensing of Atmospheric Pollutants, Ghent Belgium) June 1981", DPWB, Brussels.

SIMULATED POLLUTANT TRANSPORT OVER THE GHENT INDUSTRIAL AREA

C. Cerutti [*], G. Clerici [**], and S. Sandroni [*]

[*] Commission of the European Communities
Joint Research Centre, Ispra Establishment
21020 Ispra (Va) - Italy

[**] A.R.S. Milano - Italy

INTRODUCTION

Pollution dispersion and transport are problems usually tackled in two different ways, namely (1) extensive field measurements or (2) modelling. Apparently, measuring and modelling are two different aspects of the air pollution monitoring dilemma (Van Egmond et al., 1982) ; in practice a reasonable combination of both procedures is the ideal solution. Measuring is a realistic but very expensive and time-consuming procedure ; on the other hand, the model should be physically credible and its results should be comparable with the measured data available before it is used for predictive purposes.

Dispersion problems are usually studied by a model or experimentally by conventional monitoring networks and remote sensors which are concerned with single sources. Four of the six European Campaigns of Remote Sensing of Air Pollution so far performed under the sponsorship of the Commission of the European Communities have focused on the measurement of sulphur dioxide emission from an isolated source and on the dispersion of the effluent in the near field up to about 15 km from the plant. In these studies, a large contribution has been made by the Barringer Correlation Spectrometer (COSPEC), operating from a mobile van and used to map the gas-burden distribution.

Even though experimental methods and modelling cannot yet give a complete assessment in real cases (see for instance Hoff and Millan 1981 ; Millan, 1983), efforts to extend both techniques to more complex problems are noticeable. In EEC countries frequent pollution problems are found in industrial or urban areas (isolated sources are rare cases), and a particular question is the pollutant mass transported from the areas concerned to neighbouring urban or rural areas. Remote measurements by groundbased and airborne Cospec have been used for the assessment of a long-range transport model above the Netherlands (Van Egmond et al., 1982) as well as for the optimisation of the national ground monitoring network (Onderdelinden, 1983). In a previous study (Sandroni et al., 1982) we compared a measured SO_2 burden with a burden simulated by an analytical model around the Milano urban area in typical meteorological conditions and we deduced the corresponding mass balance.

In June 1981 we took part in the 5th CEC Campaign on Remote Sensing of Air Pollution, held at Ghent, Belgium, as part of the Belgian R and D Environmental Programme. The main objective of that exercise was the evaluation of the SO_2 mass balance for the industrial area close to the city of Ghent. According to the working programme agreed among the teams invited to participate in this exercise, the mobile units (equipped with Cospecs and ground monitors) travelled simultaneously along different roads upwind and downwind of the sources and mapped the SO_2 burden and ground concentration. The complete set of data collected from the different teams was analysed in terms of emission of the individual sources as well as of the entire area and compared with the emission inventory claimed by the factories (Final Report of the 5th CEC Campaign, Ghent, 1981, in press).

This paper compares some data gathered by our mobile unit with those computed by two models based on a bigaussian and a Fourier expansion solution of the transport equation. Attention is focused in particular on specific comparisons, namely of (1) computed vs measured distributions of burden and ground concentration, (2) computed vs experimental mass-flows and finally, (3) proposed vs bigaussian model.

EXPERIMENTAL APPROACH

The area under investigation is sketched in Fig. 1 : the letters indicate the locations of the strongest SO_2 sources while the numbers are reference points for the trajectories along which surveys have been performed during the considered days : 901 - 904 on 17 June, 139-916 on 18 and 929-931-914 on 22 June. The above mentioned trajectories have been chosen as perpendicular to the dominant wind direction for the day considered. The area includes the city of Ghent, with approximately 200.000 inhabitants and an industrial area along both sides of the sea channel Ghent-Zelzate up to the Dutch frontier over a distance of 15 km. Several pollution sources of different types are present (power plants, refineries, chemical factories) ; 19 out of the 31 stacks are higher than 90 m. According to the emission inventory based on hourly data

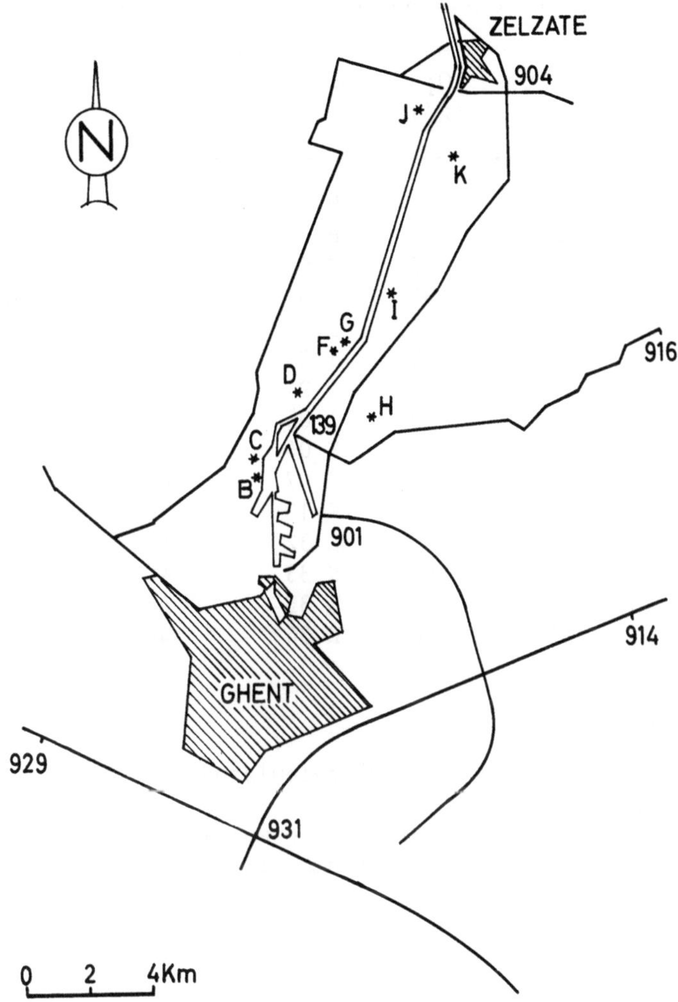

Fig. 1. Map of the industrial area of Ghent with the strongest SO_2 sources (indicated by letters) and the road marker points (indicated by numbers).

and made available by the Belgian Ministry for the Environment, the real SO_2 emission was of the order of 3.2 kg/s (the detailed emissions claimed for 17 June are listed in Table 1). Under particular meteorological conditions, the overall pollution level might be increased by long-range transport either from the industrial area of Antwerp, 60 km to the North East, or from a power plant

Table 1. Claimed SO_2 emission in the Ghent area on 17 June 1981

Source	Height (m)	Q_i (g/s)	Source	Height (m)	Q_i (g/s)
A 01	102	79	H 04	92	81
A 02	102	35	H 05	92	86
B 03	51	27	H 06	92	86
B 06	72	57	I 01	102	0
C 01	86	40	I 02	102	276
D 01	127	180	I 03	102	0
D 02	100	49	I 04	150	430
E 02	100	0	I 05	150	110
F 01	38	29	K 04	95	95
F 02	39	52	K 07	60	150
F 03	39	0	K 10	50	98
F 04	100	43	K 32	40	1
G 01	120	260	K 33	40	0
H 01	92	530	K 35	46	10
H 02	92	240	K 36	46	38
H 03	92	99			

Total emission 3235 g/s

Equivalent height (weighted) calculated for the 17 June trajectory :

205 m (by Briggs' formula)
190 m (by the Stumke II formula)

located at Ruien, 30 km to the South-West from Ghent. An evaluation of the pollutant mass coming from more distant regions was not quantified at that time.

Measurements were performed by our mobile unit equipped with (a) a Cospec III for SO_2 vertical burden, (b) a Bendix 8302 monitor for SO_2 ground concentration and (c) an odometer connected to a wheel of the vehicle to measure the distance travelled. A desk computer HP 9825 installed on board sampled data from the sensors together with the time from a clock at preselected distance intervals of 50 (or 100) m and recorded them on a tape. In order to reduce the effect of time delay between Cospec and Bendix signals, the speed of the van was kept constant at 40 km/h. A detailed description of the van as well as of the experimental procedure is given elsewhere (Cerutti et al., 1982 ; De Groot et al., 1982).

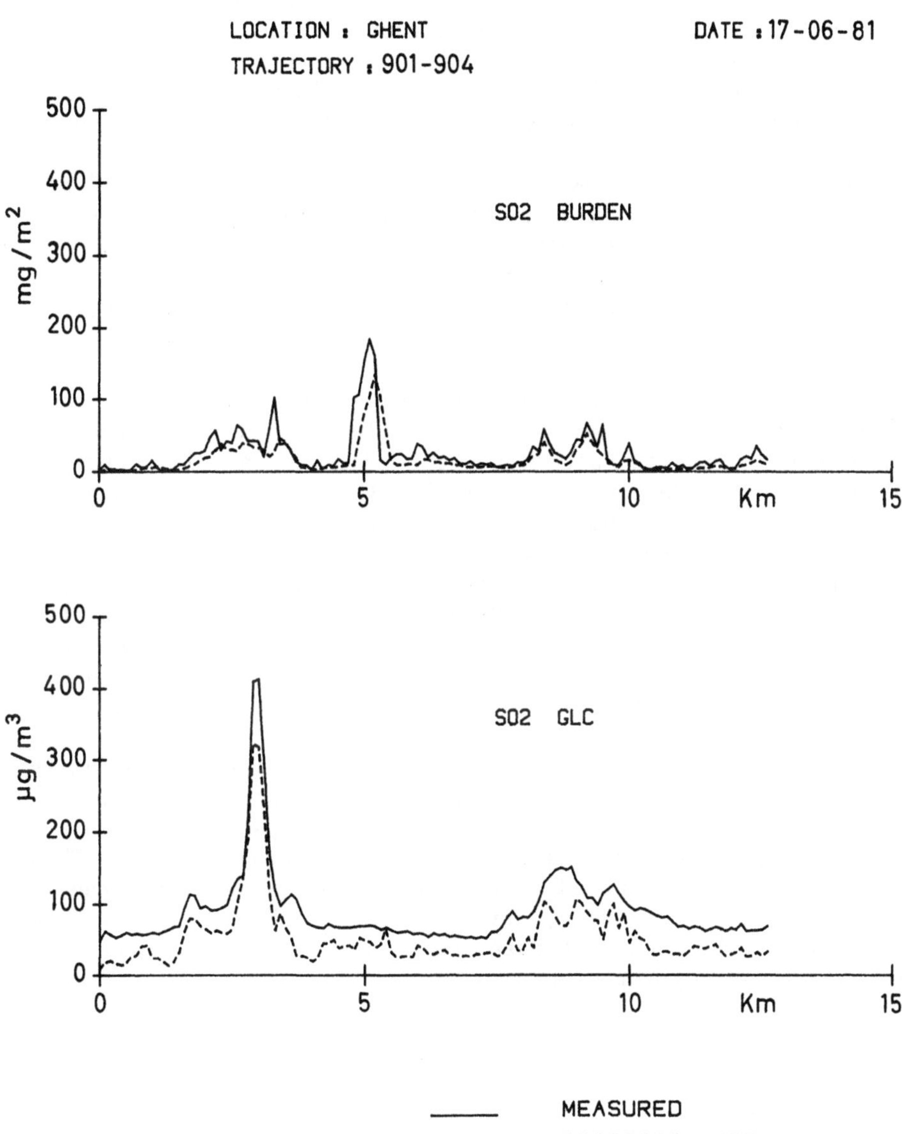

Fig. 2. Comparison of experimental vs calculated (by the proposed model) data for 17 June (time interval 10.13 - 15.20 LMT).

The burden data recorded on tape were processed afterwards according to the following procedure :

1) the raw data were plotted on a linear scale as a function of the distance travelled, then corrected for the baseline and for AGC fluctuations. These corrections are difficult to define and are a potential source of error ;
2) the individual profiles of burden distribution gathered consecutively along the same trajectory (6 up to 10 traverses) were averaged ;
3) the average profile was plotted on the map and projected perpendicular to the "average" wind direction (defined by the meteorologists) for that time interval ;
4) the projected data were integrated, then multiplied by the "average" wind speed and finally converted to mass flow, expressed in g/s.

The meteorological information coming from the several stations active during the Campaign have been collected and summarized by the team of the Université Catholique de Louvain. Generally, during the measurement time concerned (3-4 h), the wind remained nearly constant in direction while its speed changed considerably with the height. According to the meteorologists, the correct wind speed has been chosen from the vertical distribution curve in correspondance to the effective height of the "equivalent plume", defined by the Briggs or Stumke II formulae applied to the individual plumes and weighted over the source concerned. In our cases calculations by the Briggs and Stumke II formulae have led to close values (see Table 1).

As emphasized in the Final Report of the Campaign, the emissions from single sources deduced from experimental data agree well with the claimed emissions ; conversely large discrepancies (up to a factor 2) are observed for mass flow across long trajectories. These differences may be ascribed to various factors such as (i) the limited number of traverses, (ii) different methods used for data processing, (iii) uncorrect baseline assignment, (iv) inaccurate instrumental set-up, or others.

MODELLING APPROACH

Proposed model

The evaluation of the gas burden along the z-direction requires the knowledge of the pollutant concentration c(x,y,z). In order to obtain this knowledge the classical atmospheric diffusion equation

$$u.\frac{\partial c}{\partial x} = \frac{\partial}{\partial y}(K_y \frac{\partial c}{\partial y}) + \frac{\partial}{\partial z}(K_z \frac{\partial c}{\partial z}) - \lambda c \qquad (1)$$

must be solved with the following boundary conditions :

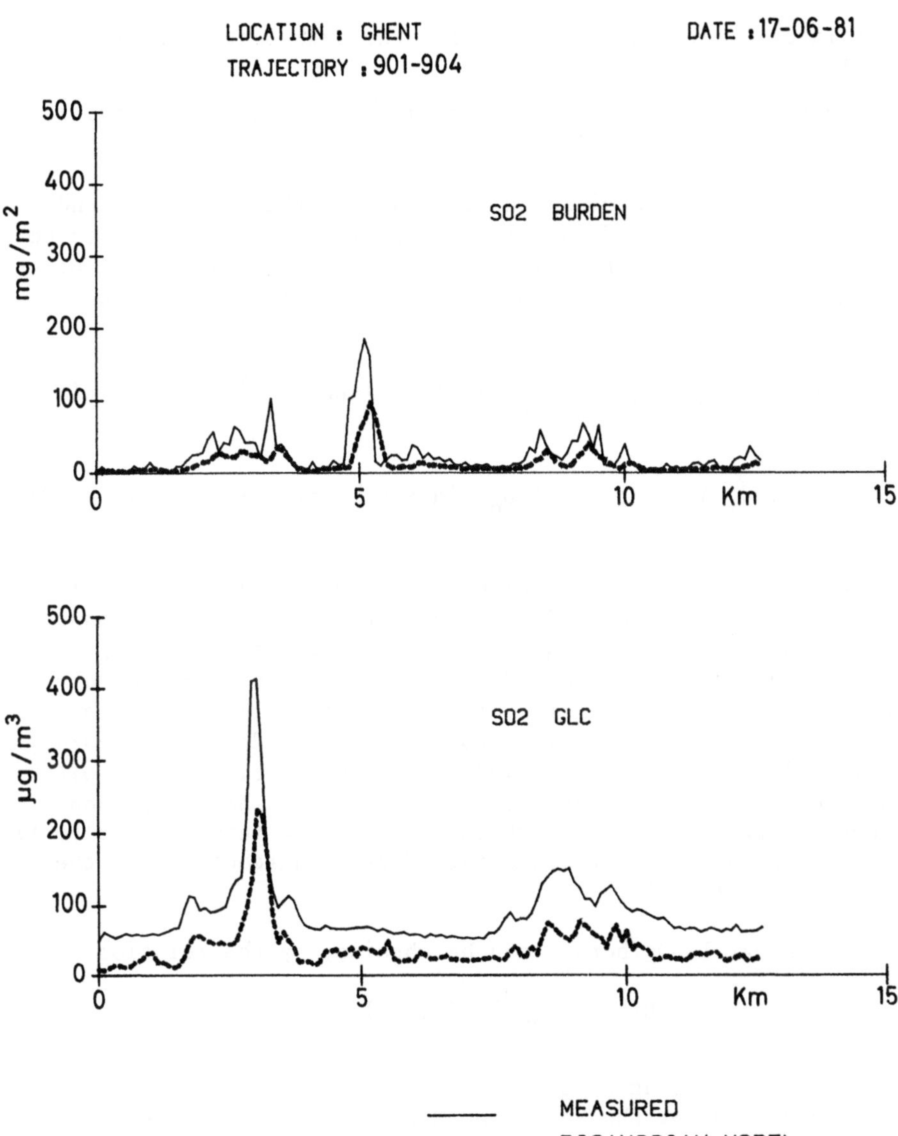

Fig. 3. Comparison of measurements vs calculations (by the bigaussian model) for 17 June.

a) the point-sources M_i with intensity Q_i and known localization x_i, y_i, z_i may be described by δ-functions of the type $Q_i/u.\delta(x-x_i).\delta(y-y_i).\delta(z-z_i)$;

b) independent of y, the solution must remain regular ;

c) a layer localized at $z=H_m$ limits the turbulent vertical mixing (for $z=H_m$, $\delta c/\delta z=0$) ;

d) the pollutant may be removed from a layer z_a at a deposition velocity v_d, i.e. $K_z(\delta c/\delta z) = v_d.c$;

where u is the average wind speed, K_y and K_z the horizontal and vertical coefficients of turbulent diffusion and λ the kinetic constant of a first order reaction including the removal of pollutant by rainout or washout. The solution of eq. (1) is as follows :

$$c(x,y,z) = \frac{1}{2\sqrt{\pi\beta}} \sum_{i=1}^{M} \frac{Q_i}{(x-x_i)^{1/2}} U(x-x_i) \exp\left\{-\left[\frac{(y-y_i)^2}{4\beta(x-x_i)} + \lambda(x-x_i)\right]\right\}$$

$$\cdot \sum_{n=1}^{\infty} A_n \{\cos \gamma_n z + B_n \sin \gamma_n z\} . \exp\left[-C_n(x-x_i)\right] \qquad (2)$$

where U is the Heaviside function (equal to 0 if $x < x_i$ or 1 if $x > x_i$), β is the ratio K_y/K_z, γ_n the eigenvalues of the transcendental equation tg $\gamma_n = N/\gamma_n$ ($n = 1, 2, \ldots$), and A_n, B_n, C_n are coefficients whose values are determined by the boundary conditions. A detailed description of the procedure leading to this formula is given elsewhere (Sandroni et al., 1982). The accuracy of this solution has been checked for a single point-source with a numerical solution of the complete equation. The comparison between the two solutions shows that only a few terms of the Fourier expansion are needed to reach a satisfactory agreement. So the use of eq. (2) is fully justified.

The integral SO_2 burden can be obtained by the relation :

$$B(x,y) = \int_0^{H_m} c(x,y,z)dz \qquad (3)$$

where H_m is the mixing height. It can be obtained for instance experimentally or by the Herlofson's diagram : once the value of H_m is known, the integration along the z-direction can be performed and the burden B can be analytically obtained. Such a procedure is possible only in particular cases (i.e. when the accurate mixing height is available) ; in that case the mass-conservation law is fully satisfied.

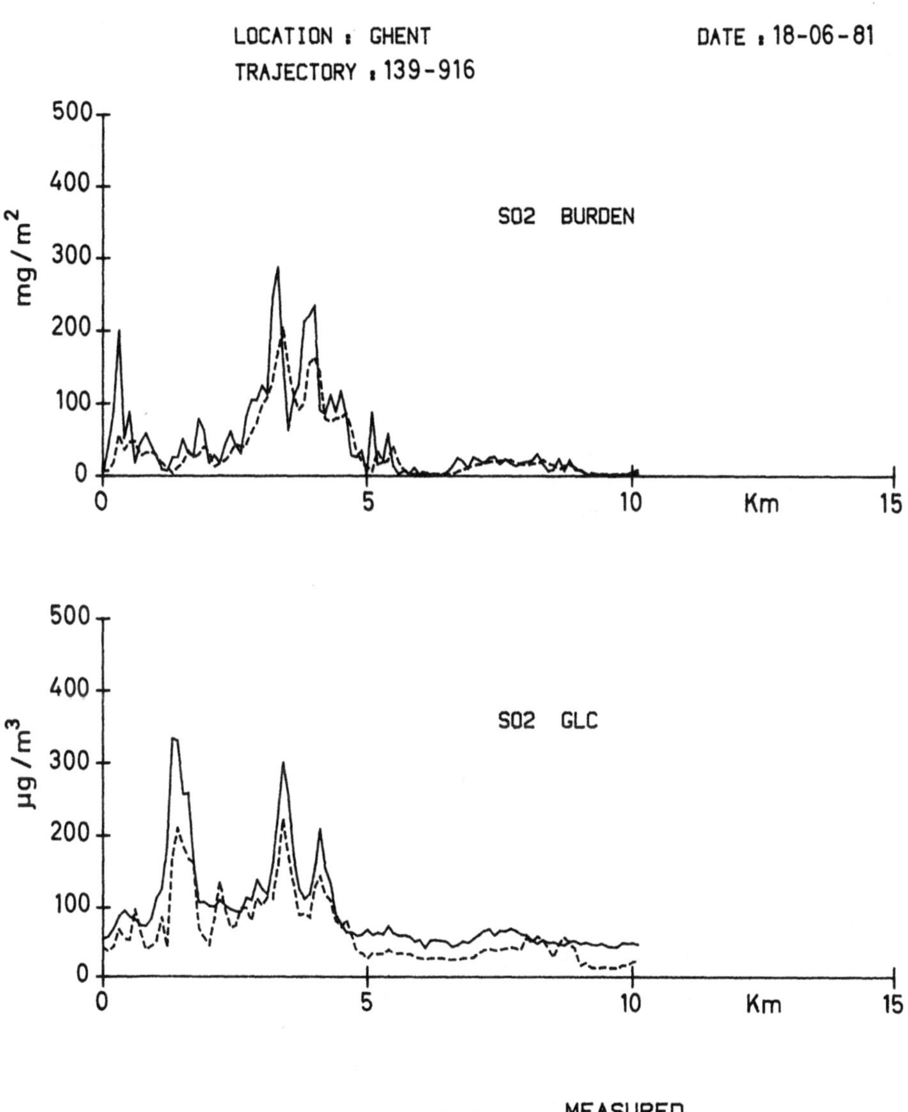

Fig. 4. Comparison between measurements and calculations by the proposed model for 18 June (time interval 13.10 - 15.27 LMT).

Bigaussian model

If we assume a gaussian cross-wind distribution and a perfect reflection at the ground, the solution of the advective-diffusion equation gives the following expression for the pollutant concentration :

$$C(x,y,z) = \frac{Q}{2\pi\bar{u}\sigma_y\sigma_z} \exp\left\{-\frac{y^2}{2\sigma_y^2}\right\}\left\{\exp\left[-\frac{(z-h)^2}{2\sigma_z^2}\right] + \exp\left[-\frac{(z+h)^2}{2\sigma_z^2}\right]\right\} \quad (4)$$

Q is the source strength, h is the effective plume height and $\bar{u}$ is the average wind speed defined by

$$\bar{u}(h) = \bar{u}(h_0) \,.\, (h/h_0)^p \quad (5)$$

where h_0 is a reference height at which the wind speed is measured, p is a numerical coefficient depending on the atmospheric stability class and σ_y and σ_z are the lateral and vertical standard deviations. They are defined by the expressions

$$\sigma_y = A_y \,.\, x^\alpha \qquad\qquad \sigma_z = A_z \,.\, x^\beta$$

in which A_y, A_z, α and β depend on the stability class and are reported in Table 2, together with p. The given stability classes refer to Bultynck-Malet (1972) classification.

Table 2.

Stability class	A_y	α	A_z	β	p
1	.235	.796	.311	.711	.53
2	.297	.796	.382	.711	.40
3	.418	.796	.520	.711	.33
4	.586	.796	.700	.711	.23
5	.826	.796	.950	.711	.16
6	.946	.796	1.321	.711	.10
7	1.043	.698	.819	.663	.33

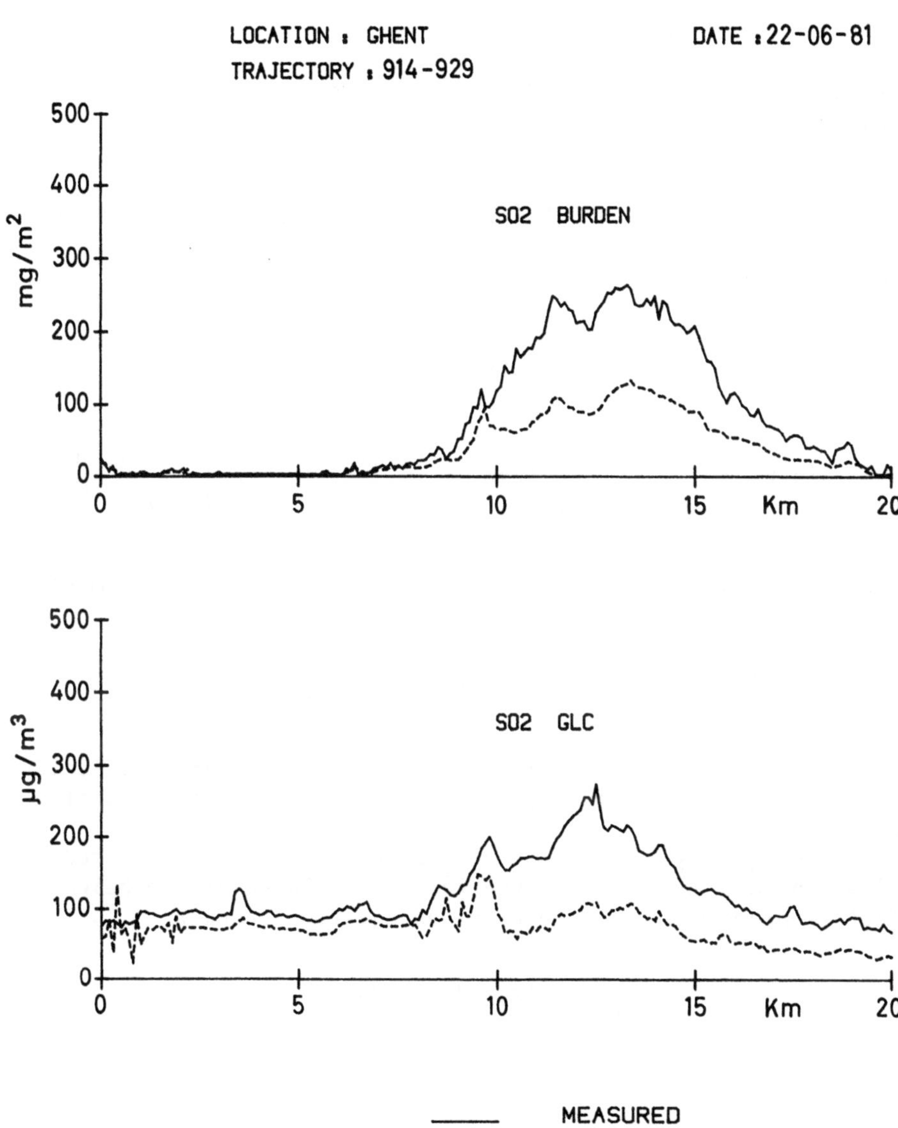

Fig. 5. Comparison between measurements and calculations (by the proposed model) for 22 June (time interval 10.20 - 14.17 LMT).

The gas burden B downwind a source characterized by an emission rate Q is defined by the equation

$$B(x,y,z_m) = \int_0^{z_m} C(x,y,z)\,dz \tag{6}$$

where z_m is the depth of the diffusion layer.
For the bigaussian model, one has

$$B(x,y,z_m) = Q.f_1(x,y).\int_0^{z_m} f_2(x,z).dz \tag{7}$$

where

$$f_1(x,y) = \frac{1}{\sqrt{2\pi}.\sigma_y(x)}\,.\,\exp\left\{-\frac{y^2}{2\sigma_y^2(x)}\right\}$$

$$f_2(x,z) = \frac{1}{\sqrt{2\pi}.\bar{u}\sigma_z(x)}\left\{\exp\left[-\frac{(z-h)^2}{2\sigma_z^2}\right] + \exp\left[-\frac{(z+h)^2}{2\sigma_z^2}\right]\right\}$$

Let us write the general conservation law for the pollutant mass

$$\int_0^{z_m} f_2(x,z).\bar{u}(z)dz + V_d\int_0^{x} f_2(s,0)ds + V_R\int_0^{x}\int_0^{z_m} f_2(s,z).ds.dz = 1 \tag{8}$$

where V_d is the deposition velocity and V_R is the removal velocity from the atmospheric medium by rainout or washout or a chemical process.
In our case $V_d = V_R \equiv 0$. If we also assume that the wind speed remains constant and equal to the value calculated at the height h (effective plume height), the previous equation becomes

$$\int_0^{z_m} f_2(x,z)dz = 1/\bar{u} \tag{9}$$

Consequently, eq. (7) is simplified as follows

$$B(x,y) = \frac{Q}{\bar{u}(h)}\,f_1(x,y) = \frac{Q}{\sqrt{2\pi}\sigma_y\bar{u}(h)}\,\exp\left\{-\frac{y^2}{2\sigma_y^2}\right\} \tag{10}$$

in which the dependence on z_m has been replaced by a dependence on h, defined as $h_g + \Delta h$. The plume rise Δh must be calculated by a suitable model. At present available models have been produced by Briggs, Stumke (II) and Berlyand. We emphasize that the evaluation of the gas burden is strictly connected with the model used for the calculation of Δh.

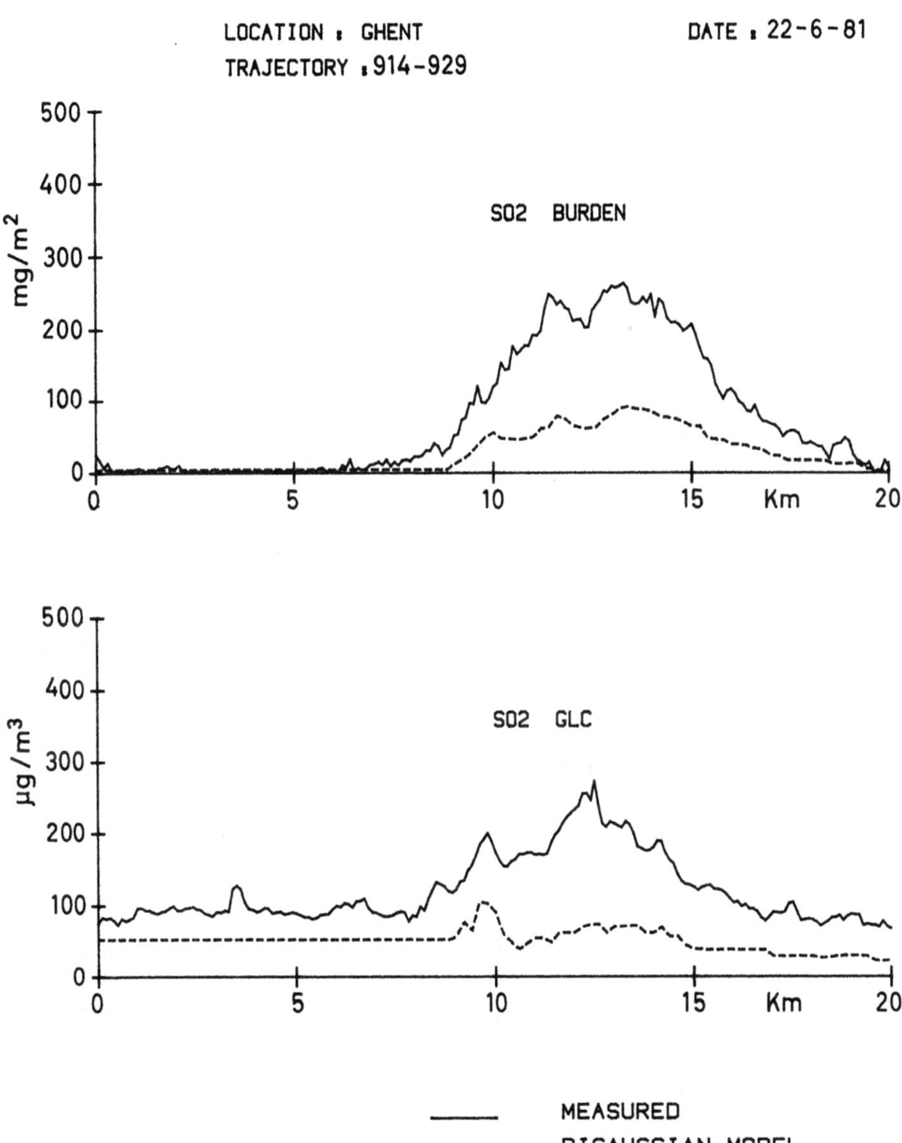

Fig. 6. Comparison of bigaussian vs measured data on 22 June (time interval 10.29 - 14.17 LMT).

The ground-level concentration is calculated from eq. (4) in which z = 0, i.e.

$$C(x,y,0) = \frac{Q}{\pi\bar{u}\sigma_y\sigma_z} \exp\left\{-\frac{y^2}{2\sigma_y^2}\right\} \exp\left\{-\frac{h^2}{2\sigma_z^2}\right\} \quad (11)$$

The ratio of ground-level concentration to burden is given by

$$\frac{C}{B} = \sqrt{\frac{2}{\pi}}\,\frac{1}{\sigma_z} \exp\left\{-\frac{h^2}{2\sigma_z^2}\right\} \quad (12)$$

which is only a function of σ_z and h.
Consequently, if the meteorological conditions (wind speed, stability class and therefore σ_z) are given and a proper model for calculating Δh is defined, the ratio C/B is only a function of σ_z, and h. Once the gas burden is calculated, the ground concentration can be deduced by eq. (12).

RESULTS AND DISCUSSION

Our analysis is focused on three important days, 17, 18 and 22 June ; for these cases we compare burden and ground-concentration computed both by the proposed and by the bigaussian model with the measured data. As measured data we have taken the data averaged over the consecutive surveys made along the same trajectory. For both models we used as input the claimed emissions (which might be lower than real emissions) and the meteorological parameters (wind speed and direction, stability class) as given by the meteorological team of the Université Catholique de Louvain.

On 17 June the input data were : trajectory 901-904 (see Fig. 1), wind direction 325°, wind speed 9.2 m/s, stability class 3 (according to the Bultynck-Malet classification equivalent to D for Pasquill-Turner). The data calculated by our model are compared to the measured data in Fig. 2 ; a similar comparison with the bigaussian model is shown in Fig. 3. As shown in Figs. 2 and 3, both models reproduce a shape for burden and ground-concentration distributions which is similar to that of the measured data.

A similar behaviour is observed for the other cases. On the 18th, the input data were : trajectory 139-916, wind direction 340°, wind speed 7.3 m/s, atmospheric stability class 4 (according to Bultynck-Malet ; C for Pasquill-Turner classification).

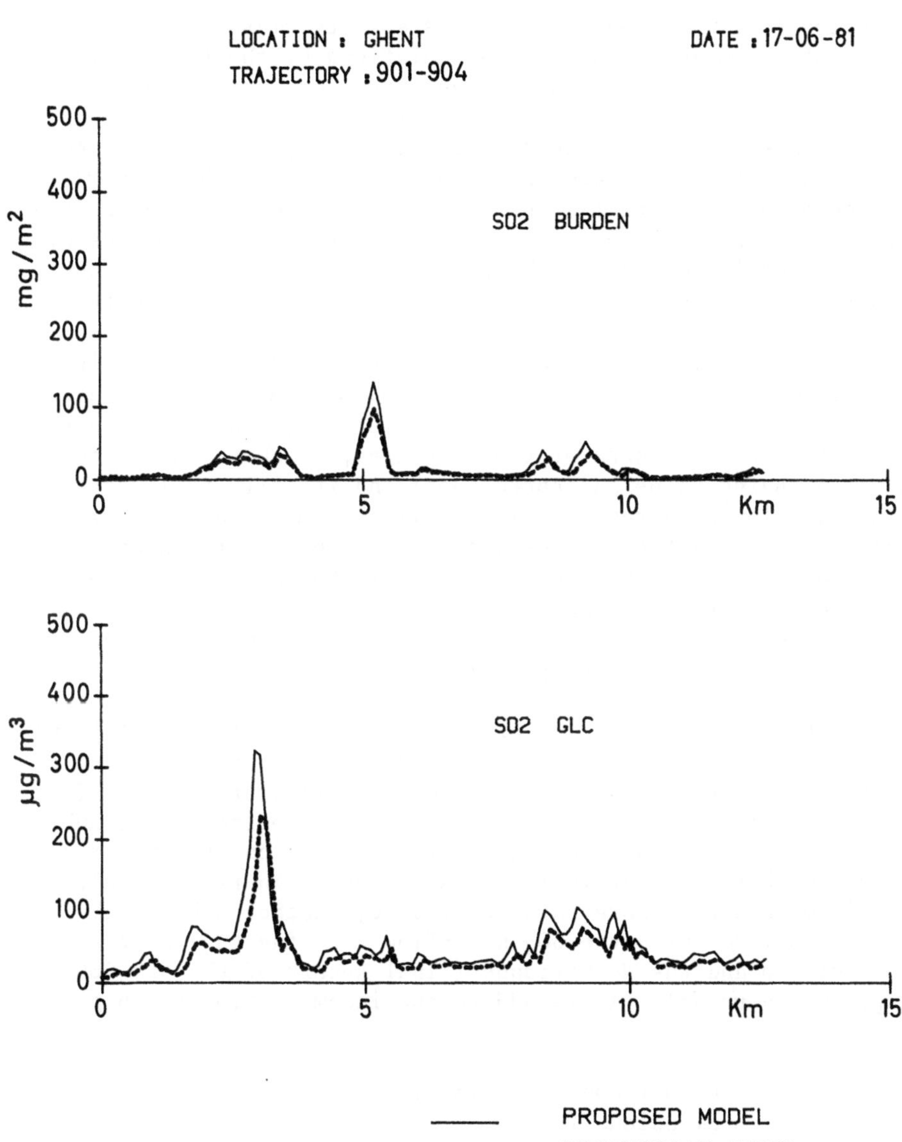

Fig. 7. Comparison between the proposed model and the bigaussian model for 17 June.

The comparison between measured and calculated data by the proposed model is shown in Fig. 4.

On 22 June the input data were : trajectory 914-931-929, wind direction 18°, wind speed 5.5 m/s, stability class 4. The comparison between measurements and calculations by the two models are shown in Figs. 5 and 6. In all the cases considered, the shape of the data distribution is almost maintained, the calculated data being remarkably lower than the measured data, particularly on 22. These differences are shown in Table 3 in which the corresponding mass-flows are listed. By comparing the two models (see for instance Fig. 7), one can see the underestimation of the bigaussian model compared to the proposed model ; the second one fits the claimed emissions used as input data (Table 3) better.

The large overestimation of measured burdens in comparison to those calculated is astonishing if we remember the good agreement observed for the Milano area (Sandroni et al., 1982). That area is located at the centre of the Po valley, which is encircled by mountain chains (Alps and Appennines) and its climate is characterized by a low wind regime. The available emission inventory includes not only the sources located inside the area under consideration but also those located upwind in the surrounding regions. The transboundary pollutant transport is negligible for the Po valley because of the barrier of the Alps ; consequently the mass-flow is defined only by regional sources.

In the Ghent industrial area, the situation is basically different : a strong wind regime and no orographic barrier allow a potential pollutant transport from distant industrial areas (Antwerp in Belgium, the Rijnmond area in the Netherlands, the Ruhr area in Germany). As a consequence, it is logical to expect a higher SO_2 burden , particularly on 22 June. In fact, simultaneously with our measurements on 22 June, another mobile laboratory (IHE) measured the SO_2 mass-flow entering the Ghent area from Antwerp and the Netherlands (Final Report of the 5th CEC Campaign, in press). If we add to the incoming flow (1726 g/s),

Table 3. Comparison of claimed, measured and calculated emissions (g/s)

	17 June	18 June	22 June
Claimed (inventory)	1853	2569	3194
Measured	2566	3050	6283
Proposed model	1719	2344	3128
Bigaussian model	1225	1817	2228

the emission of the Ghent urban area (50 g/s), the correct areal emission becomes 4507 g/s. The relative error (41 %) is comparable with the experimental accuracy : for a 20 km trajectory and a 5.5 m/s wind speed, an accuracy of 10 ppm metre corresponds to a flow of 1100 g/s.

On 17 and 18 June, the errors were considerably lower.

The hypothesis of an additional mass transported from distant regions is also confirmed by the higher ground concentration, for which the measurement technique is simpler than Cospec and does not suffer from uncontrolled factors.

The difference between the two models may be explained by investigating their dependence on the local meteorological conditions. The proposed model is influenced via a function of x and β : the ratio between horizontal and vertical turbulent diffusion parameters K_y/K_z, which are physical parameters mathematically defined. In the bigaussian model, the dependence on meteorological conditions is given only via the dispersion parameter σ_y, which can be determined only by local observations. The different dependence may explain, in our opinion, the better fit of the proposed model.

REFERENCES

Bultynck, H., Malet, L., 1982. Evaluation of atmospheric diluition factors for effluents diffused from an elevated continous point source, Tellus 24, 455.

Cerutti, C., De Groot, M., Sandroni, S., 1982. Sulphur dioxide mass-flow in the Ghent industrial area measured at the 5th CEC Campaign of Remote Sensing, Report EUR 7788.

De Groot, M., Cerutti, C., Sandroni, S., 1982. A mobile unit for mapping of atmospheric pollution, Report EUR 8260.

Final Report of the 5th CEC Campaign on Remote Sensing of Air Pollution at Ghent, June 1981, in press.

Hoff, R.M., Millan, M.M., 1981. Recent SO_2 mass flux measurements using Cospec, J. Air Poll. Control Ass., 31, 381

Millan, M.M., 1983. Effects of atmospheric interaction in correlation spectrometry. in : Optical Remote Sensing of Air Pollution, edited by P. Camagni and S. Sandroni, Elsevier.

Onderdelinden, D., 1983. Optimisation of monitoring networks by remote-sensing techniques. in : Optical Remote Sensing of Air Pollution, edited by P. Camagni and S. Sandroni, Elsevier.

Sandroni, S., De Groot, M., Clerici, G., Borghi, S., Santomauro, L., 1982. Simulation of sulphur dioxide mass-flow over Milano area, Proc. of the 13th ITM on Air Pollution Modelling and its Application, Ile des Embiez, 1982, Plenum Publish. Co.

Van Egmond, N.D., Kesseboom, H., Van Jaarsveld, J.A., 1982. Correlation spectrometry as a tool for Mesoscale modelling. Proc. of the 13th ITM on Air Pollution Modelling and its Application, Ile des Embiez, 1982, Plenum Publish. Co.

DISCUSSION

R. BERKOWICZ Is the "proposed model" not equivalent to a Gaussian model with σ_z and σ_y proportional to $x^{1/2}$?

S. SANDRONI In the general case the two models are not equivalent each other. They may be equivalent only if you substitute K_z and K_y by σ_z and σ_y (see Benarie-Urban Air Pollution Modelling - chapt. 7).

G. RESELE I didn't understand, why the two models do not yield the same values for the total flow balance (Table 3 in the Proceedings).

S. SANDRONI The mass-flow evaluations have been trusted to selected trajectories defined by end-points (for instance 929 to 914 on June 22, in Fig 1). All what is flown out of these extremes, it is not considered. For this reason, calculated mass-flows are always lower than the claimed emissions used as input data.

A NONLINEAR PROGRAMMING SEARCH TECHNIQUE TO LOCATE POSITION AND MAGNITUDE OF THE MAXIMUM AIR QUALITY IMPACT FROM POINT SOURCES

R. Joel Barnett, Glen E. Johnson, and Karl B. Schnelle, Jr.

Mechanical and Materials Engineering Department and
Chemical Engineering Department
Vanderbilt University
Nashville, Tennessee 37235

INTRODUCTION

Environmental regulations in the USA require that the effect on ambient air quality of new air pollutant emission sources be determined prior to construction of these sources. Air quality modeling is the means by which this evaluation is performed. In the process of this evaluation, the programmer carries out many similar operations entering data for various receptor locations. The objective is to locate the region of maximum concentration by constructing ever finer receptor grids narrowing down the region with each succeeding computer run. This process can be time consuming for the computer operator and may also involve a large amount of computer time. In an effort to reduce the time required by the operator and the computer, a study was conducted using modern optimization techniques that have become routine for engineering studies of this kind. This paper describes the technique of finding the location and magnitude of maximum air quality impact from point emission sources.

For this study, the USEPA computer model known as PTMTP, a Gaussian type model written by Turner (1), was used as the objective function for a conjugate gradient type nonlinear programming algorithm (2). Four case studies of different combinations of source orientation, number of sources, and meteorological data were investigated. Averaging time for the calculations was 24 hours, however, other times are feasible within the limits of the model chosen as the objective function. In all cases the optimization program was found to reliably determine the maximum impact point. The original Gaussian model with a conventional grid search strategy was used as a reference to confirm these results.

GENERAL ANALYTICAL METHOD

In performing air quality modeling for regulatory purposes in the USA, the analysis currently consists of two phases. First, a screening process is used in which, for a given source or sources, the appropriate meteorological data are selected and initial calculations determine the areas of high concentration which are of potential interest for a more detailed analysis. The screening analysis generally is performed using one or more years of meteorological data with the object of finding those shorter time periods (24-hr, 8-hr, 3-hr, etc.) which cause the highest levels of ambient concentration for the pollutant under analysis. Either a simplified model or a more sophisticated model using fewer receptors and a wider grid spacing to reduce computer time is used as the screening model. Experience has shown that although annual standards exist for various pollutants, it is the short term peak levels which are most likely to result in violation of ambient standards or allowable expansion increments. Therefore, the crucial portion of an air quality evaluation and the object of this study is to determine the location and magnitude of these short term concentrations.

Following the screening analysis, the determination of maximum air quality impact using more sophisticated Gaussian type models has generally utilized a sequential methodology which requires that the diffusion model be executed by the computer a number of times. Prior to the first execution and between subsequent executions a receptor grid containing a number of points (x, y coordinators) is input to the model. With each execution of the model, the location of the maximum impact point and the ambient concentration of that point is more accurately defined by using more closely spaced receptor grids. At some arbitrary degree of resolution (generally 0.10 km) the highest concentration present in that grid is considered to be the maximum concentration for regulatory purposes.

In contrast to this approach, the nonlinear programming technique requires only the input of the coordinates of a starting point. This starting point would generally, but not necessarily, be determined from the coarse receptor grid which is a part of the meteorological selection portion of the analysis. The maximum receptor from the screening analysis provides a starting point which should be in the vicinity of the actual maximum.

From this starting point the algorithm progresses without further input until the maximum concentration and its location are determined. The time and effort required for the input of large numbers of receptors in the conventional grid search techniques are thus eliminated.

CHARACTERISTICS OF THE CONJUGATE GRADIENT OPTIMIZATION ALGORITHM (3)

For a given set of meteorological data, source locations, and source characteristics, the predicted concentration, C, is a function of downwind coordinates (x, y). In this analysis flat terrain is assumed, so there is no variation in the vertical or z-coordinate (z = 0). The algorithm is written as a minimization procedure, therefore, the object is to find the minimum value of the quantity 1/C (equivalent to the maximum value of C) and its location. An alternative and equivalent formulation would be to determine the minimum value of the negative of the concentration C.

Functionally, the air quality model is a subroutine within the overall optimization program. Information passed from the conjugate gradient portion of the program to the air quality model consists only of the receptor coordinates for which predicted value of ambient concentration is needed. This predicted concentration is the only value passed back to the conjugate gradient routines. Therefore, in principle, most current air quality models could be utilized in an optimization program of this type.

At the initial set of coordinates provided as the starting point of the analysis, the gradient of 1/C is calculated by finite different methods (see Appendix). The negative of this gradient can be interpreted as the direction of steepest descent (direction of rate of greatest decrease in value) for the function. A univariate line search in the direction of the negative gradient is performed and the minimum along this axis is found.

The algorithm then computes a new descent direction at that point (line search minimum) composed of a linear combination of the gradient at this new point and the previous descent direction (3). This process of finding new descent directions and new minima is repeated until the gradient calculated is less than some arbitrarily small value. Reduction of the gradient to a sufficiently small value indicates that a local minimum for the function has been found. In this analysis, repeated tests of the algorithm were performed to find the value of the gradient (0.0015) below which the algorithm must reach to insure a successful determination of the maximum.

Since under some combinations of source location and meteorology it is possible for there to be local minima of 1/C (local maxima of C), it would be possible for the algorithm to converge to one of these extremes which might not be the overall or global maximum. This possibility can best be eliminated by careful examination of the output from the screening analysis. By observation of the higher concentrations from the receptor grid, it can be determined whether the possibility of multiple maxima with approximately equal magnitude exists.

If these local maxima exist, the optimization algorithm can be run with a starting point in the general vicinity of each of the candidate locations. The results of these runs can be compared to locate the global maximum. This procedure does not detract from the utility of the optimization technique since in the conventional grid search technique a full analysis would be required at each of the regions which might contain the maximum concentration.

The final procedures of the algorithm are to take the reciprocal of the minimized quantity, 1/C, giving the final maximum concentration, C, and to print this value and its coordinates. The algorithm also provides the option of printing the results at each of the line search minima referred to above, so that the trajectory of the algorithm toward the maximum can be observed.

CASE STUDIES

Four cases were chosen for study to illustrate and test the feasibility of using the conjugate gradient technique to determine the maximum concentration and location in an air quality impact analysis. Table 1 summarizes the characteristics of these case studies.

TABLE 1

Characteristics of Case Studies

	SOURCE	METEOROLOGY
Case I	single source	24 typical hours of data
Case II	single source	from screening analysis
Case III	two sources 1 km separation along predominant wind vector	from screening analysis
Case IV	three sources 0.7 km separation approx. 45 ° to predominant wind vector	from screening analysis

The first case chosen for analysis was a single source for 24 hours of typical meteorological input. Source characteristics and meteorology are shown in Tables 2 and 3, respectively.

TABLE 2

Source Characteristics

Emission rate	Height	Temperature	Emission Velocity	Diameter
50 gm/s	100 m	400 K	20 m/s	2.0 m

TABLE 3

Meteorological Data for Case I

Hour	Wind Direction Degrees from N.	Wind Speed m/s	Turner Stability Class	Mixing Height m	Temperature K
1.	260.	1.5	4	1000.	295.0
2.	261.	1.6	4	1010.	290.0
3.	262.	2.0	4	1200.	292.0
4.	263.	2.2	4	1100.	288.0
5.	264.	2.5	4	1050.	295.0
6.	265.	3.1	4	1300.	297.0
7.	266.	4.1	4	1200.	297.0
8.	277.	3.5	4	1300.	283.0
9.	268.	3.2	4	1200.	289.0
10.	269.	3.0	4	1400.	296.0
11.	270.	3.1	4	1300.	295.0
12.	271.	2.5	4	1400.	292.0
13.	272.	3.5	4	1200.	290.0
14.	273.	4.2	4	1300.	288.0
15.	274.	4.1	4	1250.	285.0
16.	275.	3.0	4	1000.	277.0
17.	276.	3.2	4	1000.	278.0
18.	277.	2.0	4	950.	275.0
19.	278.	2.5	4	960.	280.0
20.	279.	2.5	4	945.	275.0
21.	268.	2.2	4	950.	270.0
22.	269.	2.3	4	945.	272.0
23.	270.	2.3	4	940.	273.0
24.	271.	2.4	4	935.	271.0

The source location, initial point, and trajectory computed by the algorithm to termination are in Figure 1. The starting point in this case was chosen arbitrarily and did not come from a screening analysis. Also shown in Figure 1 is a portion of the (0.5 km) receptor grid used in the conventional search strategy to verify that the maximum had been found. A larger scale map of the point of termination and the region around it is shown in Figure 2. Comparison of the value as computed by the algorithm (7.77 $\mu g/m^3$) with the adjacent concentrations calculated in the final step (0.1 km spacing) of the grid search strategy shows that the true maximum value has been determined.

Other starting points for the algorithm were used and are indicated on Figure 1. Although trajectories for these cases are not shown, they also resulted in successful determination of the maximum air quality impact (termination points were same as for start point 1).

The three other cases investigated used meteorology determined by a screening analysis. This analysis was conducted with the USEPA computer model known as CRSTER (4). In these case studies different numbers of sources and source orientations and locations were used as described in Table 1. In each of these three cases, the source emission parameters for each source were the same as CASE I, and are presented in Table 2. As a result of the screening analysis the worst meteorological day was selected. For that day, Table 4 lists the meteorological data. The predominant wind vector selected by the CRSTER (4) model was 160° and the maximum concentration was found to be approximately 6 km from the source(s). Therefore, starting points in this general vicinity were chosen. However, in order to confirm the reliability of the algorithm in finding the maximum, a number of starting points were chosen at some distance from the suspected maximum.

The source locations, starting points, and termination points for these cases are given in Figures 3, 4 and 5. The region of the termination points are shown in Figures 6, 7 and 8. The 0.1 km receptor grid concentrations as determined by the conventional grid search are shown to confirm that the algorithm has successfully determined the maximum concentration and its location.

CONCLUSION

The four case studies conducted demonstrated the feasibility of using the conjugate gradient type of nonlinear programming algorithm for determining maximum air quality impact from point sources. There should be no difficulty in extending the use of this technique to area sources or to other Gaussian type of dispersion model programs as long as the dispersion program is structured to pass the required data on to the optimization program.

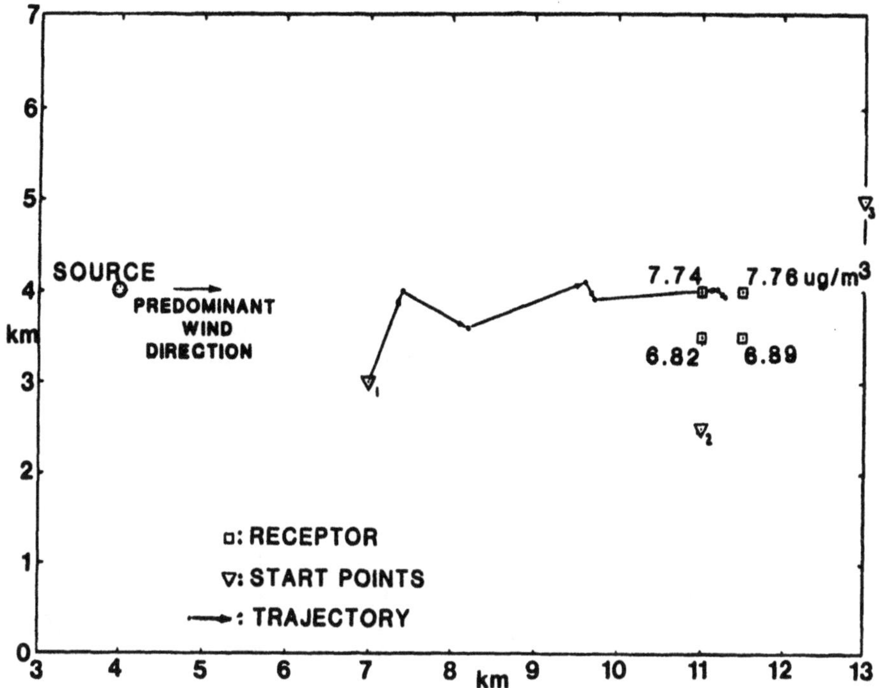

Figure 1. Case I, Trajectory of Conjugate Gradient Algorithm to Termination at Maximum Concentration

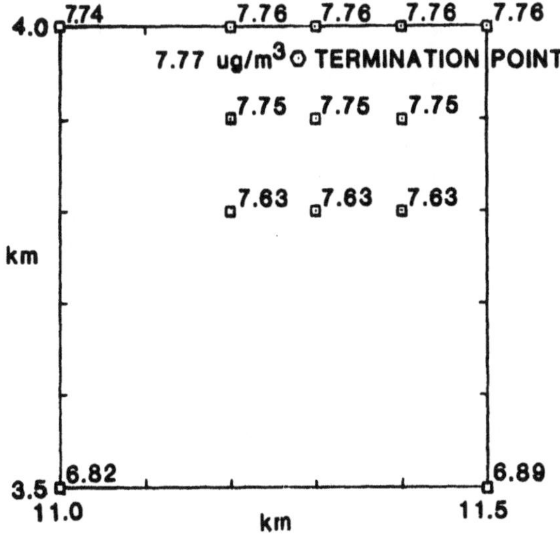

Figure 2. Case I, Detail of Termination Region of Receptor Grid for Conventional Search Strategy

TABLE 4

Worst Case Meteorological Data* Selected by the CRSTER Model for Case II, Case III and Case IV

Hour	Wind Direction Degrees from N.	Wind Speed m/s	Turner Stability Class	Rural Mixing Height m	Temperature K
1.	342.	7.2	4	937.	273.0
2.	340.	5.7	4	915.	273.0
3.	345.	6.2	4	893.	273.0
4.	337.	5.7	4	871.	273.0
5.	342.	5.1	4	849.	273.0
6.	336.	5.7	4	827.	273.0
7.	340.	6.2	4	805.	274.0
8.	338.	4.7	4	783.	274.0
9.	341.	3.6	4	760.	274.0
10.	335.	4.1	4	738.	274.0
11.	345.	4.1	4	716.	275.0
12.	336.	5.1	4	694.	275.0
13.	340.	5.1	4	672.	275.0
14.	341.	4.6	4	650.	275.0
15.	315.	5.1	4	650.	275.0
16.	319.	4.1	4	650.	275.0
17.	351.	3.6	4	654.	275.0
18.	38.	2.1	4	663.	275.0
19.	340.	2.6	4	673.	275.0
20.	26.	2.1	4	682.	275.0
21.	357.	2.1	4	691.	275.0
22.	318.	2.6	4	701.	275.0
23.	309.	2.6	4	710.	275.0
24.	339.	2.6	4	720.	275.0

* from Nashville, Tennessee Hourly Annual Data, 1964.

Central processing unit time (CPU) necessary for the conventional grid search strategy was approximately five seconds for the single source cases and 15 seconds for the multiple case. This CPU time is the total time necessary for all iterations of the procedure. Equivalent time for the complete analysis by the optimization algorithm was between 15 and 90 seconds. However, some of the longer times required by the optimization algorithm were caused in part by selection of less-than-optimum starting points. For equivalent levels of complexity of analysis, the optimization algorithm will take approximately three times as much CPU time as the conventional methodology.

Savings in personnel time cannot be easily quantified. The conventional grid search technique requires the entry of receptor location data for an extensive grid network for several computer runs. The time required for selection of an appropriate grid network and the entry of the data are dependent upon the skill and experience of the computer operator. The entry of a pair of initial coordinates required by the conjugate gradient optimization algorithm requires substantially less time. If there is a considerable computer turnaround time, the conventional grid search technique could require even a greater amount of personnel time. Total computer time and CPU time differences for the rather simple case studies reported in this paper were not significant. Additional study with more complex dispersion models is required to better determine the possible time savings.

APPENDIX

A. Nonlinear Programming

The general nonlinear programming problem can be formally stated as follows :

Minimize: $f\left[x\right] \quad x \in X^{n}$

$\left[x\right] \rightarrow$ a vector. $\quad E^{n} \rightarrow$ n-dimensional Euclidean space

subject to m linear and/or nonlinear equality constraints

$$h_j\left[x\right] = 0 \quad j = 1, \ldots, m$$

and (p-m) linear and/or nonlinear inequality constraints

$$g_j\left[x\right] \geq 0 \quad j = m+1, \ldots, p$$

The objective function for the analyses performed in this study was the EPA air quality model PTMTP. No equality or inequality constraints were placed on the objective function and, the actual quantity minimized was 1/C, the reciprocal of ambient concentration,C.

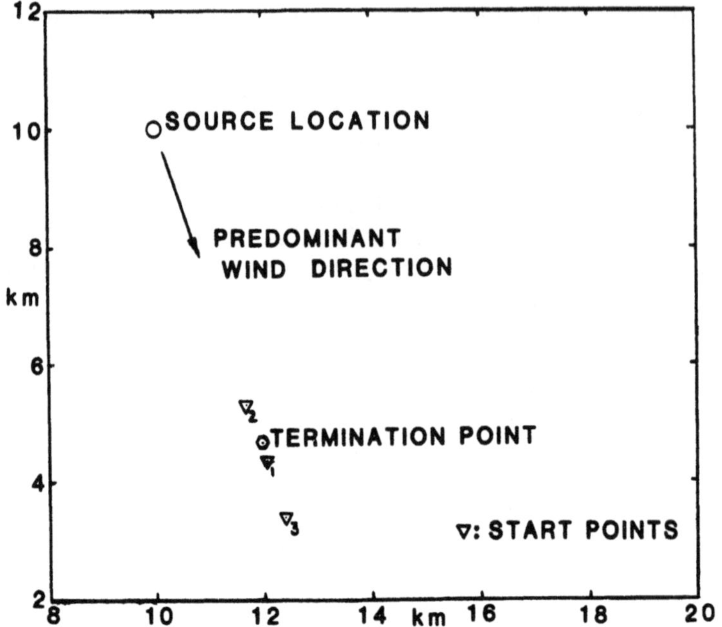

Figure 3. Case II, Source Location and Termination Point

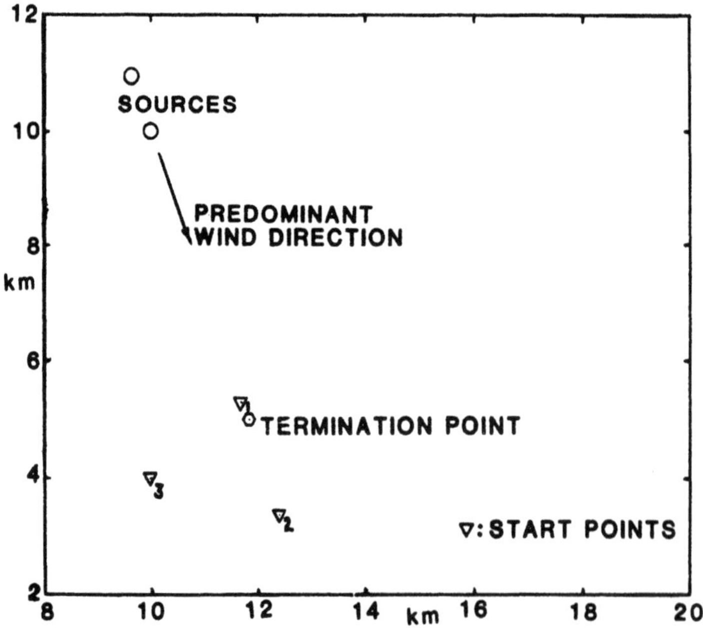

Figure 4. Case III, Source Location and Termination Point

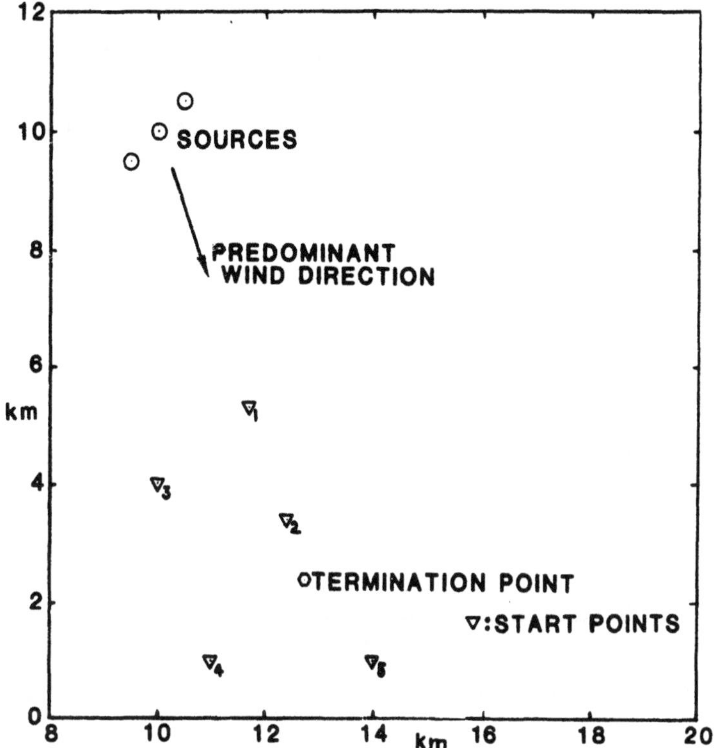

Figure 5. Case IV, Source Location and Termination Point

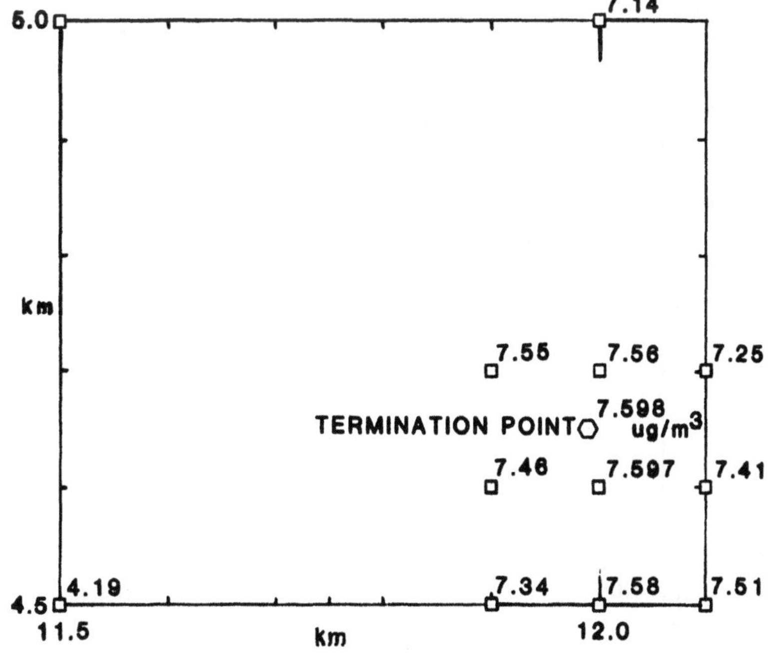

Figure 6. Case II, Detail of Termination Region

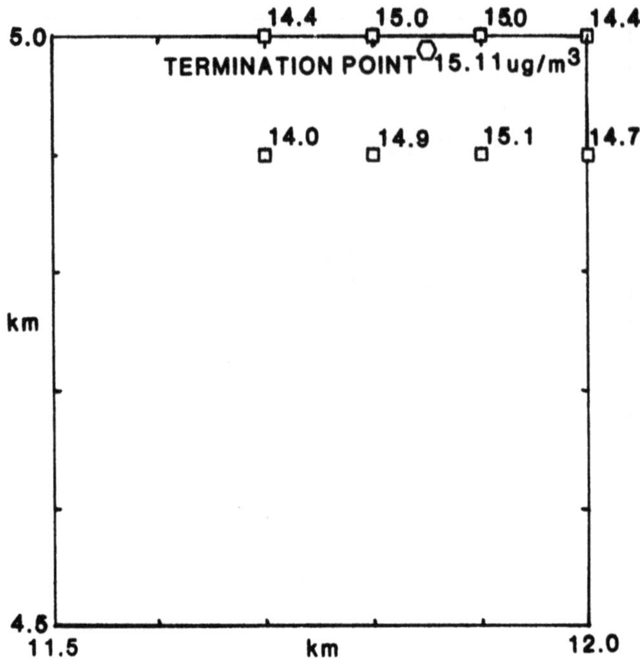

Figure 7. Case III, Detail of Termination Region

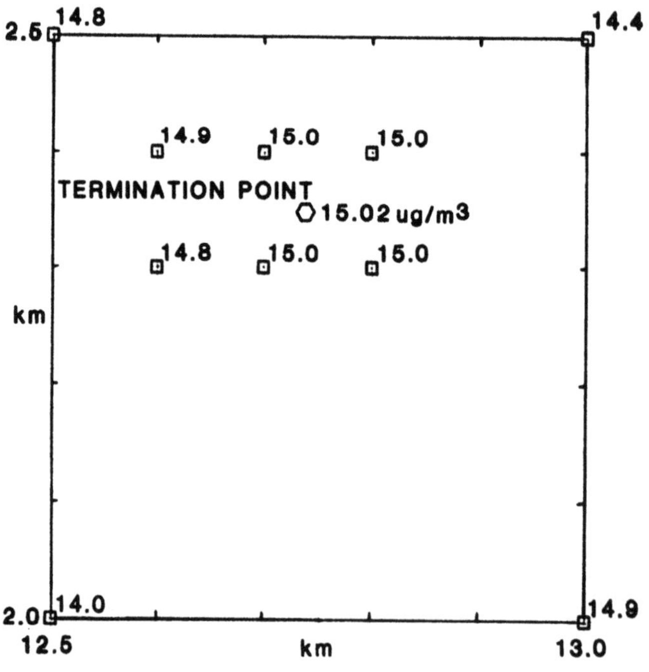

Figure 8. Case IV, Detail of Termination Region

B. Calculation of Gradient and Descent Direction

For the cases considered in this analysis the ambient concentration C is a function of receptor coordinates x and y. The gradient for the function C = f(x, y) is then

$$\nabla C = \begin{bmatrix} \frac{\partial C}{\partial x} \\ \\ \frac{\partial C}{\partial y} \end{bmatrix}$$

The partial derivatives are evaluated in the program by calculating $\frac{\Delta C_x}{\Delta^x}$ and $\frac{\Delta C_y}{\Delta^y}$, at particular values of x and y

$$\Delta C_x = C\,(x + \varepsilon) - C(x) \quad \text{y held constant}$$

$$\Delta C_y = C\,(y + \varepsilon) - C(y) \quad \text{x held constant}$$

$$\Delta_x = \Delta_y = \varepsilon = .000001$$

This negative of this gradient (the direction of steepest descent) is divided by its magnitude to determine a unit vector giving the direction for the subsequent line search.

$$|\nabla C| = \sqrt{\left(\frac{\Delta C_x}{\Delta x}\right)^2 + \left(\frac{\Delta C_y}{\Delta y}\right)^2}$$

$$\text{direction of unit vector} = \frac{-\nabla C}{|\nabla C|}$$

REFERENCES

1. D. B. Turner, PTMTP (computer program), U. S. Environmental Protection Agency, Research Triangle Park, NC

2. M. A. Townsend and G. E. Johnson, In Favor of Conjugate Directions, J. Franklin Inst., v. 306, No. 5 (1978).

3. D. M. Himmelblau, Applied Nonlinear Programming, McGraw-Hill New York (1972).

4. Unknown, CRSTER (computer program), U. S. Environmental Protection Agency, Research Triangle Park, NC.

DISCUSSION

R. STEENKIST What is the accuracy in the concentration and what in the determination of the point of max. concentration ?

K. SCHNELLE All of the current USEPA approved diffusion algorithms employ the bi-Gaussian model for dispersion and the Briggs plume rise model. The accuracy of the calculations made by any model from UNAMAP, therefore are no better than any other Gaussian model. The American Meteorological Society has reviewed these errors and the reader is referred to "Accuracy of Dispersion Models," Bulletin American Meteorological Society vol. 59, No.8, August 1978, p. 1025. My interpretation of the errors to be expected is no different than what many authors have reported in this meeting. It should be remembered that the usage of those models suggested in this paper is as a decision making tool for pollution control and not for research.

R. YAMARTINO With "local" optimization techniques, such as the conjugate gradient method, you can locate a local maximum but can't be sure that it represents the "global" maximum (i.e. there could be multiple maxima). What do you do to ensure that you have found the "global maximum.

K. SCHNELLE A screening analysis by a simplified model is applied first. This model should locate areas of possible maxima. The conjugate gradient technique would then be applied to all these areas of maximum concentration as determined by the screening analysis. To quote M. Yamartino, the conjugate gradient technique is applied as a fine tuning optimization method. It is hoped that the man hours required for application of an air quality model in a PSD analysis would thereby be greatly reduced.

AN OPERATIONAL AIR POLLUTION MODEL

R. Berkowicz, J. H. Baerentsen, A. B. Jensen*
J. S. Markvorsen**, L. B. Nielsen, H. R. Olsen, and
L. P. Prahm

Air Pollution Laboratory
Danish National Agency of Environmental
Protection
Roskilde, Denmark

INTRODUCTION

Control of present air pollution, as well as evaluation of future air quality, requires the ability to predict the level of air pollution generated by present and planned sources of emission. Mathematical air pollution models are the most common tools for this purpose. Such models have now been applied for several decades, and within this period of time, much effort has been spent on making these models more reliable. In spite of our growing understanding of the mechanisms governing the dispersion of pollutants in the atmosphere, the models are still to a high degree based on empirical methods.

Gaussian models have been the most frequently used. Here, we describe such a Gaussian model developed at the Danish Air Pollution Laboratory for the purpose of evaluation of air quality influenced by emissions from tall industrial stacks. The dispersion part of the model is similar to that reported by Weil and Brower (1982), but new operational methods for determination of the stability parameters and the mixing height have been developed to satisfy more complex conditions. The model is designed to calculate an hourly time series of pollutant concentrations from a point source.

---

*On leave from Danish Boiler Owners Association.
**On leave from Cowiconsult, Consulting Engineers and Planers A/S.

Only routine meteorological data are required for the model. Recent developments in the turbulence and dispersion theory are applied for classification of the turbulent state of the atmospheric boundary layer. Here, we describe the basic physical principles of the model and also the methods for determination of the model parameters.

DISPERSION PARAMETERS

The great majority of Gaussian models make use of dispersion parameters and classification methods proposed by Pasquill (1961) and later slightly modified by Gifford (1961) and Turner (1964). The Pasquill-Gifford-Turner (PGT) dispersion parameters were deduced from tracer experiments with near-ground sources. In spite of this, they are frequently used also for high sources. Dispersion experiments with elevated sources showed that the PGT dispersion parameters do in fact perform quite poorly. In an effort to prescribe meaningful dispersion parameters for elevated releases, Briggs analyzed the PGT curves, as well as those determined for tall sources - the Brookhaven National Laboratory (BNL) curves, described by Singer and Smith (1966), and the Tennessee Valley Authority (TVA) curves, described by Carpenter et al. (1970). Briggs deduced a new set of dispersion parameters (Gifford, 1975, 1976) which gave proper weight to the BNL and TVA data when these deviated from the PGT values. These empirical parameters are used in the present model.

STABILITY CLASSIFICATION

In order to assign the dispersion parameters to a given meteorological situation, one must asses the turbulent state of the atmospheric boundary layer. Different methods have been proposed and used. Again, the most widely used was the approach proposed by Pasquill (1961) and Turner (1964). In spite of the simplicity of this method, it displays in fact the basic features of the atmospheric turbulence. The instability of the atmospheric boundary layer increases with increasing insolation and decreasing wind speed, and this is exactly what the Pasquill-Turner scheme prescribes, at least qualitatively. However, this method is not sufficient for classification of dispersion from tall stacks in unstable conditions. The turbulence in the middle of the unstable boundary layer is dominated by the large convective eddies. The size of these eddies is determined by the thickness of the boundary layer - the mixing height z_i. Theoretical investigations by Deardorff (1972) and laboratory models by Willis and Deardorff (1974, 1976, 1978, 1981) have shown that the turbulence and dispersion in a convective boundary layer are controlled by two important parameters : the mixing height z_i and the convective velocity scale w_*,

$$w_* = \left(\frac{g}{T\rho c_p} H z_i\right)^{1/3} \tag{1}$$

where H is the surface sensible heat flux, g the earth's gravitational acceleration, ρ the air density, T the air temperature and c_p the specific heat of air at constant pressure.

The turbulence in the middle of the convective boundary layer, expressed by the standard deviation, is proportional to w_* (Kaimal et al., 1976),

$$\sigma_w \sim w_*$$

$$\sigma_v \sim w_* \tag{2}$$

where the indices w and v refer to vertical and lateral velocity components, respectively. Because w_* depends on z_i, the turbulence and dispersion depend on z_i as well. This feature is not included in the Pasquill-Turner method.

In the daytime atmospheric boundary layer, the turbulence is generated by both the convective motions and wind (mechanical turbulence). The total turbulent energy can be expressed as a sum of two parts :

$$\sigma^2 = \sigma_M^2 + \sigma_C^2 \tag{3}$$

where σ_M^2 is the mechanically generated turbulent energy and σ_C^2 is the convective part.

In the lower part of the boundary layer ($z<0.1z_i$) the vertical convective turbulence scales with $(z/z_i)^{1/3}$ (Wyngaard et al., 1971),

$$\sigma_{wC} \sim w_* (z/z_i)^{1/3} \tag{4}$$

Because w_* is proportional to $z_i^{1/3}$, the mixing height dependence cancels out. The mechanically generated turbulence is proportional to the friction velocity u_*, which in turn is mainly determined by the wind speed u. In the lower part of the boundary layer, the vertical turbulence is thus solely determined by the sensible heat flux, which is primarily governed by the insolation and by wind speed. Thus, the method of Pasquill and Turner can be expected to work reasonably well here.

To relate the dispersion parameters to the turbulence, Weil and Brower (1982) use the following procedure : According to Taylor's (1921) classical results, the dispersion coefficients within the limit of short travel time or distance can be written

$$\sigma_z = (\sigma_w/u)x$$

$$\sigma_y = (\sigma_v/u)x \tag{5}$$

Briggs's short distance limits for σ_y and σ_z (see Table 1) exhibit the same dependence on x. For the velocity variances, Weil and Brower (1982) use :

$$\sigma_w = ((0.56w_*)^2 + (1.26u_*)^2)^{1/2}$$

$$\sigma_v = ((0.56w_*)^2 + (1.6u_*)^2)^{1/2} \tag{6}$$

These expressions are only valid in the upper part of the boundary layer (above $0.1z_i$) where the turbulence is almost homogeneous, while below $0.1z_i$, the convective contribution (in the case of σ_w) varies with height according to (4).

According to (5) and (6), σ_z and σ_y are functions of w_*/u and u_*/u. For an unstable boundary layer above $0.1z_i$, the wind velocity is known not to vary much with height. For the ratio u_*/u, a typical value of 0.05 is substituted. The remaining parameter is now the ratio w_*/u, so there is a one-to-one correspondance between w_*/u and σ_z (or σ_y). Eqs. (5) and (6) are used to introduce a classification according to the stability parameter w_*/u, so each class corresponds to one of Briggs' dispersion curves. These classes, together with expressions for σ_z and σ_y, are shown in Table 1.

The method described above can only be used for unstable conditions (positive surface heat flux). For stable conditions, $w_* = 0$, but the ratio u_*/u depends strongly on the stability. Also, the assumption that the value of the wind velocity is constant with height can not be used in this case. In the stable boundary layer, the turbulence is mainly determined by the temperature gradient and wind speed. The temperature gradient is dependent on cloudiness and wind speed. Because cloudiness and wind speed are included in the Pasquill-Turner scheme, this scheme appears to be appropriate for turbulence classification in stable conditions and is thus used in the present model.

The most significant feature of the w_*/u classification method is that the unstable classes occur more frequently than according to the PGT-method. This is evident from Fig. 1 where the distribution of the stability classes computed by the two classification methods is shown. We use here 5 years of hourly data for the months of January and June from Kastrup Airport, Copenhagen, Denmark.

PLUME RISE

There exists a variety of methods for calculation of plume rise (Markvorsen, 1982). Experimental verification of these methods is, however, still very poor. Formulae for the final plume rise suggested by Briggs (1975) are used in the present model. These formulae are believed to be more physically justified than the earlier, empirical methods by Briggs (1970). In unstable conditions, the rise is assumed to terminate when the turbulence dissipation rate inside the plume has decayed to that in the surrounding turbulent air (break-up model). For convective conditions, it is furthemore assumed that the rise can terminate when the plume is brought down to the ground by large-scale convective downdrafts (touch-down model). We are not going to present here the derivation of the formulae, but only quote the final results. For more details, the reader is referred to the original paper by Briggs (1975) or to the report by Weil and Brower (1982).

For neutral (windy) conditions, the final rise is given by :

$$\Delta h = 1.3 \frac{F}{uu_*^2} \left(1 + \frac{h_s}{\Delta h}\right)^{2/3} \tag{7}$$

where h_s is the stack height, and F, the buoyancy flux, is given by :

$$F = \frac{V}{\pi} \frac{g}{T_e} (T_e - T_a)$$

where V is the volume flux, T_e the plume exit temperature and T_a the ambient air temperature. When the turbulence is dominated by convection, the final rise is given by :

$$\Delta h = 4.3 \, (F/u)^{3/5} \, H_*^{-2/5} \tag{8}$$

where $H_* = H \, g/ \, (T_a \rho c_p)$

Table 1. – Briggs' dispersion parameters σ_y and σ_z as a function of downwind distance x(m) and the corresponding range of w_*/u. The PGT-classification is used when the surface heat flux is negative.

Class	σ_y(m)	σ_z(m)	w_*/u
A	$0.22x(1+0.0001x)^{-1/2}$	$0.20x$	$0.286 < \frac{w_*}{u}$
B	$0.16x(1+0.0001x)^{-1/2}$	$0.12x$	$0.168 < \frac{w_*}{u} \leq 0.286$
C	$0.11x(1+0.0001x)^{-1/2}$	$0.08x(1+0.0002x)^{-1/2}$	$0.072 < \frac{w_*}{u} \leq 0.168$
D	$0.08x(1+0.0001x)^{-1/2}$	$0.06x(1+0.0015x)^{-1/2}$	$\frac{w_*}{u} \leq 0.072$
E	$0.06x(1+0.0001x)^{-1/2}$	$0.03x(1+0.0003x)^{-1}$	PGT
F	$0.04x(1+0.0001x)^{-1/2}$	$0.016x(1+0.0003x)^{-1}$	PGT

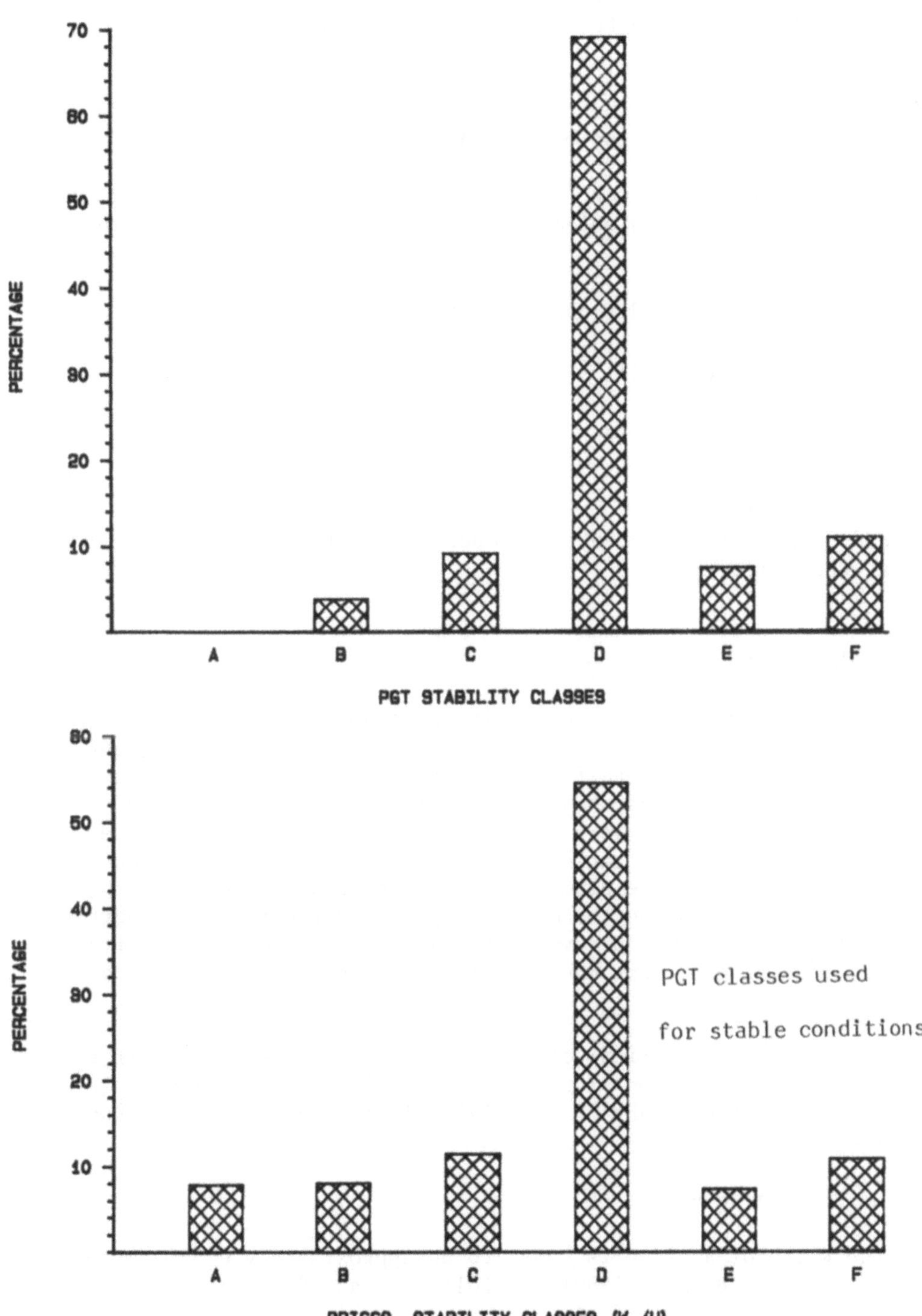

Fig. 1. - Stability classification by two methods. Five years of hourly data from Kastrup Airport, January and June, 1974-78.

In the case of strong convection, where the downdrafts can bring the plume down to the ground, the final rise is :

$$\Delta h = (F/(uw_d^2)) \; (1 + (2h_s/\Delta h))^2 \tag{9}$$

where $w_d = 0.4w_*$ is the mean downdraft velocity.

Of the expressions (7) to (9), the one which results in the lowest plume rise is selected for use in the computations for unstable conditions.

For stable conditions, the final rise is given by :

$$\Delta h = 2.6(F/(us))^{1/3} \tag{10}$$

where

$$s = \frac{g}{T_a} \frac{\partial \theta_a}{\partial z} \tag{11}$$

and $\partial\theta_a/\partial z$ is the potential temperature gradient of ambient air at the stack top.

PENETRATION

In many Gaussian models, it is assumed that when the effective plume height is predicted to be above the mixing layer height, the plume is decoupled from the ground resulting in zero ground level concentrations. Measurements show, however (Weil and Brower, 1982), that very often, in just such situations, the ground level concentrations might be very large. This is due to the fact that the plume might only partially penetrate the stable layer capping the mixing layer, and the remaining portion of the plume is mixed down to the ground. In combination with low wind speeds, this can give rise to high ground level concentrations.

Simple geometrical considerations can be used in order to compute the fraction of the plume that is above the mixing layer. Let z_i' be the height of the mixing layer above the stack top, i.e.

$$z_i' = z_i - h_s \; . \tag{12}$$

The width of a plume is known to be approximately equal to the plume rise (Briggs, 1980). It follows that if :

$$1.5 \; . \; \Delta h \leqq z_i' , \tag{13}$$

the whole plume is below the top of the mixing layer.

On the other hand, if

$$0.5 \cdot \Delta h \geqq z_i' , \tag{14}$$

the plume is totally above the mixing layer. For "in-between" cases, the ratio of the plume above the mixing layer to the total plume width, the penetration factor P, is given by :

$$P = 1.5 - (z_i' / \Delta h); \text{ for } 0.5 \leqq (z_i' / \Delta h) \leqq 1.5 \tag{15}$$

When the penetration factor P is different from zero, an effective source strength Q' is computed and used to calculate the ground level concentrations :

$$Q' = Q(1 - P) \tag{16}$$

Furthemore, when penetration occurs, the effective source height for the part of the plume remaining in the mixing layer is given by (Briggs, 1980) :

$$h_e = h_s + (0.62 + 0.38P)z_i' \tag{17}$$

Note that the effective source height h_e is in this case not equal to the height of the plume center line, which is given by $h_s + \Delta h$.

When the plume penetrates into the elevated stable layer, the further rise is limited by the stable stratification. Briggs (1980) suggests using Eq. (10) in this case to compute the final plume rise to be used in (15). This is equivalent to assuming that the lapse rate in the mixing layer is the same as in the stable layer above. In fact, the temperature profile in the mixing layer is close to adiabatic, i.e. $s \simeq 0$. In the present model, we use a procedure that is more realistic. If the plume rise computed with Eqs. (7) to (9) is larger than $z_i'/1.5$, the further rise is computed according to the formula for stable stratification but taking into account that the plume loses a part of its buoyancy when it reaches the height $z_i'/1.5$; the final rise is given by :

$$\Delta h = ((2.6)^3 (F/(us_i)) + (z_i'/1.5)^3)^{1/3} \tag{18}$$

where s_i corresponds to the lapse rate in the stable layer. Δh computed by (18) is substituted into (15) for determination of the penetration factor. A preliminary test of the model shows that for a high source ($h_s \simeq 200$ m) with a significant buoyancy ($F \simeq 600$ m^4/s^3), the highest concentrations occur under conditions when plume rise is terminated by the elevated stable layer (Eq. (17)). A proper evaluation of the penetration ability is thus very important in this case.

DETERMINATION OF THE MODEL PARAMETERS

Application of the w_*/u as the stability parameter requires the knowledge of the sensible heat flux H and the mixing height z_i. Also in the plume rise formulae, use is made of the values of w_*, u_* and z_i. None of these parameters is measured on a routine basis. At the Danish Air Pollution Laboratory, special methods were developed for evaluation of the mentioned parameters from routine meteorological data such as hourly synoptic observations from airports and radiosoundings. In the following, only a short description of these methods is given, for details the reader is referred to the original reports (Nielsen et al. 1981a; Berkowicz and Prahm, 1982a).

Sensible heat flux

The method of resistances is used for evaluation of hourly values of surface fluxes from routine meteorological data (Berkowicz and Prahm, 1982a). The governing relationship is the equation of the surface energy balance :

$$R_n = \ell E + H + G \tag{19}$$

where R_n is the net radiation, ℓE the latent heat flux and G the soil heat flux.

In Nielsen et al. (1981a), a method is presented for estimation of net radiation from hourly cloud observations and solar elevation. This method is used in the present model. In order to estimate the partitioning of the net radiation energy between the remaining three terms of (19), the method of Penman-Monteith (Penman, 1948; Monteith, 1965) is used. The latent heat flux is given by :

$$\ell E = \frac{(R_n - G)\, r_a(\Delta/\gamma) + D q \rho c_p/\gamma}{r_s + (1 + (\Delta/\gamma)) r_a} \tag{20}$$

where Δ is the gradient of the saturated vapour pressure with respect to temperature (function of temperature only), γ is the psychrometric constant (function of temperature only) and D_q is the humidity deficit in the air defined as :

$$D_q = q_s\ (T) - q \tag{21}$$

where q_s is the saturated vapour pressure at the temperature T for given reference height (usually 2 m) and q is the actual vapour pressure at the same height. The aerodynamic resistance r_a is the resistance of the atmospheric layer between the reference height (f.ex. 2 m) and the surface. This resistance is mainly determined by the wind speed (Berkowicz and Prahm, 1982a).

The surface resistance r_s is the resistance of the surface to water vapour transport. From extensive study of several sets of experimental data, it was found that r_s can be related to the humidity deficit in the air :

$$r_s = (Dq/f)\ (\rho c_p/\gamma) \tag{22}$$

where F is an empirical function of the surface moisture conditions. The main parameter governing the empirical function F is the accumulated net radiation since last precipitation (Berkowicz and Prahm, 1982a). The soil heat flux is modeled as a certain fraction of the sensible heat flux :

$$G = \alpha_g H \tag{23}$$

where the constant α_g is found to be 0.3 for Danish conditions. From (19), (20) and (23), we obtain :

$$H = \frac{R_n\ (r_a+r_s)\ -\ Dq\ (\rho c_p/\gamma)}{r_s + (1 + (\Delta/\gamma))\, r_a + \alpha_g\ (r_a+r_s)} \tag{24}$$

By means of (24), the sensible heat flux H can be estimated from routine meteorological observations only. The required information is : temperature, humidity and wind speed at only one height and also information on precipitation events and grass cover for determination of r_s.

The resistance method is capable of taking the local weather conditions into account. In Berkowicz and Prahm (1982a), tests are presented on data from several locations (Denmark, Sweden, Holland) and it appears that performance of the model is very good. In their model, Weil and Brower (1982) use the assumption that H is proportional to the global, short wave radiation, and in a later version (1983), they relate the sensible heat flux to the net radiation R_n. This procedure might be reasonable when the moisture conditions of the surface are not changing. For weather conditions such as those in Denmark, the ratio between the heat flux and net radiation exhibits a significant variation due to frequently changing moisture conditions of the surface. In the case when meteorological masts are available, the sensible heat flux can be estimated by means of the so-called "profile method" (Berkowicz and Prahm, 1982b). This method requires, however, measurements of temperature difference between two levels. High quality, ventilated thermometers must be used for such measurements (Nielsen et al., 1981b).

Friction velocity

Friction velocity, u_*, is another important parameter which is required in the model. It is used for determination of the wind speed at plume elevation and also in evaluation of the final plume rise in neutral conditions.

Friction velocity can be estimated from wind speed measurements using the profile method (Berkowicz and Prahm, 1982b). We use the well-known results from the similarity theory (Monin and Obukhov, 1954) :

$$u(z) = \frac{u_*}{\kappa} \left[\ell n \frac{z}{z_o} - \psi_m \left(\frac{z}{L}\right) + \psi_m \left(\frac{z_o}{L}\right) \right] \qquad (25)$$

where u(z) is the wind speed measured at the reference height z (usually $z \simeq 10$ m), κ is the von Karman constant (here κ = 0.35) and L is the Monin-Obukhov length :

$$L = - \frac{T}{g} \frac{u_*^3}{\kappa H} \rho c_p \qquad (26)$$

For the stability correction function ψ_m, we use the results of Businger et al. (1971). The roughness parameter z_o can be estimated by inspection of the terrain features (Davenport, 1960; Brutsaert, 1975).

Because u_* appears also in the expression for the aerodynamic resistance r_a , Eqs. (24) and (25) must be solved iteratively. Usually no more than 2-3 iterations are required to get a reasonable accuracy.

Mixing height

Mixing height is the parameter which is most difficult to determine on a routine basis. The daytime mixing height is usually defined as the height to the first stable layer and can be determined from the temperature profiles of the atmosphere. In Denmark radio soundings of the atmosphere are performed twice daily, at 00 GMT and 12 GMT (i.e. 01 and 13 hour local time). Only the latter sounding can be used for direct determination of the daytime mixing height. However, the boundary layer is still evolving at this time, and thus the mixing height estimated from 12 GMT is not representative for the daytime boundary layer. In USA, the soundings are performed in early morning and mid-afternoon. The mid-afternoon mixing height usually reaches its maximum value. In the recommended EPA-models, such as CRSTER model (EPA, 1977), the hourly values of the mixing height are evaluated by linear interpolation between the values obtained from the early morning and mid-afternoon.

For the reasons mentioned above, such procedure is not applicable for the Danish conditions. The height of the daytime convective boundary layer can be predicted from thermodynamic considerations. Several models exist with slightly different formulations, but they all originate from the work of Carson (1973) and Tennekes (1973).

The growth of the convective boundary layer is governed by the upward heat flux. The basic assumptions made when deriving models for mixing height are : a) the potential temperature is constant with height (adiabatic) in the mixing layer, b) the heat flux decreases linearly with height, c) the temperature profile in the stable layer above the mixing layer does not change with time. All these assumptions are more or less satisfied when the turbulent structure of the boundary layer is clearly dominated by convection and advective effects are absent. Jensen (1981) tested several models for mixing height on data from the Wangara experiment (Clarke et al. 1971). It was found that, for typical convective days, even the simplest models gave quite good results, since the fundamental assumptions for the models were fulfilled under such typical convective conditions. On all other occasions, the performance of the models was much worse.

The model based on Tennekes' (1973) fundamental differential equations showed the best results and was selected for further use. The model was combined with information obtained from the two radio soundings and a procedure for operational use was designed (Olesen et al., 1983). The 00-GMT sounding is used to start the integration. The daytime evolution of the mixing height is assumed to start when the surface sensible heat flux becomes first time positive (upward). The resistance method described previously is used for estimation of the heat flux. At 12 GMT, the calculated mixing height is compared with that evaluated from the temperature profile. If a substantial difference occurs, all the previous hourly mixing height values are consequently corrected. The further integration proceeds from the 12-GMT profile. This procedure is used when conditions are not too unfavourable for the application of a convective model (e.g. advection must not be extreme). A neutral boundary layer height ($z_i = 0.25\ u_*/f$, where f is the Coriolis parameter) is used when conditions do not permit the use of a convective model.

The thickness of the nighttime boundary layer is very difficult to determine. This is not just the consequence of observational difficulties, but different definitions of nocturnal boundary layer height exist as well. At present, we use a neutral boundary layer height (with a lower bound of 150 m) to estimate nocturnal boundary layer height, but work is in progress to implement an improved method. This will probably be based upon the work by

Stull (1983), who defines a length scale given by the ratio of the time integral of the downward heat flux to the temperature decrease at the surface. This length scale is simply related to the depth of the surface inversion.

CONCLUDING REMARKS

An operational air pollution model for point sources is developed. The model makes use of the Gaussian dispersion formulae and calculates hourly values of pollutant concentrations. Briggs' dispersion parameters are applied with the stability classification scheme based on the values of w_*/u for daytime and PGT-classes for nighttime. Briggs' (1975) formulae for plume rise are used with special emphasis on calculation of the ability of the plume to penetrate the elevated stable layer.

New methods are developed for estimation of the surface parameters such as heat flux and friction velocity. The mixing height is estimated from twice-daily radio soundings and a theoretical model based on Tennekes (1973) equations. The model requires only routinely available meteorological data, such as e.g. hourly synoptic data from local airports.

ACKNOWLEDGEMENT

This work was partially sponsored by the Danish Technical Council and the Danish Power Plant Association. The Institute of Mathematical Statistics and Operations Research, Technical University of Denmark is appreciated for their collaboration. G. Omstedt from Swedish Meteorological and Hydrological Institute has participated in the part of this work during his stay at the Danish Air Pollution Laboratory. We wish to thank N. Brown for his technical assistance and M. Bille and S. Mølmark for typing the manuscript.

REFERENCES

Berkowicz, R. and Prahm, L. P., 1982a. Sensible heat flux estimated from routine meteorological data by the resistance method. J. Appl. Met. 21, 1845-1864.

Berkowicz, R., and Prahm, L. P., 1982b. Evaluation of the profile method for estimation of surface fluxes of momentum and heat. Atm. Env. 16, 2809-2819.

Briggs, G.A., 1970. Some recent analyses of plume rise observations. Proc. 2nd International Clean Air Congress, Washington, D.C.

Briggs, G.A., 1975. Plume rise predictions. In Lectures on Air Pollution and Environmental Impact Analyses, American Meteorological Society, Boston, MA.

Briggs, G.A., 1980. Plume rise and buoyancy effects. In Atmospheric Science and Power Production, Atmospheric Turbulence and Diffusion Laboratory, NOAA, Oak Ridge, TN.

Businger, J.A., Wyngaard, J.C., Izumi, Y., and Bradley, E.F., 1971. Flux profile relationships in the atmospheric surface layer. J. Atmos. Sci. 28, 181-189.

Carpenter, S.B., Leavitt, J.M., Colbaugh, W.C. and Thomas, F.W., 1970. Principal plume dispersion models at TVA power plants. Proc. 63rd Annual Meeting of the Air Pollution Control Association, St. Louis, MO.

Carson, D.J., 1973. The development of a dry inversion-capped convectively unstable boundary layer. Quart. J.R. Met. Soc. 99, 450-467.

Clarke, R.H., Dyer, A.J., Brook, R.R., Reid, D.G. and Troup, A.J., 1971. The Wangara experiment : Boundary layer data. Division of Meteorological Physics Technical Paper No. 19, Commonwealth Scientific and Industrial Research Organization, Australia.

Davenport, A.G., 1960. Rationale for determining design wind velocities. J.Am. Soc. Civ. Eng. (Struc. Div.) 86, 39-68.

Deardorff, J.W., 1972. Numerical investigation of neutral and unstable planetary boundary layers. J. Atm. Sci. 29, 91-115.

EPA, 1977. User's manuel for single source (CRSTER) model. EPA-45012-77-013. U.S. Environmental Protection Agency, Research Triangle Park, NC.

Gifford, F.A., 1961. Uses of routine meteorological observations for estimating atmospheric dispersion. Nuclear Safety 2, 47-51.

Gifford, F.A., 1975. Atmospheric dispersion models for environmental pollution applications. In Lectures on Air Pollution and Environmental Impact Analyses. American Meteorological Society, Boston, MA.

Gifford, F.A., 1976. Turbulent diffusion typing schemes : A review. Nuclear Safety 17, 68-86.

Jensen, A.B., 1981. Some simple models for estimation of the mixing height in convective boundary layer. MST LUFT - A48, National Agency of Environmental Protection, Air Pollution Laboratory, DK-4000 Roskilde, Denmark (in Danish).

Kaimal, J.C., Wyngaard, J.C., Haugen, D.A., Coté, O.R., Izumi, Y., Caughey, S.J. and Readings, C.J., 1976. Turbulence structure in the convective boundary layer. J. Atmos. Sci. 33, 2152-2169.

Markvorsen, J.S., 1982. Plume Rise. MST LUFT - A35, National Agency of Environmental Protection, Air Pollution Laboratory, DK-4000 Roskilde, Denmark (in Danish).

Monin, A.S. and Obukhov, A.M., 1954. Dimensionless characteristics of turbulence in the atmospheric surface layer. Dokl. Akad. Nauk. SSSR. 93, 223-226.

Monteith, J.L., 1965. Evaporation and environment. In The State and Movement of Water in Living Organisms, 19th Symposium, Soc. Exp. Biol., 205-235.

Nielsen, L.B., Prahm, L.P. Berkowicz, R. and Conradsen, K., 1981a Net incoming radiation estimated from hourly global radiation and/or cloud observations. J. Climat. 1, 255-272.

Nielsen, L.B., Conradsen, K. and Prahm, L.P., 1981b. Analysis of Measurements from a Mast and a nearby Synoptic Station. MST LUFT - A53, National Agency of Environmental Protection, Air Pollution Laboratory, DK-4000 Roskilde, Denmark.

Olesen, H.R., Jensen, A.B. and Brown, N., 1983. Operational procedure for estimation of mixing height for air pollution models (in preparation).

Pasquill, F., 1961. The estimation of the dispersion of windborne material. Meteorol. Mag. 90, 33-49.

Penman, H.L., 1948. Natural evaporation from open water, bare soil and grass. Proc. Roy. Soc. London A194, 120-145.

Singer, I.A. and Smith, M.E., 1966. Atmospheric diffusion at Brookhaven National Laboratory. Int. J. Air Water Pollut. 10, 125-135.

Stull, R.B., 1983. A heat-flux history length scale for the nocturnal boundary layer. Tellus 35A, 219-230.

Taylor, G.I., 1921. Diffusion by continuous movements. Proc. London Math. Soc., Ser. 2, 20, 196-202.

Tennekes, H., 1973. A model for the dynamics of the inversion above a convective boundary layer. J. Atm. Sci. 30, 558-567.

Turner, D.B., 1964. A diffusion model for an urban area. J. Appl. Met. 3, 83-91.

Weil, J.C. and Brower, R.P., 1982. The Maryland PPSP dispersion model for tall stacks. PPSP-MP-36, Martin Marietta Corp., Environmental Center, Baltimore, MD 21227.

Weil, J.C. and Brower, R.P., 1983. Estimating convective boundary layer parameters for diffusion applications. PPSP-MP-48, Martin Marietta Corp., Environmental Center, Baltimore, MD 21227.

Willis, G.E. and Deardorff, J.W., 1971. A laboratory model of the unstable planetary boundary layer. J. Atm. Sci. 31, 1297-1307.

Willis, G.E. and Deardorff, J.W., 1976. A laboratory model of diffusion into the convective planetary boundary layer. Quart. J. Roy. Met. Soc. 102, 427-445.

Willis, G.E. and Deardorff, J.W., 1978. A laboratory study of dispersion from an elevated source within a modelled convective planetary boundary layer. Atm. Env. 12, 1305-1311.

Willis, G.E. and Deardorff, J.W., 1981. A laboratory study of dispersion from a source in the middle of the convective mixed layer. Atm. Env. 15, 109-117.

Wyngaard, J.C., Coté, O.R. and Izumi, Y., 1971. Local free convection similarity and the budgets of shear stress and heat flux. J. Atm. Sci. 28, 1171-1182.

DISCUSSION

R. YAMARTINO Why did you choose w_*/u, rather than σ_w/u, as your continuous variable describing stability ?

R. BERKOWICZ σ_w can be used as well as w_* because σ_w is determined by w_*.

F. DESIATO The Briggs formula for the plume rise is valid for hot gases. Has the model that you presented also a formula for a cold gas emitted from the stack with a vertical velocity ?

R. BERKOWICZ No, but it can be included.

G. DUMONT Could you comment on the differences between your sensible heat flux estimation by resistance method and the Holtslag and de Bruin approach.

R. BERKOWICZ The main difference is the way in which the moisture conditions of soil are treated. In Holtslag and de Bruin it is done by varying the parameter α, while in our procedure, the surface resistance accounts for the soil moisture conditions.

THE RELATION OF URBAN MODEL PERFORMANCE TO STABILITY

D. Bruce Turner * and John S. Irwin *

Meteorology and Assessment Division
U.S. Environmental Protection Agency
Research Triangle Park, NC 27711

INTRODUCTION

Previous studies (Turner and Irwin, 1983, and Ruff, 1981) have made use of data collected in the Regional Air Pollution Study (RAPS) in St. Louis to evaluate the urban model RAM (Turner and Novak, 1978). RAM estimates short-term (hourly) dispersion using the Gaussian steady-state modeling concepts (Novak and Turner, 1976). Concentrations from point sources are calculated by evaluation of a single equation for each source receptor pair. Concentrations from area sources are computed using the narrow plume simplification (Gifford and Hanna, 1971), with the two-dimensional integral that expresses the total area-source contribution to concentration being approximated by a one-dimensional integral that only involves knowledge of the distribution of the area-source emissions along the line in the direction of the upwind azimuth from the receptor location. Emphasis was upon averages and extremes found from the data set for the year 1976.

These studies indicated that although the model overestimated concentrations for some measurement stations and underestimated concentrations for other stations, for the network as a whole, there appeared to be little bias for both annual means and second-highest concentrations. This was not to say that the model simulated the pollutant transport and dispersion well for there could be compensating errors to arrive at a reasonable annual value. And although the proper magnitude was obtained for the second-highest values, it had not been determined that the second

---

* On assignment from the National Oceanic and Atmospheric Administration, United States Department of Commerce

highest occurred in the atmosphere under similar conditions that caused the modeled second-highest concentrations. In fact, examination of the measured and model estimated concentrations suggested that the highest values produced by the model occurred with different stability classes than the stability classes corresponding to the highest measured concentrations. Here we will stratify the modeled and measured concentrations by stability to see if there is any systematic bias.

ANALYSIS PERFORMED

Several measures related to data quality were available for the SO_2 measurements. Both the percent of zero drift and the span drift between successive calibrations (approximately every 24 hours) were available. Also the number of minutes of acceptable values for each hour was available. For this study a timeliness criteria of greater than or equal to 45 minutes was used. Quality criteria of span drift of no more than 10 percent and zero drift of no more than 1 percent of full scale were used. Full scale was 1 ppm (about 2600 $\mu g/m^3$). Use of these criteria resulted in a minimum of 58.1 percent usable data for station 106 to a maximum of 68.9 percent usable data for station 103.

A single wind speed and a stability class representative of the urban area is available for each hour. The Pasquill stability classes were determined using hourly cloud cover and ceiling height at the St. Louis Airport (Lambert Field) and an hourly urban average wind speed determined from the wind measurements made at the air quality sampling locations. There are eight classes : A, B, C, D (day), D (night), E, and F based on the Pasquill classification and G which is more stable than Pasquill's F. Using the wind speeds from the data in each stability class, a median speed was selected for each class so that the data within that class would be partitioned into two nearly equal groups. The dividing wind speeds, and the number of periods in the resulting 16 stability - wind speed categories are given in Table 1.

For each stability - wind speed category the average concentration and the maximum concentration for both the model estimates and the measurements were selected for further analysis. (Data meeting the previously discussed criteria were used for the measurements ; all data were used for the model estimates.) Table 2 gives the average concentrations and Table 3 gives the maximum concentrations for the measurements and the model estimates for the 16 categories for each measurement station. The geographic position of these stations is given by Myers and Reagan (1975), and Turner and Irwin (1983).

Table 1. Number of One-Hour Periods in Analysis Categories

Stability Class	A	B	C	DD	DN	E	F	G
Dividing Wind Speed	2.0	2.7	3.6	5.0	4.4	3.3	2.4	1.4
Number in High Wind Speed Class[a]	47	312	574	902	905	740	652	187
Number in Low Wind Speed Class	47	330	555	944	920	711	734	224

[a] cases with wind speeds greater than or equal to dividing wind speed.

Table 2. Average 1-Hour SO_2 Concentration Model Estimates and Measurements ($\mu g/m^3$) for Each Stability-Wind Speed Category. (The Number of Values in Each Category is Given in Table 1 for the Estimates on the Third Line for each Station for the Measurements).

STABILITY CLASS		A		B		C		DD		DN		E		F		G[a]	
WIND SPEED CLASS		L	H	L	H	L	H	L	H	L	H	L	H	L	H	L	H
STATION																	
101	est	34	22	38	26	56	35	57	35	76	48	130	112	134	125	202	121
	meas	79	32	70	37	63	45	51	33	58	42	57	79	73	67	47	53
	no.	16	12	193	174	342	319	610	655	559	600	425	442	421	330	120	101
103	est	26	17	25	20	30	21	28	19	35	19	69	51	67	55	98	64
	meas	72	20	56	36	67	36	34	29	38	28	31	39	30	31	23	20
	no.	29	19	221	208	405	408	698	697	645	613	494	489	466	392	146	120
104	est	45	30	35	28	50	31	46	33	57	34	101	83	100	89	157	101
	meas	92	27	111	49	177	102	94	71	132	77	161	146	127	142	144	91
	no.	25	10	194	173	358	351	659	698	590	622	454	470	424	344	133	97
105	est	49	62	50	44	50	30	38	15	38	17	70	43	73	53	114	74
	meas	68	50	65	46	63	48	47	27	50	29	48	30	48	35	40	55
	no.	30	17	211	190	387	379	706	672	611	643	481	506	457	389	135	102
106	est	23	15	28	18	28	18	32	15	36	19	65	44	70	47	117	58
	meas	72	37	68	42	66	53	56	39	52	43	58	52	65	48	41	57
	no.	23	16	200	183	330	343	600	590	524	488	431	403	403	345	127	95
108	est	16	11	19	20	25	17	22	16	29	18	52	44	51	42	77	50
	meas	52	24	51	43	51	43	36	30	38	40	34	46	23	34	15	20
	no.	22	14	194	185	360	373	707	731	617	646	462	483	443	379	133	114
113	est	16	12	19	18	18	14	20	11	23	13	43	32	39	32	67	36
	meas	41	26	51	51	50	35	35	22	37	29	39	37	34	37	25	40
	no.	26	14	198	178	344	332	649	630	560	600	421	452	431	323	110	98
114	est	22	16	21	20	18	14	23	14	26	16	44	29	39	26	61	43
	meas	65	17	45	33	42	40	38	29	43	34	33	35	29	28	19	27
	no.	29	13	196	202	364	360	689	681	613	671	478	498	456	394	129	124
115	est	26	16	18	17	23	17	18	20	25	24	44	52	41	33	47	51
	meas	77	30	45	38	40	27	24	21	32	31	21	41	15	21	11	12
	no.	28	20	194	210	382	368	700	651	627	635	471	440	460	389	127	107
116	est	13	11	11	12	14	11	12	12	15	14	24	31	23	20	34	24
	meas	40	16	42	24	31	22	22	21	36	21	21	23	20	19	17	18
	no.	29	23	220	209	370	363	633	309	562	552	468	435	453	364	142	110
120	est	10	11	11	15	12	11	13	9	12	9	26	12	22	14	31	27
	meas	33	35	36	47	30	27	31	23	30	22	25	22	23	19	19	19
	no.	24	24	202	208	366	375	669	662	590	627	457	475	472	369	125	113
121	est	11	21	17	20	16	16	18	8	18	9	28	14	27	15	36	29
	meas	47	36	58	39	40	31	33	22	32	23	32	27	31	25	22	25
	no.	18	14	167	145	319	327	618	665	584	638	440	434	380	349	131	92
122	est	13	11	11	12	12	9	10	11	12	14	18	19	18	16	24	17
	meas	42	32	41	26	33	34	27	25	27	27	24	32	16	24	18	13
	no.	21	15	187	165	307	305	605	620	529	596	400	431	408	329	121	90

[a] Stability classes A-F as defined by Pasquill; G are those cases more stable than Pasquill's F.

Table 3. Maximum 1-Hour SO_2 Concentration Model Estimates and Measurements ($\mu g/m^3$) for Each Stability-Wind Speed Category. (The Number of Values in Each Category is given in Table 1 for the Estimates and on the Third Line for Each Station for the Measurements.)

STABILITY CLASS		A		B		C		DD		DN		E		F		G[a]	
WIND SPEED CLASS		L	H	L	H	L	H	L	H	L	H	L	H	L	H	L	H
STATION																	
101	est	137	82	213	148	326	181	428	258	446	271	853	546	657	664	1101	547
	meas	224	214	1211	281	1175	531	1068	554	1130	503	1338	1310	1543	1250	251	258
	no.	16	12	193	174	342	319	610	655	559	600	425	442	421	330	120	101
103	est	129	100	148	111	288	180	251	236	378	263	519	314	407	320	622	459
	meas	227	67	842	263	1645	343	343	522	498	240	289	565	419	559	182	179
	no.	29	19	221	208	405	408	698	697	645	613	494	489	466	392	146	120
104	est	217	140	214	132	444	222	635	404	643	294	1250	543	1292	600	964	729
	meas	242	62	1366	291	1767	1712	1998	1796	2099	1066	2135	1807	1966	1724	1455	1221
	no.	25	10	194	173	358	351	659	698	590	622	454	470	424	344	133	97
105	est	252	432	407	392	407	467	423	305	385	291	737	431	842	355	747	391
	meas	266	137	271	341	500	944	537	325	556	325	417	244	582	279	262	411
	no.	30	17	211	190	387	379	706	672	611	643	481	506	457	389	135	102
106	est	99	85	183	163	227	217	338	268	332	328	596	363	718	306	548	294
	meas	240	102	366	429	789	734	1260	459	1268	458	919	549	772	829	249	365
	no.	23	16	200	183	330	343	600	590	524	488	431	403	403	345	127	95
108	est	91	54	129	130	229	210	300	245	306	268	575	398	493	355	856	396
	meas	138	98	295	290	387	329	354	324	464	338	2483	421	252	268	161	171
	no.	22	14	194	185	360	373	707	731	617	646	462	483	443	379	133	114
113	est	86	69	161	133	247	160	231	155	317	279	562	311	368	413	565	372
	meas	136	70	362	1516	410	307	371	267	414	405	358	496	396	737	199	290
	no.	26	14	198	178	344	332	649	630	560	600	421	452	431	323	110	98
114	est	114	61	158	157	197	205	357	203	456	228	869	395	746	479	1217	790
	meas	164	43	227	276	603	1062	1021	517	555	295	559	332	308	345	220	193
	no.	29	13	196	202	364	360	689	681	613	671	478	498	456	394	129	124
115	est	214	136	393	132	436	158	452	257	693	622	1428	937	1675	653	1075	1054
	meas	253	140	323	328	270	286	247	238	389	418	247	536	211	223	170	92
	no.	28	20	194	210	382	368	700	651	627	635	471	440	460	389	127	107
116	est	137	75	186	113	297	179	247	201	367	233	377	385	548	239	546	422
	meas	178	57	391	297	242	473	503	614	598	306	240	246	1079	323	144	231
	no.	29	23	220	209	370	363	633	609	562	552	468	435	453	364	142	110
120	est	78	69	111	203	246	216	352	247	313	278	449	419	427	297	288	333
	meas	99	126	279	856	394	294	248	1048	300	339	464	309	459	193	169	128
	no.	24	24	202	208	366	375	669	662	590	627	457	475	472	369	125	113
121	est	65	167	347	292	442	387	467	214	448	351	424	323	512	218	518	350
	meas	110	83	359	285	325	251	250	367	415	276	301	485	322	388	208	147
	no.	18	14	167	145	319	327	618	665	584	638	440	434	380	349	131	92
122	est	150	95	153	214	295	153	183	195	258	209	374	313	406	270	289	233
	meas	205	89	374	380	491	353	576	519	416	309	558	285	232	291	339	108
	no.	21	15	187	165	307	305	605	620	529	596	400	431	408	329	121	90

[a] Stability classes A-F as defined by Pasquill; G are those cases more stable than Pasquill's F

These data were graphed in various ways to examine performance by stability and by wind speed. Figures 1 through 5 are examples. Concentrations are shown on a logarithmic scale. The upper set are maxima (1-hour) for each stability - wind speed category. The lower set are averages for each stability - wind speed category. Measured concentrations (stippled) are on the left with model estimates on the right within each stability class. Concentrations for the light wind speed group are indi-

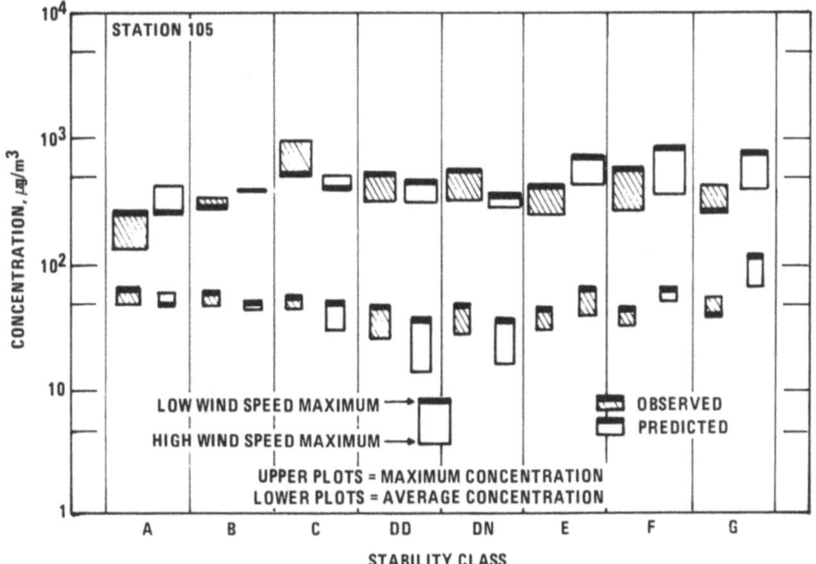

Fig. 1. Maximum and average SO_2 concentration for each stability - wind speed category for station 105.

cated by the heavy horizontal line. Graphs of this type were made for all 13 measurement locations and examined.

DISCUSSION

Comparison of Averages

For both low and high wind speed categories, the model yields lower average concentrations for the most unstable category A, with a slight trend to higher values for the stable categorys E-G. Station 108 (Figure 3) for both wind speeds and station 121 (Figure 5) for low wind speeds are typical. Station 105 (Figure 1) with low average concentration values for D (day and night) is the exception. The trend toward higher values during stable categories is more apparent in the results for the low wind speed groups than the results for the high wind speed groups.

The measurements generally exhibit higher average concentrations during A stability with lower average concentrations during stable conditions. This trend is in contrast to that seen for the model results. As with the model results, the trends are more easily seen for the low wind speed groups than for the high wind speed groups.

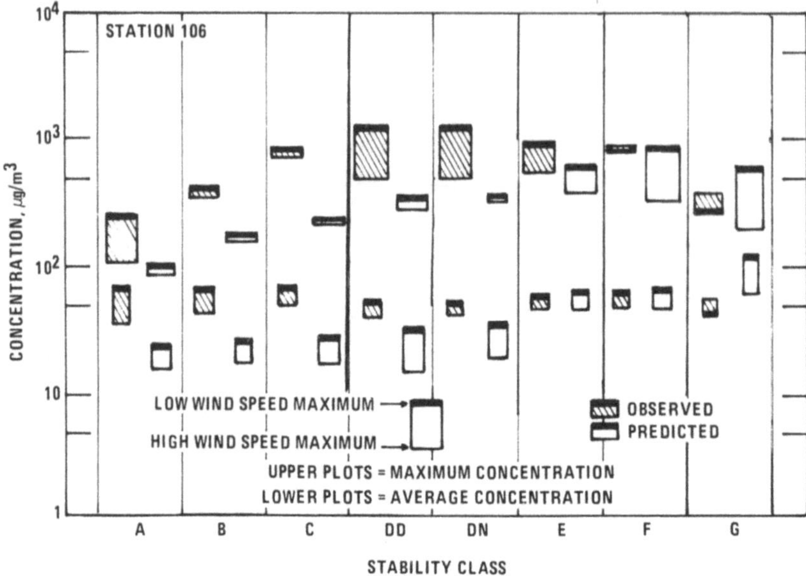

Fig. 2. Maximum and average SO_2 concentrations for each stability - wind speed category for station 106.

Comparison of Maxima

The trends seen in the model estimates for peak 1-hour concentrations are quite similar to the trends seen in the average estimated concentrations. Typically, the model yields lower peak concentrations during A stability than during stable conditions, such as at station 106 (Figure 2). Exceptions are station 121 (Figure 5) with nearly the same concentrations for classes B through G, and stations 115 (Figure 4) and 120 with lower concentrations with G stability than with F for low wind speeds.

The peak measured concentrations are generally lower for the extremes in stability (very unstable or very stable) with the highest values occurring during B through F stability. Station 105 (Figure 1) and 121 (Figure 5) display little change in the concentration values for B through G with peaks most frequently with D night stability with low wind speeds and D day or E for high wind speeds. Several stations, 105 (Figure 1) and 113, have large peaks with B and C stability.

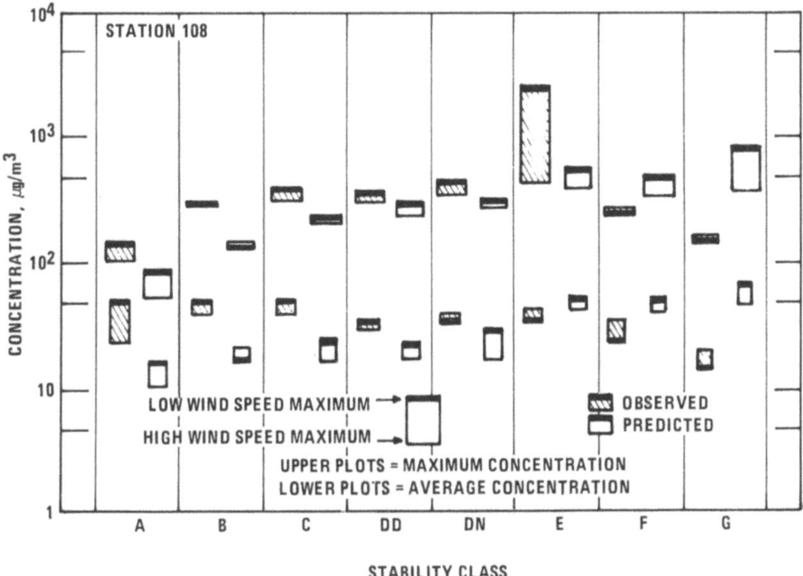

Fig. 3. Maximum and average SO_2 concentrations for each stability - wind speed category for station 108.

Comparison of Averaged Rank Order

Further calculations were made in an attempt to summarize the trends noted above over all 13 stations. For the high and low wind speed categories separately, for each station, each stability category was assigned a rank 1 to 8, according to the ranking of the concentrations. For example, for station 105 (Fig. 1), for peak measurements at high wind speeds stability category C has a rank of 1, category A a rank of 8. These ranks were then averaged for each stability category over the 13 stations. These results are displayed in Figure 6. Solid lines connect the mean ranks for low wind speeds ; dashed lines connect the mean ranks for high wind speeds. When averaged over the 13 stations some definite trends appear. Modeled concentrations for both wind speed groups and for both peak and average concentrations produce low average rank values (6 to 7) for unstable conditions and high average rank values (1 to 2) for stable conditions.

For average measured concentrations for low wind speeds, the plots of average rank suggest quite the opposite trend occurs :

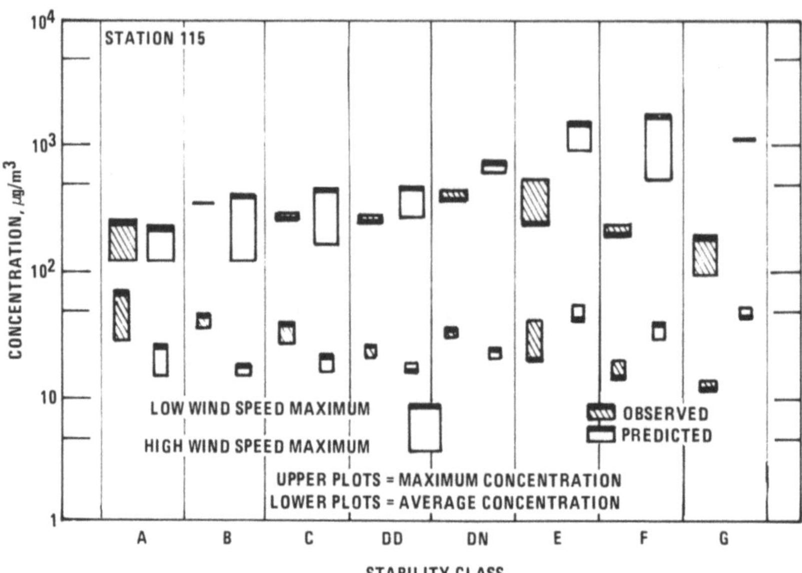

Fig. 4. Maximum and average SO_2 concentration for each stability - wind speed category for station 115.

high concentrations for unstable situations and low concentrations for stable conditions. For the high wind speed categories there is no overall trend for the measured averages. The plots of average rank for the 1-hour peak concentrations measured are very similar for both wind speed categories, suggesting low concentrations for A and G stabilities with a relatively flat distribution of average ranks for stabilities B through F.

SPECULATION ON STABILITY-RELATED PERFORMANCE

There are several possible causes that might explain the lack of agreement in the modeled and measured concentration trends with stability.

Building Downwash

The tendency to underestimate concentrations under higher wind speeds may be related, in part, to the model's rather simplistic view of dispersion. Building wake effects on dispersion and plume rise are not considered in the modeling. Observed maximum concentrations are greater for the high wind speed cases

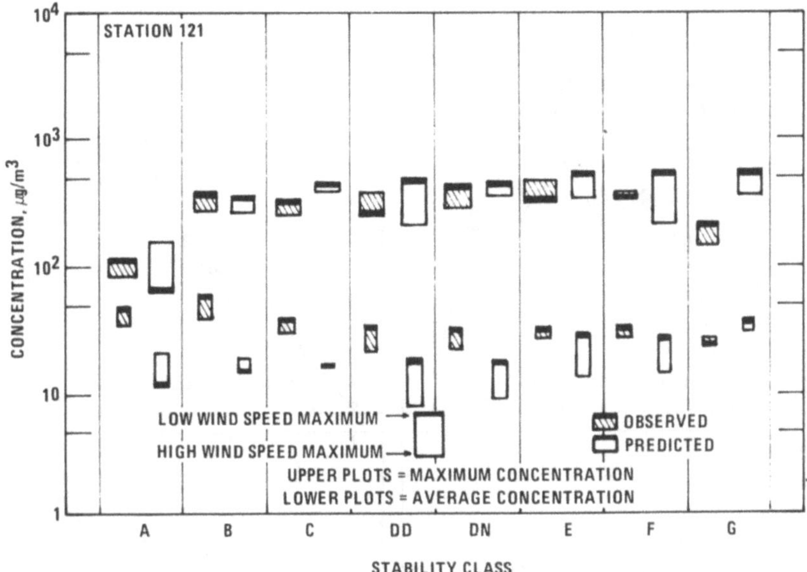

Fig. 5. Maximum and average SO_2 concentrations for each stability - wind speed category for station 121.

than for the low wind speed cases at stations 105, 113, 114, 115, 120, and 121. (The highest modeled concentration at each receptor occurs with low wind speed categories.)

Plume Penetration of Stable Layers

Underestimates during daytime may be related to the manner in which buoyant emissions are modeled. It is quite likely that the model computes full penetration of the rising plume of effluent into the elevated stable layer above the surface based mixed layer more often than really occurs. This has the effect of lowering daytime estimated concentrations. Brower (1980) and Weil and Brower (1982) have pointed out instances where a rural model employing a plume rise formulation similar to that in RAM estimates the plume above the mixing height, yet there are measurable ground-level concentrations, thus the model underestimates concentrations. This could as easily occur with urban modeling.

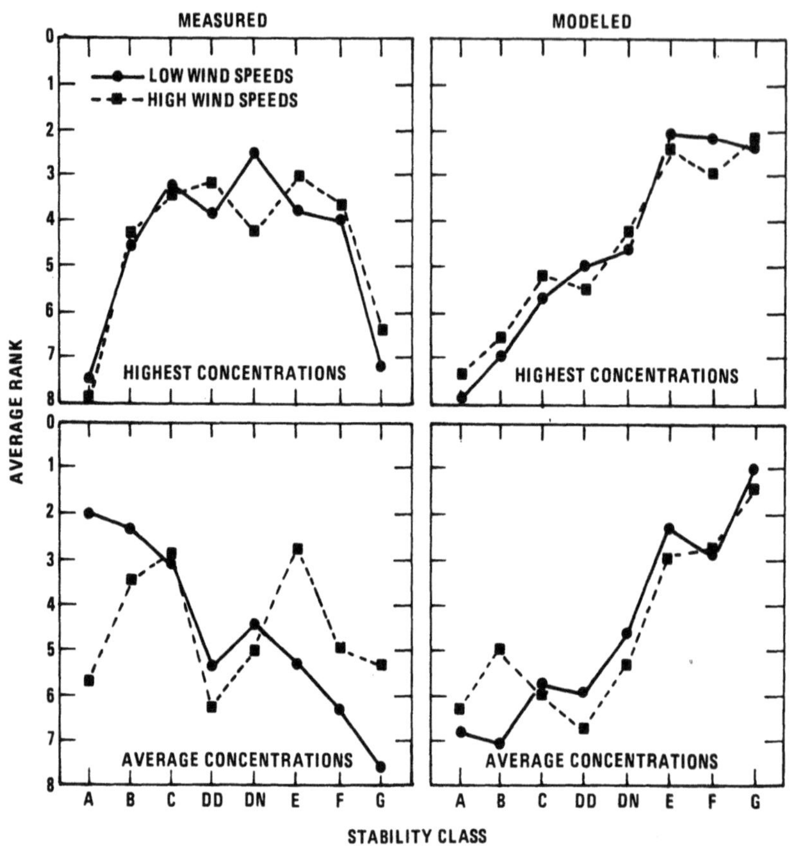

Fig. 6. Mean rankings of maximum and average concentrations over the 13 stations for each stability - wind speed category (solid lines for low wind speeds, dashed lines for high wind speeds).

Partitioning of Emissions

Of greatest importance in the authors' opinion is the strong sensitivity of the modeling results to the manner in which the emissions are characterized. The pattern of model estimates being highest under stable conditions and lower under unstable conditions is the result one might expect from modeling low level sources. The emission inventory of the St. Louis area for SO_2 for RAPS has only 208 sources considered as point sources. All other sources were included in the inventory in the area source category with relatively low emission release heights. If more of the emissions had been treated as point sources, the effect would be to decrease concentrations under stable conditions and to increase concentrations (locally near the point source) under neutral and unstable conditions since each emission would be from a point rather than a larger area. This is speculated by the authors to be a major cause of the behavior of model estimates with stability. Inclusion of point sources in the inventory as area sources was identified by Ruff (1983) as a possible cause of stability bias in the model performance.

Finally, if the modeled emission rates for either the point sources or the area sources are too large during the nighttime hours as compared to the real emissions, this would likely result in an overestimation of concentration values during stable conditions. We cannot eliminate this from the realm of possibility as the diurnal cycle of emissions for most all of the emissions was modeled in developing the emissions inventory. Only for the largest point sources were hourly records kept of the fuel usages, for use in developing emission rates.

CONCLUSIONS

Previous examinations of the performance of the urban model RAM against the RAPS data base have shown considerable station-by-station scatter but little evidence of overall bias regarding average concentrations and second-highest concentrations for 1-, 3-, and 24-hour time periods. This report has partitioned the data by stability and two wind speed groupings within each stability category. Examination of maximum 1-hour concentrations and average concentration for each category reveal quite different model behavior from that of the measurements. For model estimates for both maxima and averages, the largest concentrations occur with stable conditions and low wind speeds. In contrast to this, for measurement maxima the largest concentrations occur over a range of stabilities from slightly unstable to moderately stable with low wind speeds at some stations and high wind speeds at other stations. For measurement averages, the highest concentrations occur for the most part with A stability for the low wind speeds and have no definite trend for the high wind speeds. The authors have speculated as to possible causes for the behavior observed. Although the exact causes were not determined for the bias seen in the comparison results, the fact that the bias was seen to be an apparent function of stability is considered a significant advance. These results should help direct research activities towards discovery of the causes.

Similar analyses to those shown here will allow assessment to be easily made of any resulting improvements.

ACKNOWLEDGEMENTS

The authors thank Adrian D. Busse for furnishing measurement data containing calibration information and for writing graphics software which were very useful in exploring data displays. The secretarial assistance of Joan K. Emory and Michele Richardson is gratefully acknowledged.

REFERENCES

Brower, R., 1980, Technical Note - Comparison between predictions of two Gaussian models (CRSTER and modified CRSTER) and measured SO_2 concentrations near tall stacks under convective conditions. Environmental Center, Martin Marietta Corporation, Baltimore, MD.

Gifford, F. A., and Hanna, S. R., 1971, Urban air pollution modeling, pp 1146-1151 in Proceedings of the Second International Clean Air Congress. Edited by H.M. Englund and W.T. Beery. Academic Press, New York.

Myers, R. L., and Reagan, J. A., 1975, The regional air monitoring system - St. Louis, Missouri, U.S.A., paper 8-6 in International Conference on Environmental Sensing and Assessment, Vol. 1, Las Vegas Nevada, September 14-19, 1975.

Novak, J. H., and Turner, D. B., 1976, An efficient Gaussian plume multiple - source air quality algorithm, J. Air Pollution Control Assoc., 26 : 570-575.

Ruff, R. E., 1981, Evaluation of the RAM Using the RAPS Data Base. EPA-600/4-81-020. Environmental Sciences Research Laboratory. U.S. Environmental Protection Agency, Research Triangle Park, NC. 84 p.

Ruff, R. E., 1983, Application of statistical methods to diagnose causes of poor air quality model performance, Atmos. Environ., 17 : 291-297.

Turner, D. B., and Irwin, J. S., 1983, Comparison of sulfur dioxide estimates from the model RAM with St. Louis RAPS measurements, pp 695-707 in : Air Pollution Modeling and Its Application II, C. De Wispelaere, ed. Plenum Pub. Corp. New York.

Turner, D. B., and Novak, J. H., 1978, User's Guide for RAM. Vol. I. Algorithm Description and Use, Vol. II. Data Preparation and Listings. EPA-600/8-78-016 a and b. Environmental Sciences Research Laboratory, Research Triangle Park, NC. 60 and 222 pp.

Weil, J. C., and Brower, R., 1982. The Maryland PPSP Dispersion Model for Tall Stacks. Environmental Center, Martin Marietta Corporation, for Maryland and Power Plant Siting Program, Department of Natural Resources.

SIMULATION OF TRANSFORMATION, BUOYANCY AND REMOVAL PROCESSES BY LAGRANGIAN PARTICLE METHODS

Paolo Zannetti[(*)] and Nazik Al-Madani

Environmental and Earth Sciences Division
Kuwait Institute for Scientific Research
P.O.Box 24885, Safat, Kuwait

INTRODUCTION AND SUMMARY

Particle methods (Hockney and Eastwood, 1981) are the most recent and advanced numerical tools for computer modeling of dynamic systems. They seem particularly successful in simulating turbulent fluid dynamics, due to their capability of incorporating semi-random components. Particle modeling of air pollution diffusion phenomena has recently become the subject of a great deal of investigation (e.g., Diehl et al., 1982; Legg and Raupach, 1982; Ley, 1982, Zannetti and Al-Madani, 1983). The promising results of these studies are, however, accompanied by the persisting difficulty of properly evaluating Lagrangian velocity statistics from Eulerian measurements (see Davis, 1982). Nevertheless, particle methods provide outstanding advantages over other air pollution diffusion modeling techniques, such as Gaussian models and grid models, as discussed below.

Atmospheric diffusion processes are characterized by turbulent eddies in which the motion of different air parcels is strongly auto-and cross-correlated. With simulation particles, the computer modeling of these eddies would, therefore, require the expensive computation of the interactions between each particle and its surrounding ones. A different approach can be followed if only ensemble averages need to be computed. In this case, in fact, each simulation particle can move independently from the others and its motion can be very realistically simulated by semi-random

(*) On leave of absence from AeroVironment Inc., 145 Vista Ave., Pasadena, CA 91107, USA.

fluctuations generated by computer Monte-Carlo techniques. These Lagrangian simulations with stochastic components seem extremely useful for at least reproducing such phenomena, as atmospheric turbulent diffusion, whose physical mechanism is too complex to be simulated by deterministic techniques.

A need exists to incorporate suitable numerical tools for the simulation of atmospheric phenomena besides turbulence into the Lagrangian particle methods. Therefore, this paper, after a few introductory remarks, discusses the definition and the computer implementation of special algorithms for the simulation of dynamic plume rise, chemical decay, and deposition-resuspension effects by particle methods. These algorithms have been incorporated into a prototype computer diffusion code (MC-LAGPAR, written in APL language) whose simulation results for a few test cases are presented and discussed.

THE MODEL

In the atmospheric boundary layer, the dispersion of emitted gaseous material can be described by a suitable number of fictitious particles moving, at each time step, according to pseudo-velocities simulating (1) transport, (2) turbulent fluctuations, and (3) molecular diffusion (if not negligible). These pseudo-velocities do not intend to simulate the real trajectory of a specific pollutant parcel, but to provide realistic dynamics of the pollutant motion on an ensemble basis.

The pseudo-velocities are decomposed into two terms: the space-dependent average values $\bar{u}_x$, $\bar{u}_y$, $\bar{u}_z$ (which must be provided by a meteorological model or by an interpolation-extrapolation of meteorological measurements), plus the particle-dependent fluctuations. Different Monte-Carlo schemes have been proposed to calculate the fluctuations u', v', and w' of the pseudo-velocities (e.g., Watson and Barr, 1976; Hanna, 1981). To properly simulate the wind shear effects, Zannetti (1981) developed a scheme in which the pseudo-velocity fluctuations are auto-correlated (for all three components) and cross-correlated (between the vertical and along-wind fluctuations):

$$u'(t_2) = \phi_1 u'(t_1) + u''(t_2) \tag{1a}$$

$$v'(t_2) = \phi_2 v'(t_1) + v''(t_2) \tag{1b}$$

$$w'(t_2) = \phi_3 w'(t_1) + \phi_4 u'(t_2) + w''(t_2) \tag{1c}$$

In this scheme, the ϕ parameters and the intensities of the purely random components u", v", w" can be inferred from algebraical manipulations of known meteorological input parameters (intensities and correlations of the wind fluctuations).

In addition to transport and diffusion, particle methods can be used in a particularly effective way for providing a realistic treatment of buoyancy, chemical decay and ground deposition-resuspension effects. In the following sections, specific algorithms are proposed for the treatment of these special effects.

Dynamic Plume Rise

Particle methods provide a straightforward treatment of the dynamic plume rise. In fact, each emitted particle, tagged with its emission characteristics, can consume an increment of its initially supplied buoyancy F at each time step Δt:

$$\Delta F = \frac{\partial F}{\partial t} \Delta t \tag{2}$$

where $\partial F/\partial t$ is a function of meteorology (e.g., wind speed, temperature, stability). Each ΔF can then provide an additional vertical velocity

$$w_{pr} = f(\Delta F) \tag{3}$$

that moves each particle $w_{pr} \Delta t$ in the vertical direction, effectively simulating a dynamic plume rise.

Alternatively, a simpler computation can be performed rearranging existing semi-empirical plume rise formulas, as shown in the example below.

After the transformation x = ut, the TVA plume rise formula (Stern, 1976) can be written

$$\Delta h(t) = cF^{1/3}u^{-1/3}t^{2/3} \tag{4}$$

in which the constant c is

$$c(z) = 1.58 - 0.414 \frac{\partial \theta}{\partial z} \tag{5}$$

$\partial\theta/\partial z$ is the potential temperature gradient (^{o}C/100 m), F is the buoyancy (m^4s^{-3}) and u is the wind speed (ms^{-1}).

Both c and u vary with z. This suggests an empirical dynamic two-step computation in which the trajectory

$$z(t) = H + \Delta h(t) \tag{6}$$

of each particle is computed by (2nd step)

$$z(t+\Delta t) \simeq z(t) + w_{pr}\ \Delta t \tag{7}$$

where

$$w_{pr} = \left(\frac{dz}{dt}\right)_{t+\Delta t/2} = \left(\frac{d\Delta h}{dt}\right)_{t+\Delta t/2}$$

$$\simeq c\left[z(t+\Delta t/2)\right]F^{1/3}\ u\left[z(t+\Delta t/2)\right]^{-1/3}\ \frac{2}{3}\ (t+\Delta t/2)^{-1/3} \tag{8}$$

and (1st step)

$$z(t+\Delta t/2) \simeq z(t) + \left(\frac{dz}{dt}\right)_t \Delta t/2 \tag{9}$$

where

$$\left(\frac{dz}{dt}\right)_t = \left(\frac{d\Delta h}{dt}\right)_t \simeq c\left[z(t)\right]F^{1/3}\ u\left[z(t)\right]^{-1/3}\ \frac{2}{3}\ t^{-1/3} \tag{10}$$

Emitted particles do not need to be provided with the same buoyancy. Actually, the extra vertical diffusion produced during plume rise will be realistically simulated by releasing particles with a buoyancy defined by

$$F = \bar{F} + F' \tag{11}$$

where $\bar{F}$ is the average value and F' is a random component (particle-dependent) of suitable intensity.

Chemical Decay

An exponential decay, taking into account all removal factors except ground deposition, can be performed at each time step. If the time scale of the phenomenon is T_c (where T_c can be a function of the type of pollutant and of the meteorology) the probability of removal for each particle at each time step is

$$p_c = 1 - \exp(-\Delta t/T_c) \tag{12}$$

Consequently, $p_c n_p$ particles must be randomly cancelled from the computational domain, where n_p is the current number of active (i.e., not cancelled or deposited) particles.

Ground Deposition-Resuspension

At the end of each time step, all active particle locations need to be tested to single out those particles (say n_b) that have been moved below terrain (z<0). Some of these n_b particles will be reflected and the rest of them will be deposited on the ground. If T_d is the time constant of this partial deposition process, each of the n_b particles currently below the terrain has a probability of

$$p_d = 1 - \exp(-\Delta t/T_d) \tag{13}$$

to be deposited. Therefore, $p_d n_b$ randomly selected particles (among the previously identified n_b) will be deposited and the rest of them ($n_b - p_d n_b$) will be reflected.

Particles deposited on the ground can be resuspended back to the computational domain or permanently absorbed by the ground. If n_d is the current number of deposited particles and T_s is the time scale of the resuspension process, each of the n_b particles has a probability of

$$p_s = 1 - \exp(-\Delta t/T_s) \tag{14}$$

to be resuspended. Therefore, at each time step, $p_s n_d$ particles will be resuspended; but, at the same time, if a particle remains deposited on the ground for a period of time greater than a critical value T_{dmax}, the particle will be permanently absorbed.

T_d, T_s and T_{dmax} are functions of the meteorology (especially the surface wind speed) and the characteristics of both the pollutant and the ground surface. The proper inference of these values allows realistic diffusion simulations very difficult to obtain using other modeling techniques.

THE MC-LAGPAR CODE

A prototype computer code written in APL has been developed that incorporates, among other things, the algorithms previously described. The code simulates the diffusion of a single puff in flat terrain with non-homogeneous non-stationary meteorological conditions. The code is fully grid-free, since the meteorological variables are inputted at selected altitudes and then linearly interpolated at each particle's elevation. In this way, abrupt variations of the meteorological input parameters (causing artificial shear effects) are avoided. Moreover, since the selected altitudes do not need to be equally spaced, any degree of resolution in inputting the meteorological values can be obtained.

The meteorological variables required at each altitude at each time step are:

- the average wind components $\bar{u}_x$, $\bar{u}_y$, $\bar{u}_z$
- the standard deviations $\sigma_{u'}$, $\sigma_{v'}$, and $\sigma_{w'}$ of the pseudo-velocities (u' and v' are the along-wind and the cross-wind components; w' is along z)
- the auto-correlations $r_{u'}$, $r_{v'}$, $r_{w'}$ of the pseudo-velocities
- the cross-correlation $r_{u'w'}$
- the potential temperature gradient $\partial\theta/\partial z$.

In addition to these, T_c, T_d, T_s, T_{dmax} need to be inputted (a single value for the entire domain).

The time increment Δt must be carefully chosen. All model parametrizations are independent from Δt, but nevertheless, abrupt variations of particle elevations should be avoided and $w'\,\Delta t$ values should be less than the length scale of the vertical variation of the meteorological input. Ten seconds is probably a reasonable upper limit value for Δt.

COMPUTER SIMULATIONS

The MC-LAGPAR code has been applied for the simulation of a few test cases to provide a qualitative demonstration of the flexibility and the high degree of resolution of this numerical approach. Each simulation (150 steps of 10 seconds each) generates a puff of 100 particles at the source location. Particle locations (x,z) are plotted every three time steps thus generating a continuous plume up to a few kilometers downwind of the source. The vertical dimension chosen was twice the horizontal one.

Fig. 1 shows a slightly buoyant plume released at an altitude of 100 m during dispersion conditions of moderate vertical turbulence. The wind speed is 1 ms^{-1} at the ground and 4 ms^{-1} at the top of the domain. The $\sigma_{u'}$ and $\sigma_{v'}$ parameters start with 0.5 ms^{-1} at the ground to 1.0 ms^{-1} at 100m, remaining constant above that level. The $\sigma_{w'}$ parameter has the same behaviour, i.e., from 0.2 ms^{-1} at the ground to 0.5 ms^{-1} at 100m. The autocorrelations $r_{u'}$, $r_{v'}$, and $r_{w'}$, are constant at 0.7, 0.7, and 0.5, respectively. The cross-correlation $r_{u'w'}$ increases from -0.3 at the ground to -0.1 at the top.

Fig. 2 shows the dynamics of an elevated (150m) hot plume during conditions of low vertical turbulence. Meteorological values are similar to the previous simulation, except $\sigma_{w'}$, which now grows from 0.1 ms^{-1} at the ground to 0.35 ms^{-1} at 100m, then decreasing to 0.15 ms^{-1} at the top.

The simulation in Fig. 3 is very similar to the previous one. Now, however, an elevated inversion layer has been added by forcing $\sigma_{w'}$ equal to 0.1 ms^{-1} and $\partial\theta/\partial z$ equal 2 $^{o}C/100m$ between 500m and 550m. Most particles have enough buoyancy to penetrate the inversion layer and be trapped inside. Only a few particles perforate the inversion layer reaching the more turbulent region above. This simulation is characterized by a very unusual result, i.e., the decrease of the plume's σ_z with the downwind distance at 2 km from the source.

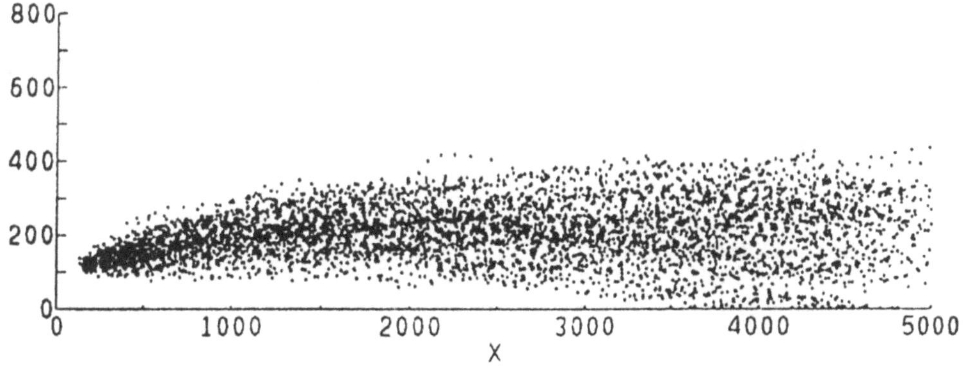

Fig. 1 - Simulation of a slightly buoyant elevated plume with moderate turbulence.

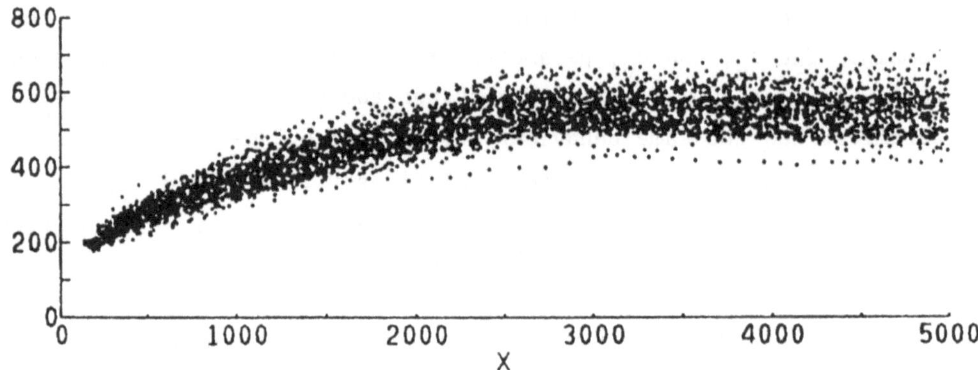

Fig. 2 - Simulation of a hot elevated plume with low turbulence.

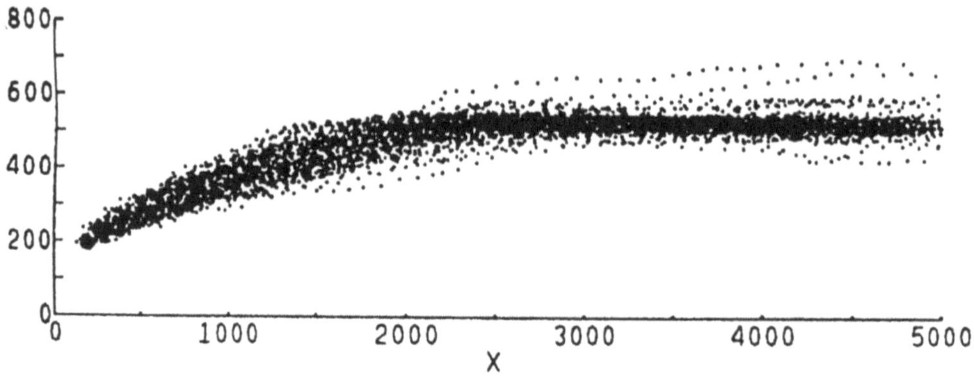

Fig. 3 - Simulation of a hot elevated plume with low turbulence and the presence of an elevated inversion layer between 500 m and 550 m.

Fig. 4 shows the release of a non-buoyant plume in moderate vertical turbulence in which $\sigma_{w'}$ reaches its maximum (0.6 ms^{-1}) at 250m, just below an elevated inversion layer between 250 m and 300 m. With a non-buoyant plume, the inversion layer acts more like a reflection barrier for the particles, even though particle trapping effects are still evident.

Finally, Fig. 5 presents a low-level release (at 10 m) with slight buoyancy in moderate vertical turbulence. Deposition-resuspension phenomena are accounted for by T_d = 10 s, T_{dmax} = 50 s and T_s = 1000 s. With these values, 23% of the emitted mass is found permanently deposited on the ground 1500 s from its release.

CONCLUSIONS

Lagrangian particle methods applied to air pollution dispersion simulations can easily provide a degree of resolution and accuracy not obtainable by other simulation techniques. This method can also incorporate a realistic treatment of such phenomena as buoyancy and deposition-resuspension. This technique can be seen as a very "natural" and effective way of simulating atmospheric dispersion processes. In fact, whereas other modeling techniques operate a questionable discretization of the atmospheric turbulence into "stability" classes or require a meteorological input (e.g., the eddy diffusion coefficients K's) not directly measurable, the particle methods require meteorological input parameters (i.e., the pseudo-velocity statistics) that seem very close to the measurable wind statistics. Nevertheless, much investigation is still required to provide a fully acceptable method of relating Eulerian wind measurements to the required pseudo-velocity (Lagrangian) statistics.

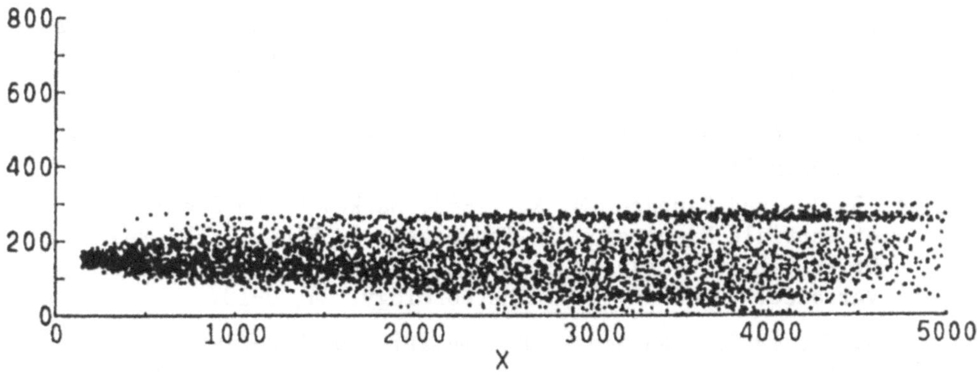

Fig. 4 - Simulation of a non-buoyant plume with moderate turbulence and the presence of an elavated inversion layer between 250 m and 300 m.

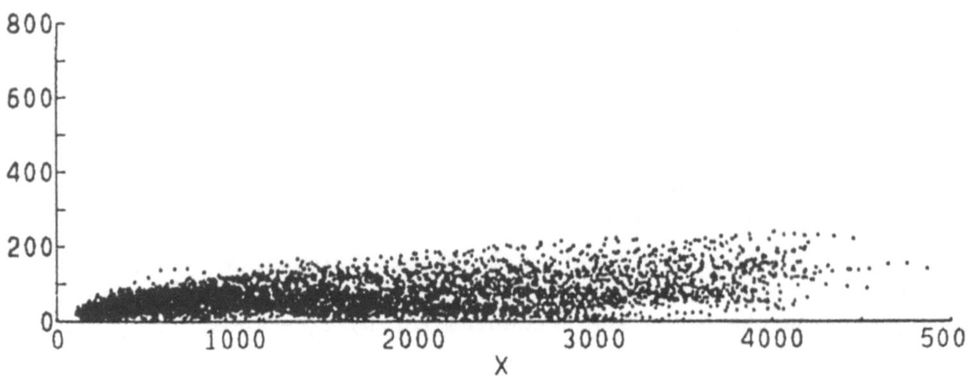

Fig. 5 - Simulation of a low-level release with slight buoyancy, moderate vertical turbulence, and deposition-resuspension effects.

REFERENCES

Davis, R. E., 1982, On Relating Eulerian and Lagrangian Velocity Statistics: Single Particles in Homogeneous Flows. J. Fluid Mech., 114, 1-26.

Diehl, S. R., Smith, D. T., and Sydor, M. 1982, Random-Walk Simulation of Gradient-Transfer Processes Applied to Dispersion of Stack Emission from Coal-Fired Power Plants. Journ. of Appl. Meteor., 21(1), 69-83.

Hanna, S. R., 1981, Lagrangian and Eulerian Time-Scale Relations in the Daytime Boundary Layer. Journ. of Appl. Meteor., 20 (3), 242-249.

Hockney, R. W., and Eastwood, J. W., 1981, Computer Simulations Using Particles. McGraw-Hill, Inc.

Legg, B. J., and Raupach, M. R., 1982, Markov-Chain Simulation of Particle Dispersion in Inhomogeneous Flows: the Mean Drift Velocity Induced by a Gradient in Eulerian Velocity Variance. Boundary-Layer Meteorology, 24, 3-13.

Ley, A. J., 1982, A Random Walk Simulation of Two-Dimensional Turbulent Diffusion in the Neutral Surface Layer. Atmos. Environ., 16(12), 2799-2808.

Stern, A. C., Ed., 1976, Air Pollution. Third Edition. Volume I p. 429-30. Academic Press.

Watson, C. W. and Barr, S., 1976, Monte-Carlo Simulation of the Turbulent Transport of Airborne Contaminants. Los Alamos Scientific Laboratory, Technical Report LA-6103.

Zannetti, P., 1981, Some Aspects of Monte-Carlo Type Modeling of Atmospheric Turbulent Diffusion. 7th Conference on Probability and Statistics in Atmospheric Sciences, AMS. Monterey, CA, Nov. 2-6, 1981.

Zannetti, P., and N. Al-Madani., 1983, Numerical Simulations of Lagrangian Particle Diffusion by Monte-Carlo Techniques. VIth World Congress on Air Quality (IUAPPA). Paris, May 16-20, 1983.

DISCUSSION

R. BERKOWICZ — Have you considered the problem that the correlation function for buoyant particles can depend on buoyancy ?

P. ZANNETTI — No, I did not. I agree that this further refinement could be useful.

R. BERKOWICZ — How do you handle the difficulties resulting from gradients in e.g. σ_w.

P. ZANNETTI — This is still an open problem. Some researchers use corrective terms associated to σ_w gradients. Perhaps, using proper values for $r_{u'w'}$, these corrective terms should not be required.

G. DUMONT — If you run the model twice with exactly the same input parameters, will the results be identical ? (because you are using Monte-Carlo methods for describing the fluctuations).

P. ZANNETTI — If you use the same random number generator routine the results will be identical.

G. RESELE — I have a follow-up question concerning the question on chemical interaction : isn't the decay of buoyancy an effect, that should also be treated by taking the position of the other particles into account. I imagine that buoyancy decays at the edge of a plume rather than in the centre of the plume.

P. ZANNETTI This effect can be incorporated into the code, if required. I agree that it may be important in some cases.

B. RUDOLF Can you take in account any chemical interactions between two ore more pollutants.

P. ZANNETTI In order to take into account chemical reactions you must superimpose a grid, count the particles, compute the concentrations in each cell, calculate the chemical reactions between concentrations and, then, go back to particle transport-diffusion. This must be done at each time step and it is very expensive.

WIND-FIELD AND POLLUTANT MASS-FLOW SIMULATION OVER AN URBAN AREA

G. Clerici*, S. Sandroni**, and L. Santomauro***

*A.R.S., Milano, Italy
**Commission of the European Communities
JRC, Ispra (Va), Italy
***Osservatorio Meteorologico di Brera
Milano, Italy

INTRODUCTION

The air quality level and its trend in large urban areas are generally deduced from a network of fixed monitoring stations, whose number and distribution depend mostly on economic and local policy factors. Local authorities may have at their disposal a source inventory from which, according to the day and the meteorological forecast, the air quality level over the next 24 hours can be predicted by an appropriate stochastic model. Although it has the well-known advantage of continuous monitoring, a monitoring network gives only part of the information about the transport of pollution.

Complete information can be obtained only by means of numerical models involving the effects of the horizontal and vertical wind fields.

Several models developed in recent years are based on a numerical integration of the mass-conservation equation for the pollutants (see for instance Randerson, 1970 ; Sklarev et al., 1972 ; Mahoney et al., 1970). Such an approach requires that the wind and diffusivity fields be known "a priori", independent of the pollutant distribution field, for instance from observations over long periods at ground level.

An alternative approach is based on the numerical integration of momentum, heat and atmospheric water content equations associated with the pollutant mass-conservation equation. All these

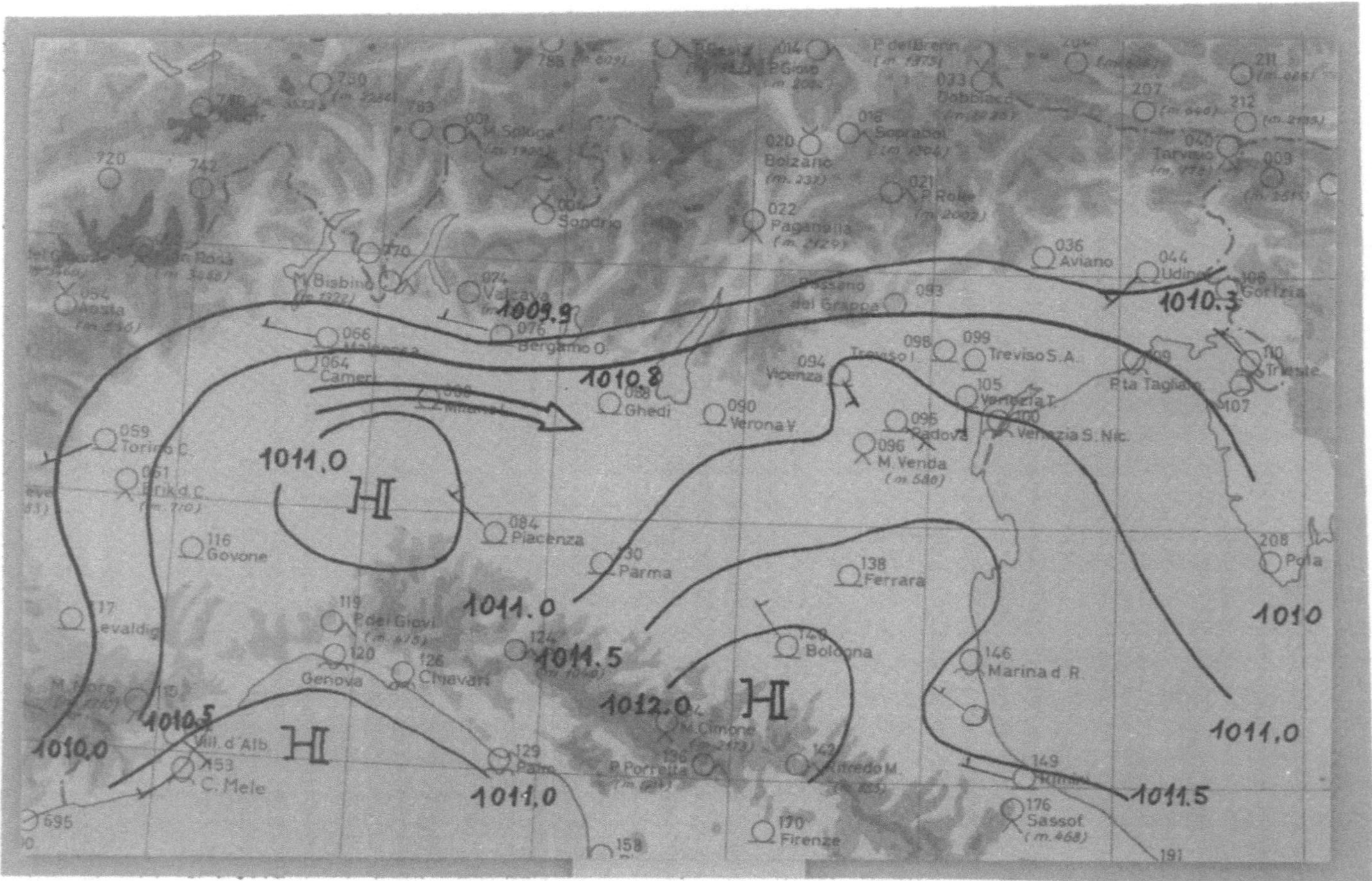

Fig. 1. Observed temperature and wind fields in the Po valley on May 20th, 1980, 1200 GMT.

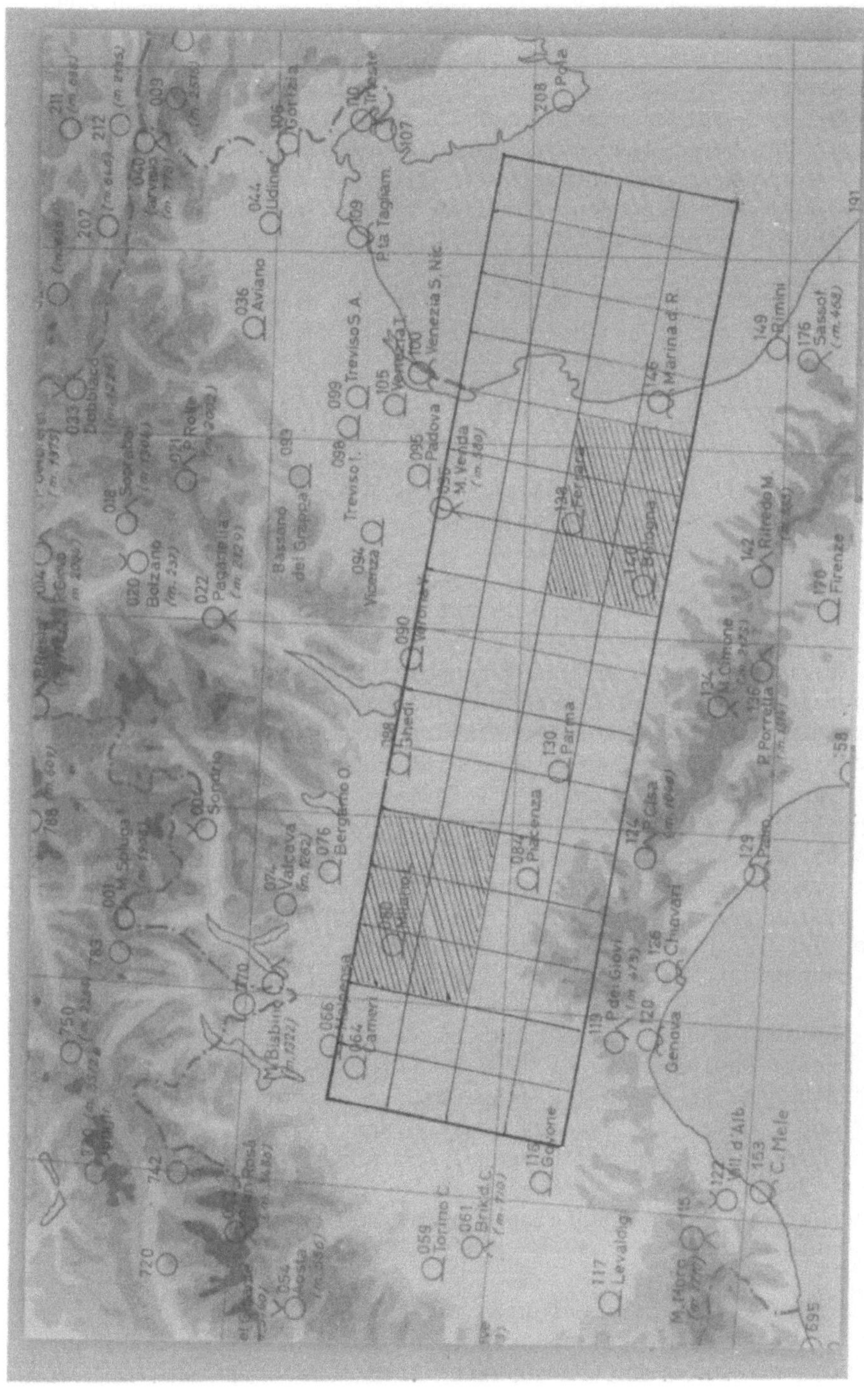

Fig. 2. Grid arrangement for the central Po valley.

equations must relate the vertical turbulent exchange to wind and temperature fields. Such a model is able to describe wind and diffusivity fields, consistent with the polluted air mass distribution ; furthermore it is applicable to any site, season of the year, hour of the day, or type of underlying surface and it is affected by a feedback action of the temperature field. Such an approach has been followed by Pandolfo et al. (1971), Pielke (1973) and by Wippermann and Yordanov (1973). All these models are three-dimensional time-dependent Eulerian models, highly sophisticated and not easy to handle.

In a previous paper (Clerici, 1982) a simplified formulation of Pandolfo's model has been described in detail. In this paper we describe some applications of the new model and in particular (1) the simulation of the wind field in the Po valley as a boundary to the Milan urban area, (2) the simulation of the sulphur dioxide distribution in relation to the emission inventories available, and finally (3) the calculation of the net mass-flow above the urban area in some typical days.

SIMULATION OF THE WIND FIELD IN THE PO VALLEY

The Po valley is a densely populated region with industrial and rural activities ; its continental climate is characterized by frequent and prolonged temperature inversions, much fog during the winter and a dominant breeze wind regime. The wind speeds rarely exceed 2 m/s. Synoptic information on wind, temperature and pressure are collected from a network of meteorological stations by Aeronautica Militare. Fig. 1 shows a map of the area considered, the location of the stations and the wind and temperature fields for a selected day (May 20th, 1980, 1200 GMT) at ground level. Twice a day (at 000 and at 1200 GMT) radiosondes are launched at Milano Linate Airport.

The region under investigation is divided into a 30 x 8 grid, each square having sides of 12.5 km (Fig. 2). Ten grid squares located on the Po Delta are considered as having an air-water interface. Some grid squares located south of the Po river (Piacenza, Parma, Bologna and Ferrara) and north of it (Brescia, Milano, Bergamo, Como, Varese) are considered as urban areas (dashed areas) ; the remaining squares are considered as rural or aquatic areas.

A more realistic simulation may be obtained if we use weighted urban and rural contributions in each grid square, but this may be considered too sophisticated a description.

In the vertical directions ten levels have been taken under consideration, not regularly spaced (50, 100, 250, 500, 750, 1000, 1500, 2000, 2500, 3000 m) in order to better describe local effects of the lowest layer.

As mentioned above, the input meteorological data have been obtained from Aeronautica Militare, Linate Airport. Wind fields and temperature were interpolated from the values of the regional network (Fig. 1) as well as of the radiosonde (Table 1).

On that day, in the presence of an easterly wind, the North, South and West boundaries are assumed as rigid reflecting walls. All initial horizontal gradients are assumed as zero. The simulation has been performed over a time interval of 9 h, from 900 to 1800 LMT.

For every grid square and for every hour we obtained the air and surface temperatures, u and v components of the wind vector (or water current), the humidity field, the salinity and the air pollutant concentration. The temperature distribution calculated across the Po valley shows that the highest temperatures are generally found in urban areas and over the upper Adriatic sea (Fig. 3). Fig. 4 shows an x-y cross section of the temperature and wind fields for the hatched urban squares : the maximum temperature is reached over the centre of Milan which is also the centre of the urban heat island top (277°K).

Table 1. Temperature and wind fields obtained from a radiosonde at Milano Airport on May 20th, 1980, 1200 GMT.

Z (m)	T(°K)	u(cm/s)	v(cm/s)
3930	247		
2300		- 340	60
1500		- 340	60
1416	265	-	-
1300		- 300	0
1030		- 300	0
750		- 200	0
500		- 150	0
300		- 100	0
104	275.8	- 100	0

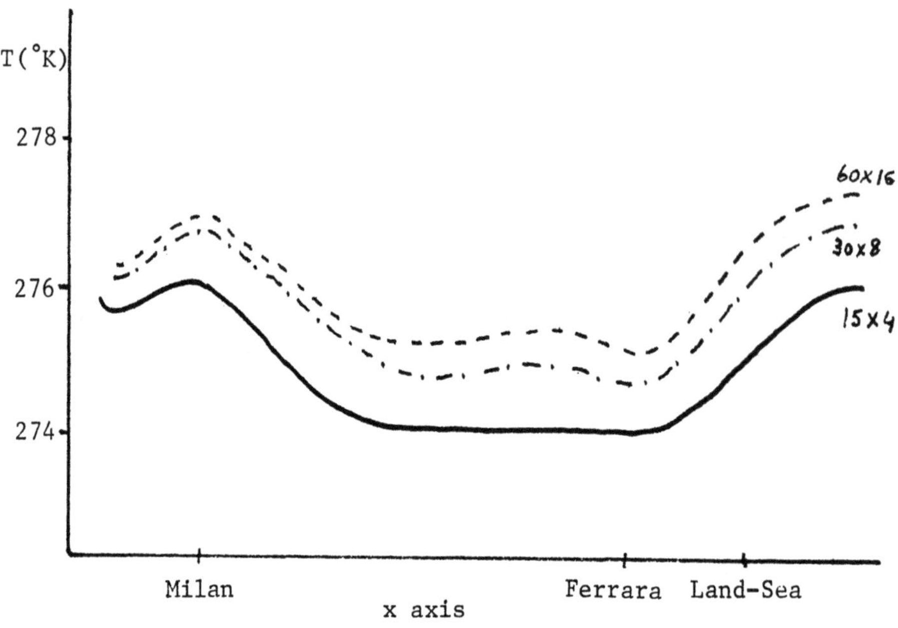

Fig. 3. Surface temperature distribution along the Po valley.

All these data suggest a pattern of a low wind regime. An accurate analysis shows that the locations of urban, rural and water surfaces play an important role in the air mass circulation. Since the urban characteristics are specified for the entire square including the urban area (and the urban area can be smaller than the square containing it), these simulations may sometimes be unrealistic. From Fig. 4 it is obvious however that the winds originate from thermally and frictionally induced circulations. Very probably such an air mass circulation is generated by the difference between urban and rural thermal characteristics. Also in this case the ascending motions over the warmer urban and water grid squares and the descending ones over the cooler rural areas are consistent with the simulated horizontal wind patterns, particularly at the lowest levels.

POLLUTANT MASS-FLOW OVER THE URBAN AREA

Once the wind field is depicted for the regional and the urban areas, the SO_2 distribution can be calculated by integration of the mass conservation equation. An input for this calculation is the emission inventory ; a time-consuming and tedious work. The mass-flow calculations have been performed for the Milan area, taking into consideration the days and emissions already analysed in a previous paper (Sandroni et al., 1982, b).

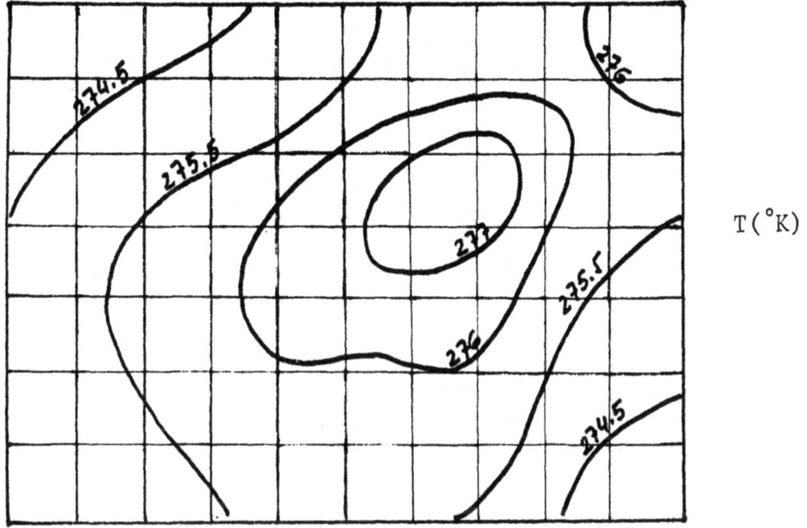

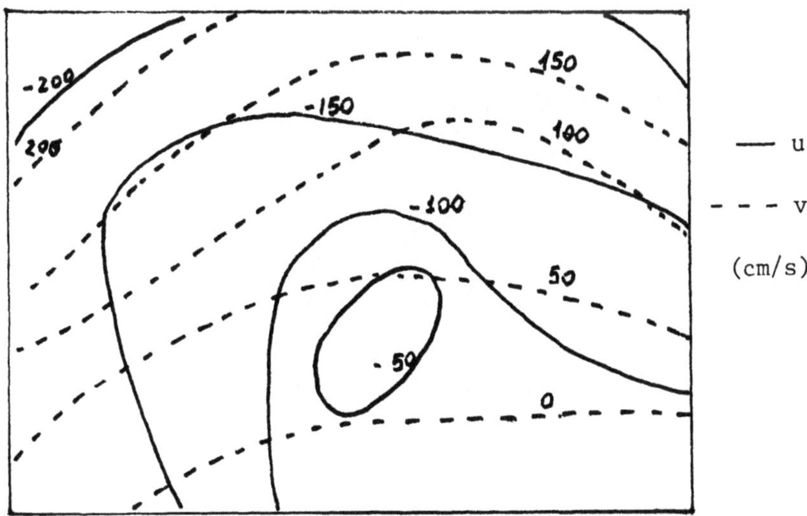

Fig. 4. Calculated temperature and wind fields for the N-W corner of the grid (dashed area). May 20th, 1980.

The ring road along which burden and ground concentration measurements have been performed by a mobile laboratory, is shown in Fig. 5. This region has been covered by a uniform grid (15 x 15) including the survey roads. Ten levels have been taken into consideration in the vertical direction, according to the criteria cited above. In total we have 2250 cells for which the parameters u, v, T, and the concentration C are calculated by the model. By assuming $\Delta x = \Delta y$ for the horizontal grid dimension and indicating the vertical one by Δz_k, the net mass flow crossing a wall of the box including the city is given by

$$\emptyset = \sum_{j=1}^{15} \sum_{k=1}^{10} \Delta x.\ z_k.\vec{V}_{jk}.\vec{n}C_{jk}$$

in which $\vec{V}_{jk}$ and C_{jk} are the wind vector and the concentration relative to the jk-th cell, $\vec{n}$ is a unit vector perpendicular to the surface $\Delta x.\Delta z_k$.

By repeating this calculation for the other walls and by summing up all the contributions, the net mass flow entering or leaving the city may be deduced.

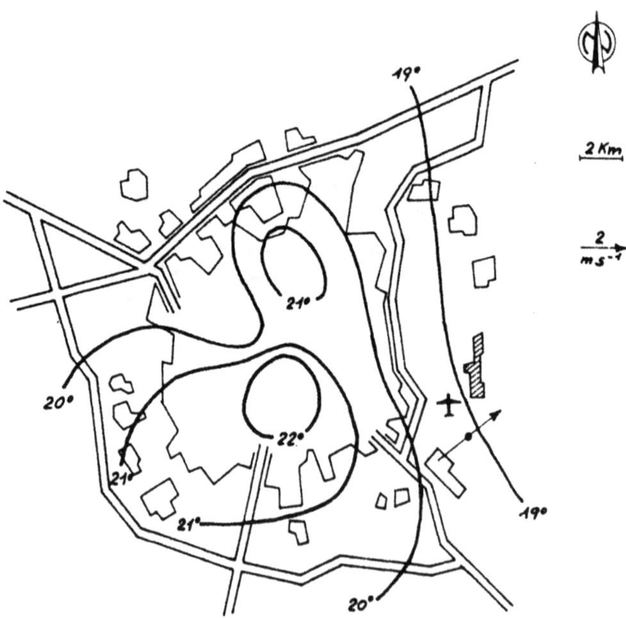

Fig. 5. Map of Milan urban area delimited by the motorways along which surveys have been performed. Temperature field deduced from ground network data on 20.05.1980.

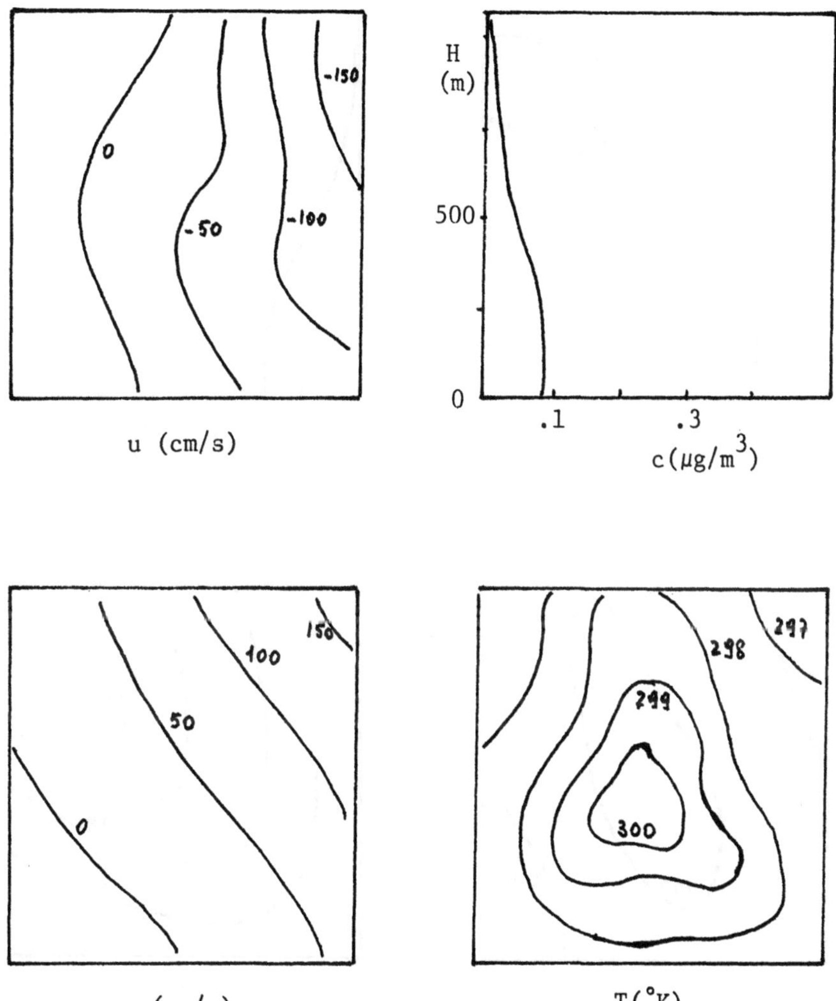

Fig. 6. Calculated wind components, temperature and SO_2 vertical distribution for August 12th, 1980.

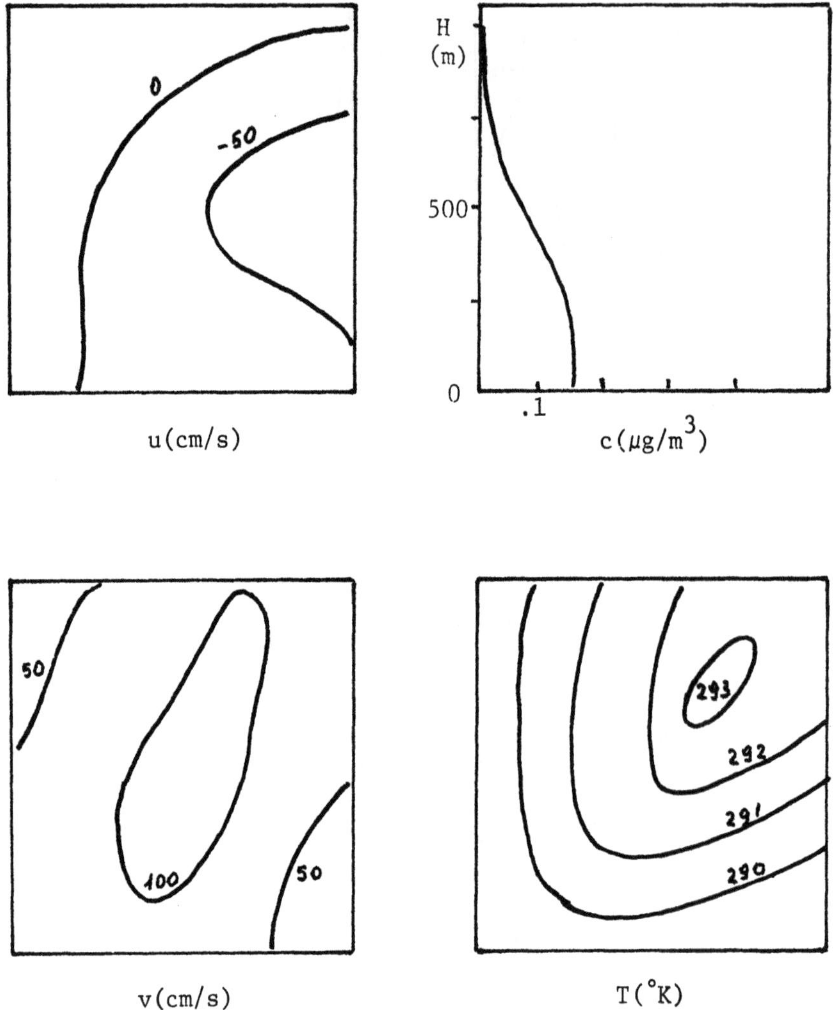

Fig. 7. Calculated wind, temperature and SO_2 vertical distribution on September 24th, 1980.

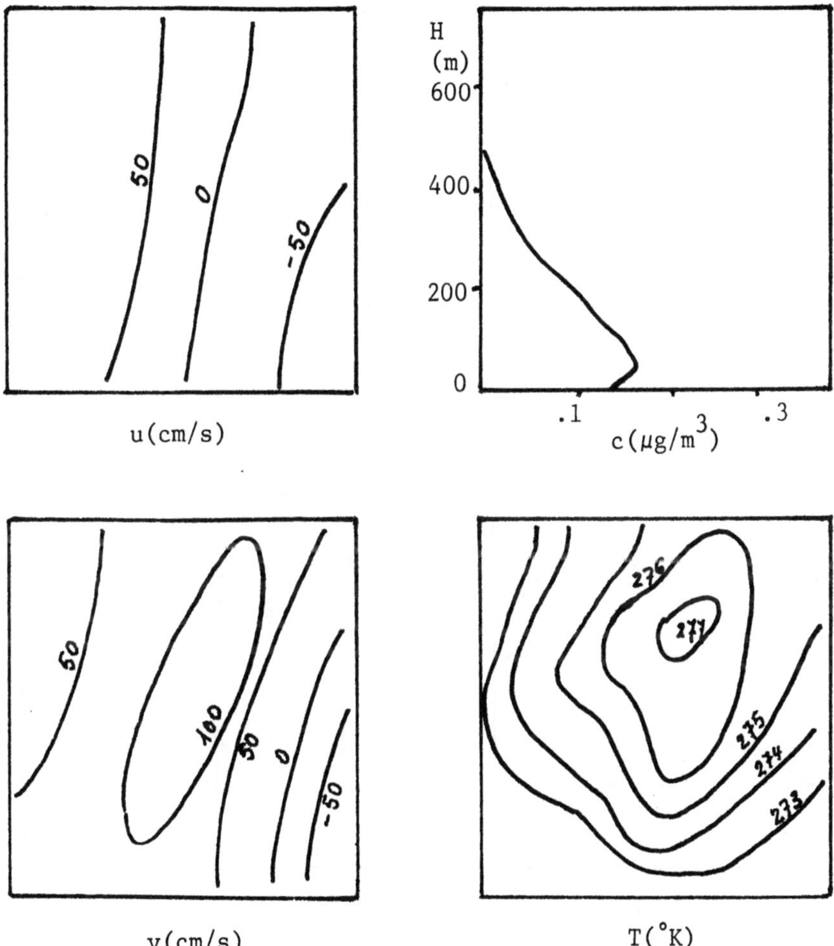

Fig. 8. Calculated wind, temperature and SO_2 vertical distribution for January, 14th, 1981.

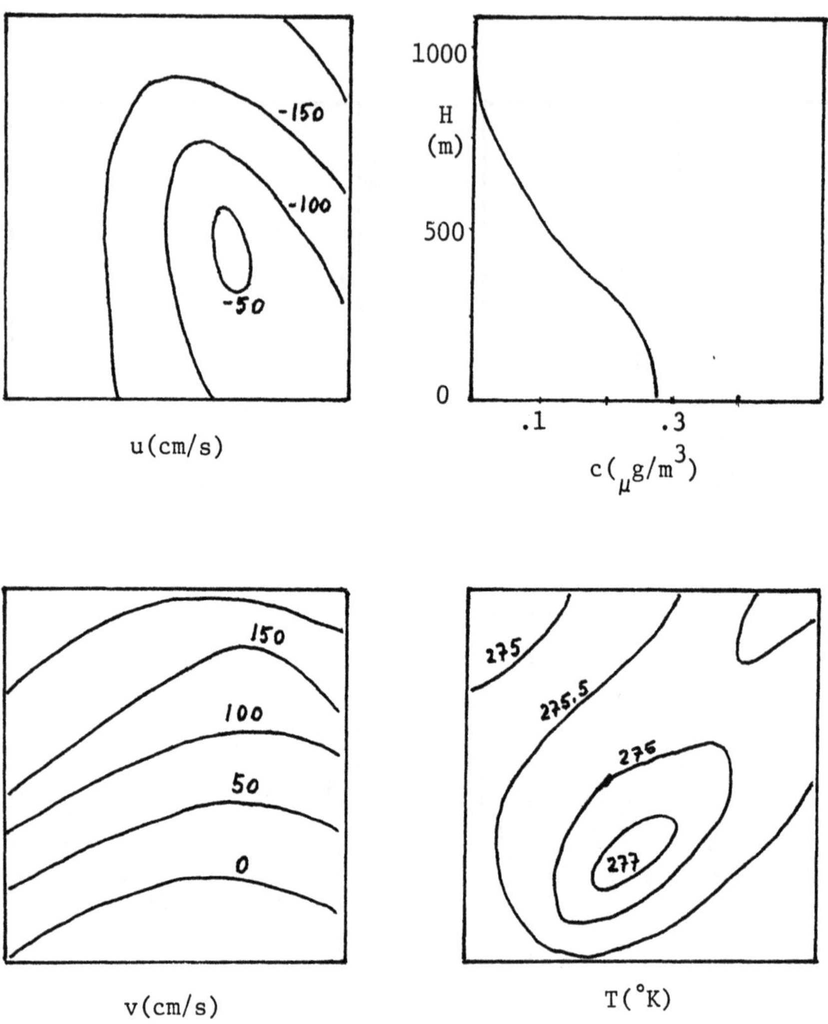

Fig. 9. Calculated wind, temperature and SO_2 vertical distribution on February 18th, 1981.

RESULTS AND DISCUSSION

Figures 6 - 9 show the horizontal components of the wind vector, the temperature field and vertical distribution of SO_2 concentration for all the days considered, as provided by the model. From the vertical profile of SO_2 concentration, one may see the possibility of estimating the mixing height (within the accuracy of Δz_k), defined as the maximum height which could be reached by the pollutant. For instance on 14th January 1981 it is estimated as 450 m.

The calculated mass flow ϕ, given in g/s SO_2 as a difference between input and output flow in respect to the wind direction are listed in Table II together with the mixing heights ; for comparison the mass-flows measured in the meantime (Sandroni et al., 1982) are listed.

As shown in Table II, the calculated flows agree satisfactorily with the measured ones. Once more it is confirmed that the urban area is a sink for SO_2 in the warm season and an emitter in the cold season.

Except for one case (August 12th) the mixing heights calculated by the model agree with the values evaluated by the thermodynamical procedure described in our previous paper (Sandroni et al., 1982). It is noticeable that the same values agree with those obtained by the experimental procedure based on burden and ground concentrations, in which a uniform vertical mixing is assumed.

Table II. Comparison between calculated and measured mass-flows and mixing heights.

Day	ϕ calc. (g/s)	ϕ meas.(g/s)	H_m calc.	H_m therm.
20.05.1980	- 1946	- 2221	1200	1300
12.08.1980	- 22	- 27	1000	1950
24.09.1980	- 834	- 1167	1000	850
14.01.1980	5137	5420	450	350
18.02.1981	8062	8952	1000	870

REFERENCES

Clerici G. (1982) Computional method for monitoring the atmospheric transport of pollutants to the Mediterranean Sea. 4th session of the Gesamt Working Group on Interchange of Pollutants between the Atmosphere and the Oceans. Monaco, 25-29 October, 1982.

Colacino M. (1982) Observations of a sea breeze event in the Rome area. Ark. für Meteo. Geophysics and Bioklimatology, Sect. B., Vol. 30 (1982).

Mahoney I.R., Egan B.A. (1970) A mesoscale numerical model of atmospheric phenomena in urban areas. Proc. 2nd Int. Clean Air Congr., Academic Press, New York.

Pandolfo J.P. et al., (1971) Prediction by numerical models of transport and diffusion in an urban boundary layer. Final Report of CEM, Contract No. 4082.

Pielke R.A. (1973) A three-dimensional numerical model of the sea breezes over South Florida. Ph. D. Dissertation, Pennsylvania State University.

Randerson D. (1970) A numerical experiment in simulating the transport of SO_2 through the atmosphere. Atmos. Environ. 4, 615-622.

Sandroni S., De Groot M., Borghi S., Santomauro L. (1982) Air pollution mass flow over Milan area, Atmos. Environ. 16 1271-72.

Sandroni S., De Groot M., Clerici G., Santomauro L. (1982) Simulation of SO_2 mass flow over Milan area. 13th Int. Meet. on Air Pollution Modelling, Ile des Ambiez (France), Sept. 1982.

Sklarew R.C. et al. (1972) Mathematical modelling of photochemical smog using the PIC method. J. Air Pollut. Control Assoc. 22, 865-869.

Wippermann F., Yordanov D., (1973) A prospective for a routine prediction of concentration patterns. Atm. Environ. 6 87.

DISCUSSION

G. RESELE — Could you comment once again on the boundary conditions you assumed to derive the wind field on the large scale grid ?

G. CLERICI — Po valley is surrounded by mountain chains on North, South and West boundaries. These three sides are considered as rigid reflecting walls. For the open side an easterly geostrophic wind is assumed.

PARTICLE SIMULATION OF DUST TRANSPORT AND DEPOSITION AND COMPARISON WITH CONVENTIONAL MODELS

L. Janicke

Dornier-System GmbH

D-7990 Friedrichshafen

INTRODUCTION

By conventional models we mean dust dispersion models based on the Gaussian plume formula; this includes the following models:

- the source depletion model,
- the surface depletion model,
- the tilted plume model.

A fourth model has been added. It is based on the solutions of the diffusion equation, where the diffusivity is computed as a function of the source distance only (it is abbreviated in the following as "diffusion model").

The German "Technical Instructions for Air-Pollution Control" (TA Luft) prescribe the application of the source depletion model for estimating the dust concentration and deposition in the vicinity of an industrial source (distance up to 50 stack heights).

In order to see whether this is actually the best model choice computations have been performed with the four models in a wide range of input parameters (stack height, stability class, particle size, deposition velocity). The results are compared with those of the particle simulation model, which is taken as a reference model, because it is best suited for modeling the dispersion properties of the atmosphere.

All models have to satisfy the following two conditions:

1. In case of small particles without deposition on the ground they follow the gas dispersion model fixed by the TA Luft.

2. The mass of the emitted material has to be conserved, i. e. the total horizontal mass flux has to decrease exactly by the amount of the material being deposited.

The particle simulation model has to be calibrated to satisfy as far as possible the first condition.

For simplicity reasons the cross-wind-integrated concentration has been considered only. All conventional models (except the surface depletion model) have a cross-wind (y-direction) concentration profile that is given by the factor exp $(-0.5y^2/\sigma_y^2(x))/(\sqrt{2\pi}\,\sigma_y(x))$.

GAUSSIAN MODELS

These models are based on modifications of the Gaussian plume formula for gas dispersion given below:

$$C_{gs}(x,z) = \frac{Q_o}{\overline{U}}\left[d(z-h,\ \sigma(x)) + d(z+h,\ \sigma(x))\right], \tag{1a}$$

$$d(z,\sigma) = \exp\ (-0.5z^2/\sigma^2)/(\sqrt{2\pi}\,\sigma), \tag{1b}$$

where Q_o = source strength, $\overline{U}$ = average transport velocity, z = height above ground, h = emission height, $\sigma(x)$ ($\equiv \sigma_z(x)$) = half width of the plume in the vertical direction. In all models it is assumed that the deposition flux is proportional to the concentration near the ground, the constant of proportionality being the deposition velocity v_d.

<u>Source Depletion</u>

Following Van der Hoven (1968) the depletion of the plume due to the deposition of material on the ground is taken into account by multiplying the source strength by a suitable factor,

$$C_{sc}(x,z) = \frac{Q(x)}{\overline{U}}\left[d(z-h,\ \sigma(x)) + d(z+h,\ \sigma(x))\right], \tag{2a}$$

$$Q(x) = Q_o \exp\left[-2\ \hat{v}_d \int_o^x d(h,\ \sigma(\xi))d\xi\right], \tag{2b}$$

$$\hat{v}_d = v_d/\overline{U}.$$

The effect of the deposition is immediately distributed over the complete vertical extension of the plume. A gravitational settling of the particles during the transport is not taken into account.

Surface Depletion

Horst (1977) proposed to treat the ground, on which material is deposited like a negative source. In this way the deposited material is taken preferably from the air near the ground, giving a more realistic picture of the actual deposition process. Horst gives the following formula for the concentration:

$$c_{sf}(x,z) = \frac{Qo}{u}\left[d(z-h,\sigma(x)) + d(z+h,\sigma(x)) - 2\hat{v}_d\int_o^x \chi_d(\xi)d(z,\sigma(x-\xi))d\xi\right] \quad (3a)$$

$\chi_d(x)$ has to be determined from the integral equation

$$\chi_d(x) = d(z_d-h,\sigma(x)) + d(z_d+h,\sigma(x)) - 2\hat{v}_d\int_o^x \chi_d(\xi)d(z_d,\sigma(x-\xi))d\xi \quad (3b)$$

This model contains as an additional parameter the deposition height z_d, which was introduced probably to guarantee the existence of the integral (3b). However, on one hand it prevents the exact mass conservation law. On the other hand no clear arguments can be presented how to choose z_d, although the result depends on it (usually z_d = 1 m is chosen).

Tilted Plume

The gravitational settling of the particles (velocity v_d positive in the negative z-direction) can be modeled by tilting the plume axis accordingly:

$$C_{ti}(x,z) = \frac{Qo}{\bar{U}}\left[d(z-h+x\hat{v}_g,\ \sigma(x))+\alpha(x)d(z+h+x\hat{v}_g,\sigma(x))\right]. \quad (4a)$$

$$\hat{v}_g = v_g/\bar{U}. \quad (4b)$$

This model differs from that proposed by Overcamp (1976) in that the plume of the mirror source is also tilted downwards with a factor $\alpha(x)$ depending on the source distance. $\alpha(x)$ is determined from the mass conservation law:

$$\int_o^\infty\left[d(\zeta-h+x\hat{v}_g,\sigma(x))+\alpha(x)d(\zeta+h+x\hat{v}_g,\sigma(x))\right]d\zeta \quad (4c)$$
$$= 1-\hat{v}_d\int_o^x\left[d(h-\xi\hat{v}_g,\sigma(\xi))+\alpha(\xi)d(h+\xi\hat{v}_g,\sigma(\xi))\right]d\xi .$$

By differentiating (4c) with respect to x we can get a differential equation for $\alpha(x)$. However, it was found that the integral equation (4c) is better suited for numerical computations.

Diffusion Model

The simple Gaussian model for gas dispersion (1) is a solution of the diffusion equation, if both the diffusion in the wind direction is neglected and the diffusivity function is properly chosen. A direct generalization to the case of dust dispersion is possible by including the gravitational settling velocity v_g in the transport term and by modifying the boundary condition:

$$\bar{U}\frac{\partial c_{di}}{\partial x} - v_g\frac{\partial c_{di}}{\partial z} = K(x)\frac{\partial^2 c_{di}}{\partial z^2} \tag{5a}$$

$$K(x) = \frac{\bar{U}}{2}\frac{d}{dx}\sigma^2(x), \tag{5b}$$

$$K\frac{\partial c_{di}}{\partial z} + v_g c_{di} = v_d c_{di} \quad \text{at } z = 0. \tag{5c}$$

An analytic solution for the case $K(x) = \text{const}$ (i. e. $\sigma^2(x) \sim x$) is given by Ermak (1977). The general case $K = K(x)$ can not be treated analytically. However, it is not necessary to solve the partial differential equation (5a) by numerical methods. Using the ansatz

$$c_{di}(x,z) = \frac{Q_o}{\bar{U}}\Big[d(z-h+x\hat{v}_g,\sigma(x))+d(z+h+x\hat{v}_g,\sigma(x)) \tag{6a}$$

$$-2\int_o^x \gamma(\xi)d(z+(x-\xi)\hat{v}_g,\tilde{\sigma}(x,\xi))d\xi\Big],$$

$$\tilde{\sigma}^2(x,\xi) = \sigma^2(x)-\sigma^2(\xi) = \frac{2}{\bar{U}}\int_\xi^x K(x')dx', \tag{6b}$$

the PDE (5a) is satisfied. From the boundary condition (5c) we get an integral equation for the unknown function $\gamma(x)$:

$$\gamma(x) = v_1(x)\left[d_+(x)+d_-(x)\right] - h\frac{\sigma'(x)}{\sigma(x)}\left[d_+(x)-d_-(x)\right] \tag{7a}$$

$$-2\int_o^x \gamma(\xi)\tilde{d}(x,\xi)\,\tilde{v}_1(x,\xi)\,d\xi,$$

$$d_\pm(x) = d(\mp h+x\hat{v}_g,\sigma(x)), \tag{7b}$$

$$\tilde{d}(x,\xi) = d((x-\xi)\hat{v}_g, \tilde{\sigma}(x,\xi)), \tag{7c}$$

$$\tilde{v}_1(x,\xi) = \hat{v}_d + \hat{v}_g\left[\frac{\partial\tilde{\sigma}(x,\xi)}{\partial x} \cdot \frac{x-\xi}{\tilde{\sigma}(x,\xi)} - 1\right], \tag{7d}$$

$$v_1(x) = \tilde{v}_1(x,o) = \hat{v}_d + \hat{v}_g\left[\frac{x\sigma'(x)}{\sigma(x)} - 1\right]. \tag{7e}$$

In the special case $v_g = 0$ the diffusion model differs from the surface depletion model only by the functional form of σ in the kernel. With $\sigma^2 \sim x$ and $v_d = 0.5\ v_g$ (i. e. $v_1(x) \equiv 0$) the tilted plume model and the diffusion model are identical.

THE PARTICLE SIMULATION MODEL

Details of the procedure to simulate gas dispersion on a computer by a stochastic particle transport have been described elsewhere (Janicke, 1983). It is extended to the case of dust dispersion by adding a gravitational settling to the particle movement. In addition each particle gets a statistical weight of 1 initially, which is diminished by the factor (1-p) with each reflection on the ground (it is assumed to be at $z_d = 6z_o$.). The reflection itself is elastic.

The probability p_d, that a particle hitting the ground is deposited can be related to the deposition velocity v_d by assuming that the distribution of the vertical velocity component of the particles hitting the ground is a drifting Maxwellian distribution (drift velocity $-v_g$, half width $\sigma_o = \sigma_w(z_d)$). The result is

$$p_d = \frac{2.5\ v_d/\sigma_o}{F_g+1.25\ v_d/\sigma_o} \tag{8a}$$

$$F_g = 1.25\ \frac{v_g}{\sigma_o} + \frac{\exp\ (-0.5\ v_g^2/\sigma_o^2)}{1+\mathrm{erf}\ (v_g/\sqrt{2}\ \sigma_o)} \tag{8b}$$

These assumptions are probably only valid for small values of v_d/σ_o and v_g/σ_o. Thus we have essentially $p_d \approx 2.5\ v_d/\sigma_o$.

The results from the particle simulation exhibit statistical fluctuations which make the interpretation more difficult. Therefore the vertical concentration profile resulting from the simulation, $c_{si}(x,z)$, is approximated using the maximum likelihood method by a smooth function $\overline{c}_{si}(x,z)$ of the form

$$\overline{c}_{si}(x,z) = \exp\ (\sum_i a_i f_i(x,z)). \tag{9}$$

The functions $f_i(x,z)$ are simple polynomials in x and z and their logarithms. Because this representation is nonlinear in the coefficients a_i, an iterative procedure has to be used. The starting values are determined from the variational principle

$$\delta \int C_{si}(x,z) \left[\ln C_{si}(x,z) - \ln \bar{C}_{si}(x,z) \right]^2 dx\, dz = 0. \quad (10)$$

Usually these starting values are already accurate enough. To estimate the statistical error of the simulation result the particles are divided into 10 groups. From the results for these groups 10 samples are drawn using the bootstrap method. Each sample is approximated as described above and from the variance of the results the 90 %-confidence interval is estimated.

Running the simulation model requires the knowledge of the turbulent velocities σ_u and σ_w, the Lagrangian correlation times T_u and T_w and the mean wind velocity u, all as functions of the height z. They are provided by boundary layer models, usually parameterized by the Monin-Obukhov-Length L, the roughness length z_o, the mixing depth z_i and the friction velocity u_*.

A test of the present models by Weber et al. (1982) has shown, that there are some uncertainties especially with respect to the correlation times. In addition the meteorological quantities being used in the simulation model have to be smooth functions of z and should be continuously dependent on the other parameters. Therefore some approximations had to be done. Making reference to the simulation results of Ley (1982) for neutral stratification as well as the discussion of the correlation times by Hanna (1981) the following boundary layer model for neutral and unstable stratification has been chosen:

$$\sigma_u = u_* \left(12 - \frac{z_i}{2L}\right)^{1/3}, \quad (11a)$$

$$\sigma_w = 1.3\, u_* \left(1-3\, \frac{z+z_o}{L}\right)^{1/3} \left[1-0.7 \left(\frac{z}{z_i-L}\right)^{2/3}\right], \quad (11b)$$

$$T_u = 0.17\, \frac{z_i}{\sigma_u}, \quad (11c)$$

$$T_w = 0.17\, \frac{z_i}{\sigma_w} \left[1-\exp\left(-1.86\, \frac{z}{z_i}\right)\right]. \quad (11d)$$

The wind profile u(z) is of the Businger type as described by Lamb (1979).

COMPARISON OF THE MODELS

For a first comparison an emission height of 100 m and neutral to slightly stable stratification (Klug/Manier III_1) is chosen, which is the most frequent stability class in Germany (about 50 %). The median of the wind velocity distribution for this class is about 5 m/s. The dispersion parameter σ_z is given by the TA Luft for this case as $\sigma(x) = 0.265\ x^{0.818}$.

The simulation model is calibrated to approach the Gaussian model for gas dispersion as close as possible. With L = -300 m, z_i = 800 m, and z_o = 1.8 m no significant differences in the predicted ground concentrations occur as shown in Fig. 1, except near the source. This is due to the neglection of the horizontal diffusion in the Gaussian model.

The high value of the roughness length used in the simulation model is appropriate to Jülich, where the dispersion parameters used in the Gaussian model have been measured.

The ground concentrations predicted by the different models for small particles (v_g = 0) with strong deposition (v_d = 8.7 cm/s) are presented in Fig. 2. As expected, the surface depletion model predicts smaller concentrations near the ground than does the source depletion model. Between both the tilted plume model nearly equals the diffusion model. This has been observed in all computations performed. The best approximation to the simulation results is given by the surface depletion model.

It is expected that the surface depletion model would give less favourable results as the mass of the particles increases. This is demonstrated in Fig. 3, which presents results for $v_g = v_d$ = 26 cm/s. The surface depletion model fails when it predicts a depletion of the portion of the plume adjacent to the surface. In reality the deposition flux is compensated by the flux of the gravitationally settling particles. Tilted plume model and diffusion model both approximate the simulation results very closely.*

Preliminary results of simulation runs for stable and unstable stratification confirm this picture with increasing differences between the Gaussian models as the stability of the thermal stratification increases: For non-gravitating particles the surface depletion model yields the best approximation. When the gravitational settling of the particles becomes important ($v_g/\bar{u} \gtrsim 0.01$), the

*For the high settling velocity used the deposition probability given by Eq. (8) probably is too low. Therefore, the concentration near the surface found with this simulation run might be somewhat too high (5 - 10 %).

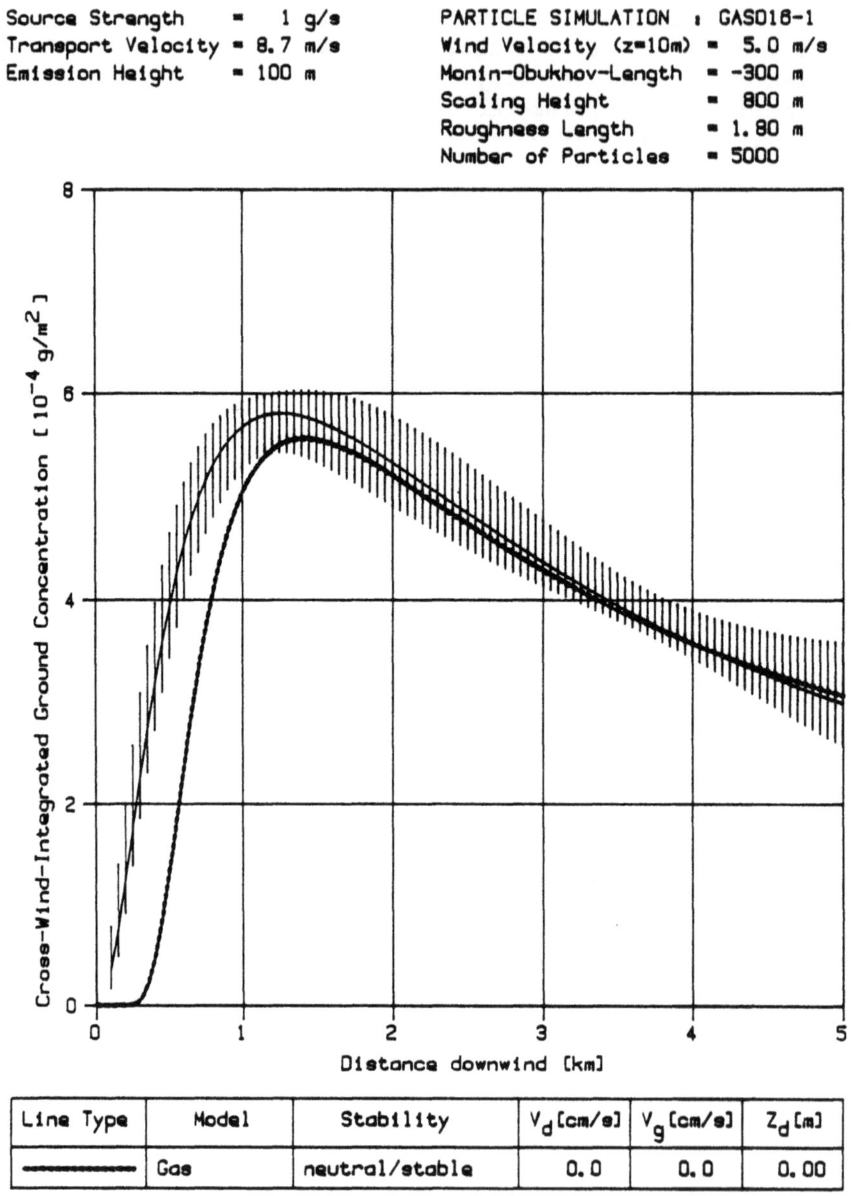

Source Strength = 1 g/s
Transport Velocity = 8.7 m/s
Emission Height = 100 m

PARTICLE SIMULATION : GAS018-1
Wind Velocity (z=10m) = 5.0 m/s
Monin-Obukhov-Length = -300 m
Scaling Height = 800 m
Roughness Length = 1.80 m
Number of Particles = 5000

Cross-Wind-Integrated Ground Concentration [10^{-4} g/m^2]

Distance downwind [km]

Line Type	Model	Stability	V_d [cm/s]	V_g [cm/s]	Z_d [m]
————	Gas	neutral/stable	0.0	0.0	0.00

Fig. 1. Comparison between the particle simulation model and the Gaussian models for gas dispersion. The vertical bars denote the 90 %-confidence interval of the simulation results.

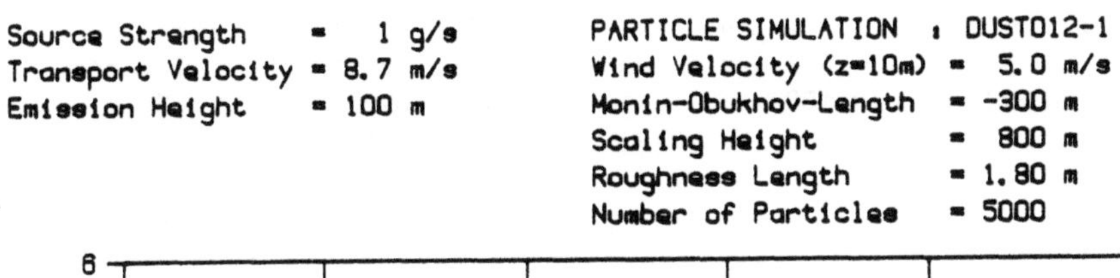

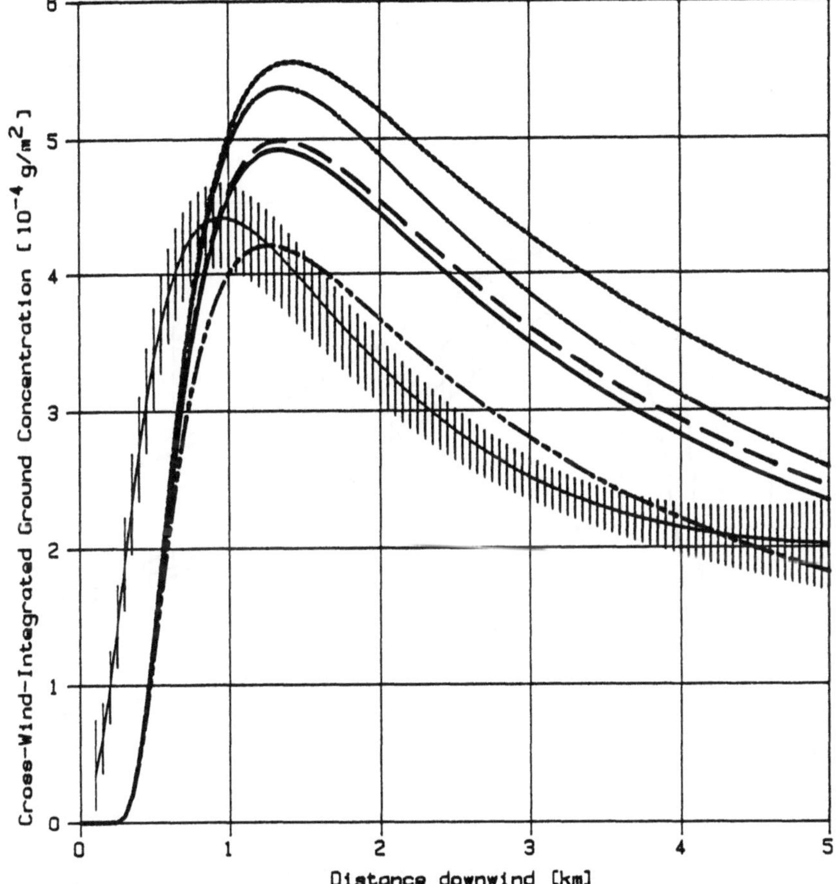

Line Type	Model	Stability	V_d [cm/s]	V_g [cm/s]	Z_d [m]
———————	Gas	neutral/stable	0.0	0.0	0.00
———————	Source-D.	neutral/stable	8.7	0.0	0.00
—·—·—·—	Surface-D.	neutral/stable	8.7	0.0	1.00
— — — —	Tilted-Pl.	neutral/stable	8.7	0.0	0.00
———————	Diffusion	neutral/stable	8.7	0.0	0.00

Fig. 2. Comparison between the particle simulation model (vertical bars) and the Gaussian models for non-gravitating particles with deposition.

Source Strength = 1 g/s
Transport Velocity = 8.7 m/s
Emission Height = 100 m

PARTICLE SIMULATION : DUST013-1
Wind Velocity (z=10m) = 5.0 m/s
Monin-Obukhov-Length = -300 m
Scaling Height = 800 m
Roughness Length = 1.80 m
Number of Particles = 5000

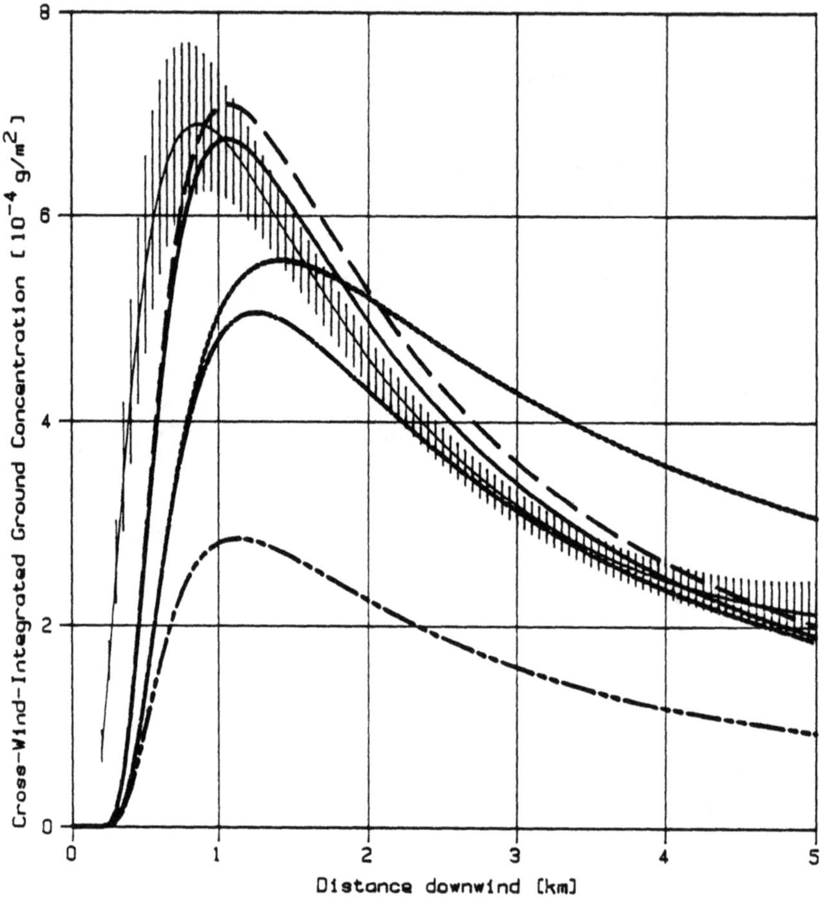

Line Type	Model	Stability	V_d [cm/s]	V_g [cm/s]	Z_d [m]
	Gas	neutral/stable	0.0	0.0	0.00
	Source-D.	neutral/stable	26.1	0.0	0.00
	Surface-D.	neutral/stable	26.1	0.0	1.00
	Tilted-Pl.	neutral/stable	26.1	26.1	0.00
	Diffusion	neutral/stable	26.1	26.1	0.00

Fig. 3. Comparison between the particle simulation model (vertical bars) and the Gaussian models for particles with gravitational settling and deposition.

diffusion model and the tilted plume model are more reliable and should be prefered.

ACKNOWLEDGEMENT

This work was supported by the Umweltbundesamt, Berlin.

REFERENCES

Ermak, D. L., 1977, An analytical model for air pollutant transport and deposition from a point source, Atmos. Environ. 11, 231:237

Hanna, S. R., 1981, Lagrangian and Eulerian time-scale relations in the daytime boundary layer, J. Appl. Met. 20, 242:249

Horst, T. W., 1977, A surface depletion model for deposition from a Gaussian plume, Atmos. Environ. 11, 41:46

Janicke, L., 1983, Particle simulation of inhomogeneous turbulent diffusion, in: "Air Pollution Modeling and Its Application II", C. De Wispelaere, ed., Plenum Press, New York

Lamb, R. G., Hogo, H., and Reid, L. E., 1979, A Lagrangian approach to modeling air pollutant dispersion, Report EPA-600/4-79-023

Ley, A. J., 1982, A random walk simulation of two-dimensional turbulent diffusion in the neutral surface layer, Atmos. Environ. 16, 2799:2808

Overcamp, T. J., 1976, A general Gaussian diffusion-deposition model for elevated point sources, J. Appl. Met. 15, 1167:1171

Van der Hoven, I., 1968, Deposition of particles and gases, in: "Meteorology and Atomic Energy", D. Slade, ed., USAEC, TID-24190

Weber, A. H., Irwin, J. S., Peterson, W. B., Mathis, J. J., and Kahler, J. P., 1982, Spectral scales in the atmospheric boundary layer, J. Appl. Met. 21, 1622:1632

ATMOSPHERIC DIFFUSION MODELLING BY STOCHASTIC DIFFERENTIAL EQUATIONS

P. Melli and A. Spirito

Centro Scientifico IBM
via Giorgione, 129 - 00147 ROMA
Rome, Italy

ABSTRACT

In the last few years Lagrangian Monte Carlo models have been proposed as a new atmospheric dispersion modelling technique overcoming the limitations and inaccuracies inherent in Gaussian models and in models based on the advection-diffusion equation. In the present paper it is shown that these models are based on the use of stochastic differential equations (SDE) describing particle movements in a random velocity field. Some cases of 2-D inhomogeneous turbulent flows are examined in order to point out the relationships between parameters appearing in the differential equations and the statistics of particle velocity distributions. The influence of the initial distribution statistics as well as that of the driving noise are also investigated. In particular the effects produced by using a random term different from the usual white noise are shown. Finally the problem of relating Eulerian and Lagrangian quantities is examined and it is shown that SDE can represent correlated Eulerian random fields as well as Lagrangian particle evolution. The possibility of using the Eulerian field as the input for the Lagrangian equation is finally considered.

INTRODUCTION

It is customary to affirm that there are three theories from which useful working models of atmospheric diffusion can be derived. These are statistical theory, gradient-transfer or K-theory, and similarity theory (see e.g. Seinfeld, 1975). Beside these approaches, a further modelling technique has been proposed and applied by several authors in recent years. The number of papers on this subject is rather large so that a complete mention is impossible and beyond the scope of this work. It will be recalled only (Gifford, 1982) that all these studies stem from Obukhov's proposal to represent the evolution of diffusing particles in the atmosphere as a Markov process. In the numerous papers available the technique has received various names: Brownian motion analogy, Langevin's model, random-force method and Lagrangian Monte Carlo model. In some of these papers the relationship with the theory of stochastic differential equations already pointed out by one of the authors (Melli, 1982) is cursorily mentioned by recalling that the schemes used are finite difference analogs of some form of Langevin's equation (Gifford, 1982; Legg and Raupach, 1982), but a coherent treatment of "Lagrangian Monte Carlo models" has not yet been carried out on the basis of the properties of the stochastic differential equations (SDE).

Moreover the four above mentioned diffusion theories are usually considered as different ways of treating turbulent diffusion, while it can be shown that, apart of the similarity theory, a unified derivation of the statistical theory, the K-theory, and the "Lagrangian Monte Carlo models" is possible starting from the same basic physical and mathematical principles underlying the theory of SDE.

This point, already detected but not exhaustively developed (see e.g. Seinfeld, 1975 , pags. 284-286), has been discussed in details elsewhere (Melli and Spirito, 1983), therefore it will be only cursorily reported in this paper whose primary aim is to show that "Lagrangian Monte Carlo models" are based on SDE and that this theory can be very useful and effective in studying diffusion in a turbulent field.

In particular, the first paragraph discusses the forms of SDE corresponding to the most used "Lagrangian Monte Carlo models" and their relationship with partial differential equations; in the second paragraph the problem of relating Lagrangian (or particle velocities) with Eulerian (or field velocities) is discussed by means of one 1-d example. Finally the third paragraph discusses how both Lagrangian and Eulerian correlations can be rigorously included in the technique.

LAGRANGIAN MONTE CARLO MODELS AND SDE

Unless otherwise explicitly stated the following discussion is restricted to the classical treatment of a point release in a 2-D idealized model of the planetary boundary layer where x is the horizontal coordinate coincident with the wind direction, z is the vertical coordinate and u and w are the corresponding components of the velocity field. This model contains all the elements needed to represent the diffusion process in a turbulent field and extension to a more complex 3-D model is straightforward.

Formulation of Lagrangian Monte Carlo models is usually achieved by means of the following finite-difference equations needed to advance in time from t to t+Δt:

$$\begin{cases} x_{n+1} = x_n+u(x_n,z_n)\Delta t & \text{(1a)} \\ z_{n+1} = z_n+w_n\Delta t & \text{(1b)} \\ w_{n+1} = w_n[1-\gamma(x_n,z_n)\Delta t]+f(x_n,z_n)\Delta t+\zeta_{n+1} & \text{(1c)} \end{cases}$$

eqs. (1a-1c) have in fact been used practically in all the studies concerning this type of approach (see e.g. Lamb, 1979; Janicke, 1982). In the above equations u is the average horizontal velocity depending on particle position, w is vertical velocity, while f is a "deterministic component" of vertical velocity whose introduction is necessary in order to achieve suitable properties of the algorithm as shown by Janicke (Janicke, 1981). Moreover γ is the inverse of the Lagrangian correlation time, ζ_{n+1} is a random number (having the dimensions of a velocity) generally assumed as Gaussian with zero mean and a given standard deviation $\sigma_\zeta(x,z)$. The above algorithm has been applied to simulate both inhomogeneous and homogeneous turbulence. As an example of the former case the application to the convective boundary layer (Lamb, 1978) must be cited, for which results provided by Willis and Deardorff's model were used to supply values of functions involved in eqs. (1a-1c). Some interesting properties of the scheme in the case of inhomogeneous turbulence but with u=const. have been derived by Janicke in the above cited paper.

A close inspection of eqs. (1a-1c) shows that they are the finite-difference analogs (based on celebrated Euler's method) of the continuous system of SDE:

$$\begin{cases} dx = u dt & \text{(2a)} \\ dz = w dt & \text{(2b)} \\ dw = (-\gamma w+f)dt+g d\beta_t & \text{(2c)} \end{cases}$$

where the same symbols as in eqs. (1a-1c) have been used while functional dependences have been omitted for simplicity's sake. Having written system (2a-2c) instead of (1a-1c) is not merely a matter of mathematical elegance and rigor: it allows, indeed, a better understanding of how the diffusion process is being represented. Before discussing this point it must be pointed out that the stochastic term ζ_{n+1} has been replaced by the product of a function g (generally depending on time and space) by the random process $d\beta_t$, which is a white Gaussian noise (formally considered as an increment of a Brownian motion), totally defined by the following properties (Jazwinski,1970):

$$\langle (d\beta_t)^n \rangle = \begin{cases} 0 & n \text{ odd} > 1 \\ 1\cdot 3\cdot 5\cdots(n-1)(dt)^{n/2} & n \text{ even} \geq 2 \end{cases} \tag{3a}$$

$$\langle d\beta_t \cdot d\beta_\tau \rangle = \delta(t-\tau)dt \tag{3b}$$

From examination of system (2a-2c) it appears clearly that the diffusion process is being represented as a 1-Markov process. A particularly appealing physical interpretation consists of considering x and z as the coordinates of a material point (representing a parcel of pollutant) moving about with velocities u and w. In particular eq. (2c) states that the material point moves under the action of a dissipative force proportional to velocity itself and of two forces depending on spatial position, the second of which is random (terms fdt and $gd\beta_t$, respectively). In the very simple case in which $u=u_0z/z_0$, f=0., and $g=\sigma$=const. system (2) can be integrated analytically to give:

$$x = \left(\frac{hu_0}{z_0}+\frac{u_0w_0}{z_0\gamma}\right)t+\frac{u_0w_0}{z_0\gamma^2}\left(e^{-\gamma t}-1\right)+\frac{\sigma u_0}{z_0}\int_0^t d\eta\int_0^\eta dv\int_0^v e^{-\gamma(v-s)}d\beta(s) \tag{4a}$$

$$z = h+\frac{w_0}{\gamma}(1-e^{-\gamma t})+\sigma\int_0^t dv\int_0^v e^{-\gamma(v-s)}d\beta(s) \tag{4b}$$

$$w = w_0e^{-\gamma t}+\sigma\int_0^t e^{-\gamma(t-s)}d\beta(s) \tag{4c}$$

provided that at time t=0 all particles are located in point S(0,h) and have a certain velocity distribution specified as w_0. From eqs.(4a-4c) all the statistical properties of the processes x, z, and w can be computed provided the initial velocity distribution w_0 is specified. In particular, if $w_0 \sim N(0,\sigma^2/2\gamma)$, then all the above processes are Gaussian with mean and standard deviations given by:

$$\langle x \rangle = \frac{u_0h}{z_0}t \tag{5a}$$

$$\langle z \rangle = h \tag{5b}$$

$$\langle w \rangle = 0 \tag{5c}$$

$$\langle x^2 \rangle = \left(\frac{u_0}{z_0}\right)^2 \left\{ h^2 t^2 + \frac{\sigma^2}{2\gamma^3}\left[\frac{2}{3}\gamma t^3 - t^2 + \frac{2}{\gamma^2} - \frac{2e^{-\gamma t}}{\gamma}\left(t + \frac{1}{\gamma}\right)\right]\right\} \tag{5d}$$

$$\langle (z-h)^2 \rangle = \frac{\sigma^2}{\gamma^3}\left(\gamma t + e^{-\gamma t} - 1\right) \tag{5e}$$

$$\langle w^2 \rangle = \frac{\sigma^2}{2\gamma} \tag{5f}$$

while if $w_0 \equiv 0$, then processes x, z, and w are still Gaussian and have the same mean as before, but standard deviations are now given by:

$$\langle w^2 \rangle = \frac{\sigma^2}{2\gamma}\left(1 - e^{-2\gamma t}\right) \tag{6a}$$

$$\langle (z-h)^2 \rangle = \frac{\sigma^2}{\gamma^3}\left[\gamma t + 2e^{-\gamma t} - \frac{1}{2}e^{-2\gamma t} - \frac{3}{2}\right] \tag{6b}$$

$$\langle x^2 \rangle = \left(\frac{u_0}{z_0}\right)^2 \left\{ h^2 t^2 + \frac{\sigma^2}{\gamma^3}\left[\gamma\frac{t^3}{3} - t^2 + \frac{t}{\gamma}\left(1 - 2e^{-\gamma t}\right) + \frac{1}{2\gamma^2}\left(1 - e^{-2\gamma t}\right)\right]\right\} \tag{6c}$$

In the former case eq. (4c) gave origin to a stationary w process, while in the latter the process had an increasing variance, which shows that in the application of Lagrangian Monte Carlo models initial velocity distribution has an influence on the characterization of the whole evolution. The influence of initial velocity distribution can also be noticed on the velocity autocorrelation function which in the two above mentioned case is respectively given by :

$$R(t,\tau) = e^{-\gamma(t-\tau)} \tag{7a}$$

$$R(t,\tau) = [e^{-\gamma(t-\tau)} - e^{-\gamma(t+\tau)}] \tag{7b}$$

Before concluding this paragraph a few words must be said about the computation of concentration. As the diffusion process has been represented by the evolution of trajectories governed by system (2) it is apparent that (ensemble) average concentration is given by:

$$\langle c(x,z,w,t) \rangle = \int_0^t \int_{-\infty}^{\infty} \int_{\Omega} p(x,z,w,t|x',z',w',t') S(x',z',w',t')\, dx'dz'dw'dt' \tag{8}$$

where function S(x',z',w',t') specifies the distribution of trajectories starting points (or the release points). For the special case where $S = Q\delta(t)\delta(x)\delta(z-h)q_0(w)$, where $q_0(w)$ is the initial velocity distribution and Q is the amount of pollutant emitted, it is apparent that eq. (8), after integration with respect to w reduces to:

$$\langle c(x,z,t) \rangle = P(x,z,t|0,h,0) \tag{9}$$

and function $P(x,z,t|0,h,0)$ is computed as:

$$P(x,z,t|0,h,0) = \frac{Q \cdot n(x,z)}{N \Delta x \Delta z} \tag{10}$$

where:

N total number of particles used
n(x,z) number of particles found at time t in the grid cell $x-dx/2<x<x+dx/2$; $z-dz/2<z<z+dz/2$

from which it is apparent that a grid needs however to be superimposed to the region where the evolution of trajectories is being computed.

It is well-known from the theory of SDE for Markov processes (Jazwinski, 1970) that evolution of the probability density in the phase space is governed by a 2nd order partial differential equation (called Kolmogorov or Fokker-Planck's equation); therefore computation of concentration field in the problem stated by system (2) could be equally well accomplished by integrating the pde:

$$\frac{\partial p}{\partial t}+\frac{\partial (up)}{\partial x}+w\frac{\partial p}{\partial z}-\frac{\partial}{\partial w}[(-\gamma w+f)p] = \frac{1}{2}\frac{\partial^2}{\partial w^2}(g^2 p) \tag{11}$$

with suitable boundary conditions and with the initial condition:

$$p(x,z,w,0) = \delta(x)\delta(z-h)q_0(w) \tag{12}$$

where $q_0(w)$ is, as usually, the chosen initial particle velocity distribution. In principle no reason suggests that the former way of proceeding is better than the latter and only a careful analysis of the computational efforts needed by the two methods can allow to discriminate between them.

LAGRANGIAN AND EULERIAN QUANTITIES

In all the previous discussion the difference between Lagrangian and Eulerian quantities has been deliberately ignored since all the treatment was purely "Lagrangian" and the only "Eulerian" or "field" quantities were represented by the several functions of the spatial coordinates introduced in the SDE. Unfortunately the most commonly measured properties are "Eulerian" and therefore the problem arises of how they can be related to the "Lagrangian" ones. In all the models up to now developed no difference has been introduced between the two based on the assumption that "Lagrangian and Eulerian quantities are simply related" (Hanna, 1978; Hanna, 1980; Pasquill, 1962). In the second reference, in particular, the hypothesis made by Hay and Pasquill that Eulerian and Lagrangian autocorrelation function are similar in shape but displaced by a scale factor is examined in the light of some turbulence measurements taken in daytime boundary layer. A ratio of 1.7 between T_L (Lagrangian time scale) and T_E (Eulerian time scale) is found and a value of T_L=70 sec. is established. From the point of view of describing diffusion in a turbulent field by means of SDE the above assumptions result in using the following system:

$$dx = udt \tag{13a}$$

$$dz = w_L dt \tag{13b}$$

$$dw_L = -\gamma_L w_L dt + \sigma_L d\beta_t \tag{13c}$$

$$dw_E = -\gamma_E w_E dt + \sigma_E d\beta_t \tag{13d}$$

where a distinction has been made between Eulerian and Lagrangian quantities by means of subscripts E an L. It is apparent that eqs.(13c-13d) describe exactly the same type of process, the only difference being in the coefficients. Moreover the first three equations are totally independent from the fourth one and therefore can be treated separately, which justifies the practice up to now followed by most researchers working with Lagrangian Monte Carlo models to neglect Eulerian quantities and work only with Lagrangian ones. Incidentally it must be emphasized that this practice has been followed both for homogeneous and inhomogeneous turbulence. From the previous discussion, however, the interesting result derives that SDE can be used to describe both particle evolution and "field" evolution, with the possibility of relating the two quantities in some way. This possibility will now be discussed with respect to a very simple one dimensional example of homogeneous turbulence stated in the following system:

$$dz = w_L dt \tag{14a}$$

$$dw_L = -\gamma_L (w_L - w_E) dt \tag{14b}$$

$$dw_E = -\gamma_E w_E dt + \sigma_E d\beta_t \tag{14c}$$

Eq. (14c) describes the random Eulerian field exactly in the same way in which eq.(2c) described the velocity of particles. This means that in every point of the field velocity is represented by a Gaussian process $w_E \sim (0, \sigma^2_E / 2\gamma_E)$ with exponential autocorrelation function whose time scale is given by $T_E = 1/\gamma_E$. Eq. (14b) establishes that particle velocity instead of assuming instantaneously the value of field velocity corresponding to its position has a delay, represented by time constant γ_L, or, if a more physical interpretation is preferred that a force is acting on the particle proportional to the difference between particle and field velocity (such a force is similar to the one acting on solid particles moving in a fluid by a laminar motion). Other types of relationships between Lagrangian and Eulerian velocities are however possible. Analytical integration of the system provides:

$$z = h + \int_0^t w_L ds \tag{15a}$$

$$w_L = w_{L_0} e^{-\gamma_L t} + \gamma_L \int_0^t e^{-\gamma_L (t-\xi)} w_E(\xi) d\xi \tag{15b}$$

$$w_E = w_{E_0} e^{-\gamma_E t} + \sigma_E \int_0^t e^{-\gamma_E (t-s)} d\beta_t(s) \tag{15c}$$

from which it can be concluded that processes z, w_L, and w_E are still all Gaussian. Computation of standard deviations is however a little cumbersome and results in very complex formulas which are not reported, but are instead plotted in fig.1, from which it can be seen that σ_z is very similar to the one supplied by the simpler system (2) represented by the dotted line. Moreover fig.2 reports the autocorrelation function $R_w(t,\tau)$.

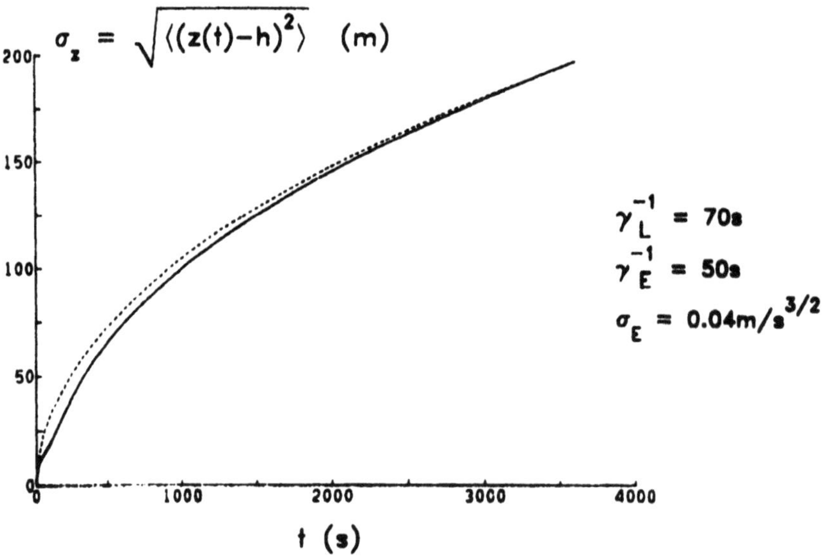

fig.1 Standard deviation σ_z for system (14) (continuous line) and system (2) (dotted line).

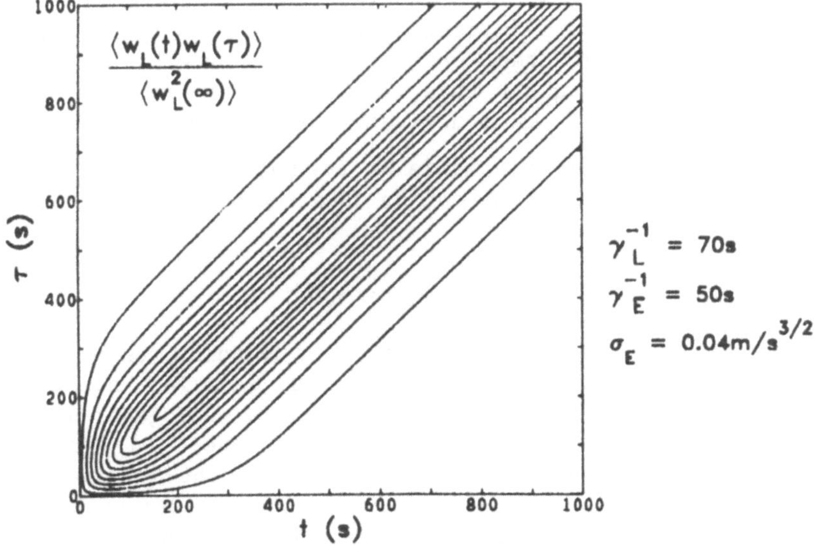

fig.2 Autocorrelation function $R_w(t,\tau)$ for system (14) (contour levels: 0.05 0.15 0.25 0.35 0.45 0.55 0.65 0.75 0.85 0.95).

It is seen that for values of t and τ sufficiently large the correlation function is coincident with the commonly used exponential function. The above discussion has pointed out that, at least in the case of homogeneous turbulence, a subtler justification of the practice of establishing simple relationships between Lagrangian and Eulerian quantities can be supplied. In the case of inhomogeneous turbulence one could use heuristically the same argument; however, better conclusions could be drawn from analysis of results supplied by simulation of system (13) with variable coefficients. Before concluding this paragraph two more remarks will be made concerning system (14). Always in the case of homogeneous turbulence process w_E can be eliminated between eqs. (14b-14c) giving:

$$\dot{z}_L = w_L \tag{14'a}$$

$$\ddot{w}_L + (\gamma_L+\gamma_E)\dot{w}_L+\gamma_L\gamma_E w_L = \sigma_E\gamma_L\dot{\beta}_t \tag{14'b}$$

which shows that using system (14) is equivalent to represent particle velocity by a 2nd order SDE, or, which is the same, as a 2-Markov process. It must be noted that in eqs. (14'a-14'b) it has been preferred to use a relationship between derivatives (indicated by dots) instead of the customary relationship between the increments, used throughout this text. Also the driving term has been indicated formally as a derivative β_t of a Brownian motion, which is actually indifferentiable by its nature. A final different interpretation of system (14) can be obtained by rewriting eq. (14b) as:

$$dw_L = -\gamma_L w_L dt + \gamma_L w_E dt \qquad (14''b)$$

this equation states that particle velocity is being represented as a 1-Markov process driven by a coloured Gaussian noise. This demonstrates that SDE allow the introduction of more general types of noise, other than the classical Gaussian one. In many cases, however, as in the one above reported, for the generation of a different type of noise one must always resort to use an additional SDE driven by a white Gaussian noise, although, of course, other methods are available.

THE TREATMENT OF LAGRANGIAN AND EULERIAN CORRELATIONS

The Lagrangian correlations

In the example considered in the first paragraph, i.e. in system (2), the horizontal velocity fluctuations were neglected as it is customary in many application where horizontal diffusion is negligible if compared to horizontal transport. If however this is not the case and if moreover correlations between horizontal and vertical velocities must be considered, as measurements in the atmospheric boundary layer show (see e.g. Kaimal et al., 1976), system (2) must be modified as follows:

$$dx = (\langle u \rangle + u)dt \qquad (16a)$$

$$dz = wdt \qquad (16b)$$

$$du = -\gamma_u u dt + \sigma_{uu} d\beta_1 + \sigma_{uw} d\beta_2 \qquad (16c)$$

$$dw = -\gamma_w w dt + \sigma_{uw} d\beta_1 + \sigma_{ww} d\beta_2 \qquad (16c)$$

System (16) differs from (2) in that a further equation was added for horizontal velocity component and moreover each velocity equation now contains two independent Gaussian white noises $d\beta_1$ and $d\beta_2$, while functions σ_{uu}, σ_{ww}, and σ_{uw} are related to correlations $\langle uu \rangle$, $\langle ww \rangle$ and $\langle uw \rangle$ respectively. In general they can be functions of both spatial coordinates and of time. In the special case when all parameters involved are constant eqs.(16c-16d) give:

$$u = u_0 e^{-\gamma_u t} + \sigma_{uu}\int_0^t e^{-\gamma_u(t-s)} d\beta_1(s) + \sigma_{uw}\int_0^t e^{-\gamma_u(t-s)} d\beta_2(s) \tag{17a}$$

$$w = w_0 e^{-\gamma_w t} + \sigma_{uw}\int_0^t e^{-\gamma_w(t-s)} d\beta_1(s) + \sigma_{ww}\int_0^t e^{-\gamma_w(t-s)} d\beta_2(s) \tag{17b}$$

with correlation functions given by:

$$R_{uu} = e^{-\gamma_u(t+\tau)} + \frac{\sigma_{uu}^2+\sigma_{uw}^2}{2\gamma_u\langle u_0^2\rangle}\left[e^{-\gamma_u(t-\tau)} - e^{-\gamma_u(t+\tau)}\right] \tag{18a}$$

$$R_{ww} = e^{-\gamma_w(t+\tau)} + \frac{\sigma_{uw}^2+\sigma_{ww}^2}{2\gamma_w\langle w_0^2\rangle}\left[e^{-\gamma_w(t-\tau)} - e^{-\gamma_w(t+\tau)}\right] \tag{18b}$$

$$R_{uw} = e^{-(\gamma_u t+\gamma_w \tau)} + \frac{2\sigma_{uw}(\sigma_{uu}+\sigma_{ww})}{(\gamma_u+\gamma_w)\langle u_0 w_0\rangle}\left[e^{-\gamma_u(t-\tau)} - e^{-(\gamma_u t+\gamma_w \tau)}\right] \tag{18c}$$

It results that in order to obtain stationary u and w processes the initial correlations should be given by:

$$\langle u_0^2\rangle = \frac{\sigma_{uu}^2+\sigma_{uw}^2}{2\gamma_u} \qquad \langle w_0^2\rangle = \frac{\sigma_{uw}^2+\sigma_{ww}^2}{2\gamma_w} \qquad \langle u_0 w_0\rangle = \frac{2\sigma_{uw}(\sigma_{uu}+\sigma_{ww})}{\gamma_u+\gamma_w}$$

The previous discussion applies of course not only to equations describing the evolution of trajectories (that we could call Lagrangian equations), but also to the equations describing the random Eulerian field thus allowing the "Eulerian velocity" to be correlated. Again this is particularly appealing as velocity cross correlations are almost always measured in an Eulerian framework.

The question now arises immediately if Lagrangian Monte Carlo models, which, on the basis of the above discussion should be more properly termed as "models based an SDE" may be further modified to represent full space-time correlations such as the ones that are encountered in real flows.

The Eulerian (space) correlations

In order to illustrate how the point raised at the end of last paragraph can be fulfilled system (14) is rewritten as:

$$dz = w_L dt \tag{19a}$$

$$dw_L = -\gamma_L(w_L - w_E)dt \tag{19b}$$

$$dw_E = -\gamma_E w_E dt + \int_{-\infty}^{\infty} \sigma(z-\xi)d\beta(\xi,t) \tag{19c}$$

as in the case previously described the third equation must be considered separately since it describes the evolution of the Eulerian random velocity field. This field therefore is only an input for the other two equations. The most interesting feature of eq. (19c) is that it contains a random force expressed by means of the 2-d white noise $d\beta(\xi,t)$, for which the following relationships hold:

$$\langle d^n\beta(\xi,t)\rangle = 0 \qquad n \text{ odd} \tag{20a}$$

$$\langle d^n\beta(\xi,t)\rangle = 1\cdot 3\cdot 5\cdot\ldots\cdot(n-1)d\xi^{n/2}dt^{n/2} \qquad n \text{ even } \geq 2 \tag{20b}$$

$$\langle d\beta(\xi,t)d\beta(\zeta,\tau)\rangle = \delta(\xi-\zeta)\delta(t-\tau)d\xi dt \tag{20c}$$

Due to the independence of the third equation it can be integrated to give:

$$w_E = w_{E_0}e^{-\gamma_E t} + \int_0^t\int_{-\infty}^{\infty} e^{-\gamma_E(t-s)}\sigma(z-\xi)d\beta(\xi,s) \tag{21}$$

It is apparent that eq. (21) generates a random velocity field with time correlation given by:

$$R_z(t,\tau) = e^{-\gamma_E(t+\tau)} + \frac{e^{-\gamma_E(t-\tau)} - e^{-\gamma_E(t+\tau)}}{\langle w_{E_0}^2\rangle}\int_{-\infty}^{\infty}\sigma^2(z-\xi)d\xi \tag{22}$$

and with space correlation given by:

$$R_t(z,\zeta) = e^{-2\gamma_E t} + \frac{1-e^{-2\gamma_E t}}{\langle w_{E_0}(\zeta)w_{E_0}(z)\rangle}\int_{-\infty}^{\infty}\sigma(z-\xi)\sigma(\zeta-\xi)d\xi \tag{23}$$

if, moreover

the velocity field is also stationary. The field thus obtained can now be used as an input in eqs. (19a-19b) to give:

$$dz = w_L dt \tag{24a}$$

$$dw_L = \left[-\gamma_L w_L + \gamma_L\left(w_{E_0}e^{-\gamma_E t} + \int_0^t e^{-\gamma_E(t-s)}\int_{-\infty}^{\infty}\sigma(z-\xi)d\beta(\xi,s)\right)\right]dt \tag{24b}$$

which is again a purely Lagrangian description of the diffusion process, but accomplished by means of a system of stochastic integro-differential equations. We notice, in fact that the evolution of the particle paths depend not only from values at the considered instant and the immediately previous one, but is linked to the whole past history of the whole field. This, of course makes the computational effort needed to integrate system (24) much higher than the one required by system (2).

CONCLUSION

The present paper has shown that a rigorous treatment of Lagrangian Monte Carlo models is possible by using the theory of stochastic differential equations. In particular it has been shown that all the models up to now used are based on the assumption that particle velocity can be represented by a 1-Markov process. In this case it is apparent that the treatment is equivalent to use a 2nd order partial differential equation in the phase space including as an independent variable particle velocity itself. The problem of relating Lagrangian and Eulerian quantities has been then examined by demonstrating that also Eulerian random fields can be described by proper stochastic differential equations. In particular the field generated by these can be used as an input in the SDE describing the motion of the particles. Finally it has been shown how the models based on SDE allow to represent Eulerian fields including all the correlations usually found in the real atmospheric flows and how consequently these can be used as an input for the equations describing particle trajectories evolution.

REFERENCES

Hanna, S.R., 1978, Some statistics of Lagrangian and Eulerian wind fluctuations, Journal of Applied Meteorology, 18, 518-525.

Hanna, S.R., 1980, Lagrangian and Eulerian time-scale relations in the daytime boundary layer, Journal of Applied Meteorology, 20, 242-249.

Gifford, F.A., 1982, Horizontal diffusion in the atmosphere: a lagrangian-dynamical theory, Atmospheric Environment, 12, 505-512.

Janicke, L., 1981, Particle simulation in inhomogeneous turbulent diffusion, Proc. of 12th NATO/CCMS Techn. Meeting, Palo Alto.

Jazwinski, A.H., 1970, "Stochastic processes and filtering theory", Academic Press.

Kaimal J.C., Wyngaard, J.C., Haugen, D.A., Cote', O.R., Izumi, Y., Caughey, S.J., and Readings, C.J., 1976, Turbulence structure in the convective boundary layer, Journal of Atmospheric Sciences, 33, 2152-2169.

Lamb, R.G., 1978: A numerical simulation of dispersion from an elevated point source in the convective planetary boundary layer. Atmos. Environ., 12, pp. 1297-1304.

Lamb, R.G., Hogo, H., and Reid, L.E., 1979, A Lagrangian approach to modeling air pollutant dispersion, EPA Report, EPA-600/4-79-023.

Legg, B.J., and Raupach, M.R., 1982, Markov-chain simulation of particle dispersion in inhomogeneous flows: the mean drift velocity induced by a gradient in eulerian velocity variance, Boundary Layer Meteorology, 24, 3-13

Melli, P., 1982, Lagrangian modelling of dispersion in the planetary boundary layer of particulate released by a line source. Proceedings of 13th Nato/CCMS ITM, Ile des Embiez, France.

Melli, P., and Spirito, A., 1983, Pollution episodes in situation of weak winds: an application of the k-model. Proceedings of the 14th Nato/CCMS ITM, Copenhagen, Denmark.

Pasquill, F., 1962, Atmospheric diffusion, Van Nostrand, London.

Seinfeld, J.H., 1975, "Air Pollution - Physical and Chemical Fundamentals", McGraw-Hill

DISCUSSION

J. KNOX — Have you tested your model over various values of the lag in the stochastic model ?

A. SPIRITO — We have done up to now just a theoretical study about the possibility of imbedding Lagrangian Monte Carlo Models in the theory of stochastic differential equations.

POLLUTION EPISODES IN SITUATIONS OF WEAK WINDS:

AN APPLICATION OF THE K-MODEL

P. Melli and A. Spirito

Centro Scientifico IBM
via Giorgione, 129 - 00147
Rome, Italy

G. Fronza

Dipartimento di Elettronica, Centro Teoria dei Sistemi
via Ponzio 34/5 - 20133
Milan, Rome

ABSTRACT

The theories of atmospheric diffusion (statistical or Gaussian model, k-theory and Lagrangian Monte Carlo model) are reviewed in order to show that they are not separate approaches, but are all amenable to the same basic physical principles and the same mathematical treatment based on the theory of stochastic differential equations (SDE). In particular it is shown how the k-theory can include either rigorously or heuristically some of the features of the statistical theory and of the Lagrangian Monte Carlo models. An application of the k-theory is then developed to describe summer pollution episodes caused by the emission of a power plant situated in the Po Valley. Different shapes for the diffusion coefficients are chosen on the basis of the previous discussion and values of parameters involved are estimated by least square fitting of the experimental concentration data.

INTRODUCTION

Modelling atmospheric diffusion in rural or urban airsheds is generally accomplished by using either the statistical theory, also called Gaussian formulation, or the well-known atmospheric diffusion equation. Although both approaches are unsatisfactory, from a theoretical point of view, to represent dispersion in a turbulent field as the atmospheric one, several applications of these models have been developed in the last fifteen years under a variety of circumstances and for the solution of a number of different problems. In order to overcome the limitations of both Gaussian and k-theory, some authors (see e.g. Lamb, 1980) have suggested to use another type of approach, usually termed "Lagrangian Monte Carlo diffusion modelling". Some work on this topic has been developed by several authors (Reid, 1979; Janicke, 1981; Legg and Raupach, 1982) but it has been essentially limited to theoretical studies. To the authors' knowledge, in fact, only two practical applications are worthy to be mentioned (Lamb et al., 1979; Hanna, 1981).

The three above mentioned diffusion theories are usually considered as different ways of treating turbulent diffusion, while it can be shown, as it will be done in the first paragraph of this work, that a unified derivation of the three above recalled atmospheric diffusion models is possible starting from the theory of stochastic differential equations. In particular, the physical meaning of the diffusion coefficients present in the atmospheric diffusion equation will be examined in the light of the above theory. Moreover it will be shown how the properties of Gaussian models, confirmed by experimental measurements, may be imbedded in the k-theory by taking a suitable mathematical structure for the diffusion coefficients. The following paragraphs report an application of the advection-diffusion equation to simulate summer pollution episodes caused by the high emission of a thermal power plant due to plume breakdown by an enhancing unstable convective layer in presence of weak winds, a situation in which the use of the k-theory is particularly critical. Two different vertical diffusion coefficients are tested: a) the usual exponential profile used by several authors (see e.g. Shir and Shieh, 1973; Wyngaard et al., 1974; Robins, 1978), b) a power law of the distance from the release. In both cases the coefficients of the profiles adopted are not assigned a priori, but they are estimated by best least square fitting between model results and observations. Results obtained are then discussed in the light of the two different hypotheses on atmospheric turbulence underlying the two different structures chosen for the vertical diffusion coefficients.

STOCHASTIC MODELLING OF ATMOSPHERIC DIFFUSION.

In this paragraph a unified approach to atmospheric diffusion by using the theory of stochastic differential equations (SDE) will be developed. First the advection-diffusion equation will be considered, subsequently attention will be put on the Gaussian formulation.

The advection-diffusion equation

The commonly used form of the atmospheric diffusion equation is given by:

$$\frac{\partial c}{\partial t}+u\frac{\partial c}{\partial x}=\frac{\partial}{\partial z}\left(k_z\frac{\partial c}{\partial z}\right)+\frac{\partial}{\partial x}\left(k_x\frac{\partial c}{\partial x}\right)+Q\delta(x-x_s)\delta(z-h) \quad (1)$$

Discussion is restricted to the 2-D case for a continuous point source located in (x_s,h) without sinks since this models contains all the elements needed to simulate the case of an inert release in the planetary boundary layer. Obviously in eq.(1) u is average wind speed parallel to x-axis, z is the vertical coordinate, and K_z and K_x are the vertical and horizontal diffusion coefficients respectively. Eq. (1) is generally derived by ensemble averaging of the continuity equation for an inert chemical species transported by a stochastic field and by expressing the turbulent fluxes as:

$$\langle u'c'\rangle = -k_x\frac{\partial c}{\partial x} \quad (2)$$

$$\langle w'c'\rangle = -k_z\frac{\partial c}{\partial z} \quad (3)$$

where u' and w' are the zero-mean horizontal and vertical stochastic component of the turbulent wind field and c' is the zero mean concentration stochastic fluctuation. In the practical applications of eq. (1) to atmospheric diffusion modelling (see e.g. Shir and Shieh, 1973) K_z is taken as a function of the vertical coordinate thus simulating the vertical inhomogeneity of the planetary boundary layer. Another derivation of eq. (1) is however possible by considering the system of SDE:

$$\begin{cases} dx = u\,dt + \sigma_x d\beta_x & (4a) \\ dz = f\,dt + \sigma_z d\beta_z & (4b) \end{cases}$$

where x and z may be interpreted as the coordinates of a material point (particle), representing a parcel of pollutant, released at a certain initial point (x_s,h) and moving in a velocity field, whose "deterministic components" are u and f, while the stochastic components are represented by the Gaussian independent white noises $d\beta_x$ and $d\beta_z$ (see Jazwinski, 1974), multiplied by the functions σ_x and σ_z which may in general depend on x and z. Direct integration of eqs. (4) provides a Lagrangian model of atmospheric diffusion. It is well known that model (4) can be also described in terms of the probability density function (pdf) $p(x,z,t|x',z',t')$ that a material particle released at the point (x',z') at time t' will be found at (x,z) at time t. The theory of SDE states (Jazwinski, 1974) that the pdf is solution to the partial differential equation (Kolmogorov equation):

$$\frac{\partial p}{\partial t}+\frac{\partial}{\partial x}(up)+\frac{\partial}{\partial z}(fp)=\frac{1}{2}\frac{\partial^2}{\partial z^2}(\sigma_z^2 p)+\frac{1}{2}\frac{\partial^2}{\partial x^2}(\sigma_x^2 p) \tag{5}$$

and that particle concentration can be then computed by:

$$\langle c(x,z,t)\rangle = \int_0^t \int_\Omega p(x,z,t|x',z',t')S(x',z',t')\,dx'dz'dt' \tag{6}$$

It is apparent that for a point source release, i.e. for $S(x',z',t')= Q\delta(x-x_s)\delta(z-h)$, by recalling the properties of Green's functions, eqs. (5-6) are equivalent to eq. (1) provided that:

$$\frac{\partial u}{\partial x} = 0 \tag{7a}$$

$$f = \frac{1}{2}\frac{\partial \sigma_z^2}{\partial z} \tag{7b}$$

$$k_x = \frac{1}{2}\sigma_x^2 \tag{7c}$$

$$k_z = \frac{1}{2}\sigma_z^2 \tag{7d}$$

The reasons requiring relationship (7b) have been discussed elsewhere in detail (Janicke, 1981; Melli, 1982) therefore they will not be recalled here. The above discussion shows that using eq.(1) to represent atmospheric diffusion is equivalent to assuming that the atmospheric turbulent field can be represented by a stochastic field of independent Gaussian noises as the ones reported in eqs.(4-5).

The Gaussian model

If the assumption is made that the vertical component of the atmospheric field can be represented by a coloured noise, i.e. a Gaussian noise with an exponential autocorrelation function, and moreover that horizontal stochastic transport may be neglected with respect to the more important contribution due to "deterministic transport", we get the "Lagrangian model":

$$dx = udt \tag{8a}$$

$$dz = wdt \tag{8b}$$

$$dw = -\frac{1}{T_L}wdt + \sigma d\beta_t \tag{8c}$$

where T_L is the characteristic time of the autocorrelation function and σ is a constant. System (8) is a little more complex than system (4) since one more SDE has been introduced for vertical turbulent velocity. Most of "Lagrangian models" developed in the literature are based on finite difference analogs of system (8). It is apparent that in order to integrate system (8) for a point release we need to specify the initial distribution of particle velocities. If such a distribution is assumed to be Gaussian with zero mean and standard deviation $\sigma^2T_L/2$ then the stochastic field generated by eq. (8c) is homogeneous and stationary and the pdf (already integrated over the w-distribution) is given by:

$$p(x,z,t|0,h,0) = \frac{\delta(x-ut)\ Q}{\sqrt{2\pi}\Sigma_z(t)} \exp\left[-\frac{(z-h)^2}{2\Sigma_z^2(t)}\right] \tag{9}$$

where:

$$\Sigma_z^2(t) = \sigma^2T_L^3\left[\frac{t}{T_L}+\left(e^{-t/T_L}-1\right)\right] \tag{10}$$

Eq. (10) is the basis of the Gaussian models and has become very popular in atmospheric diffusion modelling as it reflects the experimental property of plume dimensions to grow initially at a rate proportional to t and after a while (depending on the Lagrangian time scale T_L) at a rate proportional to $t^{\frac{1}{2}}$. As a further consideration it can be noted that the pdf in eq. (9) is the same as that governing the system:

$$dx = udt \tag{11a}$$

$$dz = \Sigma_z(t)d\beta \tag{11b}$$

whose Kolmogorov equation is:

$$\frac{\partial p}{\partial t}+u\frac{\partial p}{\partial x} = \frac{1}{2}\frac{\partial}{\partial z}\left(\Sigma_z^2\frac{\partial p}{\partial z}\right) \tag{12}$$

which shows that system (10), and therefore system (8) is equivalent to the advection-diffusion equation (1), provided that the term considering horizontal diffusion is neglected and the vertical diffusion coefficient K_z is supposed to depend on the time of release $\tau=(x-x_s)/u$.

On the basis of the above discussion two important conclusions can be drawn:

1) the k-theory is a Lagrangian Monte Carlo model in which the stochastic part of velocity is represented by a white Gaussian noise;

2) the Gaussian model is a Lagrangian Monte Carlo model in which the stochastic part of velocity is represented by a coloured Gaussian noise with exponential autocorrelation function;

3) the k-theory can imbed rigorously in the case of constant horizontal average wind speed the properties of the Gaussian model confirmed by field measurement in some atmospheric conditions. This is obtained by taking a vertical diffusion coefficient depending on the downwind distance, which is of course in contrast with the strict Eulerian conception of the traditional k-models.

These considerations must be kept in mind as reference to them will be made in the following when the practical application of the k-model reported in this paper will be discussed with more detail.

DESCRIPTION OF THE PRESENT APPLICATION

This paragraph is devoted to describing the pollution problem investigated, the area of application and the methodology employed.

The pollution problem and the area of investigation

The application developed in the present study concerns the use of a k-model to study summer sulphur dioxide pollution from a power plant in the Po Valley. The case has been studied from a completely different viewpoint in a number of previous works (Finzi et al., 1978; Bacci et al., 1981; Melli and Fronza,1981). In the last work, in particular, a pollution episode predictor was investigated based on the k-model and the Kalman filtering technique. In the present paper the k-model alone is applied to see at what extent suitable modifications of the diffusion coefficients in terms of both space and time variability are able to represent correctly the summer fumigation episodes in conditions of strong instability, namely in the presence of plume breakdown by an enhancing unstable convective boundary layer (see e.g. Slade,1968). The area under investigation is shown in fig. 1, reporting the position of the polluting source and the concentration measurement points. The network shown in the figure is not the actual one as actual measurements have been spatially interpolated in order to get measurements in the gridpoints. The area is flat so that there is no significant orographic effect to be accounted for. The network provides hourly concentration values, hourly meteorological measurements (wind speed

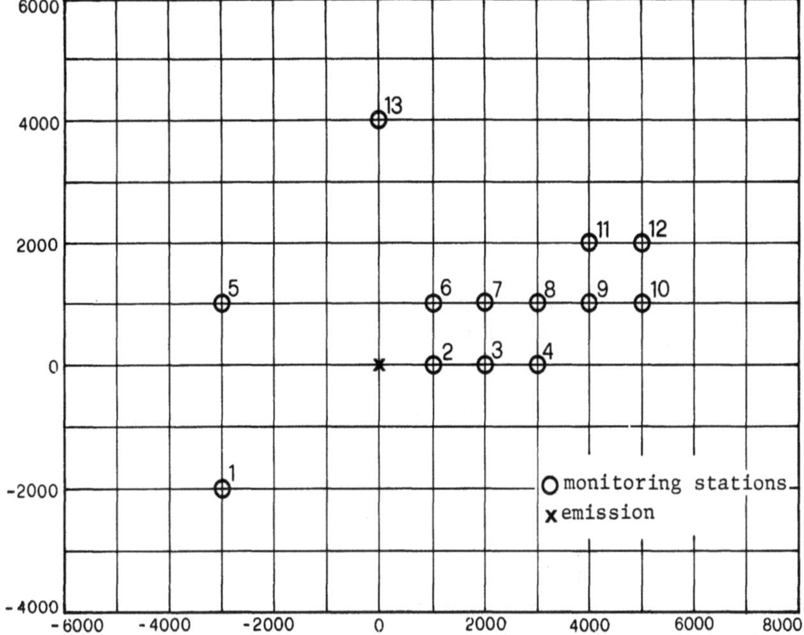

fig.1 Area under investigation and integration grid.

and direction) at ground level and at 120m close to the source and hourly sulphur dioxide emission rates. The source consists of three stacks (respectively tall 110, 120 and 100m), whose target power in the year of investigation was of 320 MW and is practically the only sulphur dioxide emission present in the area. From the analysis of data gathered in a whole year (1973) it resulted that most severe pollution episodes (with concentration reaching at ground level values as high as .5 ppm) occurred in the summer during late morning and early afternoon warm hours especially in conjunction with calm or very low wind condition. This is a situation in which application of the k-theory is most critical since all its underlying assumptions are practically violated. A correct way of representing the phenomenon would consist of applying a boundary layer model describing the evolution of the mixed layer developing over the ground since sunrise. This cannot be done in routine computation, therefore it was decided to investigate if inclusion of peculiarities of other diffusion models (Gaussian and "Lagrangian" models) in the k-theory may improve its performance in the considered case.

The k-model and its parameters

The k-model used in the present work is given by:

$$\frac{\partial c}{\partial t}+u\frac{\partial c}{\partial x}+v\frac{\partial c}{\partial y} = \frac{\partial}{\partial x}\left(k_x\frac{\partial c}{\partial x}\right)+\frac{\partial}{\partial y}\left(k_y\frac{\partial c}{\partial y}\right)+\frac{\partial}{\partial z}\left(k_z\frac{\partial c}{\partial z}\right)+S \qquad (13)$$

where:

c	pollutant concentration
u, v	wind components in the horizontal plane
k_x, k_x, k_z	diffusion coefficients
S	source term

By recalling the above discussion on stochastic modelling of atmospheric turbulence we point out that using eq.(13) is equivalent to assume that the following equations of motion hold for "particles" representing the pollutant release:

$$dx = udt+\sqrt{2k_x}d\beta_x \tag{14a}$$

$$dy = vdt+\sqrt{2k_y}d\beta_y \tag{14b}$$

$$dz = \frac{\partial k_z}{\partial z}dt+\sqrt{2k_z}d\beta_z \tag{14c}$$

i.e. atmospheric turbulent field is represented by means of three independent Gaussian noises multiplied by functions specifying spatial inhomogeneities of the flow. As far as wind field is concerned direction was assumed to be constant in space and equal to the hourly average value supplied by the anemometer at stack height, while speed was assumed to vary by a power law with height up to the stack height and then to be constant. Parameters of the power law were derived by using the two hourly average values of wind speed supplied by the two anemometers at different heights. Before discussing in more details the problem of choice of the diffusion parameters a few words are to be said on the problem of integrating numerically eq.(12). A lot of work has been developed on this subject and several schemes have been proposed based on different numerical approaches. Here it will be only specified that the integration has been carried out by a fractional step procedure analogous to the one described in previous works (see, e.g. Runca et al., 1979). The integration grid is reported in fig. 1 which shows also the values of grid spacings in the horizontal; the grid extends vertically up to 1200m and vertical grid spacing is assumed uniformly equal to 100m.

The diffusion coefficients

The limitations inherent in the formulation represented by eq.(13) have been already pointed out, but one more crucial point has to be faced in applying a k-model. This concerns the lack of definite mathematical expressions for the diffusivity coefficients. Some power law profiles have been proposed by several authors both for the surface layer and for the whole planetary boundary layer (see Seinfeld, 1975) and other semiempirical profiles were proposed by F.B. Smith for all stability conditions (Pasquill, 1974). For the neutral atmosphere an exponential profile expressed as:

$$k_z = k_0 \frac{z}{H} \exp\left(-\rho \frac{z}{H}\right) \tag{15}$$

where:

$K_0 = u_* k$
u_* = friction velocity
k = von Karman constant
ρ = 4
H = PBL height

has been proposed and used in 3-D modelling of atmospheric dispersion (Shir and Shieh, 1973; Shir and Shieh, 1974). This type of formulation has also received some experimental support (see e.g. Robins, 1978).

A final results must be cited concerning a work where diffusion coefficients were evaluated by using a numerical model of the convective planetary boundary layer (Lamb and Durran, 1978). In this paper the authors fitted the concentration field provided by Willis and Deardorff model by the concentration field supplied by the k-model and determined the "best" vertical eddy diffusion coefficient (in the least square sense). It turned out that this coefficient was a function of the source position, which is again a contradiction with the usually assumed "Eulerian nature" of the k-theory. Moreover the coefficient was related to convective velocity scale w_*. Due to the fact that no element allows to discriminate between the above reported results, it is generally suggested (e.g. Seinfeld, 1975) to assume a constant value reflecting the average diffusivity in the planetary boundary layer. In the present paper a little different approach was undertaken aiming at establishing which one of the previous suggestions could represent in the best way the pollution episodes above recalled. More precisely the diffusion coefficients were not assigned a priori and inserted into eq. (13) in order to simulate the pollutant concentration field, but an estimation technique was adopted to evaluate the values of K_x, K_y, and K_z providing the best least square fit between measured and computed concentration values. The technique will be described more in details in the following, while for the moment the different shapes chosen for the diffusion coefficients will be discussed. Horizontal diffusion coefficients K_x and K_y were supposed to be constant and equal over the whole integration region so that only one coefficient had to be estimated for each hour. As to vertical diffusion coefficient three different hypotheses were made:

a) K_z constant over the whole integration region.
b) K_z expressed by the same expression used for the neutral layer

as in eq. (15), but with K_0 and ρ considered as two coefficient to be estimated at each hour; H has been taken as the maximum integration height.

c) K_z expressed as a power law of the distance from the release i.e. as:

$$K_z = K_0(d/d_0)^n$$

with K_0 and n considered as coefficients to be estimated at each hour.

It is apparent that choices made at point b) and c) reflect respectively the trend to account for vertical inhomogeneities of the planetary boundary layer and to introduce into the k-model the features of the Gaussian formulation.

Method of determining the optimal diffusion coefficients

The optimal values for the coefficients defining the shape of diffusion coefficients as defined in the previous section have been determined by minimizing the index:

$$P(k_z) = \sum_{l=1}^{J}(c_{ml}-c_{cl})^2 \tag{16}$$

where:

J	number of concentration measurements points
c_m	measured concentration
c_c	computed concentration by eq.(13)

The method adopted to seek minimum of (16) is exactly the classical one described by in Lamb's above cited paper (Lamb and Durran, 1978) and therefore no further information on it will be given here.

RESULTS

Five different pollution episodes recorded in the summer of 1973 (the year to which the available data set referred) were examined, occurring under the conditions described in the first section. The episodes generally started in the late morning when ground level sulphur dioxide concentration steadily began to increase and reached its maximum at different hours (between noon and 17.00). In general wind speed was very low (0.5-2.0 m/s) both at ground level and at stack height and in several hours the anemometer recorded a calm condition. Moreover a high insolation occurred for all the hours of all the episodes (out of the five three occurred in September and two in May).

As it is impossible to report all the concentration evolution plots for all measurements points and for all the episodes it has been chosen to evaluate them in term of two statistical indexes: the correlation coefficient (r) between measured and computed values and the average root-mean-square error (E) between the two, defined as:

$$r = \frac{\sum_{l=1}^{J}(c_{ml}-c_m)(c_{cl}-c_c)}{\sqrt{\left[\sum_{l=1}^{J}(c_{ml}-c_m)^2\right]\left[\sum_{l=1}^{J}(c_{cl}-c_c)^2\right]}} \qquad E = \frac{\sqrt{\sum_{l=1}^{J}(c_{ml}-c_{cl})^2}}{\sum_{l=1}^{J}c_{ml}}$$

The distribution of the two indexes are reported in figg. 2-3-4 for the three different choices of the dispersion coefficients made at points a), b) and c) respectively of previous section.

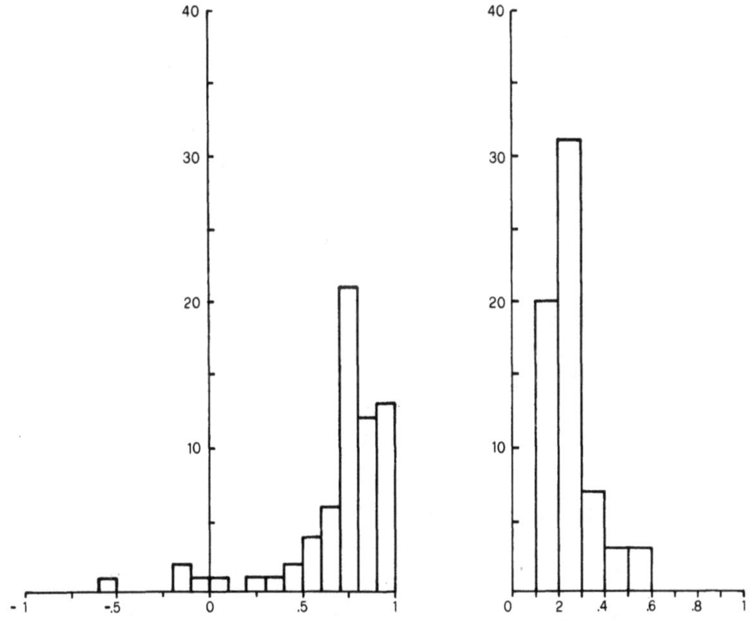

fig.2 Distributions of correlation coefficients (left) and root mean square errors (right) for k_z=const.

It is apparent, as could have been expected, that the homogeneity hypothesis behaves quite badly for both indexes, while much better and comparable results are obtained for the coefficients as in points b) and c). From the above mentioned plots it is also apparent that the improvement obtained by introducing the features of the Gaussian formulation into the k-model are not such to suggest that this choice is superior to the usual vertical inhomogeneity assumption. In order to show how the model represents the spatial

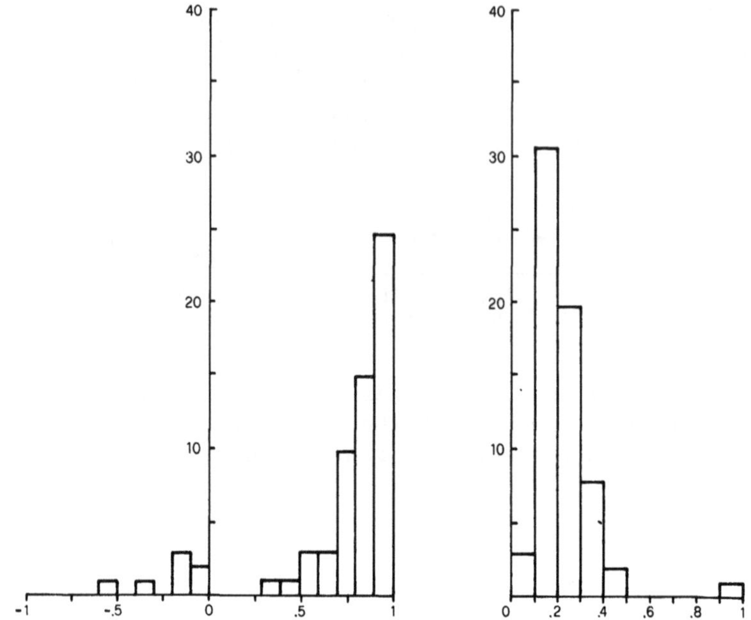

fig.3 Distributions of correlation coefficients (left) and root mean square errors (right) for $k_z=k_z(z)$.

variations of pollution levels the three tables I-II-III report, exactly in the same order than the figg. 2-3-4 the maximum, minimum, and average values of r and of E for each of the measurement points. It is seen that even for the extreme values the same considerations as above hold. As a final information it must be said that a very small variability was found for the values of K_x and of K_y which always oscillated around the value 200 m^2/s for all the episodes, a much larger variability was observed for the vertical diffusion coefficient K_z. Extreme values for the different coefficients are reported in table IV. It must be noted that maximum values were always obtained in the hours in which maximum concentration values were recorded.

CONCLUSION

The present work has shown that a unified treatment of the atmospheric diffusion is possible by using the theory of stochastic differential equations. In particular it has shown that Gaussian formulation, k-models and Lagrangian Monte Carlo models are all amenable to the same physical principle and the same basic governing equations. It has also been pointed out that even the features of the Gaussian models can be either rigorously or heuristically

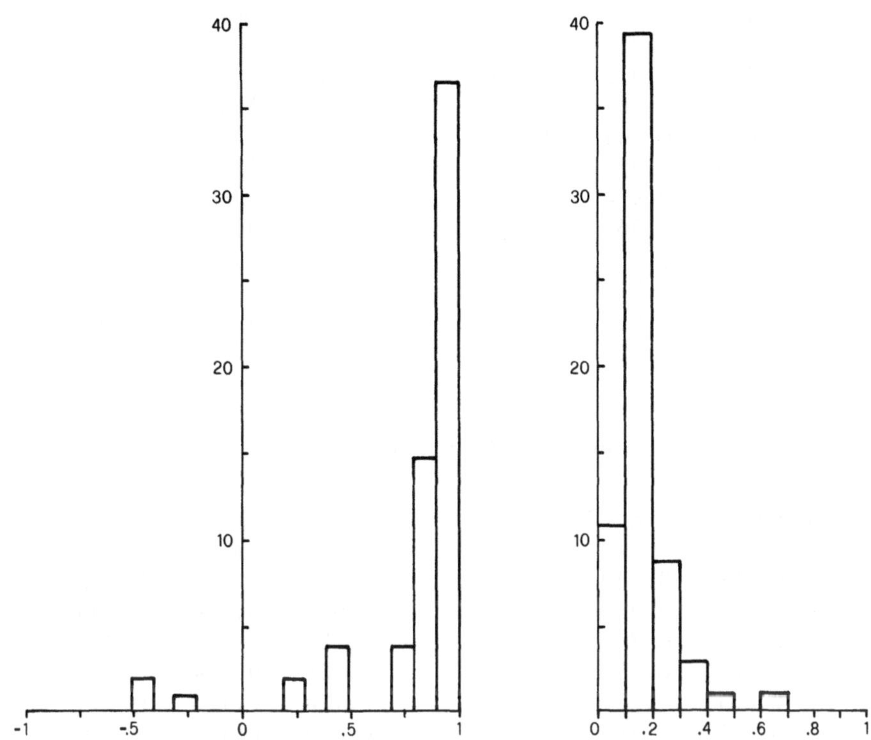

fig.4 Distributions of correlation coefficients (left) and root mean square errors (right) for $k_z = k_z(d)$.

included in the k-models. On the basis of these results application of a k-model to describe summer pollution episodes caused by a power plant in the Po Valley has been considered. The key problem of defining the diffusion coefficients has been tackled by estimating them by means of a classical least square technique. Results point out obviously that inhomogeneity must be accounted for, but they do not provide sufficient ground to assert that inclusion of Gaussian features into k-models definitely improves their performance although some improvement is present almost in any episode and in every measurement point. Ranges of variability for the diffusion coefficients have been computed and reported.

TABLE I (k_z = const.)

st.	correlation coefficient			root mean square error		
	min	max	mean	min	max	mean
2	0.565	0.858	0.740	0.161	0.528	0.306
3	0.771	0.943	0.841	0.163	0.345	0.234
4	0.750	0.961	0.828	0.178	0.347	0.236
6	0.697	0.941	0.855	0.161	1.044	0.375
7	0.739	0.956	0.846	0.143	0.273	0.193
8	0.755	0.974	0.848	0.116	0.389	0.215
9	0.671	0.968	0.760	0.120	0.432	0.233
10	0.480	0.966	0.658	0.141	0.529	0.296
11	0.560	0.932	0.785	0.162	0.354	0.242
12	0.613	0.952	0.758	0.186	0.515	0.285
13	0.359	0.826	0.652	0.177	0.422	0.283

TABLE II ($k_z = k_z(z)$)

st.	correlation coefficient			root mean square error		
	min	max	mean	min	max	mean
2	0.569	0.962	0.805	0.080	0.276	0.161
3	0.709	0.973	0.850	0.068	0.348	0.206
4	0.794	0.964	0.871	0.125	0.308	0.211
6	0.860	0.955	0.921	0.128	0.934	0.322
7	0.861	0.967	0.920	0.120	0.178	0.144
8	0.773	0.982	0.909	0.089	0.193	0.146
9	0.663	0.972	0.833	0.106	0.241	0.179
10	0.472	0.967	0.749	0.144	0.321	0.236
11	0.579	0.952	0.831	0.137	0.283	0.189
12	0.637	0.950	0.801	0.158	0.325	0.229
13	0.346	0.929	0.704	0.106	0.420	0.256

TABLE III ($k_z = k_z(d)$)

st.	correlation coefficient			root mean square error		
	min	max	mean	min	max	mean
2	0.820	0.990	0.931	0.050	0.164	0.108
3	0.934	0.965	0.950	0.089	0.177	0.131
4	0.896	0.961	0.937	0.106	0.201	0.146
6	0.778	0.984	0.887	0.111	0.629	0.269
7	0.899	0.996	0.935	0.112	0.190	0.149
8	0.913	0.991	0.945	0.074	0.160	0.119
9	0.875	0.987	0.927	0.059	0.152	0.112
10	0.736	0.980	0.887	0.086	0.210	0.140
11	0.802	0.942	0.895	0.112	0.193	0.148
12	0.827	0.968	0.899	0.116	0.191	0.157
13	0.243	0.893	0.569	0.171	0.432	0.272

TABLE IV

	k_z		ρ	
k_z=const.	min	max		
	0.1	4933.2		
	k_{z0}		ρ	
$k_z=k_z(z)$	min	max	min	max
	0.1	4999.8	1.0	20.0
	k_{z0}		n	
$k_z=k_z(d)$	min	max	min	max
	0.001	200.0	0.001	2.5

REFERENCES

Bacci, P., Bolzern, P., and Fronza, G., 1981, A stochastic predictor of air pollution based on short-term meteorological forecast, J. Appl. Meteor., 20, 121-129.

Finzi, G., Fronza, G., Rinaldi, S., and Spirito, A., 1978: Prediction and real time control of SO_2 pollution from a power plant, Proceedings of APCA Annual Meeting, Houston (USA).

Hanna, S.R., 1981, Effects of release height on σ_z and σ_x in daytime condition. Air Pollution Modelling and its Application, C. de Wispelaere, ed., Plenum Press, New York.

Janicke, L., 1981, Particle simulation in inhomogeneous turbulent diffusion, Proc. of 12th NATO/CCMS Techn. Meeting, Palo Alto.

Jazwinski, A.H., 1970, "Stochastic processes and filtering theory", Academic Press.

Lamb, R.G., 1980, Mathematical principles of turbulent diffusion modelling, in: "Atmosperic planetary boundary layer physics", Proc. of 4th Course of the International School of Atmospheric Physics, A. Longhetto, ed.

Lamb, R.G., and Durran, D.R., 1978: Eddy diffusivities derived from a numerical model of the convective planetary boundary layer, Il Nuovo Cimento, vol.1C, n.1.

Lamb, R.G., Hogo, H., and Reid, L.E., 1979: A Lagrangian Monte-Carlo model of air pollutant transport, diffusion, and removal processes. 4th AMS Symposium of Turbulence, Diffusion and Air Pollution, (Reno,Nevada)

Lamb, R.G., Hogo, H., and Reid, L.E., 1979, A Lagrangian approach to modeling air pollutant dispersion, EPA Report, EPA-600/4-79-023.

Legg, B.J., and Raupach, M.R., 1982, Markov-chain simulation of particle dispersion in inhomogeneous flows: the mean drift velocity induced by a gradient in eulerian velocity variance, Boundary Layer Meteorology, 24, 3-13

Melli, P., 1982, Lagrangian modelling of dispersion in the planetary boundary layer of particulate released by a line source. Proceedings of 13th Nato/CCMS ITM, Ile des Embiez, France.

Melli, P., and Fronza, G., 1981, An Application of Pollution Episodes Predictor Derived from a K-theory Model, in: "Air Quality Modelling and its Applications", C. de Wispelaere, ed., Vol. 1, Plenum Press.

Pasquill, F., 1974, "Atmospheric Diffusion", 2nd Ed. Horwood, Chichester.

Reid, J.D., 1979, Markov-chain simulations of vertical dispersion in the neutral surface layer for surface and elevated releases, Boundary-Layer Met., 16, 3-22.

Robins, A.G., 1978, Plume dispersion from ground level sources in simulated atmospheric boundary layers. Atmos. Environ., 12, 1021-1032.

Runca, E., Melli, P., and Spirito, A., 1979, Real time forecasting of air pollution episodes in the Venetian region; Part I: The advection-diffusion model, Appl. Math. Modelling, 3, 402-408.

Seinfeld, J.H., 1975, "Air Pollution - Physical and Chemical Fundamentals", McGraw-Hill

Shir, C.C., and Shieh, L.J., 1973, A preliminary numeric study of atmospheric turbulent flows in idealized planetary boundary layer, J. Atmos. Sci., 30, 1327-1339.

Shir, C.C., and Shieh, L.J., 1974, A generalized urban air pollution model and its application to the study of SO_2 distributions in the St. Louis Metropolitan Area. J. Appl. Meteor., 13, 185-204.

Slade, D.H., 1968, "Meteorology and atomic energy", USAEC Div. of Technical Information Extension Oak Ridge, Tenn.

Wyngaard, J.C., Cote', O.R., and Rao, K.S., 1974, Modeling the atmospheric boundary layer. Adv. Geoph., 18B.

DISCUSSION

R.M. VAN AALST — Did you include dry deposition of SO_2 in your model. Could you comment on the conseguences of this to your results ?

A. SPIRITO — We did not include dry deposition in our model. Dry deposition can be included in the k-model as a boundary condition on the ground boundary.

PARTICIPANTS

The 14th NATO/CCMS International Technical Meeting on Air Pollution Modeling and its Application
Copenhagen, September 27-30, 1983 Denmark

AUSTRIA

Kaiser A.	Zentralanstalt f. Met. v. Geodyn Hohe Warte 38 A-1190 Wien
Whiteman D.	Institut f. Meteorologie und Geophysik Schoepfstrasse 41 A-6020 Innsbruck

BELGIUM

Berger A.	Institut d'Astronomie et de Géophysique 2, Chemin du Cyclotron B-1348 Louvain-La-Neuve
De Wispelaere C.	Diensten van de Eerste Minister Programmatie van het Wetenschapsbeleid Wetenschapsstraat 8 B-1040 Brussels
Dumont G.	I.H.E. J. Wytsmanstraat 14 B-1050 Brussels
Gallee H.	Institut d'Astronomie et de Géophysique 2, Chemin du Cyclotron B-1348 Louvain-La-Neuve

Kretzschmar J.	Nuclear Energy Research Center SCK/CEN Boeretang 200 B-2400 Mol
van Der Auwera L.	Meteorological Institute of Belgium Ringlaan 3 B-1180 Brussels
Vanderborght B.	Nuclear Energy Research Center SCK/CEN Boeretang 200 B-2400 Mol

CANADA

Den Hertog G.	Atmospheric Environment Service 4905 Dufferin St. Downsview M3H 5TY
Scholtz T.	The MEP Company 7050 Woodbine Avenue, Suite 100 Markham, Ontario L3R 4G8

DENMARK

Jensen A.B.	Danish Boiler Owners Association EPA Air Pollution Lab. Gladsaxe Møllevej 15 DK-2860 Søborg
Berkowicz R.	National Agency of Environmental Protection Air Pollution Laboratory Risø National Laboratory DK-4000 Roskilde
Baerentsen H.	National Agency of Environmental Protection Risø National Laboratory DK-4000 Roskilde
Christensen O.	Københavns Magistrats 5. Afd. Sekretariatet Stormgade 20 DK-1555 København V

Corlin A. — Københavns Magistrats 5. Afd.
Sekretariatet
Stormgade 20
DK-1555 København V

Fenger J. — Air Pollution Laboratory
Risø National Laboratory
DK-4000 Roskilde

Flyger H. — Air Pollution Laboratory
Risø National Laboratory
DK-4000 Roskilde

Gislason K.B. — National Agency of Environmental Protection
Air Pollution Laboratory
Risø National Laboratory
DK-4000 Roskilde

Gryning S.E. — Risø National Laboratory
Meteorology Section
DK-4000 Roskilde

Heidam N.Z. — Luftforureningslaboratoriet
Risø
DK-4000 Roskilde

Høg J. — Københavns Magistrats 5. Afd.
Sekretariatet
Stormgade 20
DK-1555 København V

Larsen P.A. — Cowiconsult
Consulting Engineers & Planners
Teknikerbyen 45
DK-2830 Virum

Larsen S. — Risø National Laboratory
Physics Dept.
DK-4000 Roskilde

Lyck E. — National Agency of Environmental Protection
Air Pollution Laboratory
Risø National Laboratory
DK-4000 Roskilde

Markvorsen J.S. — Cowiconsult
Teknikerbyen 45
DK-2830 Virum

Mikkelsen T.	Risø National Laboratory Physics Dept. DK-4000 Roskilde
Nielsen L.B.	National Agency of Environmental Prot. Air Pollution Laboratory Risø National Laboratory DK-4000 Roskilde
Olesen H.R.	Air Pollution Laboratory Risø DK-4000 Roskilde
Jensen F.P.	Air Pollution Laboratory Risø DK-4000 Roskilde
Prahm L.	Danish Meteorological Institute Lyngbyvej 100 DK-2100 København
Thykier-Nielsen S.	Risø National Laboratory P.O. Box 49 DK-4000 Roskilde
Torp U.	National Agency of Environmental Prot. Strandgade 29 DK-1401 København K

FRANCE

Bernard P.	Dept. Mathématiques Appliquées Les Cézeaux BP 45 F-63170 Aubiere
Bessemoulin P.	Centre National de Recherches Météoro- logiques Avenue Eisenhower F-31057 Toulouse Cedex
Huguet P.	Remtech S.A. 2 et 4 Avenue de l'Europe Velizy F-18360 Velizy
Saab A.E.	Electricité de France 6, Quai Watier F-78400 Chatou

FEDERAL REPUBLIC OF GERMANY

Bott A.	Max-Planck-Institut für Chemistry Airchem. Dept. P.O. Box 3060 D-6500 Mainz
Herbert M.	Dornier System GmbH P.B. 1360 D-7990 Friedrichshafen
Janicke L.	Dornier System GmbH Postfach 1360 D-7990 Friedrichshafen
Kampe W.	German Military Geophysical Office Mont Royal D-5580 Traben Trarbach
Kardels D.	Industrieanlagen Betriebsgesellschaft mbH Einsteinstrasse 21 D-8012 Ottobrunn
Löbel J.	VDI-kommission Reinhaltung der Luft Graf-Recke-Str. 84 D-4000 Düsseldorf
Ludwig C.	Umweltbundesamt Bismarckplatz 1 D-1000 Berlin 33
Moellmann M.	Kernforschungsanlage Juelich KFA Juelich Abteilung Ass. UW Postfach 1913 D-5170 Juelich
Rudolf B.	Deutscher Wetterdienst Zentralamt Frankfurter Strasse 135 D-6050 Offenbach
Stern R.	Inst. für Geophysikalische Wissenschaften Freie Universität Berlin Thiehlalle 50 D-1000 Berlin 33

Vogt S.
Kernforschungszentrum
Karlsruhe GmbH, Hauptabt.
Sicherheit
Postfach 3640
D-7500 Karlsruhe 1

Wilcke F.
Freie Universität Berlin
Thiehlallee 50
D-1000 Berlin 33

Zimmerman P.
Max-Planck-Institut f. Chemie
Saarstrasse 23
D-6500 Mainz

ISRAEL

Itzchack A.
Israel Institute for Biological
Research
P.O. 19 Nessziona
ISRAEL

ITALY

Anfossi D.
Istituto Cosmogeofisica
Del CNR
Corso Fiume 4
I-10133 Torino

Bonino G.
Istituto Cosmogeofisica CNR
Corso Fiume 4
I-10133 Torino

Desiato F.
ENEA-Disp
Viale Regina Margherita 125
I-00198 Rome

Hasenjager H.
Commission of the European Comm.
JRC, Ispra Establishment
I-21020 Ispra, Varese

Longhetto A.
Istituto di Fisica Generale
Universita Torino
Corso Massimo D'Azeglio 46
I-10125 Torino

Melli P.
IBM-Scientific Center
Via Giorgione 129
I-00147 Roma

Sandroni S.	Joint Research Center European Communities I-21020 Ispra, Varese
Spirito A.	IBM-Scientific Center Via Giorgione 129 I-00147 Roma
Stingele A.	Commission of the European Comm. JRC, Ispra Establishment I-21020 Ispra, Varese

KUWAIT

Zannetti P.	Kuwait Institute for Scientific Research P.O. Box 24885 Safat

MALAYSIA

Inouye R.	University Pertanian Malaysia 5-6 Pantai Towers Lorong Bukit Pantai Kuala Lumpur

THE NETHERLANDS

Asman W.	Institute for Meteorology and Oceanography 5, Princetonplein NL-3584 CC Utrecht
Builtjes P.	MT-TNO Postbox 342 NL-7300 AH Apeldoorn
de Jong T.	Gravelandseweg 565 NL-3119 XT Schiedam
de Leeuw F.	National Inst. of Public Health Antonie van Leeuwenhoeklaan 9 NL-3721 MA Bilthoven
Steenkist R.	KEMA Utrechtseweg 310 NL-6812 AR Arnhem

van Aalst R.M.	Division of Technology for Society TNO NL-2600 AE Delft
van Ham J.	Study and Information Center TNO for Environmental Research P.O. Box 186 NL-2600 AD Delft

NORWAY

Eliassen A.	Norwegian Meteorological Inst. P.O. Box 320 Blindern N-Oslo 3
Hov O.	NILU Box 130 N-2001 Lillestrøm
Lehmhaus J.	Norwegian Meteorological Inst. P.O. Box 320 Blindern N-Oslo 3

SWEDEN

Enger L.	Meteorological Dept. of Uppsala University Box 516 S-751 20 Uppsala
Omstedt G.	The Swedish Meteorological Institute Box 923 S-60119 Norrköping
Widemo U.	Studsvik Energiteknik AB S-61182 Nyköping

SWITZERLAND

Resele G.	Motor Columbus Ing. AG Parkstr. 27 CH-5401 Baden
Schneiter D.	Swiss Meteorological Institute Jolimont 6 CH-1530 Payerne

U.S.A.

Clarke J.F.	U.S. Environmental Protection Agency Research Triangle Park North Carolina 27711
Gillani N.	Washington University Campus Box 1185 St. Louis, Missouri 63130
Gudiksen P.	Lawrence Livermore Laboratory P.O. Box 808 Livermore, California 94550
Knox J.B.	Lawrence Livermore Laboratory P.O. Box 808 Livermore, CA 94550
Lavery T.	ERT Inc. 696 Virginia Road Concord, Mass. 01742
Liu M.K.	Systems Applications Inc. 101 Lucas Valley Road San Rafael, CA 94903
Ludwig F.	SRI International 333 Ravenswood Ave. Menlo Park, California 94025
Mermall S.	University of Toledo Dept. of Civil Engineering Toledo, Ohio 43606
Nagler L.	U.S. Environmental Protection Agency 345 Courtland Street Atlanta, Georgia 30340
Reynolds S.	Systems Applications 101 Lucas Valley Road San Rafael, CA 94903
Schiermeir F.A.	U.S. Environmental Protection Agency Environmental Sciences Research Laboratory Research Triangle Park North Carolina 27711

Schnelle K.B. Jr. — Vanderbilt University
Box 1683 Station B
Nashville, Tennessee 37235

Seigneur C. — Systems Applications Inc.
101 Lucas Valley Road
San Rafael, California 94903

Shannon J.D. — Argonne National Laboratory
ER Bldg. 181
Argonne, Illionois 60439

Venkatram A. — ERT Inc.
696 Virginia Road
Concord, Mass. 01742

Yamartino R. — ERT Inc.
696 Virginia Road
Concord, Mass. 01742

UNITED KINGDOM

Clark P.A. — Central Electricity
Research Laboratories
Kelvin Avenue
Leatherhead, KT22 7SE Surrey

Ghobadian A. — Imperial College of Science and
Technology
Exhibition Road
London SW7 2BX

Jones C. — Chemical Defence Establishment
Porton Down
Salisbury SP4 OJQ Wiltshire

Jones J.A. — National Radiological Protection
Board Chilton
Didcot, Oxon

Thomson D. — Meteorological Office
London Road
Bracknell RG12 2SZ Berkshire

Timmis R. — Warren Spring Laboratory
Gunnels Wood Road
Stevenage SE1 2BX Hertfordshire

Todd	UK Atomic Energy Authority Safety and Reliability Dir. Wigshaw Lane Culcheth WA3 4NE Warrington

<u>YUGOSLAVIA</u>

Loncar E.	Hydrometeorological Institute of Croatia Gric 3 41000 Zagreb
Sinik N.	Hydrometeorological Institute of Croatia Gric 3 41000 Zagreb
Vidic S.	Hydrometeorological Institute of Croatia Gric 3 41000 Zagreb

AUTHOR INDEX

SUBJECT INDEX

[illegible] have any [illegible] about [illegible]
[illegible] contact us on
[illegible]@springer[illegible]
[illegible] established outside the EU,
[illegible] EU authorized representative [illegible]
[illegible] Customer Service Center [illegible]
[illegible]platz 3, 69115 Heidelberg, Germany
Printed by [illegible]
in Hamburg, Germany

MIX
Papier aus verantwortungsvollen Quellen
Paper from responsible sources
FSC® C105338

If you have any concerns about our products,
you can contact us on
ProductSafety@springernature.com

In case Publisher is established outside the EU,
the EU authorized representative is:
Springer Nature Customer Service Center GmbH
Europaplatz 3, 69115 Heidelberg, Germany

Printed by Libri Plureos GmbH
in Hamburg, Germany